AF677981

Self-Healing Structures, Machines, and Systems

This book describes the behavior, underlying principles, and design of self-healing materials, structures, machines, and systems. Self-healing is a ubiquitous phenomenon. It appears in many systems ranging from the molecular scale up through to large macroscopic systems and in domains ranging from materials such as self-healing polymers to self-sealing tires, water distribution networks, and information systems, including control systems for damaged aircraft. Self-healing extends performance and endurance in ways that are just not possible otherwise. This book presents a unifying holistic approach to the operation and design of self-healing systems. It acts as a valuable reference for students, researchers, and engineers who are interested in understanding self-healing mechanisms and acquiring techniques to extend the performance and endurance of the structures, machines, and systems that they build, design, and study.

Key Features:

- Describes the design, operating principles, manufacture, and performance assessment of self-healing materials, structures, machines, and systems.
- Presents a unique holistic approach to the engineering and inclusion of self-healing into structures, machines, and systems.
- Topics covered include materials, machines, vessels, structures, networks, and systems, with detailed discussions of polymers, concrete, machinery, pressure vessels, fuel tanks, knives, clothing, lasers, biohybrids, networks, and information systems.

Dryver R. Huston has been an engineering faculty member at the University of Vermont since 1987. His research interests include subsurface sensing, structural health monitoring, and electromechanical systems design. Recent research applications include the use of ground-penetrating radar, acoustic sensing, and augmented reality to map and assess urban subsurface infrastructure. Dr. Huston has authored the book *Structural Sensing, Health Monitoring and Performance Evaluation* (2010), He has a Ph.D. (1986) and MA (1982) from Princeton University and a BS (1980) from the University of Pennsylvania.

Self-Healing Structures, Machines, and Systems

Dryver R. Huston

CRC Press
Taylor & Francis Group
Boca Raton London New York

CRC Press is an imprint of the
Taylor & Francis Group, an **informa** business

Designed cover image: Shutterstock

First edition published 2026
by CRC Press
2385 NW Executive Center Drive, Suite 320, Boca Raton FL 33431

and by CRC Press
4 Park Square, Milton Park, Abingdon, Oxon, OX14 4RN

CRC Press is an imprint of Taylor & Francis Group, LLC

© 2026 Dryver R. Huston

Reasonable efforts have been made to publish reliable data and information, but the author and publisher cannot assume responsibility for the validity of all materials or the consequences of their use. The authors and publishers have attempted to trace the copyright holders of all material reproduced in this publication and apologize to copyright holders if permission to publish in this form has not been obtained. If any copyright material has not been acknowledged please write and let us know so we may rectify in any future reprint.

Except as permitted under U.S. Copyright Law, no part of this book may be reprinted, reproduced, transmitted, or utilized in any form by any electronic, mechanical, or other means, now known or hereafter invented, including photocopying, microfilming, and recording, or in any information storage or retrieval system, without written permission from the publishers.

For permission to photocopy or use material electronically from this work, access www.copyright.com or contact the Copyright Clearance Center, Inc. (CCC), 222 Rosewood Drive, Danvers, MA 01923, 978-750-8400. For works that are not available on CCC please contact mpkbookspermissions@tandf.co.uk

Trademark notice: Product or corporate names may be trademarks or registered trademarks and are used only for identification and explanation without intent to infringe.

ISBN: 9781032488493
ISBN: 9781032499208
ISBN: 9781003396048

DOI: 10.1201/9781003396048

Typeset in Times
by KnowledgeWorks Global Ltd.

Contents

Preface

This book originally began as a short summary of the state-of-the-art and future directions of self-healing structures, machines, and systems. Upon filling in a couple of chapters, the author realized that the subject matter is broad, deep, and rapidly growing, largely fueled by inventions and advancements from enthusiastic and creative researchers worldwide. Complete coverage of the field in a single manuscript is not practical. The flow of content begins in the first couple of chapters from physically small systems to big, then surfaces and vessels, followed by multiscale, controlled, and networked systems. Specific chapters cover electronics, optics, design, and speculation about future directions. The appendices include listings of patents and a high-level overview of underlying related theoretical topics. During these discussions, the author takes a very broad and inclusive view of what constitutes self-healing. Hopefully, the reader will find the organization and content useful and informative. The author welcomes comments for improvement and corrections of errors and omissions.

The original motivation and a portion of the material stems from research by collaborators with the author: Mohammad Abdul Qader, Saleh Al-Ghamdi, Josh Allen, Brent Boerger, Owen Brandriss, Dylan Burns, Mandar Dewoolkar, Brian Esser, Robert Farrell, Alireza Fath, Anthony Gervais, Kenneth Gollins, Diarmuid Gregory, David Hochman, Shannon Hughes, David Hurley, Ezra Kahn, Mathew Kaplita, Eric Kim, Patrick Lee, Yi Liu, Patrick O'Connor, Daniel Orfeo, Scott McNulty, Steve Pearson, Jonathan Razinger, Frederic Sansoz, Daniel Savin, Robert Seal, Graham Spencer, Ting Tan, Bernie Tolmie, Robert Worley II, Tian Xia, Bismark Yeboah, and Timothy Ziegler. The author would like to express his appreciation to Xiaoyan Sun for many suggestions regarding healing of civil structural systems, Jihong Ma for advice on an early draft of the manuscript, Ronald Huston for continued suggestions on finishing the manuscript, and Danny Kielty of Taylor & Francis Group for patience and support during the final stages of production.

Dryver R. Huston
Burlington, VT, USA, 2025

1

Introduction

1.1 Healing, Autonomy, and Taxonomy

Self-healing is essential to life and to many human-built structures, machines, and systems. While some methods date back to antiquity, others represent the latest advances of modern technology [Martin 1997] [Fischer 2010a] [Fratzl 2007] [Frei 2013]. Extending and expanding self-healing enables fabricating structures, machines, and systems with extraordinary durability and resilience, perhaps even exceeding that of living systems.

1.1.1 Definitions

To clarify the discussions, a few definitions:

Health: The World Health Organization defines health as "a state of complete physical, mental and social well-being and not merely the absence of disease or infirmity" [World Health Organization 2009]. While this definition aims at humans, it translates to human-built structures, machines, and systems.

Self-healing: This book uses a broad definition. Without healing or maintenance, all living biological systems as well as human-built structures, machines, and systems eventually fail. Structures, machines, and systems with self-healing capabilities improve their own condition in response to damage processes. There is often ambiguity about whether the system boundary includes active healing agents. Figure 1.1 shows an extrinsic system that is not able to heal itself

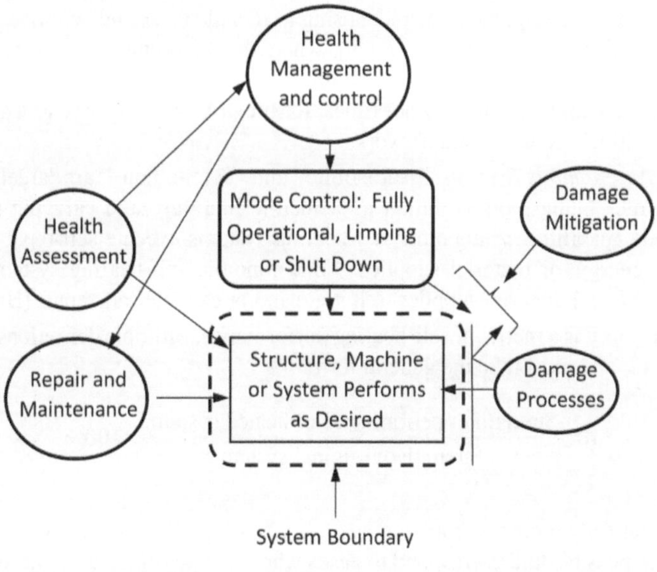

FIGURE 1.1 System maintained by nonautonomous healing agents external to the system boundary, that is, extrinsic healing.

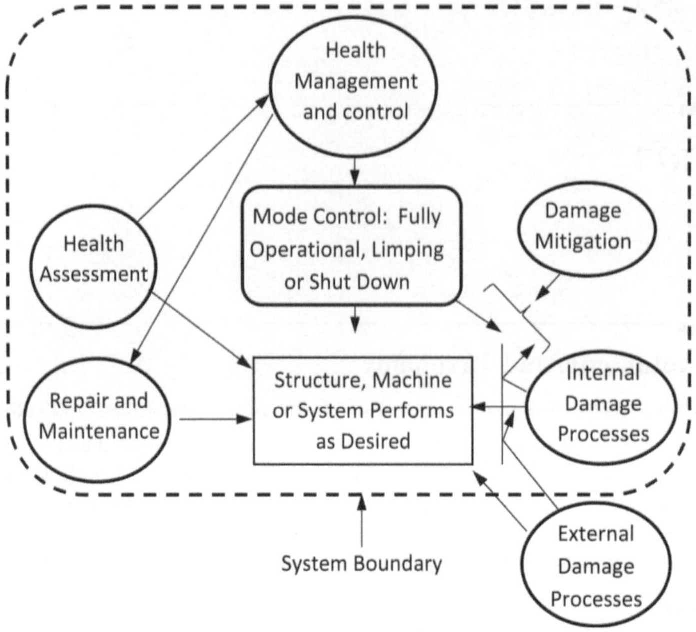

FIGURE 1.2 Autonomic system with healing supplied by agents located within the system boundary, that is, intrinsic healing.

but has external healing and maintenance agents tending to it. Figure 1.2 represents an intrinsic self-healing system. It contains similar sets of events and actions as in Figure 1.1, but now the system boundary includes healing agents.

Structures, Machines, Systems, and Networks: These are human-built. Structures transmit mechanical force with small relative motions between elements and sustain long-term stability in geometric form. Machines transmit force to do work, often with large relative motions between internal elements. Systems are collections of interacting entities with distinguishable boundaries. The entities can be physical components, as in structural girders, connected networks, or perhaps more ephemeral, for example, organized information. The system boundary may be dynamic – opening, moving, and closing to include or exclude entities. Closely related are networks, which can be viewed as systems where the interconnections between subsystems are a dominant characteristic. When needed for clarity, the discussions distinguish structures, machines, systems, and networks. Other times, for brevity, *systems* serve as a generic term for structures, machines, systems, and networks.

Autonomic and Autonomous Systems: "Autonomic" and "autonomous" are largely synonymous, with "autonomic" being more common in academic literature and carrying implications of self-contained capabilities. Autonomic self-healing systems execute self-repair without external stimulus, energy, or material supplies. Nonautonomic self-healing systems use external stimulus, energy, and material supplies to initiate and/or execute self-repair [Hager 2010].

Healing Efficiency η: It is a metric of self-healing performance. Multiple definitions appear for this (see Chapter 12) and most take the general form:

$$\eta = \frac{\text{Strength or performance of healed system}}{\text{Strength of original system}} \times 100\% \tag{1.1}$$

A system that fully recovers from damage has a healing efficiency of 100%. Values greater than 100% are possible and correspond to cases where the healing leaves the system in a state that is superior to the original undamaged state.

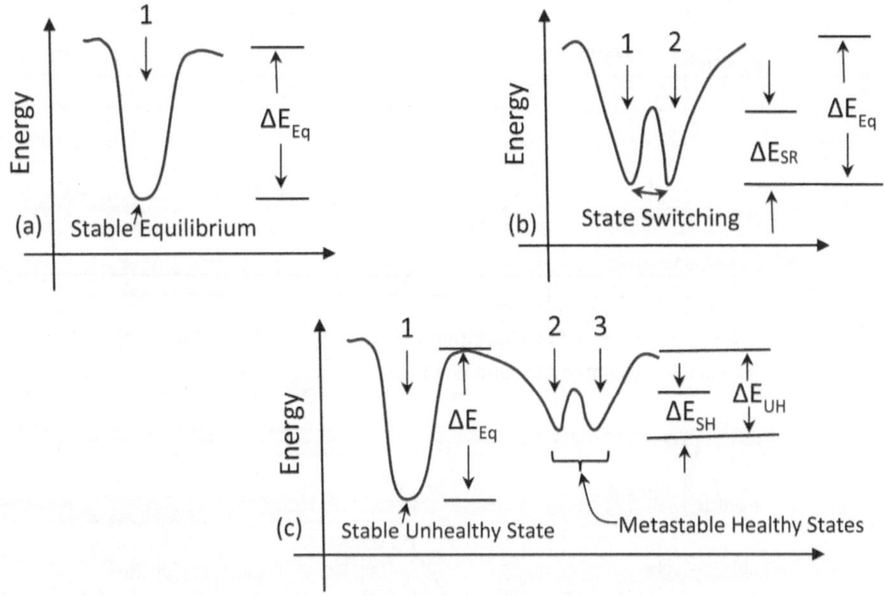

FIGURE 1.3 Energy equilibrium states with energy available for repair. (a) System with a single stable equilibrium state (1) and energy barrier of ΔE_{Eq}. (b) System with two stable states (1 and 2) that switches between the states with stimulus-response behavior and energy ΔE_{SR}. (c) System with stable low-energy unhealth state (1), and two metastable healthy states (2 and 3) with self-healing switching through energy ΔE_{SH}, and injury-causing energy of ΔE_{UH} that moves system to unhealthy state (1). (Adapted from [Urban 2009].)

Smart, Intelligent, Cognitive, and Functional: These are adjectives for systems with stimulus-response behaviors that exceed passive systems. Smart generally confers sharpness or snappy stimulus-responsive behavior. Intelligence implies that thinking is associated with behavior. Cognition is a higher form of intelligence that includes memory, reasoning, adaptation, and self-formulation of new ways of thinking based on the specific circumstances presented to the system. Functional materials respond by using available physical effects, usually with specialized arrangements of internal constituents and components.

Self-referencing: Systems that observe and act on themselves introduce a wide variety of interesting and often paradoxical behaviors. These concepts extend beyond that of self-healing systems to include contexts of logic, music, philosophy, and art [Hofstadter 1999]. In the context of self-healing, this appears as feedback control to sense and direct healing to a self-referenced state of healing.

Antagonism and Heterogeneity: Autonomic self-healing systems need available energy and material to perform a repair. Antagonistic effects, such as prestress, and heterogeneous distributions of materials in high free-energy states often increase the performance of self-healing, at the expense of supporting the prestress (Figure 1.3).

1.1.2 Advantages of Self-healing

Figure 1.4a shows the lifetime health history of an unmaintained system without healing. This is throw-away and then replace operation – a technique that can be quite effective if the loading is not too severe and the systems reliably sustain functionality over desired time frames. Often better are systems with self-healing that endure and even take advantage of loads and attacks that normally render passive non-healing systems inoperable (Figure 1.4b and c).

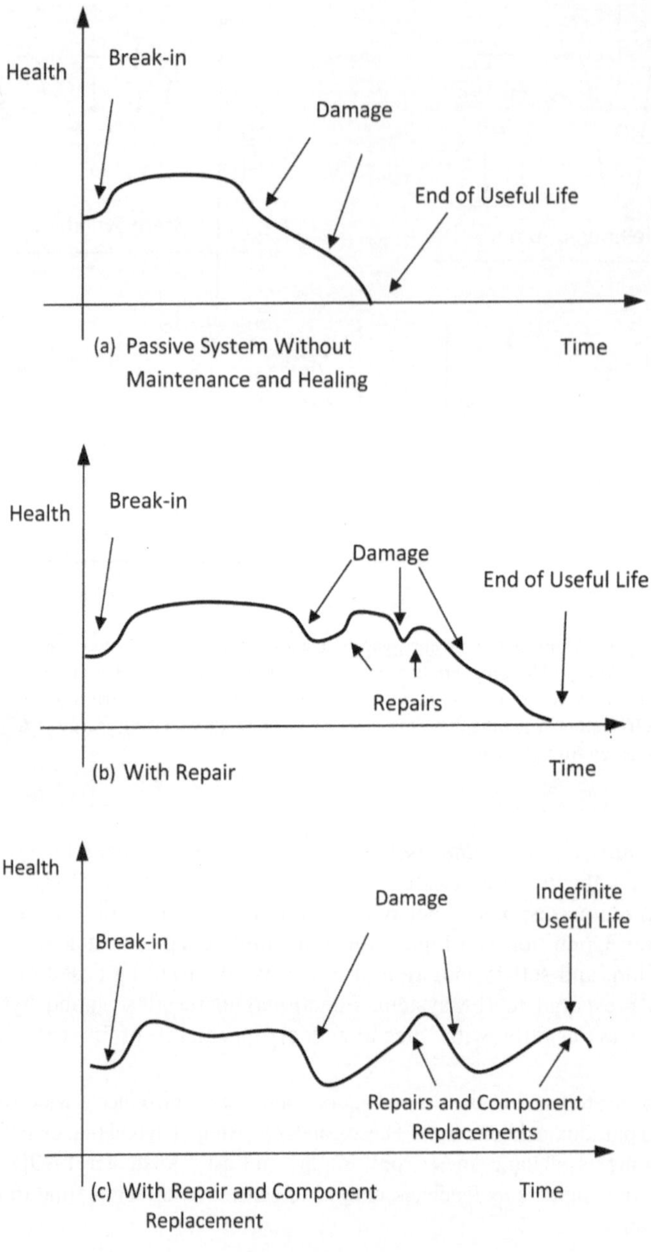

FIGURE 1.4 Lifetime health histories of structures, machines, and systems. (a) Without maintenance and repair, possibly with an early end of life. (b) With maintenance and repair that extends lifetime. (c) With component replacement and possible indefinite lifetime.

1.1.3 Fundamental Questions

Some questions with partial answers:

Are there fundamental principles that govern the design and action of self-healing systems?

Yes, fundamental principles underlying self-healing include health metrics, self-awareness, cognition, and the role of pain.

Can something last forever?

A reactive answer is "No. Nothing can last forever." However, nuanced answers provide some wiggle room. One is "forever." The long-term fate of the universe and the laws of physics as we know them – barring supernatural intervention – appear to be headed for multi-billion years of future existence. This may be a good approximation to forever. Complementing the laws of physics is the notion of information. It may be that certain types of information, such as Plato's mathematical truths, are also immutable. If information includes the structure of a system, then the endurance of a system with replacements of parts can be indefinite, as with the fabled ax that endures by having the handle and blade replaced as needed, or with long-lasting bridges and airplanes [Azizinamini 2014].

Living biological systems are masters at self-repair. What lessons can be learned?

Self-healing in biological systems is complicated and only partially understood. Biological systems recognize damage to themselves and then coordinate resources to execute repairs. Repairs occur over virtually all spatial scales ranging from the molecular to ecological systems that span the globe. In addition to making proteins, gels, new tissue, and organs, rebuilding inevitably involves the selective removal of diseased, dead, or damaged tissue, while leaving healthy tissue intact. The repair may be complete or may be partial and leave a scar.

What are effective measures of health?

If the probable failure modes are known and well understood, it may be possible to use simple quantitative measures, such as the remaining wall thickness on a pipe, to quantify health. More complicated failure modes often require more complicated measures of health.

What are the general attributes that favor self-healing?

Adding self-healing normally adds costs with increased weight, exotic materials, controls, and manufacturing details. Increased complexity introduces additional costs, such as reduced performance, operational compensation for antagonistic healing and restraining of healing effects, and unintended consequences. Excellent opportunities arise when there is a confluence of conditions, including a means to self-heal at relatively low cost and the high cost of failure without healing. Examples are self-sealing fuel tanks and tires.

What are the roles of antagonism and complexity?

Antagonism often stabilizes systems, but at a cost (Figure 1.5). Self-healing involves feedback mechanisms, often at multiple temporal and spatial scales, leading to complexity. Guidance may come from Western and Eastern medical practices that nominally follow opposite approaches of reductionist evidence-based methods and holistic approaches. A suggestion by Alfred North Whitehead is "The only simplicity that can be trusted is that which lies beyond complexity," that is, the best simplifications are those based on an understanding of the underlying complexity [Lytton 2000].

The notion of antagonism further appears in materials that separate into two phases at the microscopic scale in a thermodynamically metastable configuration. Examples are bitumen as a binder in asphalt, and certain ionic polymers. Both have self-healing capabilities. Atomic force microscope images of bitumen reveal "bumblebee" features that are phase-separated waxy crystalline inclusions within a hydrocarbon polymer matrix [Kringos 2011]. These waxy inclusions aid in healing microscale damage and deformation by a phase transition back into a solution of the bitumen.

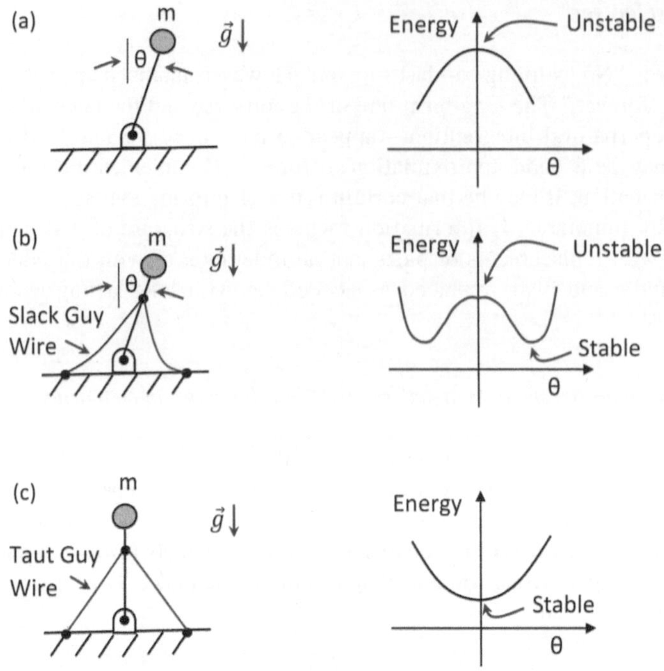

FIGURE 1.5 Energy landscape of inverted pendulum stabilized by loose and taut guy wires indicating stabilization of system by antagonism at a cost of maintaining tight wires. (a) No guy wires and unstable. (b) Slack guy wires without antagonism tighten and stabilize at tilted position. (c) Taut guy wires with antagonism stabilized at upright position.

What are practical techniques for implementing self-healing?

Some of the answers are tricks of the trade. For example, certain mixes of asphalt pavement can crack and heal multiple times [Zou 2012]. Introducing self-healing capabilities raises issues of safety, compatibility, and cost. Self-healing must work, be cost-effective, and do no harm.

When is the best time to activate self-healing?

The timing and location of self-healing actions can be particularly important in enhancing the performance of the self-healing system. For example, failures in many systems are the culmination of damage-dependent actions. Failure follows initiating events and conditions that progress through a cascading series of events. Intervening in a timely manner with mitigation and/or self-healing can prevent more damage.

What amount of restoration is possible?

It is possible to recover original strength and functionality – including self-healing capabilities. It is difficult but possible to repair a structure without leaving a scar.

Is there a role for limping?

Limping arises when systems have redundancy and excess performance capability. When damaged, a limping-enabled system reorganizes operations and load paths to retain overall function, often with reduced capability and performance.

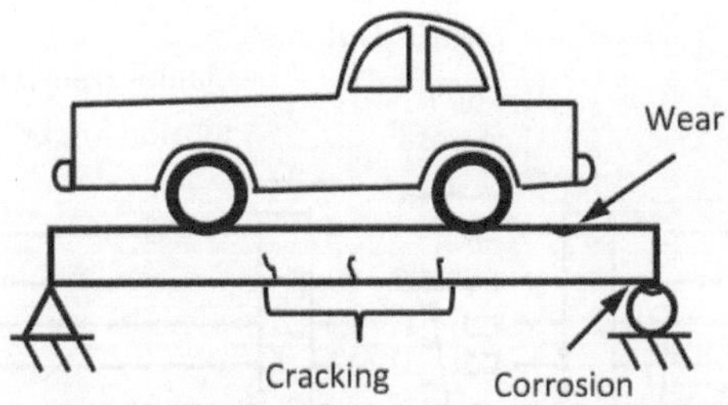

FIGURE 1.6 Level 0 passive structure with no self-healing capabilities inevitably suffers degradation.

1.1.4 Taxonomy

A six-level classification scheme ranging from simple to complicated:

1.1.4.1 Level 0: No Self-healing

Most engineered structures do not self-heal. They are well-built, perform the desired tasks as needed, and tolerate modest to moderate levels of damage. Nonetheless, such structures eventually wear out and fail (Figure 1.6). Survival requires maintenance, repair, and component replacement by external actions.

1.1.4.2 Level 1: Health and Performance Improvement with Break-in and Use

Break-in and wear improve many structures, machines, and systems. Examples are leather shoes and journal bearings in machinery. It is economically prohibitive to manufacture mating parts in machinery without geometric imperfections. A more practical approach uses break-in to modestly damage the shoe and bearing to improve the shape and structure. Bearings autonomically smooth geometric imperfections by breaking and grinding asperities as the parts slide over one another (Figure 1.7).

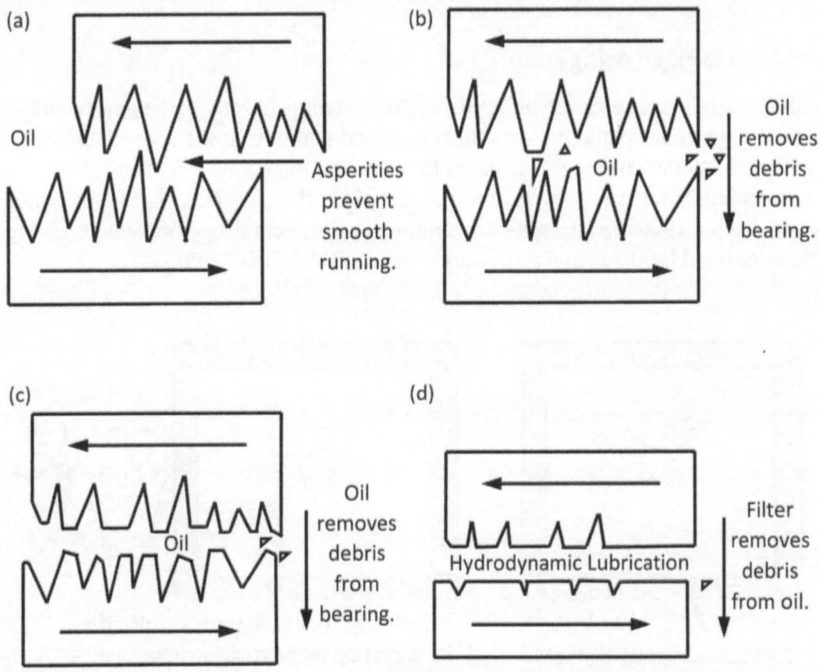

FIGURE 1.7 Level 1: Break-in process smooths machine bearings by mutual grinding of asperities to create a healthier machine component. (a) Machine bearing parts that slide past one another as initially manufactured. (b) Early break-in process generates significant levels of debris. (c) Bearing surfaces smooth out. (d) Smooth running machine.

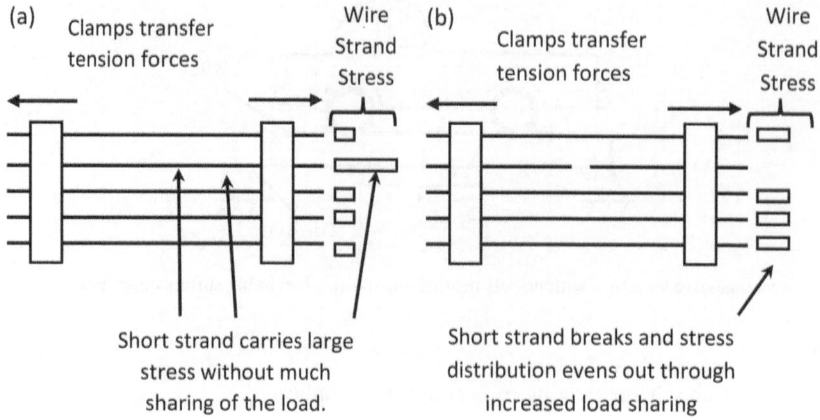

FIGURE 1.8 Simplified representation of parallel strand suspension bridge cables with statically indeterminate mechanics that distributes loads of failed shorter strands onto healthy population of strands with similar lengths. (a) Before. (b) After.

Modern anti-wear lubricants improve the process by acting at the sliding nanocontacts of single asperities and removing the debris [Gosvami 2015].

Statically indeterminate structures may also undergo break-in processes through the favorable redistribution of internal loads. An example is the use of multiple strands of parallel wire in a suspension bridge cable. Clamping, interstrand friction, and varying lengths of the strands combine to produce a resilient configuration [Waisman 2011]. Initially, the shorter strands carry excessive loads, then break and redistribute the loads for a more even distribution. The spreading of the load distribution reduces the excess loads on individual strands and prevents further breakage (Figure 1.8).

1.1.4.3 Level 2: Damage Mitigation

Mitigation reduces damage progression by acting and intervening directly in the processes by (1) altering the structural topology to decouple the structures and load paths from the active damaging forces and (2) taking actions that deflect or otherwise degrade the damaging action. Mechanical fuses, shear pins, and feathering windmills are examples. Reactive armor alters the specific action of the damaging force. Figure 1.9 shows a circa 1908 reactive armor technique that protects strong boxes with an explosion that both deters unauthorized intrusion and acts as a loud alarm [Vaughan 1908].

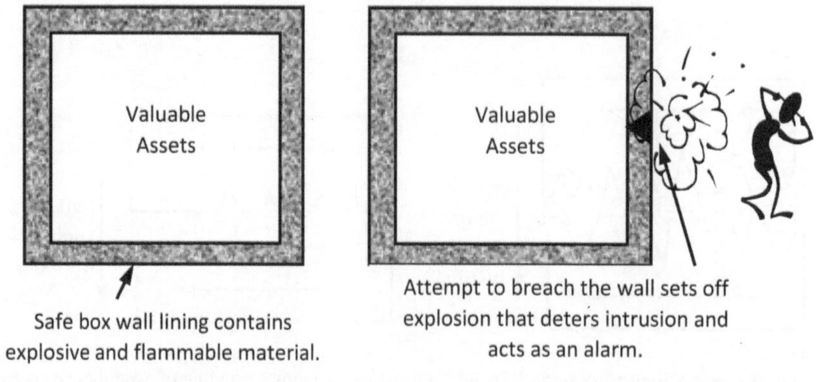

FIGURE 1.9 Active damage mitigation technique – a safe box with reactive armor resists penetration through the walls with explosion and fire. (From [Vaughan 1908].)

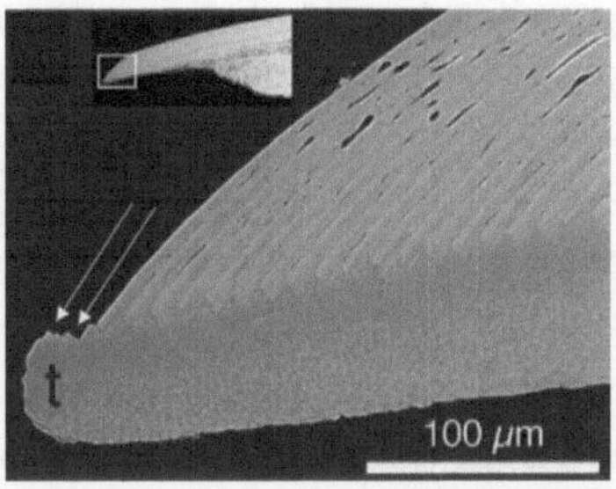

FIGURE 1.10 Regenerative self-sharpening sea urchin tooth. (From [Killian 2011].)

1.1.4.4 Level 3: Regeneration with Redundant Components

Regeneration replaces damaged components with prepositioned functionally equivalent components. Many cases of regeneration are topological reconfigurations that transfer duties from a damaged to an undamaged component. In many cases, damage initiates replacement with minimal sensing and control. The sea urchin uses a regenerative self-sharpening tooth in that shed shards as they become dull to uncover a new sharp edge (Figure 1.10) [Killian 2011]. A human-built example is the regenerative tire tread. Damage and wear cause the outer tread layer to shed and expose subsurface features with a fresh tread.

1.1.4.5 Level 4: Local Damage Repair with Local Material Transport

Certain cases of damage, for example, tight cracks, only require localized movements of small amounts of material for repair. Such processes benefit from prepositioning self-repair capabilities throughout the structure. An example is the self-sealing wood barrel. Water from small leaks causes the wood to swell and seal the leak. Another example is the healing of a tight crack in a thermoplastic polymer (Figure 1.11). The molecular structure of these polymers is that of long molecules that move due to thermal processes. These mobile long-chain molecules diffuse and flow across the crack to heal in a process known as reptation.

Asphalt pavement on roads is a ubiquitous successful example of local damage self-healing. Asphalt is a conglomerate of small stones held together by tar binders. Vehicular and environmental loading causes cracking and large deformations of the asphalt. The sticky polymer nature of the tar binder enables

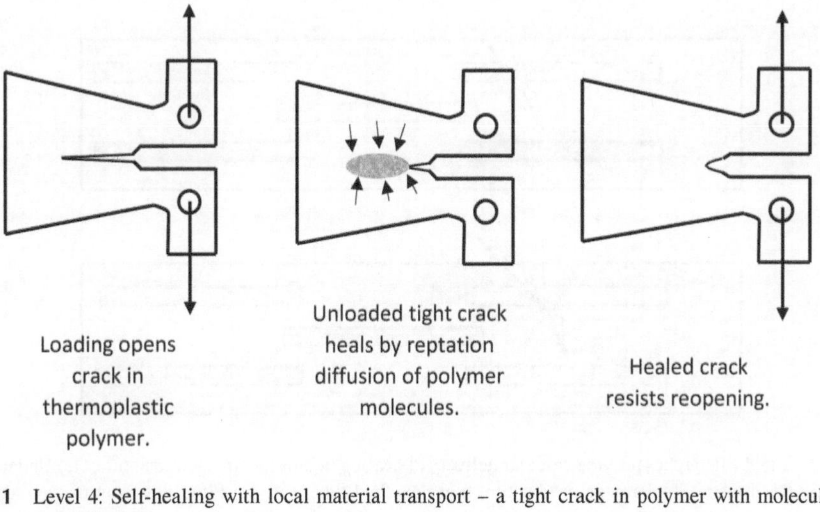

FIGURE 1.11 Level 4: Self-healing with local material transport – a tight crack in polymer with molecular mobility repairs by reptation diffusion.

the asphalt to recover from damage and endure as a low-cost pavement of choice for multiple years. Alternatives include nonhealing asphalt that fails by crumbling after only a year or two, and significantly more expensive concrete pavements.

1.1.4.6 Level 5: Damage Induces Reactive Processes with Significant Levels of Material Repair

Using encapsulated healing fluids for repair is a reactive repair. Cracking damage to the material ruptures the capsules, allowing the fluid to exit the capsule, flow into the crack, bridge the crack, and solidify to heal the structure (Figure 1.12).

1.1.4.7 Level 6: Coordinated Repair

Damage induces coordinated follow-up actions that heal the system, generally following these steps:

1. Sense damage
2. Assess damage
3. Marshal resources and execute repair processes
4. Assess the state of repair
5. Taper and terminate the repair process when the healing is complete
6. Remove damaged material

Coordinated healing may follow an algorithmic or rule-based approach, that is, if certain damage conditions are met, then specific healing actions are executed. More complicated situations may benefit from adaptive approaches, possibly with cognitive reasoning. Damage avoidance and mitigation is an effective alternative.

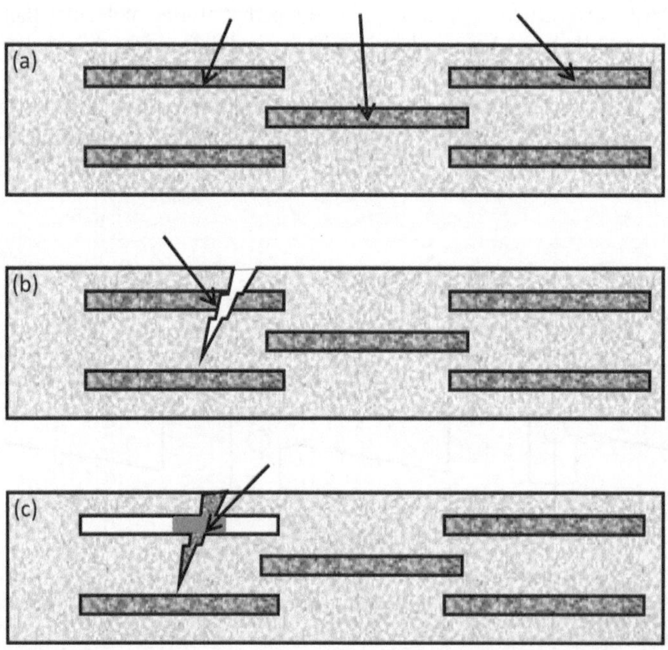

FIGURE 1.12 Fluid-filled tubes provide vascular delivery of healing liquids to damaged structures. (a) Undamaged structure with fluid-filled tubes. (b) Damage to structure breaches fluid tube wall. (c) Fluid flows out of tube, congeals, and repairs crack.

1.2 Material Repair and Damage Processes

Macroscopic manually applied repair techniques, that is, welding, epoxy injection, and fiber-reinforced polymer (FRP) patches, are a starting point for the design of self-healing systems [Petrie 2000] [de Smet 2008]. Self-repairing systems may be an automated version of the typical manually applied patch and repair methods, especially if the damage mode is known a priori, for example, a tire puncture.

1.2.1 Damage Processes

Corrosion is a common, somewhat complicated, and pernicious damage process resulting from the oxidation of metals. A variety of electrochemical processes cause metals to oxidize. Sometimes oxidation forms a protective coating. Other times oxidation runs unabated and leads to significant damage [Fontana 1978]. A thermodynamic view is that oxidation occurs because the metallic state has more free energy than the metal oxide. Corrosion processes involve a self-organized interplay of surface chemistry, bulk fluid chemistry, electric currents, confined spaces, surface treatments, and several other effects. Some effects slow corrosion. Others accelerate. An example is the formation of an Al_2O_3 skin on the surface of aluminum. The skin self-heals to form a tight bond that prevents oxygen diffusion and further oxidation. Stress corrosion cracking is a combined-effect acceleration of corrosion. The localized electrochemical environment combines with deleterious metallurgy of stress at the crack tip surface to produce rapidly growing cracks.

1.3 Early Developments

Some main themes of development:

1. *Pre-industrial Era:* Some self-healing techniques date back to antiquity. Most exploit inherent physical properties. Self-sealing wooden fluid containment systems are an example (Figure 1.13). Wooden ships, barrels, and water pipes use leaking water to swell wood in a conformal manner that tightly seals against leaks. An important detail is to control the shape of the swelling by

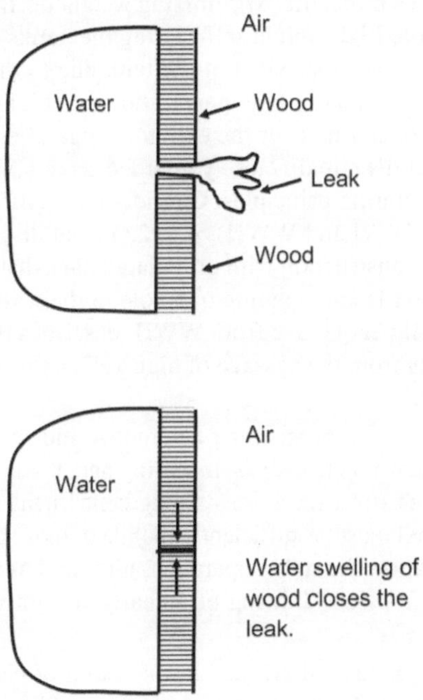

FIGURE 1.13 Water leak swelling and sealing of leak in wood vessel.

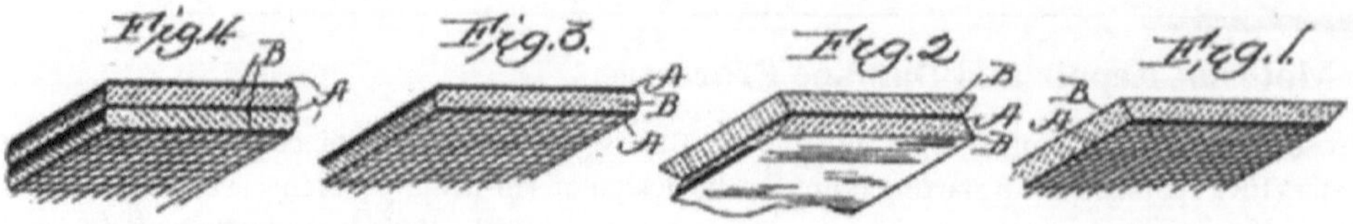

FIGURE 1.14 Sandwich of rubber, reinforcing fabric and petroleum jelly, produces an early self-sealing composite material. (From [Mercier 1896].)

using boards cut with a proper grain pattern. A radial rift sawn cut can produce isotropic and heterogeneous swelling [Twede 2005].

The ancient practice of using natural fibers, such as straw, in bricks and mortar mitigates damage by reducing the size of cracks [Snoeck 2015]. Cements and mortars, including those used by the Romans, autogenously heal tight cracks through the hydration of free lime. Break-in self-improvement techniques, for example, wear-in to form smoother running machines, also date back to antiquity.

2. *Industrial Era:* Self-healing developments arise largely to solve new challenges posed by the new technologies in the industrial revolution. Industrial and military entities conducted much of the research, with a minimum of public disclosure. Notable examples are self-sealing tires and fuel tanks.

A precursor to many self-sealing tires and fuel tanks was a nineteenth century invention of a composite material panel that autonomically sealed puncture holes. The construction was a sandwich of partially vulcanized rubber, cloth mesh, and petroleum jelly (Figure 1.14) [Mercier 1896]. Through-wall perforations by nails, projectiles, and so on, initially create a hole, but also comingle the rubber and petroleum jelly in the surrounding region. The diffusion of petroleum jelly into partially cross-linked rubber, followed by rubber swelling with in-plane directional constraint from the cloth mesh seals the hole.

Self-sealing fuel tanks in military aircraft represent one of the more successful self-healing structural systems in the twentieth century. The goal is to suppress fuel tank leaks and fires due to penetration by a projectile. Minimizing weight during flight restricts using heavy armor as passive protection. Lightweight self-sealing fuel tanks are a competitive alternative. The design requirements include survival of the tank due to the intense mechanical shock, known as hydrodynamic ram, that follows penetration by a high-speed projectile and quickly stopping leaks, all while operating over the extreme range of environmental conditions during flight. While many details remain closely guarded secrets, the patent literature describes some of the important operating principles. Considerable activity clustered around the First and Second World Wars (WWI and WWII). A 1922 patent discloses a self-sealing fuel tank as a combined multilayer construction with articulated flaps that swing open and shut following the passage of a projectile and opening of a hole in the wall (Figure 1.15) [Friant 1922]. Eckelmeyer, in a historical perspective from WWII, describes the development of fuel tanks that survive and seal leaks from the passage of high-velocity projectiles with diameters up to 50 mm [Eckelmeyer 1946].

Similarly, the demands of the burgeoning automotive industry in the early twentieth century prompted a flurry of developments in self-sealing pneumatic tires. An important technical constraint is that a rotating tire experiences strong centrifugal and vehicle to road dynamic forces. The sealing material must be sufficiently mobile to move and seal a leak as needed, but sufficiently stationary to allow for stable operation in normal undamaged but highly dynamic conditions. Figure 1.16 is a patent drawing of an early version with a sponge rubber sealing layer [Cochrane 1900].

Another track from the industrial era lies in the realm of electric power transmission. A chronic issue is the difficulty with containing concentrations of large amounts of electrical energy. High-voltage containment requires sturdy insulators, switches, and fuses. Dielectric insulators

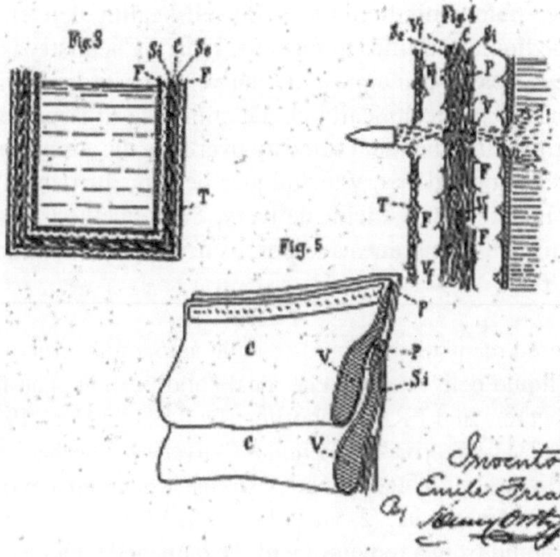

FIGURE 1.15 Early design of a self-sealing fuel tank using a combination of multilayer construction with articulated flaps. (From [Friant 1922].)

tend to fatigue and break down when exposed to high voltages. A standard containment method dating back to the early days of electric power uses liquid insulators that flow in a vascular manner to replenish the potentially damaged dielectric.

3. *Academic Interest and Developments:* Many early threads of academic activity examined the healing of tight cracks. As early as 1836, the French Academy of Science studies autogenous carbonation-based crack healing of concrete [Kenneth 1956]. Another early report of concrete crack healing research dates to 1918 [Maddocks 1925]. In 1959, Kolsky described the healing of tight cracks in polymers [Kolsky 1959]. Outwater and Gerry in 1969 experimentally quantified the energies of self-healing polymers [Outwater 1969]. de Gennes provided a viable theoretical model of polymer healing based on wiggling snake-like reptation motions of polymer molecules, leading in part to the award of Nobel Prize in Physics [de Gennes 1971]. The academic community began to examine self-healing more earnestly in the late 1980s and early 1990s with a focus on healing cracks in materials [Xu 2007] [Aïssa 2012] [Murphy 2009]. Subsequent research and development covers virtually all aspects of engineering that require high performance from the system [Ghosh 2009b] [Zako 1999].
4. *Thermoplastic Polymer Crack Healing:* Some polymers, notably thermoplastic polymers, heal tight cracks by bridging and infilling [Volynskii 2009]. The molecular structure of thermoplastic polymers is that of long polymer chains held together in intertwining conglomerations by weak hydrogen bonds. At a sufficiently high temperature (often the glass transition temperature, T_g), the molecules become mobile and reptate, that is, slither in a snake-like manner [de Gennes 1971]. Statistical mechanical models based on reptation can explain many aspects of the behavior of thermoplastics, including autonomic bridging and healing of tight cracks.

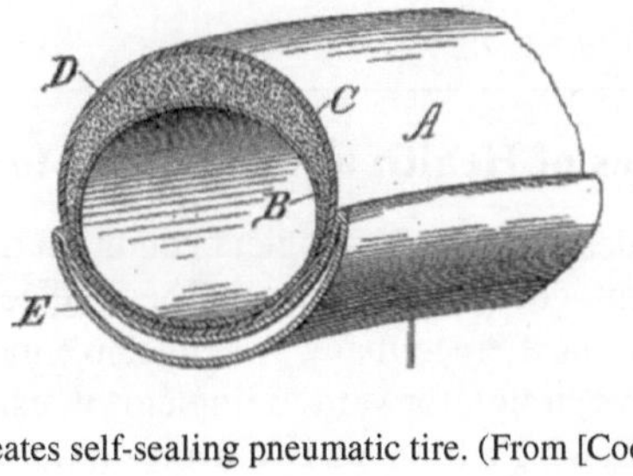

FIGURE 1.16 Sponge rubber layer creates self-sealing pneumatic tire. (From [Cochrane 1900].)

5. *Liquid to Solid Healing:* Autonomic liquid-to-solid self-healing delivers liquid to the damaged region and then the liquid solidifies to form the repair. Chemistry is the primary means of solidification. While simple in principle and relatively easy to implement manually, autonomic liquid-to-solid healing is technically challenging due to issues of liquid delivery to the proper location on time, and chemical stability over long durations and wide temperature ranges. Early solid in situ liquid delivery techniques became available in the middle of the 1970s for applications, including insecticide delivery. These methods used erodible containers, diffusion release through membranes and porous networks, and hollow fibers filled with healing agent [Paul 1976].

In the 1990s, C. Dry pioneered many of the liquid-to-solid autonomic structural healing methods by combining controlled in situ liquid delivery with chemical solidification. The focus was on the insertion of healing liquids into concrete and FRP composites [Dry 1992a] [Dry 1993a] [Dry 1996a]. This work introduced the following: (1) *Encapsulated Liquid Delivery* – The healing system stores liquid in capsules throughout the structural or material matrix so that when damage occurs, the liquid has to flow only a short distance to the damage site. (2) *Vascular Liquid Delivery* – A prepositioned plumbing system transports the healing liquid from a remote site to the damaged area.

White et al. further advanced liquid-to-solid healing by storing low-viscosity healing liquids in micrometer-scale capsule and embedding catalysts into the polymer matrix. The effect is to activate curing only after fluid flows out of the ruptured capsules [White 2001]. Miniaturizing the liquid capsules permits finer-scale dispersion in solids – an important detail in FRP composites. In many respects, this development was a turning point that motivated a broad spectrum of self-healing research and development.

In the twenty-first century, research, development, and deployment of self-healing structures continued at a rapid pace with the increased sophistication and complexity of active materials, along with systems-based healing approaches based on sensing and quantification of damage and control/coordination of the repair actions [Carlson 2006a] [Balazs 2007] [Trask 2007a] [Hurley 2011a] [Song 2011].

Vascular systems provide higher speeds of internal material transport and have the potential for accelerating and accentuating self-healing [Bond 2007]. Vasculature appears in both biological and human-built situations that benefit from high rates of material transport. Most of these human-built systems flow material in an open-loop one-way fashion to the damaged site. The technology of material removal by analogous combined circulatory and lymphatic approaches in human-built systems appeared in the early twentieth century with self-sealing fuel tanks.

Self-cleaning surfaces and fabrics are additional threads of development. The accumulation of unwanted material degrades the appearance and performance of many surfaces. Autonomic resistance to the buildup and shedding of debris has advantages [Youngblood 2008]. Many self-cleaning surfaces use micro- and nanoscale surface textures to effect hydrophilic and hydrophobic cleaning actions. Other surfaces harvest ambient energy from the environment to create chemicals with free energy that break down contaminants, such as through photocatalysis [Wang 1997]. Applications include destruction of bacteria [Rtimi 2013], and surface tension of small water droplets on glass surfaces for hydrophilic anti-fogging action [Rampaul 2003]. Incorporating these techniques into fabric makes self-cleaning clothing [Wu 2011].

1.4 Quantitative Measures of Health and Capacity for Self-healing

Health is a familiar qualitative description of biological condition that readily transfers to engineered systems. Health can be "excellent, good, fair, or poor." These adjectival descriptions lack specificity, yet have easily understood meanings. Since many systems have multiple failure modes, quantitative scalar descriptions are not always straightforward to implement, especially as the level of complexity rises. Being healthy requires following a life path that avoids failure modes, as shown conceptually in

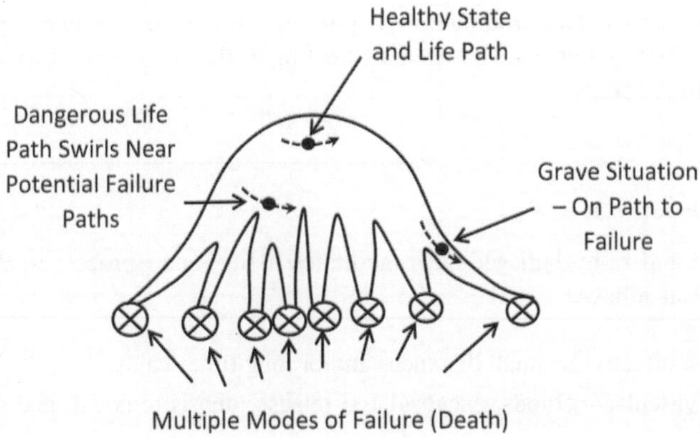

FIGURE 1.17 Avoiding danger in life paths sustains good health.

Figure 1.17. Hybrid qualitative and quantitative health ratings are common for multicomponent systems with even modest levels of complexity.

Living biological systems proactively promote their own states of health. Imbuing human-built structures, machines, and systems with similar health-promoting capabilities is complicated. A conservative approach identifies failure modes, possible paths to failure, and maximizes the distance of the system from failure based on available information. An example is a machine that fails when specific parts, such as bearings, wear out. Bearing failure is diagnosed and predicted by detecting an increase in particular vibration signatures. Measuring bearing vibrations is a simple, but effective, measure of health. The situation becomes more complicated when remedial steps prevent a singular wear event from being a primary mode of failure. What remains is secondary, often unexpected, failure modes. For example, if a machine uses long-endurance bearings to minimize bearing failures, the ultimate failure mode then may become unpredictable. This leads to the ironic situation of taking steps to improve safety, and reliability by eliminating failure modes may actually result in a seemingly less reliable machine.

Since the health of biological or human-built systems is inherently complicated, a single-scalar health metric may not provide sufficient information. An alternative approach combines components of measurand vectors.

Reliability, *redundancy*, *resilience*, and *recoverability* are overlapping interrelated concepts. Reliability is a description of survival and failure that combines estimates of random variations in strengths of the individual components of the system with estimates of the random variations in the loads acting on the systems [Liu 2001]. Redundancy arises when multiple structural load paths or systemic configurations carry the loads or perform desired tasks. Damage to a component causes the parallel undamaged components to carry the load. Redundancy may take on more complicated forms where the overall topology of the system rearranges to carry the desired loads into either fully functional or partially functional limping configurations. In these cases, network and graph theory configurations may be useful. Resiliency is the ability to survive and then recover from damaging loads and extreme events. Recoverability is the ability of a system to recover a particular state, often as a specific measurable quantity.

1.5 Scaling Aspects

Length scales with important self-healing activity run from subatomic to macroscopic dimensions. Similarly, time scales range from the smallest measurable instances up through historical epochs. Key interactions often remain confined to narrow temporal and spatial scales. There are many important cases, however, where damage and healing interactions span and even skip multiple space and time scales.

Multiple scale interactions create the opportunity for many interesting and difficult to predict occurrences, along with some great opportunities for innovation. An example is an earthen dam. A small leak at the base can lead to internal dam-eroding flow, beginning with the formation of pipes and ending with catastrophic collapse. Intervening with autogenous healing at the small-scale leak stage can be a low-cost means of preventing collapse.

1.6 Material Transport

Material transport to and from damaged sites facilitates many repair processes. Possible methods of material transport are as follows:

1. *Diffusion:* Most effective at small distances and/or long time scales
2. *Liquid Flow:* Vascular methods, encapsulated release, debris removal, and surface cleaning runoff
3. *Dendritic and Filamentary Growth:* Effective directed method with minimal material use over extended distances
4. *Swelling, Contractions, and Other Shape Changes:* Good for bulk movement of material coupled with mechanical action
5. *Accretion and Deposition:* Deposits material from liquid onto solid surface

1.7 Thermodynamic Considerations

Table 1.1 lists some thermodynamically intensive self-healing actions. Classical thermodynamics asserts that macroscopic systems tend to move to equilibrium states. The processes minimize available energy and increase entropy [Fermi 1956] [Callen 1960]. While movement toward equilibrium is an inexorable universal process, it is not instantaneous or homogeneous. Processes that favor maintaining and/or establishing nonequilibrium conditions are ubiquitous and lead to many interesting and complicated features in our universe [Prigogine 1981]. Biological life is a prime example [Schrödinger 1944] [Ho 2008]. An additional feature of thermodynamics is the relation to information theory [Shannon 1949]. At present, it is not clear if this relationship is a coincidental consequence of phenomena with similar organizational structure or is a more fundamental relation.

Simpler molecular scale coordination techniques arise when a state of minimum free-energy thermodynamic equilibrium is the desired state of health. Annealing processes can nudge damaged states back

TABLE 1.1
Thermodynamic Phenomena and Self-healing

Thermodynamic Phenomenon	Self-healing Action	Reference
Stable equilibrium	Recovery of strength and length of byssal thread	[Holten 2011]
Release constraint, move to new equilibrium	Liquid solidification for healing	[White 2001]
Phase change	Shape memory alloy wound closing	[Burton 2006] [Jefferson 2010]
Cycle	Reversible thermal cycling of DA reaction to heal polymer	[Kavitha 2009]
Diffusion	Diffusion reptation healing of thermoplastic polymers	[de Gennes 1967] [Wool 1981]
Diffusion-reaction	Asphalt with segregated bumblebee inclusions for crack healing	[Kringos 2011]
Irreversible Onsager reciprocity	Chemically selective membranes that support healing	[Katchalsky 1965]
Irreversible dissipative structure	Dissipative actions heal dissipative structure, SUSY laser stabilization	[Belkin 2015] [El-Ganainy 2018] [Qiao 2021]

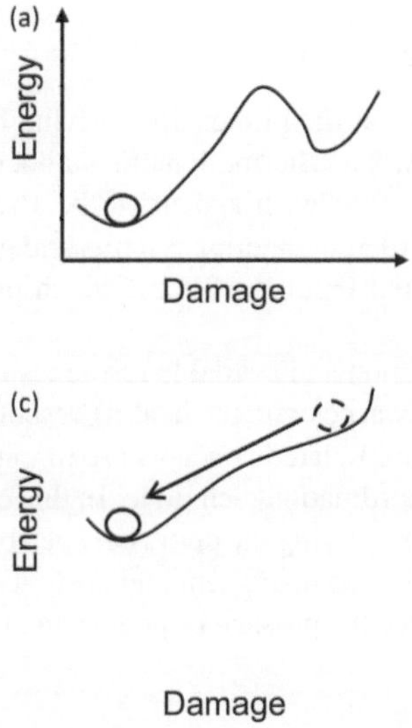

FIGURE 1.18 Thermodynamic-driven movement to healthy structures in equilibrium state with application of heat, as in annealing. (a) Original heathy. (b) Damage increases internal energy, but state is stable. (c) Heating changes energy landscape and material moves to lower energy healthy state. (d) Material remains in healthy state upon cooling.

into healthy equilibrium states through the injection of activating energy that eventually dissipates as the system heals (Figure 1.18). Many of these processes can repeat through numerous cycles of damage and repair without long-term degradation. The following are examples of simple system movements to healing:

1. *Damage Raises the Free Energy:* Movement to equilibrium is healing.
2. *Damage State Is Frozen in a Nonequilibrium Configuration:* Energy injection processes enable movements for healing.
3. *Hierarchical, Complicated, and Living Systems:* Many material systems are nonhomogeneous, with nonhomogeneous distributions of free energy. This is especially the case of biological life, where packets of stored free energy available for release on cue to perform healing can be released upon triggering by damage.

Hierarchical organization explains some of the thermodynamics of self-healing systems. Damage increases entropy defined as an increase in configurational disorder, such as microcracking. If there are components in the system, such as microcapsules filled with healing liquid stored at a lower entropy, then damage may initiate movement to a nonequilibrium state, such as the release of a healing liquid. The healing liquid flows into the cracks, congeals, and solidifies in processes that increase entropy. The result is that the healing liquid reduces the configurational entropy of the microcracking, but the overall entropy of the system increases [Nosonovsky 2009].

Open systems with energy and material flowing in and out often have stable dynamic nonequilibrium states that exhibit high degrees of organization. External events that force temporary deviations from the stable states induce dynamical processes that heal the system and drive it back to the stable nonequilibrium state, often with maximum entropy production being the physical mechanism [Belkin 2015]. Limit cycles, stable vortices, and nonhomogeneous patterning of materials are examples. Lying in-between static equilibrium and stable self-organized dynamical systems are static self-assembled systems, such as folded proteins, nematic crystals, or brick-like block assemblies. In these cases, the assembled static system has a lower entropy, but the overall system has a higher total entropy [Frenkel 2015].

1.8 Self-assembly and Coordinated Healing

Self-assembly places components in the proper positions at the proper time. Living biological systems and ecosystems are extreme examples of self-assembly. The thermodynamics underlying the process is interesting and only partially understood [Whitesides 2002a]. It appears to be a case of antagonism between order and disorder with crystals representing ordered material systems and glasses disordered. Component interaction is a combination of attractive and repulsive forces, which promotes heterogeneous distributions of free energy.

Coordination uses information to bring the healing action and available resources to bear on damage when and where needed. Such techniques simultaneously accelerate the healing response and reduce the need for distributed and parasitic stores of repair material. Related concepts are limping and mitigation.

Topological structural reconfiguration is a primary coordination technique. In this context, reconfiguration is more than redundancy. Ships at sea are examples. Closing the doors between bulkheads during a severe leak event stops water passing from compartment to compartment and prevents sinking. Keeping the doors open during normal operations is convenient for the passage of people and material.

1.9 Biomimetics, Living Systems, and Biohybrids

Biomimetic and *bioinspired* are adjectives describing the use of natural biological systems as guides for the conceptual design of engineering systems (Figure 1.19) [Bar-Cohen 2006] [de Mestrai 1961] [Trask 2007b] [Meyers 2013]. It is natural to examine healing in biological systems as a source of ideas and concepts for the design of artificial self-healing systems. Biological systems are faced with an extraordinary task – healing by regeneration of living tissue actions ranging over length scales from molecules to ecosystems. In mammals, the healing of a relatively small cut is a multistage process that begins with blood clotting and scab formation followed by inflammation, infilling of cells, and tissue structures, often aided by mechanical actions that close the wound, and finally scar maturation with matrix remodeling [Martin 2010] [Singer 1999]. Similar wound-covering and healing processes also occur at smaller length scales, with the formation of protein patches over lesions in cellular membranes [Roostalu 2012].

Interesting possibilities for bioinspiration come from blood clotting that uses complex antagonistic cascades of clotting and dissolution biochemical reactions to form macroscopic clots, only when needed [Mann 2010] [Davie 1964] [Lam 2011] [Pleydell 2002] [Diamond 1999] [Hougie 2004] [Macfarlane 1964] [Velnar 2009]. A characteristic feature of blood clotting is nonlinear amplification with chain reactions. The products of one reaction catalyze and amplify another reaction. Healthy blood clotting strikes a remarkable balance between clotting and dissolution processes. Creatures thrive for decades without bleeding to death or dying from clot-induced strokes or heart attacks. Similar in scope, but different in detail, actions occur in plants [Bauer 2009] [Shigo 1984]. Antagonism also plays active roles in removal of tissue components that are dead, injured, or in intermediate healing states, such as a clot or scab, leading to the notion that the effective management of death is essential to life.

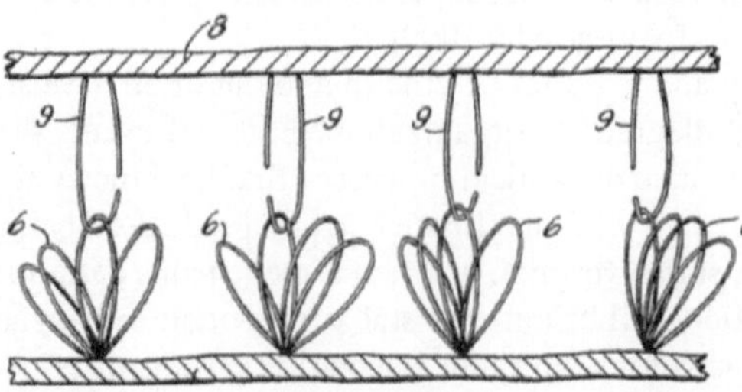

FIGURE 1.19 Original biomimetic concept of cocklebur-mimicking Velcro fastener. (From [de Mestrai 1961].)

Accommodating wide ranges of operating temperatures is a challenge in the biomimetic design of self-healing systems. Biological systems thrive and heal via a sophisticated interplay of chemical reactions that operate in narrow temperature ranges. Operating outside of the preferred temperature range leads to degraded performance, injury, and even death. Engineered systems typically operate over much wider temperature ranges than those tolerated by biological systems.

The following are some of the approaches to imbuing life and life-like properties into inanimate objects [van der Zwaag 2009]:

1. *Inanimate Living Properties:* Living polymers grow by the addition of molecular units, that is, monomers, to an end [Szwarc 1956]. The number of available monomers regulates the growth of the polymer so that growth stops when the "food" of available monomers runs out, and then resumes upon replenishing the supply.
2. *Biohybrids:* Merging of living biological tissue with engineered systems.
3. *Engineered Biological Living Systems:* These are advanced versions of breeding and domestication, along with the development of living biosystems from low-level constituents [Gibson 2010] [Kriegman 2020].

2

Molecular Scale Healing

2.1 Introduction

The *molecular scale* is the realm where molecular bonds and atomic interactions dominate the overall interactions. Molecular length scales generally run from subnanometer to multi-micrometer. Slightly larger is the *supramolecular scale*, which encompasses large molecules and/or conglomerations of smaller molecules. Many important self-healing interactions occur at molecular length scales – a concept not lost on notable twentieth-century visionaries [Feynman 1960] [Drexler 1981].

The following are forms of molecular scale healing:

1. *Individual Molecules:* Healing of damage.
2. *Bulk Materials:* Healing of bulk materials at the molecular scale, typically large molecules, many with a biochemical origin. This includes annealing of metals and fluid recovery.
3. *Molecular Movement:* This is for tight crack repair.
4. *Repair of Large Molecular Structures:* Examples include gels and supramolecular assemblies.

2.1.1 Repair of Molecules

Molecules lose utility if damaged. Sometimes the damage is a simple scission break of a polymer chain. Other times the damage is more complicated – as many atoms and subatomic particles are stable and not easily damaged. Perfect healing at the molecular scale is possible by merely rearranging atomic configurations. A convenient situation is when thermodynamic equilibrium minimizes available energy at a preferred healthy state. More complicated situations arise when the desired healthy state is not at thermodynamic equilibrium.

2.1.1.1 Polymer Chain Repair

Stress, radiation, and chemical attack break polymer chains and degrade macroscopic polymer properties. Repairing broken polymers often restores the integrity of the bulk polymeric material. Some examples follow.

Chemical Repair: Polycarbonate prepared by an ester exchange method (instead of the standard bisphenol-A and phosgene method) self-repairs following scission using Na_2CO_3 as a repairing agent. Reported healing efficiencies for material strength approach 100% [Takeda 2003]. Polyphenylene ether exhibits a somewhat more complicated behavior. Oxygen catalyzed by copper feeds a reaction that repairs broken polymer molecules while excreting water as a by-product [Imaizumi 2001]. Favorable chemical thermodynamics induces repairs in large molecules, such as using carbon monoxide to reduce epoxide, hydroxyl, and ketone pair defects in graphene oxide sheets [Narayanan 2013].

Fluorescein Repair: Fluorescein molecules have active optical properties that absorb and reemit photons. The cycling of optical and chemical energies damages fluorescein molecules, by leaving them in a high-energy triplet or radical ion state that no longer absorbs and reemits photons. Macroscale manifestations of damage are bleaching (irreversible) and blinking (reversible)

DOI: 10.1201/9781003396048-2

[Dave 2009]. A two-step reduction oxidation process heals the molecules: First is to transfer an electron to a reducing agent and then transfer one from an oxidizing agent while lowering the energy back to a fluorophore-capable state, that is, a healed molecule. Fluorescein healing requires the diffusion of reducers and oxidizers at concentration levels that can be inconvenient for biological studies [Tinnefeld 2012]. An effective and more biocompatible alternative places stabilizing molecules, such as cyclooctatetraene, 4-nitrobenzyl alcohol, or 6-hydroxy-2,5,7,8-tetramethylchroman-2-caboxylic acid (Trolox – a vitamin E derivative), close to the cyanine fluorophore Cy5 by covalent binding [Altman 2012] [van der Velde 2013]. The stabilizing agents then execute a similar healing energy reduction without the need for high concentrations and diffusion as required by the two-step process.

Catalytic Repair: Molecular catalysts lower energy barriers that nominally hinder thermodynamically favored chemical reactions. Catalysts help with healing when the desired molecular structures have a lower energy state than the damaged or imperfect state. Metal catalysts heal defects in growing carbon nanotubes [Yuan 2012].

Catalyst Repair: Sometimes catalytic molecules themselves need healing. Catalysts operate by exchanging energy between the reactants and themselves in a sequence of intermediate reactions that temporarily alter and then restore the state of the catalyst. The reactions may move the catalyst into altered undesirable configurations that are no longer catalytic. Healing that revitalizes the catalytic state is sometimes possible. Of note is cobalt–phosphate (Co-Pi), a water-oxidizing catalyst used for thin films on anodes in electrolytic splitting of water molecules. Separation of the cobalt and phosphate erodes the thin film. Adding phosphate ions replenishes the Co-Pi thin film catalyst [Lutterman 2009].

2.1.1.2 Large Biomolecule Repair

Repairing DNA comprises some of the more sophisticated and widely studied molecular repair techniques. Damage to DNA molecules is common, and if uncorrected can lead to disease, cancer, and death. DNA damage is also an issue when using DNA molecules for engineered nanotechnology applications. Living cells use multiple techniques to detect and repair damage to DNA molecules [Sancar 2004]. The following are some of the defects, detection, and repair techniques for DNA:

Stalled Replication: DNA replicates by unwinding the double-strand helix and building complementary strands onto the newly formed single strands. The interruption and stalling of this process prevents DNA replication.

DNA Base Damage: DNA consists of the four base molecules that appear in complementary pairs – adenine and thymine, and guanine and cytosine – in a double-stranded helix. Damage to these molecules adversely affects DNA. Healing proteins recognize the presence of a damaged base molecule, excise the base molecule, and then use the remaining complementary base molecule to guide a new undamaged base into position.

Single-strand Gaps: These defects occur with the removal of multiple base pairs on a strand. The repair recognizes the missing gap, synthesizes a patch using the remaining strand for complementary base pairs, and guides the patch into place.

Mispairing: DNA replication induces conformational bending deformations in the molecule that lead to mispairing. MutS proteins detect this bend and then initiate series of reactions that cut out and replace the damaged portion [Hura 2013].

Double-strand Breaks: Repairing complete scission of DNA molecules is a multistep process that begins with recognition of broken ends by MRX complexes, followed by trimming of the ends with a combination of MRX and Sae2, and then a passive resection by Sox1 and/or Sgs1 (Figure 2.1) [Gobbini 2016] [Mimitou 2008].

Cross-link Defects: These are excess binding between the strands in the DNA helix. The repair is complicated and requires excising both strands near to the cross-link and then executing a double-strand repair.

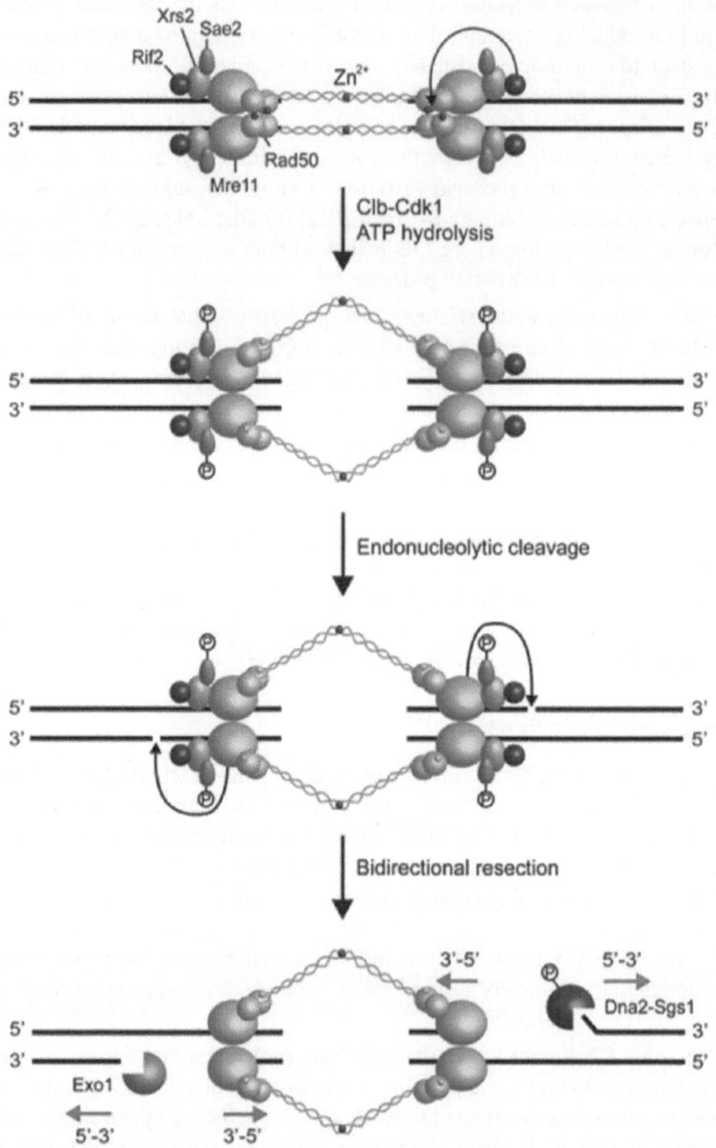

FIGURE 2.1 Repair of DNA double-strand break as a two-step process facilitated by DSB molecular complex. (From [Gobbini 2016].)

Photosynthesis converts light energy into chemical energy. The process transfers energy from individual photons into photosynthetic molecular complexes of lipids and proteins and then to secondary molecules. The intermediate molecules normally revert to the original state following energy conversion. Sometimes the photon energy alters the state, that is, damages, the photosynthetic molecules. Biological photosynthetic systems repair molecular damage with a process that moves the damaged molecules away from the light-harvesting region, breaks down the molecules into base constituents of damaged and undamaged components, and reassembles functional molecules for reuse. A molecular system that mimics this process uses an array of lipid bilayers that aggregate on the surface of a carbon nanotube (Figure 2.2) [Ham 2010]. Light-harvesting proteins assemble on the lipid bilayers to form photosynthetic complexes. This assembly is stable under most conditions and then disassembles in the presence of surfactants. Repair and reassembly of the photosynthetic complex occurs upon removal of the surfactant.

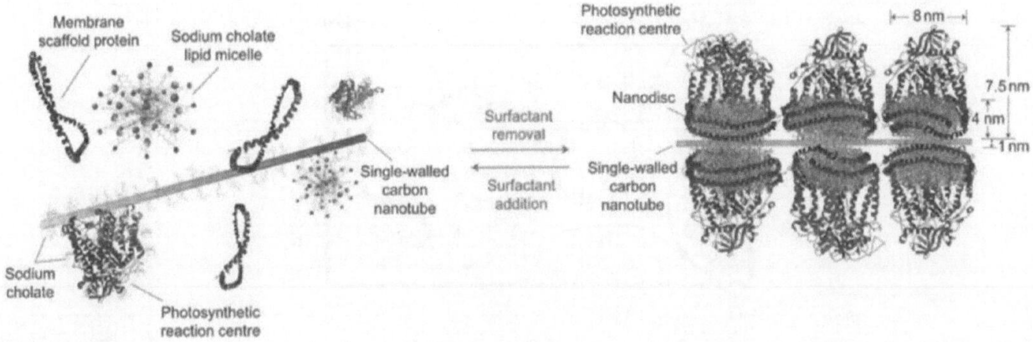

FIGURE 2.2 Lipid bilayer molecular system with repair process that mimics the repair of photosynthesis molecules. (From [Ham 2010].)

2.1.2 Supramolecular Assemblies

Supramolecular assemblies are large molecular structures held together by combinations of strong and weak chemical bonds. Special geometric arrangements give specific assemblies unique and facile functional properties [Brunsveld 2001]. The number of supramolecular assembly variations is virtually limitless. Many have self-healing capabilities [Hart 2013]. Some of the techniques are as follows:

Chemistry: Chemistry offers an increased ability to control molecular activity beyond the random action available with simple thermal methods. Chemical bonding drives many, if not most, molecular scale interactions. Chemical methods, while not immune to thermal fluctuations, direct the flow of energy to individual atoms and molecules to control specific bonds. Biological systems offer examples of exquisite control of molecular interactions, such as the use of DNA to encode the amino acid sequences into proteins that fold into 3D shapes. Coordinating protein–protein interactions at the molecular scale up through meso- and macroscale produces life. Figure 2.3 shows a synthetic self-healing system that combines soft and hard protein segments into folding molecular chains [Sariola 2015]. The molecules fold along the soft segments. The hard segments bind to one another to provide a combination of molecular healing and bulk strength.

Mechanical Ultrasound and Mastication Damage and Healing: The supply of external mechanical energy raises the free energy in a material, which then mobilizes molecular degrees of freedom for repair. Scission by ultrasound, that is, sonication, induces reversible damage as well as repairs polymer molecules by separating the molecules that subsequently reform, often in favorable configurations, such as the high-molecular-weight coordination polymers diphosphane in dilute solutions with palladium dichloride [Paulusse 2004]. Masticating rubber is a similar process.

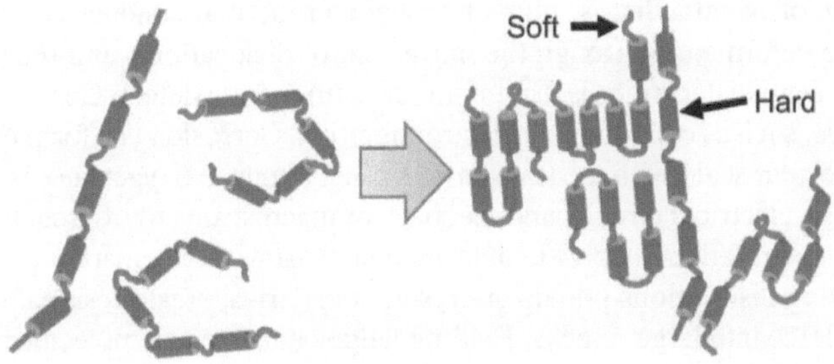

FIGURE 2.3 Synthetic self-healing material based on strong bond chains of soft and hard proteins that fold and link with weak bonds. (From [Sariola 2015].)

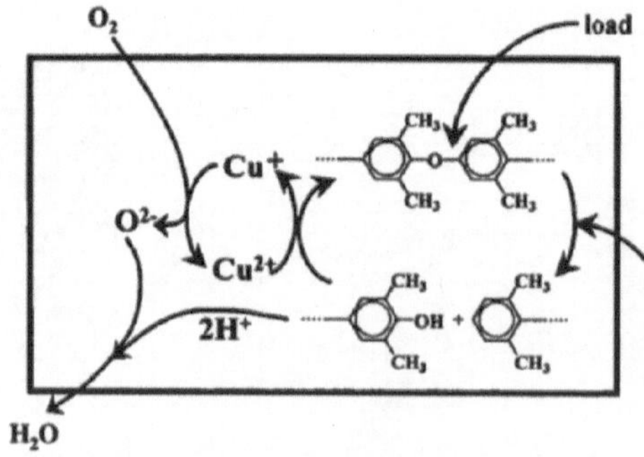

FIGURE 2.4 Self-repair reaction in polyphenylene ether using oxygen catalyzed by copper. (From [Takeda 2003] [Imaizumi 2001].)

2.2 Molecular Scale Healing of Bulk Materials, Surfaces, and Cracks

Molecular scale damage moves atoms and molecules from favorable into less favorable configurations. Materials of various classes tend to use distinctive mechanisms for repair.

2.2.1 Fluids

The molecular structure of fluids is that of mobility. Molecules slide past one another and produce macroscopic properties of unlimited shear deformation. Simple fluids are homogeneous at the molecular scale and normally have simple macroscale physical properties. Complex fluids achieve superior and functionalized performance with the heterogeneous inclusion of microscale, nanoscale, and molecular scale constituents.

Mechanical, thermal, and electrical stresses damage molecules. Damage to fluids breaks molecules into smaller pieces and/or cross-links them into unwanted larger complicated molecular structures. Macroscale manifestations of molecular scale damage include the loss of viscosity in lubricating oils and the breakdown of dielectric fluids in high-power electric devices. Thermally driven processes that reform the fluid molecules may also heal the molecules. The repair of heterogeneous inclusions in complex fluids requires specialized repair and/or replenishment techniques.

2.2.2 Metals

The molecular and atomic structure of metals differs from nonmetals and requires specialized self-healing techniques [Manuel 2008]. Metals are atomic aggregates held together by metallic bonds, often in crystalline arrangements. A key feature of metals is the free movement of electrons in the outermost atomic orbitals from atom to atom. Collectively, the free moving electrons give rise to many properties that are characteristic of metals, that is, high electrical and thermal conductivity, shiny metallic optical properties, plastic deformation through the movement of dislocations, and the tendency to oxidize. Deformation derived from dislocations is a common molecular scale defect. Coalescence of dislocations leads to larger damage, such as cracking and electromigration. Corrosion is a form of multiscale damage beginning at the molecular scale with the transfer of free electrons to oxygen atoms and organizing over multiple scales through electrical circuits and feedback of macroscale environmental conditions.

A typical case for metal fatigue is plastic deformation causing the formation and migration of crystal dislocations. Moving dislocations pile up at crystal boundaries, create residual stresses, and initiate microcracks that coalesce into larger cracks. Healing fatigue damage at a molecular scale stage prevents progression to larger scales and the formation of cracks.

Annealing reforms crystalline structures metals to more homogeneous states without melting. Annealing removes dislocation damage by evaporation at grain boundary edges, consolidates smaller crystal grains into larger ones, reduces fatigue damage, and relaxes residual stresses. Damage evaporates by dislocation motion in thin films [Bucher 1994]. The controlled application of heating above annealing and below melting temperatures, combined with judicious application of mechanical pressing, can repair large bends and buckles in structural steel [Hirohata 2008]. Since annealing usually requires the application of heat, many annealing healing actions are more of an extrinsic repair than true self-healing.

Sometimes dislocations evaporate at crystal boundaries without the application of external heat energy and effectively heal the crystal (Figure 2.5). Some ductile metals with large, exposed surfaces, such as gold plating or nanoscale rods of aluminum, evaporate dislocations at room temperatures [Wang 2015d]. Very thin sheets of aluminum or gold evaporate dislocations transversely through the exposed grain boundaries, leading to metallic structures that sustain unusually large plastic deformations without cracking [Rajagopalan 2007]. Shape and damage recovery mechanisms also appear in thin membranes made of polymer composites containing arrays of gold nanoparticles [Jiang 2004]. Upon overstretching membranes, with thicknesses on the order of 25–70 nm and spans approaching 1 mm, the damage evaporates, and the membranes return to their original shape and mechanical performance. Evaporative dislocation healing also occurs between grain boundaries and nanovoids in radiation-damaged copper [Bai 2010] [Chen 2015e]. Dislocation resorption occurs in silver nanowires with pentatwinned crystal structure, producing an anomalous strong Bauschinger effect [Bernal 2015].

Like dislocations, disclinations are linear lattice defects in crystals, but they are larger and in many respects superpositions of arrays of dislocations [Romanov 2003]. Numerical simulations indicate that disclination motions close small nanocracks by moving crystal boundaries and leaving dislocations on the boundary [Xu 2013a].

Intrinsic elastic stresses in epitaxially grown thin films due to atomic lattice mismatches cause nanocracks and delaminations. The growth of disclinations can heal excessive intrinsic elastic stresses that form in epitaxially grown thin films formed by atomic lattice dimension mismatches, thereby mitigating the potential for nanocracks and delaminations.

Stainless steels are iron alloys with high levels of chromium (greater than 10%) that resist corrosion. The three main classes are austenitic phase, a ferritic or martensitic phase, and a mix of the two. Austenitic stainless steels have a face-centered cubic (FCC) crystal structure and are not ferromagnetic. Ferritic and martensitic stainless steels have a body-centered cubic (BCC) structure and are ferromagnetic.

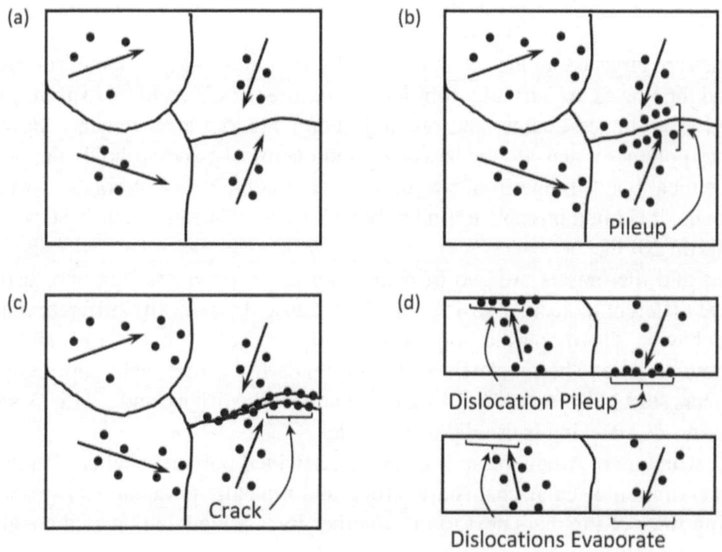

FIGURE 2.5 Dislocations (a) form under load, (b) pile up at crystal boundaries, and (c) initiate the formation of cracks, (d) elongated geometries allow for dislocations to evaporate at free surface.

The packing density of FCC crystals is higher than that of BCC. Some stainless steel alloys switch from austenitic to martensitic phases when mechanically deformed. This phase transformation gives the stainless steel enhanced deformation-tolerant and crack-arresting properties [Zackay 1967].

Ostwald ripening is the growth of crystals in polycrystalline materials by the absorption of smaller crystals into bigger crystals in a thermodynamically favored process. The growth tends to occur in specific temperature ranges and is similar to annealing. Heating a material above a certain temperature causes the crystals to grow bigger. If bigger crystals are a favored state, Ostwald ripening is a form of thermodynamically driven healing that occurs in bulk 3D materials or at 2D interfaces between different material phases, such as a solid–liquid interface [Samorí 2004].

Certain alloys self-heal by a mechanism similar to thermal annealing – instead of heat, the healing uses energy available from internal sources and from the deformation itself. Such an effect appears in underaged alloys of Al-Cu-Mg and Al-Cu-Mg-Ag [Somoza 2007] [Lumley 2002]. The degree of aging in the alloy is a critical factor. Fully aged Al-Cu-Mg alloys use a heat treat cycle to precipitate secondary-phase particles from the alloy matrix. The particles immobilize crystal dislocations and toughen the material against plastic deformation. Underage alloys have an incomplete precipitation of particles that provides a potential reserve capacity and enables diffusion and clustering of defects near to retained solute Cu atoms [Hautakangas 2008]. Stainless steels with Cu and Ce-Ti additions exhibit similar behavior. Creep and damage begins by the formation of nanoscale and microscale defects. The defects form points of mechanical weakness that coalesce into bigger points of damage. The defects also promote diffusion and intrusion of deleterious chemicals. In Cu-added alloys, Cu crystals precipitate out of the steel and fill in the defects. In Ce-Ti-added alloys, Ce and Ti precipitate into sulfides and block damage by S. Adding B-N accelerates the precipitation of both Cu and Ce-Ti and promotes localized healing for creep resistance [He 2010a] [Shinya 2006].

2.2.3 Diffusion Repair of Semiconductors

Diffusion repairs some semiconductor devices, such as GaAs nanowires with a zinc blende structure. Numerical simulations indicate that healing is size dependent. The effect causes larger diameter nanowires to be less capable of healing, with a reduction of 46% as the diameter increases from 2.3 nm to 9.2 nm [Wang 2012b].

2.2.4 Polymers

Polymer molecules are large assemblies of smaller molecules, known as monomers or mers. The variety of geometries and topologies are virtually limitless – mixtures of different mer species; folded, branching, looping, and cross-linked chains; and ordered and/or random arrangements. Polymer molecules conglomerate into polymer materials. Intertwining and chemical bonding holds the polymer material together. The chemical bonds range from strong with irreversible covalent bonds to weak with reversible hydrogen bonds. It is often possible to find polymers with self-healing capabilities within this large design space [Martin 2013b].

Thermoplastics and thermosets are two principal classes of polymers. Thermoplastic and thermoset materials have different molecular structures which cause dramatically different macroscopic thermomechanical behavior. Thermoplastic polymers solidify at cooler temperatures and soften and melt with increased temperature. Thermoplastic softening tends to be reversible with temperature cycles. Thermoset polymers start in an initial liquid state that congeals, stiffens, and solidifies with the application of heat. Thermoset stiffening is usually irreversible.

The molecular structure of thermoplastics is long intertwined polymer chains of mers. The chemical bonds holding mers into molecular chains are strong and typically covalent. Conversely, weak hydrogen bonds holding sides of the mers next to one another form folded and tangled conglomerations. At low temperatures, the hydrogen bonds are strong enough to limit molecular mobility and form a solid. At higher temperatures, the molecular mobility increases which softens the polymer. A further rise in temperature liquifies the polymer. Since the polymer molecules are not all the same size, the melting is gradual and does not correspond to a distinct melting point as is characteristic of molecularly homogeneous materials. Above a glass transition temperature, the molecules have sufficient energy and mobility to slither past one another in a snakelike motion known as reptation. Below the glass transition temperature, the molecules remain frozen in place. Sometimes the molecules arrange themselves into ordered crystalline or semicrystalline structures [Cannon 1976]. If the melt temperatures are not high enough to damage the polymer molecules, melting is reversible through multiple cycles. Both melting and reptation heal thermoplastic polymers at macroscopic scales.

In thermoset polymers, strong covalent bonds cross-link individual molecules together to form a network that prevents the easy movement of molecules. Increasing the number of cross-link binding reactions by thermal or chemical activation fixes the polymers in place and increases the solidification stiffness by fixing the polymers in place. Thermoset polymers do not have a glass transition temperature, generally do not melt upon heating, and resist self-repair by reptation and thermal cycling techniques.

Generally, thermoplastic polymers are amenable to self-healing but are relatively weak. Thermoset polymers have difficulty with self-healing but are relatively strong. Hybrid polymers combine both thermoplastic and thermoset properties with potentially superior mechanical performance and self-healing capabilities.

Ultraviolet (UV) light damages polymers by breaking molecular chains in a process known as photoscission. Methods of mitigating UV light damage include adding photo absorbers – often at the cost of mechanical strength and performance – and through additives that scavenge the free radicals arising in photoscission [Wiles 1980].

2.2.5 Oxidation Healing

Thermal barrier coatings (TBCs) are thin layers of thermally resilient materials that limit the flow of heat and high temperatures to underlying layers. Many TBCs find use in high-performance applications by being bound to sturdy and high-temperature-tolerant substrates, such as turbine blades. The endurance of the TBC depends in large part on strong adhesive bonds between the ceramic layer and the substrate. The diffusion of S and C atoms degrades the adhesion of TBC Al_2O_3 TBC layers grown on Ni/Al substrates [Bennett 2006]. Imbuing the TBC with reactive elements, such as Y, Zr, and Hf, remediates the diffusion of S and C atoms by a scavenging process that ultimately preserves the integrity of the TBC bonding.

2.2.6 Molecular Diffusion Healing

Everything appears to be in motion at the molecular and atomic scales. At thermal equilibrium, each dynamic degree of freedom has an average kinetic energy of $\frac{1}{2} k_B T$. Here $k_B = 1.380649 \times 10^{-23}$ J/K is the Boltzmann constant and T is the temperature (K). These dynamical degrees of freedom include modes of molecular vibration. The result is that atoms and molecules are always in motion. This motion sometimes promotes diffusion healing.

2.2.6.1 Diffusion and Reptation Repair of Thermoplastics

Some thermoplastic polymers autonomically heal tight cracks in a thermodynamically driven process, as noted in 1959 by Kolsky while studying stress waves in pentamethylflavan [Kolsky 1959]. A macroscopic explanation is that there are circumstances when a crack has a higher free energy than the solid and thermodynamics favors a solid material. A molecular scale description of the thermoplastic healing process lies in the mobility of polymer molecules at the crack interface. Strong covalent bonds individually hold long polymer molecules together in chains, and weak hydrogen bonds and intertwining collectively hold the polymer matrix together. This arrangement allows for thermal kinetic energy to cause the molecules to wiggle and move. The Edwards model describes the short-time elastic mechanics with molecules moving relative to a set of fixed obstacles [Edwards 1967]. This model led de Gennes to propose a statistical mechanics-based *reptation model* of thermoplastics with collections of individual molecules wiggling inside constraining tubes formed by a matrix of surrounding molecules [de Gennes 1967]. The rms displacement of the molecule along the length of the tube, $\langle s(t)^2\rangle^{1/2}$, follows the Einstein description of the Brownian random walk motion of small particles or isolated molecules.

$$\langle s(t)^2\rangle^{1/2} \sim t^{1/2} \tag{2.1}$$

Since the tube for an intertwined molecule follows a contorted random path, the same random walk statistics apply again. The result is that the rms motion of the end of the molecule relative to the solid mass, $\langle r(t)^2\rangle$, and consequently the strength of the bond, is proportional to the quarter power of time:

$$\langle r(t)^2\rangle^{1/2} = \left(\langle s(t)^2\rangle^{1/2}\right)^{1/2} \sim t^{1/4} \tag{2.2}$$

Experiments with thermoplastics, such as poly(ether ketone) (PEEK), confirm this prediction [Cho 1995].

When a thermoplastic polymeric material with reptating molecules has a closed crack, the molecules wiggle and bridge the gap and heal the damage (Figure 2.6) [Wool 1981] [Wool 1982] [Ageorges 2001]. The de Gennes model assumes that the strength of the healed crack depends on the length of the molecules that bridge across the gap [de Gennes 1983]. This approach provides quantitative predictions of polymer–polymer diffusion healing behavior in a variety of timescales and geometries [Kausch 1983]. Specifically, the amount of work done pulling a bridging molecule out of the reptation tube, and the fracture energy, G(t), is proportional to the length of the bridged tube section; that is, it is proportional to the square root of time.

$$G(t) \sim \langle s(t)^2\rangle^{1/2} \sim t^{1/2} \tag{2.3}$$

At some point, the number and length of bridging reptated molecules saturates and the fracture energy reaches a plateau.

The *recovery rate*, R(t), is a time-dependent dimensionless measure of healing, defined as the ratio of the present state of healing and long-term healed state. The recovery rate predicted by the reptation diffusion model for healing of tight cracks that instantly wet upon contact is a half-power function of time:

$$R(t) = \frac{G_{Ic}(t)}{G_{Ic\infty}} = C(T)t^{1/2} \tag{2.4}$$

Here, $G_{Ic}(t)$ and $G_{Ic}\infty$ are the critical strain energy release rates for a specimen healed over time t and infinite time, respectively. C(T) is a material- and temperature-dependent parameter proportional to the polymer self-diffusion coefficient [Loos 1994].

It is common for reptation diffusion to occur at temperatures above the glass transition temperature T_g for a polymer and cease when the temperature drops below T_g. Exceptions occur. For example, bonding tests in polystyrene indicate that surfaces behave as if the material is above T_g, even though the bulk material is

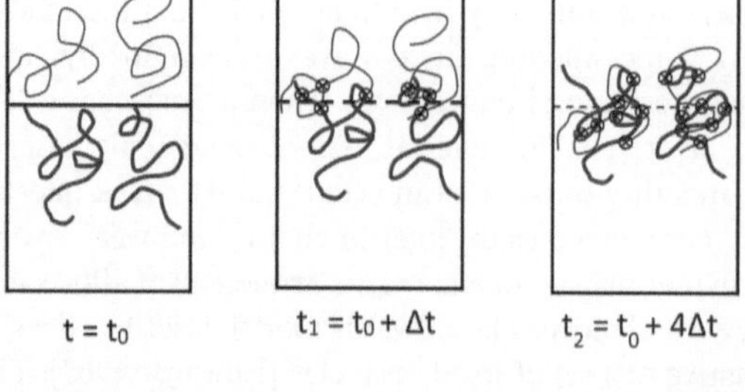

FIGURE 2.6 Reptation and interpenetration of molecules across crack surface. (Adapted from [Jud 1979].)

below T_g [Boiko 2004] [Guérin 2003]. Solvents, such as ethanol for poly(methyl methacrylate) (PMMA) and carbon tetrachloride for polycarbonate, and chemical reactions at the rejoined surfaces also promote healing with diffusion at temperatures below T_g [Avramova 1993] [Wang 1994] [Wu 1994] [Liu 2008a] [Lin 1990] [Hsieh 2001]. Interesting counterexamples are crystalline polymers, such as polyethylene or polypropylene, that do not reptate above T_g and only do so near the melting temperature [Shimada 2004].

The ability of a reptation-healed surface to resist re-cracking depends on the amount of diffusion and the molecular weight of the polymers [Sperling 1994]. Polymers with low molecular weight are not long enough to penetrate deeply across the crack surfaces and get a good grip. Short molecules tend to pull out from the opposing matrix. Longer molecules with higher molecular weights, typically 150,000 g/mol, penetrate deeper. When loaded, these longer molecules tend to break, rather than pull out.

The time dependence of crack healing is complicated. Experiments on PMMA determined that crack healing occurs quickly with near full-strength recovery after 5 min [Jud 1979]. Fresh crack surfaces re-healed quicker than stale surfaces. Subsequent experiments using deuterium isotopes of hydrogen confirmed many of the details of the polymer-diffusion reptation model [Whitlow 1991] [Wang 1993] [Russell 1993] [Kim 1994] [Welp 1999]. Kim and Wool extended the reptation model of diffusion healing by considering the molecular weight, M, of the polymer molecules, leading to a fractional power relation for the ratio of the fracture stress $\sigma(t)$ of the partially healed material to the long-term healed fracture stress $\sigma\infty$, [Kim 1983].

$$\frac{\sigma}{\sigma_\infty} \sim \left(\frac{t}{M^3}\right)^{1/4} \tag{2.5}$$

Competing mechanical effects produce different healing rates, with molecular weight distributions of the polymer affecting different stages of healing. The Griffith theory of cracking predicts that fracture corresponds to $\sigma^{1/2}$, which indicates that re-healing of σ based on a $t^{1/2}$ scaling gives a $t^{1/4}$ behavior [Prager 1981]. The minor chain model considers scission of polymer molecules at the fracture interface and further distinguishes different phases of interpenetration and timescales switching between $t^{1/4}$ and $t^{1/2}$ as the penetration depth increases [Zhang 1989] [Kim 1996].

2.2.6.2 Craze Healing

Crazes are line-like opening defects that form inside polymeric solids. A common cause is dilatational stress. Reptation heals crazes in a manner like thermoplastics with some differences. Crazes tend to form at nucleation sites that persist and promote new crazes along the same lines, even after healing [McGarel 1987]. Polycarbonate exhibits a similar healing process when microcracks form due to the thermal stresses of boiling water [Narkis 1982].

2.2.6.3 Dual Material Diffusion Healing

Reptation-type diffusion healing occurs between two different thermoplastic materials, even though the materials may have significantly different molecular weights, such as polystyrene and poly(vinyl methyl ether) [Jabbari 1993]. Bilayers of glassy deuterated and hydrogenated polystyrene with the same molecular weight and held together at sub-T_g intermingle in a manner that leaves a saturated 1–3 nm thick interfacial zone. This indicates that the healing mechanism does not strictly follow the statistical mechanics of the reptation model [Yuan 2010].

2.2.6.4 Functionalized Crack Surface Healing

Solids suffering from cuts or cracks self-repair by a molecular scale process. The surface becomes sticky and the material bonds when brought into tight contact. Surfaces may initially not be sticky but later become sticky through functionalization. One example functionalizes polyacrylate and polymethacrylate materials with mussel-inspired catechols. Low-pH water solution activates hydrogen bonds that heal cut surfaces [Ahn 2014].

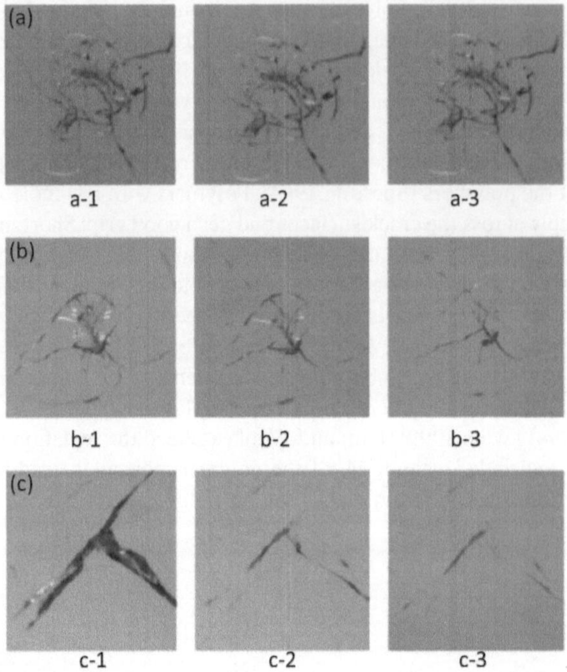

FIGURE 2.7 Visual inspection of thermal remendability of cured DGFA polymer. First treatment was at (a) 100°C, (b) 119°C, and (c) 125°C for 20 min, respectively. The next treatment was in an oven preset at 80°C for (1) 0 h, (2) 12 h, and (3) 72 h, respectively. (From [Tian 2009].)

Active chemical bonding at the interface is another approach to the healing of tight cracks. Mixing furan and epoxide groups creates a solid with two types of intermonomer linkages – thermally reversible DA bonds and thermally stable epoxide and anhydride bonds [Tian 2009]. This mixture combines the mechanical properties of epoxies with an annealing-type thermal remendability (Figure 2.7).

2.2.6.5 Glass

Solid glass has an amorphous atomic structure that strikes a balance between brittle ceramic and viscous liquid behavior. Tight cracks in glasses sometimes self-heal by diffusion and rearrangement. Observed cases include metaphosphate Nd-doped laser glass [Crichton 1999] and microcracks in barium titanate glass ceramic thin films with high dielectric constants undergoing annealing [Yao 2002].

2.2.6.6 Dynamers with Hybrid Weak and Strong Bond Healing

Dynamers are polymers capable of dynamic reorganization, adaptation, and self-healing at the molecular level via hybrid interactions of reversible covalent and noncovalent bonds (Table 2.1) [Roy 2015] [Wojtecki 2011]. Dynameric behavior appears in both polymers and gels. The goal is to achieve thermosetting performance, for example, stiffness and resilience, with thermoplastic annealing, melting, and strength recovery [Wouters 2009]. Combinations of ionic, dipole, covalent, and hydrogen bonds make

TABLE 2.1
Dynamer Polymer Systems that Operate at Room Temperature

Bond Type	Material	Thermal Nature	Citation
Ionic	Cross-linked poly(*n*-butyl acrylate) (pBuA) using a very low amount of a specific tin oxo-cluster	Room temperature	[Potier 2014]
Hydrogen ureidopyrimidinone (UPy)	Polyester–polyurethane networks	Below T_g	[Wietor 2009]
Hydrogen peptide or ionic	Silicone	Room temperature, solvents accelerate healing	[Harreld 2004]
Disulfide	Polyimide	Room temperature solvent step followed by cycle above 300°C	[Tesoro 1993]
Disulfide	Polyurethane	Room temperature with UV radiation	[Chang 1983]
Coumarin	Polyurethane	Room temperature with UV radiation	[Ling 2011]
Hemiaminal dynamic covalent networks	Elastic organogels from poly(ethylene glycol) (PEG) diamine monomers	Below 49°C melting point	[García 2014]
Metal–protein coordination covalent	Zinc–histidine	Room temperature with mediation by Zn^{2+} ions	[Degtyar 2015]
Noncovalent metal–ion ligand	Rubbery, amorphous poly(ethylene-*co*-butylene)	Room temperature with UV radiation	[Burnworth 2011]
Hydrogen and click cycloaddition	Four-arm star functionalized polyisobutylene with dual bonds	Increased activity with temperature in the range of 20–40°C	[Döhler 2015]
Tri-*n*-butylphosphine catalysis	Air-insensitive disulfide metathesis	Room temperature	[Lei 2014]

materials that sustain large deformations. During loading, the weak reversible bonds break and reform as needed. The strong bonds do not break and instead form a skeleton with memory that guides the recovery process during unloading (Figure 2.8). A challenge with this hybrid dynamer approach, especially with the thermoreversible variants, is that specific bonds in the polymer need to dissociate at specific temperatures, while the rest of the bonds remain intact [Wagener 1993].

2.2.6.7 Dynameric Elastomers

Rubbers are a class of elastomeric solids with cross-linked molecular network structures that enable large reversible elastic deformations, usually in the form of volume-preserving distortions. The stiffness of the rubber increases with the degree of cross-linking. Actively modifying the cross-links modifies the mechanical properties of the rubber accordingly.

Polyisobutylene (PIB), also known as butyl rubber, is a common synthetic polymer known for low T_g, ease of reformability, gas impermeability, and sticky behavior. PIB is the base ingredient of inner tubes in the tire and appears in applications ranging from self-healing vascular layers in electrical wire insulation to chewing gum. The molecular structure is that of a copolymer of isobutylene and a small amount of isoprene that facilitates cross-linking. It is possible to functionalize the ends of the PIB molecules through the addition of bifunctional barbituric acid [Herbst 2012]. This creates a dynamic

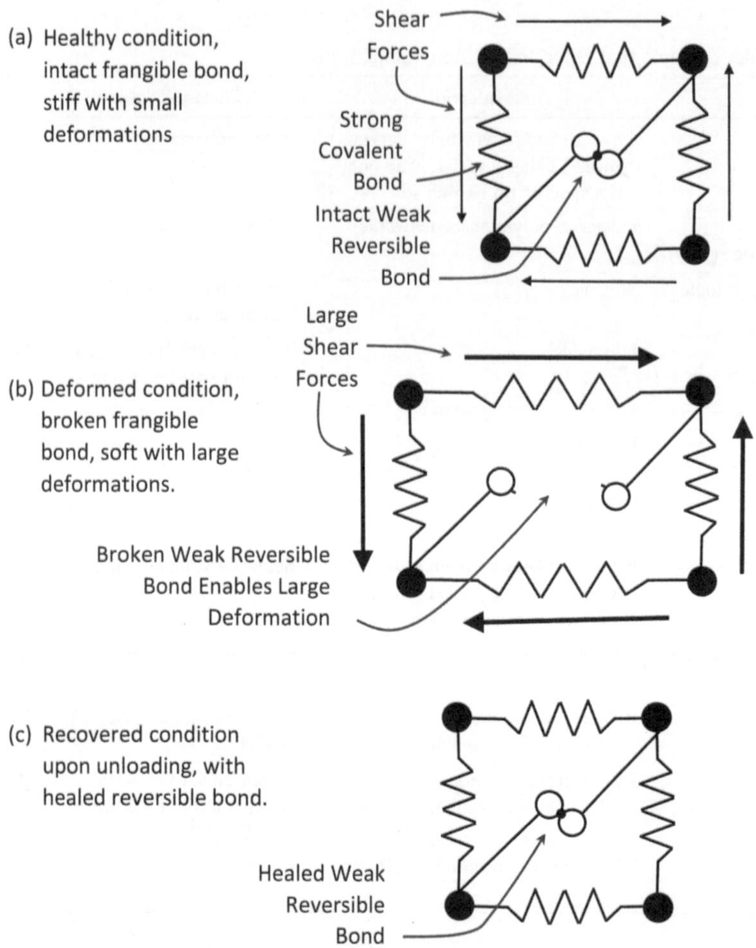

FIGURE 2.8 Notional representation of combination of weak reversible bonds and strong covalent bonds creates a material that heals from large loads and deformations. (a) Original condition, with intact reversible bond stiff over small deformation range. (b) Large loads break reversible bond and material becomes soft and easily undergoes large deformation. (c) Reversible bond heals with unloading; the material regains original shape and stiffness.

supramolecular polymer where functionalized PIB groups dynamically join supramolecular clusters of other PIB molecules in a living manner (Figure 2.9). Material made of this functionalized supramolecular PIB readily heals at the macroscopic level from cuts and tears. Similarly, adding dibromocyclopropane mechanophores to the polymer backbones functionalizes the polybutadiene with mechanophores that react to shear distortions, become activated, and promote more cross-linking of the polymer. The result is a significant increase in material modulus in response to applied loading – a potential damage-mitigating property [Ramirez 2013].

An alternative approach to severe loading resilience occurs in natural rubber near to a crack tip which forms a crystalizing soft–hard double molecular network, resulting in an increase in the fracture toughness by three orders of magnitude [Zhou 2014b]. This contrasts with synthetic rubbers, which do not crystalize as readily, presumably due to a difference in the balance of chirality, that is, left- and right-handed isomers. Natural rubbers have molecules with the same chirality, which readily crystalize. Synthetic rubbers have equal distributions of chirality and do not crystalize as easily.

The relative weakness and sensitivity to external stimuli of the metal–ligand bonds offers another set of possibilities for mixed-band strength supramolecular polymers with potential for self-healing, such as

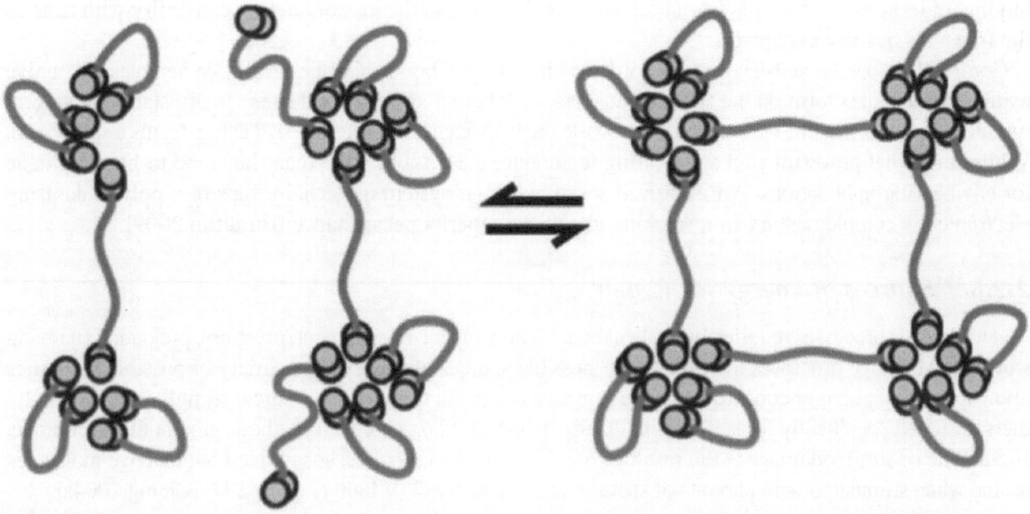

FIGURE 2.9 Formation of a dynamic supramolecular network with functionalized PIB. (From [Herbst 2012].)

the use of UV light to stimulate and disengage metal–ligand bonds in damaged material to produce an annealing-type healing [Kersey 2007].

2.2.6.8 Covalent Cutting Repair with Danglers

Functionalized dangling molecules repair tight cuts and cracks. In the basic technique, the cut or crack surface becomes covered with active dangling molecules that repair the cut or crack by tightly contacting and bonding to another cut or crack surface (Figure 2.10). Dangling repair generally requires molecules with easily reversible bonds.

Models of the statistical mechanics of dynamers account for a combination of diffusion motion of polymer molecules and the binding kinetics of labile bonds formed due to damage. The model assumes that dynamers consist of molecules made with reversible stickers at one end and tethered by irreversible covalent bonds at the other end [Stukalin 2013]. These models account for the healing of cuts by the

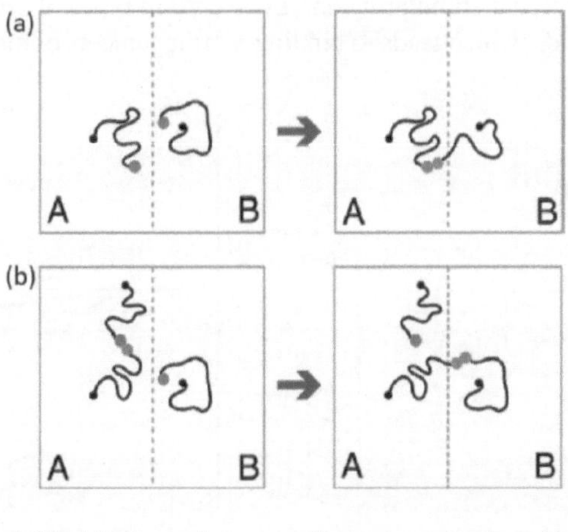

FIGURE 2.10 (a) Movement and (b) rebinding in dynamers of polymer molecules with reversible bonds at one end that are free to move and bind, and tethered by irreversible bonds at the other end. (Reprinted with permission from [Stukalin 2013]. Copyright 2013, American Chemical Society.)

binding of molecular free ends from both sides of the cut, and the loss of healing capability with time as the free ends become occupied.

Cordier developed a self-healing elastomer with hydrogen bonds [Cordier 2008]. When cleaved, active hydrogen bond sites form on the freshly cut surfaces. These hydrogen bond sites remain stable for several minutes. Recontacting the cleaved surfaces while the hydrogen bond sites are still active forms a solid bond. While somewhat powerful as a self-healing technique, these rubber-like materials tend to be too pliable for engineering applications. An improved supramolecular system uses chain folds of a polyimide using electronically complementary π–π stacking to achieve superior performance [Burattini 2009].

2.2.6.9 Sources of Energy for Repair

Material repair processes require energy to execute. The ambient environment, prestress, preloaded chemical stores, and damage processes themselves are possible sources of energy. The variety of possible properties allow for tuning polymer constituents to be sensitive to certain types of mechanical stimuli, including polymers designed specifically for self-healing [Caruso 2009a]. Photochemical methods are another technique. If cracking or similar damage is the result of breaking molecular bonds, specialized photoactive molecules rebind when stimulated with photons of specific energy, such as UV light (Figure 2.11) [Chung 2004a].

2.2.6.10 Reversible Stimuli-Responsive Materials

Stimuli-response behaviors can be reversible. The stimulation can be intrinsic or extrinsic. Intrinsic reversibility uses mechanical loads and damage to provide both the stimulus and energy for internal reconfiguration and healing. Extrinsic behaviors use external stimuli and energy for reconfiguration and healing, with externally applied thermal cycling being the most common. Several molecular mechanisms exhibit this behavior [Montero 2015]:

1. *Hydrogen Bonds:* Hydrogen bonds are relatively weak but can cross-link molecules with sufficient strength to alter mechanical properties yet dissociate and rebind upon thermal cycling. Polybutadienes modified with urazole units that cross-link by hydrogen bonds create thermally reversible thermoplastic elastomers [de Lucca 1987]. Partially cured thermosets contain molecules that attach to the molecular network using a subset of the available bonds. This leaves some of the molecules dangling with loose ends available for intertwining with other dangling molecules, including those on the other side of a crack or cut. This method can recover nearly all of the strength of cut sections of partially cured polyurethanes [Yamaguchi 2007].
2. *Covalent Bonds:* Covalently cross-linked polymers are particularly difficult materials for self-healing reactions because the molecules are tightly bound and lack mobility. Applying the appropriate level of external stimulating energy to certain types of cross-linked polymers can break the covalent bonds, which leads to binding rearrangements of molecules across damaged

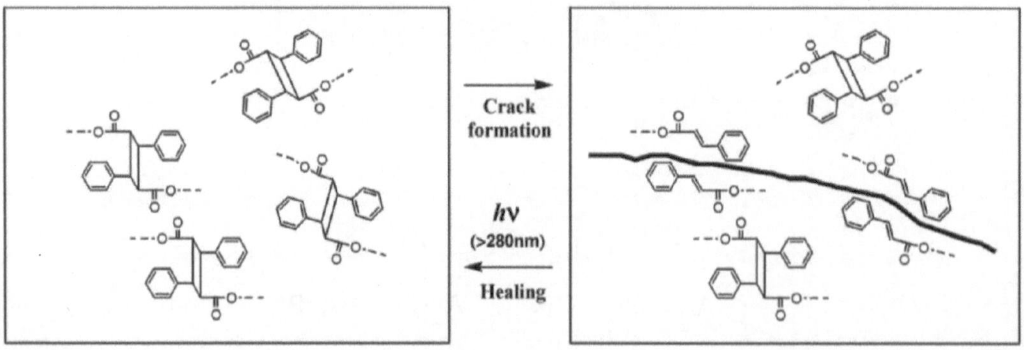

FIGURE 2.11 Healing of crack damage by photochemical [2 + 2] cycloaddition of cinnamoyl groups. (Reprinted with permission from [Chung 2004]. Copyright 2004, American Chemical Society.)

scissions of the material. Networked polymers that are not tightly cross-linked can be suitable because scising of polymer chains running between the cross-links creates dangling chains with sufficient mobility to intertwine with neighboring molecules for healing. This conceptual system also requires that polymers form suitable sites for binding at the dangling ends of the chains. Such a material system reforms based on reshuffling reactions of trithiocarbonate (TTC) units subjected to UV radiation (Figure 2.12) [Amamoto 2011].

3. *Diels–Alder* (DA) *Reactions:* These have a versatility that enables developing room temperature dynamer materials. Thin films when cut and then overlaid heal to form a sufficiently solid bond to enable stretch testing. It is noted that the healing occurs not at freshly cut surfaces, but at overlaid surfaces. The explanation is that the bulk solid of this dynamer is in an active energetic state that permits healing [Reutenauer 2009].
4. *Covalent Bonds that Reversibly Dissociate and Re-bond at Room Temperatures:* Among these are the aromatic disulfide bonds. Fabricating rubber and polyurethane elastomers with dynamic disulfide cross-linking bonds that when cut or otherwise damaged, heal closed wounds through the dynamic covalent re-bonding and the conversion of stored elastic energy [Canadell 2011] [Rekondo 2014] [Yoon 2012]. Another set of reversible covalent bonds are those based on boronate ester formation to heal hydrolytic damage to polymer chains [Niu 2005].
5. *Light:* This can stimulate and activate reversible molecular reconfiguration in polymers with suitable combinations of photosensitivity and internal molecular degrees of freedom [Fiore 2013].

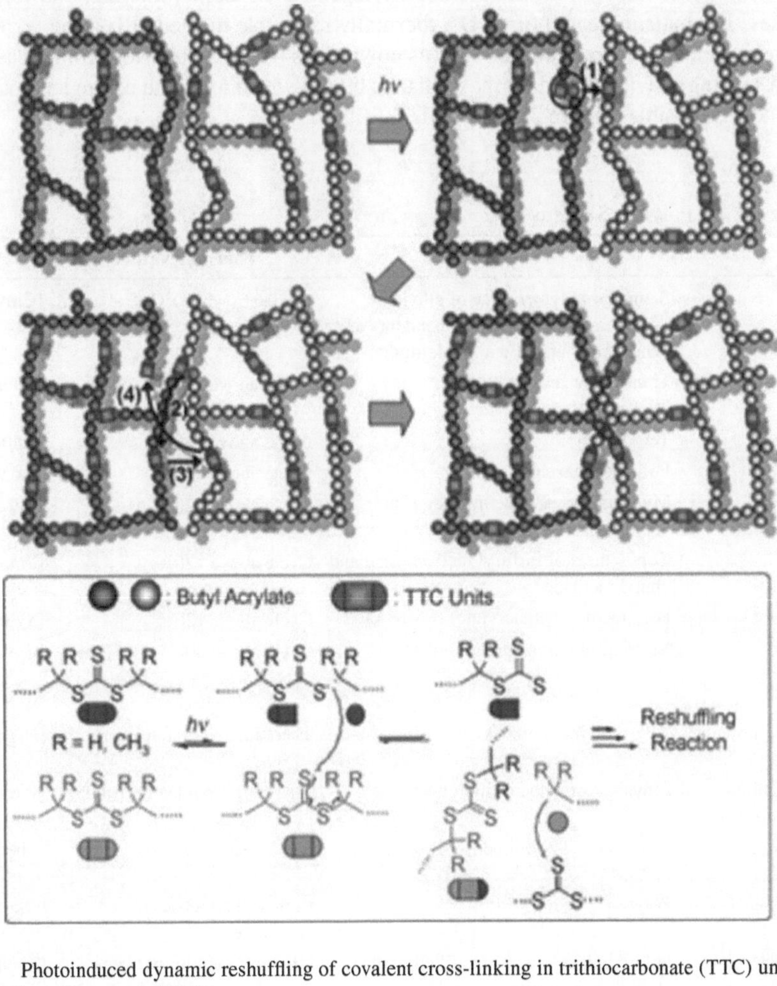

FIGURE 2.12 Photoinduced dynamic reshuffling of covalent cross-linking in trithiocarbonate (TTC) units subjected to UV radiation. (From [Amamoto 2011].)

2.2.6.11 Thermally Reversible Materials

Chemical thermal cycle reversibility is the basis for a variety of molecular scale material self-healing actions. A spectrum of reversible strong covalent and weaker hydrogen and supramolecular bonds all contribute to thermal cycle self-healing. Thermal cycle healing requires heating and cooling, which can come from extrinsic or intrinsic sources. Included in this chapter is a discussion of molecular scale thermal healing. Table 2.2 lists selected thermally reversible polymer systems.

Covalent bonds that reverse with thermal cycling often have tunable functional properties with potential self-healing capabilities. A typical approach uses thermally reversible DA reactions with polymerizing combinations of furan and maleimide monomers [Wudl 2005] [Wang 2007b]. The reversibility and versatility of DA reactions are largely due to the energy levels between the various states being small and easily reversed with a change in temperature [Kuramoto 1994] [Bergman 2007] [Murphy 2008]. The adaptation of tools for creating polymer molecules with controlled molecular weight, such as atom transfer radical addition applied to the DA reaction, create thermally amendable polymers with well-controlled click chemistry properties [Kavitha 2009].

An interesting counterpoint to the thermoset–thermoplastic polymer paradigm is the development of polymers that exhibit both thermoset cross-linking and thermoplastic melting properties with remendable cross-linking bonds [Zhang 2009b]. The polymer uses furan-functionalized, alternating thermosetting polyketones (PK-furan) and bis-maleimide by using the DA and retro-DA reaction sequence (Figure 2.13). The result is a reversible sequence of thermosetting at 50°C and melting at 150°C. The mechanical properties of the cross-linked solid-phase material are unaffected by repeated shredding, melting, and reforming recycling processes. The maleimide and furan DA thermally reversible molecular healing technique can be extended to robust molecular frameworks, such as polyimides [Liu 2007b]. More complicated thermally reversible DA healing activity, including the exchange binding of adducts and entanglement intertwining of polymer ends, is possible (Figure 2.14) [Yoshie 2010].

TABLE 2.2
Thermally Reversible Polymer Systems

Bond Type	Material	Thermal Nature	Reference
Diels–Alder	Saturated polymers free of ethylenic unsaturation with furanic pendants or ends cross-linked with maleimide	Cycle above 120°C	[Craven 1969]
Diels–Alder	Transparent furanic polymer cross-linked with maleimide	Cycle above 120°C	[Chen 2002b]
Diels–Alder	Terpolymer	Cycle above 120°C	[Kötteritzsch 2015]
Diels–Alder	Polyphosphazenes	Cycle above 150°C	[Salamone 1988]
Diels–Alder	Multifunctional maleimide and furan compounds	Cycle above 170°C	[Liu 2006b]
Diels–Alder	Copolymers of furfuryl methacrylate and butyl methacrylate	Cycle above 120°C	[Pramanik 2015]
Diels–Alder	Poly(methyl methacrylate) (PMMA)	Cycle above 120°C	[Syrett 2010]
Supramolecular hydrogen-bonding networks	Vulcanized polyisoprene rubber	Cycle above 80°C and higher depending on mix ratios	[Chino 2001]
De-cross-linking polymer system	Alkoxyamine moieties	Thermally reversible	[Higaki 2006]
Diamines and amino-alcohol cross-links	Ethylene/propylene copolymers	Thermoreversible cross-links	[van der Mee 2008]
Anionic equilibration	Crosslinked poly(dimethylsiloxane) (PDMS)	Anionic equilibration	[Zheng 2012a]
Disulfide bonds	Polyrotaxane network	Reversible thiol-catalyzed cleavage	[Oku 2004]
Cross-linking molecules and scising cross-links	Poly(ethylene glycol) modified with 9-anthracenecarboxylic acid	UV light cycling between wavelengths of 365 nm and 254 nm	[Zheng 2002]

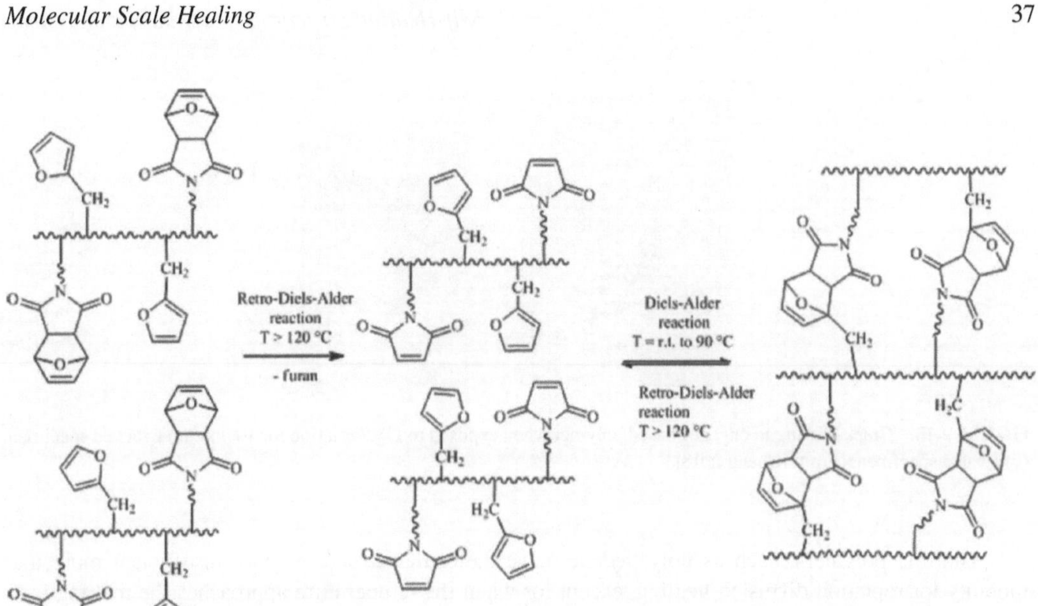

FIGURE 2.13 Retro-Diels–Alder reaction for thermally reversible dynamer. (From [Kötteritzsch 2015].)

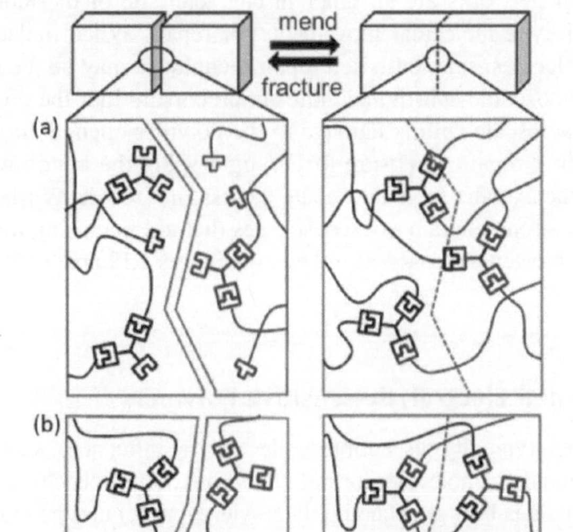

FIGURE 2.14 Three modes of reversible Diels–Alder healing. (a) Sacrificing dissociation of adducts. (b) Exchanging furan (maleimide) groups between adducts. (c) Chain entanglement. (From [Yoshie 2010].)

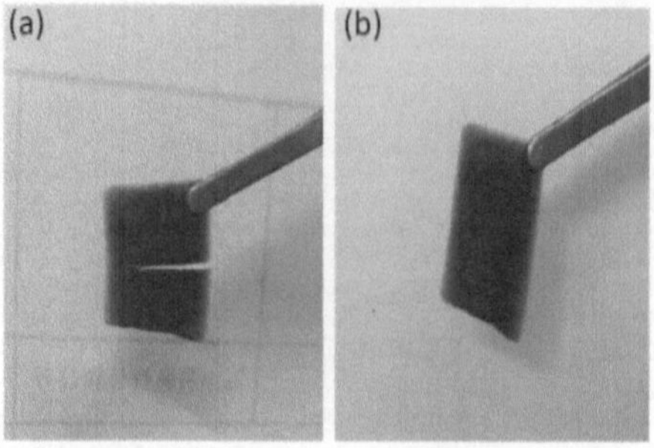

FIGURE 2.15 Crack healing in eugenol-based polymer when exposed to UV radiation for 10 min. (a) Cracked specimen. (b) Healed specimen. (From [Cheng 2018].)

Crystalline polymers, such as polyethylene, have molecular structures with insufficient molecular mobility for reptation diffusion healing, except for when the temperature approaches the melting temperature. Mechanical damage, such as puncturing, moves the molecules into crystalline arrangements, such as from orthorhombic to monoclinic forms, that are in principle reversible, but remain in distorted shapes. Grafting mobile molecules (poly(hexyl methacrylate)) onto the ends of crystalline molecules (polyethylene) produces a molecular system that maintains many of the structural properties of the crystalline polymer, but has an enhanced ability to flow at temperatures well below melting temperatures [Shimada 2004]. Grafted polyethylene achieves closures that leave only 30% of the hole open after 1 hour versus neat polyethylene opening 80% of the area of the original pinhole.

Cross-linked molecular networks are an inherent characteristic of thermoset polymers. The cross-linking hinders reptation-type molecular movements for repair. When molecular damage consists of scission of cross-link molecules, scission-based repair techniques may be the solution. A technique for polyurethanes intermixes oxetane-substituted chitosan precursors into the cross-linked molecular network [Ghosh 2009a]. Macroscale cutting damage to the polymer opens four-member oxetane rings as possible sites for cross-link repair. Exposure to UV light splits the chitosan molecules to open sites that bind to those from the oxetane rings to execute a cross-link repair. While the repair mechanism is molecular in nature, it is strong enough to exert stresses that aid in closing macro- and mesoscale cuts and scratches for further molecular cross-linking repair(Figures 2.12 and 2.15). Table 2.3 lists radiation-reversible polymer systems.

2.2.7 Water-soluble and Electrolytic-sensitive Polymers

Water-soluble polymers are typically susceptible to electrolytic influences, such as pH or the presence of ions in the solution. Antagonistic molecular formulations with oppositely charged polyelectrolytes bound together in complexes, such as bulk branched poly(ethylene imine) and poly(acrylic acid) (BPEI/PAA) complexes, when activated with water, will self-heal cracks and tears at the macroscopic scale. Salt ions and variations in pH have a strong influence on healing efficiency [Zhang 2016a].

TABLE 2.3

Radiation-reversible Polymer Systems

Bond Type	Material	Thermal Nature	Reference
Disulfide	Polyurethane	Room temperature with UV radiation	[Chang 1983]
Coumarin	Polyurethane	Room temperature with UV radiation	[Ling 2011]
Noncovalent metal–ion ligand	Rubbery, amorphous poly(ethylene-*co*-butylene)	Room temperature with UV radiation	[Burnworth 2011]

2.2.8 Molecular Conglomeration Repair and Living Polymers

Living polymers grow by the controlled addition of mers from a surrounding fluid. As such, living polymers support self-healing applications, including the repair of polymer molecules subjected to radiation-induced scission damage that creates open sites for living polymer growth [Chipara 2005].

2.3 Solid Shape and Strength Recovery

Shape memory materials (SMMs) undergo large inelastic deformations and return to an original shape, often with the help of thermal cycling. Large loads produce damage in the form of permanent deformation. This may be scratches, dents, gouges, or more gross shape distortions. Shape recovery is both a macroscopic and molecular scale effect. Large deformations corresponding to molecular rearrangements heal the unwanted deformation. Stainless steels, proteins, and polymers have molecular rearrangement capabilities that enable structural elements to sustain large energy-absorbing deformations before suffering permanent damage and then to recover shape and functional capabilities.

SMMs differ from elastomers and other large deformation-tolerant materials by specific nonlinear thermomechanical behaviors. Some materials use the stored energy in the deformed molecular structure to drive a return recovery, typically over a timescale that is much longer than the original loading cycle. Other materials require extrinsic stimuli, such as thermal cycling. Metallic and polymer SMMs are readily available, as well as some ceramic and gel variants [Lendlein 2002.] At the molecular scale, SMMs effectively retain a template of the original atomic arrangements.

Metallic SMMs typically use switchable crystal structures at the molecular scale to change shape. The shape and volume difference between FCC and BCC crystals and various twinning arrangements is part of the thermomechanically driven switching between the phases. The behavior is hysteretic, nonlinear, thermodynamically irreversible, yet often mechanically reversible. Following large deformation, the retained molecular template guides the recovery back to the original shape, once the thermomechanical configurations favor it.

The rearrangement of atomic crystalline structures in the presence of deformation and temperature changes can explain much of the behavior of metallic SMMs, that is, SMAs. In nickel–titanium, that is, nitinol, alloys, shape memory appears as a change between austenitic and martensitic phases. Austenitic crystals are FCC. Martensitic crystals are BCC. The FCC configurations occupy more volume than the BCC. BCCs also allow twinning configurations and associated shear-type change in shape. The FCC and BCC phases can coexist with a strain energy difference. A 3D stress–stress–temperature graph illustrates this behavior. The instantaneous state of the SMA is that of a point in the 3D stress, strain, and temperature space. The state moves along various paths dictated by the thermomechanical loading [DesRoches 2004]. Figure 2.16 shows a scalar 1D

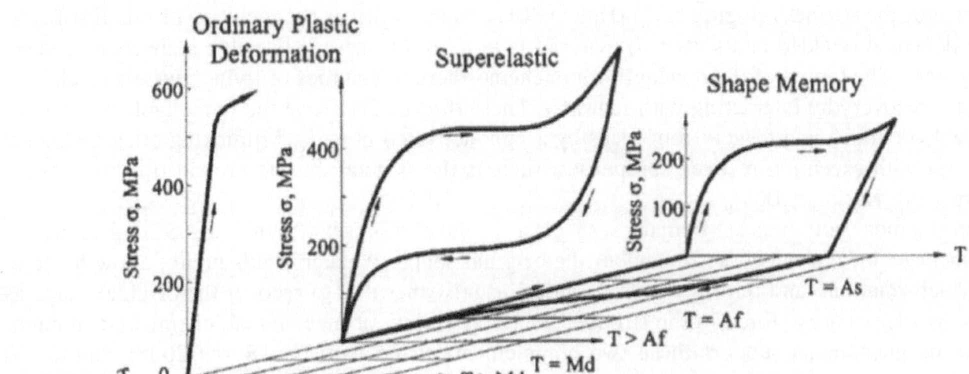

FIGURE 2.16 Stress, strain, and temperature behavior of SMA, showing nonlinear history-dependent behavior. (From [DesRoches 2004].)

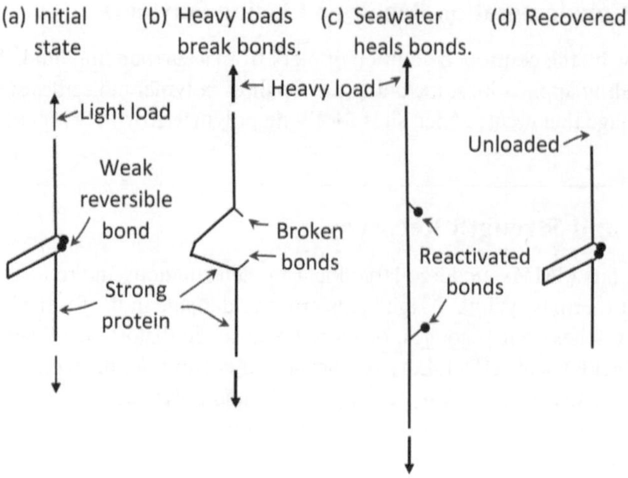

FIGURE 2.17 Unfolding molecular mechanism for polymer shape recovery, as employed by byssal thread proteins. (a) Initial state carries light loads and holds mussel onto rocks in light surf with light loads. (b) Heavy loads from heavy surf break weak reversible bonds allowing thread to stretch and absorb energy without breaking. (c) Seawater chemistry heals broken reversible bonds. (d) Upon unloading in light surf, thread recovers initial short length.

model with behavior that readily adapts into 3D constitutive models for numerical simulations [Popov 2007]. The paths designated superelastic and shape memory have practical implications in self-healing. Superelastic behavior at constant temperature corresponds to large deformation with shape recovery. Shape memory is a large deformation-driven thermodynamic engine.

Biology provides inspiration for polymer mechanisms of shape recovery. Many biological structures thrive under repeated small loads, yet survive and quickly recover from occasional large loads. These structures benefit from materials with damage-mitigating large deformation recovery. Examples include the muscle protein titin, polymers that hold bone and abalone nacre together, and the byssal thread used by mussels and other mollusks to attach to rocks [Thompson 2001] [Vaccaro 2001]. Rough surf introduces occasional large loads. Stretching and not breaking with reversible recovery is a viable strategy. Internal protein molecules stretch and unfold by breaking weak molecular bonds, leaving the broken bonds in a reversible heightened energy state. This is a sophisticated version of the process shown in Figure 2.8. The recovery process is complicated with supramolecular interactions of metals, disulfides, and noncovalent interactions aided by seawater chemistry to restore molecules to a state that restores the stiffness (Figure 2.17) [Holten 2011]. Entropy drives the recovery of initial stiffness where damaged (yielded and softened) material is in a higher energy state that recovers to a lower energy state. The tuning of the nonequilibrium chemo-thermodynamics of living mussels mechanics promotes recovery by interacting with seawater. These threads also have the remarkable property of being able to attach and detach from rocks by a light-activated process. Mimicking other biological molecules with exception recovery properties, such as the skeletal muscle protein titin, is possible (Figure 2.18) [Kushner 2009].

Shape memory polymers are in many respects a specialized subset of dynamers. Strong, relatively permanent molecular networks remember the original shape. Weaker bonds break, allow for temporary deformation, and reform with a stimulus, usually thermal, to recover the original shape as guided by the memory stored in the stronger bonds. A variety of mechanical, chemical, and nanocrystalline phenomena support these two complementary mechanisms [Ratna 2008]. Numerical simulations indicate that dynameric cross-linked nanogels recover up to 90% of the original shape from small scratches, such as from an atomic force microscope (AFM) tip, with only 10% of the bonds being labile [Duki 2011].

Multiple thermally triggered molecular bindings combine into a single material to exhibit multiple temperature-dependent shape memory effects. Combinations of chemical cross-linking and crystalline

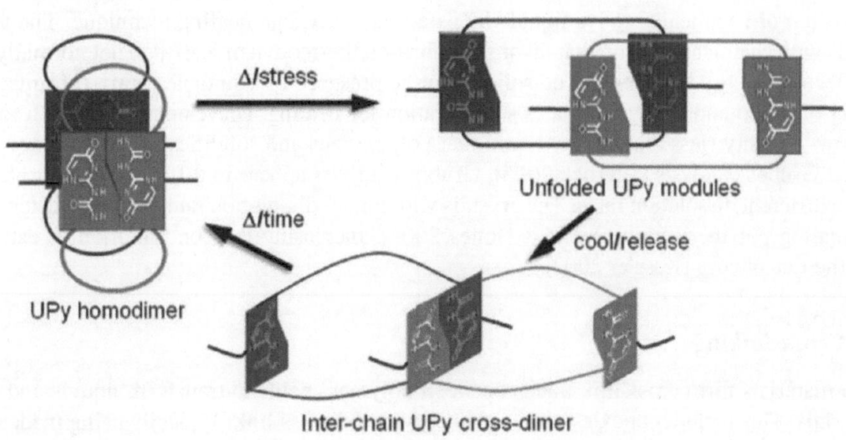

FIGURE 2.18 Proposed unfolding mechanism of titin-mimicking shape memory polymer with an interwoven arrangement of hydrogen 2-ureido-4[1*H*]-pyrimidone (UPy) bonds that are weaker than surrounding covalent bonds and can reversibly detach and reattach. (From [Kushner 2009].)

chemical folding activate at different temperatures in a three-step shape memory polymer [Behl 2010]. Some epoxies have a higher temperature SMP behavior in what appears to be a mixture of covalent bonds, which hold a molecular network together for shape recovery, and noncovalent bonds, which govern intermediate configurations [Rousseau 2010]. SMPs with crystalline and semicrystalline folding–unfolding molecular arrangements have sharper temperature threshold behaviors than noncrystalline variants [D'Hollander 2009].

2.4 Liquid-to-solid Healing Transitions

2.4.1 Chemistry-driven Liquid Solidification

2.4.1.1 Catalyzed Solidification

Liquid solidification is a thermodynamically driven phase change process. The liquid has a higher free energy than the solid. An energy barrier stabilizes the liquid state (Figure 2.19). Solidification requires an activation energy to cross the barrier, penetration of the barrier, or removal of the barrier. Ideally, activation should only occur when and where needed, that is, at a damage site following damage formation.

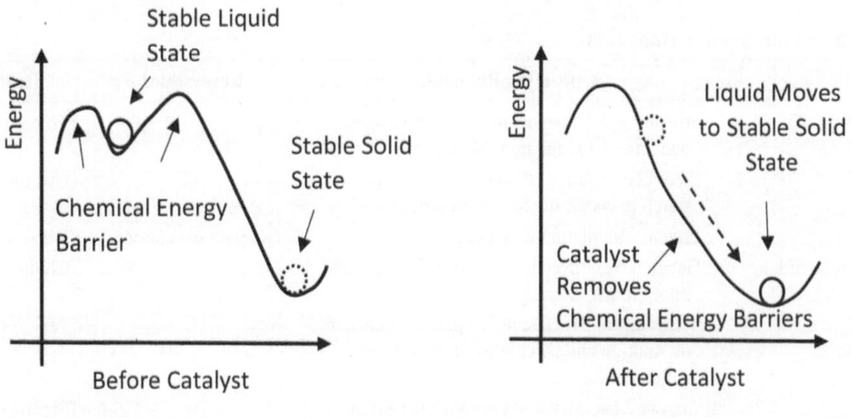

FIGURE 2.19 Notional thermodynamics of catalyst promoting liquid solidification by removing energy barrier.

Solidification of chemically active liquids by catalysis is a viable healing technique. The technique uses liquids with chemical makeup that favor solidification thermodynamically, do not normally solidify due to molecular energy barriers, but do solidify in the presence of a suitable catalyst. Damage-driven comingling of liquid and catalyst initiates solidification for healing. Dicyclopentadiene is a room temperature low-viscosity (less than water) liquid that polymerizes and solidifies in the presence of ruthenium-based Grubbs catalysts [Schrock 2006]. Grubbs catalysts appear in different polymorph crystals, which have different dissolution rates. The crystals with higher dissolution rates produce faster and more efficient healing polymerization reactions [Jones 2006]. Increasing the concentration of catalyst also increases the rate of cure [Kessler 2002].

2.4.1.2 Cross-linking

Polymeric materials form cross-link bonds between polymer chains to transform liquids and gels into solid materials. This includes epoxies and urethanes with the cross-links typically being made of strong and irreversible covalent bonds.

2.4.1.3 Multi-parameter Triggering

Effective liquid-to-solid repair requires that the formative ingredients remain stable in an unreacted state until the need for solidification arises. Simple triggers that activate upon the change of a single-state parameter lack specificity and long-term stability. Combined multi-parameter triggering may be better. Stimulus-initiated liquid solidification appears in many self-healing materials. Most liquid solidification methods rely on molecular domain interactions, with many using the growth and cross-linking of polymer chains to form gels then solids (Table 2.4). Some of these reactions are reversible. Most are irreversible.

2.4.2 Mechanical Stimulation of Liquid-to-solid Transitions

2.4.2.1 Molecular Stretching and Cross-linking

2.4.2.1.1 Thixotropic Shear Thinning Recovery

Some materials switch from a stiff gel or solid state to virtually fluid-like when deformed and then switch back to recover stiffness. Such thixotropic or shear thinning material behavior typically has a molecular or supramolecular mechanism. Control of lipophilic and hydrophilic interactions in alginate hydrogels with supramolecular inclusions tunes the rheology of thixotropic gels to create materials that quickly recover stiffness following shear thinning [Miao 2015].

TABLE 2.4

Methods of Liquid-to-solid Transition

Stimulus	Operating Principle	Reversible?	Reference
Shear deformation	Stretching of polymers to induce hydrogen bonding and jamming with platelets	No	[True 2012]
Electricity	Electrocoagulation of material, such as corn starch in water, by flow of electricity	No	[Huston 2011]
Chemical	Protein coagulation with acids	No, with exceptions	[Watson 1946]
Chemical cross-linking agent	Gelation by conversion between disulfide and thiol groups	Yes	[Chujo 1993]
Thermal	Temperature-dependent competition between dissolution and percolation-type sol–gel solidification	Yes	[Petit 2007]
Ultrasound	Activate a latent catalyst to promote chemical solidification	No	[Piermattei 2009]

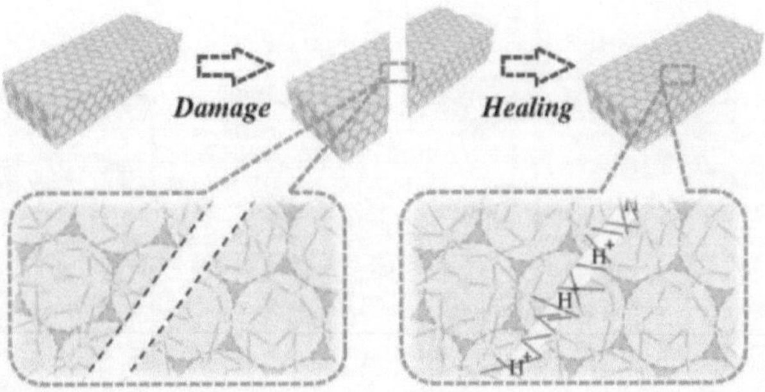

FIGURE 2.20 Healing of structural color gel. (From [Fu 2017].)

2.5 Gels

Gels are materials with cross-linked polymer chains, known as gelators, that form molecular networks with sufficient internal porosity to sustain the presence of fluids in between the links. Excessive chemical, mechanical, or thermal stress damages the structure of the gelator network. Gels recover original properties by internal reorganization following the application of external stimuli of sufficient intensity to cause damage. Recovery occurs at the molecular level with rejoining of cross-linking bonds in the gelator. The possible bonding mechanisms of reversible chemical bonds range from hydrogen to covalent to supramolecular, including the healing of structural color gels (Figure 2.20) [Fu 2017] [Wei 2014]. Table 2.5 lists dynamic self-healing gel formulations.

TABLE 2.5
Dynamic Self-healing Gels

Bond Type	Material	Thermal and/or Thermodynamic Nature	Reference
Covalent Schiff-base linkage	Chitosan and biocompatible telechelic difunctional poly(ethylene glycol)	Room temperature	[Yang 2012]
Covalent poly(vinyl alcohol) or a catechol-functionalized copolymer	2-Acrylamidophenylboronic acid (2APBA)	Room temperature cycling of pH	[Deng 2015]
Covalent acid-sensitive imine bonds	PIB with side-chain amino acid–based polymers	Room temperature cycling of pH	[Haldar 2015]
Hydrogen bonds on dangling covalent polymers	Acryloyl-6-aminocaproic acid with balance of hydrophobic and hydrophilic interactions	Room temperature cycling of pH	[Phadke 2012]
Covalent	Tetrahedral borate ester with catechol end groups	Room temperature cycling of pH	[He 2011]
Covalent H–N	Acylhydrazone	Room temperature cycling of pH	[Deng 2010]
Hydrogen	Supramolecular PIB gel with inclusions of functional nanoparticles	Liquifies at 60°C, stimulated by superparamagnetic nanoparticles	[Binder 2007]
Covalent	Diarylbibenzofuranone dimer of arylbenzofuranone	Room temperature dynamic bonds	[Imato 2012]
Covalent host--guest redox	Poly(acrylic acid) with cyclodextrins and ferrocene side chains	Room temperature active for 24 h and reactivated with NaClO	[Nakahata 2011]
Metal–ion	Acryloyl-6-amino caproic acid with flexible dangling side chains	Room temperature activated with $CuCl_2$	[Varghese 2006]
Schiff base cross-links	Mixture of modified chitin and alginate	25°C	[Ding 2015]
Entangled cholesterol-capped polyethylene glycol	Liposome gel	Room temperature and thermal cycling 4–37°C	[Rao 2011]

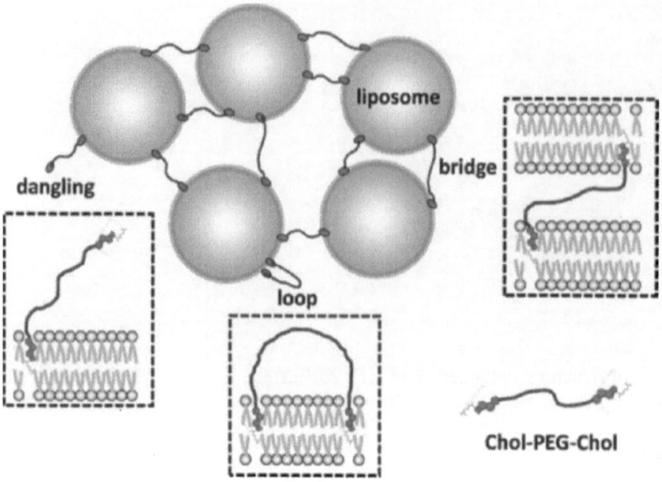

FIGURE 2.21 Reversible liposome gel. (From [Rao 2011].)

Protein-based gels made with synthetic (as opposed to natural) polypeptides often have superior mechanical properties, presumably due to improved consistency. A synthetic gel made from self-assembling low molecular weight diblock copolypeptide amphiphiles remains stable at temperatures up to 90°C – a temperature that dissolves natural protein gels. These synthetic gels recover mechanical properties within seconds from strong mechanical agitation that breaks the gel structure. Credit goes to the relatively low molecular weight of the polypeptides [Nowak 2002].

Assembling nanoscale and microscale particles with the appropriate binding network is an alternative means of making gels. Liposomes are small, usually submicrometer, encapsulating structures with lipid bilayer walls. Joining the liposomes with polymer links makes a gel, such as cholesterol-end capped polyethylene glycol (Figure 2.21) [Rao 2011]. The polymers bind to the liposomes with sufficient strength to form a gel but are reversibly weak. Mechanical agitation disrupts the bindings without breaking the liposomes, which then rejoin once the agitation stops. This process creates a macroscopic stiff gel, which when stirred aggressively liquefies and then reversibly heals back into a stiff gel.

Zwitterions are net-zero charge molecules with separate functional groups containing positive and negative ions. Attaching zwitterions to hydrogels creates a material with spontaneous healing properties [Bai 2014a].

3

Micro- and Nanoscale

The dimensions of micro and nano length scales run from those of molecules and atoms up through small visible features. The nano length scale is approximately 0.1–100 nm. The micro length scale runs from 0.1 μm to 100 μm. Physical effects at these dimensions often promote the action of particles, surface effects, and occasionally quantum phenomena. Important to self-healing are microencapsulation of healing agents and the mechanical actions of nanoparticles, dendrites, and tendrils [Thakur 2015].

3.1 Nano- and Microscale Interactions for Healing of Larger Objects and Systems

3.1.1 Crack Filling Methods

Many cracks have widths too wide for healing by molecular diffusion alone. The healing of wide cracks requires methods that either close or bridge across larger gaps. Filling cracks with liquids, bridging tendrils, and growing solids is a viable method. The filling material comes from fluids in the crack or the crack subsurface. Tight cracks with widths on the order of micrometers or smaller have small overall volumes that need to be repaired and may take advantage of a favorable chemical mini environment in the crack (Figure 3.1).

3.1.2 Concrete Crack Filling

Cracks in concrete weaken the structure and reduce durability. Some cracks portend major structural issues. Other cracks are more benign, but nonetheless permit ingress and water, salt, and other material that adversely affect durability. Repairing cracks counters these effects and improves structural performance and durability.

Nonautogenic crack repair techniques for concrete generally use the external application of sealers and patches. Calcium silicate is a common sealant that flows into cracks and solidifies to form a water-stopping seal [Martin 1996] [Connell 2007].

Autogenic crack repairs use concrete chemistry and heterogeneity. Most modern concrete uses Portland-type cement to bind the aggregate together. A primary hydrating ingredient is calcium silicate ($CaO{\cdot}SiO_2$), often modified with performance-enhancing additives. Pozzolanic concrete based on volcanic ash hydrations is an alternative that dates to the Romans and has a renewed interest in the modern era due to lower carbon footprints. The inclusion of aggregate-scale lime clasts in Roman concrete provides reactive calcium for crack infilling and may be a primary factor in millennia-scale durability of the material [Seymour 2023]. Modern variants use coal plant fly ash for hydration, often as an add-on to calcium silicate mixes. Fly ash increases the strength of the concrete. Calcium silicate and fly-ash-laden cements cure into not fully hydrated states, leaving residual capacity for hydration that promotes autogenous healing. Upon formation of cracks, a chemical microenvironment draws calcium-based mineral products out of the cement and solidifies in the crack [Şahmaran 2008] [Neville 2002]. The healing effect is more active in early age concrete. Experiments on one-day old concrete with compressive stress (1 M/mm^2) applied across the crack found virtually complete healing with nearly ideal 100% efficiency [Schlangen 2006] [Qian 2009]. Additives, such as fly ash, promote hydration-based healing, even after 28 days of curing [Termkhajornkit 2009].

DOI: 10.1201/9781003396048-3

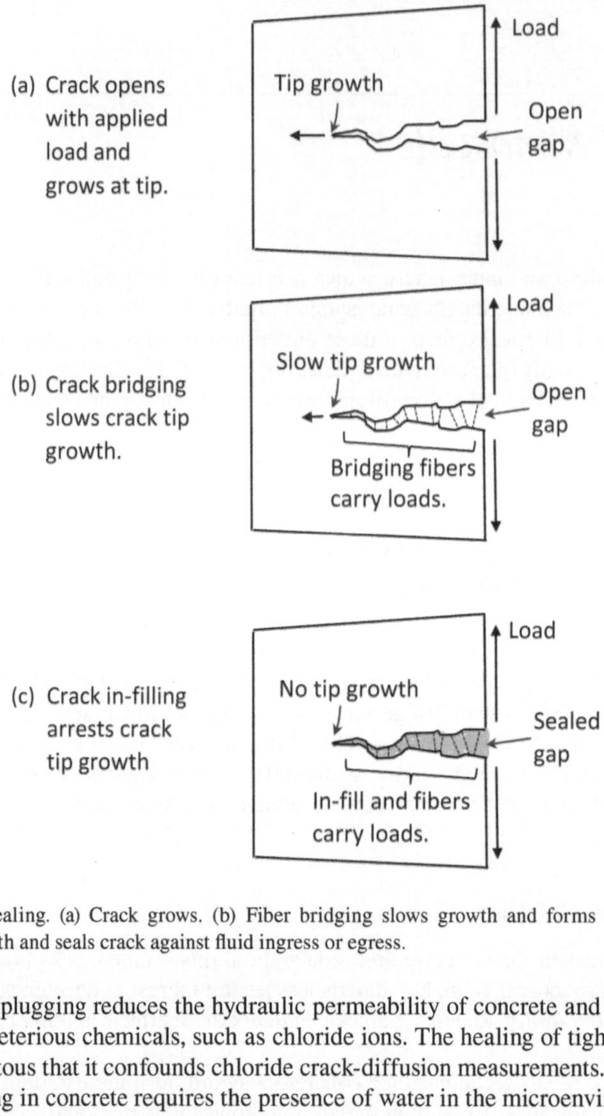

FIGURE 3.1 Crack healing. (a) Crack grows. (b) Fiber bridging slows growth and forms scaffolding for infilling. (c) Infilling arrests growth and seals crack against fluid ingress or egress.

Autogenous crack-plugging reduces the hydraulic permeability of concrete and prevents the penetration of water and deleterious chemicals, such as chloride ions. The healing of tight sub-1-mm cracks in concrete is so ubiquitous that it confounds chloride crack-diffusion measurements.

Autogenous healing in concrete requires the presence of water in the microenvironment of the crack, but is relatively insensitive to external moisture, such as high humidity. Soaking in lime-laden water promotes partial crack healing following freeze-thaw damage measured as recovered ultrasonic pulse velocities and reduced chloride diffusion rates, but not in compressive strength [Jacobsen 1996]. Reinforcing steel and/or fibers may hold the cracks in a tight configuration and promote effective autogenous healing [Mangat 1987].

Somewhat like concrete are the lime-based mortars used in masonry. CaO, $Ca(OH)_2$, and $CaCO_3$ bind the mortar together and provide a residual chemical capacity for healing small cracks, through dissolution and deposition of lime-based mineral products in the cracks [Lubelli 2011]. Studies of a wide variety of lime mixes indicate that the healing process is robust and ubiquitous in most mortars [Dhir 1973]. Ancient and historic mortars tended to use more lime than modern mixes and retain healing capacities that to this day continue to be active.

3.1.3 Microencapsulated Liquid Crack-filling Healing

Healing cracks by liquid solidification combines suitable liquids, means of delivery, and solidification on cue. Frangible microcapsules place the healing liquid throughout a solid matrix where subsequent damage induces release and activation followed by activation for healing (Figure 3.2).

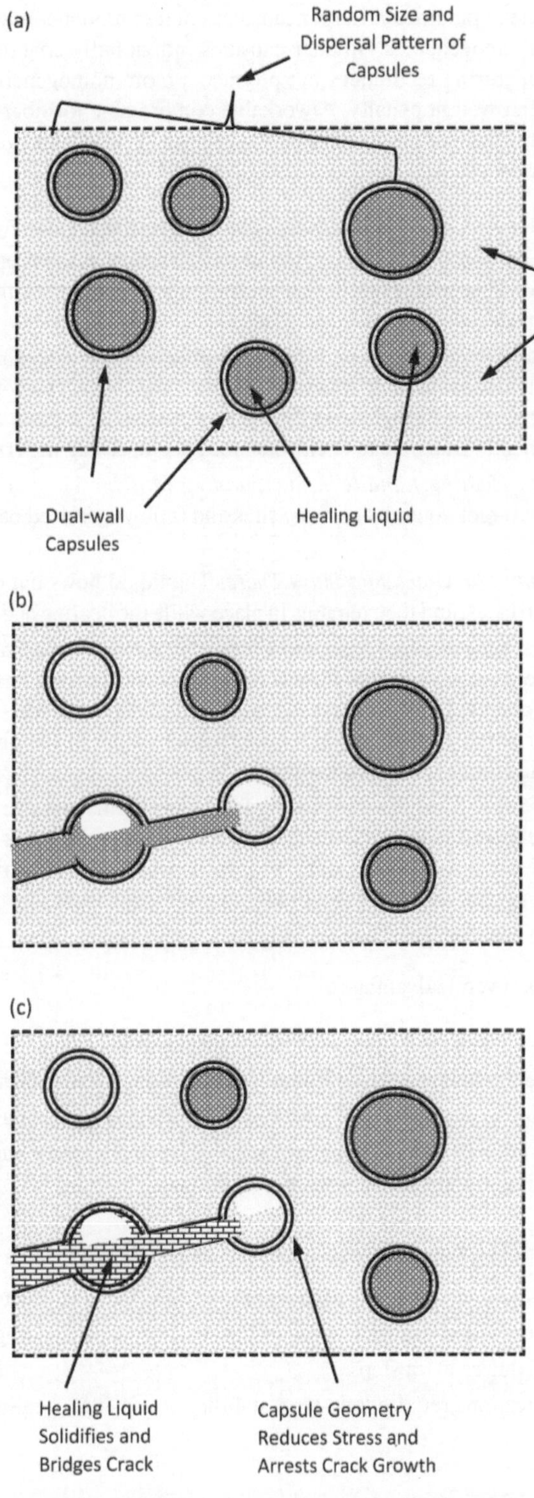

FIGURE 3.2 Encapsulated liquid healing, with pre-dispersed capsules – crack ruptures frangible capsules, liquid flows into crack, solidifies, and heals crack. Spherical capsule shell reduces stress at crack tip and arrests propagation. (a) Microcapsules containing healing liquid pre-dispersed in solid matrix. (b) Damage ruptures capsules and releases healing liquid to flow into cracks. (c) Liquid solidifies and heals cracks. (Adapted from [White 2003].)

Capsules size is an important design parameter. Larger capsules contain more healing liquid but introduce larger parasitic voids and inhomogeneities. Smaller capsules individually contain less liquid and require more sophisticated manufacturing techniques but produce a more homogeneous material with smaller voids and possibly a smaller weight penalty. A workable compromise for fiber-reinforced polymer composites is to use microcapsules with diameters of 10–100 μm. While some variations in technique exist, most require the following steps:

1. *Suitable Healing Liquid:* The viscosity and surface tension must be low enough for the liquid to flow out of the capsule into damaged regions. The liquid must solidify at the proper time. This includes remaining stable for long periods of time prior to damage. The repair must be strong enough to serve the needs of the application.
2. *Suitable Capsules:* The encapsulation system must be compatible with the mechanics and chemistry of the healing liquid and surrounding structural matrix. The capsules should minimize weakening of the material, be strong enough to withstand long periods of normal structural use, without rupture yet be sufficiently frangible to open when the encompassing matrix cracks.
3. *Placement of Capsules with Healing Liquids in Surrounding Matrix:* The placement of the healing liquid capsules must target potential damage sites and follow a desired pattern, such as avoiding clumping together.
4. *The Healing Liquid Flows into the Crack and Stays There:* The liquid flows out of the capsule into cracks, wets the crack surfaces, and then remains in place while the healing action takes hold.
5. *Post-damage Stimulus Causes the Healing Liquid to Congeal and Repair the Crack:* A stimulus triggers the liquid to congeal once it has flowed into the crack, but not before. Possible stimuli include a catalyst embedded in the matrix, a two-part liquid system when the stimulus is also encapsulated, and/or a matrix that partially dissolves into a gel in the presence of the healing liquid and resolidifies into a healed configuration.
6. *Lifetime Stability and Durability:* Capsules and healing liquid must remain stable until damage requires activation. A key concern is manufacturing. In FRPs, the matrix starts as a polymer liquid that undergoes a cure cycle for solidification. Placing capsules requires mixing into the matrix without rupturing and then surviving the thermomechanical stresses of matrix curing and the stresses of normal structural use.

Microcapsule liquid healing has several advantages:

1. *Technique Works:* Healing efficiencies in excess of 100% are possible.
2. *Minimal Coordination:* The method requires minimal coordination, control, or maintenance.
3. *Healing Variety:* It is possible to accommodate a wide variety of loads and damage.
4. *Multiple Functionality:* Additional functionalities are possible, such as capsule rupture causing color changes, that is, bruising to indicate the presence of damage.

Disadvantages of microcapsule healing:

1. *Nonrenewable:* The healing is nonrenewable. Once the capsules rupture, they are difficult to replenish.
2. *Weakening:* Embedding liquid-filled capsules weakens the structure while adding weight.
3. *Extra Steps:* Manufacture requires extra steps.
4. *Instability:* Long-term applications require long-term stability of capsules, healing liquid, and structural interfaces.

3.1.3.1 Early Developments in Microencapsulation

Healing with microencapsulated liquids is conceptually simple, but technically sophisticated. Microencapsulation dates back to 1902 with a serendipitous discovery by Wolfgang Pauli (father of the famous physicist with same name) and Peter Rona [Lednicer 2003] who noticed that a combination of neutral salts with gelatin separates into two layers with a differentially prestressed double layer forming at the boundary. Upon mechanical agitation, the dual-layer boundary breaks into small pieces. Differential stress across the membrane thickness curls the small pieces into micrometer-scale liquid-containing capsules. H.G. Bungenberg de Jong named the process *coacervation* from the Latin *acervus*, with an English translation being heap or mass. Industrial applications based on controlled rupture of liquid-containing microcapsules began to appear in the 1930s and 1940s.

Forming a stable capsule wall is essential. The wall needs to be compatible with both internal liquids and external material matrices. This led to the development of dual-layer walls, with the inner layer being compatible with the inner liquid, the outer layer being compatible with the surrounding matrix, and both layers compatible with each other. One tract of early development was to use sugar to encapsulate food [Shirrif 1930] [Scherer 1941]. Sugar is fairly stable, nonpoisonous, and dissolves in water. Dual-wall methods enable encapsulating a wider range of liquids, such as oil-based vitamins with water-soluble outer shells [Taylor 1939] [Brynko 1961].

Storing multiple different reactive liquids in separate sets of microcapsules is a method of increasing functional performance. A stimulus, such as damage or deformation, breaks the capsules and releases the liquids for mixing and reacting. Barrett K. Green, while working for the National Cash Register company in Dayton, OH, wrote at least 13 U.S. patents (ca. 1942–1957) using microencapsulated fluid to make "carbonless carbon paper" for multilayer paper copying of typewritten and handwritten documents (Figure 3.3) [Green 1942]. The technique embeds microcapsules of color-functionalized liquids into paper. Local stress due to writing with a ball pen or typewriter ruptures the capsules and releases the liquids, which change color when mixed. Formaldehyde is a key ingredient in solidifying the capsule walls for long-term stability. A further set of developments occurred in the 1950s with the use of microencapsulated styrene and carboxylic acid–based polymers in easy-to-use color photographic films [Godowsky 1955].

Early developments of dual-liquid microencapsulated systems with mechanically active structural healing implications appeared in the late 1950s and through the 1960s. These include adhesives and adhesive tape [Eichel 1959] [Jensen 1962] [Capozzi 1958] [Meier 1962] [Washburn 1968].

Perhaps the first application of microencapsulation to structural self-healing was a corrosion-inhibiting coating with encapsulated dichromate, nitrite, borate, and molybdate quaternary ammonium corrosion inhibitors in a deactivated cellulose ether outer shell [Bailin 1987]. When damaged, for example, by a scratch or gouge, the shell ruptures and releases the corrosion inhibitors. About a decade later, Jung reported a microencapsulated material system for healing mechanical cracking damage to structures [Jung 1997]. The method successfully repaired polyester matrix composites with a *para-tert*-butylcatechol-stabilized styrene monomer liquid healing agent and polyoxymethylene urea microsphere shells. The liquid styrene remained stable in the microcapsules until flowing out through a capsule crack into the cure-initiating polyester matrix.

Most self-healing microencapsulation systems use mechanical breakage of the capsules to trigger the release of liquids. Mechanical breakage has the advantages of sensitivity and specificity but limits the choice of capsule wall material. Other methods of triggering capsule release include photo triggering or pH changes due to damage [Pastine 2009]. More sophisticated multi-cue triggers are possible, such as combinations of pH and temperature, or compression cycling with magnetic fields acting on nanoparticles embedded in the shell wall [Peleshanko 2007] [Long 2015].

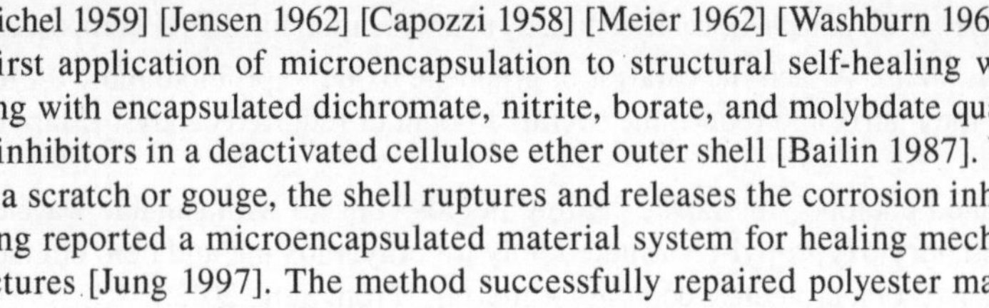

FIGURE 3.3 Carbonless carbon paper invented by Barrett Green in 1942 using two types of microencapsulated materials that rupture, comingle, and change color under the mechanical stress of a typewriter or pen tip. (From [Green 1942].)

3.1.4 Single-component Intrinsically Activated Liquid Healing

Single-component catalyst-free healing liquids are conceptually simple – just release the liquid and it solidifies. These liquid healing methods are attractive for situations where a single liquid provides a palliative action. Some liquids solidify when exposed to specific environmental conditions, such as liquid isocyanates in wet environments [Yang 2008]. Other liquids provide a protective anticorrosion coating for scratching damage. Epoxy-coated steel reinforcing bars for use in concrete placed in high-salt environments are a potential application. Epoxy coating protects steel rebars from corrosion, as long as the coating remains undamaged. Routine construction activities scratch and gouge the coating, often leaving corrosion hot spots known as holidays. Encapsulating liquid epoxy in a sublayer that releases when damage occurs to the top epoxy layer produces an active material system with superior corrosion resistance capabilities (Figure 3.4) [Enos 1998].

A potentially viable single-component healing liquid for epoxies is glycidyl methacrylate (GMA) [Meng 2010]. It is a low-viscosity liquid at room temperature that binds to epoxy fracture surfaces through both covalent and hydrogen bonds, and readily encapsulates inside poly(melamine formaldehyde) walls.

3.1.5 Single-component Liquids Activated by Catalysts and Other Extrinsic Agents

Catalysts extend the functional range of single-component healing liquids. The configuration stores the catalyst separately from the liquid and then mixes with the healing liquid following damage. Effective approaches include the following: (1) Disperse ruthenium-based Grubbs catalyst particles throughout a polymer matrix to initiate solidification of dicyclopentathiene that flows from a microcapsule into a crack [White 2001] [Grubbs 2004]. (2) Wax-protected tungsten chloride combined with phenylacetylene and nonylphenol is also suitable [Kamphaus 2008].

Custom geometric placements increase catalytic performance:

1. *Capsule Wall:* Attach the catalyst particles directly to the outside of the capsule wall to promote comingling of the catalysts and healing liquid (Figure 3.5) [Skipor 2006].
2. *Fiber Reinforcement:* Attach the catalyst directly to fiber reinforcing in FRP composites [Yang 2007].
3. *Uniform Dispersal:* Attach the catalyst to graphene oxide to promote more uniform dispersal throughout the matrix and reduce the overall amount of required catalyst [Mariconda 2015].

Microencapsulated photopolymerizable healing liquids congeal with suitable wavelength light, such as mixture of methacryloxypropyl-terminated polydimethylsiloxane and benzoin isobutyl ether in a 49:1 mass ratio as part of a protective coating for mortar. [From Song 2013.]

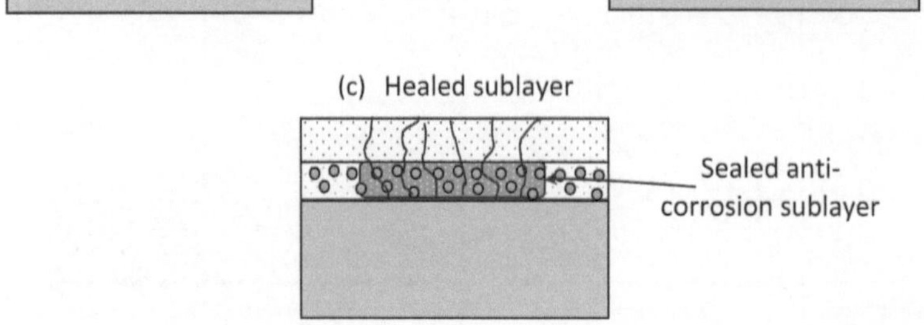

FIGURE 3.4 Sublayer of encapsulated single-component epoxy produces high-performance corrosion protection following scratches and gouges by sealing potential holidays. (a) Healthy configuration. (b) Damage creates potential holiday. (c) Microencapsulated sublayer releases sealing liquid that prevents holiday and local corrosion. (Adapted from [Enos 1998].)

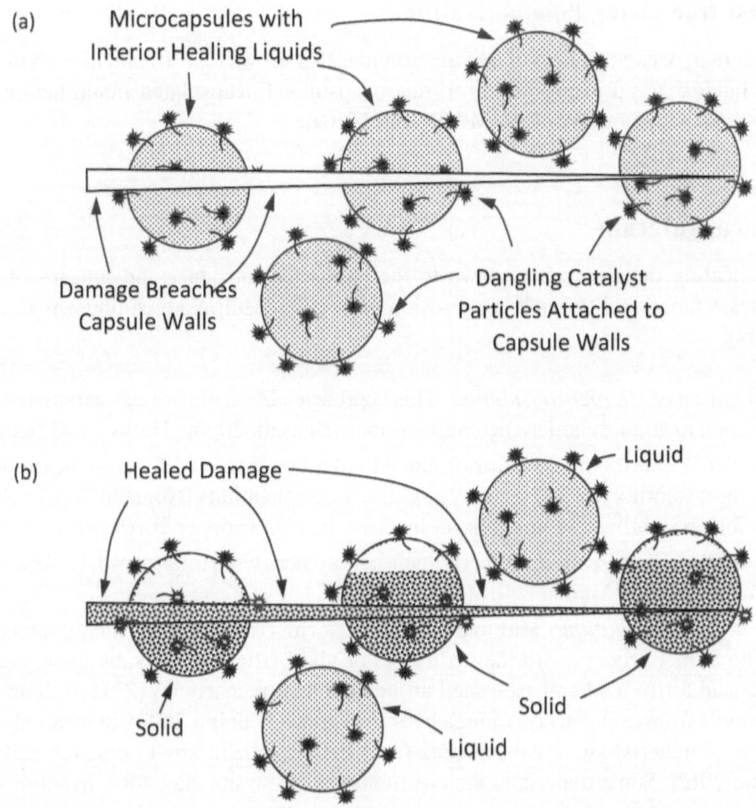

FIGURE 3.5 Attaching catalyst particles directly to the outside of the capsule wall promotes comingling of the catalysts and healing liquid. (a) Damage breaches microcapsule wall and releases healing liquid. (b) Wall-attached catalyst particles solidify liquid to heal damage. (Adapted from [Skipor 2006].)

3.1.6 Dual-component Liquid Systems

Microcapsules containing multiple liquids that synergistically heal materials are practical for specific material combinations. One combination embeds microcapsules into a combination of liquid anisole solvent core and a small amount of linear poly(methyl methacrylate) (PMMA) to heal cracks in PMMA [Celestine 2015]. Epoxy adhesives are usually two-part systems with a liquid resin and hardener. Mixing sets off a covalent cross-linking polymerization reaction that solidifies the resin. Co-embedding separate capsules of resin and hardener heals damage sufficiently severe to rupture the capsules and comingle the liquids [Xiao 2009]. A variant for anticorrosion coatings uses 10 wt% epoxy resin capsules and 2 wt% catalyst in an epoxy matrix [Yang 2011].

3.1.7 Catalyst-free Liquid–Solid Matrix Interaction

Specialized healing liquids solidify when contacting damaged material. An example is aqueous sodium silicate that gels and then solidifies by hydration when contacting damaged cement [Pelletier 2011].

For appropriately matched polymers, solvents act as catalyst-free healing liquids. The solvent softens and mobilizes damaged regions so that the material reforms and flows to fill and repair cracks, for example, through the enhancement of reputation-type healing. Conversely, thermoset polymers tend to resist dissolution and softening by mild solvents. Nonetheless, microencapsulated healing systems based on solvents for thermoset epoxies are viable [Caruso 2007]. A difficulty is that the solvents have to be sufficiently aggressive to soften a thermoset polymer, yet not degrade the microcapsule wall. A compromise is to use a mild solvent such as chlorobenzene for the healing liquid and then use a thermoset matrix that is not completely cross-linked and is more soluble.

3.1.8 Catalyst-free Living Polymerization

A solid polymer matrix can be made of a living polymer that grows to infill cracks with monomers accumulated from liquids stored in pre-dispersed microcapsules. Encapsulated liquid healing agent GMA monomer combines with a matrix of PMMA [Wang 2010a].

3.2 Nanoencapsulation

Miniaturizing healing liquid capsules down to the nanoscale introduces advantages of better dispersal, and nanoscale functionality, such as corrosion protection. Multiple techniques make nanocapsules [Shchukin 2007]:

1. *Miniaturization of Microencapsulation:* This approach adjusts the process parameters for microencapsulation to attain submicrometer dimensions [Blaiszik 2008] [Tiarks 2001] [Kirk 2009].
2. *Stiffen Lipid Vesicles:* Noncovalently bonded lipid bilayers readily form nanocapsules but lack the long-term stability needed in many engineering applications [Noguchi 2001] [Meier 2000]. It is possible to stabilize the lipid bilayer by cross-linking [Förster 2002] [Peyratout 2003].
3. *Nanocontainer Ensnarement:* Hyperbranched polymer dendrimers ensnare nanoparticles to form nanocontainers [Manna 2001] [Sunder 1999].
4. *Nanocontainer Formation:* Multiple processes form containers at submicrometer dimensions. The Stöbel process with the hydrolysis of alkyl silicates and subsequent condensation of silicic acid forms a silica-based shell around compatible particles [Zoldesi 2006]. Capsules self-assemble from cyclic arrays of urea hydrogen bonds [Ebbing 2002]. Sonication induces the formation of material-laden nanospheres from functionalized amylose starch and magnetite [Mauricio 2012]. Some materials, such as titania or halloysite clay, form nanotubes that convert into nanocontainers with further processing for applications, such as corrosion protection [Arunchandran 2012] [Abdullayev 2013].
5. *Layer-by-layer Buildup:* Nanoparticles that build up in a layer-by-layer combination of encapsulating and active layers provide opportunities for functionalized applications, such as anticorrosion coatings. Nanocontainers impregnated with corrosion inhibitor, using a silica–zirconia-based hybrid film is a corrosion inhibitor for 2024 aluminum alloy [Zheludkevich 2007]. A synergistic aspect of this approach is that the encapsulating walls include benzotriazole corrosion inhibitor layers along with polyelectrolyte layers, which then surround silica nanoparticles.
6. *Biological Methods:* Many biological processes use nanocapsules that hijack the molecular machinery of biosystems, including viruses, builds and loads nanocontainers [Graff 2002].

Multiple methods control the release of materials from inside nanocontainers. Many of the early developments focused on controlled drug release agents [Barbe 2007]:

1. *Polymeric Systems that Trap Chemicals Inside and Then Release:* Diffusion, tortuous nanochannels, chemical dissolution, or solvent-induced swelling of a hydrogel.
2. *Liposomes:* Lipid bilayer capsules, release material in response to stimuli, such as pH changes.
3. *Micelles:* Single-layer lipid capsules release quickly in response to stimuli, but have limited encapsulation ability and limited thermodynamic stability.
4. *Bioactive Glass:* Encapsulates bioceramics.
5. *Polymer Spaghetti Mesh Attached to Nanoparticle:* Swells and shrinks in response to stimulus, such as pH changes [Ballauff 2007].
6. *Cascades with Autocatalytic Behavior:* The release of reagents from nanocapsules promotes additional release of reagents [Hof 2002].
7. *Labile Nanocages:* Heating of superparamagnetic nanoparticles by tuned radio frequency electromagnetic waves releases chemicals from thermally sensitive nanocages [Derfus 2007].

Nanocages made of photosensitive peptides degrade and release chemicals with light stimulus, usually at UV wavelengths [Fay 1999].

8. *Thermal Cycling:* Thermal cycling drives the release of healing liquids by storing material inside the capsules in a solid form that only liquefies when the material cycles to a higher temperature. Low earth orbit satellites are a potential use case. The temperatures swing from –150°C to 150°C per orbit, which fractures the material while at a low temperature state and then heat up and melt the encapsulated material into a healing liquid. Low-melting-point metals, such as solders, as well as epoxies, heal with this extreme thermal cycling [Rohatgi 2013] [Merle 2014].

Simple liquids are isotropic and homogeneous at molecular to macro length scales. Complex liquids have inhomogeneities and/or anisotropies at intermediate length scales that create functionalized behavior, including stimulus-responsive healing. An example is microencapsulated carbon nanotubes suspended in nonpolar chlorobenzene and ethyl phenylacetate [Caruso 2009b]. The nanotubes remain stable inside the microcapsules until mechanical forces rupture the capsule walls and disperse the nanotubes for applications, including the repair of electrical conductors.

3.3 Solid Healing and Damage Mitigation with Nano- and Microscale Features

3.3.1 Nano- and Microcomposites for Damage Mitigation

Many forms of structural damage start at the atomic scale and cascade up through multiple length scales to emerge as severe macroscopic damage. Arresting cascading damage at early stages reduces the total amount of damage. A material that yields ahead of the crack tip can arrest crack growth. Homogeneous ductile materials generally resist cracking, but often lack the yield strength required of many applications. Conversely homogeneous high-strength materials tend to be brittle and lack toughness. Nonhomogeneous materials effectively arrest crack growth through actions at the nano and micro length scales [Ritchie 1988]:

1. *Crack Deflection and Meandering:* Local inhomogeneities, such as tough nanoparticle inclusions in a ductile material, deflect and pin crack growth through processes that act on nascent shear bands. Important details are particle size, stiffness, and dispersion pattern and adhesion to the surrounding matrix. Metal matrix composites often use this technique, as do a variety of nanoparticle-doped polymers [Zuiderduin 2007] [Carey 2011]. Insertion of microscale glass particles as reinforcing for epoxy composites shields and mitigates against fatigue cracking [Kawaguchi 2004].
2. *Zone Shielding:* Local yielding around the crack tip relaxes the stress field. Formation of shear bands ahead of the crack tip is effective, as with glassy amorphous metallic palladium [Demetriou 2011].
3. *Contact Shielding:* Mechanical action in the crack reduces tip stress and growth. Examples are crack bridging, infilling, and shear friction.
4. *Combinations of Zone and Contact Shielding*: Synergistic combinations of zone and contact shielding provide enhanced crack arresting performance.

3.3.2 Dendrites

Dendrites are small-scale filamentary structures that extend and protrude from larger solid features, often with branching and hierarchical patterns. Internal growth, accretion, or attachment of preformed elements produce dendrites. Self-healing applications extend dendrites from a solid element and grip onto something else, possibly other dendrites or other structural elements. Gripping techniques include molecular bonding, mechanical intertwining, intergrowth, and accretion.

3.3.3 Functionalized Microgels

Assembling and conglomerating microcomponents with functional properties produces functionalized gels. Numerical simulations using labile bonds between nanoparticles covered with a sticky gel find that such functionalization leads to self-healing materials [Kolmakov 2009]. An example is a gel assembly made of small soft microspheres with thermally responsive sticky coatings [Lyon 2009]. Attaching sticky macromolecular dendrites to clay particles creates a self-healing hydrogel with noncovalent bonds [Wang 2010b]. Dendrites also form scaffolding meshes for infilling with supplemental materials.

Shake gels transform from a liquid to a gel upon shaking, that is, upon being sheared. Such materials show potential for use in coagulation-type healing. Mixing clay particles with polymer fluids produces shake gels. Shearing combinations of laponite clay and polyethylene oxide liquid increases the number of molecular binding sites between the clay and polymer. The increase in binding dramatically stiffens the material system by enabling clay particles to clump together in a percolating network [Zebrowski 2003]. The process reverses when the shear stops. The number of polymer binding sites decreases, and the network of conglomerated clay particles relaxes. Increasing the level of shear reaches a point of diminishing returns in which the polymers start to self-bind to the clay particles and prevent particle-to-particle glomming.

Light-sensitive binders control the glomming of nanoparticles to create a light healable composite gel system. UV light controls ureidopyrimidone hydrogen bonding to functionalize telechelic poly(ethylene-*co*-butylene) and cellulose nanocrystals so that irradiation heats the material, releases the bonds to liquify the material, and then initiate a reversible healing process [Biyani 2013].

Colloidal crystals assembled from microparticles provide insight into the healing and crack arresting of molecular and atomic crystals. Curvature and topological frustration in curved colloidal crystal surfaces produce interesting effects when subjected to the insertion of equilibrium-disturbing interstitial particles [Irvine 2012]. Unlike flat crystals, curved crystals provide an impetus for self-healing through localized actions that collectively rearrange particles.

3.3.4 Crack Bridging and Filling

Bridging across crack openings by actively growing and fixing micro- and nanostructures controls the flow of stress and deformation around the crack, especially near the tip. Bridging also provides scaffolding for infilling, sealing, and functional activities.

Nanoparticles dispersed into polymer composites fill and bridge cracks by migrating through the polymer matrix in an entropy-driven action. Placing thin brittle layers between nanoparticle-laden polymers creates a sandwich geometry. Cracking of the thin layers induces the particles to move into the cracks and serve as band-aids for healing [Lee 2004a] [Tyagi 2004]. In the case of a thin film sandwich of nanoparticle-doped PMMA on top of a silicon wafer and covered with a glassy SiO_2 layer, thermal and process cycling cracks the brittle top layer (Figure 3.6). Ethylene oxide–covered particles migrate through the PMMA and fill the crack. Cooling and closing of the crack tightly binds the particles in place as crack bridges [Gupta 2006]. Nanoparticles with different coatings, such as tri-*n*-octylphosphine oxide, will not migrate into the cracks under the same conditions. This observation lends credence to the hypothesis that thermodynamic forces drive the motion of the nanoparticles.

More complicated nano-structuring has advantages. An optically clear material made of glassy-rubbery layered block copolymer nanostructures absorbs large amounts of energy from high-velocity projectiles and then heals back into a solid structure in the wake [Lee 2012a]. Superparamagnetic nanoparticles migrate into and repair tight cracks when jostled by external oscillating magnetic fields (Figure 3.7) [Corten 2009].

3.3.5 Healing of Tribology

Minimizing sliding friction and wear in machinery are important considerations that fall into the engineering field of tribology. Bearing and sliding surfaces of machines are high-performance components, with comparatively small volumes, yet can lead to costly failures. These traits are the ingredients of a good opportunity for the application of self-healing effects. Wearing surfaces need to be replenished, regenerated, or replaced, as does the lubricating fluids and low-friction nonlubricated wearing surfaces.

Designing low-friction long-lasting wearing surfaces requires satisfying competing considerations. Being slippery relies on low chemical affinity between the sliding materials. Being strong implies that the material binds strongly with itself so that it resists mechanical deformation. PTFE resists van der Waals and other types of chemical binding and does not readily cross-link with itself to form mechanically robust

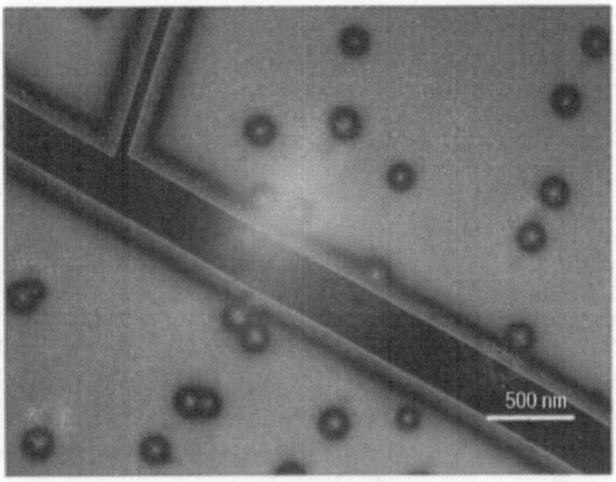

FIGURE 3.6 Nanoparticle motion into cracks. (From [Gupta 2006].)

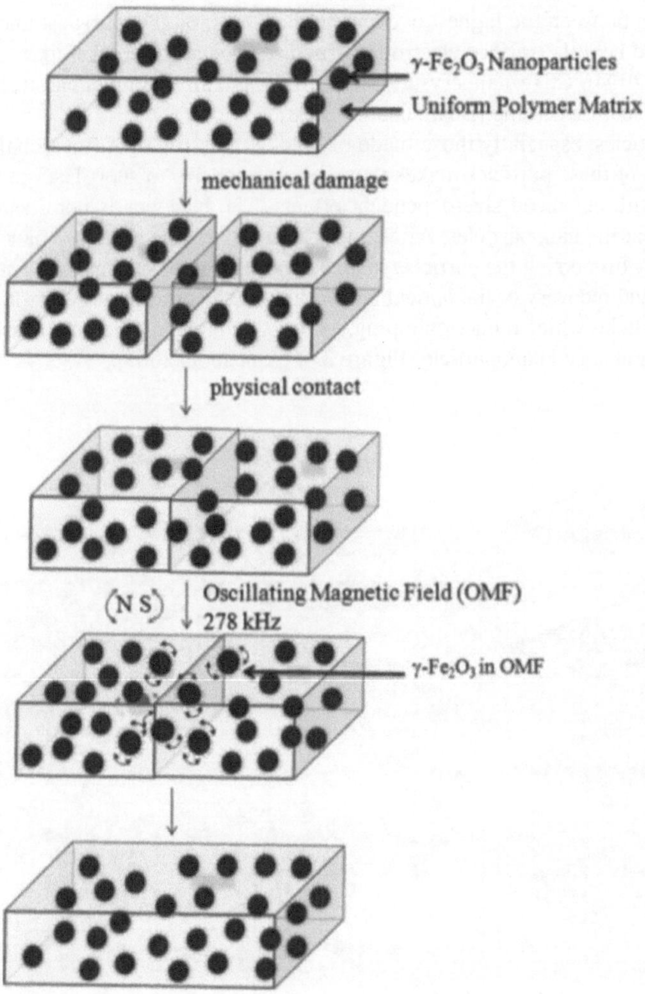

FIGURE 3.7 Superparamagnetic nanoparticles agitate in the presence of oscillating magnetic fields and seal a crack in thermoplastic polymer. (From [Corten 2009].)

structures. Creep and gouging-induced wear are common maladies of PTFE. Doping with sub-100 nm particles of carbon, alumina, ZnO, and other hard materials reduces surface deformation and gouging [Lee 2007] [Sawyer 2003]. Arresting shear bands, preventing crack propagation, and controlling the size and shape of debris contribute to mitigation [Burris 2006]. Nanoparticle doping generally requires an activation or functionalization step to the surfaces for nonclumping uniform dispersal of particles. Control of ligand chemistry on the nanoparticle surfaces is an effective functionalizer [Glogowski 2006].

3.4 Self-repair of Micro- and Nanodevices

Micro- and nanosized objects provide functionality in multiple domains [Vincenzo 2009]. Possible actions include the migration and movement of nanoparticles, molecular scale conglomerations, molecules, and even subatomic particles. The accumulation and conglomeration of material fills gouges and similar surface damage defects in nanodevices [Smith 2005].

Semiconductor quantum dots have small crystalline cores and an outer shell that relaxes from the crystalline form into a more amorphous layer. Interactions between core and shell lead to functional behavior, such as optical reactivity. In CdSe nanocrystals with a wurtzite structure, the relaxed outer layer widens the gap between the highest occupied molecular orbital and lowest unoccupied molecular orbital to reform and heal the surface electronic structure, which preserves bulk optical and electronic properties [Puzder 2004]. CdTe nanocrystal quantum dots exhibit similar passivation protection and surface electronic structure healing [Bhattacharya 2008].

Metallic nanoparticles, especially those made of gold, exhibit multiple functional optical behaviors. The metallic nature of these particles makes them very responsive to short laser pulses, largely due to surface plasmons with enhanced size-dependent effects. The behavior is nonlinear. Energetic pulses damage and fragment the nanoparticles. After a few laser pulses, the shattered nanoparticles lose their optical functionality. Immersing the particles in an activating liquid, such as zinc phthalocyanines, promotes self-healing and recovery of the optical performance. The process activates liquid charges on the fragmented nanoparticles which induce clumping, melting, and fusing following a laser pulse. The result is a larger optically enhanced nanoparticle (Figure 3.8) [Amendola 2009].

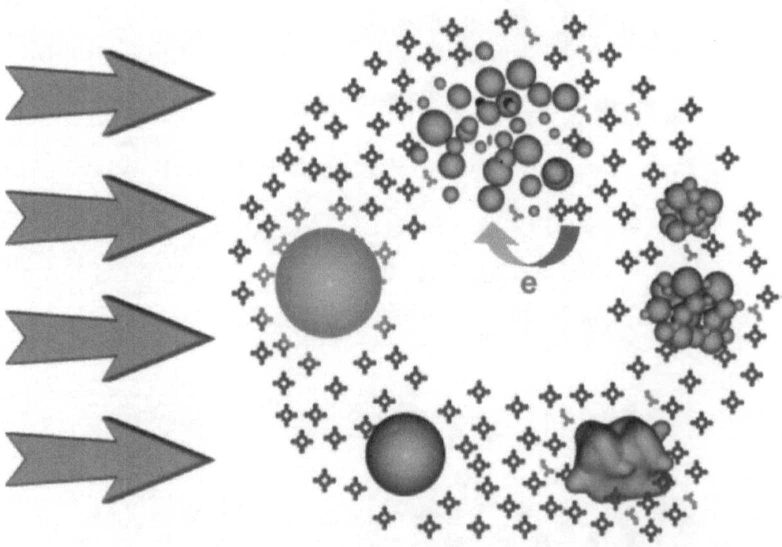

FIGURE 3.8 Self-healing of gold nanoparticles subjected to intense laser pulses while immersed in zinc phthalocyanines. The particles fragment, aggregate, melt, and then fuse into a larger optically enhanced nanoparticle. (Reprinted with permission from [Amendola 2009]. Copyright 2009, *American Chemical Society*.)

4

Meso- and Macroscale

Macro length scales are features big enough to be visible to the human eye. The mesoscale bridges between the macroscale and microscale with dimensions nominally running from 0.1 mm to 1.0 mm. Many mesoscale self-healing techniques take advantage of granular homogeneities and include the sloughing off of material for regeneration of damaged surfaces, the insertion of mesoscale fibers and features of macroscale composite structural elements, and flow control of granular material. Macroscale healing includes patching, connecting structural continuity, and sealing large leaks.

4.1 Regeneration

Regeneration replaces damaged material with undamaged, usually by transport from interior regions. Regeneration is a staple of biological systems. Animals routinely regenerate skin, intestines, fur, and/or scales by shedding and sloughing off dead outer tissue followed by growth from below. Regenerative wearing surfaces on structures and machines appear where wear is an important concern, is predictable, and the competing alternative is an expensive replacement.

An example is a pneumatic tire that must withstand extreme wearing loads while providing good traction, a smooth and quite ride, and operate over a wide range of temperatures. Treads are short cantilevers that meet the road to transmit combined axial and shear loads. Tread stiffness is a compromise between being soft for improved traction and being stiff to minimize rolling resistance and maintain stability during maneuver. As the tread wears, the geometry changes to shorten the cantilever, which increases the effective stiffness, decreases traction, and promotes susceptibility to hydroplaning. Preplacing cavities in the tread that expose upon wear and create new tread geometries with the proper stiffness and traction characteristics is a regenerative method of extending the useful life of a tire. Filling the cavity with a water-soluble elastomer further improves the performance of the regenerative system by both stabilizing the original tread structure and introducing a more aggressive change in tread structure stiffness upon wear and cavity exposure (Figure 4.1) [Corvasce 2006].

The flagging brush bristle is another commercially successful regenerative self-healing system [Collichio 1957]. The flagging bristle has a frayed end that improves dust, debris, and paint pickup. The frayed ends wear off with use. The construction of the flagging bristle is that of a set of fine fibers bundled and held together by a polymer. Wear causes the binding polymer to release at the end and regenerate a new flagging end (Figure 4.2).

4.2 Self-healing Composites

Composite materials attain superior performance by a synergistic integration of the properties of different materials. Composites include fiber-reinforced polymers (FRPs), asphalt pavements, and reinforced concrete structures. The manufacture of composites assembles multiple materials of smaller dimensions. These steps provide opportunities for inserting self-healing components. A practical issue is that many manufacturing processes fuse the composite together with heat cycles and/or chemical reactions, which require provisions to prevent these energetic processes from triggering the activation of the self-healing materials.

DOI: 10.1201/9781003396048-4

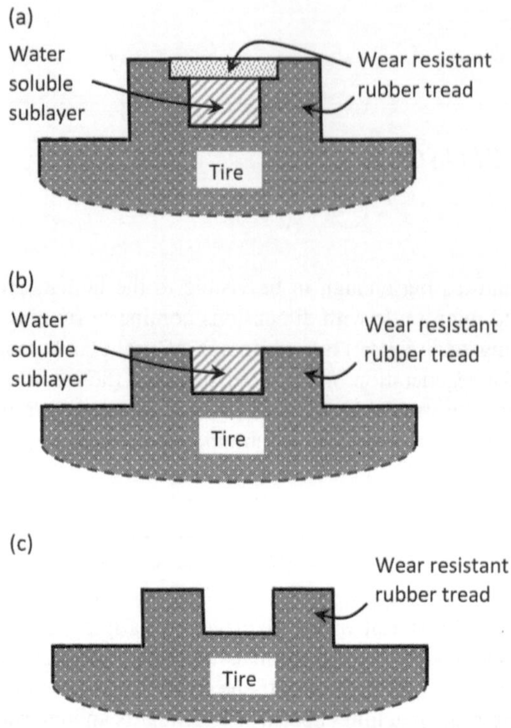

FIGURE 4.1 Regenerative tire tread with water-soluble elastomeric cavity that dissolves when exposed to environment by wear to regenerate tread. (a) Original healthy tire tread. (b) Tire wears to expose sublayer. (c) Regenerated tread. (Adapted from [Corvasce 2006].)

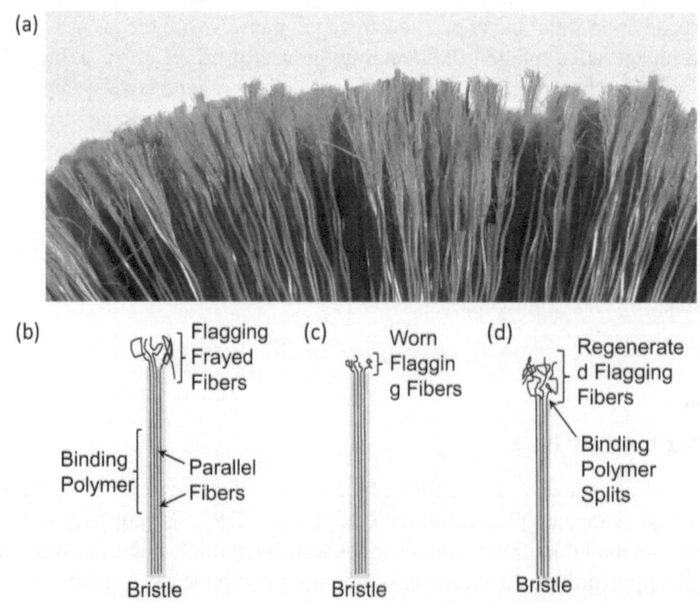

FIGURE 4.2 Flagging brush bristles self-heal by regeneration. (a) Brush with flagging bristles. (b) Initially intact bristle with frayed flagging end. (c) Flagging end wears with use. (d) Binding polymer splits and releases new flagging fiber end.

4.2.1 FRP Composites

FRP composites take advantage of the unidirectional strength of individual fibers and the toughness of the polymer, while eliminating buckling and separation issues with the fibers and the relatively weak mechanics of the polymer. The internal heterogeneity leads to failure modes that normally do not occur in homogeneous materials. Both high-intensity transient loading, such as from impacts, and low-intensity long-term fatigue loading are concerns.

4.2.2 Unbonded Granular Materials

Macroscopic assemblies of mesoscopic granular materials, such as piles of sand, can behave as solids or as fluids depending on the packing configuration, which depends on loading and recent mechanical history. Collective solid behavior occurs when particles interlock to sustain large shear loads without flowing. Fluid-like behavior occurs when the interlocking relaxes and the grains flow in bulk. Reversible and rapidly switchable interlocking solidification creates possible opportunities as components for use in self-healing structural systems. Vacuum loading switches a compliant flowable state into a rigid granular solid mass that gains strength from antagonistic prestress. Releasing the vacuum reversibly returns the pile to an easy flowing state. An example application is a variable stiffness core for a self-healing sandwich composite structure [Phillips 2010]. Vacuum loading stiffens the core. Heavy loading damages the macroscopically stiff core by shearing the pile, without damaging the mesoscopic granules (Figure 4.3). Releasing the vacuum allows for the core to reform to an original unsheared state with a minimum of effort. Reapplying the vacuum heals the composite structure by forming a new stiff core.

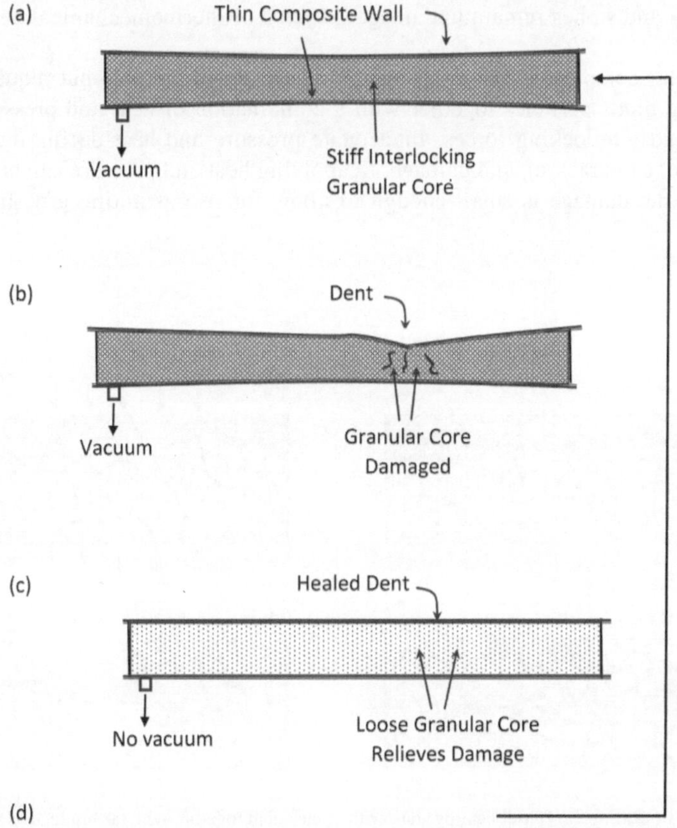

FIGURE 4.3 Vacuum healing granular core in unbonded composite panel. (a) Healthy condition with vacuum preload. (b) Damage dents composite wall. (c) Releasing vacuum frees granules and heals dent. (d) Reapply vacuum and return to healthy state (a). (Adapted from [Phillips 2010].)

4.3 Bonded Granular Materials

Binding granular particles together forms composites with strength and durability. Most applications derive useful properties from the particles with collective interlocking of compression forces to provide shear stiffness, as with asphalt pavement or as individual agents, such as abrasive particles in a grinding wheel. Sand, stones, metallic particles, and small nonmetallic crystals are common granules. The shape, size, and position of the particles are individually random within tightly controlled distributions and ranges.

An important detail is the nonlinear contact force behavior between particles. Collectively, particles transmit large, combined forces, as long as the particles remain in contact. If the stress and kinematic state causes the particles to lose contact, then the transmitted force drops to zero. The amount of compressive stress required to sustain contact and enable force transmission is relatively small when compared to the force transmitting capacity. Often, only a soft enveloping binder material, possibly combined with fibers, or vacuum loading is necessary to sustain the contact. Binder systems often provide good opportunities for self-healing applications.

Asphalt pavement is a matrix of petroleum-based tar binders that hold small stones together in a durable composite. While much of the strength derives from the interlocking of the aggregate particles, the binder is also a significant contributor. Microcracking of the tar binder is an early-stage failure mechanism that leads to large cracks and potholes [Little 2001]. Tar being thermoplastic heals with the application of heat and/or mechanical loading (Figure 4.4) [Carpenter 2006] [Kim 2006a] [Liu 2011b]. Temperature-dependent damage and healing phases acting over multiple spatial scales complicate the process [Qiu 2011] [Lytton 2001] [García 2012]. As long as the tar molecules and stones remain undamaged, cycles of thermomechanical healing may repeat indefinitely.

Healing granular composites that do not use binders is also possible but requires heat or other stimuli. Sintering binds particles together with a combination of heat and pressure that activates van der Waals and interlocking forces. Inadequate pressure and heat during manufacture leaves sintered parts prone to cracking and damage. Reapplying heat and pressure can heal sintered parts, especially when the damage is small enough to allow for reconstituting a desired overall shape [Luding 2008].

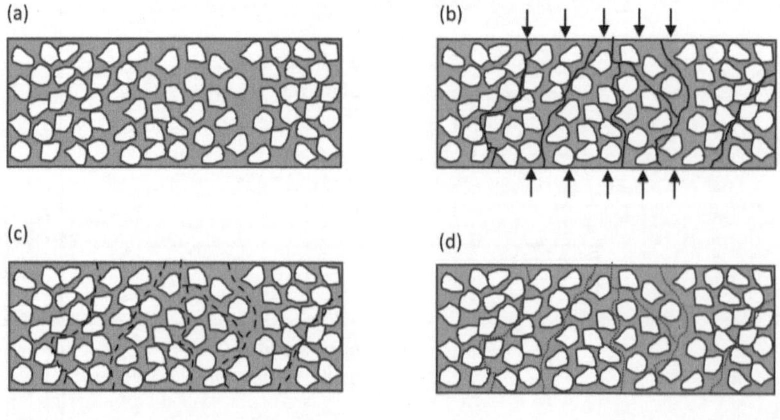

FIGURE 4.4 Asphalt healing. (a) Initial healthy state with stones held together with tar binder. (b) Mechanical and environmental loading damages tar. (c) The tar heals during rest periods, largely through diffusion. (d) Healed state almost as good as initial state.

4.4 Patching, Bridging, and Filling of Cracks

Cracks are principal forms of macroscopic damage to structural elements. Cracks alter the load paths inside the element, which may promote more crack growth. Effective healing of cracks restores and promotes a healthy flow of loads inside the element. Some of the primary methods of healing macroscale cracks are as follows:

1. *Externally Bridge with a Patch:* This usually works best if the patch gets loaded in tension. The ultimate strength depends on the weaker tensile strength of the patch, the shear strength of the patch, and the shear strength of the bond [Soutis 1999].
2. *Internally Bridge:* Fibrous material moves internally from the crack walls and bridges across the crack to connect to the wall on the other side. The bridging itself may be strong enough to transfer loads and stress across either by shear or by tension. An alternative is that the contact is weak but strong enough to act as a scaffolding to support more material infilling and promote interlocking of asperities.
3. *Externally Fill:* Insert material that fills the crack. The material is typically elastomeric and prevents seepage while being compliant.
4. *Internally Fill:* Concrete with unhydrated cement infills cracks exposed to water. A primary mechanism is deposition of $CaCO_3$ crystals in the cracks. Crack infilling decreases water permeability and strength, while increasing the strength of the cracked material [Granger 2007].

4.5 Heat-sensitive Functional and Multicomponent Materials

Thermal cycling affects material properties in both reversible and irreversible manners, some of which promote healing. An advantage of reversibility is the ability to sustain multiple bouts of damage and heat-cycle healing. The disadvantages of reversible processes are that the material deformations tend to be small and that the materials weaken when held above a critical healing threshold temperature. Irreversible thermomechanical processes often have advantages of larger material deformations and material toughness at temperatures above the healing activation threshold value but cannot heal over multiple cycles.

Autonomic thermomechanical healing within a structural system requires suitable heat sources. Electric or magnetic heating has the convenience of easy embedment and the ability to heat local regions under control. Electricity that drives the heat is easy to control but requires power. Oscillating electromagnetic fields induce electric currents, or direct power, such as from batteries, to drive Joule heating in electrically conductive elements. A related but different technique uses oscillating electromagnetic fields to flip the magnetization of ferromagnetic solids back and forth, producing magnetic hysteresis heating. Some "inductive" electric household stoves use magnetic hysteresis for heating but require using ferromagnetic cookware. The following are the demonstrated self-healing uses of thermomechanical healing: (1) A multifunctional composite materials with functional electromagnetic metamaterial properties that incorporate thermally reversible DA reaction polymers [Plaisted 2003]; (2) Anneal healing of specialized electrically conductive polymers based on complexes formed between *N*-heterocyclic carbenes and transition metals, and aided by solvent vapors of dimethylsulfoxide [Williams 2007c]; (3) Asphalt pavements with conductive fibers of graphite and steel wool inserted into binders to establish conductivity by a percolation effect [Garcia 2009] [Dai 2013b]; (4) Epoxy resins with embedded microparticles, such as polyethylene-*co*-methacrylic acid, that melt at relatively low temperature, reflow across cracks, and solidify to heal upon cooling [Meure 2009].

Large-scale macroscopic material deformations promote healing. Heat shrink tubing is a geometrically irreversible material used to cover wiring and electrical connection assemblies [Nyberg 1977]. Shrink tubing is a composite with a radially expanded polymer bonded to an elastomer tube. Upon heating, the expanded polymer shrinks radially, but only minimally in longitudinal directions (Figure 4.5).

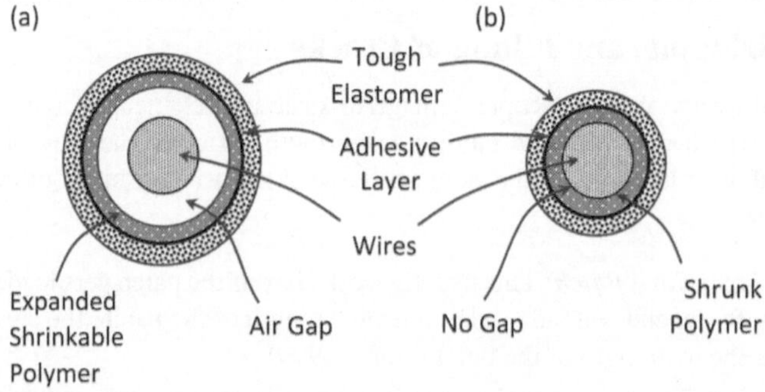

FIGURE 4.5 Cross section of shrink tubing as a composite with a tough elastomeric layer bonded to a layer that shrinks irreversibly with application of heat. (a) Expanded composite tubing loosely slipped over wire. (b) Composite tubing shrinks with heat.

The shrinkage is irreversible. The elastomer layer props open the expanded polymer for insertion of wires and then guides the shrinkage to envelope underlying wires upon application of heat.

4.6 Active and Robotic Repair

Robots and similar machines perform many repair processes. Cost, safety, and technical sophistication are factors favoring the use of robots over humans for repairs. The following are some situations to consider:

1. *Dangerous:* The repair environment is dangerous, confined or otherwise for humans.
2. *Cost-effective:* Robots are less expensive to use than humans.
3. *Simple Repairs:* The repairs are simple enough to not require human cognition or dexterity to execute.
4. *Performance:* The repair capabilities of robots outperform those of humans, both in physical capability, such as dexterity and strength, and in cognition, including intelligence, judgment, condition assessment, and planning.
5. *Robots improve with improved programming:* A process that may continue indefinitely.

The boundary delineating when robots outperform humans continues to shift. Semiautonomous repair robots presently appear in deep sea oil rig repair, buried pipe repair, spaceborne repair applications, and medical surgery. Fully autonomic repair integrates sensing damage conditions, deciding on a repair plan, executing the plan, and evaluating the result.

Metallic structures with localized damage can lend themselves to robotic repair:

1. *Crack Tip Management:* Most cracks grow at the tip. Managing the stress field around the tip arrests the growth. For isolated large cracks, it is possible to drill a hole through the crack tip to increase the tip radius and reduce the stress concentration. For microscopic surface cracks grinding, polishing and peening reduces and even reverses surface tensile stresses.
2. *Material Change:* Annealing and heat treatment with lasers, induction heaters, and other heat sources [Blackshire 2004].
3. *Crack Bridging:* Injection of material, such as epoxy, or welding to bridge and seal a crack.
4. *3D Printing:* 3D printing by extrusion and droplet deposition are becoming mainstream manufacturing and prototyping techniques. Printers and extruders can attach to robot arms to execute repairs [Brodbeck 2012].
5. *Places Not Suitable for Humans:* Many places are dangerous, geometrically awkward, or otherwise unsuitable for human repair. Sophisticated robots can execute repairs in some of these situations. Examples include deep underwater structures, buried pipes, large fuel tanks, and spacecraft [Caron 2012].

5

Surfaces: Healing, Sealing, and Cleaning

Maintaining the cleanliness, functionality, and geometry of surfaces has advantages. The small volume, cost, weight, and overall importance makes surfaces excellent candidates for applications of self-cleaning, healing, and sealing techniques.

5.1 Self-cleaning Surfaces

Self-cleaning surfaces are an attractive alternative to those that require manual cleaning [Parkin 2005] [Bayer 2013]. Applications include circumstances where accumulation of unwanted surface material degrades both appearance and performance, and manual cleaning is expensive, dangerous, and/or ineffective. An example is dirt and debris that accumulate on the skin of an airplane. The dirt increases noise and fuel consumption and is unsightly [D'Angelo 2010]. Manual cleaning requires removing the airplane from service, is dangerous to both personnel and the aircraft, and is at best only intermittent. Self-cleaning airplane skins alleviate these issues.

Figure 5.1 shows possible schemes for self-cleaning surfaces. These range from surfaces that actively prevent debris from accumulating to those that actively remove, breakdown, or consume unwanted material. A wide range of liquid, solid, and chemical phenomena, often as performance-enhancing combinations of effects promote self-cleaning [Genzer 2008].

Fluid-based cleaning methods clean both fluidic and particulate surface contamination. Particulate-based cleaning methods have an advantage in cases where there is a minimum of available cleaning fluid, or fluids cause problems, such as smearing. Methods that prevent debris and contamination buildup may be even more attractive since they eliminate the need for cleaning.

5.1.1 Debris-repelling Surfaces

The following are methods of preventing and repelling debris buildup on surfaces.

5.1.1.1 Electrostatic Methods

Static electric charges produce forces large enough to move debris and particles. Opposing electric charges attract particles to surfaces for fluid cleaning and filtering applications. Like charges repel particles from a surface for cleaning. Dipole charges produce more complicated effects, but also spawn forces that move particles. Airborne oil droplet removal systems in commercial kitchens use electrostatic methods to charge, attract, and remove droplets from the air.

5.1.1.2 Repelling Debris by Sweating, Eluting, and Slipping

Self-cleaning slippery surfaces shed particles and prevent particle buildup on lubricated surfaces. Liquids seep through the surface and cause contaminants to slide off quickly (Figure 5.2). Many practical examples of lubricated systems remove surface debris as a secondary characteristic: (1) *Porous Brass Bearings Laden with Oil and Grease:* The primary use is in bushings and bearings with a minimal supply of lubricant. Articular cartilage and synovial fluid lubricate in similar manners. (2) *Porous Air Bearings:* These are often made of porous graphite pressurized on the backside to push air out through

DOI: 10.1201/9781003396048-5

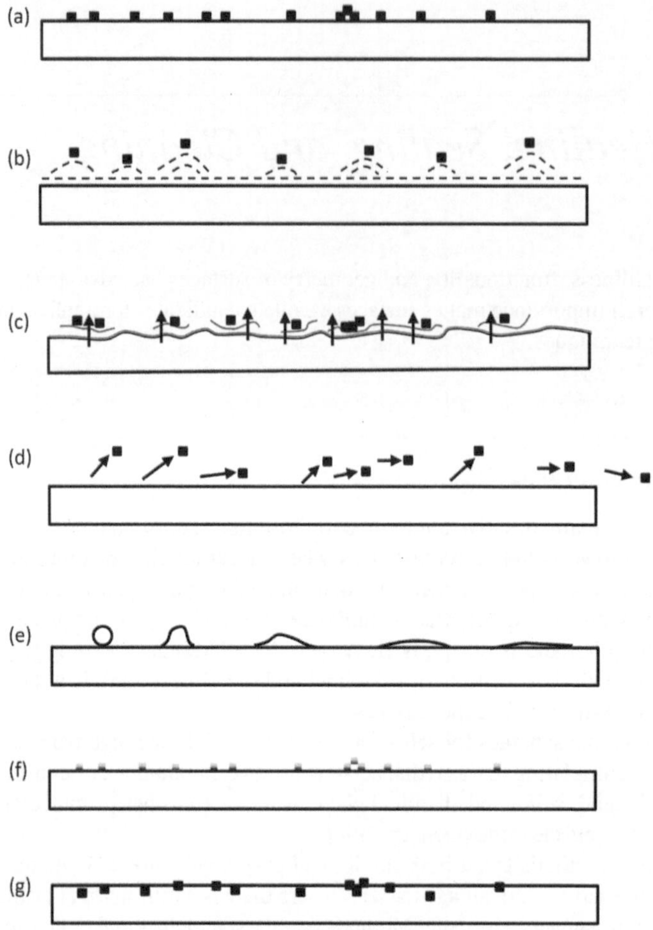

FIGURE 5.1 Operating principles of self-cleaning surfaces. (a) Debris accumulates on ordinary surface, sometimes synergistically, as with films and clumps. (b) Debris does not adhere, or surface repels. (c) Outer layer sloughs off to regenerate clean surface. (d) Mechanical, fluid-mechanical, and electromechanical forces remove debris. (e) Surface flattens liquid droplets. (f) Surface causes accumulated debris to break down. (g) Surface consumes debris.

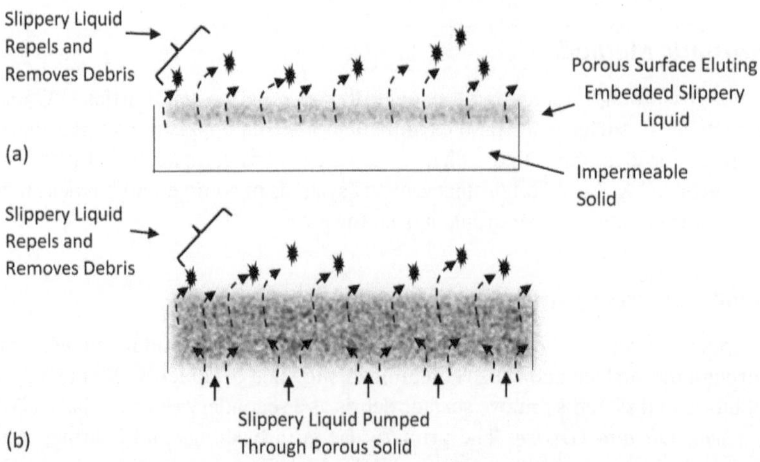

FIGURE 5.2 Sweating surface repels fluid to promote self-cleaning. (a) Slippery liquid embedded and eluting from micro- and nano-textured porous surface. (b) Slippery liquid pumped through porous surface.

the surface. Primary uses are low-friction and high-precision motion control applications. More aggressive approaches design slippery surface behavior as a primary mechanism.

Embedding slippery liquids in microstructure, nanostructure, and colloidal structure creates omniphobic surfaces [Vogel 2013]. The slippery liquid produces a self-healing effect by flowing into damaged regions. A method inspired by *Nepenthes* pitcher plants is the slippery liquid-infused porous surface (SLIPS) [Kim 2012b] [Bohn 2004]. Being slippery has broad applicability to a variety of situations, including repelling immiscible liquid droplets [Smith 2013].

5.1.2 Mechanical Debris Removal

5.1.2.1 Self-cleaning by Droplet Control

Moving liquid droplets along and off surfaces can clean surfaces. Autonomic droplet control is common in nature. Water rolls off on duck feathers or combined water roll-off and debris removal on lotus flowers are examples [Cassie 1944] [Extrand 1995]. Surface tension drives many of the actions. Possible droplet movements are creeping, spreading, conglomeration, and roll-off. Bigger drops move and creep more readily than smaller ones. Spreading and conglomeration converts drops into a film. Creeping and spreading tend to leave debris adhered to the surface (Figure 5.3). Rolling drops clean by picking up debris along the way (Figures 5.4 and 5.5) [Gould 2003] [Barthlott 1997] [Solga 2007].

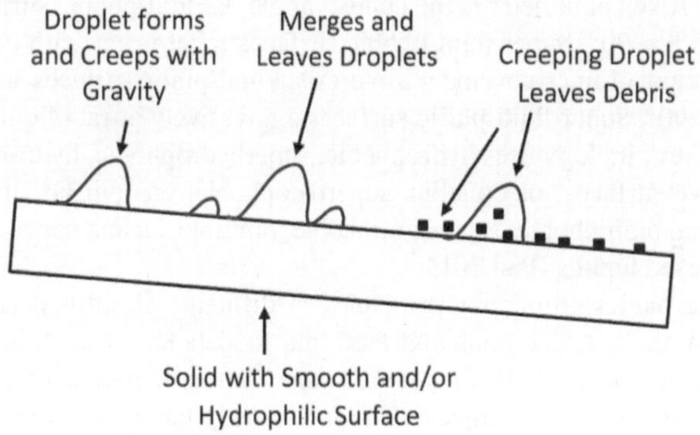

FIGURE 5.3 Smooth and/or hydrophilic surface forms droplets that move and merge with other droplets. Creeping movement of droplets leaves debris adhered to surface. (Adapted from [Barthlott 1997].)

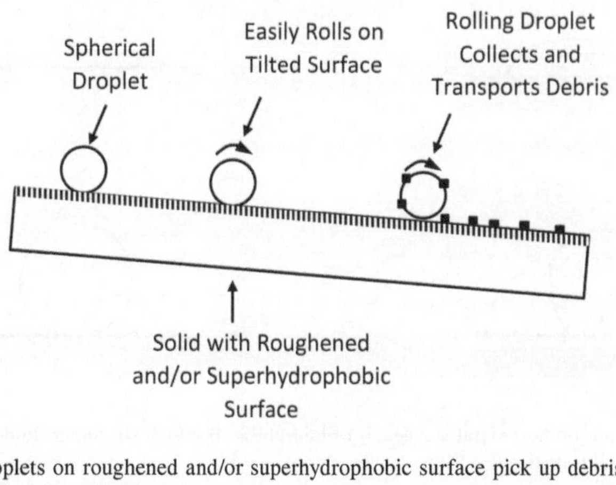

FIGURE 5.4 Rolling droplets on roughened and/or superhydrophobic surface pick up debris, often by droplet surface adhesion. (Adapted from [Barthlott 1997].)

FIGURE 5.5 Dust removal by water droplets rolling down superhydrophobic paint. (From [Solga 2007].)

Shape affects droplet movement and control. Figure 5.6 shows the general shapes of droplet that form on a horizontal surface. A key parameter is the contact angle θ. Fluid-phobic surfaces repel liquid fluids and have contact angles θ > 90°. Super fluid-phobic surfaces repel water with contact angles θ > 150° and usually have a minimum of line-movement hysteresis. Fluid-philic surfaces attract liquids by wetting with contact angles θ < 90°. Super fluid-philic surfaces aggressively attract liquids with contact angles θ < 150°. Fluid-specific terminology uses hydrophobic, superhydrophobic, hydrophilic, and superhydrophilic for water-sensitive surfaces; oleophobic, superoleophobic, oleophillic, and superoleophillic for oil-sensitive surfaces; and omniphobic, superomniphobic, omniphilic, and superomniphilic surfaces act on both water and oil-based liquids [Pan 2013].

Modeling droplet mechanics from first principles is difficult. Simplified semiempirical methods incorporate thermodynamic considerations and feed into models based on Newtonian and continuum equilibrium mechanics [Tadmor 2004]. *Surface Texture:* Smooth, rough, or patterned is a primary consideration. The smooth surface is simplest. The model only has to account for intrinsic properties of liquid, solid, and air, which are macroscopic manifestations of molecular scale phenomena. Rough and patterned surfaces have more complicated interactions. Microstructures with large local curvatures

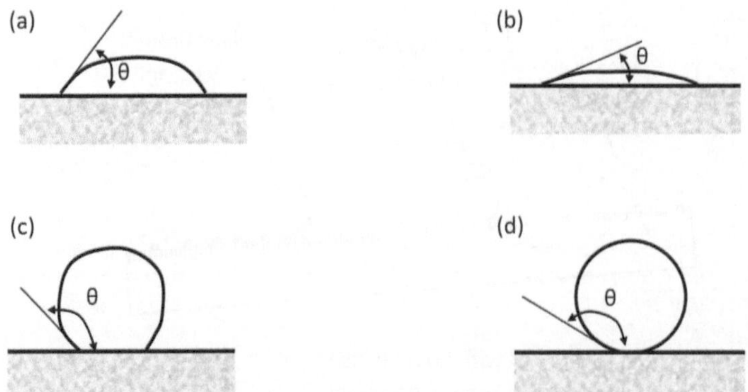

FIGURE 5.6 Droplet shapes on horizontal surface. (a) Fluid-philic, θ < 90®. (b) Super fluid-philic, θ < 30®. (c) Fluid-phobic, θ > 90®. (d) Super fluid-phobic, θ > 150®.

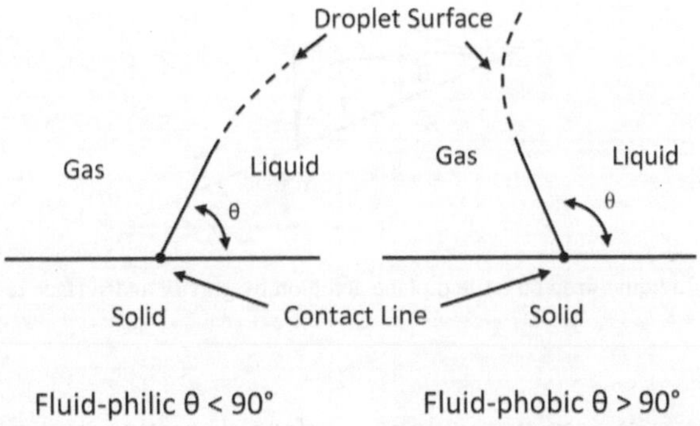

FIGURE 5.7 Two-dimensional geometry of solid–liquid–gas interfaces in a droplet resting on a smooth surface.

introduce behaviors not normally observed in smooth surfaces, that is, super fluid-phobicity, and effects that depend on the history of previous droplet movements.

The interfaces between the gas, liquid, and solid phases dominate droplet mechanics [de Gennes 1985]. In 3D, the three material phases meet in pairs forming surfaces, that is, solid–gas, solid–liquid, and liquid–gas, or as a triplet forming a contact line. Figure 5.7 shows a 2D slice of the droplet interface geometry in both fluid-philic and fluid-phobic configurations. The lines correspond to 3D surfaces. Figure 5.8 shows the intersection point corresponding to the contact line. The tension forces acting at the three-phase contact point appear with θ_0 = the equilibrium contact angle, γ_{SL} = solid–liquid tension, γ = gas–liquid tension (aka surface tension), γ_{SV} = gas–solid tension, and γ_{CL} = pinning contact line force. A consideration is whether the contact line is free to move along the surface or is pinned in place. Figure 5.8a represents the case of a negligible pinning force γ_{CL}. Neglecting the pinning force and summing the interface forces for equilibrium in the horizontal direction acting at the contact point leads to a static force balance equation, known as the Young equation:

$$\gamma_{SL} + \gamma\cos\theta = \gamma_{SV} \tag{5.1}$$

Including the pinning contact force leads to a modification of the pinning force, as shown in Figure 5.8b, with

$$\gamma_{CL} = \gamma_{SL} + \gamma\cos\theta - \gamma_{SV} \tag{5.2}$$

The pinning force grows to a maximum and then breaks away as the droplet moves. The magnitude of the breakaway force often depends on whether the droplet is advancing or receding.

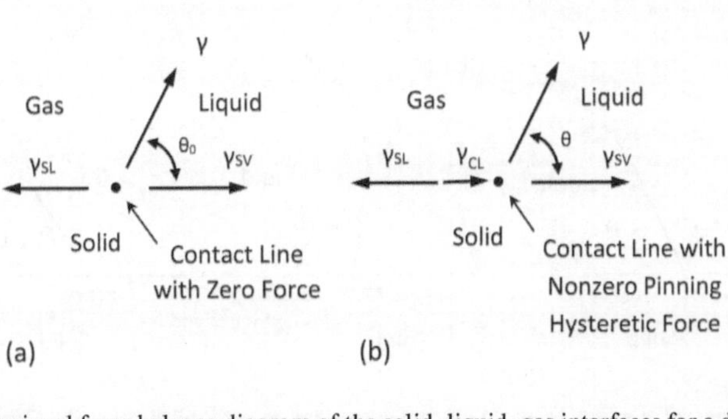

FIGURE 5.8 Two-dimensional force balance diagram of the solid–liquid–gas interfaces for a droplet. (a) Young's model resting on a smooth surface without contact line forces surface. (b) Inclusion of a pinning contact line force.

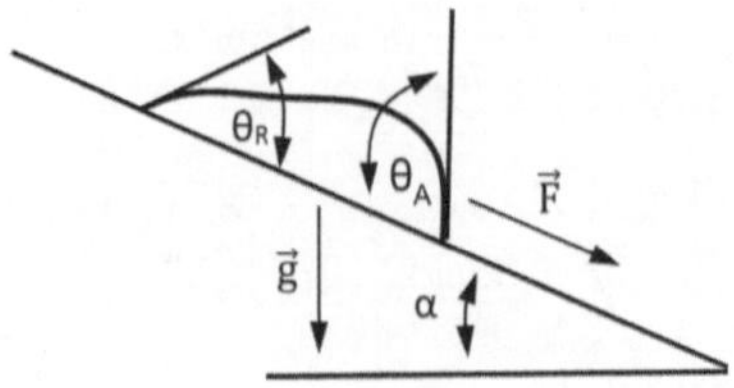

FIGURE 5.9 Side view of a liquid drop on a tilted plane acted on by gravity and surface tension forces. (Adapted from [Extrand 1995].)

Standard tests to measure droplet wettability, surface absorption, and surface adsorption place a droplet of specified volume on a surface, and observe the resulting geometry, including contact angles (Figure 5.9) [ASTM 2008a] [ASTM 2008b]. These experiments find that the advancing angle θ_A is usually bigger than the receding angle θ_R and confirm the presence of a nonzero pinning contact force γ_{CL} and droplet movement hysteresis (Figure 5.10) [Gao 2008].

Droplets move when

$$F = \Sigma \text{ Forces inducing movement} > \Sigma \text{ Forces resisting movement} \tag{5.3}$$

The forces driving movement include gravity, fluid drag, inertia, electrostatics, and surface effects. Hysteretic fluid–solid contact line effects are primary forces resisting movement. The critical force F_c required for the droplet to move depends on how strongly the drop pins to the surface. A semiempirical relation for the critical force is as follows [Extrand 1995]:

$$F_C = k\gamma w(\cos\theta_R - \cos\theta_A) \tag{5.4}$$

Here, k is an empirical constant and w is the half-width of droplet footprint. Experiments on circular droplets generally confirm the theoretical value of $k = 4/\pi$, with droplets of other shapes producing similar values.

An examination of (5.4) indicates possible means of reducing F_C and enabling easy roll-off of droplets for self-cleaning. Reducing the size of the contact footprint w is a primary factor. The pinning force depends on material properties and the length of the wetting perimeter – smaller perimeters correspond to smaller net pinning forces. The small wetting perimeters explain much of why droplets on fluid-phobic and, even more so, super fluid-phobic surfaces have small net pinning forces and easily rolls off surfaces. Reducing the contact line hysteresis, as represented by the difference in θ_A and θ_B, is another path to promoting droplet roll-off.

Electronic structure explains many of the mechanisms underlying hydrophilic and hydrophobic surfaces at the molecular scale [Azimi 2013]. Materials with surface electrons arrayed so that the OH

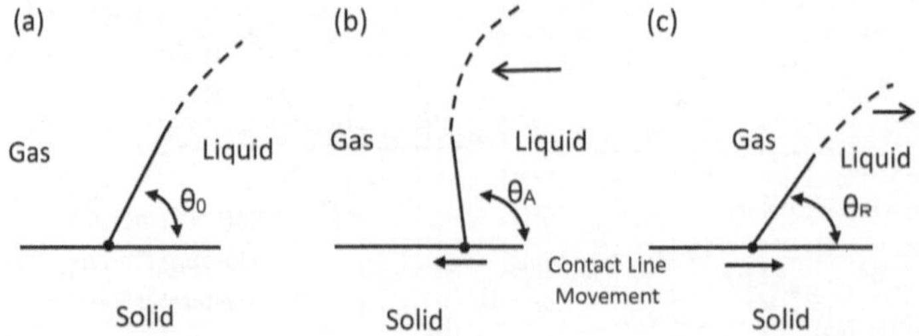

FIGURE 5.10 Geometries of contact lines. (a) Droplet stationary in equilibrium. (b) Droplet advances to the left. (c) Droplet recedes to the right.

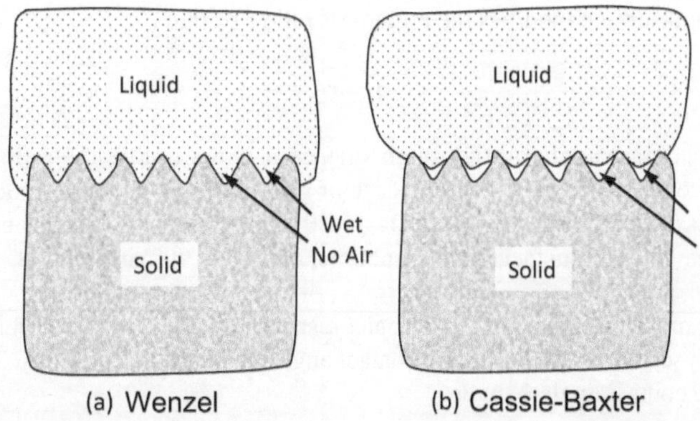

FIGURE 5.11 Two states of a liquid on a rough surface. (a) Wenzel state – the liquid completely wets the surface. (b) Cassie–Baxter state – the liquid partially wets the surface with air pockets in the valleys of the rough surface.

portion of water molecules fit nicely in a coherent polarized arrangement set up a hydrogen bond network leading to hydrophilic surface, as in alumina (Al_2O_3) crystals. Conversely, if the electronic structure prevents attaching the OH legs of water molecules, the surface becomes hydrophobic. Rare earth crystals, such as neodymia (Nd_2O_3), exhibit such hydrophobic behavior.

Micro- and nanoscale features that roughen surfaces affect wettability [Blossey 2003]. A consideration is whether the droplet completely wets the rough surface and forms the Wenzel state, or only partially wets the surface and forms the Cassie–Baxter state (Figure 5.11) [Wenzel 1936] [Cassie 1944]. The energy level of the Cassie–Baxter state is higher than the Wenzel state. Both states are stable with an energy barrier separating the two (Figure 5.12) [Johnson 1964] [Dettre 1964]. Transitioning between the Cassie–Baxter and Wenzel states requires energy to break pinned lines of contact sitting on the crevice walls to wet the entire microtextured surface. Mechanical forces may provide the energy impetus, but other sources, such as electrowetting of saline solutions, also work [Heikenfeld 2008]. Moving from the Cassie–Baxter state to the Wenzel state is an irreversible process that breaks the pinned contact lines [Ren 2014]. Switching back from the Wenzel state to the Cassie–Baxter state is very difficult and often requires evaporation or boiling to remove liquid bound in the crevices. Edge pinning and droplet mobilities are primary macroscopic properties resulting from the differences between the Wenzel and Cassie–Baxter models. The Wenzel state strongly pins the droplet in place. The Cassie–Baxter state is much more amenable to droplet movement by unpinning followed by creep or roll-off [Lafuma 2003].

Some simplified cases of surface roughness lend themselves to modeling with reasonable quantitative results. The Wenzel model begins by considering equilibrium on a smooth surface and a rearranged version of Young's equation (5.5):

$$\cos\theta = \frac{\gamma_{SV} - \gamma_{SL}}{\gamma} = A \tag{5.5}$$

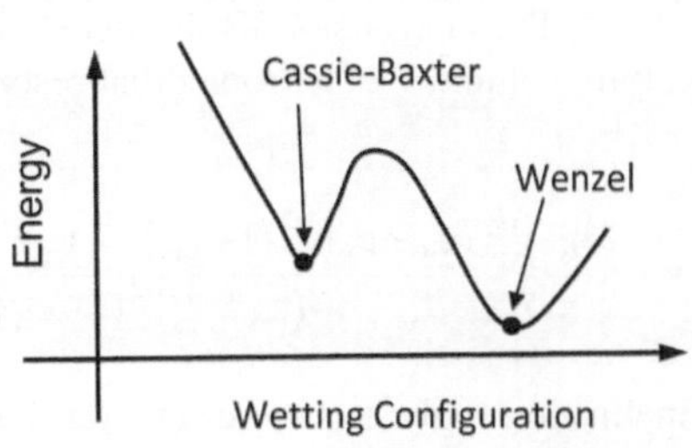

FIGURE 5.12 Energy versus wetting configuration of Cassie–Baxter and Wenzel states of liquid on a rough surface.

Wettability of a rough surface modifies the roughness ratio r so that

$$\cos\theta^* = rA \tag{5.6}$$

θ^* is the observed contact angle on the rough surface. r is nominally the ratio of the rough surface to smooth surface areas so that $r \geq 1$. The case of $r = 1$ corresponds to a smooth surface. When the smooth surface is hydrophilic, then $\theta < 90°$ and $\theta^* < \theta$, that is, roughness enhances wettability. Conversely, if the smooth surface is hydrophobic, then $\theta > 90°$ and $\theta^* > \theta$, that is, roughness decreases wettability.

The Cassie–Baxter model assumes that the droplet rests on a surface of two different materials with area fractions f_1 and f_2 such that $f_1 + f_2 = 1$ and contact angles θ_1 and θ_2 [Yoshimitsu 2002]. Energy considerations predict the contact angle θ^* to be

$$\cos\theta^* = f_1\cos(\theta_1) + f_2\cos(\theta_2) \tag{5.7}$$

When area 2 is filled with air, a droplet is spherical, and $\theta_2 = \pi$, then

$$\cos\theta^* = -1 + f_1[1 + \cos(\theta_1)] \tag{5.8}$$

An examination of (5.8) confirms that textured fluid-phobic surfaces in the Cassie–Bassie state with $\pi > \theta_1 > \pi/2$ have increased fluid droplet mobility with increased contact angles, smaller wetting perimeters, and reduced pinning. Noting that $0 < f_1 \leq 1$ in (5.8) leads to a condition of increased fluid-phobicity for the textured state, that is,

$$\theta^* > \theta_1 \tag{5.9}$$

The development of (5.8) applies to both advancing and receding contact angles with the difference

$$\cos(\theta_R^*) - \cos(\theta_A^*) = f_1[\cos(\theta_R) - \cos(\theta_A)] < \cos(\theta_R) - \cos(\theta_A) \tag{5.10}$$

Comparison of (5.10) with (5.4) leads to the conclusion that the textured-surface Cassie–Baxter state has smaller pinning forces than a smooth surface. This may be expected because there is a smaller amount of solid surface available for pinning [Quéré 2003].

The critical radius above which a drop rolls off is

$$R_c = \left(\frac{3\Delta\theta\varepsilon^2}{4\sin\alpha}\right)^{1/2} \kappa^{-1} \tag{5.11}$$

Here $\Delta\theta$ and θ are the difference and mean of the advancing and receding contact angles, respectively; $\epsilon = \pi - \theta$; α is the tilt angle; ρ is the mass density; γ is the surface tension; g is the acceleration due to gravity; and κ^{-1} is the capillary length = $[\gamma/\rho g]^{1/2}$ [Callies 2005].

The above description of droplet mechanics is largely 2D. Droplets are 3D objects. Interactions with solids occur along curved lines. The analysis of 3D droplet movements requires more detailed analysis. For example, the maximum volume V of a droplet that rests on an inclined surface without moving is

$$\left(\frac{\rho g \sin\gamma}{\sigma}\right)^{3/2} V \sim \left(\frac{96}{\pi}\right)^{1/2} \frac{(c_R - c_A)^{3/2}(1+c_A)^{3/4}\left(1 - 1.5c_A + 0.5c_A^3\right)}{(c_A+2)^{3/2}(1-c_A)^{9/4}} \tag{5.12}$$

Here ρ is the density; γ is the inclination angle; σ is the surface tension; c_A is the cosine of the advancing angle of inclination; c_R is the cosine of the receding angle of inclination; and g is the acceleration due to gravity [Dussan 1985] [Nguyen 1987].

5.1.2.2 Hierarchical and Regenerative Fluid-phobic and Fluid-philic Surface Structures

Hierarchical ganging of effects from different length scales creates surfaces with fluid-phobic performances superior to those produced by any single effect using. Close-up examinations of superhydrophobic materials found in nature, such as the Lotus leaf, find hierarchical combinations of nanoscale and microscale textures. Experiments with artificial materials made of similar hierarchical textured arrangements confirm that such geometries do produce super fluid-phobic surfaces (Figure 5.13) [Koch 2009]. Combining materials with durable hydrophobic molecular surfaces, such as with a rare earth oxide, and then patterning the materials with a suitable fluid-phobic microtexture creates an abrasion-resistant hydrophobic surface system [Azimi 2013]. Plants with water-repellant self-cleaning surfaces typically use surface regeneration for long-term durability [Neinhuis 1997].

Custom hierarchical arrangements of micro- and nanoscale textures produce specialized fluid-phobic behaviors. A notable example is when nanoscale texture pins into place a droplet formed by superhydrophobic behavior. An example is the rose petal. Like the lotus, the rose petal has a microtexture with superhydrophobic behavior. Water droplets form with large contact angles in Cassie–Baxter states. Unlike the lotus, droplets pin to the surface of the rose, even when inverted. An explanation of the difference lies in nanotexturing of the tips of the microbumps on the rose. The nanotexture on the rose is sharp and pins the droplets while the microtexture still promotes Cassie–Baxter droplets [Law 2014]. The reason for roses and lotuses to have such different droplet pinning behaviors is not clear, but may lie in a complicated competitive ecological landscape where water droplets on roses attract pollinators on a dry day and lotuses growing in wet conditions do not gain such an advantage. Another micro–nano specialization places a polymer layer on top of a microroughened fluoropolymer to enhance Cassie–Baxter effects for omni-phobicity [Coulson 2000].

A somewhat more complicated situation arises with hairy surfaces and flexible hairs. A droplet interacts with hairs at contact angles of <90°, yet due to the flexing and clumping of the hairs, a large macroscopic contact angle appears. Superhydrophobic hair plant leaves, such as the Lady's Mantle, use this effect [Otten 2004]. These configurations remove droplets that form by condensation – a microscopically driven process that wets most textured superhydrophobic surfaces into the Wenzel state. Water striders are insects that can walk on water. Their legs have specialized hairs and hair patterns that remove condensed water in a three-step process: (1) Conical hairs collect condensed water into droplets at the tips. (2) The water drops deflect the hairs to contact neighboring hairs and to conglomerate the drops into bigger drops. (3) The arrangement of hairs into anisotropic spiral patterns moves the bigger drops down and off the legs [Wang 2015c].

The surface textures of many hydrophobic surfaces have spatial pitches on the order of wavelengths of visible light. Such surfaces can have bright colors due to optical interference while also being superhydrophobic for self-cleaning, as with butterfly wings [Gu 2003].

5.1.2.3 Fluid-phobic Applications

Super fluid-phobic surfaces appear in multiple applications. An example is superhydrophobic coatings on solar cells that promote self-cleaning for optical performance [Zhu 2010] [Lin 2016]. A biofluid control example is hydrophobic surfaces on engineered nanoparticles that prevent the buildup of tough protein

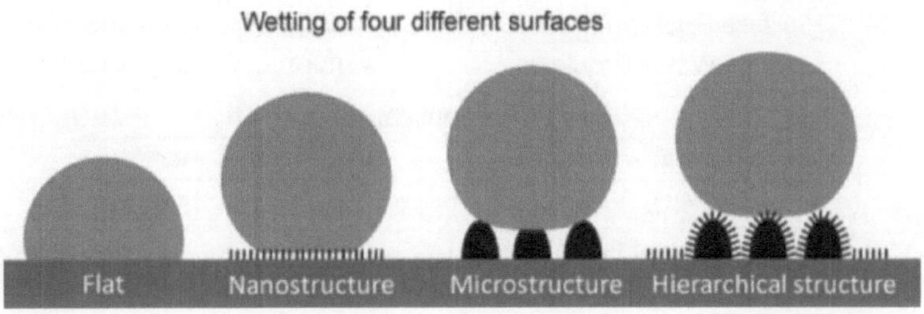

FIGURE 5.13 Droplet shape control with smooth, nanostructure, microstructure, and hierarchical superhydrophobic surfaces. (From [Koch 2009].)

coronas and enable realizing functional properties [Moyano 2014]. Water-repellant organosilane coatings combine antireflectant, wear-resistant, self-cleaning optical coatings [Nair 2014].

Much of the above discussion assumes that the super fluid-phobic surfaces behave in a quasi-static manner. Droplet dynamics introduce additional behaviors, such as hydrodynamics that favor droplets that normally stick to but instead bounce off surfaces [Bird 2013].

5.1.2.4 Superhydrophilic Applications and Antifogging Systems

The buildup of micrometer-scale water droplets that scatter light causes fogging of optical surfaces. Evaporation, wiping, and/or merging of droplets are effective defogging.

Forced evaporation defogs by forced convection of low-humidity air or by effects that alter surface temperature. Heating dehumidifies humid ambient air prior to convection. Forced convection of cool air cools the surface and alters the dewpoint in a manner that promotes evaporation. Electrical resistance heating of the surface with embedded or surface-mounted wires is another option. All three techniques are common options for defogging vehicular windshields.

Merging droplets by hydrophilic effects eliminates fogging. Superhydrophilic coatings aggressively wet droplets so that they flatten upon contacting a wetting surface. The flat drops merge with neighboring flat drops to form a smooth fluid film that allows for coherent light image transmission and reflection. This approach mitigates fogging of optical surfaces by micrometer-scale water droplets, as in antifogging rear-view mirrors in cars. A superhydrophilic method uses a titania surface that activates with incident light to free oxygen and leave binding sites for the oxygen in water [Fujishima 2000]. In a time-dependent process, the binding of water reduces the surface tension contact angles to a nearly perfect superhydrophilic state. Surface wetting occurs by growing water droplets that spread over the entire surface (Figure 5.14a). Durability of the hydrophilic surface can be an issue. Many are prone to washing off, flaking off, scratching, and/or degrading with water absorption. Transparent hydrophilic cross-linked polyurethane surfaces resolve many of these issues [Rädisch 1986]. An alternative and potentially more durable approach uses a layer of gold nanoparticles sandwiched between two layers of titania [Haechler 2023]. The nanoparticle density is sufficient to permit electrical conduction by percolation and broadband absorption of near-infrared light, but thin enough to be optically transparent to visible light. In sunlight, the infrared heats and defogs the glass (Figure 5.14b).

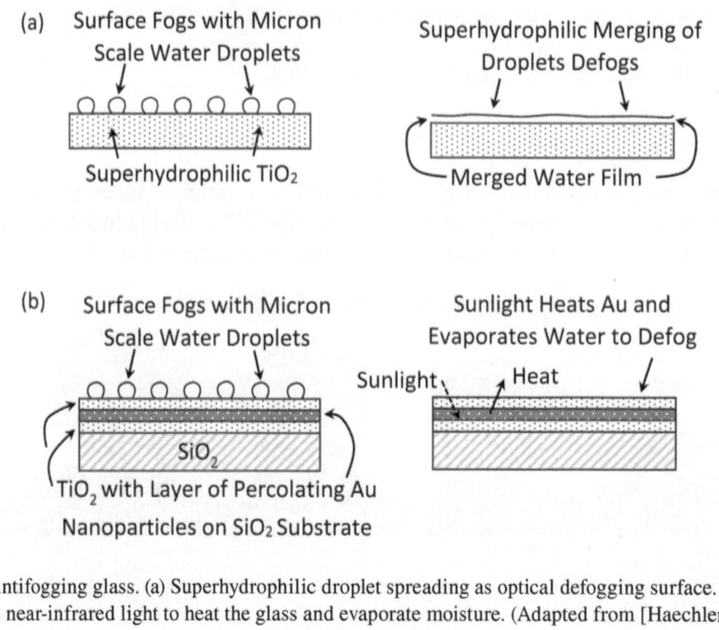

FIGURE 5.14 Antifogging glass. (a) Superhydrophilic droplet spreading as optical defogging surface. (b) Gold nanoparticle layer absorbs near-infrared light to heat the glass and evaporate moisture. (Adapted from [Haechler 2023].)

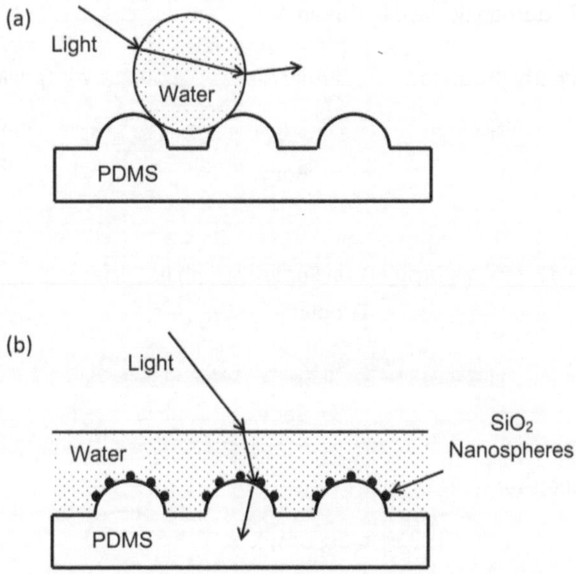

FIGURE 5.15 Hierarchical surface mimics antifogging capability of insect eye. (a) Uncoated and hydrophobic textured PDMS forms micrometer-scale water droplets that scatter light and fog optics. (b) Hierarchical surface of textured PDMS coated with hydrophilic SiO_2 nanospheres forms transparent non-fogging water layer. (Adapted from [Gao 2007].)

Superhydrophobic surfaces are typically not very effective for optical defogging applications as micrometer-scale fog droplets foul and prevent droplet roll-off. A biomimetic mosquito-eye-inspired workaround combines texturing at nano- and microscales to prevent fogging fouling of superhydrophobic surfaces (Figure 5.15) [Gao 2007].

5.1.2.5 Switching and Patterned Super Fluid-phobic and Super Fluid-philic Systems

Actuating hierarchical texture switches between hydrophilic and hydrophobic states. An example is stearic acid atoms that are hydrophobic on one end and hydrophilic on the other end. Electrically stimulating copper surfaces immersed in stearic acid forms flower-like hierarchical micro- and nanostructures and creates a superhydrophobic surface, with the hydrophobic tails pointing outward [Wang 2013c]. Thermal annealing reverses the effect and causes the surface to become superhydrophilic. Reapplying stearic acid converts the surface back into a superhydrophobic state. Placing and then removing ink from titania nanopillar arrays produces similar switchable wetting behavior [Lai 2013]. Elastomeric materials with microtextured surfaces switch between phobic and philic states, and from optically opaque to transparent, by gross mechanical deformations that alter the pitch distances and fluid infilling of surface details [Yao 2013]. Rapid and spatially patterned switching is possible. The reversible switching of flat thin films of ZnO, TiO_2, and photochromic azobenzene from hydrophobic to superhydrophilic with UV illumination opens additional possibilities for hierarchical texturing and patterning [Sun 2001] [Liu 2004] [Ichimura 2000].

Patterning a surface with patches of fluid-philic and fluid-phobic regions creates a material with properties capable of spatially controlling fluid droplet formation and movement. One effect found in nature is the harvesting of water from morning fog by the desert-living Namib beetle. Small superhydrophilic patches collect and pin small airborne water droplets. The pinned microdroplets are points of coalescence for more droplets and grow into larger droplets. As the droplets grow, the boundary moves into a superhydrophobic region. Upon reaching a critical size, the droplet mobilizes and rolls off the surface

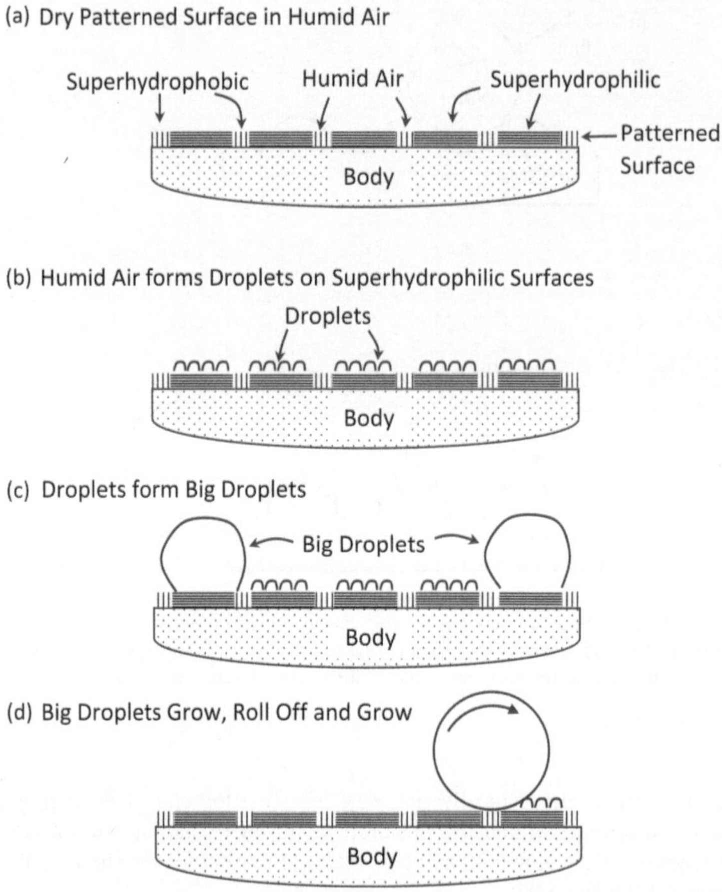

FIGURE 5.16 Patterned superhydrophobic and superhydrophilic surface harvests water drops from the atmosphere. (a) Patterned surface in humid air. (b) Droplets form at superhydrophilic patches. (c) Droplets grow. (d) Droplets grow, roll off, and grow while rolling.

and into the beetle's mouth. Artificial materials patterned with fluid-phobic and fluid-philic regions provide a variety of means of droplet control, ranging from channels, roll-off, and mixing (Figure 5.16) [Zhai 2006] [Feng 2015]. Impregnating the surface with a lubricant promotes the roll-off of small (100 μm) droplets that form by condensation [Anand 2012].

5.1.2.6 Active Mechanical and Electrodynamic Debris Removal

Surfaces self-clean with mechanical motions and electrodynamic actions that propel debris along and away from the surface. At the molecular length scale, functionalized surfaces dislodge and push debris away from the surface with active structures, such as polymer brushes (Figure 5.17) [LeMieux 2007]. At the meso length scale, synchronized wave motions of actuated synthetic cilia readily move particles [Masoud 2011].

Mechanically active surfaces manipulate multiple fluid phases to produce useful damage mitigation. One example is the trapping of a thin layer of air or vapor in a Cassie–Baxter state on superhydrophobic surfaces. The thin layer prevents deleterious effects of direct liquid contact, such as excessive heat transfer due to boiling [Vakarelski 2012].

Electrostatic and electrodynamic forces propel particles along and off surfaces. Charged surfaces projecting electrostatic fields repel like-charged and attract opposite-charged dust particles. Simple electrostatic repulsion is largely ineffective due to practical issues, including the need for particle

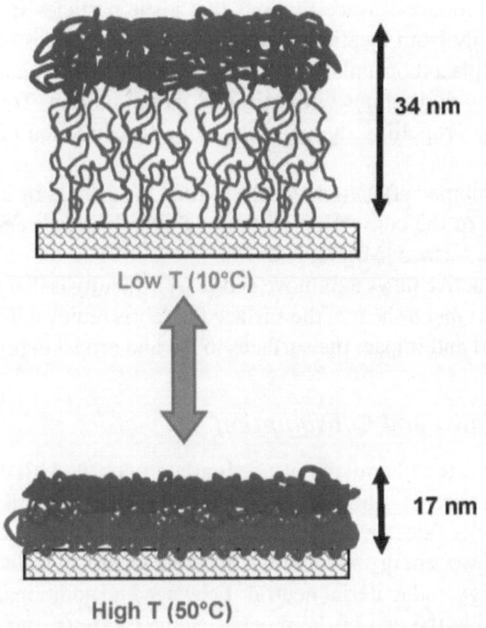

FIGURE 5.17 Temperature-sensitive active molecular brushes with potential to dislodge and push debris away from a functionalized self-cleaning surface. (Reprinted with permission from [LeMieux 2007]. Copyright 2007, *American Chemical Society.*)

removal, not just repulsion. Properly timed relatively low-frequency (~60 Hz) quasi-static electric fields configured to act as traveling waves repel and move particles to desired locations. Early applications of quasi-static electrodynamic effects removed charged airborne particles with traveling waves [Masuda 1973]. Later applications extend the technique to self-cleaning surfaces [Mazumder 2005] [Mazumder 2007]. An issue is that most dust particles are only lightly charged and do not react strongly to electric fields. Slight dragging motions of dielectric particles along the surface introduces additional static electric charges by dielectrophoretic-triboelectrification. Similarly, induction–triboelectrification adds charges to conductive particles. The result is a positive feedback cycle that favors electrodynamic particle motions. A three-phase coordination of electrostatic waves is convenient for moving electrically charged dust particles across a surface, along with meshing nicely into standard three-phase power supplies (Figure 5.18). Estimated power requirements for solar panel cleaning operations run in the range of 10 W/m^2. Combining superhydrophobic surface textures with controlled electric fields offers the possibility of enhanced control of droplet motions.

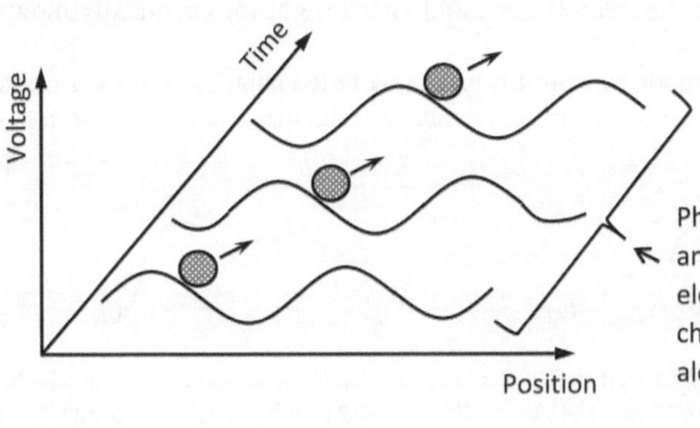

FIGURE 5.18 Three-phase traveling wave quasi-static electric fields propel charged particles along a surface.

Traveling elastic waves produce surface motions that jostle particles sitting on the surface and propel them somewhat randomly from locations with large surface motions to quieter locations with relatively small motions. Sand placed on a plate vibrating in resonance with a mode shape forms distinctive Chladni patterns that follow nodal lines with minimal modal motion. Traveling elastic surface waves produce a similar effect by propelling the particles along the direction of the wave propagation [Bar 2012] [Kalkowski 2015].

A thermomechanical combined effect reduces the fouling of surfaces by condensation. Liquid droplets jump from the surface due to the conversion of excess thermal surface energy and then minimize the return of the droplets to the surface [Miljkovic 2013].

Surfaces can use fluid in active flows to remove debris. A difficulty is that steady flows build up boundary layers which do not exert much shear at the surface for debris removal. Flow past macroscale features generates vortices that swirl and impact the surfaces to lift and propel debris [Slocum 2013].

5.1.2.7 Debris Breakdown and Consumption

Active surfaces break down the chemistry of accumulated debris with catalytic and photocatalytic processes. Many materials are in a high-energy state, yet are nominally stable due to energy barriers (Figure 5.19a). Catalysts sidestep the barriers, destabilize the chemical configurations, and enable the material to move to a lower energy state with a different chemical state (Figure 5.19b). Nonetheless, catalytic processes are energy and material neutral. They need no additional energy to proceed and then leave the catalyst unaltered at the end of the process. Some catalysts, such as platinum, act on a wide variety of chemical species. Others are more specialized.

Photocatalytic cleaning uses energy harvested from incident photons as a driver for cleaning favored chemical reactions (Figure 5.19c). The necessary ingredients are as follows: (1) Surface debris is amenable to photocatalytic breakdown, that is, many organic chemicals. (2) Surface contains photocatalysts. (3) Light energy is available. (4) Contaminant material breaks down under the action of photogenerated chemicals. The primary mechanism of photocatalysis is to split water molecules into a higher energy state of molecular hydrogen and atomic oxygen using energy from an incident photon with sufficient energy, usually in the UV range [Fujishima 1972]. Atomic oxygen being a strong oxidizer breaks down many organic compounds, along with inorganic compounds, such as nitric oxide [Hüsken 2009]. Stains, oil-spills, dangerous organic chemicals, microbes, and the like fade away when attacked by atomic oxygen.

Titania (TiO_2), usually in the anatase crystal form, is the most common photocatalytic substance, but others, such as polyoxometalates, the semiconductors TiO_2, SnO_2, WO_3, ZnO, CdS, ZnS, $SrTiO_3$, and Fe_2O_3 are viable photocatalyzers [Kim 2004a] [Luo 2001]. Hybrid nanoclusters of MoS_2, WS_2, or gold nanoparticles with TiO_2 enhance performance, such as extending the useful input light from UV into the visible range [Ho 2004] [Subramanian 2004] [Li 2001] [Lorret 2009]. 2-Anthraquinone carboxylic acid has potential for facile insertion into cotton cloth that promotes the photoinduced production of reactive oxygen species and hydrogen peroxide from water [Liu 2011a]. The most effective light wavelengths are usually the ultraviolet components of sunlight with wavelengths less than 400 nm for neap titania, and visible light for chemically modified variants [Yuranova 2007].

Geometric packaging is an important concern for titania-based photocatalytic self-cleaning systems. A submicrometer size is good for titania particles to interact with material and to absorb light. Favorable

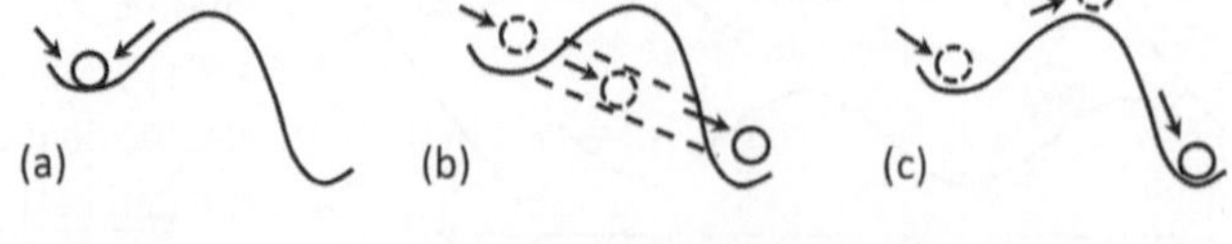

FIGURE 5.19 Energy diagrams for material stability and degradation. (a) Stable material – potential barrier stabilizes system in high-energy state. (b) Catalytic degradation – catalyst provides path from high-energy state to low-energy state. (c) Photocatalytic degradation – photocatalyst provides energy to overcome potential barrier with move to low-energy state.

spatial distributions are near to the surface for absorbing the light, yet at a sufficient depth to be protected from wear and abrasion or distributed throughout the outer depths to achieve a regenerative capability with surface wear. The placement of titania in carbon nanotubes improves the packaging in cotton and clothing with organic stain-cleaning and antimicrobial capabilities [Qi 2006] [Lee 2004b]. Placement of titania on the outer surface of glass requires provisions for preventing sodium in the glass from interacting with and fouling the titania [Greenberg 2000]. Titania photoinduced superhydrophilic wetting may occur simultaneously with titania photocatalysis used in self-cleaning surfaces.

An example is the breakdown of organophosphorous pesticides on clothing with an embedded organophosphorous hydrolase (OPH) enzyme. While simple in principle, the technique requires careful packaging for the enzyme to stick to the cloth, survive through repeated wet–dry cycles, and remain active. A viable technique is with polyelectrolyte multilayers (PEMs) that encapsulate the OPH with a polystyrene sulfonate layer acting as a first layer primer [Singh 2004]. Another possibility is a Cu-containing catalyst attached to single-wall carbon nanotubes to promote breakdown [Bailey 2014].

5.1.2.8 Surface Anti-icing Systems

Operating machinery in subfreezing water-laden atmospheric conditions runs the risk of icing external surfaces. The mechanics of ice buildup on surfaces are complicated. Icing is a combination of water impact, subfreezing temperatures, and nucleation sites. Active anti-icing prevents and removes ice buildup.

Bare hydrophobic surfaces repel liquid water and are effective for anti-icing due to supercooled liquid water droplets, as in freezing rain [Mishchenko 2010]. Once ice forms on the surface, the texture changes and hydrophobic properties degrade. Ice formation requires a nucleation site. The critical radius for nucleation is on the order of 22 nm, which lies below the range of many superhydrophobic coating textures (Figure 5.20) [Cao 2009]. Hierarchical micro–nano textures effectively blunt the growth of ice through a combination of minimizing nucleation sites and the insulating effect of air trapped within the textured surface [Shen 2015]. The effect is strongest at moderate supercooled temperatures and mostly disappears when near to the homogeneous nucleation temperature [Alizadeh 2012].

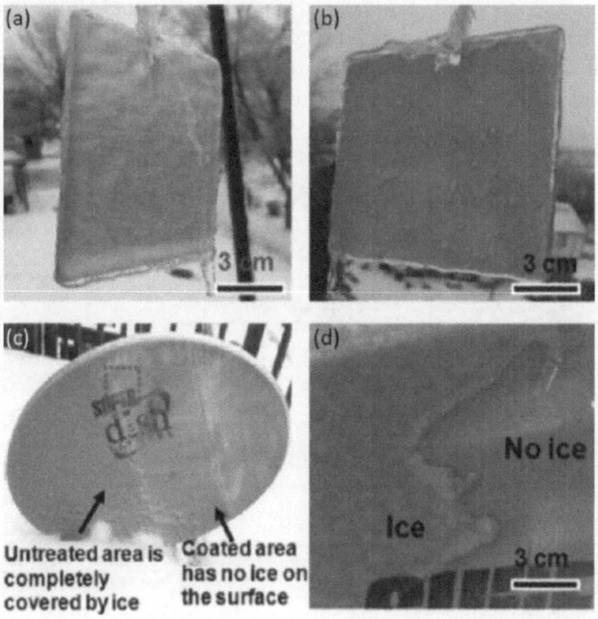

FIGURE 5.20 Superhydrophobic anti-icing coating performance. (Reprinted with permission from [Cao 2009]. Copyright 2009, *American Chemical Society*.)

Methods of reducing existing ice built up on external surfaces:

1. *Mechanical:* Flexing of a surface breaks off ice buildup. Inflatable and deformable surfaces on the leading edges of airplane wings flex to break off ice buildup.
2. *Application of Deicing Chemicals:* Halite and brine are roadway deicers.
3. *Thermal Melting:* Concrete and composites with conductive fibers that produce electrical resistance Joule heating melt ice on the surface [Chung 2004b] [Tuan 2008].
4. *Change Atmospheric Conditions:* Move the system to more favorable conditions, such as changing the elevation of an airplane, or cycling through a defrosting cycle in a refrigerator freezer.
5. *Slippery Surfaces:* Slippery fluid-infused surfaces provide anti-icing functionality. Slippery liquids need to remain liquid at subfreezing temperatures to mitigate icing [Subramanyam 2013]. Hydrophilic behavior, such as produced by a conjugate of poly(acrylic acid)–dopamine or water–glycerin mixtures, produces a slippery surface at subfreezing temperatures [Chen 2015a]. Sublimation from solid carbon dioxide (dry ice) avoids freezing of the fluid [Antonini 2013].

Applications of autonomous self-deicing systems include airplane deicing systems, frost-free refrigerators, windmill blades, overhead electric power lines, ships at sea, and automated bridge deck anti-icing systems. During winter weather, highway bridge decks can ice up before the approaching roadways, leading to treacherous driving conditions. Deicing salts are expensive, environmentally hazardous, promote corrosion, and require timely application. Self-deicing bridge decks are an attractive alternative, with Joule heating of specialized conductive concretes [Tuan 2008].

5.1.3 Self-healing Self-cleaning Surfaces

Self-cleaning surfaces, especially those that rely on surface texture, are vulnerable to damage. Self-healing of self-cleaning surfaces is the remedy. Replenishment of hydrophobic behavior brings active surfaces up from underneath as the top surface wears. For example, methacrylate-terminated dangling chains of perfluorinated poly(ethylene glycol diacrylate) replenishes hydrophobic functionality following wear of the top surface layers [Zhang 2015f].

Colloidal particles migrate to liquid–gas surfaces to produce a superhydrophobic surface upon cooling and solidification. When damaged by scratching and scraping, remelting allows more of the colloidal particles to migrate and heal the superhydrophobic surface (Figure 5.21). Perfluorinated wax with silane-modified silica particles exhibits this behavior [Puretskiy 2012].

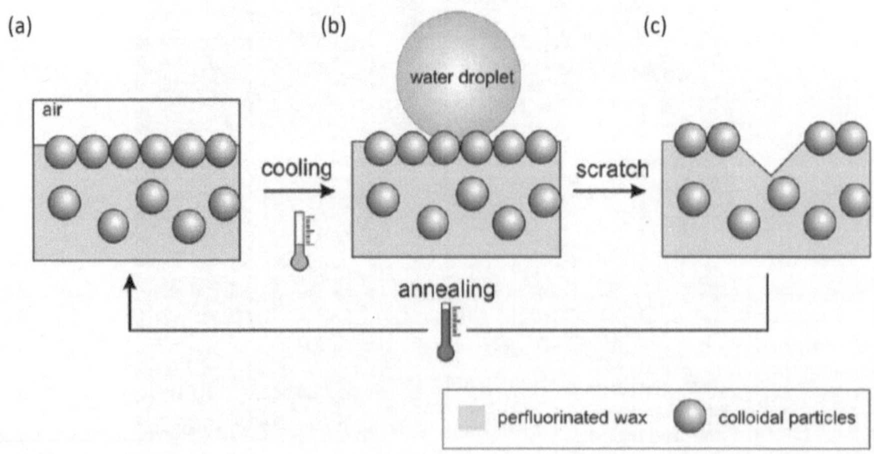

FIGURE 5.21 Migration of particles to colloidal surfaces creates a self-healing superhydrophobic surface. (Reprinted with permission from [Puretskiy 2012]. Copyright 2012, *American Chemical Society.*)

5.2 Surface Crack, Scratch, Puncture, and Gouge Repair

Cracks, punctures, and scratches are common forms of surface damage that affect both aesthetics and functional performance of structural elements. The techniques available for surface self-repair are well-developed for organic coatings, primarily due to the relative ease of activation [García 2011]. Inorganic surface materials, that is, metals and ceramics, are more difficult to activate and have more limited choices for self-repair.

5.2.1 Material Remodeling

Polyurethanes are a versatile family of tough organic polymers with a wide range of functionalities. Much of the original discovery of useful polyurethane compounds was empirical. More recent developments use molecular and nanostructure analysis. Surface coatings made of functionalized polyurethane compounds self-repair from small scratches and punctures through a remodeling process where the molecular structure of damaged material induces reflow to fill the defect. Patent literature dating back to the 1980s discloses how to make scratch-healing polyurethanes, but it does not disclose the underlying healing mechanisms [Orain 1980] [Musil 1987]. More recent efforts using molecular designs achieve superior healing performance with complete scratch-healing at room temperature in 2 hours (Figure 5.22).

Microencapsulated liquid healing systems located near the surface provide an opportunity of self-healing. Surface damage ruptures the microcapsules and initiates a healing reaction. Size and placement of the capsules, along with chemical durability and timely activation, are key design challenges. Thin surface coatings increase the effective parasitic size of the capsules. Harsh environmental loadings, including temperature swings and UV radiation while waiting for the time of repair, challenge the use of microencapsulated self-healing systems. Several systems provide surface self-repair with microencapsulated healing liquids, including two-part siloxane systems [Braun 2010].

Additional heterogeneous material techniques combine the strengths of the individual components, such as crack-healing capabilities of thermoplastics and load-bearing capacity and shape-recovery capabilities of thermosets. A phase-separated mix of 10–30% crack-healing thermoplastic polycaprolactone inside a matrix of shape-recovery thermoset polyurethane produces a durable wood coating system [Ou 2014]. The phase separation occurs during an initial reaction process and produces separate components with submicrometer dimensions for the material to be transparent. Adding clay nanoparticles toughens the thermoplastic.

Polymer networks, such as cross-linking of poly(vinyl butyral) and hexamethylene diisocyanate, recover from surface scratches by a combination of solvent and thermal shape memory effects [Bai 2014b]. The underlying mechanism is that the network of polymers retains a memory of the original unscratched profile to guide the reconstruction of a smooth surface. Similarly, the introduction of

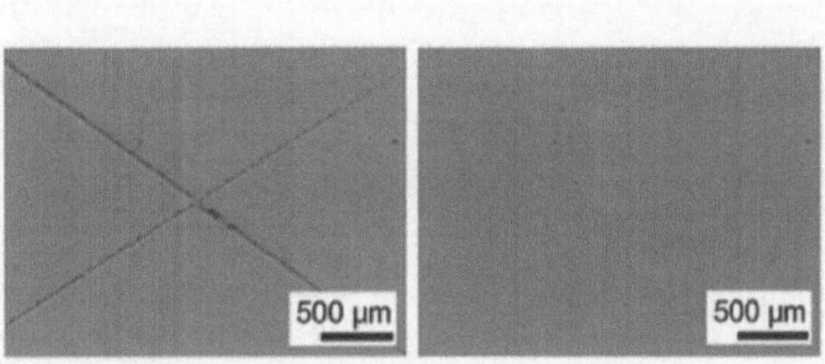

FIGURE 5.22 Polyurethane scratch-healing at room temperature in 2 hours using formulation based on a heterogeneous structure with soft segments containing polytetramethylene ether glycol and hard segments containing isophorone diisocyanate. (From [Kim 2018].)

poly(aryl ether ketone) (12F-PEK) into epoxy resin produces a material that recovers from scratches, with the underlying mechanism presumably explained by the 12F-PEK competing with the epoxy for cross-linking sites at proportions that promote scratch-healing action [Brostow 2002].

5.2.2 Passivation and Accretion Healing of Damaged Surfaces

The deposition of material from a fluid onto surfaces can fill, seal, and heal damage. Such deposition processes are often entropy-driven, self-limiting, and require the following:

1. *Fluid in contact* with the damaged surface
2. *Material in fluid* with sufficient mobility for deposition onto surface
3. *Thermodynamic processes* that favor deposition or reaction onto damaged area
4. *Limiting process* that prevents excessive accumulation.

Passivation results from chemical deposition of a durable barrier on the outside of a solid (Figure 5.23). The outer molecular layers of the solid react with the fluid. The reaction products form an impermeable barrier that protects the solid from further penetration. An example is the corrosion-preventing alumina (Al_2O_3) layer that forms on the outside of solid aluminum. Damage, such as scratches and gouges, expose pure aluminum which quickly initiates a new passivation reaction and heals the barrier. Titanium exhibits similar behavior. Passivation tends to be moderately robust and active only in specific ranges of temperatures and chemistry. Other metals, such as plain steel, may or may not form an impermeable barrier to corrosion. "Black rust" in the form of Fe_3O_4 exhibits anticorrosion behavior, while "brown rust" in the form of Fe_2O_3 does not. Aggressive environments, such as saline water, break down nominally impermeable oxidation corrosion barriers.

Situations that promote the swelling of the top surface layer seal and close tight cracks and small punctures. The uptake of environmental material promotes the top surface layer swelling. A high-temperature application of the technique is spray-on ceramic coatings that protect steel from harsh thermal loading, such as is needed in a nuclear fusion reactor [Gao 2010a]. Multilayer coatings of TiC + mixture (TiC/Al_2O_3) + Al_2O_3 plasma-sprayed on martensitic steel have oxidation-induced swelling in the TiC layer. This swelling produces an overall healing response in a coating system that survives 50 thermal cycles, or more, at over 800°C with minimal damage, not even pinholes.

There are fortuitous situations where passivation acts favorably in aggressive environments, such as high-temperature refractory materials that self-heal by surface remodeling. For example, AlN, CaO, or Y_2O_3 acts as self-healing electrical insulators in liquid-lithium-cooled fusion reactor wall blankets, with operating temperatures ranging up to 500°C [Gohar 1995]. Similarly, thermodynamics favors the formation of an electrically insulating barrier on V–Cr–Ti alloys from calcium dissolved in the lithium. A damaged barrier reforms autonomically [Park 1996] [Park 2003b].

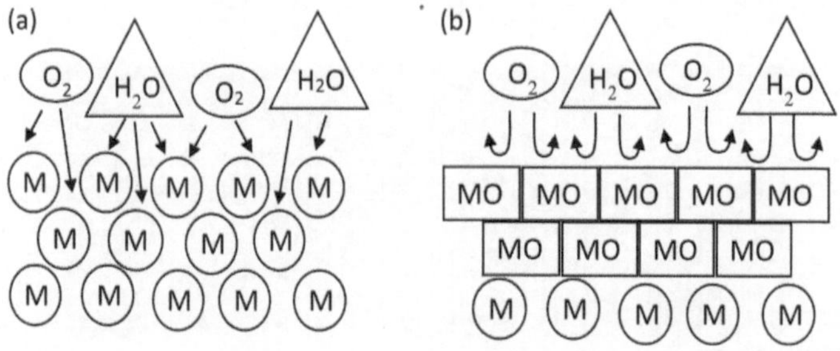

FIGURE 5.23 Oxidation of metal creates impermeable passivating layer that prevents further oxidation. (a) Metal (M) surface in the presence of oxygen (O_2) and water (H_2O) is prone to oxidation and damage that penetrates into the solid. (b) Metal oxidizes to form passivating impenetrable metal oxide (MO) layer.

Surfactants, that is, surface-active agents, are molecular structures that cling to surfaces and provide functionality, usually to reduce surface tension, often with molecules that are hydrophobic at one end and hydrophilic at the other end. Surfactants deposit from a bulk fluid onto either a fluid–solid or fluid–fluid interface. Surfactants are ubiquitous in everyday life and appear in consumer products such as soaps, detergents, and cosmetics. Many surfactants spread and cover broad surface areas, sometimes with long-range organized structures, such as micellar crystal layers. When damaged or disrupted, the surfactant layers reconstruct through entropy-driven deposition from the bulk fluid, sometimes at a fast rate. Measured timescales of the reconstruction of micellar surfactant crystals are on the order of 6 ms. A hypothesis is that surface molecular distribution drives high-speed reconstruction rather than bulk fluid deposition [Schniepp 2006]. Lipids are a subset of surfactants based on naturally occurring fats and waxes. Lipids form bilayers of hydrophilic ends on the outside and hydrophobic ends on the inside that exhibit robust healing of scratches by entropy-driven deposition [Creczynski 2009].

5.3 Surface Texture, Shape, and Reflow Repair

5.3.1 Surface Texture Control

Surface texture is an important functional property. Some surfaces should be smooth, others rough, and others patterned. Mechanical loading wear and chemical erosion degrades surface texture.

Electromigration smooths wrinkled metallic surfaces by the forced movement of atoms due to momentum transfer from the wind of flowing electrons. High current densities, weak metallic bonding (as in aluminum versus copper), and other available energy sources (heat, surface, and elastic) promote electromigration. Electromigration normally moves only small amounts of material but is a major source of failures in microelectronic circuits [Ho 2022]. Surface smoothing comes from disrupting the Asaro–Tiller–Grinfeld (ATG) instability that occurs when elastic strain energy and surface energy compete to form wrinkles as a preferred low-energy state. In a somewhat unusual role, electromigration is beneficial and disrupts the destabilizing competition to heal and smooth the surface [Tomar 2008].

5.3.2 Surface Reflow Repair

Many surfaces are active, with material that flows, that is, reflows, along the surface. Surface reflow healing requires the following:

1. *Surplus material* that flows to the damage without jeopardizing integrity of undamaged regions.
2. *Forces that drive* material to flow across the damaged region.
3. *A templating or shape memory action* that guides the reflow to form a smooth surface.

Shape memory drives many polymer-based repairs. Pinholes produced by mechanical puncturing leave residual stresses that drive and guide reflow sealing of the pinhole damage. A simple version is stress relaxation that moves the walls to close the hole. Some crystalline polymers exhibit this behavior, possibly because the crystalline structure contains a templated memory of pre-damaged configurations. Raising to elevated, but submelting, temperatures relaxes the prestress left by a puncture pinhole to close holes in crystalline photografted polyethylene-*g*-poly(hexyl methacrylate) [Shimada 2004]. An empirical three-term exponential relaxation relation describes the hole closure:

$$\frac{S(t)}{S(0)} = A_{vf}\exp\left(-\frac{t}{B_{vf}}\right) + A_f\exp\left(-\frac{t}{B_f}\right) + A_s\exp\left(-\frac{t}{B_s}\right) \tag{5.13}$$

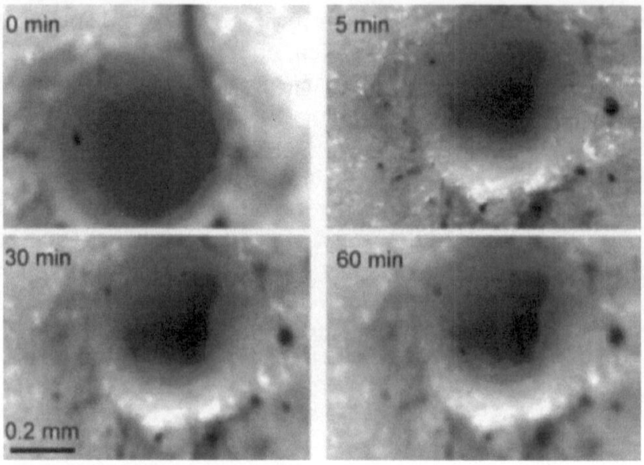

FIGURE 5.24 Time evolution photographs of the shape of a pinhole closed by shape memory recovery in PE-*g*-PHMA (GR 61%) at 335 K. (From [Shimada 2004].)

Here S(t) is the area of the hole as a function of time; A_{vf} and B_{vf}, A_f and B_f, and A_s and B_s represent the very fast, fast, and slow relaxation constants, respectively (Figure 5.24).

Some variants of polyurethanes heal surface scratches by a mechanism attributed to molecule slippage. Adjusting the mix and measuring the damping factor, tan(δ), while the curing polymer is in the glass transition region tunes the scratch-healing action [Ho 1998] [González 2011]. Damping factor values less than 0.4 correspond to materials that do not self-heal scratches. Materials with larger damping factors, a value between 0.7 and 0.9 being optimal, self-heal. Higher damping values correspond to more rapid healing, but the material may be too soft and prone to scratching. Presumably, the same molecular slippage behavior gives rise to self-healing, softness, and larger damping values. Similarly, adding ionic liquid to an epoxy mix creates a material that is softer and more prone to scratching. When scratched, the ionic-liquid-modified epoxy reflows back to form a smooth surface, while the neat epoxy being initially more difficult to scratch does not heal and remains scratched [Saurín 2015].

5.3.2.1 Molecular Scale Surface Reflow

Molecular monolayers and bilayers heal scratches by entropy-driven reformulations of the surface. The process is sensitive to the relative entropies of the coating and substrate. Octadecylphosphonic acid self-assembled monolayers built with a drip coating method heal scratches from an atomic force microscope, but do not heal when created by crystal melting [Neves 2001].

5.3.2.2 Micro- and Nanoscale Surface Reflow

Colloidal hydrogels formed on surfaces as thin film assemblies of submicrometer-scale gel particles undergo an initial fabrication step with entropy-driven self-limiting processes placing the gel particles in a smooth tightly packed surface-covering configuration. These hydrogels are normally soft and susceptible to mechanical damage. Changing the environment, such as adding water, reactivates forces that drive the gel particles so that they reflow back into a smooth surface configuration [South 2010].

Preloading microcapsules of healing liquids into paints and surface coatings facilitates recoating and healing of scratches, prevents corrosion, and inhibits biofouling [Sarangapani 2007] [Samadzadeh 2010] [Sauvant 2008].

5.4 Active Anticorrosion and Antioxidation Coatings

Oxidation and related electrochemical processes corrode metals. Corrosion transfers electrons from the metal to oxygen which induces ionic bonds that form metal oxides. Electrical conductivity of the metal is essential to the process, as it allows long-distance transfers of electrons. Uninterrupted, the flow of electrons leads to progressively larger amounts of corrosion. Impeding the flow of electrons impedes corrosion. The electrical flow can serve as trigger and power source for smart anticorrosion coatings that act only when corrosion is active.

Some key issues to consider for mitigation and healing of corrosion:

1. *Corrosion is not a single process*: Corrosion is a family of related processes that cause damage by oxidation.
2. *Oxidation is a thermodynamically favored state for metals*: Oxidizing metals lowers the free energy.
3. *Various corrosion processes differ* in complexity, aggressiveness, and mitigation techniques.
4. *Corrosion processes require multiple conditions* to be in place for the process to move forward.
5. *Interventions* stop corrosion at an early stage and prevent major damage.

5.4.1 Physical and Electrochemical Corrosion Protection

Electrochemical reactions comingle electromotive forces, electrically conductive paths, and fluid-borne mobile ions, often in competing and/or cooperating combinations. While some electrochemical reactions drive toward corrosion damage, others, such as cathodic protection (CP), mitigate and even heal damage [Perez 2004]. Passive CP methods use sacrificial metals, such as zinc, to drive the reaction. Sacrificial anode methods can be autonomous but require periodic replenishment of the sacrificial material. Active CP methods use external electric power to drive the beneficial reactions.

The controller for active CP must compensate for nonlinear percolative aspects of corrosion, especially when protecting surfaces on subsurface structures, such as steel pipes buried underground or steel reinforcing bars in concrete. The system may initially lack conductive paths and require high voltages and low currents to percolate the system. The applied voltage drives changes in the electrochemical structure of the system to create conductive paths, eventually leading to a percolated state with dramatic drops in electrical resistance and the ability to sustain a healthy corrosion-free state with lower voltages and currents. Excess currents produce hydrogen, which embrittles steel and can be an explosion hazard in confined environments. Pulsed current methods using square waves with duty cycles on the order of 50% and frequencies in the kilohertz range may be more effective at achieving the low-resistance conduction paths while avoiding deleterious side effects of continuous-current CP [Koleva 2009].

For CP in concrete applications, specialized conductive admixtures provide a conductive path without the need for a percolating break-in period [Yehia 2010]. Less common, but still useful, are a similar set of electrophoretic processes where mobile dipoles replace mobile ions. Retrofit implementations of cathodic protection typically occur following severe corrosion. Including cathodic protection into the repair is often expedient, especially for aggressive corrosion environments, such as in fiber-reinforced polymer wraps that reinforce underwater concrete columns [Sen 2010].

Electrical antiscaling devices for water pipes prevent the buildup of calcium carbonate and similar materials on the inside of water pipes. Scale buildup is a chemical precipitation process. Calcium and carbonate ions dissolved at unsaturated concentrations bond and adhere to existing scale layers on the pipe surface. Applied electromagnetic forces disrupt scale buildup by agitating the water-borne calcium and carbonate ions, breaking them free from surrounding water molecules, and enabling the formation of molecular calcium carbonate that is free and not bound to the pipe surface. Flushing of water removes the calcium carbonate from the system. Tuning and then varying the frequency and amplitude of the electromagnetic field is often effective [Jefferson 1998].

5.4.1.1 Physical Separation Barriers

Epoxy coatings form tough physical separation barriers to protect steel from corrosive environments. Additives, such as alkoxy silanes, formulate an epoxy anticorrosion coating that is both hydrophobic and capable of reflow self-repair of scratch damage [Shah 2008].

5.4.1.2 Enhanced and Active Passivation Barriers

Releasing and electrochemically activating passivating reagents enhances passivation as corrosion initiates. Conductive polymers have a potential role in this regard since they combine polymer and electrochemical functionality. Polyaniline in inherently conducting polymer films releases beneficial anions when a breach in the coating causes it to become a cathode on aluminum [Kendig 2003] [Wessling 1996]. Polyaniline doped with phosphonic acid forms a passivating iron-dopant salt barrier on the surface of mild steels [Kinlen 2002]. Double-stranded polyaniline molecular complexes of noncovalently bound side-by-side pairing of polyaniline and a polyanion is electrically conductive, electrochemically anodic with respect to aluminum with layering that promotes oxidation passivation of aluminum surfaces [Racicot 1997].

Aggressive environments, such as those prone to saltwater spray, easily overcome oxide passivation in aluminum. More effective are multilayer and multicomponent active protective coatings. One system begins with surface mechanical attrition (SMAT) to harden the surface of the aluminum leaving it in a dense nanocrystalline state, followed by placing an outer microarc oxidation coating (MAC) [Wen 2011]. An aggressive environment may breach the outer MAC layer to expose the underlying SMAT layer. The edges of the SMAT and MAC layers interact to form a resilient oxide layer on top of the SMAT, which heals the system and impedes further corrosion (Figure 5.25). An alternative uses anodic porous alumina as a base layer on top of aluminum followed by an outer layer of zinc, aluminum-containing layered double hydroxide (ZnAl-LDH) film grown in situ. This active coating heals scratches and gouges, even in a saline environment. A dissolution/recrystallization mechanism leaches ZnAl-LDH out of the neighboring healthy crystals and grows new crystals onto the bare aluminum (Figure 5.26) [Yan 2013].

Functionalized surfaces inhibit corrosion. Many of these surfaces release a chemical inhibitor upon receiving a corrosion signal. Mechanical damage that exposes bare metal triggers the inhibitor to leach out of the coating onto the bare metal region, and then bind to form a healed impermeable barrier. Electrochemical processes induced by the bare surface drive the binding process, along with covering and completely healing the damage. Some schemes are as follows:

1. *Chromate-laced Coatings:* Chromates are excellent corrosion inhibitors, even when applied with nanometer-scale thicknesses, but are environmentally toxic [Buchheit 2003].
2. *Chromate-free Environment-friendly:* Sol–gel-covered porous titania layer loaded with cerium nitrate and benzotriazole corrosion inhibitors to protect titanium [Lamaka 2007].

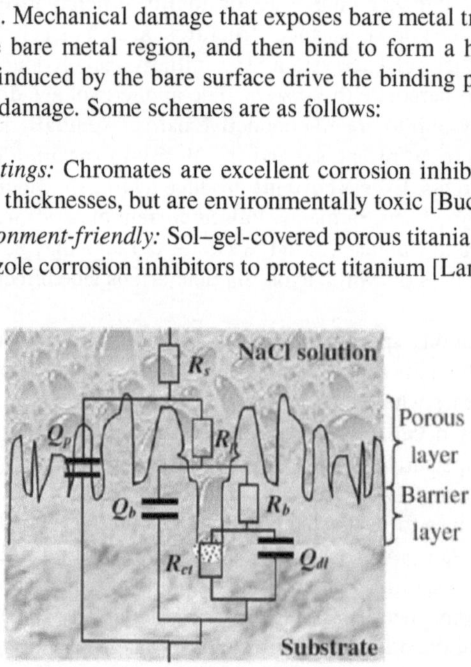

FIGURE 5.25 SMAT and MAC form active anticorrosion surface on 2024 aluminum immersed in 3.5% NaCl solution. (From [Wen 2011].)

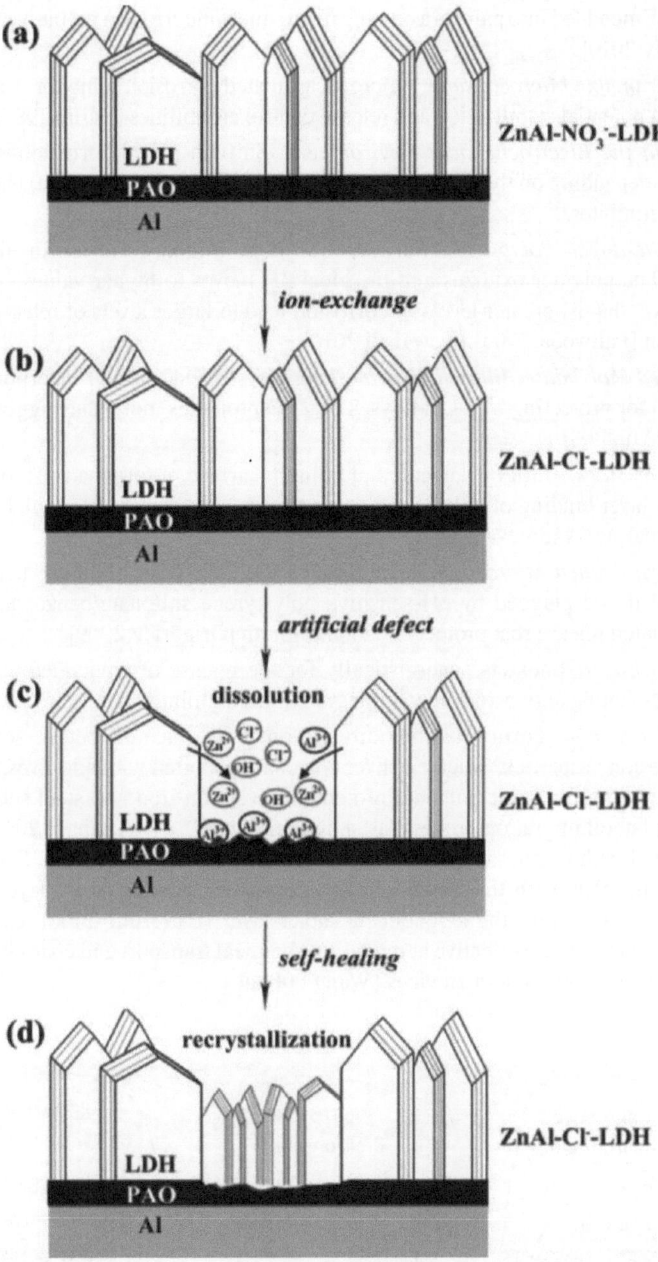

FIGURE 5.26 Dissolution/recrystallization anticorrosion self-healing mechanism where ZnAl-LDH leaches out of the neighboring healthy crystals in a saline environment and grows new crystals onto the bare aluminum. (From [Yan 2013].)

3. *Micrometer-scale Cage-like Particles:* Cages hold anticorrosion particles in reserve as part of a protective layer. Damage to the outer layer causes the particles to swell and open the cage pores, thereby releasing the anticorrosion particles, which deposit on the metallic object and form a new protective layer [He 2009].
4. *Coated Steel:* First with a layer of Al-Zn followed by second layer of phosphoric acid salt laden into an insoluble chromic organic coating [Matsuzaki 2003].
5. *Magnesium and Vanadium Oxide:* Nanometer-thick coatings form nanoscale flowering crystals that heal pitting damage and restore corrosion resistance [Hamdy 2011].

6. *Linseed Oil:* Embedded into paint is a coating repair and anticorrosion palliative [Suryanarayana 2008] [Jadhav 2010].
7. *Embedded Frangible Microcapsules:* Microencapsulated corrosion inhibitor is an alternative to leaching with potential stabilization and release control capabilities [Bailin 1987] [Mehta 2009].
8. *Alterations in the Electrochemical Environment:* Shift in pH to corrosion-initiating values (high or low, depending on the details of the specific corrosion environment) trigger the release of corrosion inhibitors.
   a. *Molybdate-doped Polypyrrole Polymer Coatings:* Inhibit corrosion in iron by releasing anions. The polymer oxidizes and the local pH moves to higher values. The release rate is reactive, that is, greater levels of corrosion lead to larger levels of release and corrosion inhibition [Paliwoda 2005] [Kowalski 2010].
   b. *Nanosized Molybdate Pillared Hydrotalcite:* (HT-MoO42–)/ZnO incorporates composite (HTMZ) for protecting Mg–Li alloys. The ZnO promotes molybdate aggradation on alloy surface [Yu 2008].
   c. *Multilayer Stack:* Built on sonicated aluminum surface, it forms a fresh oxide that facilitates top-layer binding of polyelectrolytes stack laden with corrosion inhibitors and a sol–gel top coating [Andreeva 2008].
   d. *Mesoporous Silica Microcontainers:* Loaded with corrosion inhibitor benzotriazole and surrounded and plugged by pH-sensitive polystyrene sulfonate/benzotriazole for corrosion-initiated release that protects 2024 aluminum [Grigoriev 2009].
   e. *Chitosan Layer:* Interacts synergistically for the release of embedded cerium nitrate to form self-healing anticorrosion coating system for aluminum [Carneiro 2012].
9. *Chemical Conversion:* Forms coatings directly on the surface of metals, some have excellent anticorrosion properties. Silicate conversion coatings, often with additives, such as cerium nitrate, placed on top of zinc galvanic protection layers on iron and steel self-repair scratch damage by a lateral migration process, as shown in Figure 5.27 [Aramaki 2002] [Yuan 2011]. Primers loaded with magnesium protect aluminum in a two-stage process, first with cathodic polarization and then with the formation of a porous magnesium oxide layer, driven by the leaching of material from the neighboring intact layer [Battocchi 2006]. Continuum-based finite element methods are effective at modeling chemical transport and coating buildup related to corrosion protection, even in crevices [Wang 2004a].

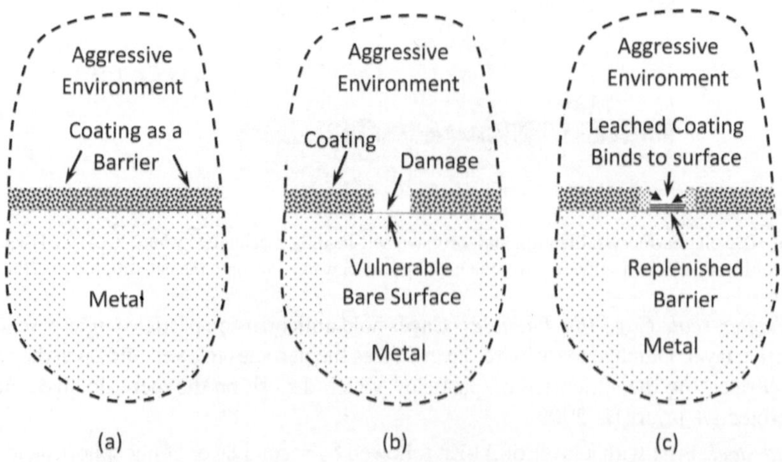

FIGURE 5.27 Active anticorrosion barrier. (a) Coating forms a barrier to block aggressive environment from reaching metal surface. (b) Damage breaches protective barrier and exposes bare metal to aggressive environment. (c) Anticorrosion compounds leach out of coating, bind to bare metal, and regenerate barrier. (Adapted from [Yuan 2011].)

Oxidation also attacks nonmetallic materials such as semiconductors. Diamond oxidizes at high temperatures (~1,000°C). Boron carbide coatings block the oxidation and self-heal by fluidic flow to fill pinholes and improve overall compressive strength of diamond [Sun 2016]. This may be superior to boron doping, which also limits oxidation, but distorts the crystal lattice of the diamond.

5.5 Lubrication and Wear

Smooth-running long-lasting machinery depends on wear-minimizing well-lubricated joints. Failure of wearing surfaces disables machines. Self-healing lubrication systems and wearing surfaces are a remedy.

5.5.1 Lubrication Replenishment

Lubricating fluids fill tight gaps between the moving parts of machines. The fluid separates the parts, allows low-friction and wear-minimizing movements, and removes debris. Breakdown and loss of lubricating fluid damages and disables machinery.

Hydrodynamic and hydrostatic processes pressurize fluids that keep parts separated. Hydrodynamic lubrication uses the motion of the moving parts to create viscous drag forces and pressures that pull lubricating fluid into the joint. Hydrostatic lubrication pumps fluid under pressure into the joint, largely independent of the motion. Oil-filled journal bearings are often hydrodynamic, while air bearings are typically hydrostatic. Start-up and intermittent motions are concerns. Machinery that moves intermittently suffers from the problem of stiction where solid parts stick together, often unpredictably. A lubricant that remains in the gap between the parts while they are not moving minimizes stiction. An ordinary fluid with Newtonian viscosity bleeds out and causes solid-to-solid contact and stiction. Many modern lubricating fluids prevent stiction with a non-Newtonian formulation that has a portion of the fluid bind to the surface and not bleed out. Cartilage, brass bearings, and some plastic bearings are self-lubricating. The lubricating fluid flows through pores on the solid-bearing surface, usually in response to motion and pressure across the joint (Figure 5.28).

Microelectromechanical systems with sliding parts are especially prone to stiction and other friction problems. This is largely a consequence of length scaling. Stiction and friction forces tend to scale as L^2, with a characteristic length L as a critical dimension. Inertia forces scale as L^3. At micrometer length scales the stiction and friction forces can dominate inertia forces. The materials (silicon, PMMA, etc.) commonly used in MEMs devices are a contributing factor. An obvious solution is to insert a lubricant between the moving parts and keep the solid surfaces separated. However, length scaling again is an issue.

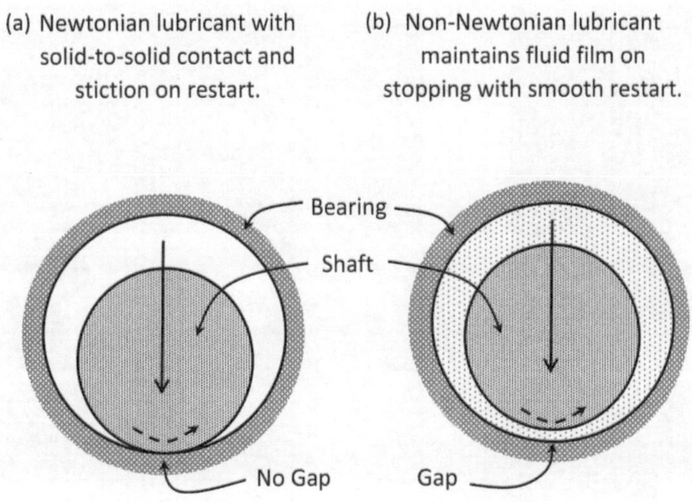

FIGURE 5.28 Hydrodynamic fluid-bearing settles under load. (a) Newtonian fluid flows out of gap and leads to solid-to-solid contact. (b) Non-Newtonian fluid remains in gap and prevents solid-to-solid contact and easy start-up without stiction.

Viscous forces scale as L^1 and become effectively quite large at the microscale. A workaround is to bind a thin layer of lubricant to the wearing surface. Surface-bound lubricants are vulnerable to scraping removal during wear cycles and typically need replenishment. Prepositioned surface-chemistry-bound lubricants can free the movement of silicon-based MEMs devices. Fluoropolymers and hydrocarbons act as autonomic mobile-to-bound replenishing lubricants [Eapen 2005] [Hsiao 2011]. Supramolecular techniques provide enhanced capabilities, such as with quadruple donor–acceptor moieties based on 2-ureido-4[1*H*]-pyrimidone (UPy) [Li 2013b].

5.5.2 Seals for Lubricants

Seals on bearings allow relative movement of machine parts while retaining lubricating fluid inside and keeping dirt and debris outside. Most bearing seals use tight gaps, labyrinths, and compliant materials as barriers that allow relative motion. Thermal expansion, damage, and wear to moving parts change shape and affect seal performance. A self-healing seal adapts to the shape changes and still performs as required. One possibility is leaf seal assemblies that use hydrodynamic lubrication to maintain the gap while actuating as needed to accommodate shape changes and maintain the seal [Grondahl 2013].

5.5.3 Controllable Sealing Surfaces

Controlling surface permeability facilitates self-healing by regulating the flow of material through a structural system. Closing of pores reduces permeability and material transport. Biological systems maintain stasis by regulating flow-through pores. Plants control internal hydration with the opening and closing of stomata pores on the bottom side of leaves. Internal physiological signals promote internal swelling of guard cells against asymmetric constraints on the side of the stomata. Curling due to asymmetric loading opens and closes the pore.

Pore closing in engineered materials combines pore closing mechanics with stimulus-response behavior. Pore wall and material matrix swelling is one of the primary methods of pore closing with potential to seal and open upon command. Numerical simulations confirm that artificial lipid bilayer membranes create controllable pores with hydrophilic–hydrophobic Janus particles. Upon formation of a pore with excess tension, the Janus particles migrate to the perimeter with the hydrophilic tail pointing inward. As the tension releases, the pore closes, but the Janus particles remain in position. A smaller tension quickly reopens the pore due to the presence of the nanoparticles (Figure 5.29) [Alexeev 2008]. An irreversible

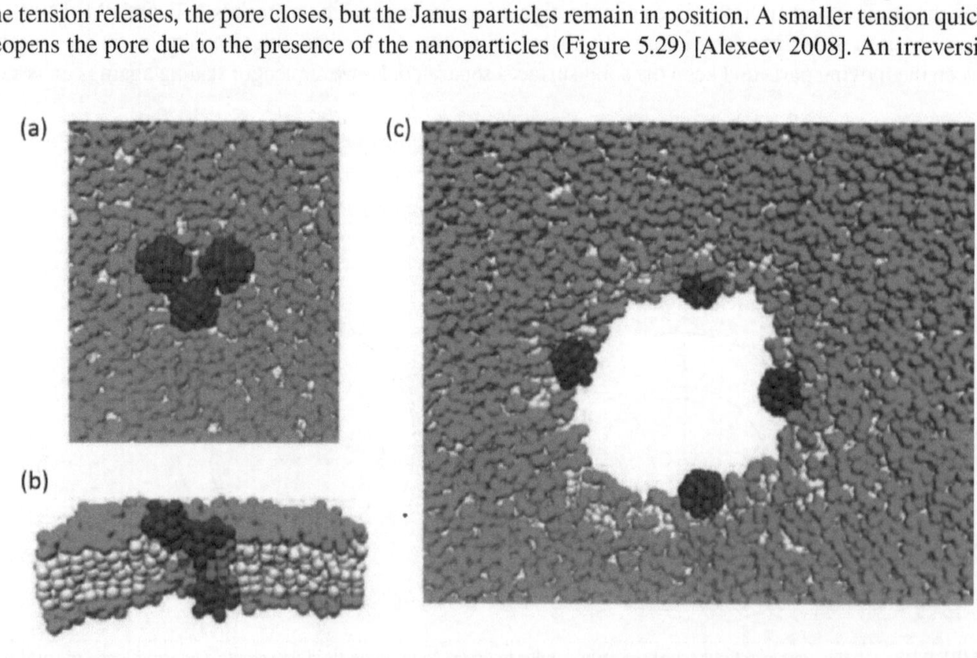

FIGURE 5.29 Janus particles breach and then form controllable pores in lipid bilayer membrane. (From [Alexeev 2008].)

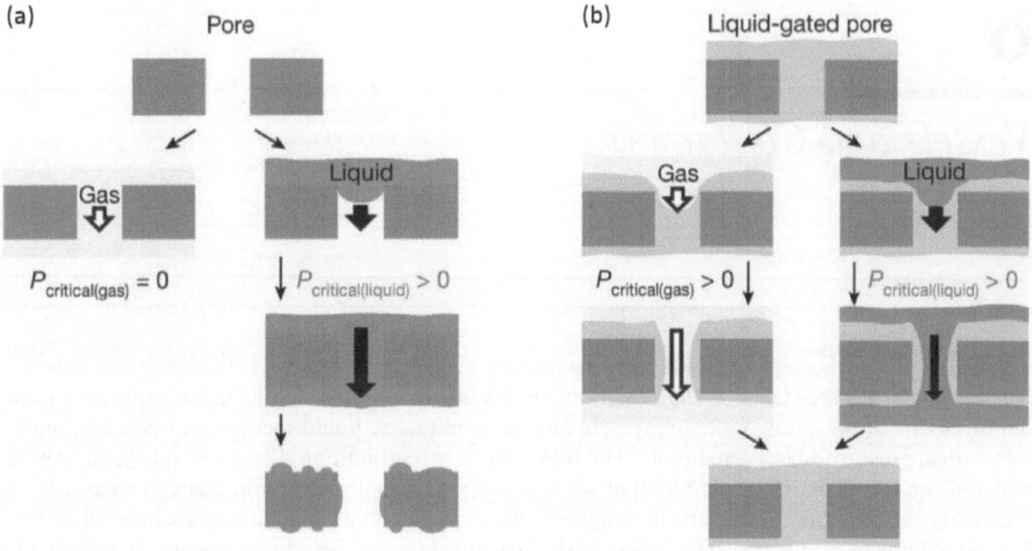

FIGURE 5.30 Liquid control of gating of pore. (a) Uncontrolled fluid with uncontrolled pore. (b) Controlled fluid with controlled pore. (From [Hou 2015].)

one-use activation approach for controlled permeability is a dual-layer surface where one layer is porous and allows the flow of liquids and the other is impermeable. As long as the impermeable layer remains intact, the fluid does not pass through the surface. When environmental conditions change, a triggering action disintegrates the impermeable layer exposing the underlying permeable layer and allowing for the flow of healing liquid.

Infusion of a porous surface with a controlling liquid dramatically affects the way that fluids – both liquid and gas – traverse the pores. Without the controlling liquid, gases flow directly through the pores and liquids penetrate when the differential pressure overcomes surface tension produced by the meniscus of the fluid. Length scaling favors surface tension forces over pressure differentials for small orifices. Tuning the surface tensions of the controlling fluid allows for the selective passage of gas, liquid, and/or a combination through the pores (Figure 5.30) [Hou 2015].

5.6 Optical Surfaces

Self-healing and reconfigurable surfaces in optical instruments, in particular diffraction gratings, benefit from control and reformation of surface texture. An example is azoaromatic polymer films [Rochon 1995]. High-intensity interference patterns from an argon laser produce a localized sinusoidal diffraction grating texture that is long-term stable. Heating the film above the glass transition temperature of approximately 95°C smoothens the surface texture. Upon cooling, the surface rejuvenates and is capable of forming another diffraction grating.

6

Vessels and Containers

6.1 Introduction

Keeping materials separate is a vital task of many engineering systems. Among the applications are bulkheads in ships, dams, food storage containers, fuel tanks, liquid medicine dispensers, pneumatic tires, pressure vessels, and roofs. The following are the conditions that favor self-healing with material-separating barriers: (1) Small physical damage to a material barrier leads to a cascade of secondary failures that disable an entire system, for example, a nail puncture in an automobile tire forces the vehicle to stop. (2) Intervening early to arrest damage cascade scenarios is effective. (3) The material separation barrier is only a small part of the overall weight and cost of the system. (4) The barriers naturally set up a difference of free energy on opposite sides that may initiate and power the repairs.

6.1.1 General Problem of Material Separation

Material separation barriers often divide sealing and structural support into separate tasks performed by different components. It is common to place a relatively thin impermeable sealing layer on the outside of a thicker, structurally stiff, but perhaps not as impermeable, supporting layer. Damage to the sealing layer causes leaks. Material penetrates the barrier and damages the underlying structural layers. The loss of localized structural support exacerbates the leaks with positive feedback that cascades into a complete collapse of the wall. An example is the phenomenon known as piping at the base of earthen dams. Small leaks form self-reinforcing water passages, that is, pipes, that grow by erosion and lead to large-scale structural failures (Figure 6.1). Stopping piping early in the process with autonomous self-sealing techniques has advantages.

Material separation structures and tasks vary widely but have common issues. Unwanted material flows through a breach in the separation wall. The relative pressure across the barrier is a consideration. Low differential pressures are generally less than 1 MPa (150 psi). The failure consequences of low-pressure systems are usually relatively mild, but occasionally are dangerous. High differential pressures, that is, those larger than about 1 MPa (150 psi), generally store large amounts of mechanical energy in both the contained materials and the vessel walls and have more severe failure consequences.

The intent of most containment vessels is to maintain strict material separation. Sometimes the walls are functional and allow material to flow through under control. The control mechanism depends on the size of the fluid paths, the interconnection of the paths, and material properties, such as fluid viscosity and wetting. This flowing fluid can heal other structural components and may help to heal or clean the holes, pores, and fluid regulation processes themselves.

6.2 Methods of Self-repair to Stop Leaks

Multiple methods for self-stopping of leaks are available (Figure 6.2). Most rely on a vessel wall remaining sufficiently intact to support the patch, plug, or new seal.

DOI: 10.1201/9781003396048-6

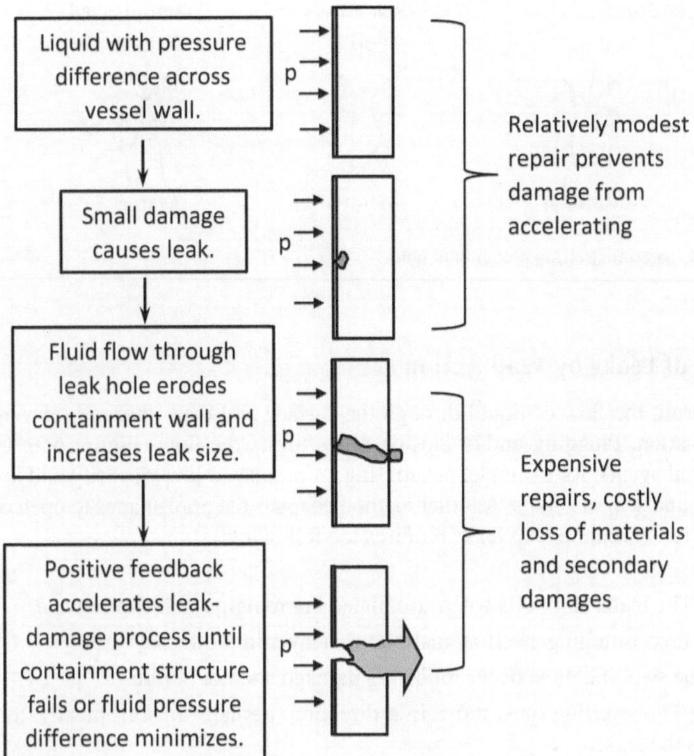

FIGURE 6.1 Advantage of early leak sealing prevents cascading failure.

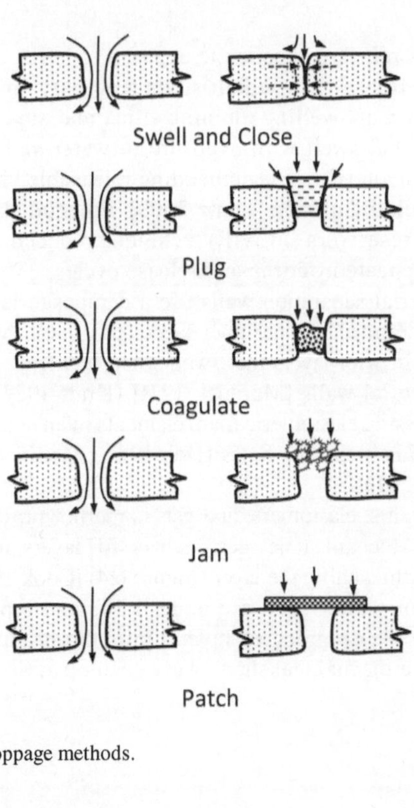

FIGURE 6.2 Hole-closing leak stoppage methods.

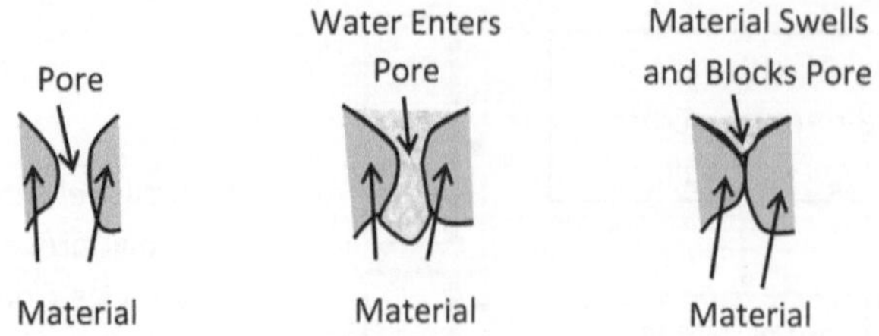

FIGURE 6.3 Bulk material swelling plugging of holes.

6.2.1 Closing of Leaks by Wall Action

Porous solids permit the flow of liquid through the surface and then internally through the bulk solid, largely by percolation. Plugging and/or closing pores limits the flow (Figure 6.3). One method plugs pores with external agents, for example, penetrating, or preplaced low-viscosity and low-surface-tension liquids that enter and seal the holes. Another method changes the pore shapes to open or close as needed.

Bulk volumetric swelling to stop leaks requires the following:

1. *Material:* The material swells when stimulated but remains stable otherwise.
2. *Location:* Prepositioning swelling material as a layer in the vessel wall.
3. *Timing:* The swelling must occur following damage and not before.
4. *Direction:* The swelling must move in a direction, perhaps anisotropically, to close the hole that causes the leak.
5. *Limited:* Swelling is spatially and temporally limited to the needed healing action.
6. *Strength and Durability:* Swelling material is tough enough before and after damage to serve as a structural material or can be packaged in a suitable stiffening composite structure.

An important application is reinforced concrete exposed to harsh environmental conditions. The penetration of water leads to freeze-thaw, alkali-silica reactivity, and chloride-ion-induced corrosion vulnerabilities. Mixing water-swelling alumina silica and superabsorbent particles into concrete creates pore structures that swell with exposure to water and prevent further water ingress [Kishi 2012] [Jensen 2013]. An alternative is embedding releasable viscosity modifiers into the concrete, which decreases the bulk diffusivity [Bentz 2009]. Embedded polymers using the fine lightweight aggregates as internal reservoirs (FLAIR) technique reduced chloride penetration depths by up to 40%. FLAIR supports repeated wetting and release cycles.

Many swelling-sealing material separation walls use a composite layered structure. The outer layers are normally stiff and impermeable. The inner layers can swell. Many fuel tanks use elastomers and cross-linked polymers for interior layers that swell when fuel leaks through nonswelling, yet compliant and burst-resistant structural walls [Murdock 1921] [Kraft 1927] [Sullivan 1946] [Smith 1947] [Underwood 1970]. Precompressing elastomeric foam elements in an antagonistic configuration provides supplemental expansion that aids in closing holes [Holt 1947]. Modified sol–gels act as absorbent and swellable materials [Edmiston 2013].

Isotropic bulk swelling of foams, elastomers, and gels is inefficient. Much of the material movement does not help to close the hole. Mechanical devices, such as stiff layers, use swelling to urge the material directly into the gap instead of thickening the layer (Figure 6.4) [Cook 1972].

Sometimes plugged holes in a bulk solid are undesirable and represent a form of damage. An example is an oil well where the plugging of holes limits the production of oil. The insertion of downhole devices that produce mechanical shock waves can open the holes and increase productivity of the well [Ageev 2014].

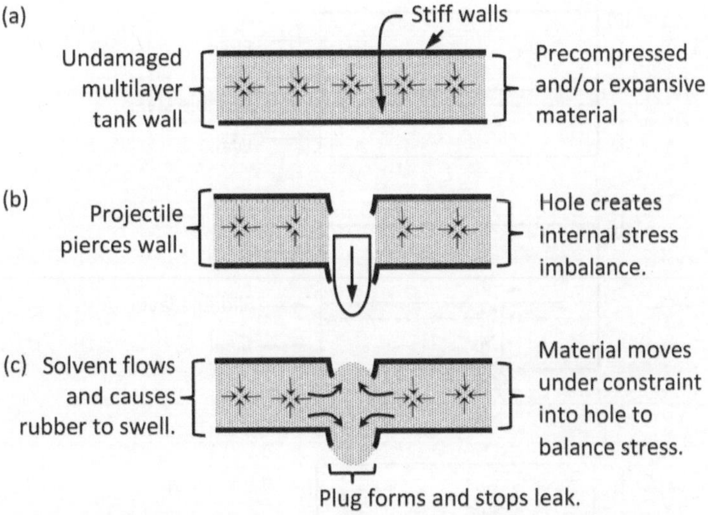

FIGURE 6.4 Swelling and movement due to relief of precompression plugs hole in fuel tank.

A similar autonomic version combines a swellable material with a reticulated skin that opens and closes "scales"' as it swells and contracts with the uptake or release of fluid (Figure 6.5). This includes nanocrack-regulated self-humidifying membranes [Park 2016b]. Swelling-plugging smaller holes is also possible. Nanoclay particles in concrete induce ettringite dendrite swelling that closes pores [He 2008].

Cellular materials have microscale degrees of freedom available for use in swelling sealing. Cork and stiffer forms of wood have sealed bottles, jugs, and barrels for millennia. As early as 1900, Cochrane patented a compressed sponge rubber layer to self-seal puncture leaks in tires, with pneumatic pressure providing the prestress [Cochrane 1900]. A patent from 1919 disclosed a self-sealing fuel tank with a multilayer wall with a precompressed rubber layer that expands to seal projectile-induced leaks (Figure 6.6a) [Thacher 1919]. A patent from 1972 uses a layered structure with pockets of encapsulated material that expand into a sealing foam upon breaching of the pocket wall by the evaporation of a volatile solvent [Olevitch 1972].

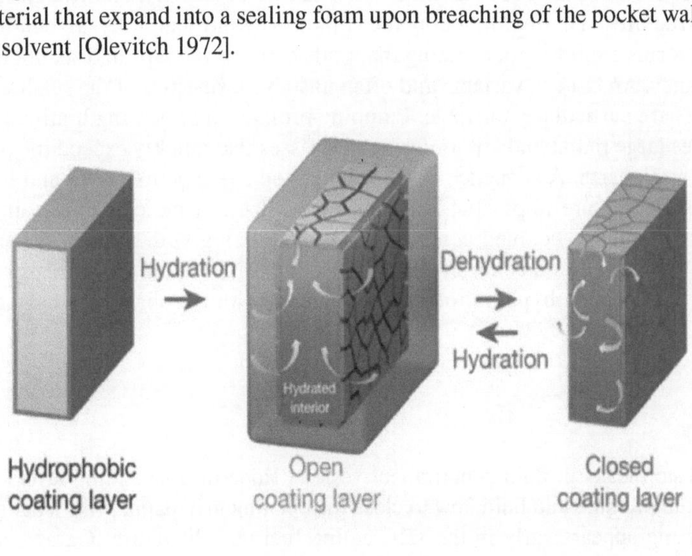

FIGURE 6.5 Reticulated skin on swellable bulk material for autonomic fluid-level regulation. (From [Park 2016b].)

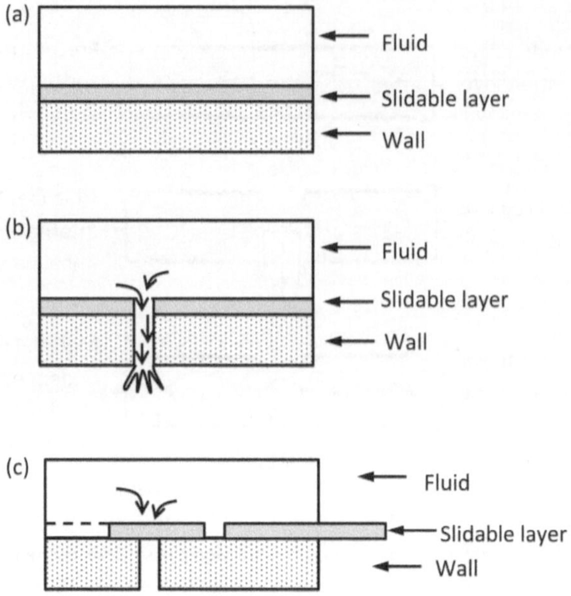

FIGURE 6.6 Sliding and covering layer that forms a patch to seal leaks. (a) Intact wall seals against leaks. (b) Through-wall puncture causes leak. (c) Layer slides and fluid pressure seals leak.

An example of swelling-induced sealing is downhole-drilling petroleum mining which requires a good seal between a steel pipe drill string and geologic formations, usually with cement mix that has density-modifying foam. Leaks of oil and water through the cement seal are undesirable. Compounds that swell in the presence of water and/or oil plug the leaks. One formulation uses dual oil/water swellable particles made of an oil-swellable material (latex rubber) and a water-swellable component (sodium bentonite) [Roddy 2011].

Foams that expand to fill voids and then stiffen have a broad potential for self-healing. Concrete mixes that include foaming nitrogen stabilizes deep sea well holes drilled through spongy and unstable geological structures. The foaming concrete fills the cylindrical hole surrounding the pipe. The filler sets up and forms a solid support casing that seals between the pipe and seafloor, while maintaining proper buoyancy. An issue is variable and often unstable properties of the seafloor rock structures. Methane hydrates are particularly unstable. Pumping ordinary nonfoaming liquified concrete into the drillhole produces large radial and circumferential stresses that quickly exceed the pressure tolerance of a borehole in weak earth. A sequence of short-length concrete pours set up and limit destabilizing hydrostatic pressures but are impractical and expensive due to time constraints and the risk of gas migration bypass problems. Foamed concrete is self-stabilizing with a much lower specific gravity and with a volumetric compliant compressibility that controls the hydrostatic pressure and counteracts shrinkage and gas migration bypass problems associated with the shrinkage-curing of normal concrete [NAE 2010] [Olevitch 1972].

6.2.2 Patching

Patches seal and stop leaks in fluid containment vessels. Impermeable sliding layers that move over a hole use differential pressure and fluid flow to close the opening in a manner like a flat rubber sink drain. Sliding and covering appears early in the self-sealing fuel tank literature [Garagnani 1925] [Pescara 1944] [Wilson 1956]. A variation for tires has the covering layer moving normal to the leaking surface [Wilson 1941]. The initial placement of the sealing layer of foam was offset at a smaller radius from the tread to prevent damage and overheating during routine driving conditions.

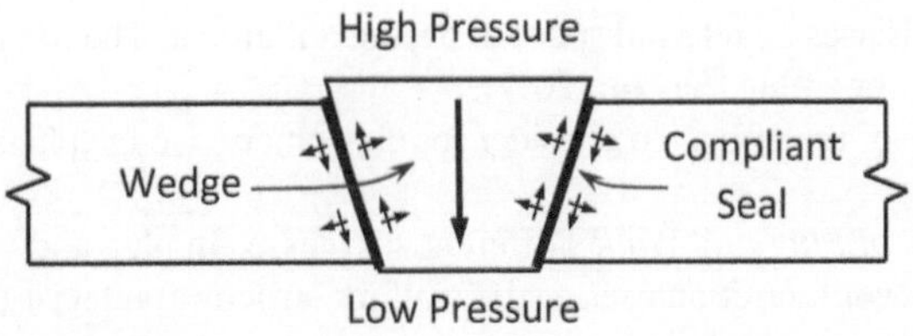

FIGURE 6.7 Compliant wedge seals leak when jammed by pressure differential.

6.2.3 Plugging and Jamming

Plugging stops leaks with material placed inside the leak hole. Effective plugging requires that the plug be of sufficient strength to hold the pressure difference across the hole, attain a good seal between the plug and the leak hole, and be made of material that is impermeable and insoluble (Figure 6.7). Most plugging sealing systems rely on mechanical fluid forces to transport and jam plugs into the leaking holes. Plugging and jamming forces become more aggressive as the plug moves into the hole and jams, and then relax as the flow stops. Autonomous plugging requires prepositioning or transporting plugging material in the structure, such as with tethered plugs (Figure 6.8). The plugging material may be solid, conglomerate, or a liquid that solidifies in the hole. Wedging, compliance between the plug and the wall, and the possible insertion of gummy fluids all help to ensure a good seal. Convective fluid forces act strongly on plugs operating near leaks.

It is possible to use a payout device to dispense and/or disperse plugging material into a leak. An advantage of controlled plugging material payout is that it can cause a gradual slowdown of flow, which may reduce the mechanical shocks and fluid hammer produced by rapid plugging. A method for plugging

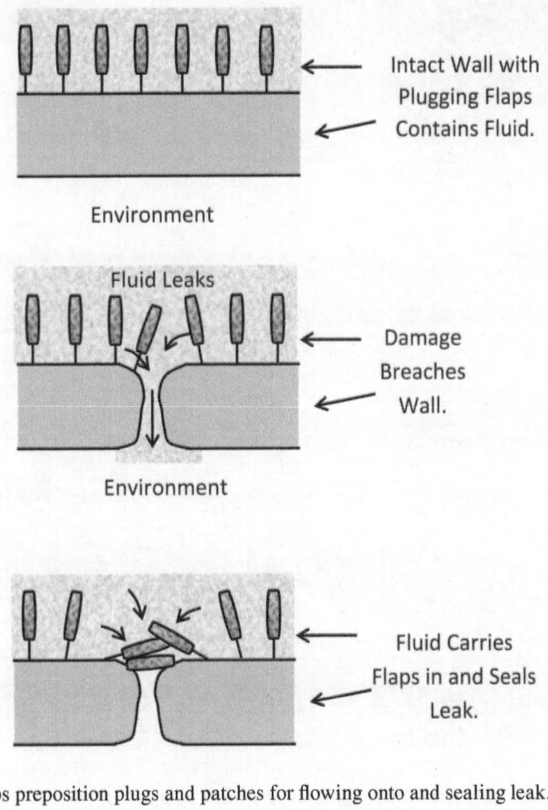

FIGURE 6.8 Tethered flaps preposition plugs and patches for flowing onto and sealing leak.

downhole leaks in oil wells uses an internal spool to payout a filament. The filament tangles into a clump and forms the scaffolding for a plug [Slocum 2017].

The following are some of the effects to consider for elastomeric wedge plugs:

1. *Elastomeric Nature of Plug and Wall:* Different stiffnesses in plug and wall material promote plugging. The usual configuration uses a stiff wall for structural integrity and a highly compliant plug that fits into irregular holes.
2. *Differential Pressure:* Pressure differences on opposite sides of the wall force wedging of the plug.
3. *Location:* The plug must be placed near the hole so that fluid motion pushes and sucks it in.
4. *Coagulation and Conglomeration:* Particles can form a plug.

Floating and coagulating particles are generally smaller than wedges or tethered plugs. A flowing fluid leak drags particles into the confined quarters of the leak hole where they jam and conglomerate to form plugs in holes that are big enough to pass individual particles (Figure 6.9). Jamming particles forms plugs that stop leaks. Particles move with the flow into the leaking hole, conglomerate, and jam the hole. The following are some of the requirements for jamming with particles:

1. *Size:* Particles must be smaller than the diameter of the hole, otherwise they will not fit into the hole and jam; also, they cannot be too small, otherwise they will not jam and stop the leak.
2. *Sticky:* The particles must be sticky enough to stick to one another and jam. Heterogeneous mixes of particles and nonspherical shapes enhance jamming. When not plugging a leak, the particles should remain dispersed in the fluid, not conglomerate into large clumps, and otherwise cause trouble.

FIGURE 6.9 Plug formed by coagulation with tire inner tube sealant compound.

3. *Dispersal:* The particles must mix in and remain dispersed within the fluid so that when the leak occurs, the fluid carries the particles into the hole.
4. *Compatibility:* The particles must be compatible with the rest of the system. They cannot clog or damage other parts of the system.

Conglomeration enhances jamming, such as with inherently sticky particles, shear activation of sticky particles leading to coagulation, and heterogeneous particle jamming (Figure 6.10). Sticky particles significantly enhance particle conglomeration plugging. Conglomeration plugging differs from particle jamming plugging in that weak chemical and molecular intertwining bonds hold particles together rather than compressive stresses. Conversely, granular nonsticky materials lose shear strength and flow in a liquid-like manner in the absence of compressive forces. Following jamming and bonding, polymerization can permanently solidify the plug.

Reports of particle-conglomeration self-sealing date back to at least 1919 with a saw-dust-based stop leak formulation for aircraft fuel tanks [Murdock 1919]. Barton developed stop leak formulations in 1952 for engine water cooling systems using ginger root ground into a fine powder dispersed in oil as the plugging agent [Barton 1952]. An alternative radiator stop leak formulation is a combination of organic pulp and sodium borate [Wynn's 2008]. Similar products are in routine use today, such as for stopping leaks in tire inner tubes. Leak stoppage depends on the relative size of the particles and hole diameter. Some compounds flow into an inner tube through Schrader valves with an 8 mm outer diameter but clog smaller Presta valves with a 6 mm outer diameter. Presta valves require different formulations with lower viscosity and smaller particles, such as 0.01–0.5 mm diameter mica flakes mixed with hydrated bentonite clay particles and propylene glycol [True 2000].

Coagulation changes a liquid into a solid by microscale, nanoscale, and molecular scale effects. Possible transition mechanisms include non-Newtonian fluid mechanics, chemical reactions, and solvent concentration changes. Both non-Newtonian thixotropic and anti-thixotropic fluids plug leaks. Thixotropic fluids become less viscous when subjected to shear deformation. Anti-thixotropic fluids become more viscous when sheared. The basis of thixotropy and anti-thixotropy lies in interactions of

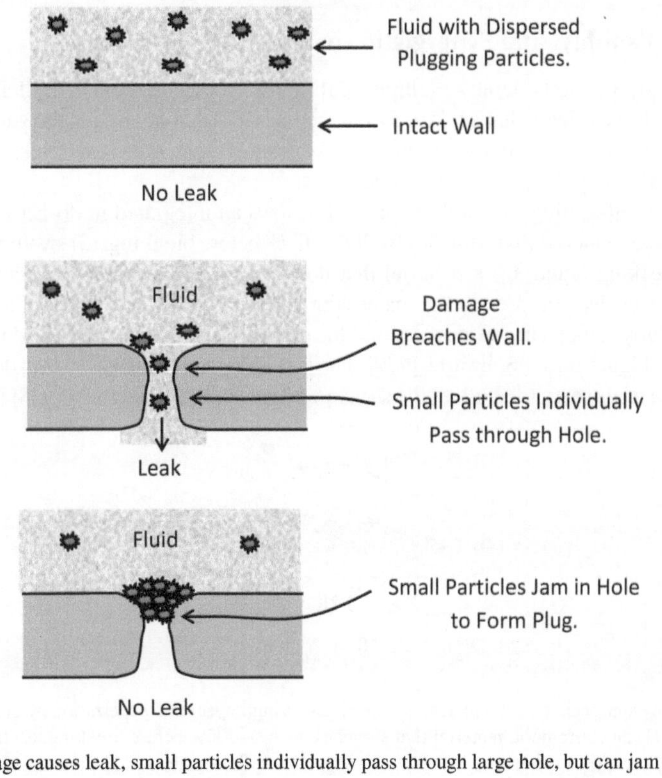

FIGURE 6.10 Damage causes leak, small particles individually pass through large hole, but can jam together to form plug.

micro to molecular length scale structures that organize and/or disorganize under shear loading. The Herschel–Bulkley model represents non-Newtonian fluid stress–shear rate behavior [Morariu 2009].

$$\tau = \tau_0 + K\dot{\gamma}^n \tag{6.1}$$

Here τ is the shear stress; τ_0 is the yield stress; K is the consistency factor; and n characterizes the pseudoplasticity of the system.

Coagulation converts a fluid into a solid or gelatinous state. Localized coagulation after liquid flows into a hole forms a leak-stopping plug. A three-step coagulation process is as follows:

1. *Initial Triggering Stimulus:* Usually by either fluid shear rate or chemical signals
2. *Coagulation:* Molecular entanglement and cross-linking, particulate conglomeration, and hybrid processes
3. *Termination:* Removal or consumption of the triggering stimulus

Effective coagulation requires that the clotting occurs reliably when and where a leak occurs, but not elsewhere. This follows a biomimetic analogy. Cuts in vascular plants and animals should clot at the wound site, but not elsewhere. Clotting in the wrong location causes ischemia, strokes, and heart attacks. Some methods of controllable coagulation are as follows:

1. *Solvent Removal:* Certain materials reside in a flowable state due to the presence of a solvent. Removal of the solvent by evaporation, or by another fluid, such as aircraft fuel, causes the material to solidify [Cook 1974].
2. *Thixotropic Non-Newtonian Behavior:* The introduction of micro- and nanoscale particles into a fluid induces non-Newtonian behavior. Synthetic silicate nanoplatelets mixed into gelatin-based hydrogels produces a shear-thinning effect that reduces the propensity for excessive clotting [Gaharwar 2014].
3. *Anti-thixotropic Non-Newtonian Behavior:* Dispersed silica particles in polyethylene glycol produce a shear-thickening effect that promotes clotting [Decker 2007].

6.2.4 Systemic, Panoply, and Synergistic Leak-sealing Methods

Systemic and panoply methods deploy multiple sealing and repair methods simultaneously. A simple approach inserts multiple effects that act largely independently. Total healing is the sum of the individual healing actions. More complicated and sophisticated systemic approaches synergize healing effects to produce actions that exceed the sum of individual effects.

In the context of a self-sealing fuel tank, Figure 6.11 shows an integrated multi-active system that combines (1) an elastomeric material that stretches by 300–400% before breaking, (2) swelling of imbiber beads with contact with leaking liquid, (3) a material that flows and coagulates upon contact with the leakage liquid, and (4) top layer that compresses the underlying layers to guide the material flow to close the hole [Monk 2007]. Similarly self-sealing spacesuits combine prestress, lightweight filler, and vacuum-solidifying binder to seal leaks (Figure 6.16). [Schwartz 1970]. The leak-stoppage materials are made to be soft and pliable, while other materials carry the bulk of the structural fluid load [McLaughlin 1948] [Underwood 1970].

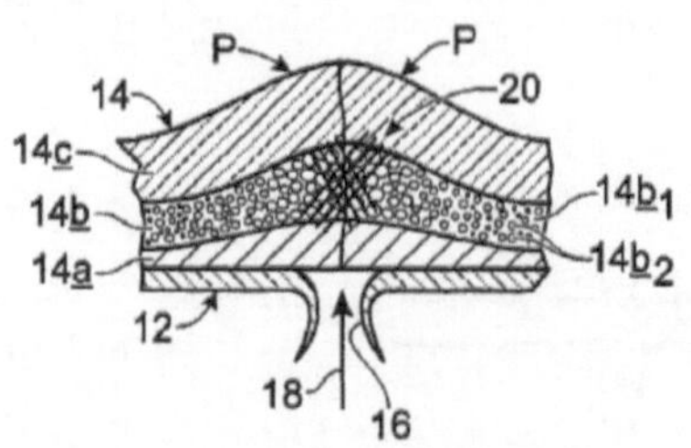

FIGURE 6.11 Panoply approach to fuel tank leak stoppage following impact and penetration by high-velocity projectile. The system combines (1) an elastomeric material that stretches by 300–400% before breaking, (2) leakage-liquid contact swelling with imbiber beads, (3) a material that flows and coagulates upon contact with the leakage liquid, and (4) top layer that compresses the underlying layers to guide the material flow to close the hole. (From [Monk 2007].)

Electric batteries have undesirable failure modes. The amount of stored electrochemical energy makes for particularly dangerous fires [Peskar 2015]. Panoply approaches can counter dynamic damage events, such as projectile penetration. Skins made of self-healing ionomers quickly reseal an outer container following projectile impact. Resealing contains battery chemicals inside the battery and restricts the flow of fire-feeding oxygen from the outside. Internal phase change materials, such as microencapsulated wax, absorb heat, and prevent fires from spreading.

6.3 Applications

6.3.1 Fuel Tanks

Leak-induced swelling is a principal method of stopping fuel tank leaks. Partially vulcanized rubber swells upon contact with petroleum-based fuels. The mechanism is like hydrogel swelling. The fuel penetrates the interstitial spaces between the elastomeric molecular network.

Fluid-activated swelling foams seal leaks in perforated fluid containment vessels. A specific application for self-sealing aircraft tanks is an elastomeric polyurethane foam that swells up to 300% within 5 minutes when the foam encounters jet fuel [Rosthauser 1988].

Sealing holes due to the penetration and passage of high-speed projectiles requires additional accommodations (Figure 6.12). A primary issue is the formation of sharp folded edges, known as petals, on the exit side of the hole. Petals impede the movement and placement of materials for covering and plugging the hole. Workarounds include inserting petalling-resistant laminates or a separating toughness-tuned foam layer between the tank wall and the sealing layer (Figure 6.13) [Baker 1971] [Harr 1971]. Another issue is that the impact of a high-speed projectile into liquids contained inside a vessel sets off a mechanical shock wave. Known as *hydrodynamic ram*, this shock wave causes rupturing and other damage (Figure 6.14) [Ji 2021].

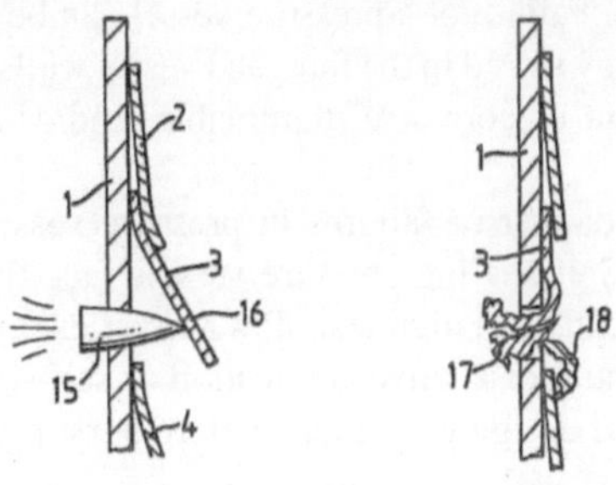

FIGURE 6.12 Prepositioned flaps serve as plugs to stop leaks in tank following penetration with a projectile. (From [Grosvenor 1983].)

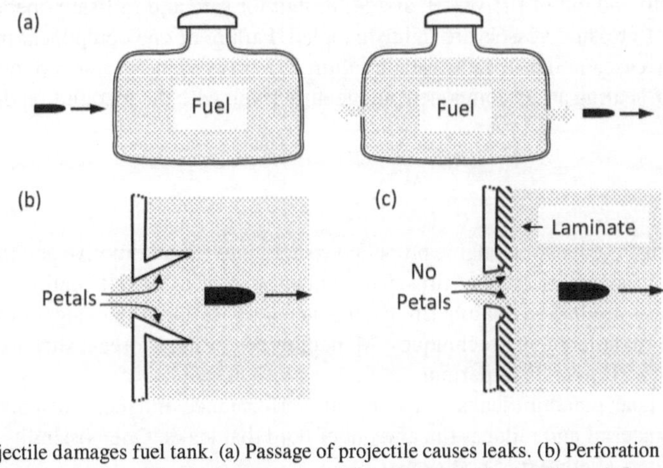

FIGURE 6.13 Projectile damages fuel tank. (a) Passage of projectile causes leaks. (b) Perforation of tank wall by projectile leaves petal deformation in the wall, which aggravates self-sealing methods. (c) Fiber-reinforced elastomeric laminate suppresses petal formation.

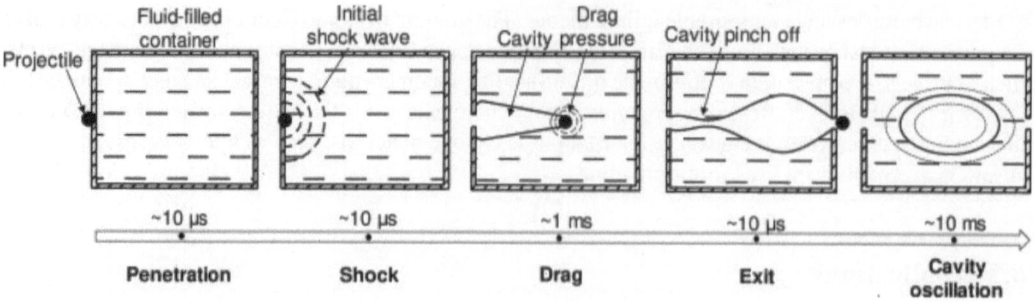

FIGURE 6.14 Hydrodynamic ram due to projectile penetrating and passing through fluid-filled container. (From [Ji 2021].)

6.3.2 Tank and Pipe Sealing, and Multi-active Methods

Self-sealing fuel lines use many techniques similar to that of self-sealing fuel tanks [Heitz 1978] [Hall 1978] [Wilkinson 1943].

6.3.3 Pressure Vessels

Pressure vessels contain or exclude fluids at high differential pressures across the walls of the vessel. Boilers, compressed-gas fuel tanks, pressurized cabins in airplanes, hydraulic and pneumatic power system lines, and air-filled tires are examples. In most cases, the high-pressure fluid is inside of the structure, but many important applications, such as submarines and vacuum chambers, keep the fluid with high pressure on the outside. Failure of a pressure vessel can be catastrophic. The rapid release of large amounts of mechanical energy stored in the fluid and vessel walls causes injuries and damage. High temperatures and the containment of corrosive, flammable, and/or explosive fluids exacerbate many pressure vessel failures.

The high cost and associated danger of failures in pressure vessels prompts consideration of self-healing and sealing techniques. Self-healing pressure vessels face the technical challenges of sealing leaks with high-pressure differentials and high leak flow velocities, including those that are supersonic. A complicating issue is the potential of a naive application of self-sealing compromising the safety of the overall system. Pressure vessel design uses a leak before burst principle. Leaking is less dangerous than bursting.

Typical features of pressure vessels are as follows: (1) minimal and controlled leaking; (2) contain the pressures through suitable means of carrying the forces; (3) suitable means of controlled movement of pressurized fluid into and out of the vessel; and (4) design for safe and fail-safe operation. The components and features of pressure vessels are tightly coupled. Failure of one component induces a failure in another, which may cascade into a catastrophic failure. There is value in uncoupling the functionality and introducing self-healing at the component and feature level, with the recognition that systemic issues must be considered.

6.3.4 Tires

As soon as the internal combustion engine propelled society into the automotive era, there was an urgent need to plug leaks in pneumatic rubber tires. This spawned a series of self-sealing tire inventions dating back to as early as 1900. Self-sealing tire patents generally fall into the categories of plugging goo, plugging flaps, and manufacturing techniques. Managing centrifugal forces, such as with macro- and mesoscale cellular constraints, is important.

Preplacing plugs near potential leaks is a stop leak performance enhancer. It reduces the amount of parasitic plugging material and reduces the amount of fluid that leaks. One possibility is a shag rug-like array of plugging fibers (Figure 6.15) [Dirienzo 1920].

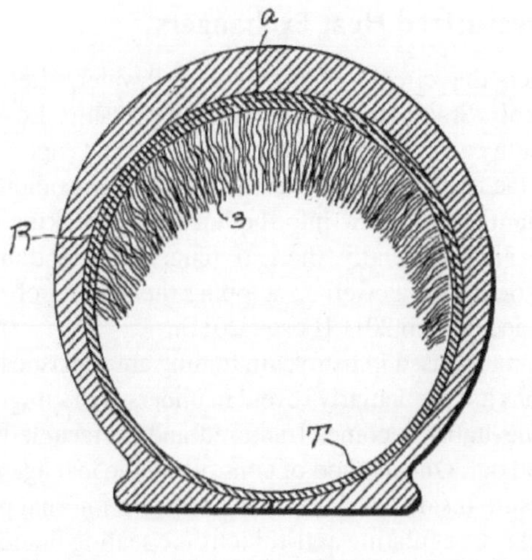

FIGURE 6.15 Early (ca. 1920) shag rug arrangement of prepositioned plugs for self-sealing pneumatic tire. (From [Dirienzo 1920].)

Mucilaginous goo flows into holes and forms plugs [Landis 1915]. An issue is that compliant gels or liquids move under the action of centrifugal forces of rotating tires and vibration that arise with driving. This requires special formulations, that is, non-Newtonian fluids that liquefy and spread uniformly under centrifugal force when the tire spins but congeal in place and resist gravitational creep when the tire is stationary [Yamagiwa 2008].

6.3.5 Space Suits

Space suits are pressure vessels that protect astronauts from the near vacuum of outer space. Small leaks caused by high-velocity micrometeorites or space debris are a constant concern. Space suits that autonomously seal leaks have advantages. Figure 6.16 shows a space suit concept with a prestressed wall construction with an outer layer in tension and a flowable inner layer in compression. Perforation by a micrometeorite causes the inner layer to relieve the prestress by flowing to seal the load.

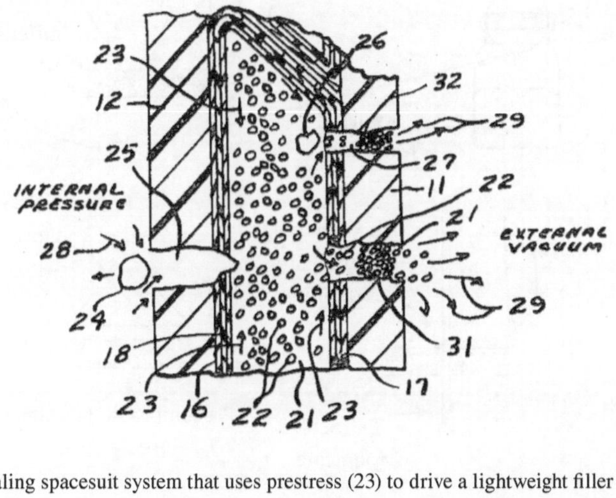

FIGURE 6.16 Self-sealing spacesuit system that uses prestress (23) to drive a lightweight filler (22) and resin (21) into a hole. The external vacuum of space causes the resin to outgas leaving a solid plug (31). (From [Schwartz 1970].)

6.3.6 Radiators and Pressurized Heat Exchangers

Radiators are heat exchangers that circulate hot-pressurized water, often mixed with additives, through grids cooled by convective air. Small leaks lose fluid and pressure. Loss of fluid reduces heat transfer performance. Loss of pressure causes boiling and further reduced capacity for heat transfer.

Self-sealing additives to the cooling liquid such as liquid glass sodium silicate stop leaks. Dissolved sodium silicate and potassium silicate flow into the hot wall of leaking container. Heat evaporates the water and leaves a glassy solid. Eventually, the leak plugs. Glass sodium silicate appears in stop leak compounds for automobile radiators, as well as stopping the leaking of radioactive water following the Fukushima nuclear reactor accident in 2011 [Foster 2011].

The drilling fluids, that is, muds, used in petroleum mining are sophisticated liquids with anti-thixotropic sealing properties. Conditions are particularly severe in undersea drilling. As the drill penetrates through rock strata, the drill hole inevitably becomes fractured and permeable through which water and other unwanted liquids flow in and out. One method of controlling the leakage is to pressurize the shaft with a counterbalancing fluid pressure inside the shaft hole. Excessive internal pressure runs the risk of further fracturing soft soils and rocks. Coagulating anti-thixotropic sealing liquids added to the drilling mud are common sealing mechanisms.

6.3.7 Self-sealing Medicine Jars

Storing and dispensing liquid medicines for hypodermic injection are excellent opportunities for self-sealing systems. Container systems must meet conflicting constraints of holding the medicine separate from the external environment and biological contamination while allowing extraction of the liquid into the hypodermic needle through multiple penetration and extraction cycles. A 1933 patent for a valve stemless self-sealing pneumatic tire inner tube foreshadowed self-sealing medicine bottles [Richardson 1933]. The invention is to inflate a tire through an inner tube by piercing with a hypodermic needle and avoid using a valve with the tube resealing itself upon needle removal of the needle. Figure 6.17 shows a similar concept for self-sealing medicine bottles that dispense sterile liquids through a hypodermic needle.

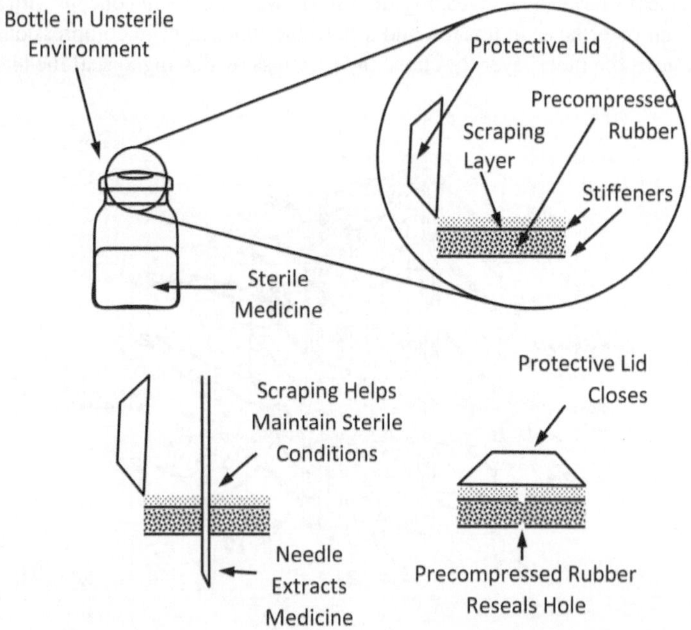

FIGURE 6.17 Self-sealing medicine bottle dispenses liquids into hypodermic needles and maintains a sterile seal.

6.3.8 Landfill Liners

Modern landfills use impermeable barriers to prevent hazardous pollutants leaching into the earth. Breaches to these barriers defeat pollution containment and can be difficult to repair. Self-sealing impermeable membranes mitigate these problems. Design issues include geometric scale and cost. A barrier with physically separated layers of lime and pozzolanic grains is effective. Mechanical breaches comingle the lime and pozzolanic material, which in the presence of water solidify into an impermeable mass [Shi 2005]. Another design for a landfill liner is a multilayer structure that includes an impermeable clay layer. Cracking due to shrinkage and earth movement breaches the container wall and allows for the outflow of dangerous pollutants. Calcium-bentonite clays swell and plasticize under the weight of a confining layer of 0.75 m of earth. The combination of stress and confinement causes the clay to flow and seal cracks as they form [Egloffstein 2001] [Madsen 1998].

6.3.9 Joints, Seams, Seals, and Gaskets

Fluid containment vessels inevitably have seams and seals, often with gaskets. The presence of seams, seals, and gaskets provides advantages, such as ease of assembly and material handling. Gaskets are sources of leaks and other failures. The relatively small size and importance makes seams, seals and gaskets potentially attractive options for self-sealing methods.

Often inserting compliant materials between mating solid tank components helps to block fluid transport in seams and seals. These inserted materials include putties, gaskets, swelling compounds, and so on. A typical gasket serves as a compliant layer between two stiff surfaces. Tight fluid seals require that the solid surfaces mate with minimal gaps. The mating of surfaces is statically indeterminate contact. Squeezing the parts together uses compliance to form the seal (Figure 6.18).

Maintaining leak-free seals has competing requirements. The seal material must be sufficiently compliant or fluidic to flow and fill holes and gaps that cause leaks. Simultaneously, the seal must be strong enough to sustain the mechanical loads due to fluid pressure, container wall movement, and repeated seal–unseal actions. One solution is self-compensating seals with functional task separation into multi-component, multilayer material (Figures 6.19 and 6.20). Some components and layers provide compliant

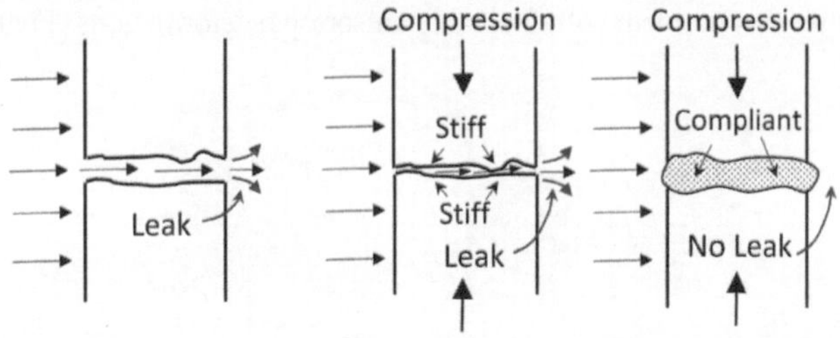

FIGURE 6.18 Compliant gasket stops leak. (a) Loose joint allows fluid to flow freely. (b) Tightening joint slows flow, but stiff walls prevent complete closure. (c) Compliant gasket completely fills gap and stops fluid flow.

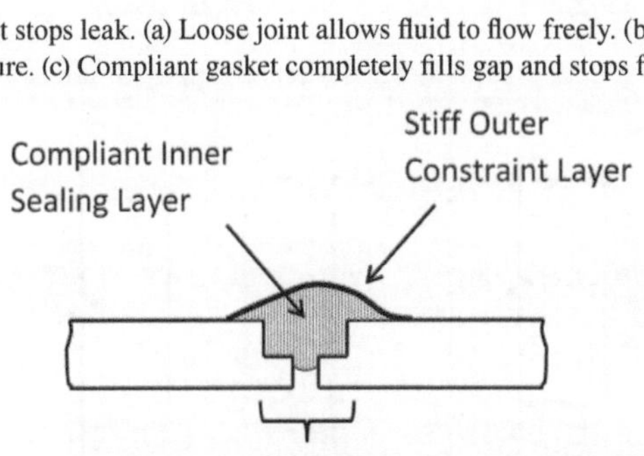

FIGURE 6.19 Layered tape using stiff outer constraint layer to guide inner compliant layer into self-compensating seal. (From [Karrfalt 1998].)

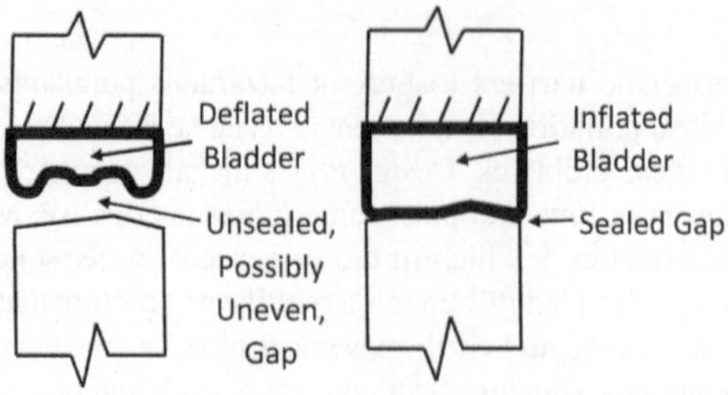

FIGURE 6.20 Self-compensating pneumatic bladder gasket seals uneven gaps.

material that flows and fills leaks, while the other components are stiff enough to sustain the pressure and associated mechanical loads. Including thick layered tapes and inflatable pneumatic bladders inside the gasket produces a compliant, yet mechanically robust, self-sealing interface [Karrfalt 1998] [Paugh 2006]. Self-relief seals in earth-boring drill bits expand an elastomeric seal with pressure from the fluid to get a good fit and then release when the pressure drops off [Neville 2008].

Compliant and rigid are varieties of mechanical seals. Compliant seals use an elastomeric material, typically in a gasket configuration, held in place by compressive stresses. Compliant seals offer resilience in the face of mechanical distortion but require a resilient elastomer. High temperatures, high pressures, and aggressive chemicals challenge the successful use of compliant seals. Rigid seals can offer a sealing solution for such demanding applications. However, rigid seals require matching of thermal expansion properties, tend to be brittle, and suffer from cracking. Borosilicate glasses satisfy rigid sealing requirements for fuel cells with high-temperature cycling and exhibit spontaneous healing of tight microcracks during high-temperature cycles that soften the glass [Liu 2008c] [Lu 2013].

A drawback to using ordinary elastomers as compliant sealing methods is that ordinary elastomers are isotropic and incompressible, that is, when squeezed in one direction, they expand in another. This may lead to awkward geometric behaviors, such as the swelling out of a gap into the travel surface of elastomeric gasket seals for expansion joints in bridge decks. Anisotropically constrained preprogrammed shape memory polymers resolve this problem with favorable anisotropic deformations (Figure 6.21) [Li 2012b].

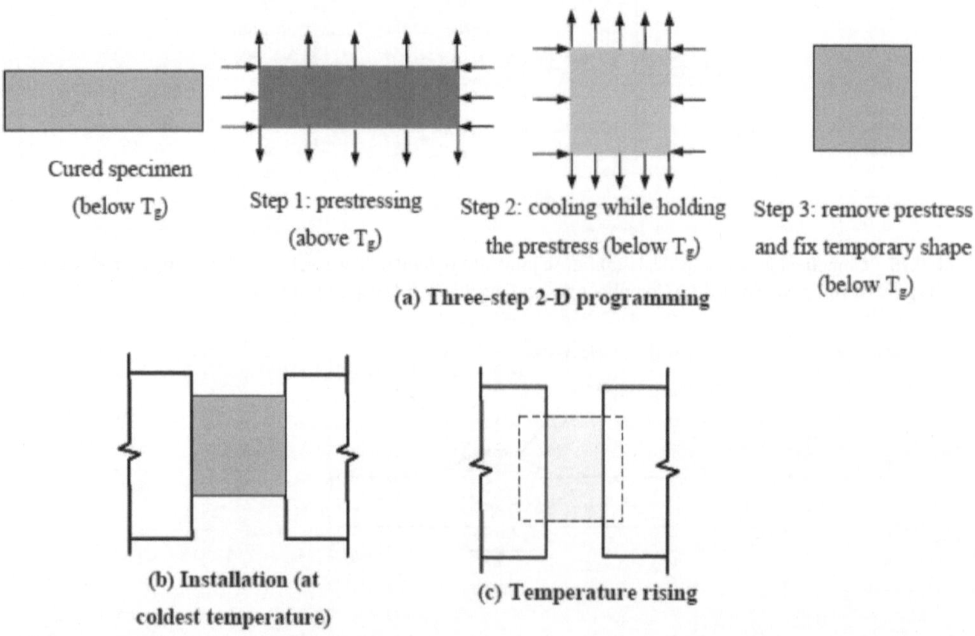

FIGURE 6.21 Anisotropic programmed shaped memory polymer. (a) Prestressing above T_g creates anisotropic shape memory. (b) Installation at cold temperature below T_g. (c) Anisotropic contraction as temperature rises above T_g. (From [Li 2012b].)

6.3.10 Machinery Boots and Covers

Moving machinery uses boots to separate the moving parts from the external environment. Boots are made of tough fabric, rubber, or a composite of the two. Failure of the boot allows contamination into the moving parts and exposes pinch points that risk ensnaring objects and limbs. Self-healing boots using a layer with microencapsulated healing liquids mitigate some of these issues by healing cracks as they grow [Kwon 2012].

6.3.11 Reinforced Concrete

Concrete is an excellent opportunity for active pore closing methods. Hearn lists several proposed mechanisms for self-sealing in concrete, primarily through activity in the hydrated cement paste (HCP) [Hearn 1998].

1. Air in the HCP matrix – (a) incomplete saturation of the test specimen, and (b) dissolution of air under pressure, into permeating water
2. Swelling of HCP
3. Chemical interaction of water and HCP – (a) continued hydration of residual clinker, (b) dissolution and deposition of soluble hydrates, such as $Ca(OH)_2$, along the flow path, and (c) carbonation of dissolved $Ca(OH)_2$
4. Osmotic pressure
5. Physical clogging caused by downstream movement of loose particles in the HCP matrix in a process analogous to sediment transport and formation of a filter cake.

Lining iron water pipes with concrete or mortar allows for the use of autogenous cementitious crack healing to stop minor leaks. Stagnant water conditions are more favorable than flowing. Observations find that cracks with widths up to 0.1 mm will heal shut over one year [Wagner 1974].

6.3.12 Particle Filters

Fluid filters remove unwanted components, such as solid particulates, with a selective barrier that catches the particles but allows the fluid to pass. The performance requirements of fluid filters tend to be severe and often impose conflicting design requirements. The filter must pass large quantities of fluid with minimal pressure drop, while removing small particles with a fine-holed mesh. Failure modes include leaking, crushing, plugging, improper bypass, and insufficient particle removal – all of which may cause systemic failures. Filters with damage mitigating, fail-safe, self-cleaning, and self-healing features reduce the need for expensive maintenance shutdowns for filter replacements and machinery repairs.

Methods for self-healing filters:

1. *Active Damage Mitigation with a Bypass:* A clogged filter with impeded flow causes severe problems, including stopped flow and crushing of the filter. Bypassing the clogged filter maintains the flow with unfiltered fluid – often a worthwhile trade-off. Pressure-difference-driven bypass valves are common additions to many filters (Figure 6.22).

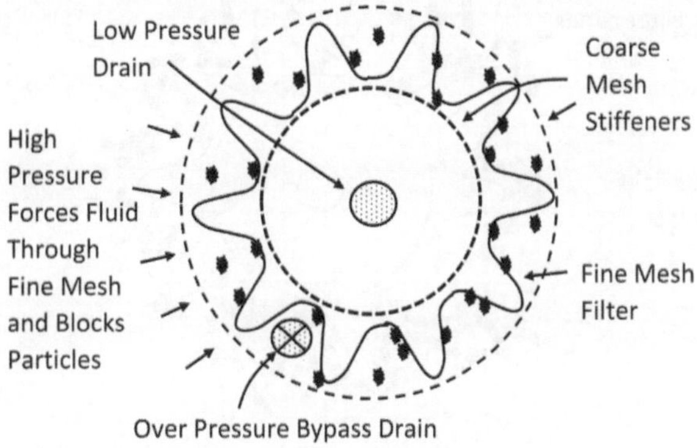

FIGURE 6.22 Fluid filter with overpressure bypass drain.

2. *Backflow and Supplemental Fluid Motions:* As a filter collects particles, the particles accumulate. Continued accumulation clogs the filter. One method of unclogging the filter is to replace it with a clean one. Backflow is an alternative method that reverses the flow of the fluid and removes particles from the filter mesh (Figures 6.23 and 6.24). Transverse motions then remove the loose debris as secondary step [Villares 2004]. Similarly, macroscale features generate vortices and nonsteady flows with swirls that impact the surface to lift and propel debris away [Slocum 2013].
3. *Re-conglomeration:* Many filters use conglomerates of particles embedded and attached to the filter mesh to filter additional particles. Moderate amounts of particle buildup produce a positive effect as part of an active filtering process. A common pre-conglomeration method is swimming pool filters that use diatomaceous earth. Initially, the diatomaceous earth forms a primer conglomeration layer on the filter paper. The primer layer helps to filter particulates in

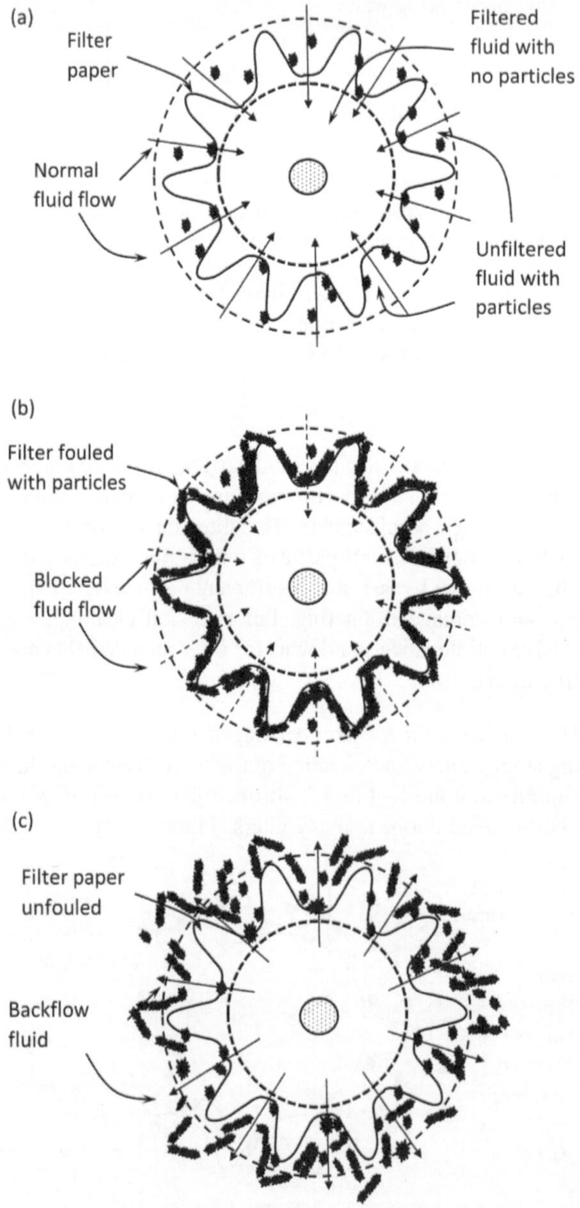

FIGURE 6.23 Filtering of particulates. (a) Normal operation. (b) Clogging and fouling by debris. (c) Backflow cleaning.

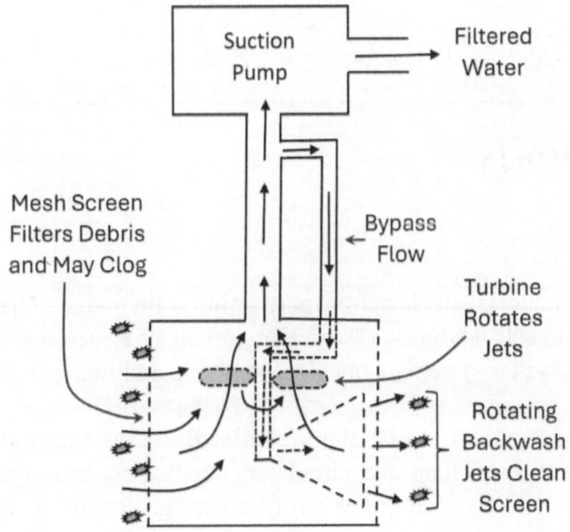

FIGURE 6.24 Water pump intake screen with integrated self-powered backwash self-cleaning system. (Adapted from [Hosford 2003].)

the water. Subsequently, the primer layer aids cleaning filters once they are clogged. Reversed or tangential hosed flow of water easily dislodges the particles from the filter mesh. The cycle repeats with a new layer of diatomaceous earth.

4. *Self-cleaning Filter Media:* A method of enhancing backflow cleaning and defouling of filters uses active media that pushes the clogging particles away from the filter surface. Thermally triggered shape memory nanobrushes when heated above a transition temperature, the nanobrushes lengthen and push particles off the clogged filter surface [NASA 2012]. Lowering the temperature causes the filter brush surface to recover its original shape (Figure 6.25).

Closely related to fluid filters are gas separation membranes that act at molecular level to selectively allow transport of different gas species. These membranes often operate under harsh conditions, pressure differentials, mechanical stresses, and inevitably sustain damage. Self-healing is a key enabler of this technology. Ionic polyamides demonstrate high CO_2 selectivity, while being able to heal upon molecular scale and macroscale damages [O'Harra 2020b]

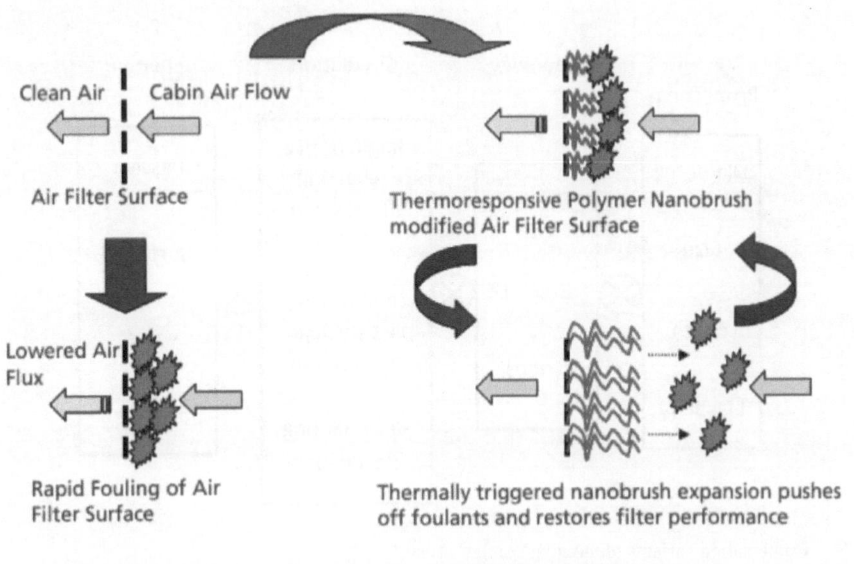

FIGURE 6.25 Lynntech's self-cleaning particulate air filter surface. (From [NASA 2012].)

7

Vascular Methods

Vascular systems enhance material and heat transport within solid bodies. The following are the primary roles of vascular systems in self-healing – (1) facilitate repair of structures through enhanced material transport, and (2) repair vascular systems, typically the piping and tubing used in the operation of engineered systems.

Many configurations of engineered vascular systems mimic those found in nature. Configuration variables include vessel size, branching and circulatory topologies, and control methods (Figure 7.1). Closely related are vascular-like techniques using wicking, porous media, planar channels, and open channels.

7.1 Biomimetics

Biological vascular systems are generally either internal systems that circulate material or external systems that transfer material in and out of the organism.

Some questions:

1. Are there biomimetic or bioinspired principles that aid in the conceptualization and design of vascular systems for the support of self-healing?
2. How to fabricate and operate vascular self-healing engineering systems?

Diffusion is a macroscopic means of transferring mass and energy from regions of high concentration and temperature to regions of lower concentrations and temperatures. The mechanics of diffusion tie directly to the random nature of molecular motions. Diffusion is a primary material transfer process supporting the molecular-level metabolism of living biological systems.

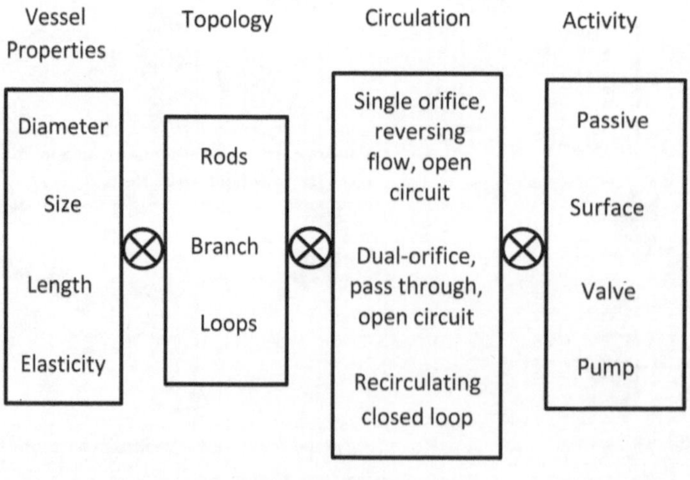

FIGURE 7.1 Combination variants of vascular configurations.

DOI: 10.1201/9781003396048-7

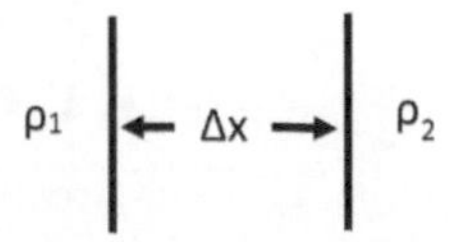

FIGURE 7.2 Geometry of 1D diffusion.

A 1D model of diffusion geometry appears in Figure 7.2.

Material diffuses across a material at a rate q according to

$$q = -kA\frac{\rho_2 - \rho_1}{\Delta x} \tag{7.1}$$

Here k is the diffusivity; A is the cross-sectional area; ρ_1 and ρ_2 are the concentration of the material on the left side and right side, respectively. Material flows from high to lower concentration. Altering the parameters in (7.1) alters the flow rate. Changing the diffusivity and geometric parameters of area and distance are often not easy. Equation (7.1) indicates that diffusion rates decrease as the distance Δx increases. Transfer of material depletes the gradient between ρ_1 and ρ_2, leading to a decrease in q.

Larger organisms cannot thrive on diffusion alone due to the slow rate of diffusion transport over large distances. Most of the higher order plants and animals use vascular systems for the transport of materials within their bodies, especially for applications that require material transport over distances and at rates that are not practical by diffusion. Vascular systems replenish the concentrations ρ_1 and ρ_2 on opposite sides and maintain a high diffusion rate.

Biological systems have wide varieties of vasculature. The different architectures depend largely on the metabolism rate, the distances covered, and the need for redundancy. The following are the two main classes:

1. *Circulatory Diffusion:* Transport of chemicals to provide a macroscale down through molecular scale efficient transport of material. Blood circulation, lymphatic system, and xylem and phloem in plants
2. *Bulk Consumption or Excretion of Material:* Lungs and gills, digestive tracts, stomata in leaves, kidneys and urinary tracts

Most vascular systems are scale-dependent solutions to transport problems. The larger and more complex life forms, for example, trees and vertebrates, tend to have more sophisticated vasculatures that support their vital operations. An example is respiration, that is, the transport of gases to and from the atmosphere through the body of a creature or plant. Higher metabolic rates require higher respiration rates. Animals with high metabolic rates exchange oxygen and carbon dioxide with ambient air through lungs and gills using forced convection and then forced circulation of blood. Animals have higher levels of metabolism, the cold-blooded animals having lower rates than the warm-blooded animals. Comparatively, high transport appears in gills in fish. Water contains less oxygen than air, but the metabolism is relatively slow. Gills use an open-circuit pass through design that is aided by current flow and swimming. Lungs in mammals have an in–out structure with hierarchical branching. In a sense, this is suboptimal because the reversal of flow reduces the diffusive transfer rate. Lungs in birds combine in–out macro-aspiration with flow-through meso- and microscale aspiration to achieve higher rates of transfer than possible with mammal lungs [Wrenn 2018]. Plants with lower metabolic rates use less-active means of gas transport. Leaves use small-scale stomata, with respiration driven at low rates.

There are numerous constraints and design considerations with vasculature. One is the issue of leaks and open circuits that accelerate the delivery of material. Additional considerations are the ability to respond to damage by inflammation and swelling.

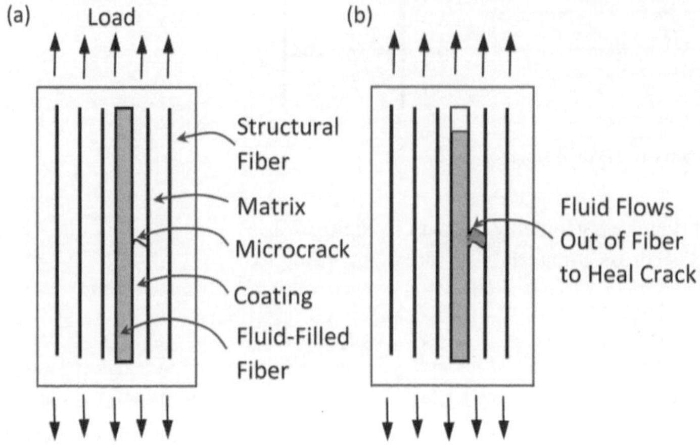

FIGURE 7.3 Cracking time release system. (a) FRP composite forms microcrack. (b) Vascular system heals crack. (Derived from [Dry 1993b].)

7.2 Non-branching Tubular and Nodal Vasculature

A fluid-filled non-branching tube embedded in a structural element is one of the simpler engineered vascular topologies. Much of the early development and conceptual underpinning of self-healing non-branching tubes arises from pioneering efforts of Carolyn Dry in the 1990s (Figure 7.3) [Dry 1992b, c; 1993b; 1994; 1996b, c, d; 1997; 2001b; 2003b]. The single tube has conceptual simplicity and provides control over the flow. In many respects, this technique is a precursor to the micro- and nanoencapsulation techniques discussed in Chapters 3 and 4. A difference is that the elongated shape of the non-branching tube acts as both a reservoir of healing fluid and a vasculature enabling the long-distance transport of material.

Primary steps of implementation:

1. *Embed Tubes:* Fill with healing fluid into the structural element, usually during fabrication.
2. *Damage Scale:* The structural element suffers damage of sufficient severity to require repair, but not large enough to destroy the structural element. Virtually any structural material that suffers from cracking damage, is repairable by healing liquids at ambient temperatures, and tolerates the embedment of tubes is a good candidate for this method. Concrete and fiber-reinforced polymer composites are candidate applications.
3. *Damage Transfer:* Damage to the structural element propagates into the tube, causes it to crack, and then dispense healing liquids by leaking.
4. *Healing Fluid:* Flows into the crack.
5. *Fluid Solidifies:* Forms an effective repair in the crack.

This process is conceptually straightforward, but technically nontrivial. Some items of concern are as follows:

1. *How to Get the Fluid to Flow?* Most simple vascular configurations do not have pumps and instead rely on other means of forcing the healing fluid to flow out of the tube and into the damaged region.
2. *How to Miniaturize?* The Hagen–Poiseuille law finds that viscosity dominates flow through small diameter tubes. The pressure drop due to viscosity increases as an inverse power of

4 relative to tube diameter. Questions arise as to how to miniaturize the vasculature to reduce parasitic loads, while maintaining adequate flowability. For example, viscosities on the order of 0.10 Pa-s are adequate for liquid to flow out of borosilicate glass microcapillary pipets with inner diameters on the order of 130–380 μm [Motuku 1999].

3. *How to Get the Vessel to Crack and Release Healing Fluid at the Appropriate Time?* This requires that the vessel wall material be tough enough to survive embedment and routine operational usage, yet readily crack open in the presence of nearby structural damage. Brittle glass tubes crack in complicated patterns, but are generally effective at releasing healing fluids into the correct location at the correct time [Li 1998a]. An alternative is vessel walls that leak when stretched.
4. *How to Minimize and Offset Parasitic Loads So As To Not Degrade the Overall Performance of the System?* There are competing design objectives. Hollow fibers with diameters compatible with the free flow of most fluids in vascular systems are much larger than the load-bearing fibers in most FRPs, leading to stress/strain mismatches and the formation of damage initiation sites. While smaller diameter vascular tubes are readily available, filling the tubes and then inducing outflow when needed becomes increasingly difficult as the diameter shrinks. One possibility is to cut the viscosity of the healing liquid with solvents, for example, acetone for epoxy or cyanoacrylate liquids [Bleay 2001]. Combinations of dilution with geometric optimization enable fabrication of vascular self-healing FRP composites that recover up to 97% of the original strength with parasitic weight penalties as low as 1–2% [Pang 2005]. Miniaturization extends down to the nanoscale and molecular scale with the use of carbon nanotubes as vessels, containers, and structural filaments [Lanzara 2009]. Electrospinning with a functionalized fluid-filled core is a method of mass-producing vascular wadding [Kotrotsos 2021]. The tube can also act as a structural reinforcing filament.
5. *How to Maintain Healing Fluid Stability?* The fluid must exert the desired healing action once deposited in the damaged region but remain liquid for long durations prior to the damage. The fluid must congeal once it flows into the crack, not before and not too much afterward. Possibilities are single-part materials that solidify when exposed to the air or the structural matrix. Solvent evaporation and air-induced polymerization are common solidification mechanisms. Ethyl cyanoacrylate (commercially available as Super Glue) polymerizes in the presence of moisture [Li 1998a]. Two- and three-part systems are possible [Dry 1996b].
6. *The Amount of Healing Liquid Must Match the Crack Size* [Li 1998]. Bigger cracks require more healing liquid. Mechanical wound closing reduces the amount of healing liquid required.

7.3 Non-branching Vascular Cables

Cables are elongated bundles of components that individually conduct electricity, light, fluids and/or forces. The addition of tubes for conveying healing fluids is straightforward. Figure 7.4 shows an early vascular system used to replenish dielectric fluid in high-power electric transmission lines.

Vascular electric cable systems are often called "flooding cables" [ASTM 2013]. Applications of flooding wire repair are as follows: (1) Maintain and replenish dielectric layers surrounding conductors in high-voltage applications [Bennett 1932] [Bennett 1934], and (2) Self-sealing, primarily against water intrusion (Figure 7.5).

Another non-branching vascular application protects cables that transmit mechanical forces, that is, parallel strand steel cables in tension. Corrosion is a problem, particularly corrosion stress cracking. In the absence of corrosion, excessive loads, and fatiguing loads, steel cables generally survive indefinitely. The traditional approaches of coating, sealing, and painting all prevent corrosion reasonably well, but

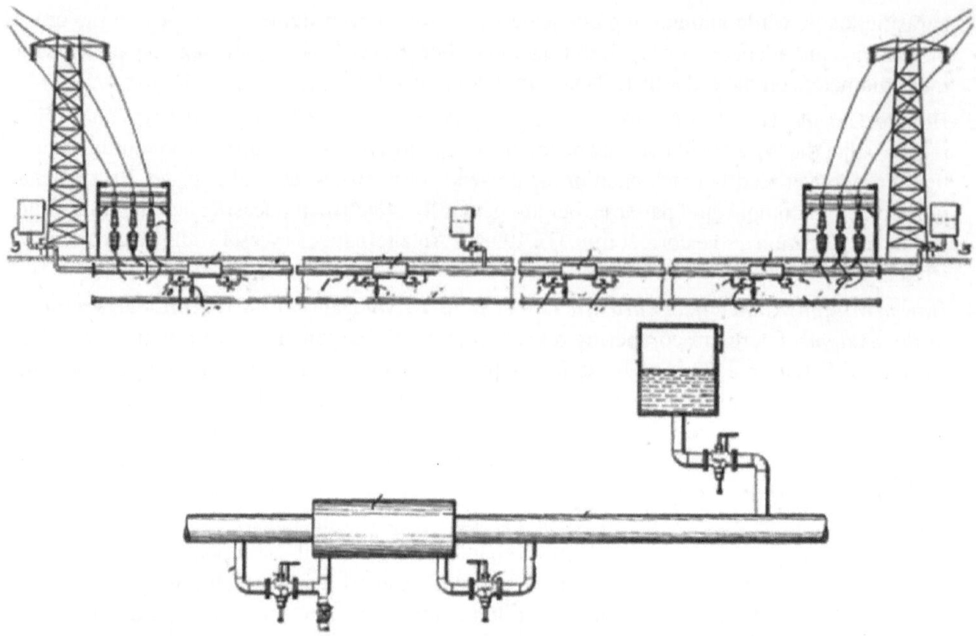

FIGURE 7.4 Early (ca. 1932) self-replenishing liquid dielectric system for high-power electric conductors. (From [Bennett 1932].)

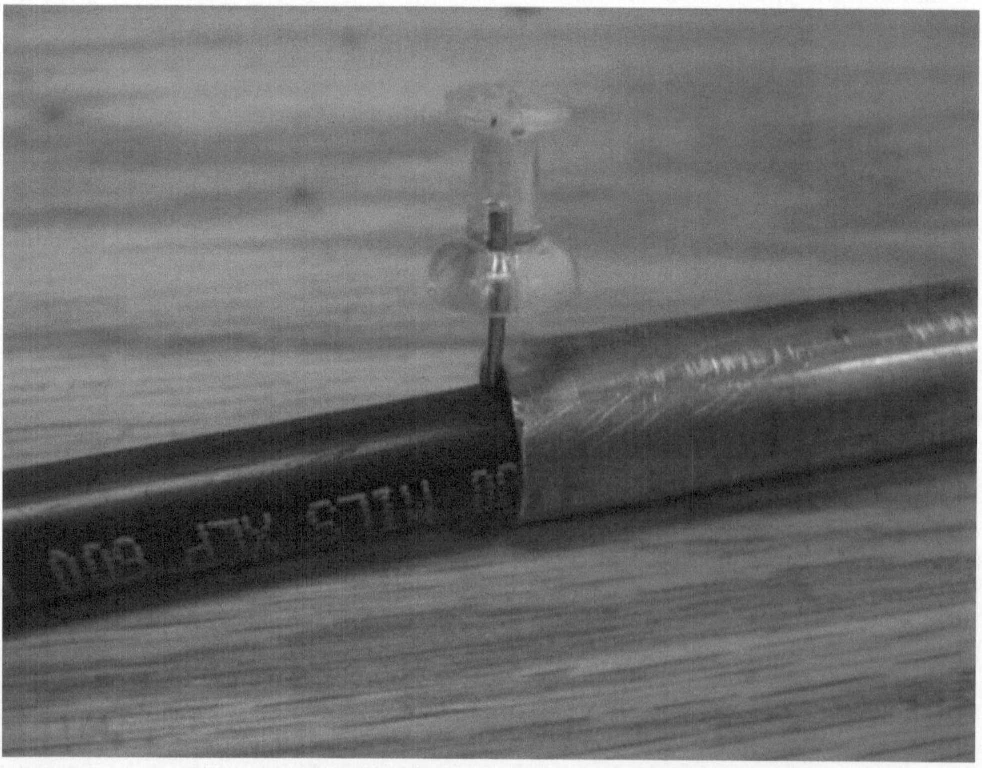

FIGURE 7.5 Flooding cable with viscous layer that flows and prevents water flowing to inner conductor when damage breaches outer jacket.

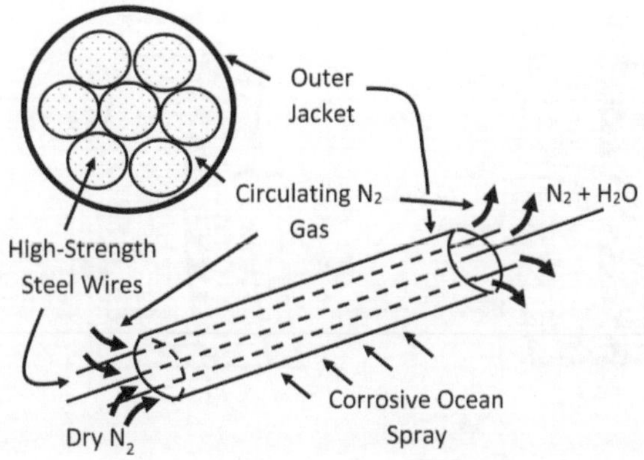

FIGURE 7.6 Circulating dry nitrogen gas prevents corrosion in steel cables.

nonetheless eventually break down. An example is the cables in cable-supported bridges. Siting these bridges near marine environments with salt-laden ocean spray accelerates corrosion. Corrosion occurs in the subsurface, and is difficult to detect and mitigate at early stages. An effective vascular corrosion-prevention application uses circulating oxygen-free dry nitrogen gas in a vascular configuration that prevents corrosion in bridge stay cables (Figure 7.6) [Doe 2005].

7.4 Non-branching Layered Vascular Configurations

Increasing the diameter of the coaxial channel eventually produces a planar vascular configuration, with the fluid flowing in a confined layer. The effect is to provide a broad area of coverage for healing. Figure 7.7 shows a self-healing liquid container with a layered wall construction.

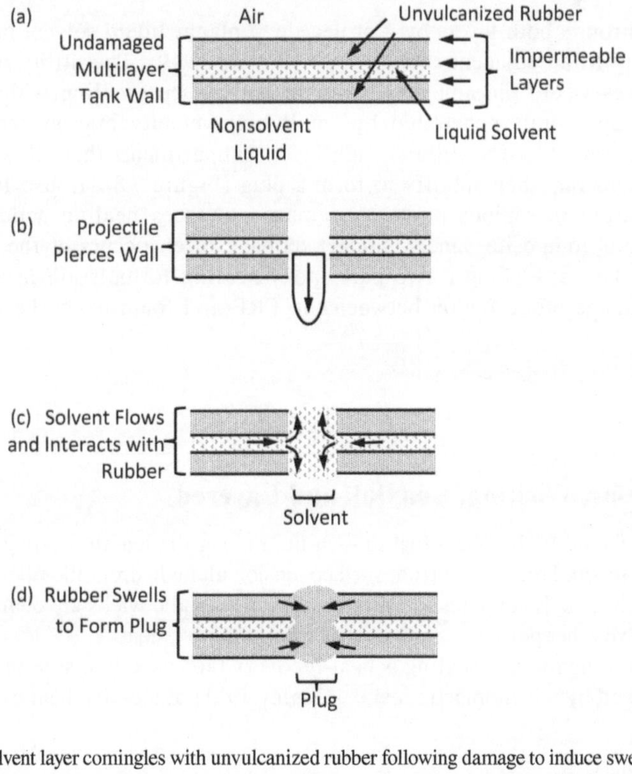

FIGURE 7.7 Liquid solvent layer comingles with unvulcanized rubber following damage to induce swelling that plugs hole.

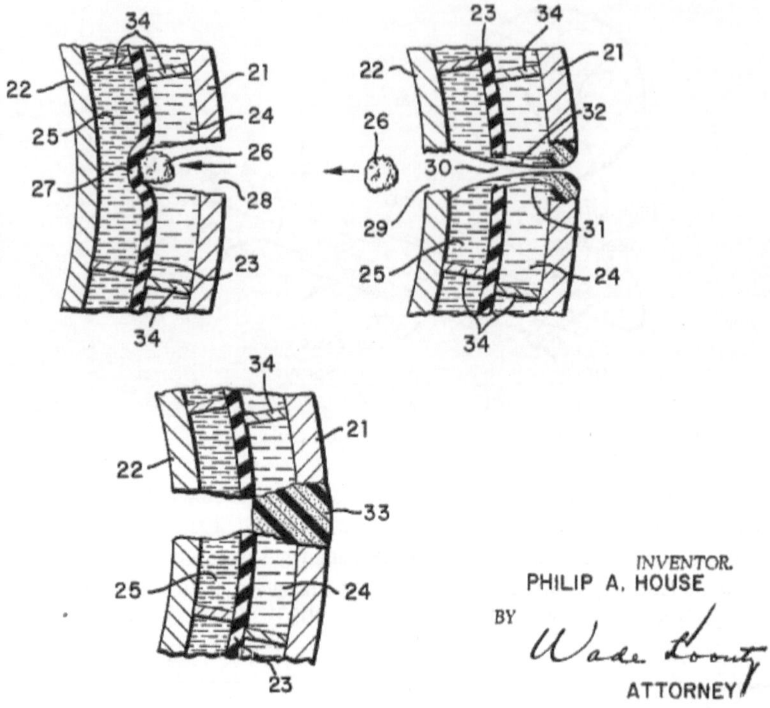

FIGURE 7.8 Cellular layer structure for two-part mixing coagulation plugging of spacecraft wall following penetration by micrometeorites. (From [House 1966].)

Upon penetration through both layers by a projectile, a planar liquid solvent layer induces swelling and then sealing in an adjacent rubber layer [Potters 1949]. The utility of solvent-sensitive layers arises in cases where the ambient fluids do not induce swelling with sufficient promptitude for sealing. Spacecraft penetrated by small high-velocity micrometeorites may benefit from similar configurations using cellular multilayer compartments that allow for the mixing of two-part fluid systems that then solidify to form a plug (Figure 7.8) [House 1966]. Strategically placing the vasculature in regions prone to damage enhances healing action. An example is fiber-reinforced foam composite panels. Impact damage often occurs at the interface between the FRP and foam layers. Placing a two-part epoxy-healing formulation in a parallel vascular structure in the damage-prone region between the FRP and foam heals the damage as needed [Chen 2013d].

7.5 Quasi-vascular, Wicking, Coaxial, and Layered

Wicks are bundles of fluid-philic fibers that draw a fluid along the length, leading to a quasi-vascular fluid transport mechanism. Free energy from surface tension along hydrophilic fibers drives the motion, while viscosity and fluid body force resist. Paint brushes and candle wicks are common examples. The fiber bundles can serve other purposes, such as structural reinforcement.

An example of wicking for self-healing is heat-recovery tape used in insulating wire with a fiber-based wadding backed by thermoplastic resins [Hanley 1941]. Excessive heat causes thermoplastic

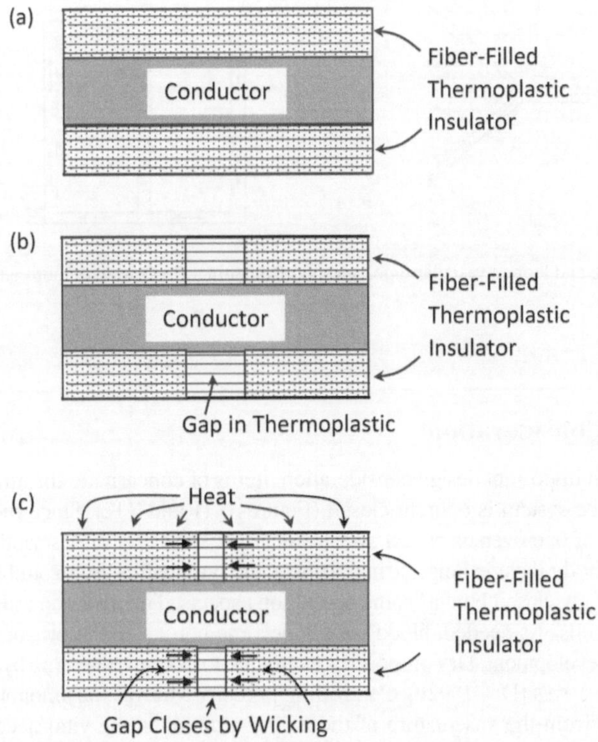

FIGURE 7.9 Wicking promotes flow of healing liquid. (a) Healthy electrical cable. (b) Heat, damage, or manufacturing forms gap in insulation. (c) Heat melts and fibers wick thermoplastic to close gap.

insulation to melt, flow, and expose electrical conductors. As a self-healing mechanism, fiber-based wadding wicks insulating liquids along the cable and heals the melted regions to prevent electrical shorts (Figure 7.9). A similar wicking approach with Kevlar™ strands transports and combines a two-part epoxy-healing system for wire insulation in cable bundles [Esser 2005]. Combinations of hydrophobic and hydrophilic fiber patterns confine the flow into predetermined channels with forces that are strong enough to counteract gravity (Figure 7.10) [Xing 2013]. Textured superhydrophilic surfaces operate as 2D wicking surfaces where drops spread as a quarter power of time [Kim 2013a].

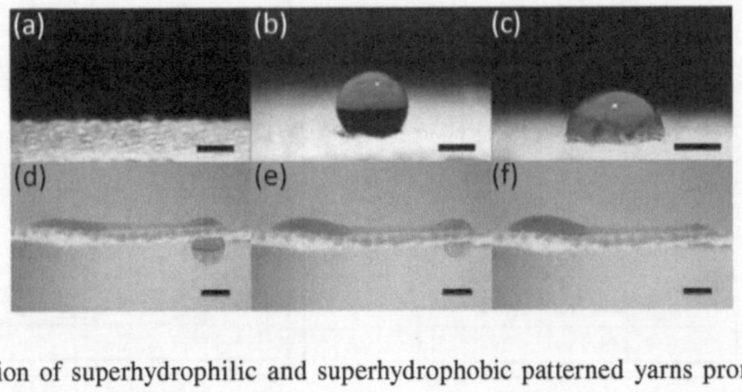

FIGURE 7.10 Combination of superhydrophilic and superhydrophobic patterned yarns promoting controlled wicking flow. (a) Unpatterned hydrophilic yarn fabric. (b) Unpatterned superhydrophobic yarn fabric. (c–f) Patterns of hydrophilic and hydrophobic yarns promote controlled movement of drops. (From [Xing 2013].)

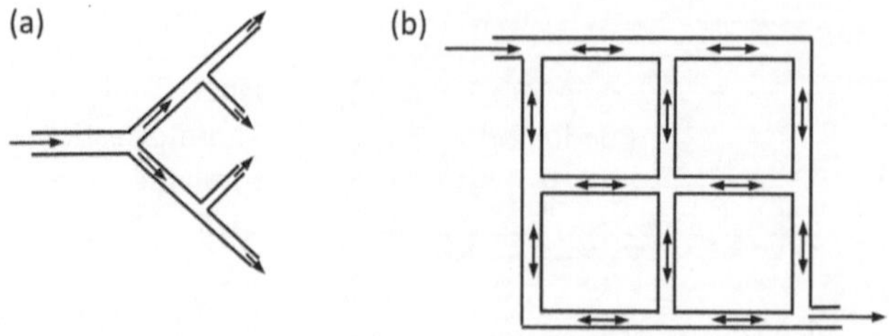

FIGURE 7.11 Branching and looped vascular topologies. (a) Branching. (b) Grid with loops and redundant flow paths.

7.6 Topological Considerations

Vascular topology is an important design consideration. Items of concern are the amount of branching and looping and whether the system is open or closed (Figures 7.11 and 7.12). Since most systems have open aspects, the definition of open versus closed systems is largely a matter of descriptive convenience. Open systems in the human body include lungs, urinary tracts, the lymphatic system, and leaves. The circulation system for blood is largely closed but has some open-loop aspects. The liver and kidneys cleanse the blood and remove old dead cells. As needed, blood vessels become porous and allow for through-wall material transport. In an early development, Dry proposed a technique for vascular healing by through-wall material transport with porous vessels [Dry 1992b]. Controlling porosity controls the amount of material transport.

An interesting item from the vasculature of the higher animals is the vital necessity of an open-loop lymphatic system in combination with the closed-loop blood-carrying hematic system. Principal tasks of the lymphatic system are immunological and the removal of waste tissue products. The number of reported applications to date of lymphatic-type vasculature in engineered self-repairing systems is limited. Most use systems to drain or otherwise remove unwanted liquids. One example is the use of fume hoods to remove noxious gases and particulates from buildings. Another is the use of a draining layer to remove salt-laden water from the top of bridge decks to protect the underlying reinforced concrete deck slab (Figure 7.13)

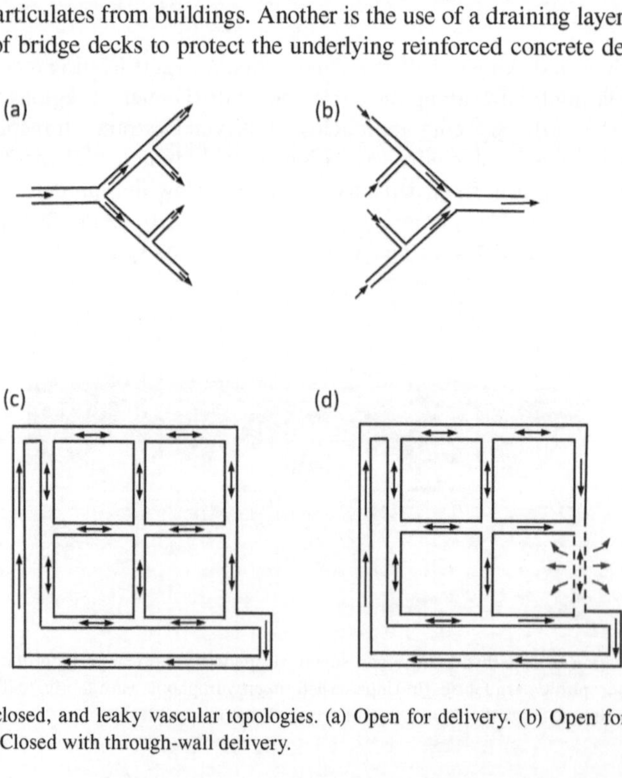

FIGURE 7.12 Open, closed, and leaky vascular topologies. (a) Open for delivery. (b) Open for drainage or lymphatic removal. (c) Closed. (d) Closed with through-wall delivery.

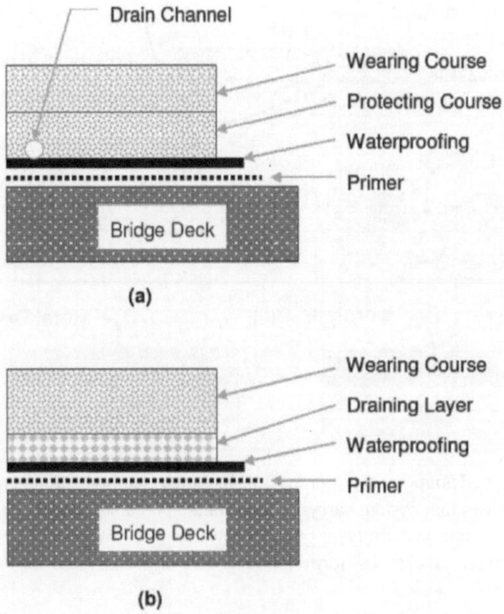

FIGURE 7.13 Drainage systems in bridge deck pavements. (a) Traditional system with drain channel. (b) System with draining layer. (From [Kim 2009].)

[Lee 2008]. A third example is the removal of leaking fuel and combustible fluids to prevent fires and collateral damage. A gravity-fed lymphatic-type vascular system embedded into the wall of a fuel tank in the form of corrugated cardboard drains fuel from leaks [Ericsson 1921]. Another technique uses multilayered walls. When the leak penetrates through only one of the layers, an interior layer captures and drains the fluid into a location suitable for handling. Double-wall fuel tanks, double-wall container ships, and building façades use these methods. As early as World War I, Mougey and Jacobs used such methods to mitigate leaks in aircraft fuel tanks (Figure 7.14) [Mougey 1919]. Preconstructed berms that surround large chemical and fuel storage tanks are variants on the theme of fluid containment with drainage.

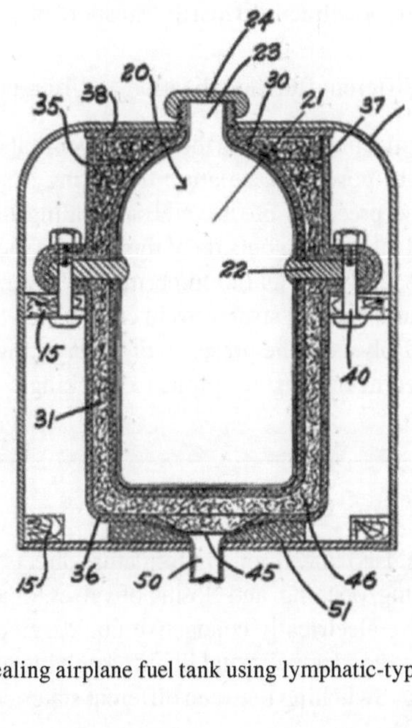

FIGURE 7.14 Early (ca. 1919) self-healing airplane fuel tank using lymphatic-type drainage system to dispose of leaking fuel. (From [Mougey 1919].)

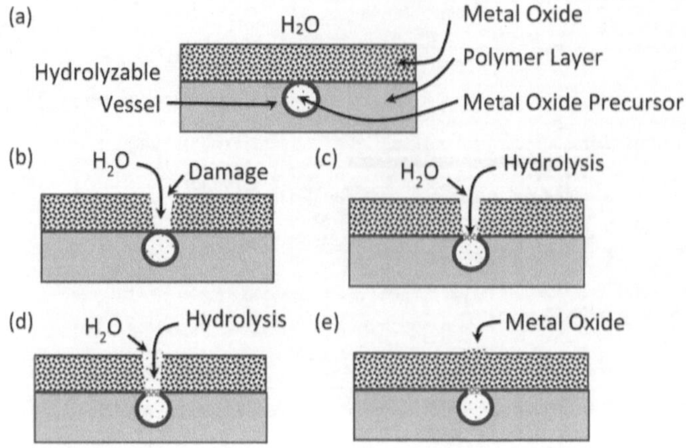

FIGURE 7.15 Irreversible triggered nonporous-to-porous transition that enables healing. (a) Metal oxide layer on top flexible polymer layer with hydrolyzable vascular system carrying metal oxide precursor. (b) Water flows into damage that breaches the metal oxide layer. (c) Water penetration hydrolyzes polymer to cause porous release of metal oxide precursor. (d) Precursor flows into the crack. (e) Second hydrolysis process forms a metal oxide layer that heals the crack. (Adapted from Liu 2008a].)

Branches and loops provide options for enhanced material delivery. Non-branching tubes are good for the high-volume transport of materials over long distances but are not very effective at distributing material over surfaces or throughout volumes. Branching topologies offer potential advantages for material transport into volumes and over areas. Branching comes in five primary varieties: tubes, branches, loops, openings for release of material in and out of the vessels, and combinations of the above. Some underlying concepts are as follows:

1. *Branching of Flow:* Distributes flow in a large channel to multiple smaller channels.
2. *Branching of Diffusion:* Increases the ratio of wall surface area to internal volume. The increased area increases the potential for through-wall diffusion and reduces the effective length of small-diameter, high-viscosity resistance runs.
3. *Hierarchical Layering of Branches:* Balance the needs of high-volume long-range transport over large channels with the localized diffusive transport of small amounts of material over short length scales.
4. *Functional Grading:* Use different functionalization at different length scales.

A selectively dissolvable impermeable barrier as the outer layer that covers a porous vessel wall implements an irreversible nonporous-to-porous vasculature triggering system (Figure 7.15). The impermeable layer does not dissolve in the presence of the internal healing liquid, but it does dissolve when a different damage-signaling liquid attacks the outside of the vessel. One scheme uses $TiCl_4$ as the healing liquid and poly(lactic acid) (PLA) polymer as the impermeable layer [Liu 2008b]. The $TiCl_4$ acts as a precursor for the formation of titania healing structures in composites. The PLA layer does not react with $TiCl_4$ but does break down by hydrolysis in the presence of water or high humidity. This vascular system heals brittle materials when placed near a surface prone to cracking.

7.7 Active Fluid Pumping and Control

Actively forcing and controlling fluids to flow through vasculature directs material transport when and where needed. Possibilities include pumping, opening, and closing of valves, dilation of vessels, and surface effects.

Electrodynamic techniques move electrically conductive liquids. Eutectic indium gallium, also known as Galinstan, is a room temperature conductive metal liquid. Applying an electric field changes the surface tension from wetting to nonwetting. Switching between different states causes the fluid to flow [Tang 2014].

7.8 Local Material Delivery and Resolution

Local material delivery may be most effective with local control of delivery, including dispensing with through-wall diffusion, with orifices and leaky vessel walls, and porous networks of orifices. Through-wall diffusion is common in biological systems, as is open channel in the lymphatic system. For example, the pumping of fluids through porous layers increases the level of heat transfer to create an active thermal insulation system that protects against intense long-duration thermal loads [Maruyama 1989].

Alternative means of stimulated fluid release also promote fluid healing [Dry 1996a]. Concrete damage, such as reinforcing bar corrosion, cracking due to shape changes during curing, cracking due to alkali silica reactions (ASR), and freeze-thaw degradation benefits from liquid sealants, such as linseed oil and sodium silicate; anticorrosion liquids, such as calcium nitrite; antifreeze agents, such as polypropylene glycol; crack-filling epoxies and urethanes; and solvents that soften or dissolve material. Alternative fluid release methods use coatings that degrade in the presence of particular external stimuli, such as heat, pH, specific ions, acoustic waves, low-frequency waves, hydrostatic pressure, photosensitivity, electric currents, voltages, electrorheological excitation, and radiation.

7.9 Support of Bio-based Healing Systems

Vasculature provides nutrients to bio-based healing systems. Figure 7.16 shows vasculature used to provide nutrients for bacteria healing of reinforced concrete.

FIGURE 7.16 Vascular system for providing nutrients to bacteria-based self-healing of concrete in diagonal tubes placed in formwork. (From [Paine 2016].)

7.10 Healing of Vasculature

Vasculature is an important component in many engineered systems. This includes piping for cooling, lubrication, hydraulic power, pneumatic power, ventilation, and firefighting. Piping is often buried underground or located inside machinery and walls in inaccessible locations. The self-repair of vasculature, that is, piping, in engineered systems is an important concern.

Fluid-carrying pipes have multiple failure modes ranging from small leaks to big leaks and blowouts, cracks, erosion, corrosion, clogs, pump failure, loss of fluid, and breakdown of fluid. Vascular healing techniques:

1. *Damage Mitigation with Blowout Valves:* Excessive pressure damages piping, valves, and other devices. Pressure release devices let the fluid escape upon receiving excess pressures. These include pressure relief valves, membranes that burst with excess pressure, freeze-out plugs in water-cooled engine blocks, and high-pressure transients in hydraulic systems mitigated by accumulators.
2. *Damage Mitigation with Check and Cutoff Valves:* Automated sense, more sophisticated two-way cutoff.
3. *Stop Leak Compounds:* Many of the leak-plugging methods for pressure vessels discussed in Chapter 6 also work for piping. These generally use floating particles that jam and conglomerate in the hole. A concern is to select the sizes of the particles so that they are big enough to plug the leak, but do not clog the pump or foul ancillary equipment, such as pumps, valves, and sensors.
4. *Patches and Annular Walls:* External patches on piping must contain the pressure of the leaking fluid. Tension and shear with adhesives can get a grip onto the pipe, but a circumferential grip provides better mechanical force transfer and ease of positioning, such as annular wall patches for leaks in the casing string of downhole well pipes [Allison 2012].
5. *Systemic Reconfiguration:* A significant leak at one location drains the working fluid and reduces pressure to levels that may disable an entire piping system. Closing the fluid flow to the piping with the leaks mitigates large leaks. This disrupts the local fluid flow on the cutoff branches of piping but leaves the rest of the system operable. Figure 7.17 shows a leak

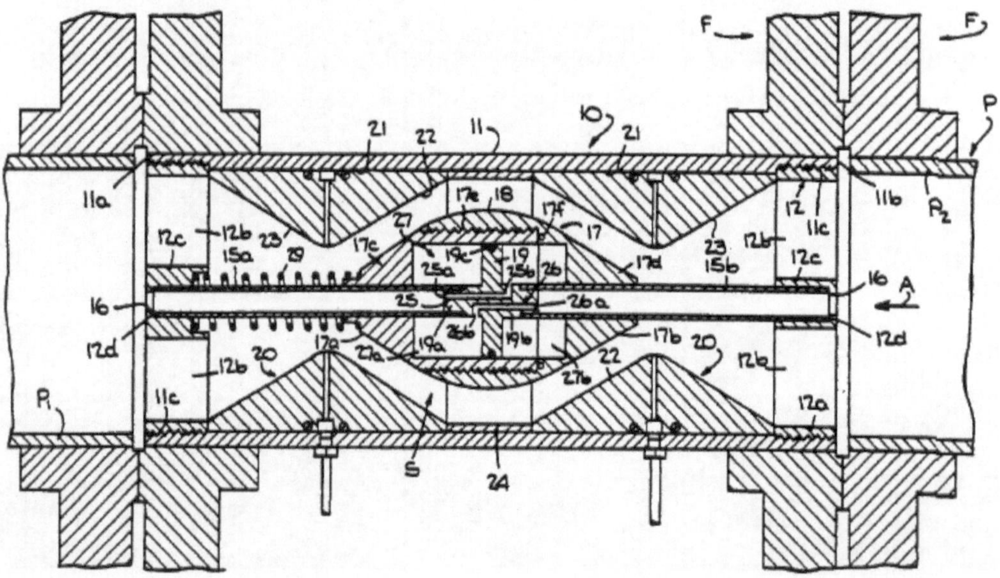

FIGURE 7.17 Two-way check valve for use in high-voltage liquid dielectric vascular insulation system closes in the event of a significant leak in the piping on either side. (From [Sarro 1994].)

mitigating cutoff valve for a liquid dielectric insulating system used in high-voltage power distribution [Sarro 1994]. This system uses an automated two-way check valve to close the pipe via secondary hydraulic connections that actuate in the event of a pressure drop due to a significant leak.

6. *Trenchless and Inside-out Methods:* Trenchless methods repair buried pipe from the inside out without digging. Placing a new internal liner seals leaks, but also reduces the internal diameter. It is possible to increase the cross section along with relining on unreinforced pipes, first with pressurized bursting and then relining [Lueke 2001].
7. *Pipe Repair Robots:* Use of robots to crawl inside, inspect, and repair pipes. The relatively simple geometry and confined spaces favor the use of robots.
8. *Futuristic:* Angiogenesis, growth of new vasculature largely remains to be developed into a viable technology.

8

Multiscale and Multifunctional Methods

8.1 Introduction

This chapter addresses self-healing systems with interactions that occur over multiple length and time scales, often through the use of multiple functionalities. These multi-X approaches lead to better performing damage mitigation and self-healing systems. Scale-dependent effects selectively allow for some physical processes to dominate at some scales, while other processes at other scales. Mixing processes can scale these effects to act over multiple scales. Sometimes, the mixing cascades layer by layer up and down length and time scales, in a manner analogous to fully developed turbulence (Figure 8.1). Other times, mixing effects leapfrog over multiple length and time scales, as in molecular scale interactions producing macroscopically observable phase changes. Timing is especially important for materials with states that change significantly during their lifetimes, such as concrete that solidifies as it cures [Dry 1998].

8.2 Continuum Damage and Healing

Many self-healing materials heal at small length scales, yet exhibit macroscopic behavior describable by models derived in the framework of continuum mechanics. Constitutive theory asserts that constitutive models of material behavior must follow the laws of physics, including thermodynamics [Eringen 1980]. Damage modeling describes the degradation of materials with observable macroscopic properties, such as deformation and temperature in terms of internal variables, such as energy, entropy, and damage.

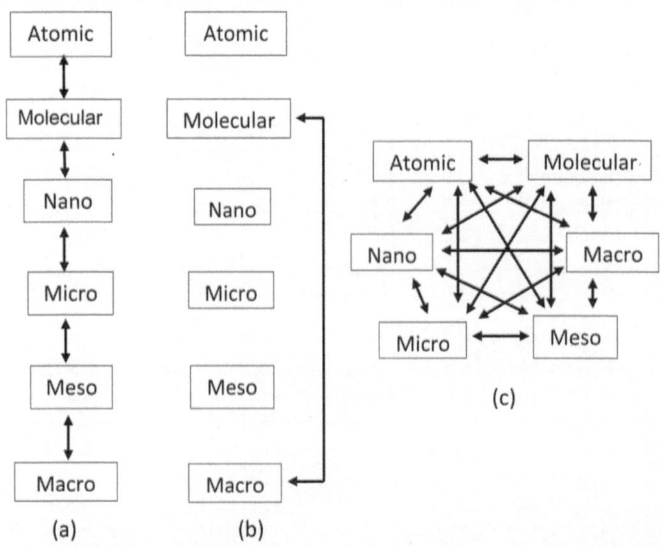

FIGURE 8.1 Interactions over multiple scales. (a) Cascade. (b) Leapfrog. (c) Fully mixed interactions.

DOI: 10.1201/9781003396048-8

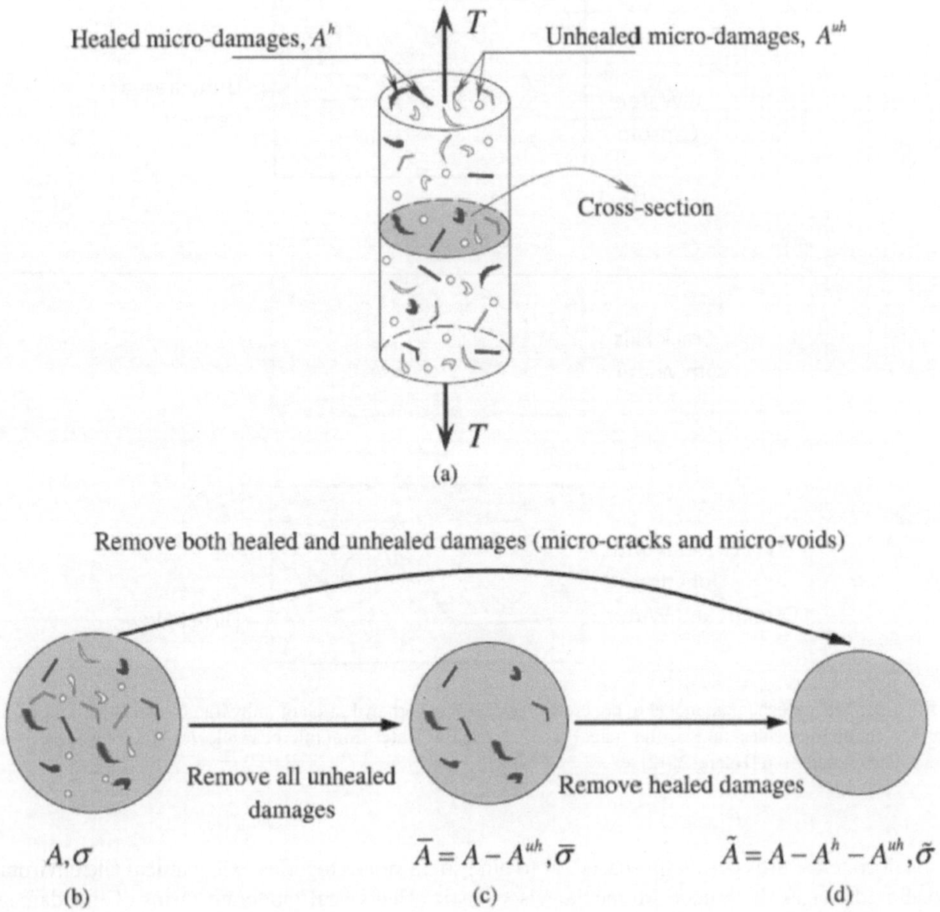

FIGURE 8.2 Healing of microdamage modeled as a continuum process. (a) The bar in tension. (b) Damaged (nominal) configuration. (c) Healing configuration. (d) Effective (undamaged) configuration. (From [Abu 2012].)

Adding healing as an internal state variable develops constitutive models that are consistent with the laws of physics and describe macroscopic damage and healing behavior (Figure 8.2) [Barbero 2005] [Voyiadjis 2011] [Miao 1995] [Abu 2012].

8.2.1 Viscoelastic and Creep Models

Asphalt pavements sustain repeated loading due to the weight and traction of passing vehicles. Rest periods between loading allow mildly damaged asphalt to heal [Bazin 1967]. Even though asphalt is a composite of granular stones and tar, continuum models describe much of the macroscopic behavior. A pseudo-stiffness model derived from viscoelastic material properties provides a reasonable semiempirical framework for explaining fatigue crack healing [Si 2002].

8.2.2 Micromechanics

Simplified models of micromechanical interactions in solids describe the healing effects of embedded fibers and inclusions. The models largely draw on the framework established for damage to micromechanical constituents with healing substituted for damage [Pugno 2011] [Bosia 2015]. Figure 8.3 shows the geometry of a micromechanical model of using embedded capsules of water to promote the autogenous healing of cracks in concrete by hydrating unhydrated cement [Huang 2012].

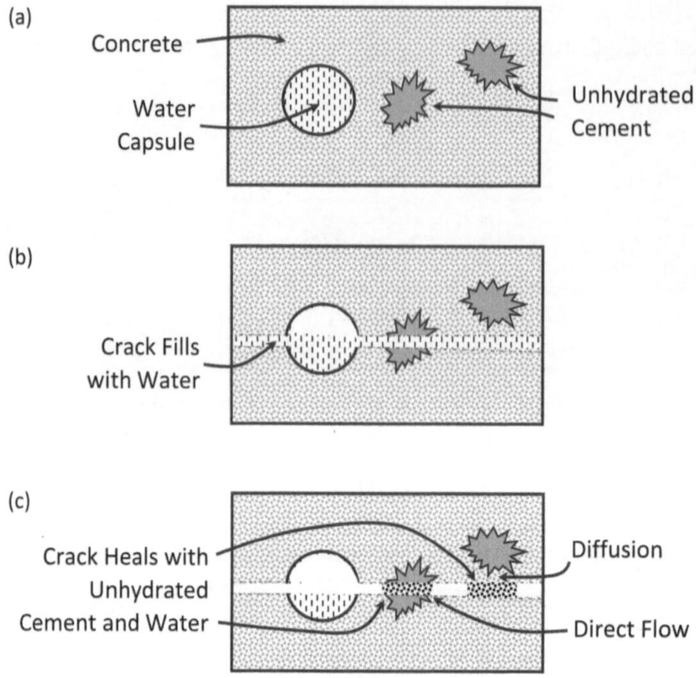

FIGURE 8.3 Micromechanical model of concrete crack healing with hydration of unhydrated cement by internal release of water from microcapsules. (a) Pristine state. (b) Crack releases water from microcapsule. (c) Water hydrates cement to heal crack. (Adapted from [Huang 2012].)

Asphalt concrete uses petroleum-based tar to bind small stones together. Mechanical and environmental loading damages the binder. Molecular viscoelastic effects heal moderate forms of this damage. A memory model based on a convolution integral can model the damage and healing processes [Carmona 2007]. Equation (8.1) expresses the quantity p(t) as a von Mises-type damage for a beam specimen. The first term contains the effect of axial strain, ε, with respect to a threshold value ε_{th}. The second term contains the effect of bending through the end rotations θ_i and θ_j, with respect to a threshold value θ_{th}. Equation (8.2) captures the healing effect with a fading memory in a convolution integral. The characteristic time τ governs the rate of healing. Failure occurs when $q(t) > 1$.

$$p(t) = \left(\frac{\epsilon}{\epsilon_{th}}\right)^2 + \frac{\max\left(|\theta_i|,|\theta_j|\right)}{\theta_{th}} \tag{8.1}$$

$$q(t) = p(t) + f_0 \int_0^t e^{-(t-t')/\tau} p(t')\,dt' \tag{8.2}$$

Syntactic foams are composite materials made by mixing small hollow spheres and a binding matrix. The result is a multiphase foam material with tunable properties. Combining a shape memory polymer with hollow glass microspheres creates customized thermally active self-healing foam [Xu 2010b].

Halite is a rock crystal form of NaCl. It has inherent self-healing capabilities that makes it useful for large-scale sealing layers. A typical halite installation begins with rock salt granules that merge and heal into a solid mass by diffusive transport that largely squeezes out the open gaps and cracks. This is a multiscale process acting at molecular scale as well as micro-, meso-, and macroscales. Fabric tensors are second-order symmetric tensors that describe statistical properties of granular media as well as the micro- and mesoscale mechanics of these structures [Zhu 2015].

Large earthquakes often coincide with the fracture of subterranean rack layers. High temperatures and pressures underground promote self-healing. Water permeability measurements in wells find that fault healing is a punctuated process with exponential time constants on the order of 0.6–2.5 years [Xue 2013].

8.2.3 Nonrational Models

Rational mechanics is an approach to continuum and related applied mechanics problems based on formal derivations from first principles. The rational mechanics approach provides important insights, such as the role of thermodynamics in continuum models, but often has difficulty in capturing the behavior of complicated material systems. Nonrational methods combine experimental observations with heuristics and other subjective inputs to form models of expected behaviors [Arson 2012]. Fuzzy reasoning and, more recently, some artificial intelligence methods, are examples. Fuzzy methods can provide insights into which parameters are important to describe the healing of asphalt mixtures [Huang 2016].

8.2.4 Percolation

Percolation in heterogeneous material systems occurs when localized connections coalesce to form long-range connections. Small changes in the rate of local connectedness sometimes cause large long-range changes, with the change having many characteristics of a phase change. Aside from a few simple cases, most percolation analyses use numerical methods based on simulations of random fields. Percolation describes some aspects of fatigue crack growth, suppression, and self-healing. Fatigue cracks grow through the connecting coalescence of smaller cracks. Bigger cracks connect to smaller cracks until a threshold is reached in which the part fractures. Embedded self-healing agents, such as liquids in microcapsules, heal and suppress the growth of small fatigue cracks before they start to percolate into bigger cracks [Dementsov 2008].

More complicated percolation dynamics arise with active materials. Gels with arrays of nematic tubules form hierarchical dynamic patterns of percolating bundled active networks comprised of microtubule bundles [Sanchez 2012]. Such bundles are active, typically moving, fracturing, and self-healing in nonequilibrium processes.

Percolation modeling describes the strength of polymer–polymer bonds with reptation interdiffusion in terms of the number of molecules that entangle across a bond through reptation diffusion [Wool 2006]. If there are enough bonds of sufficient strength, the interface is strong enough to withstand the applied load. If the bond strength and density are insufficient, the macroscopic interface cannot hold the load and fails.

8.3 Multiscale Thermomechanical Healing

It is possible to combine thermal and mechanical loading to induce multiscale self-healing interactions. Thermoplastic polymers when heated above their glass transition temperature T_g heal tight cracks by reptation. Being above T_g also weakens the material. Thermoset polymers are tougher, but do not readily heal by diffusion. The differences in thermomechanical properties of thermoplastic and thermoset polymers result in big differences in damage processes and the ability to self-heal. Hybrid polymers combine thermoset and thermoplastic properties to produce unique energetic mechanical and thermomechanical loading properties.

Custom networks of bonded polymer molecules change stiffness in response to a stimulus. A primary technique is to control the amount of polymer cross-linking. More cross-linking increases stiffness and less cross-linking softens the material.

Antagonistic mixes of strong and weak bonding within the same material form highly tunable internal molecular structures. Stimuli-responsive behavior percolates across the material in gels and elastic solids. An example is a combination of ring-opening metathesis bonds with weak hydrogen bonding molecular motifs [Nair 2008].

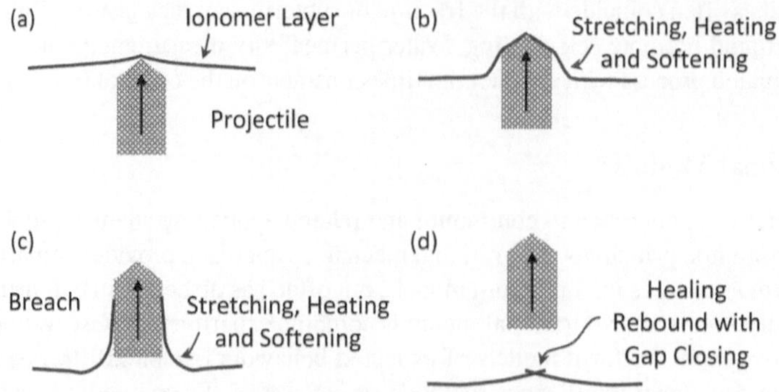

FIGURE 8.4 Elasto-plastic rebound hypothesis for thermoelastic healing with ionomers. (a) Initial contact with projectile. (b) Projectile stretches ionomer. (c) Projectile breaches ionomer. (d) Ionomer rebounds to close gap. (Adapted from [Varley 2010].)

8.3.1 Ionomers with Nanoscale Thermomechanical Functionalization

Adding sodium and other molecular species to polymers forms ionic bonds that alter the material at nanoscale and molecular scale. The result is hierarchical actions up through macroscales, including self-healing compatible combinations of elastic rebound plus plastic reformability.

Ionic polyamides demonstrate remarkable macroscale healing through nonbonding interactions. Observations of the response of the ionic polyamide Surlyn™ to penetrating ballistic impacts produce an impressive near-recovery of initial shape [Fall 2001] [Wool 2008] [Varley 2008]. Projectile penetration and pass-through stretches heat and tear the polymer. A combination of elastic rebound and viscous material effects, including a thermoelastic transition from hard to soft material acting at molecular scale as well as meso- and macroscales, causes the hole to close. At the molecular scale, ionic polymers are a mix of hydrophobic polymer components and hydrophilic ionic components that form into clusters [Tadano 1989]. Highly energetic damage, such as ballistic penetration or rapid saw-cutting, elastically distorts the material on a macroscale and melts the material in the immediate vicinity of the damage (Figure 8.4). The first healing stage is an elastic rebound that closes much of the wound. The second stage is the thermal-melt-driven sealing of the hole. Less energetic damage, such as slow sawing and drilling, does not energetically excite the material and cause rapid healing [Varley 2010]. Measurements on the ionic polymer poly(ethylene-*co*-methacrylic acid), that is, Surlyn, and a nonionic copolymer with the trade name Nucrel™ exhibit similar behavior. Figure 8.5 shows an ionene-polyamide film that recovers

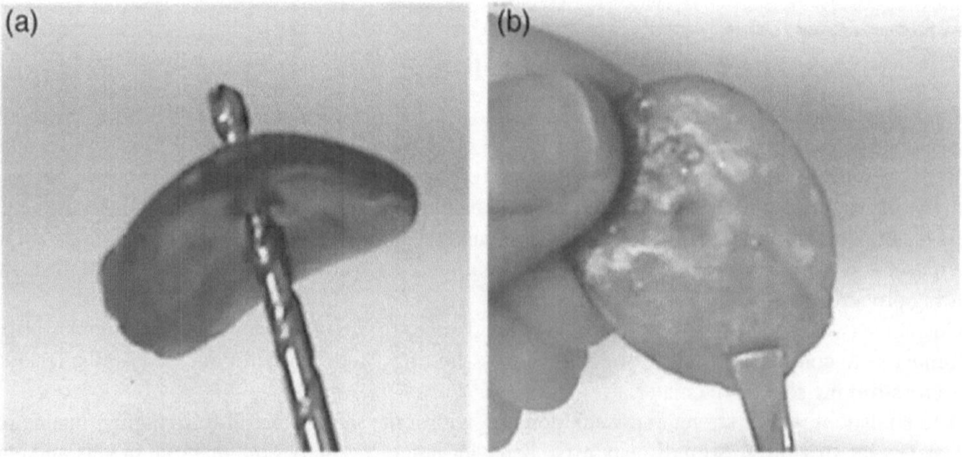

FIGURE 8.5 Macroscale healing with ionene-polyamide film. (a) Perforation with drill. (b) Resealed hole following thermal cycling at 60 °C for 30 s. (From [Kammakakam 2019].)

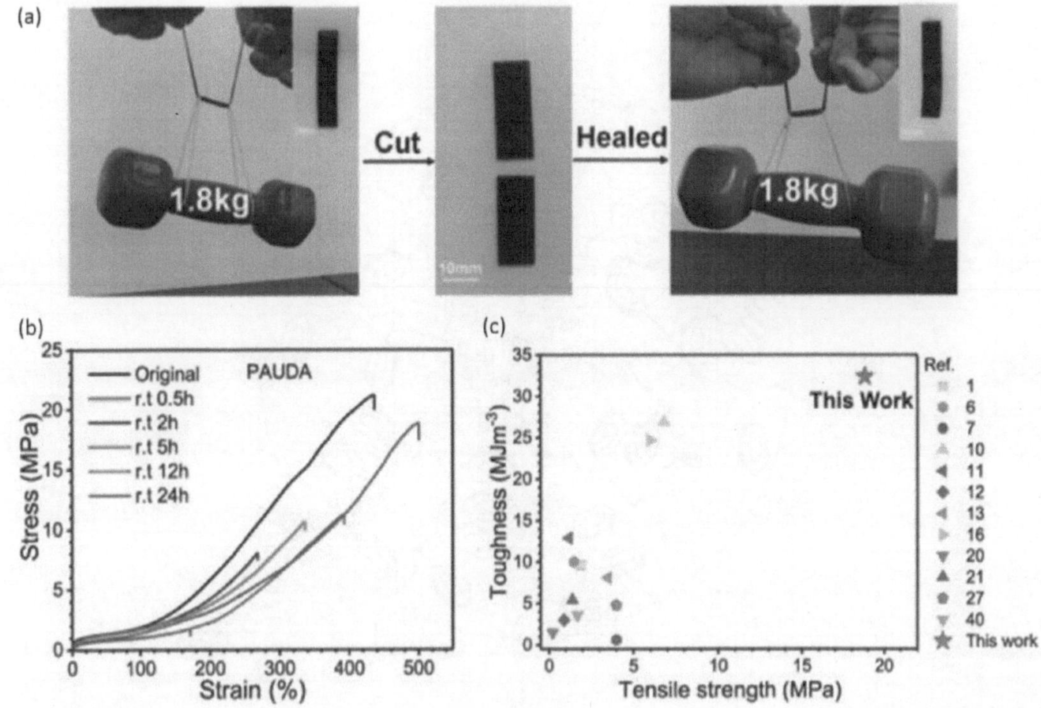

FIGURE 8.6 High-strength polyamide elastomer that recovers from a cross-sectional cut. (Reprinted with permission from [Wu 2020]. Copyright 2020, American Chemical Society.)

from perforation by a drill with 30 s of heating at 60 °C [Kammakakam 2019] and Figure 8.6 shows a high-strength polyamide elastomer that recovers from a cross-sectional cut [Wu 2020]. Another application is noise-suppression systems on firearms where a self-healing ionic polymer captures particulates from each discharge to improve the performance and durability of the silencer [Oliver 2014]. Isolated mechanical or isolated thermal loading does not produce the same effect with these materials.

Molecular scale descriptions (density functional theory, density functional tight binding theories [Vuong 2019], and classical molecular dynamics simulations) of the mechanics of healing with ionic polymers suggest that healing combines dissociation of ionic clusters and supramolecular backbone relaxation. The process has three timescales – the terminal relaxation time t_d; the supramolecular relaxation timescale τ; and the macroscopic healing timescale t_{heal}. Good healing seems to require a separation of timescales with terminal relaxation being fastest, followed by supramolecular relaxation and macroscopic healing. This separation enables dangling sticky molecules to form on freshly cut surfaces and promote adhesion, but also to fade with time so that normal exterior surfaces are not sticky. Experiments with poly(butyl acrylate-*co*-acrylic acid) ionomers using cobalt counterions in temperature ranges that promote self-healing found terminal relaxation times on the order of tens of seconds, supramolecular relaxation times of hundreds of seconds, and macroscopic healing times of thousands of seconds [Bose 2015].

The Eisenberg–Hird–Moore model combines multiplet clusters to form a more detailed explanation of some of the behavior of ionomers (Figure 8.7) [Eisenberg 1990]. The model considers ionomer molecules as long chains of nonionic mers with a small number of interdispersed ionic mers. Ionic components attract other ionic components and form ionic bonds as well as clusters of ionic bonds. Clustering leads to nanoscale heterogeneity with localized regions of ionic bonds restricting the mobility of polymer molecules interspersed in a matrix of thermoplastic nonionic regions. The heterogeneity combined with hierarchical clustering and the reversibility of the ionic cluster bonds explains the macroscale thermoelastic behavior of ionomers. Adjusting the level and type of ionic bonding affects performance.

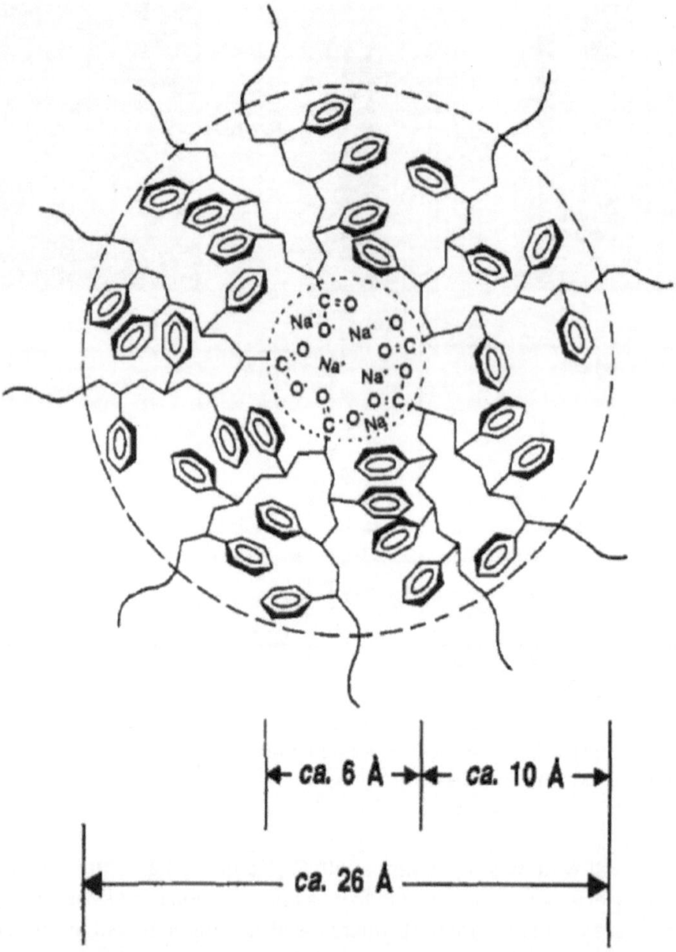

FIGURE 8.7 Eisenberg–Hird–Moore model of multiplet clusters in ionomers. (From [Eisenberg 1990].)

Too small of an ionic dose and insufficient linking occur. Too large of a dose overly stiffens the material. An ionic loading of 15% methacrylic acid ionized with Na or Zn in an ethylene matrix produces a material with high impact resistance [Akimoto 2001].

Hybrid compositions of elastomers and ionomers produce material systems with both the large reversible deformation typical of elastomers and the elasto-plastic rebound of ionomers for self-healing. A combination of *n*-butyl acrylate as the elastomer with acrylate acids as the ionomer promoter produces a self-healing elastomer material system with adjustable properties [Hohlbein 2015]. A 5% molar ionic load may be optimal.

Inserting nanoparticles into hydrogels produces composite material systems with tunable shear thinning and self-healing properties that are superior to neat hydrogel [Appel 2015]. Polymer molecules bind to the nanoparticles. Upon macroscopically stressing the material, the polymer molecules stretch and detach from the nanoparticles, as a toughening effect. The polymer molecules reattach to the nanoparticles in a subsequent self-healing process. Key features are to have the polymers noncovalently link to one another and to the nanoparticles. To promote percolation, the particles need to have average diameters smaller than the length of the polymer molecules, with an average number of interactions between molecular and nanoparticle >2. Similarly, the introduction of pyrene-functionalized gold nanoparticles into a mixture with supramolecular pyrene-functionalized polyamide and polydiimide polymers produces recognition between the three components leading to a doubling of tensile strength versus control without recognition, and the ability to heal with solvent and thermal cycling [Vaiyapuri 2013]. Numerical simulations indicate that the nature of the bond between the polymer and nanoparticles has a large effect on overall self-healing material performance. Catch bonds, that is, bonds that grow in strength with time and load, produced more robust material networks than slip bonds [Iyer 2015]. The technique requires polymer backbones with sufficient mobility and flexibility to migrate to the crack surfaces [Schäfer 2015].

Annealing for healing requires the following:

1. *Homogeneous State:* The material has a nominal homogeneous state with no damage or stress.
2. *Minimal Energy:* The homogeneous state has a minimum of intrinsic energy.
3. *Damage Raises Energy into a Stable State:* Damage, cracking, distortion, residual stress, magnetization, and so on alter the internal energy, homogeneity, and entropy of the system. This altered state freezes the damage into a stable nonequilibrium configuration.
4. *Mobility:* Adding energy that is just enough to mobilize the frozen degrees of freedom allows the material reconfigure to a lower energy healthy state.

External heating as a means of promoting healing is the mainstay method of annealing. Additional techniques using other materials in multicomponent functionalized arrangements that activate with heating began to appear in the 1990s, such as materials with embedded epoxy particles in a state of localized high free energy, that repair damage in glass-fiber-reinforced polymer (GFRP) composites with external heat [Zako 1999].

Some metal alloys solidify in a nonequilibrium state with excess free energy. Damage to the alloys, such as large mechanical deformation, raises the energy state to a level that releases the free energy, resulting in release of the free energy and phase changes, leading to precipitation at damage nucleation sites that heals the metal [Nosonovsky 2012]. Similarly, some materials remain stable in out of equilibrium conditions by multiphase effects that prevent decomposition. An example is gas hydrates which exhibit a self-preservation effect by forming an ice-like solid outer layer that prevents decomposition at temperatures above the melting point [Bai 2015].

Point defects arise in germanium crystals due to a variety of manufacturing processes. These defects degrade the performance of electronic devices, such as leakage currents from on/off action of n+/p junction diodes. Annealing between 550 °C and 650 °C heals point defects in germanium crystals [Shim 2013].

8.4 Granular and Cementitious Materials

Asphalt, mortar, and concrete are granular composite materials that routinely self-heal, typically by infilling tight cracks. The self-healing behavior is not very aggressive, but nonetheless leads to long-term endurance, even with harsh environmental and loading conditions.

Asphalt cement is a composite mix of small stones and tar-like polymer binders. A common damage mode is fatigue and cracking at the microscopic scale of the binder that appears on macroscopic scale as a reduction of stiffness and visible cracks. Most of the healing techniques use rest periods between loads. The healing phenomena depend on viscoelastic reflow and molecular reptation diffusion healing of cracks in the binder matrix. The time constants of these thermodynamically driven healing processes are often several times longer than the damage-causing loads. The rate of healing activity increases with temperature.

Many tests of healing performance measure stiffness, which damage reduces and healing increases. Testing applies mechanical loads of sufficient magnitude to damage the material, followed by rest periods with no loading. Subsequent testing evaluates the healing by measuring the material stiffness and the number of load cycles to failure.

Concrete is a composite of stones, sand, and reinforcing elements held together by cementitious material. The precipitation of calcium carbonate to fill tight cracks is a common self-healing mechanism.

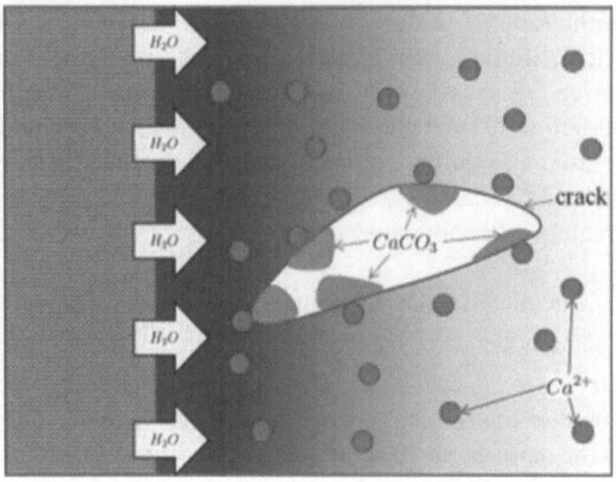

FIGURE 8.8 Diffusion of H_2O and Ca^{2+} forms $CaCO_3$ to fill crack in concrete. (From Aliko 2015].)

Figure 8.8 shows a simplified model of the process where water diffuses through the cement, accumulates Ca^{2+} ions, flows into the crack, and precipitates $CaCO_3$ to fill the crack. A continuum model of the process describes the rate of $CaCO_3$ formation. An ordinary differential equation (8.3) describes the formation rate. A partial differential equation (8.4) describes the spatial variation. Here, k is the kinetic rate constant; D is the diffusivity; H[x] is the Heaviside step function; and Ca^{2+} is the concentration of calcium ions (Figure 8.8) [Aliko 2015].

$$\frac{d}{dt}[CaCO_3] = -\frac{d}{dt}\left[Ca_3^{2-}\right] = k\left[Ca^{2+}\right]\cdot H\left[Ca^{2+}\right] \tag{8.3}$$

$$\frac{d}{dt}\left[CO_3^{2-}\right] = D\cdot\nabla^2\left[CO_3^{2-}\right] - k\cdot\left[CO_3^{2-}\right]\cdot H\left[Ca^{2+}\right] \tag{8.4}$$

8.5 Phase Change and Percolation Methods

Phase changes are macroscopic manifestations of interactions initiated and thermally driven at the molecular scale. A macroscopic characteristic is that different phases of the same material, such as ice and water, coexist at the same temperature and pressure, yet have distinctly different internal energy levels and different macroscopic properties. Adding or removing energy transforms the material from one phase to another, often crisply at specific changes of temperature, pressure, or other physical parameters. Phase changes appear at the molecular scale with atomic reconfigurations, such as transforming from being immobilized in solids to being free to move in fluids. Phase changes act as large reservoirs (or sinks) of energy. For example, the amount of heat required to raise liquid water at atmospheric pressure from 25°C to 26°C without a phase change is 4.19 J/gm. A phase change, such as the amount of heat needed to melt ice into water at atmospheric pressure, is 335 J/gm. Vaporizing water by boiling at atmospheric pressure requires 2,250 J/gm.

Thermodynamic properties associated with phase changes provide the impetus and/or store of available energy for damage mitigation and self-healing. Technical challenges arise due to hysteresis and nonlinear behavior, which favors systems with linear behavior. A notable exception is the use of a condenser by James Watt to heal steam by removing entropy. Thermodynamic healing of steam was a major innovation that continues to fuel the Industrial Revolution and sustain modern society.

8.5.1 Liquid Phase Change

The spontaneous formation of liquid phases can also play a role in healing. Methylamine gas at room temperature causes perovskite thin methylammonium lead iodide used in solar cells to heal solid defects by liquifying and then solidifying with the removal of the gas [Zhou 2015].

8.5.2 Solid-to-solid Phase Change

Void formation between crystal grains in metals is a principal mechanism of cracking. The application of external pressures closes and heals the cracks in a thermodynamically favored process [Wang 2003].

8.5.3 Shape Memory Materials

Shape memory materials are integral components of multiscale healing systems, typically as inserted or interwoven components that actuate with thermal stimulus and change phase. SMA actuator and pseudo-elastic behavior enable large deformations at constant temperature. A notable example is the expandable mesh arterial stent. Change in temperature changes the phase and creates macroscopic actuators – maybe capable of exerting 100 ksi.

8.6 Heterogeneous Material and Structural Systems

Another application of phase change for healing is a material that separates into a heterogeneous patchwork of different phases during manufacture. This occurs when thermodynamics has a nonconvex phase energy equilibrium diagram, with two or more valleys. One possible mechanism of the self-healing behavior of asphalt is a thermodynamically driven phase separation heterogeneity with a waxy buildup of bumblebee patterns (Figure 8.9) [Kringos 2011]. Reaction–diffusion effects are a means of introducing heterogeneity.

8.6.1 Foams and Cellular Material Systems

Foams and cellular solids are multiphase materials with arrays of fluid-filled compartments surrounded by stiffer solid or semisolid walls and struts. The combination of lightweight and the ability to imbue foams and cellular solids with a wide variety of customizable and functional properties makes them quite useful in a variety of sealing and healing applications.

Some authors distinguish foams and cellular solids by the phase of the original source material. Foams come from liquids and cellular solids from solids [Gibson 1997]. The manufacture of foams and many cellular solids is a gas insertion process or supersaturated gas release with the initiator being heat, chemical reaction, or pressure release. High-performance materials produce high-performance foams and cellular solids but are difficult to fabricate. Examples include PTFE, which is difficult to foam, because it does not melt, polyimide, and metals [Kolmschlag 2004] [Abdul 2008]. Material removal and assembly

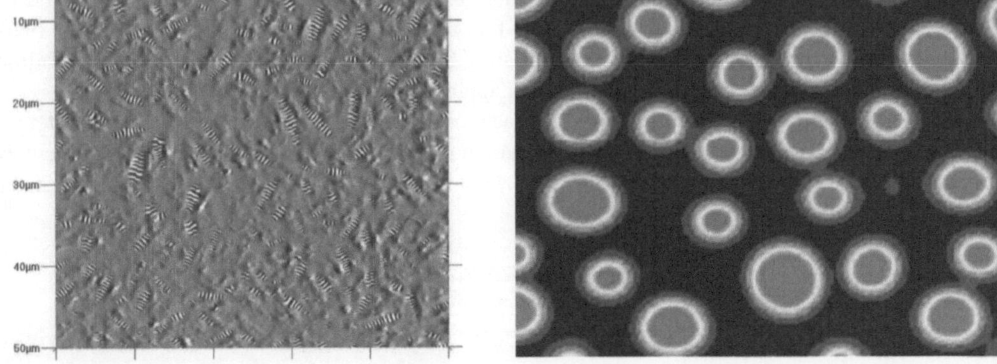

FIGURE 8.9 Microscale heterogeneities in asphalt bitumen with stores of free energy that enable self-healing. (a) Bumblebee heterogeneities observed by AFM. (b) Patchwork simulation of phase separation in thermodynamic valleys. (From [Kringos 2011].)

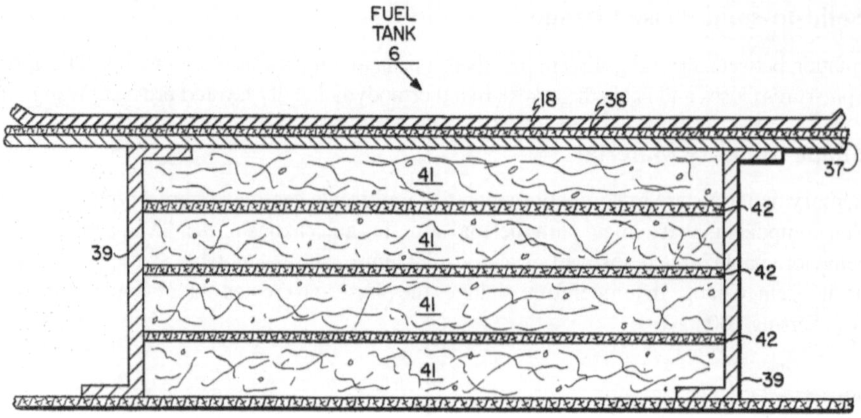

FIGURE 8.10 Hierarchical cellular fuel tank system mitigates projectile damage by absorbing shrapnel and leaking fuel. (From [Hietz 1982].)

at the cellular level are alternative methods for making cellular solids. Syntactic foams are assemblies of individual spheroids. Some syntactic foams use spheroids filled with functional fluids.

Thin wall cellular systems exhibit macroscale deformation and shape recovery through elastic buckling interactions at the cellular level. Some of these materials can be stiff, compliant, and lightweight. An extreme example is a periodic tube metallic microlattice that is capable of 50% strain and full shape recovery while sustaining bulk densities of less than 0.9 mg/cm^3 [Abdul 2008].

Hierarchical architectures that embed foams or other sealing compounds inside of structural cellular compartments provide enhanced functionality. Hierarchical cellular systems appear in a variety of self-sealing systems. Hierarchical architectures separate the functions so that the macroscale cellular structure provides mechanical support while the embedded material (usually gel or foam) provides enhanced leak-induced sealing. One possibility is a cellular swelling foam self-sealing aircraft fuel tank system (Figure 8.10) [Hietz 1982]. This system sits underneath fuel tanks, mitigates shrapnel damage, and contains fuel leaks. An embedded cellular gel structure for sealing leaks in tires appears in Figure 8.11 [Knowlton 1934]. Cellular structures that mix vulcanized and unvulcanized rubbers require specialized manufacturing techniques [Waber 1939].

Some materials swell when exposed to solvents. Unvulcanized rubber, polyisobutylene, and neoprene swell when in contact with gasoline. Such materials tend to become weak and dissolve with long-term soaks in solvents. Blending the swelling-soluble material with an insoluble material, such as vulcanized rubber, produces a material that swells in the presence of the solvent and yet retains structural resilience. Foamed variants produce enhanced properties, such as more rapid swelling, lighter weight, and the ability to sustain larger deformations. Self-sealing fuel tanks are a primary application [Dasher 1940]. Open-cell polyurethane foams act as thermal insulating layers while also producing healing capabilities in thermal storage tanks for technically demanding applications, such as in aircraft [Cline 1968].

A multilayered self-sealing fuel tank places an elastomeric foam layer between polyurethane outer and metal inner tank walls in a configuration with the foam layer being several times thicker than the polyurethane or metal layers (Figure 8.12) [Baker 1972]. Fuel-induced swelling sealing and pressure-resistant

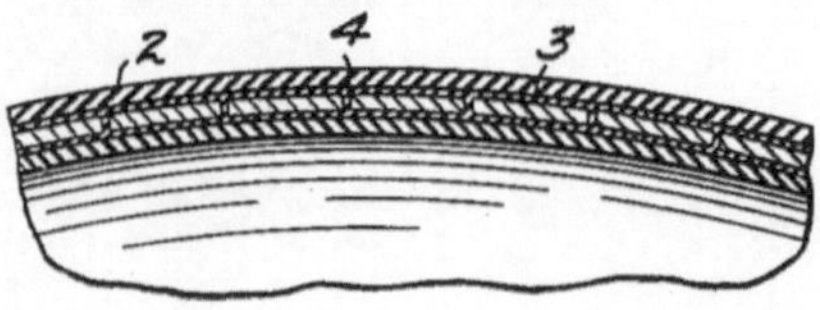

FIGURE 8.11 Early (ca. 1934) cellular self-sealing tire tube with flowable gel inside the cells that provide sealing capability and cell walls that provide a mechanical structure. (From [Knowlton 1934].)

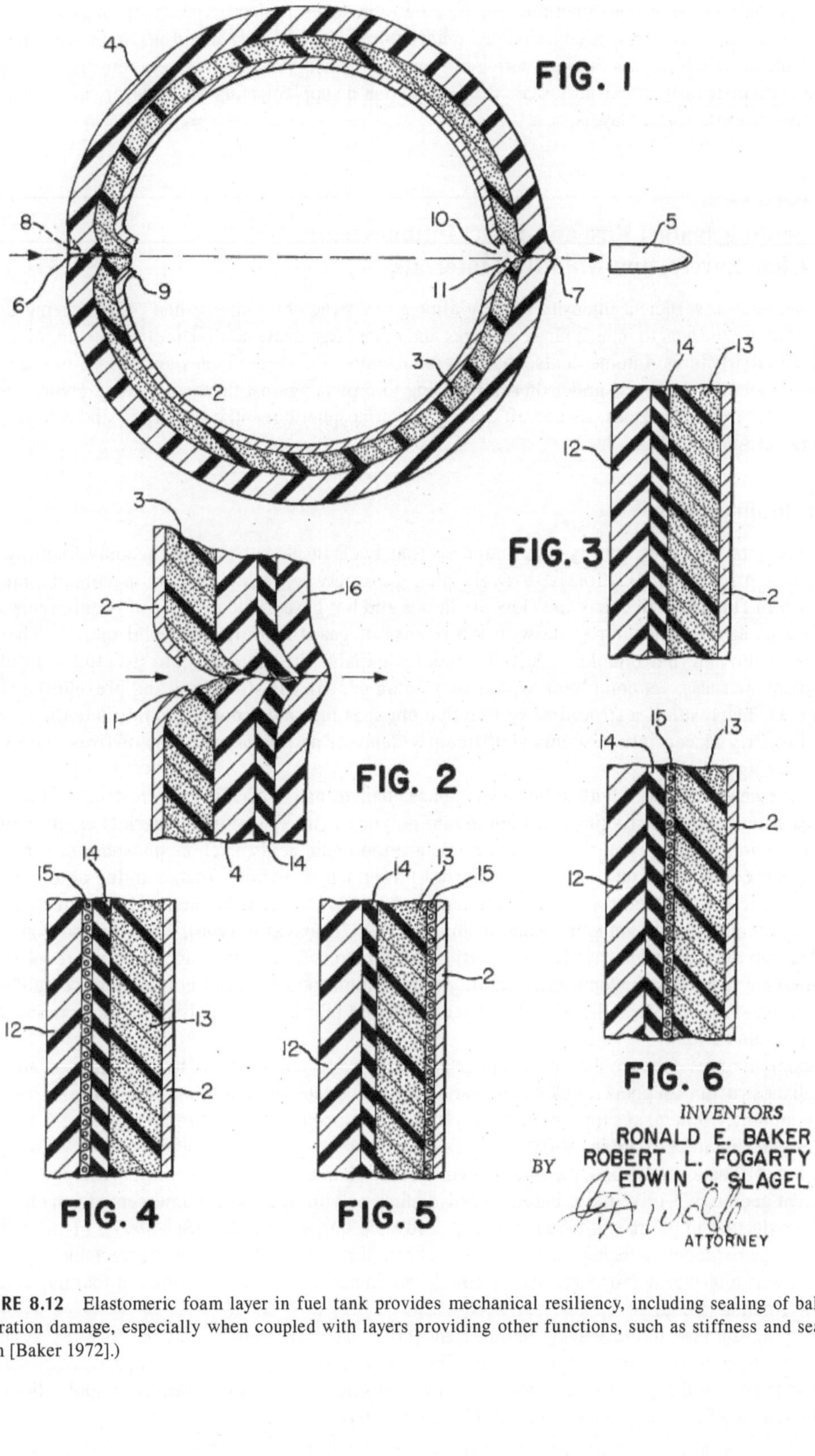

FIGURE 8.12 Elastomeric foam layer in fuel tank provides mechanical resiliency, including sealing of ballistic penetration damage, especially when coupled with layers providing other functions, such as stiffness and sealing. (From [Baker 1972].)

layers provide supplemental functionality. When a high-speed projectile passes through the tank, the elastomeric foam stretches, accommodates petals formed by the stiff metal and polyurethane walls, allows the projectile to pass through with a minimum of tearing, and then largely returns to an original shape that either fully or partially seals the hole so that a supplemental sealing layer can complete the seal fiber-reinforced cloth layers.

8.7 Sealing against Fire and Heat: Intumescents, Char Layers, and Ablative Materials

Fire and heat of sufficient intensity destroy almost any structure. One method of countering thermal loads is to use the heat to induce phase changes and other large-scale nonlinear effects to alter the physical state of structures. Intumescents, char layers, ablators, and ceramfiables are classes of specialized materials that change phase under thermal loading to protect against thermal loading. Hypersonic aircraft, rocket exhaust nozzles, spacecraft undergoing atmospheric reentry, buildings, and electric power lines are example applications.

8.7.1 Intumescents

Intumescents are solid or putty-like materials that swell in-place by spontaneously foaming when heated. A primary application is to swell, plug gaps between structural compartments, and stop the flow of fluids, particularly the flow of flames and hot gases. The process of intumescent swelling begins at the molecular scale with the release of gases from the material matrix. The gases coalesce into internal bubbles. Multiple bubbles expand the material into a cellular solid with sufficient stiffness, yet compliant expansion to plug gaps in the structures and prevent the flow of hot gases. The result is a structural system that changes internal topology to mitigate the spread of fire. Properly placed intumescents significantly delay or arrest the progress of fires and extreme heat [Hering 2001].

Intumescents that swell with solid to gas phase changes use irreversible microstructural processes to create multicellular structures in a compliant polymer matrix. Multiple materials exhibit intumescent behavior. Common intumescent materials are inorganic acids, such as phosphoric; carbon-rich polyhydric compounds, such as starch or phenol-formaldehyde resins; organic amine or amides, such as melamine; halogenated compounds; aromatic sulfonates in polycarbonate, and mixtures with inorganic compounds and synthetic resins [Camino 1989] [Kobayashi 1995]. Halogenated compounds produce noxious gases and have fallen into disfavor. It is sometimes advantageous to microencapsulate intumescent materials for long-term stability. Ammonium phosphate in polyurethane or polyether–polyurethane shells reduce the flammability of bulk and fabric-coated polyurethane materials [Saihi 2005] [Girand 2005].

Flexible fixtures enhance the fire-stopping performance of intumescents by providing an additional compliant structural support for elements that need to penetrate through floors and walls, such as wires and plumbing, while also supporting the swelling intumescent during a fire and guiding it to form a tighter seal (Figure 8.13) [Stahl 2008].

In a multifunctional method, the intumescent forms a char layer on the outer layers after swelling to a different geometry. Unlike those based on polyolefins, polyimide-based intumescents form char layers by the reduction in the rate at which material feeds and burns in the fire [Siat 1997]. Flame-retardant additives help. Modeling multiscale interactions is complicated, but there are some tractable approaches, most of which include detailed analysis of kinetic parameters and micromechanics of foaming and char-layer formation [Butler 1997].

The parasitic load introduced by intumescent-active materials in bulk polymer material systems is small, often on the order of 1–2% by weight. The reduction in size of intumescent particles down to the nanoscale enables the graceful insertion of intumescents into materials in a variety of applications, such as clothing and FRP composites [Panse 2013] [Toldy 2011].

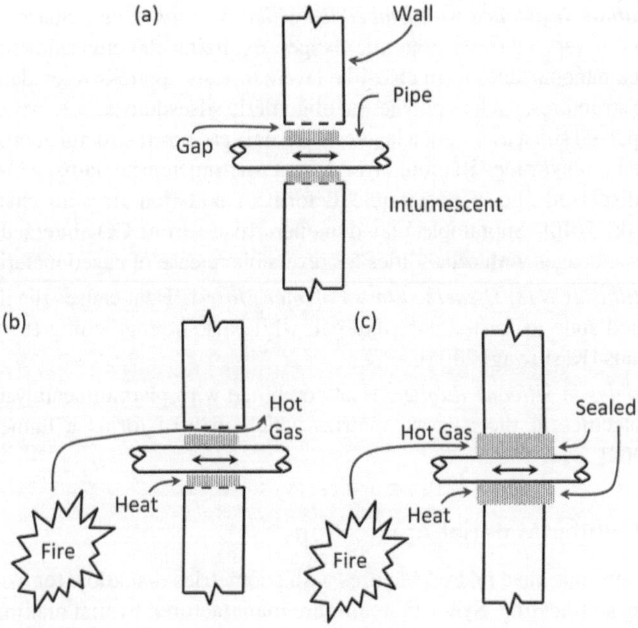

FIGURE 8.13 Self-adjusting intumescent fire stop swells and seals gap in building wall to change gas exchange topology when heated by fire and prevents a conflagration. (a) Gap allows small motions during routine use. (b) Fire heats and sends hot gases through gap. (c) Intumescent swells and plugs gap. (Adapted from [Stahl 2008.)

8.7.2 Char and Self-passivating Layer Forming Materials

Materials that form char layers are like intumescents. These materials use heat-induced swelling and material property changes to slow heat damage. The char layer forms as an impermeable outer layer that resists oxidation with an insulating porous sublayer that minimizes the flow of heat (Figure 8.14). Naturally occurring wood cork is an excellent char-forming material with multiple industrial and aerospace applications. Several techniques produce similar effects:

1. *Doping Polymers with Silica and Other Ceramic Particles:* Intense heat burns off, that is, ceramifies, the polymer matrix leaving a thin silicate layer that resists further degradation by heat [Vaia 1999]. Additives, such as phosphates, can enhance the durability of the ceramic layer following fire exposure [Alexander 2013]. Silicate-based ceramifiable insulation provides a short-term shelf-healing capability for high-power electric power lines. The electrical resistance stays intact following fires and prevents major electrical outages.

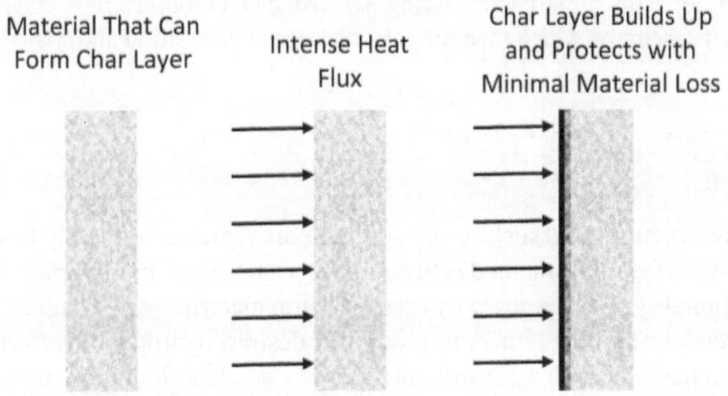

FIGURE 8.14 Intense heat flux creates passivating char layer that insulates against further heat damage.

2. *Polymer Oxidation Resistance with Embedded Silica Nanoparticles:* Spacecraft in low earth orbit experience a very aggressive atomic oxygen oxidizing the environment. Polymers with embedded silica nanoparticles form char-like layers to resist aggressive oxidation [Fong 2001]. Inorganic cage structures, such as polyhedral oligomeric silsesquioxane, serve as delivery agents of the nanodispersed silica as part of a layered external spacecraft structure consisting of an SiO_2 outer coating on a polyimide (Kapton) layer. When micrometeorite damage breaches this outer skin, the nanodispersed silica particles quickly form an oxidation-arresting char layer [Tomczak 2004] [Miyazaki 2010]. Supramolecular dynamers made from CO_2 interacting with amines assemble into nanocages with capabilities for reversible release of caged material [Xu 2003].
3. *Carbon Nanotubes (CNTs) Dispersed in a Polymer Matrix:* Heat causes the nanotubes to jam into a percolated state to form a char-like gel, while also strengthening the polymer against mechanical loads [Kashiwagi 2005].
4. *Platinum Particles in Silicone Rubber:* Heat combined with platinum catalysis induces cross-linking of molecules in the silicone matrix, which in turn forms a flame-retarding layer [Hayashida 2003]

8.7.3 Foam and Cellular Material Applications

Foams and cellular materials have microstructures with geometries that allow for insertion of functional capabilities, including self-healing. Syntactic foams are manufactured by first making hollow spheroidal particles and then conglomerating the particles into what becomes a foamed solid. The process, while generally more expensive than in situ foaming, allows for a wider range of functionalization of the cells. A combination of glass microbubbles, shape memory polymer (SMP) matrix and CNTs forms a self-healing syntactic foam [Li 2008a]. Placing the self-healing syntactic SMP foam in a sandwich composite produces a structural element that recovers from repeated impacts, with a heating cycle to activate the shape memory effect. The CNTs act both as micro-reinforcing and as a potential source of heat through external electrical or infrared optical effects.

In concrete, active aggregates made of expanding material, such as clay, readily absorb and release a healing agent. Encapsulating the aggregate with a material barrier that breaks down with damage creates an active self-healing concrete system for blast furnace slag mortars, as shown in Figure 8.15 [Sisomphon 2011]. Macroscale expanded clay aggregate particles contain and then release sodium monofluorophosphate (Na_2FPO_3) through micropores upon encountering a carbonation front. Combined interactions with calcium hydroxide released from the cement paste act to limit salt-scaling damage – a complicated process that depends on chemistry, stress, porosity, cracking state, and hydration.

Metal organic frameworks (MOFs) are crystalline solids with a molecular framework structure that has internal pores big enough to contain small molecules. The internal stores of molecules flow out of the framework when needed and bond crystals together to produce macroscopically observable healing [Mondal 2022] [Półrolniczak 2021]. Copper MOFs heal at temperatures as low as –50°C, where the overall elastic modulus of healed crystals increases from 4 GPa to 12 GPa and hardness from 400 MPa to 1,000 MPa [Chen 2013c]. This remarkable healing performance combines molecular diffusion effects to drive nanoscale crystal assembly, which then propels macroscale formation and stiffening of bulk crystals.

8.8 Regeneration

Regeneration reforms structures and surfaces by adding more material, typically by moving material up from the bottom of the surface. The method is to construct a surface with subsurface features that take on the appropriate functionality when exposed by erosion of top material. An example is a surface that uses covalently bonded pendant fluoropolymers to produce a desired hydrophobic effect [Dikić 2012]. It is possible to construct a material with a 3D structure that has pendant fluoropolymers bond to an internal network. Upon erosion of the top surface, the pendant fluoropolymers emerge to replenish hydrophobic functionality (Figure 8.16).

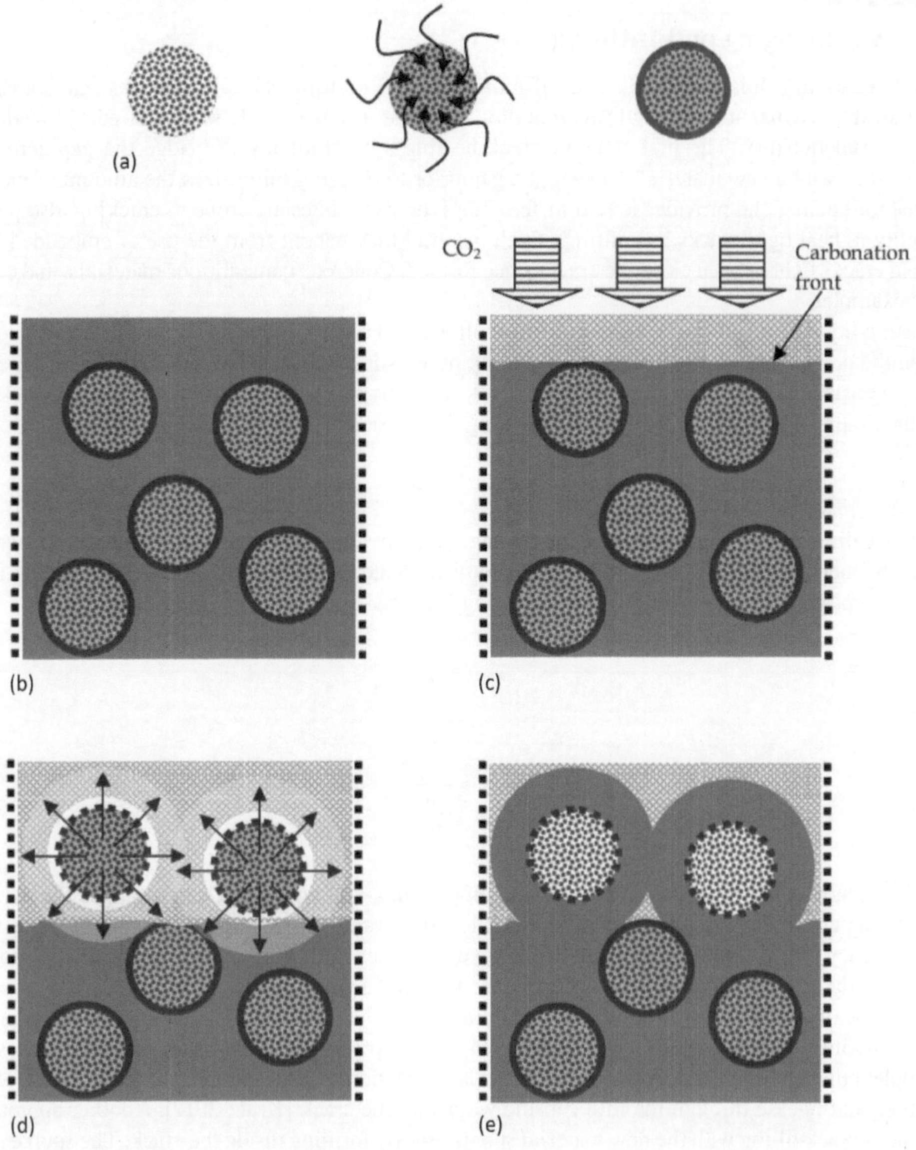

FIGURE 8.15 Encapsulated aggregate with a material barrier that breaks down with damage creates an active self-healing concrete system. (a) LWAs impregnated with Na-MFP solution is prepared under vacuum. Then, the particles are encapsulated in a cement paste layer. (b) Encapsulated LWA embedded in the matrix. (c) Carbonation coarsens the matrix porosity. (d) Carbonation damages the coating layer. (e) Healing takes place. The microstructure of the matrix is improved. (From [Sisomphon 2011].)

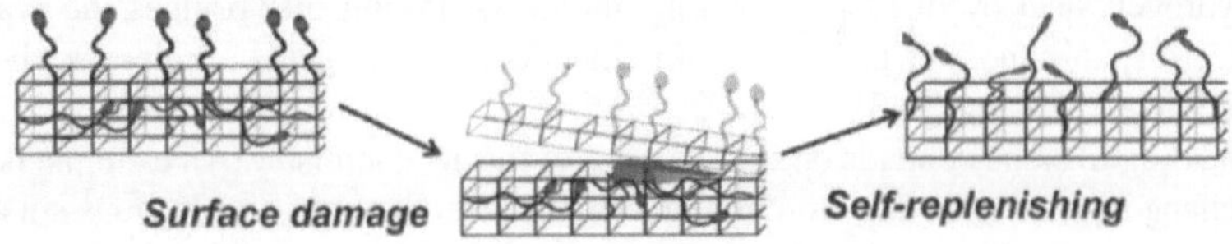

FIGURE 8.16 Surface replenishment system for functional material with dangling polymer chains. Damage removes the top layer and releases subsurface dangling molecular chains. (From [Dikić 2012].)

8.9 Crack-Bridging and Infilling

Healing cracks in solids comprises a set of multiscale and multifunctional processes that sometimes restore all the original strength and functionality, and other times provide sufficient partial healing to restore of functionality. The first steps in crack healing are commonly to bridge the gap across the crack surface with a new material. Closing the gap prior to bridging minimizes the amount of material required for healing and provides load transfer across the gap. Materials prone to cracking also possess the ability to heal tight cracks by infilling. Such material may benefit from the use of embedded fibers that hold cracks tight or even close the cracks once formed. Concrete, cementitious materials, and ceramics are examples.

Injecting liquids into cracks that subsequently solidify and form a bridge heals cracks. A complication is when the crack continues to grow as the healing process is underway. For a polymer composite, this requires consideration of both τ_{Heal}, the time it takes for the liquid to solidify and heal the crack, and τ_{crack}, the time of crack growth. If the crack-healing time is shorter than the growth time, that is,

$$\tau_{heal} < \tau_{crack} \tag{8.5}$$

then the healing process stays ahead of the damage growth. Otherwise, the crack growth outpaces the healing. Numerical modeling beginning at the atomistic scale up through mesoscales for healing liquid gelation in epoxy composites has found that healing and crack growth follow the timescales of

$$\tau_{heal} = \frac{1}{B} \tag{8.6}$$

and

$$\tau_{crack} = \frac{l_c}{\frac{da}{dn}\omega} \tag{8.7}$$

Here B depends on the specific healing liquid, temperature, and amount of catalyst; l_c is the cohesive length of the crack; da/dn is the amount of crack length advance per loading cycle; and ω is the cyclic load frequency [Maiti 2006]. High loads and high cyclic rates reduce τ_{crack}. An important issue to consider is that the healing of the liquid may introduce wedging actions on the crack tip.

Forcing healing material to migrate into tight cracks is a challenge. Encapsulated droplets of healing liquids containing nanoparticles are normally too big to penetrate tight cracks. A workaround process has the droplet adhere to the crack-opening through a combination of hydrophobic and hydrophilic effects for subsequent release through the thin capsule walls into the crack [Kratz 2012]. Another approach is autogenous crack-filling with the new material spontaneously forming inside the crack. The source of the healing material may be a fluid that fills the cracks, diffusion of material through the solid, flow of material through pores in the solid, and/or combinations of these effects. A vascular system delivers adhesive polyurethane foams inside a solid structure by mixing two different liquids in situ [Patrick 2012].

Lubricants for heavily loaded sliding and rolling machine parts, such as cams or bearings, alter the growth of fatigue cracks on the surface. Many water-based lubricants accelerate fatigue crack growth, while others, such as white mineral oil, extend the life, with a hypothesis being that these oils promote the healing of fresh fatigue cracks [Galvin 1964].

Cracks in hydrogels heal by diffusion bonding and cross-linking that bridges the crack. Inserting cross-linked electrospun nanofiber networks with embedded redox agents accelerates the healing and overall healing efficiency [Fang 2013].

Tight crack healing in stone depends on the specifics of the geochemistry. An example is the Longyou Caverns in Zhejiang Province, China. Humans carved the caverns out of argillaceous siltstone approximately 2,000 years ago. Situated near to the Qu River, the caves filled with water long ago and remained forgotten and unmolested underwater until being discovered in 1992 and subsequently pumped dry. As discovered, the walls of the caverns were largely uncracked, yet quickly cracked when exposed to air.

It is hypothesized that the slightly acidic conditions of the flooding water placed the siltstone in an autogenous self-healing state that removed any cracks that formed [Yue 2010]. Removing the acidic water removed the self-healing capability. Unlike siltstone, neutral pH or alkali conditions promote autogenous healing of tight cracks in concrete. Slightly acidic conditions prevent healing [Ramm 1998].

8.9.1 Ceramic and Crystalline Crack Filling

The molecular structure of ceramics is arrays of alternately charged ions and covalent bonds (Figure 8.17). These bonds hold the solid together and prevent free movement of atoms and molecules, causing cracking to be a principal damage mode. Healing by rearranging molecules and atoms with thermal and/or plastic deformation processes is not practical in most cases. Instead, many healing methods for ceramics combine a heat cycle with a gaseous-phase contributing material, typically through oxidation, to fill the tight cracks. Various forms of SiC, Si_3N_4, Al_2O_3, and other ceramics heal by gaseous solidification.

Oxidation-driven infilling at elevated temperatures heals tight cracks in ceramics. Oxidation in air with temperatures as low as 800°C heals silicon nitride. Most ceramic oxidation healing occurs at higher temperatures (~>1,100°C) [Choi 1992]. Ternary carbide ceramic Ti_3AlC_2 fills cracks of widths less than 1 μm with Al_2O_3 [Song 2008]. Larger width cracks fill with increased levels of TiO_2. The filling mechanism comprises the deposition of nanoscale particles at the crack walls and infilling until the crack is bridged, with observable layers appearing within minutes and filling completed within a couple of hours. Bulk SiC heals at temperatures between 600°C and 1,500°C with healing efficiencies up to 75% for fatigue in bending. Ceramics, such as zirconium diboride or silicon nitride, reinforced with silicon carbide whiskers hold the cracks tight and promote infilling healing of the cracks [Zhang 2008] [Yao 2000].

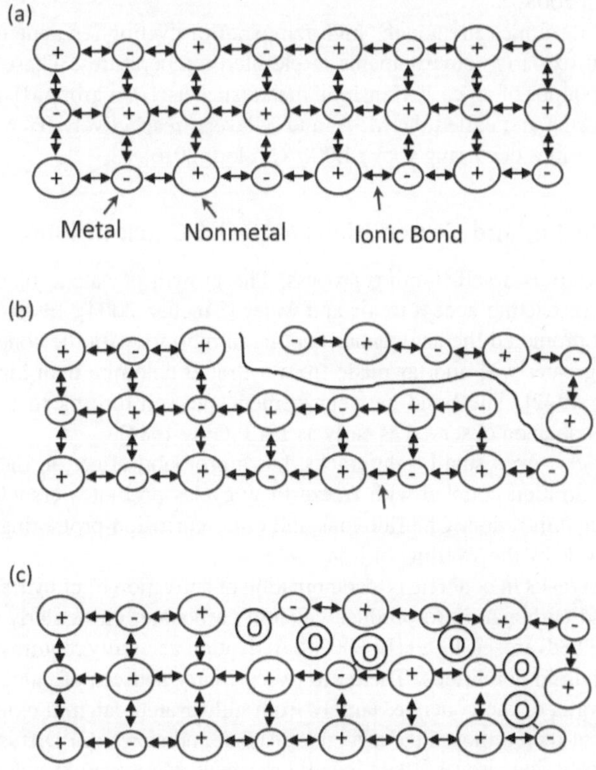

FIGURE 8.17 Ionic structure of ceramics, microcracking, and oxide healing. (a) Healthy state with no broken bonds. (b) Cracks difficult to repair with rebonding. (c) Oxide infilling heals cracks.

Mechanical actions aid ceramic healing. If the cracks in a ceramic do not run all the way through the solid and are reasonably tight, it is possible to heal the cracks by infilling. Oxygen diffuses into the crack and then reacts with the ceramic properties to fill the crack. Si_3N_4 crystals fill in by forming glassy silica (SiO_2) [Korouš 2000] [Zhang 1998]. Stress relaxation, especially in the absence of oxygen, and the inclusion of composite materials and SiC whiskers aid in tight crack healing. The self-healing effect of tight cracks restrained by composite whiskers, such as SiC in Al_2O_3, significantly increases the endurance of the ceramic component by preventing small cracks from growing into large cracks [Sugiyama 2008]. Metamorphic ceramic surface layers with an inherent compressive stress protect underlying ceramic crystals, such as NiAl/α-Al_2O_3 composites oxidized to form $NiAl_2O_4$ and Al_2O_3 grains. Oxidation at high temperatures recovers the strength of damaged surface layers better than in nonoxidizing Ar environments, up to 531 MPa versus 343 MPa as a recovery from a damaged 150 MPa [Abe 2006].

More complicated ceramic materials also heal by oxidation infilling. Mullite is stoichiometric mix of alumina (Al_2O_3) and silica (SiO_2) ceramics. Cracks heal with a heat treatment of 1 h in air at 1,300°C [Chu 1995]. Composites of mullite reinforced with SiC whiskers exhibit similar healing behavior [Ono 2005]. Oxidation of intergranular media in liquid-phase sintered silicon carbide ceramics provides healing of cracks through heat cycles in excess of 1,100°C [Kim 2003b].

TBCs are often thin ceramic layers bonded to a sturdy and high-temperature-tolerant substrate, such as a turbine blade. TBC durability depends in large part on strong adhesive bonds between the ceramic layer and the substrate. The diffusion of S and C atoms degrades the adhesion of the TBC. Al_2O_3 TBC layers grow on Ni/Al substrates [Bennett 2006]. Imbuing the TBC with reactive elements, such as Y, Zr, and Hf remediated the diffusion of S and C atoms by a scavenging process that ultimately preserves the integrity of the TBC bonding. Crystals of pure iron heal microcracks at temperatures elevated above 1,000°C in low-oxygen environments, as oxidation prevents the healing [Gao 2001]. Radiation damage recovery in yttria-stabilized zirconia (YSZ) with aliovalent doping is possible with oxygen vacancy migration [Devanathan 2008].

MAX-phase metalloceramics are tough, high-temperature cycling-resilient materials that autonomously heal cracks in oxidative environments at elevated temperatures. These are ternary ceramics based on layered compounds of an early transition metal (such as Ti), a group IIIA or IVA element (usually Al or Si), and C or N; and called the M, A, and X layers, respectively. Ti_2AlC is an example with repeated healing capabilities occurring above 1,000°C [Sloof 2016].

8.9.2 Concrete, Mortar, and Cementitious Material Crack Infilling

Crack infilling in mortars is a self-limiting process. The growth of calcite blocks further growth by closing the crack and restricting access to air and water [Sanchez 2004]. In 1942, Anderegg reported that high lime content promoted the autogenous healing in mortars with the conclusion that the degree of healing is slightly greater with mortar made from hydrated dolomite than high-yield, high-calcium quicklime [Anderegg 1942]. The bond between cement and reinforcing in concrete can also heal autonomously – a phenomenon observed as early as 1913 [Gray 1984].

The presence of cracks, even small tight cracks, has a profound effect on the long-term endurance of concrete. Water infiltration coupled with freeze-thaw cycles promotes cracking, which feeds back to promote more water infiltration. Chloride ions and other corrosion-promoting agents accelerate the cracking-infiltration cycle by the swelling of iron oxides.

The healing of tight cracks in concrete is a combination of activation of unhydrated cement to produce calcium hydroxide and carbonation to produce calcium carbonate, particularly the stable polymorph calcite [Hannant 1983] [Edvardsen 1999] [Li 2007a]. Activation generally requires the presence of water, either by soaking or humid conditions. There are two primary sources of calcium and carbonate: (1) Leaching out of the walls of the concrete, largely from unhydrated calcium oxides. Precipitation proceeds fairly quickly, but then removes the available reagent and slows. (2) Diffusion through the walls of the crack is a relatively slow process that typically proceeds over several weeks or more. Minimizing water flowing through the crack aids in both of these precipitation processes. Fiber-reinforced concrete is a multiscale approach to crack sealing. Autogenic sealing of cracks is possible, as long as the crack

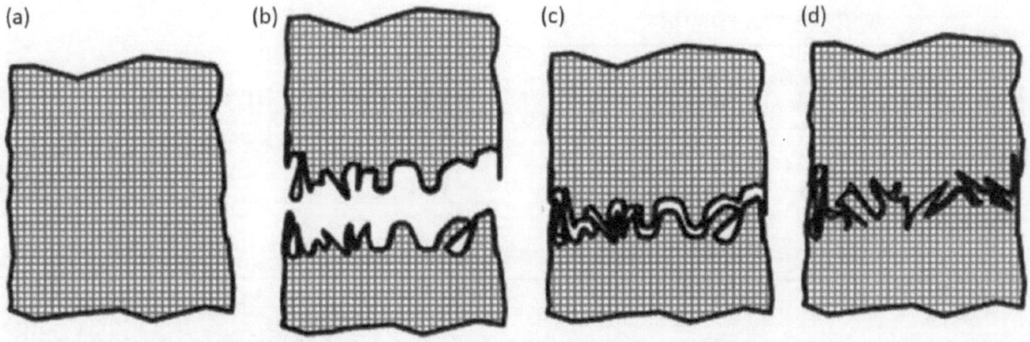

FIGURE 8.18 Interlocking, also known as topological bonding, of micro- and nanotexture followed by stress relaxation and intertwining of dangling polymer chains heals tight cracks in thermosets, such as epoxies. (a) Cross-linked epoxy–amine network. (b) Surface topology after fracture. (c) Intimate contact between crack interfaces. (d) Rearrangement of aligned free surfaces to cause mechanical interlocking. (From [Rahmathullah 2009].)

widths remain small enough, typically less than 0.2 mm, for the microenvironment in the crack to support infilling [Mangat 1987].

Engineered cementitious composites (ECC) are custom cement-based materials like concrete, but with enhancements due to fiber reinforcement and other supplements, and often with little or no aggregates. Conventional cementitious materials are generally weak in tension and fail by cracking. Fiber reinforcement prevents the cracks from opening wide by bonding to the cement, bridging across gaps that emerge and pulling across the gap with sufficient force to prevent the crack from opening wide. Polyethylene fibers are common as they bond well to cement and are mass produced at low cost [Homma 2009]. The remaining relatively tight cracks form a micro-environment where material growth from the solid faces inside the crack fills the crack, especially when subjected to wet–dry cycles [Li 2009a]. The tight crack repair can be sufficiently active to reduce the water permeability of the concrete [Ma 2014].

8.9.3 Crack Interlocking

Interlocking and intertwining of crack surfaces are methods of mechanically gripping across a closed crack. Interlocking and intertwining are effective for materials, such as thermosets and epoxies that do not readily heal by reptation diffusion because the polymer chains are bound into the molecular matrix. Experiments with tight cracks indicate that some epoxy thermoset materials achieve healing efficiencies of over 40%, even after multiple fracture-heal cycles [Rahmathullah 2009]. The intertwining of molecules, sometimes termed topological bonding, bound at one end to the matrix and dangling on the other, seems to explain some of the observed healing. The interlocking and relaxing of micro- and nanotextures at the crack surface provide additional mechanical strengthening (Figure 8.18).

8.10 Composite Materials that Heal

Composite materials contain nano-, micro-, and mesoscale heterogeneities. Collectively, these heterogeneous components act to form macroscopic materials with enhanced properties, including the possibility of self-healing materials.

Composites effectively counter expected damage with a heterogeneous distribution of self-healing capabilities. An example is FRP laminate panels that are vulnerable to delaminations caused by impacts. Placing a thin self-healing layer in a strategic position facilitates delamination damage recovery that minimizes the parasitic costs of the healing system. A damaged layer releases encapsulated epoxy and

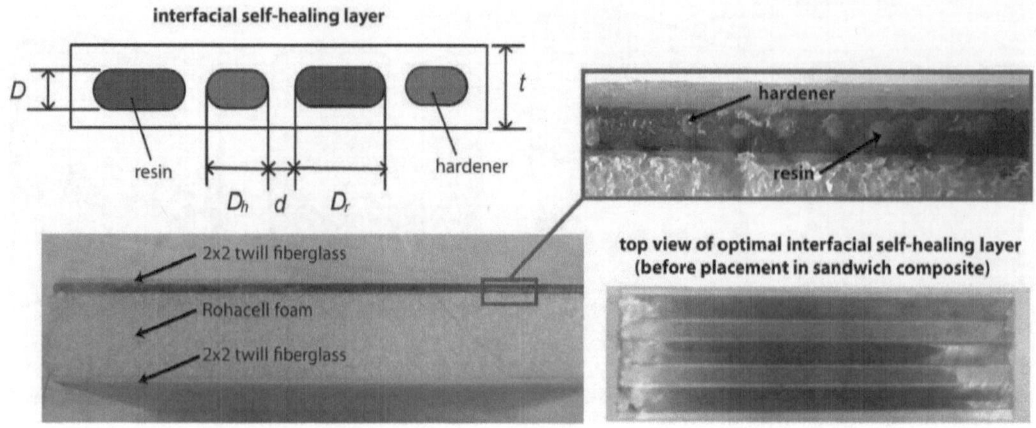

FIGURE 8.19 Interfacial vascular healing layer using two-part epoxy in a layered FRP composite panel. (From [Chen 2012].)

hardener that comingle and then solidify to heal (Figure 8.19) [Fugon 2012]. Another approach uses a healable polymer, such as one that acts with a thermally remendable Diels–Alder mechanism preplaced with a 3D printer [Fleet 2015].

8.10.1 FRP Composites

FRP composites use high-strength fibers embedded in polymer matrices to create structural elements with superior properties. The assembly is often hierarchical with fibers grouped into yarns and fabrics with macroscopic length scales. The diameters of the fiber are on the order of micrometers. The binding of the fibers to the matrix occurs over molecular, nano, and micro length scales. Figure 8.20 shows some of the principal physical interactions and associated length scales in a fiber-reinforced structural element.

Delamination between fiber layers is a common failure mode of FRP composites. The failure is a split in the polymer layer running between fiber layers. Healing this split layer heals the delamination. Possible healing actions include the following: (1) Use of microencapsulated liquid healing agents. Healing efficiencies exceed at least 80% [Kessler 2003]. (2) Molecular scale reptation-type healing of delaminations in PEEK polymer matrix layers, with the aid of external compressive forces normal to the delamination and heat [Laczynski 2002]. (3) Use of a thermally remendable DA polymer, aided with encapsulated solvents [Ghezzo 2010] [Peterson 2010].

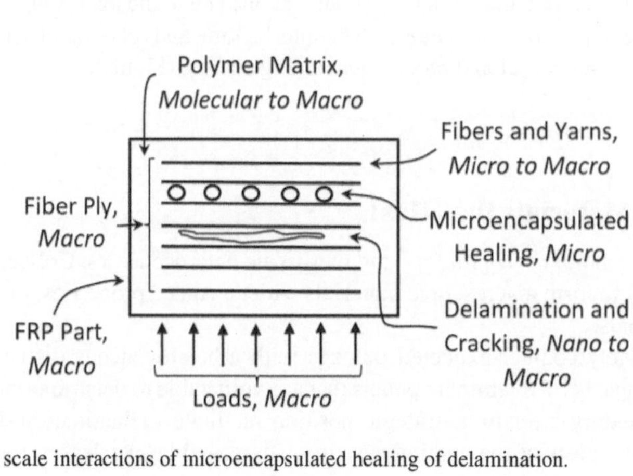

FIGURE 8.20 Length scale interactions of microencapsulated healing of delamination.

8.10.2 Asphalt

Conventional cementitious concrete and asphalt concrete are composite materials made from cement, binder, stones, sand, reinforcement, and chemical additives. The composite nature of these materials allows for the introduction of functional components to enable smartness, that is, sensors, actuators, and self-healing [Han 2015a].

8.11 Wound Closing

Open wounds in a structure expose the subsurface to the external environment. Healing begins with an active subsurface that initiates chemical and mechanical action to close the wound (Figure 8.21). The advantages for wound closing are as follows: (1) The contraction reduces the volume and area of the wound to be healed. (2) The closed wound requires less material for a more substantial healing, including subsurface structures. (3) Closing the wound reestablishes the barrier between the interior and the exterior of the system.

Plants, animals, and even single-cell organisms are quite adept at closing wounds and healing. In animals, connective tissue matrix deposition, contraction, and epithelization act in concert to heal skin wounds [Diegelmann 2004]. Closing actions range in length from cellular membranes up through multicellular tissue [Woolley 2000] [Braiman 2007].

Closing techniques vary in detail and yet have common features, including damage signaling that triggers a cascade of wound-closing mitigation and healing actions. A key feature is a compliant yet mechanically active substructure and infilling from the edges and below of healthy tissue. Sensing crack damage, closing the crack, and then healing the crack is a natural approach to building a multifunctional engineered self-healing skin system.

8.11.1 Wound Closing by Prestress and Intrinsic Passive Mechanical Loads

Prestressing predisposes a material to close wounds by passive mechanical methods. Applying prestress in a bulk heterogeneous manner across a structural member so that external regions are in compression with internal regions in tension prevents surface cracks from growing in glass and other brittle materials. In PMMA, methanol-induced swelling causes compressive stresses in a subsurface layer that first arrests and then promotes healing of cracks [Kawagoe 1997]. Prestressed fibers, yarns, and tendons in cement close cracks after damage and restore shape, largely through an elastic rebound effect. Such healing is advantageous in thin-walled concrete structures [Hull 1969]. Tuning the stretchability and recovery of the fibers and tendons to the cracking behavior of the structure is crucial to success. Closing the crack

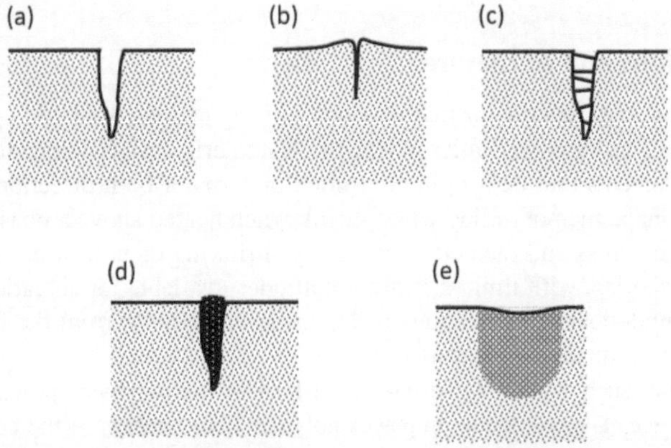

FIGURE 8.21 Wound closing heals cracks, scratches, and gouges by a combination of deformation maneuvers and infilling. (a) Solid surface with crack, scratch, or gouge. (b) Swelling closes the gap. (c) Dendrites, fibers, or stitches span the gap. (d) External agent or material transport fills the gap. (e) Material flows and fills the gap.

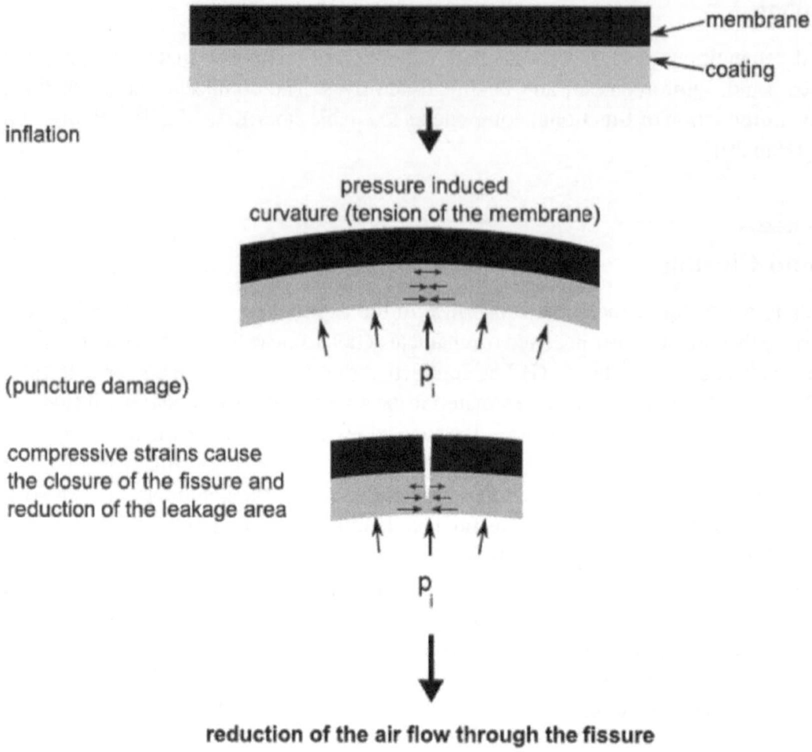

FIGURE 8.22 Internal pressure autonomically closes pressure wound due to convex curvature in the thick flexible wall of inflated structure. (From [Rampf 2013].)

generally requires the tendons to be able to stretch enough to survive as bridging elements during the initial opening and then pull all the way to crack closure. The superelastic behavior of shape memory alloys is well-suited for this application. Iron-based SMAs can be effective as post-tensioning strands to close cracks in concrete structural elements [Soroushian 2001].

Another prestressed example is a pressurized inflatable structure made of a flexible multilayer wall with appreciable thickness. Internal pressure curves the flexible wall in a way that autonomically closes puncture wounds (Figure 8.22) [Rampf 2013].

8.11.2 Active Tendons and Actuators

Active materials provide the impetus for many wound-closing actions. Embedding thermally sensitive shape memory materials as patterned microfibers in elastomeric materials creates stimuli-responsive macroscale material elements that deform in controlled patterns of both direction and location [Sakai 2003] [Mather 2014]. Shape memory alloy wires shrink when heated above a phase transition temperature. SMA wires bridge across and then close cracks by shrinking upon heating (Figure 8.23) [Burton 2006]. Piezoelectric materials with limited strain amplitudes available for actuation are best suited for small deformation wound-closing applications or for those that benefit from the introduction of stress with small associated deformations [Wang 2012e].

Embedded SMA wires, such as nitinol, Cu-Al-Mn, or heat shrink polymers, provide the tension needed to close cracks in concrete. The technique preplaces polymer fiber tendons in the concrete before curing, and then heats the concrete post cure following crack formation. It is necessary to use materials that shrink at less than 100°C to avoid steam release cracking the concrete [Isaacs 2013]. Heat-shrink polymers offer a large deformation, one-time, thermally activated crack closure method for concrete [Jefferson 2010]. Vascular dispensing of healing liquids, such as epoxies, finishes the repair (Figure 8.24) [Kuang 2008] [Pareek 2014].

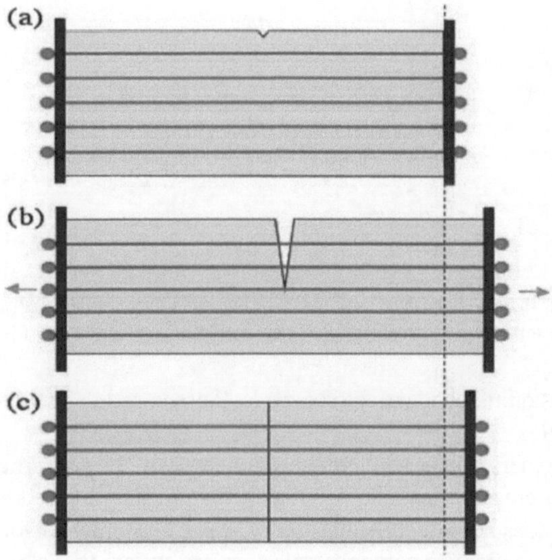

FIGURE 8.23 Closing cracks by heating and shrinking SMA wires that bridge across the gap. (a) Pre-notched initial configuration. (b) Propagating crack during loading. (c) Healed crack after heating. (From [Burton 2006].)

Closing wounds with bridging and shrinking tendons followed by tight crack healing works in thermally compatible metal systems. An example is Sn alloy that softens and melts at temperatures comparable to those that active shrinking phase transitions in nitinol SMA wires (450K approx.). Cycling through temperatures that both soften the Sn alloy and activate the SMA produces a healing system with efficiencies more than 80% [Bernikowicz 1998].

Engineered systems can use similar wound-closing techniques but require mechanical devices to execute stress redistribution and material deformation. SMAs and SMPs exert the large deformations needed for wound closing. SMAs undergo large superelastic deformations that close wounds. An issue of concern is the tolerance of temperature changes. The wider temperature hysteresis NiTiNb SMAs favors their use over NiTi SMAs in civil structural applications [Choi 2010]. Embedding SMA fibers in dentures closes cracks after fracture for bonding repairs [Hamada 2003]. An issue is the maintenance of shape following the SMA crack closure contraction. Embedded SMA wires control stress intensity factors near crack tips [Shimamoto 2007]. Three-dimensional woven SMA fabrics recover shape to heal impact dents in FRP composites [Nji 2010]. An array of stimuli-responsive fibers inspired by spider silk demonstrated a thermally induced wound-closing healing system in composite polymer panels [Li 2012a]. A detail of concern is that SMA wires undergo large lateral dimensional changes along with the longitudinal

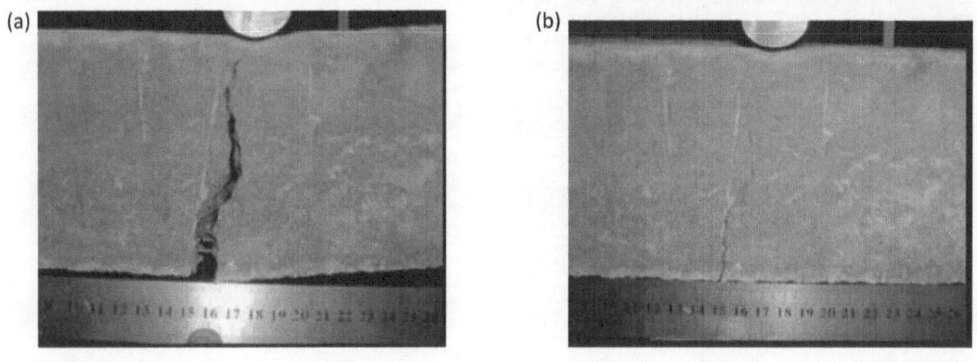

FIGURE 8.24 Wound closing and repair in concrete beam. (a) Damaged condition. (b) Crack closed by SMA wires and healed by vascular delivery of epoxy. (From [Kuang 2008].)

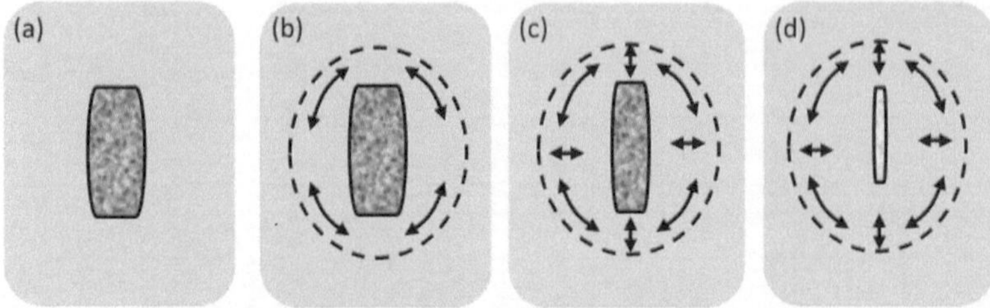

FIGURE 8.25 Purse string method of wound closing and healing. (a) Fresh wound in structural panel. (b) Purse string knittle around wound in tension. (c) Compression and swelling against knittle closes wound. (d) Ingrowth heals closed wound.

changes in martensitic–austenitic changes. This pseudo-Poisson ratio effect causes debonding of SMA wires in composite structures.

Most conventional elastomers do not heal cracks autonomously. Mixing silicone elastomers with ethylene vinyl acetate (EVA) creates an elastomer that heals cracks with heating to a temperature that activates the EVA and yet does not affect the elastomer. Since elastomers deform easily, an SMA spring actuates the wound closing to create a composite elastomeric system that closes and then seals wounds by thermal cycling [Wang 2012a].

8.11.3 Wound Closing by Purse String and External Actuation Techniques

The purse string method acts on planar structures (Figure 8.25). A tension ring, that is, knittle, forms around the wound. Inside the knittle swelling induces compression, which closes the wound. Figure 8.26 shows a test rig and results of a cellular actuated purse string method.

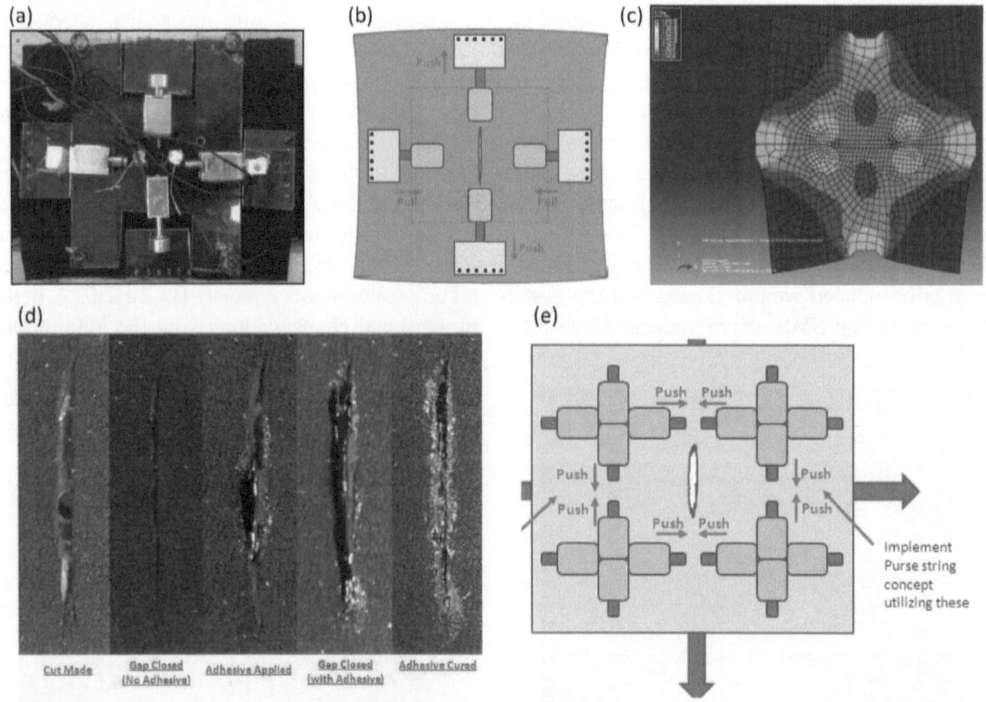

FIGURE 8.26 Cellular mechanical wound-closing test rig. (a) Backside sensing and actuation mechanism. (b) Conceptual model. (c) Numerical model. (d) Wounds closed and then healed in neoprene skin. (e) Plan for scalable multicell system. (From Huston 2013].)

The purse string effect occurs automatically in certain dynamic thermomechanical healing processes. Thermal and mechanical healing on materials introduces molecular scale and long-scale interactions that combine to close the wound. Detailed studies of wound closing in multicellular organisms indicates that a mechanical model, including homogeneous purse string tensions around the wound with material infilling, is perhaps too simple and does not adequately describe the mechanics in detail. Instead, epithelial wounds heal by a heterogeneous purse string that exerts dipole-like tension and compression forces to underlying tissue at localized anchors. The positive Poisson's ratio of the substrate tissue subjected to heterogeneous tangential forces deforms into the wound and promotes cellular migrations to close the wound [Brugués 2014].

8.11.4 Three-dimensional Out-of-plane Delamination Closing

Out-of-plane forces perpendicular to the plane of delamination close the defect. Foam core sandwich structures self-healing impact damage through out-of-plane motions actuated with SMP-based syntactic foam [Li 2008a]. Mathematical models of out-of-plane closing based on SMA wires appear in Bor [Bor 2010].

Polyethylene fishing line twisted into a helix makes for suitable gap bridging actuators. The semi-crystalline structure causes twisted helices to shrink when heated above the glass transition temperature of 31°C. Melting and flowing a thermoplastic polymer into a tight gap in an epoxy matrix with cracks bridged and then closed by the helix actuators creates an integrated thermally activated healing system [Zhang 2015c]. Piezoelectric patches close, and even provide compressive stress, across cracks and delaminations in composites [Liu 2007a] [Alaimo 2013].

8.11.5 Wound Closing by Constrained Swelling

Liquid that penetrates and then swells seals cracks. Clays, such as hectorite or montmorillonite, interact with water at the nanoscale and expand to fill small cracks or gouges [Hikasa 2004] [Cases 1997].

The geometry of a mathematical model of swelling–diffusion-based crack closing a 1D clay layer appears in Figure 8.27 [Miccichè 2008]. The water concentration versus position and time, C(x,t), following a step change in surface concentration follows a standard diffusion response:

$$C(x,t) = C_S - (C_S - C_0)\,\mathrm{erf}\left(\frac{2}{x\sqrt{Dt}}\right) \tag{8.8}$$

Here C_s is the moisture level at the surface; C_0 is the moisture level initially in the clay layer; and D is the moisture diffusion coefficient. For clays with approximately a linear relation between relative humidity moisture and volume and that $\mathrm{erf}(0.5) \approx 0.5$

$$\Delta x = \frac{k}{2}\sqrt{Dt} \tag{8.9}$$

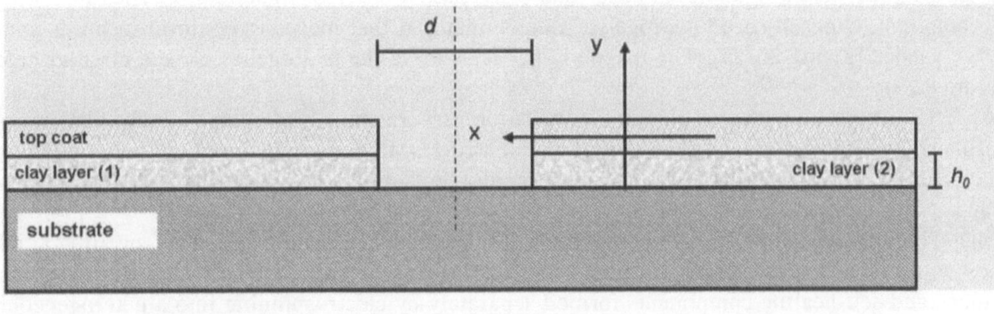

FIGURE 8.27 Geometry of 1D diffusion-driven swelling surface crack healing. (From [Miccichè 2008].)

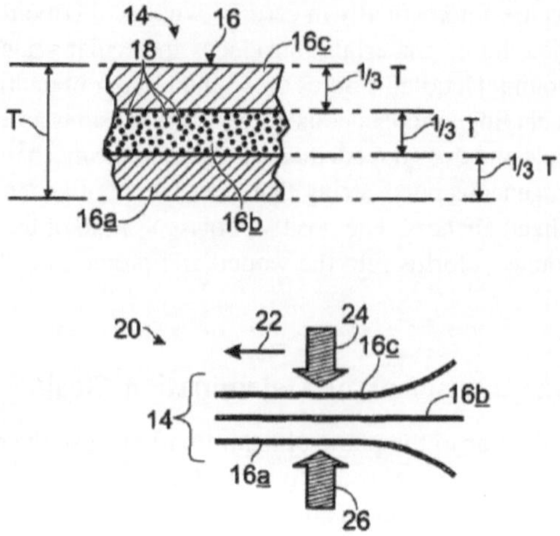

FIGURE 8.28 Sandwich structure constrains swelling layer for anisotropic motion that closes the hole. (From [Ohnstad 2011].)

Here k is the volumetric expansion.

$$k = \frac{V_{\%RH} - V_0}{V_0} \tag{8.10}$$

Here $V_{\%RH}$ is the volume based on relative humidity and V_0 is the volume when the clay is dry. Based on estimated values of D and k, an estimated closing time is 2 min for a 32 μm scratch, a value somewhat smaller than observed but sufficiently close to lend credibility to the assumptions underlying the approximate model.

The swelling of material for wound closing is much more effective when stiff layers constrain the swelling (Figure 8.28). Such a technique seals punctures in tires [Powell 1980]. Another wound-closing method is a multilayer approach used in closing a hole in a fuel tank wall. The multilayer construction uses rigid walls to constrain a sandwich of elastomeric and a swellable layer made of a combination of elastomeric and swellable imbiber beads. Upon perforation, the fuel comingles with the imbiber beads. The sandwich construction constrains the swelling so that it flows anisotropically to fill the hole (Figure 8.28) [Ohnstad 2011]. Example applications include fuel tanks puncture-sealing tires and electric wire repair patches [Nagaya 2006] [Reynaert 1990].

8.11.6 Infilling, Transient Softening, and Material Migration

Materials that close wounds by mechanical deformation must be sufficiently compliant and deform without sustaining more damage. The combination of ferromagnetic properties into stiff, yet compliant, chitosan-based hydrogels creates a material that moves to conform to lower total energy shapes [Zhang 2012b]. The magnetic particles drive the hydrogel to close a circular hole (Figure 8.29).

An FRP composite puncture-plugging application uses an embedded composite-layer microencapsulated system, based on liquid silanol-terminated PDMS resin and liquid tin catalyst. Upon puncture, the liquids flow from the capsules, comingle, solidify, and stop the leak [Beiermann 2009].

Shape memory polymers swell upon heating to close damage in surfaces. This swelling combines with embedded thermoplastic fibers to seal the closed crack [Luo 2013b]. Mixing shape memory and self-healing components formed separately by electrospinning into a heterogeneous

FIGURE 8.29 Wound closing in a chitosan-based hydrogel with embedded ferromagnetic particles. (From [Zhang 2012b].)

mesh creates a material that closes wounds by shape memory effects and heals the wounds by reptation diffusion. Using poly(ε-caprolactone) as the shape memory material and poly(vinyl acetate) with a 25% by weight mix heals large wounds with thermal cycling over 75°C with 99% efficiency [Nejad 2015].

Sometimes, it is desirable to use tougher and more resilient polymer materials in shape memory applications that close wounds. Conventional mechanisms that toughen polymers also resist large-scale deformations. Mixing nanoparticles into polymers promotes shape-memory-driven deformations, such as SMPs made of polyurethane with embedded functionalized silica nanoparticles [Jung 2010]. A 1% loading of nanoparticles may be optimal. A variety of fillers based on CNTs, clay, and SiC support noncovalent binding with shape memory behavior. Filler-particle-based noncovalent shape memory effects occur more quickly than similar covalent variants [Gunes 2010].

8.11.7 Wound Closing by Transient Material Softening

Compliant elastomeric, gel-like, or plastic materials are nominally compliant for wound-closing, but have the disadvantage of being too compliant after the repair. A biomimetic workaround uses a material that changes from compliant when closing wounds to stiff afterward. One possibility is a two-phase material made of a compliant skeleton surrounded by a thermally reversible matrix [Cheng 2014a]. Upon heating, the thermally reversible matrix softens, allowing for large deformation of the compliant skeleton. Upon cooling, the matrix stiffens, leaving a rigid structural element in a new deformed shape. The insertion of canted coil springs into elastomeric gaskets produces a similar effect due to the nonlinear spring stiffness providing a recoverable deflection over large deformations [Balsells 1993].

8.11.8 Zippers and Active Patches

Active zippers and sliding patches are direct mechanical means of closing wounds. A subsurface solid matrix with a homogeneous copolymer system containing isocyanate groups and covered by a surface fluorine-containing barrier layer heals surface cracks with a process initiated by exposure to moisture in the environment. The generation of amine groups that react with other isocyanate groups form urea crosslinks that heal the crack with a zipper-like molecular wound-closing effect (Figure 8.30) [Zhang 2012c]. Similarly, azobenzene main chain polymers with side chains open and close as molecular zippers with UV light stimulation [Weber 2015].

Composites of metals with different melting points repair cracks by selective melting. Upon heating, the material with a lower melting point flows into damage in a material with a higher melting point material. Eutectic indium tin is a low-temperature melting metal layer that heals cracks with temperature cycling in structural aluminum or titanium alloy elements [Smith 2014].

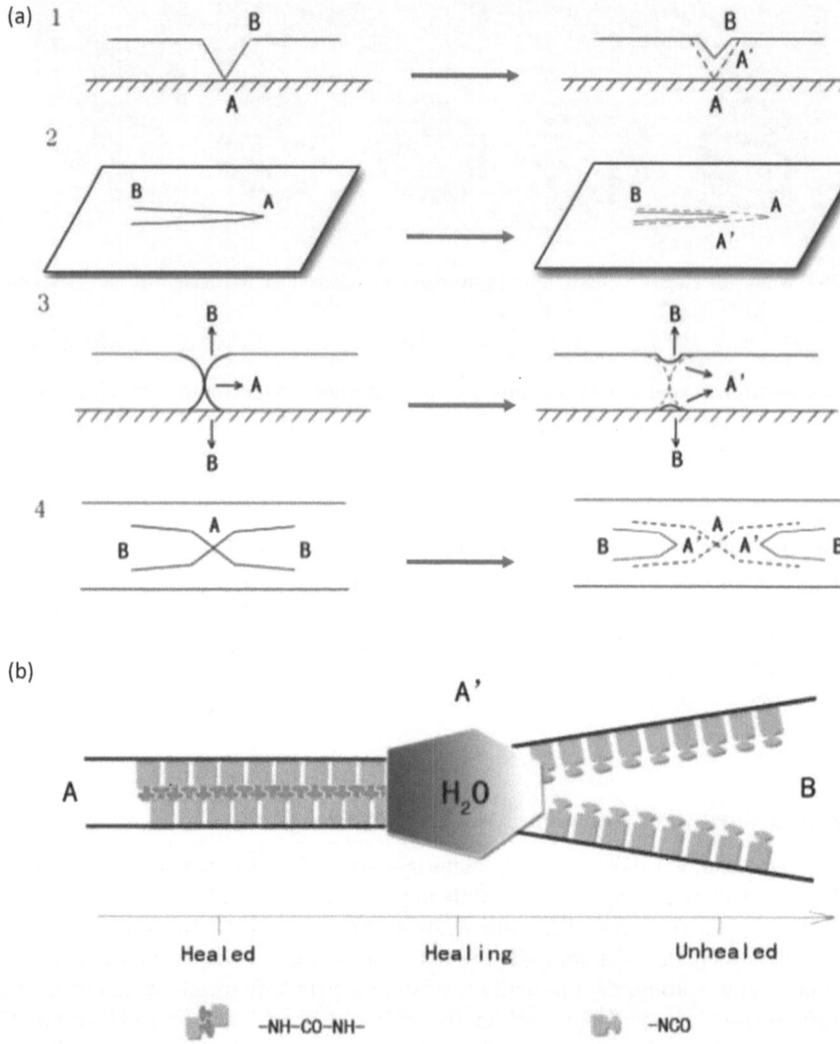

FIGURE 8.30 Molecular-based zipper wound-closing method. (From [Zhang 2012c].)

8.12 Electrochemical Processes and Systems

Electrochemical processes and systems, as appear in corrosion, batteries, and fuel cells, use coordinated, antagonistic, and synergistic effects that act over length scales running from molecular to macroscopic. These processes and systems are prime targets for multiscale and multifunctional self-healing methods.

8.12.1 Corrosion

Corrosion is a family of electrochemical processes that combine (1) electromotive force, (2) conductivity, and (3) suitable environmental conditions. Textural features in heterogeneous materials, such as pores, cracks, and wicks, provide supplemental forces acting over multiple scales to accentuate diffusive transport.

Many corrosion damage processes are a synergistic combination of effects that at times can combine to rapidly degrade the concrete. An example is a steel-reinforced concrete bridge deck subjected to winter environmental conditions with the application of deicing salts. Small cracks promote water and chloride ingress which promotes freeze-thaw and corrosion damage, which then promotes more water and chloride ingress and more damage. Autonomously arresting and recovering from these attacks, especially at early stages, significantly enhances the maintenance-free lifetime of concrete structures. Standard methods of preventing water-ingress-related damage are anticorrosion admixtures, surface sealants, and coatings on steel reinforcement bars [Cusson 2009].

8.13 Multifunctional and Multieffect Systems

8.13.1 Multifunctional Materials

Advanced high-performance multifunctional materials gain increased performance with the inclusion of self-healing capabilities [Nemat 2004]. Some of these materials take on life-like qualities. An example is the skin on a soft robot that provides proper mechanical action with embedded electronics for sensing. The skin should heal damage to both mechanical and electrical properties. An example is an artificial skin with combined healing that uses a mixture of a urea supramolecular hydrogen bond dynamic and microscale conductive Ni particles [Tee 2012]. The dynamer delivers mechanical healing, while electrical percolation through the Ni microparticles produces a crack-healable conductive material. Full mechanical healing requires about 10 min. Electrical healing efficiencies of 90% occur in about 15 s.

8.13.1.1 Thermal Multifunctional Healing

Thermally active inclusions enhance the performance of thermally healable composites. The insertion of electrically and thermally conductive polymers sets up conductive paths for thermal mending of fiber-reinforced polymers. Inserting electrically conductive graphite particles and thermally conductive BN into a polysulfide elastomer forms materials that successfully undergo multiple debond failures and recovery [Kwok 2007] [Lafont 2014]. The inclusions appear to enhance thermal conductivity of the elastomer.

Alternating magnetic fields heat magnetic nanoparticles by the hysteresis of switching the polarity of the magnetic domain [Schmidt 2006]. Mixing magnetic nanoparticles into a shape memory polymer with embedded mechano-chromic molecules makes a material system that changes color when deformed sufficiently large to cause damage and reversibly recovers shape and original color with magnetic nanoparticle heating. Ferrite MnZn nanoparticles with a Curie temperature of 230°C are well-suited for this application with the shape-memory-active polymer polyethylene-*co*-vinyl acetate [Ahmed 2015].

Foams functionalized with thermal shape memory polymer integrate into self-healing material systems containing combinations of a stiffening grid and skin. Such systems can recover multiple times from impact damage through thermal cycling (Figure 8.31) [John 2010]. Three-dimensional confinement of the foam by the grid aids the healing response. Residual antagonism with compressive prestrain at an elevated temperature of 79.4°C of the shape memory polymer enhances the healing performance with a 20% prestrain exhibiting superior tolerance to impact and the ability to heal multiple damage over specimens with a 3% prestrain.

8.13.1.2 Dynameric Multifunctional Healing

Dynamers offer opportunities for multifunctional self-healing behaviors.

8.13.1.2.1 Micro- and Nanoscale Features

Dynamers can be stiff materials that mitigate against macroscopic damage and are self-healing. Figure 8.32 shows a stiff elastomeric thermoplastic material made of polystyrene backbone molecules with pendant dynameric hydrogen-bonded polymer brushes [Chen 2012]. The polystyrene collapses into hard nanodomains with the dynamer brushes pointing outward to form a two-phase material. The dynameric matrix provides self-healing. The hard inclusions provide stiffness.

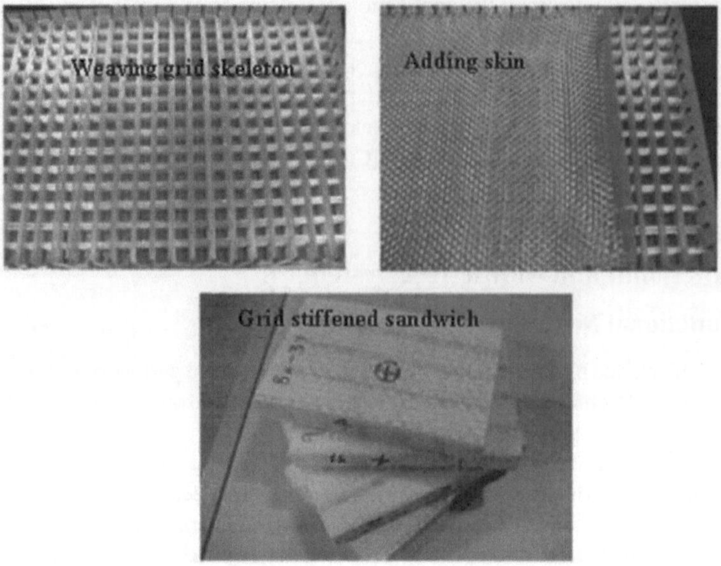

FIGURE 8.31 Thermally functionalized syntactic shape memory polymer foam integrated with stiffening grid for self-healing panel. (From [John 2010].)

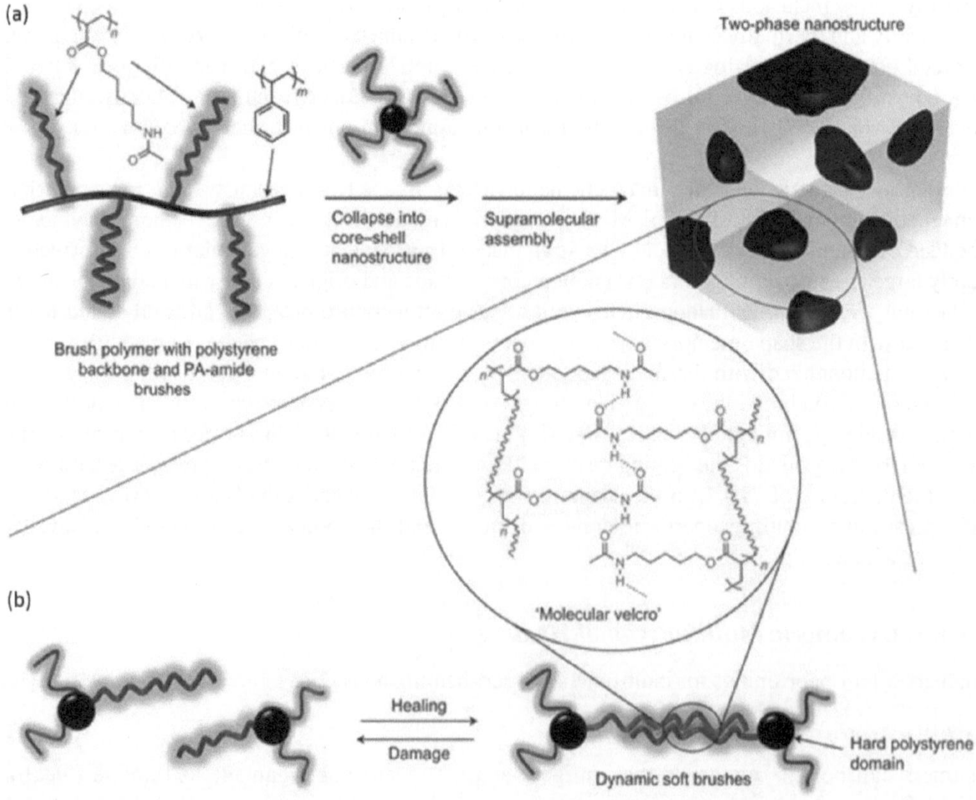

FIGURE 8.32 Self-healing stiff elastomeric material with two-phase nanostructure. (a) Dynamer matrix self-assembles into heterogeneous mix of hard nearly crystalline and soft amorphous phases. (b) Hard phase stiffens material, yet remains a dynamer with reversible deformation behavior. (From [Chen 2012b].)

Thin film hydrogels built up in a monolayer stacking process with alternating anionic and cationic polyelectrolyte process quickly heal scratches when placed in solvent. Experimental evidence suggests that swelling plays a significant role in healing [Spears 2014]. It is believed that the swelling relaxes the molecules and induces increased dynameric behavior that combines with molecular templating provided by the individual layers at the edges of the wound.

8.13.1.2.2 Optically Transparent Polymers

Dynameric processes heal optically transparent polymers. These include transparent electrically conductive ionomer elastomers with ion–dipole interactions, and transparent PDMS that uses reversible imine reactions with dynameric self-healing [Cao 2017] [Zhang 2017].

8.13.1.2.3 Scratch-healing Hydrophilic Antifogging Coating on Glass

The coatings recover from scratches by reforming both a smooth surface for optical transparency and hydrophilic properties for antifogging. Elastic polyurethanes with difunctional sulfonated polyether polyol heal surface scratches by elastic rebound, while also possessing a hydrophilic nature that is antifogging [Fock 1988]. A hydrogen-bond dynamer made of a mixture of partly cross-linked polyvinyl alcohol and polyacrylic acid exhibits both autonomous surface reforming and hydrophilicity [Zhang 2015d].

8.13.1.2.4 Scratch-healing Antibiofouling

Hydrogel surface coatings with dynamic disulfide dynamers both self-heal scratches and resist the buildup of Gram-negative bacteria [Yang 2015c].

8.13.1.2.5 Self-lubrication

Hydrogels with dynameric molecular structures contain and secrete liquids onto their surfaces to set up a feedback-stabilized equilibrium configuration with a layer of liquid sitting on top of a liquid-infused hydrogel. Removal of a surface layer of liquid induces a dynameric reconfiguration of the hydrogel to contract and release more liquid onto the surface until equilibrium is reestablished [Cui 2015].

8.13.1.2.6 Multifunctional Nanocomposites

Insertion of nanoscale structures into dynameric matrices creates composite materials with enhanced functionality. A nanocomposite formed from the hydrogen bond dynamer fatty di- and triacid and diethylenetriamine attached to exfoliated boron nitride nanosheets mends multiple cuts and maintains high dielectric breakdown strength [Xing 2016].

8.13.1.3 Photoactive Multifunctional Healing

Light activates chemical bonds in cracks for mending and also activates shape memory action in polymers, creating a multifunctional light-activated self-healing material system [Habault 2013].

UV light activates TiO_2 which alters surface wettability and self-cleaning oxidizing chemistry. This functionality combines with other antagonistic functionalities to create highly responsive self-cleaning material systems. A hierarchically textured high-density polyethylene surface with embedded TiO_2 nanoparticles reversibly switches from superhydrophobic to superhydrophilic with oxidizing self-cleaning [Xu 2013b]. A sponge imbued with hydrophobic hydrocarbon and hydrophilic TiO_2 nanoparticles initially absorbs oil and then desorbs the oil when exposed to UV light [Kim 2015b].

Inserting mechanophore molecules into self-healing polymers creates a material that changes color when overly strained and heals itself as needed. The covalent mechanophore spiropyran in copper-catalyzed azide–alkyne cycloaddition-based tridentate ligand 2,6-bis(1,2,3-triazol-4-yl)pyridine produces such a material system. A nanoscale segregated soft–hard material structure forms with the color-changing mechanophores separating out into the soft regions. Macroscopic deformation of the material amplifies the deformation of the soft regions containing the mechanophore and activates the color change. The material heals damage, such as cutting with a solvent-based relaxation of the polymer backbone [Hong 2013].

8.13.2 Materials with Self-healing Properties Organized at Molecular Scale, Nanoscale, and Microscale

Combining sensing and damage repair requires multifunctional activity. CNTs inserted into fiber-reinforced polymer composites have both sensing and repair functionalities. CNTs act as damage sensors through a change in resistance and as electrical conductors for thermal cycle repair of the thermoplastic matrix following cracking and delamination [Zhang 2007b].

8.14 Panoply Methods

Combining different healing technologies into a panoply is effective for healing, usually with the design tuned to a specific application.

Some epoxies have shape memory. A large deformation imparted at a lower temperature reverts back to the original shape by placing the material at a higher temperature. Combining shape memory healing with damage mitigation produces a superior material system that resists scratching and then has an easier time of healing the smaller scratches that do occur [Xiao 2010]. Shape memory enables these materials to recover smooth surface scratches with a heating cycle. Inserting a small amount (0.0125 vol%) of nanolayered graphene significantly mitigates scratching damage, presumably through crack pinning and similar deformation-arresting processes. Similarly, combining shape memory with reptation style healing into a single miscible polymer creates a material system that can close cracks and heals the cracks. Shape memory–assisted self-healing (SMASH) systems combines a cross-linked poly(ε-caprolactone) network (n-PCL) for shape restoration with a linear poly(ε-caprolactone) (l-PCL) as a interpenetrating network that heals the closed cracks [Rodriguez 2011].

8.15 Applications

8.15.1 Fire and Thermal Protection Systems (TPSs)

TPSs are specialized material systems that endure extreme high temperatures for high-performance applications, such as heat shields in spacecraft undergoing hypersonic atmospheric reentry. The requirements for these TPS systems are quite demanding because of the competing needs of lightweight, high strength, and thermal protection. Microcracking combined with oxidation facilitated by the microcracks is a common failure mode. Plugging the microcracks reduces oxidation and promotes endurance. Molten viscous glasses are good for flowing into and plugging the cracks but are heavy in bulk quantities. Placing glass precursors, such as B_4C, into select places inside the TPS composite matrix is a lightweight alternative (Figure 8.33) [Keller 2006a]. Upon heated in an oxidative environment, the B_4C converts into a glassy B_2O_3 and CO_2. The B_2O_3 seals the microcracks in a process that works up to 1,300°C, above which two-part systems work.

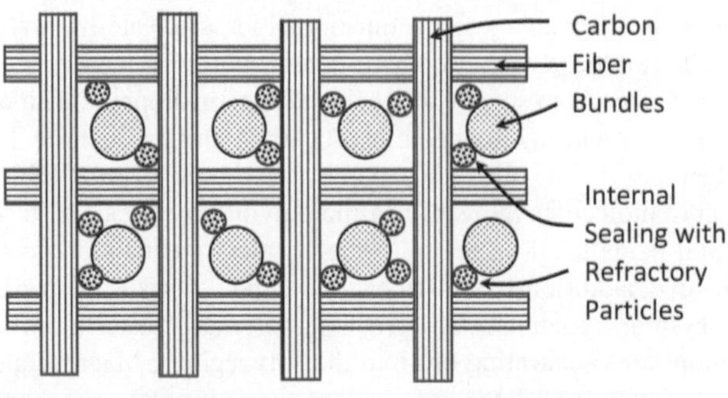

FIGURE 8.33 Ceramic B_4C glass precursor placed into TPS composite matrix plugs microcracks and limits oxidation. (Adapted from [Keller 2006].)

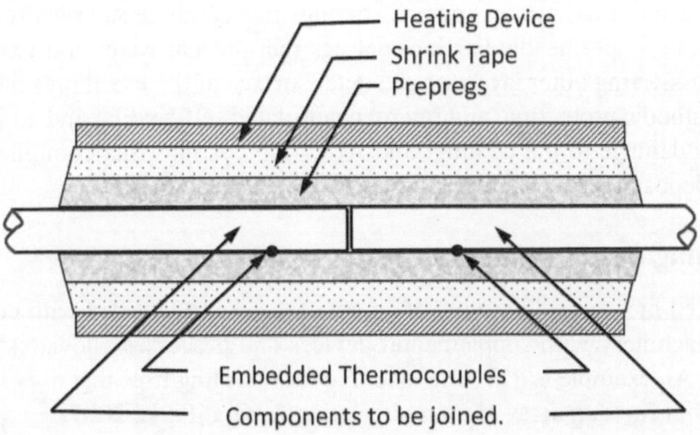

FIGURE 8.34 Autonomic joining of pipes. (Adapted from [Stubblefield 1998].)

8.15.2 Piping

Combining the molecular action of heat shrinking with other materials, such as thermosets, holds out the possibility of making superior mechanical joining and sealing systems by mechanical actions, such as squeezing the compliant uncured thermoset to fill gaps. Figure 8.34 shows a combined heater, heat shrink, and thermoset system for joining cylindrical rods in a butt joint arrangement. This is a heat-activated thermal coupling for joining composite-to-composite pipe using a combination of shrink tubing and thermoset resins for adhesion [Stubblefield 1998].

8.15.3 Multiscale Patch and Plug

Liquid leak plugging is a multiscale technique that bridges or partially closes leaking holes with large objects and elements, and then fills-in with smaller items. A plug forming system in fuel tanks uses a brush-like array of macroscale polymer fibers that bend and flow into a hole to provide scaffolding for smaller scale coagulation to occur in fuel tank leaks (Figure 8.35) [San 1978].

8.15.4 Corrosion

Corrosion is a persistent multifaceted threat to the integrity of metallic structures. Small inroads of corrosion damage can alter structural and electrochemical environments in ways that accelerate corrosion. Prevention and mitigation benefits from panoply methods that stop corrosion at multiple fronts

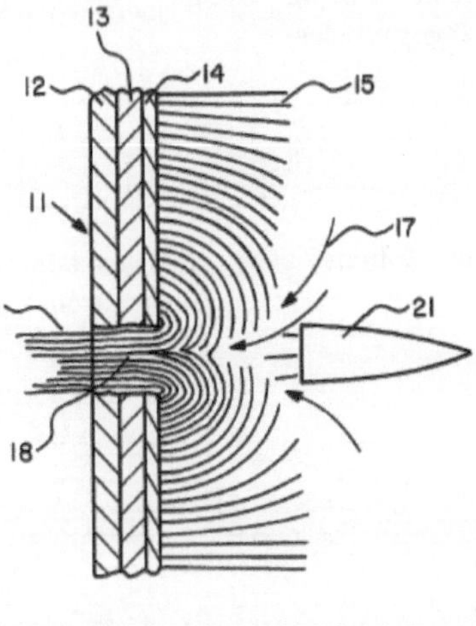

FIGURE 8.35 Prepositioned macroscale fibers backflow and fill leaking hole to provide scaffolding for microscale coagulant plugging. (From [San 1978].)

and multiple length scales. Possibilities include coatings that combine superhydrophobic outer skins to prevent water adhesion, impermeable barrier coatings that prevent water and oxygen from contacting the metal surface, passivating outer layers on the outer surface of the metal prevent further contact with water and oxygen, cathodic protection, and microencapsulated self-healing and anticorrosion agents that release upon scratch damage to the coating; and Al_2O_3-Nb coatings that combine low pinhole density with metallic layer deposition self-healing [Koene 2009] [Yasuda 2003].

8.15.5 Self-cleaning Machines and Systems

Machinery and related mechanical systems often get clogged and jammed with debris and other detritus. Self-cleaning machinery with supplemental devices and processes alleviates the need for manual cleaning operations. An example is a cutting machine for recycling tires that uses supplemental disks to remove excess material from an array of counterrotating cutters [Parke 2007].

8.15.6 Permanent Press Fabrics

Permanent press fabrics are one of the most commercially successful self-healing materials. Wrinkled clothing heals itself into a smooth pressed state, normally through a thermal tumble dry cycle. The mechanics of self-pressing derive largely from antagonistic multiple length scale effects. The fabric must be stiff enough to recover creases during normal usage of wear, storage, and washing, and yet compliant enough for comfort [Holzman 1973]. Most permanent-press fabrics use a mixture with some fibers being stiff and others compliant. When combined as a mixture, the fabric is soft enough to wear with comfort. During normal usage, the compliant fibers crease and wrinkle. The stiff fibers maintain their original unwrinkled shape, but the wrinkled compliant fibers are collectively too strong and the material wrinkles. During washing and drying, the compliant fibers soften, and the stiff fibers dominate the mechanics and reproduce the unwrinkled state (Figure 8.36).

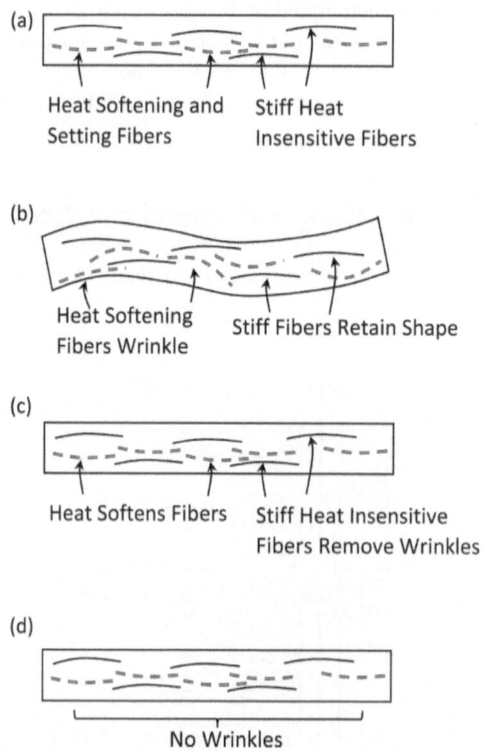

FIGURE 8.36 Permanent press fabrics. (a) Original pressed fabric. (b) Wear wrinkles fabric. (c) Heat softens fibers to remove wrinkles. (d) Cooling sets fiber and pressed fabric.

Wrinkle-free self-pressing fabrics require special handling and temperature control during spin wash and dry cycles, a feature included in modern laundry machines. More sophisticated permanent-press fabrics use heat-activated SMPs [Hu 2010].

8.16 Bioinspired Fracture and Tear Mitigation and Healing

Fracture and cracking processes comingle loads, stress, deformation, and material properties at multiple length scales. Biological systems counteract multiscale damage processes with multiscale mitigation and healing techniques. Bones, shells, wood, and other biological structures must resist and recover from routine and occasionally intense loads encountered throughout life, yet remain lightweight and minimize the consumption of resources, also while growing in place.

Damage mitigation plays a key role. Limiting cracking damage during severe loading reduces the amount of healing required for restoration and reduces the need for being overly strong and heavy. Bone exhibits exceptional fracture toughness with viscoplastic flow, the formation of unconnected microcracks, crack bridging, and crack deflection. These mitigation methods have fine-tuned optimized location and direction dependencies (Figure 8.37) [Peterlik 2006]. At the molecular and nano length scales, bone crystals form with calcium-based platelets held together by citrates in disordered arrays that promote fracture toughness [Davies 2014]. At the meso and macro length scales,

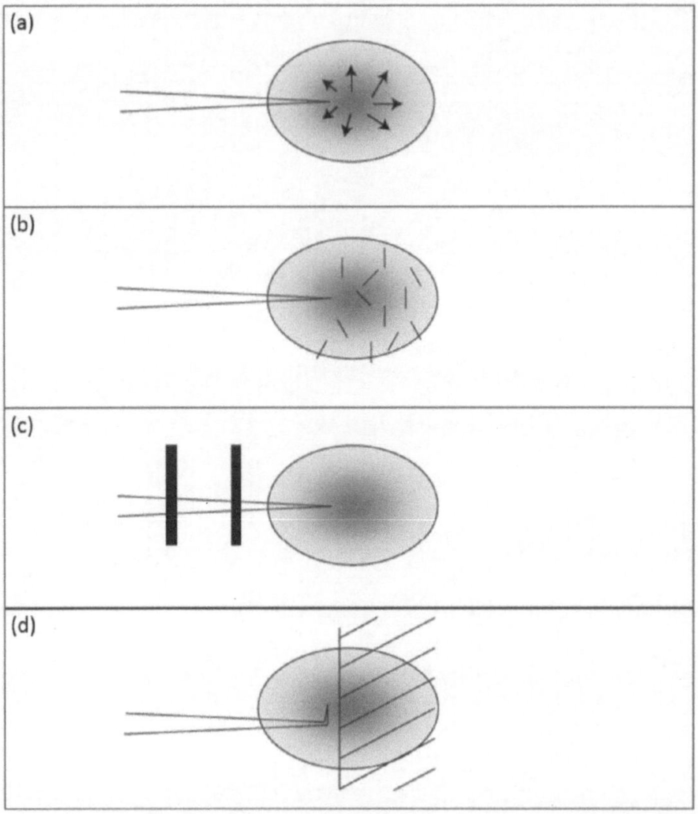

FIGURE 8.37 Microscale methods of mitigating crack propagation. (a) Viscoplastic flow (gels). (b) Microcracking and crazing (ceramics and polymers). (c) Crack bridging (ceramic fiber composites). (d) Crack deflection (composites). (From [Peterlik 2006].)

fractures running in the direction of the bone axis propagate more easily than those running across the axis, with fiber-bridging slowing the growth of larger fractures [Koester 2008].

It is possible for bone to carry reduced levels of loads while recovering from small mitigated levels of damage. Similarly, the shells that protect mollusks are impressive. They combine micro- and nanoscale brittle calcium carbonate crystals with relatively small amounts of polymers to form very tough macroscale materials. The molecular structure of the polymer adhesives is inherently tough with normally folded structures that unfold, absorb energy, and yet remain intact when severely loaded [Bettye 1999]. Soft tissue, such as skin, benefits from embedded networks of fibers, often with micro- and nanoscale diameters, that prevent tearing while allowing for routine stretching and flexibility through network-scale fiber straightening and reorientation [Yang 2015b].

Introducing antagonistic micro- and nanoscale effects into engineered material systems mitigates fracture propagation. Dispersing ductile particles of brittle ceramics bridges cracks and toughens against cracking (Figure 8.37). The toughening performance increases if the particles undergo partial debonding at the appropriate level – too much or too little debonding degrades performance [Tvergaard 1992].

9

Coordinated Repair

9.1 Introduction

Coordination enhances self-healing with efficient use and allocation of limited resources. Applications of coordination range from relatively simple feedback and control strategies up through adaptive, anticipatory, and cognitive self-aware systems (Figures 9.1 and 9.2) [Hurley 2011a]. Coordination promotes damage avoidance and mitigation, limping and reconfiguration recoveries, effective repair processes, and apoptotic behaviors that include destroying and deleting components as needed. Modern sensing and control systems offload the coordination tasks of self-healing from humans onto autonomous machines and materials. [Brandon 2011] [Mihashi 2012].

Some features of coordinated self-repair:

1. *Sense of Self:* Systems and subsystems must distinguish themselves from non-self and other items in the environment.
2. *Sense of Health:* The system has notions of an idealized self-state.
3. *Health Assessment:* The system assesses its own condition and configuration to distinguish deviations of its self-state from ideal state.
4. *Healing Plan:* The systems either have a preset plan for healing or formulate plans as needed.
5. *Healing Capability:* The system can alter its state to move toward a more ideal configuration.

Note that Step 5 can loop back onto and through Steps 1–4, sometimes spawning a multi-cycle self-referencing feedback loop.

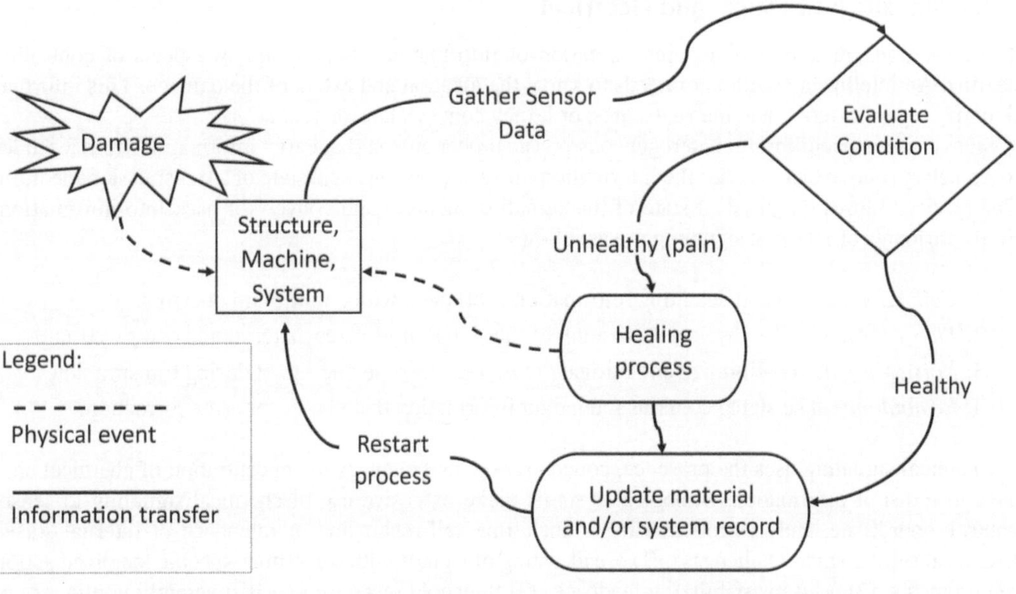

FIGURE 9.1 Coordinated self-healing of structures, machines, and systems. (Adapted from [Hurley 2011a].)

DOI: 10.1201/9781003396048-9

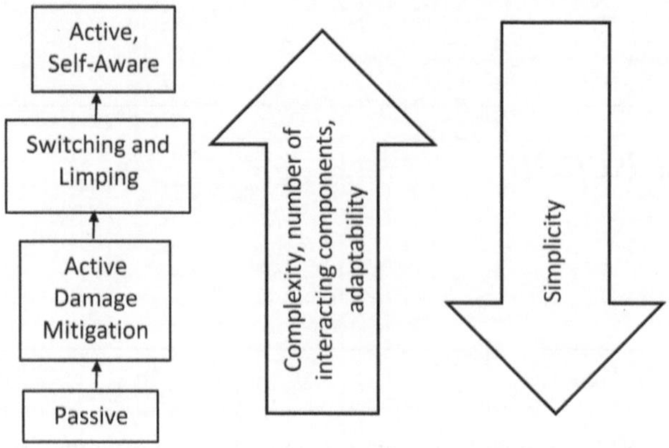

FIGURE 9.2 Active to passive levels of coordination of repair versus complexity.

9.2 Enabling Technologies

Effective coordination of healing requires tools for sensing, signaling, actuation, component switching, and decision-making.

9.2.1 Damage Sensing

Controlling, directing, and evaluating the effectiveness of self-healing actions benefits from sensors that determine the location and extent of internal damage [Thostenson 2006]. Detecting, locating, and assessing damage is often nontrivial. Many forms of damage result from phenomena acting over multiple spatial and temporal scales [Huston 2010]. Conventional health monitoring systems use networked arrays of sensors distributed throughout the structure. Nonconventional sensors and assessments, such as with chemical signaling, are also effective.

9.2.2 Signaling: Chemical and Electrical

The collection, transmission, and interpretation of information underlies many aspects of controlled healing. An intelligent coordinator needs to know the location and extent of the damage. This information may regard external and internal states, or it may convey plans of action.

Signals are nonequilibrium heterogeneous physical phenomena that carry information from a sender to a receiver. The sender encodes the information into a specific physical state of a transmission medium. The receiver senses the physical state of the signaling medium and converts it back into information. Some attributes of effective signaling are as follows:

1. *Coding:* Encoding and decoding information is efficient with a minimum of errors.
2. *Physical Layer:* The physical realization of the signal minimizes material and energy costs.
3. *Noise Mitigation:* Minimize and mitigate noise that corrupts the signal during transmission.
4. *Redundancy:* The signal contains redundant information that allows for error correction.

Chemical signaling uses the presence, concentration, and changes in concentration of chemical species, to transmit information. Biological systems make extensive use of chemical signaling to sense, control, coordinate, and control activities – including self-repair and maintenance of internal stasis. Chemical signaling has challenges – (1) broadcasting of signals with sometimes specific localized action requirements, (2) need for stabilizing feedback, (3) threshold response that is inherently nonlinear, (4) compatibility of signals, and (5) stability and ability to maintain in reserve, or to generate signals quickly.

9.2.3 Passive Switching of Loads, Load Paths, and Stiffness

Relatively simple are passive methods where damage directly drives component switching to alter load paths and structural stiffness for healing. Redundant structural systems, with indeterminate mechanics, as in suspension bridge cables, are examples. Damage to a redundant element reduces the carried load, which then transfers onto other elements. These altered load paths may or may not be robust against additional damage. Coordination of changes in load paths enhances resilience and structural performance.

Statically indeterminate systems often use prestress to aid in connectivity to ensure effective load transfer among the constitutive components. When localized damage breaks the load path continuity, it is sometimes possible to coordinate mechanical reassembly of the connection. Regenerative surfaces, such as tire treads, also heal by passive switching methods where the healthy surfaces emerge from underneath as the top surfaces wear.

Thermal expansion is a viable method of controlling the stiffness and load paths in a structure. One method uses thermal expansion to alter in-plane compression or tension and the associated nonlinear geometric stiffness in elongated structural members. Self-tuning pianos alter string tensions by thermal expansion [Gilmore 2003].

It is possible to alter the stiffness of composite structures based on observed damage. Early demonstrations used distributed sensors, such as optical fibers, to sense the damage, shape memory alloys (SMAs) to alter the stiffness in a favorable manner, and artificial neural networks to perform the supervisory control of the system (Figure 9.3) [Degang 1999] [Shiammoto 2005]. The use of SMAs to alter stiffness relies upon a combination of antagonistic storage of prestrain in the fibers, and then on a combination of geometric stiffening in the concrete with closing of cracks and stiffening through fibers.

Controlling the stiffness of structures improves resilience by engendering the ability to detune from dangerous vibrations, and to tune the mechanical properties to recover suitable overall stiffness following

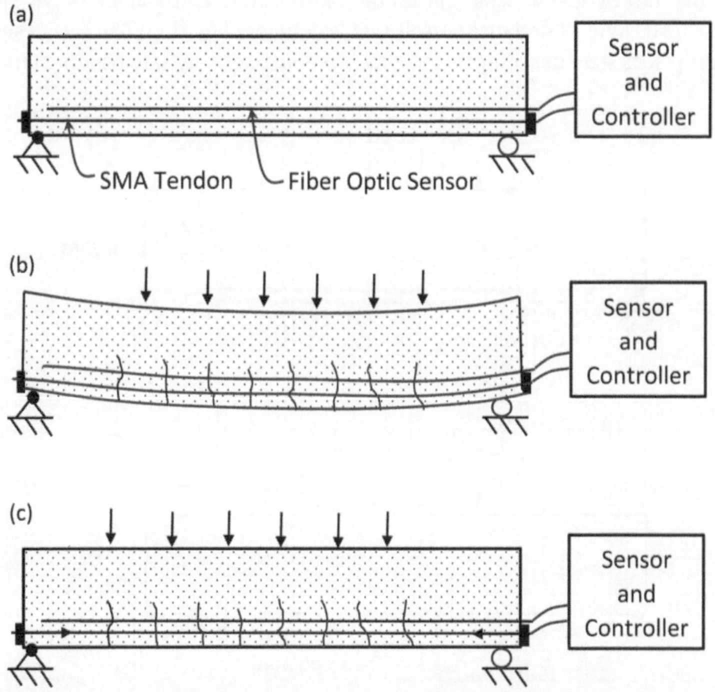

FIGURE 9.3 Distributed fiber-optic sensing combined with SMA tensioning to stiffen concrete structural elements and recover from damage. (a) Healthy concrete beam with SMA tendons and fiber-optic sensors. (b) Load cracks beam. (c) SMA tendons restores shape and stiffness.

damage. Using SMA wire in a heating-phase change mode alters preloads to induce nonlinear elastic effects resulting in a stiffness change. An example is geometric nonlinearity arising due to axial tension and compression in an elongated member.

9.2.4 Damage Avoidance and Mitigation

It is highly effective to coordinate the behavior of a system to avoid or mitigate damage-inducing processes. Mitigation reduces the extent and consequences of damage-causing loads, which eases future repairs. Mitigation actions range from moderating the rate and type of damage growth to stopping growth to avoiding damage altogether.

Cracking and delamination, being self-reinforcing phenomena, provide excellent opportunities for mitigation. The presence of small cracks and delaminations alters the mechanics to produce conditions that promote growth of cracks and delaminations. Sometimes, relatively small efforts can stop or prevent small cracks and delaminations from forming. In the case of delaminations, it may be just a small compressive force applied across the layers that prevents formation. Figure 9.4 shows a situation where delaminations tend to form at the ends of FRP layers on beams in bending due to the formation of tension stresses transverse to the layers. Piezoelectric patch layers at the ends apply differential tension and compression parallel to the layer which cause a curling effect leading to transverse compressive stresses that prevent the delaminations [Rabinovitch 2007].

9.2.5 Graceful Damage Progression

In cases where failure of a structure or system is unavoidable, graceful failure mitigates surrounding damage and costs. Graceful damage progression extends beyond mitigating against progressive damage to having the damage proceed in a manner that facilitates repairs. For example, it is possible to design automotive tires that fail by a slow leak, but do not blowout [Gramelspacher 1939]. Similar considerations apply with safe designs of pressure vessels that leak before bursting. The safe design of reinforced concrete structures places steel in arrangements that promote relatively safe bending failures where steel

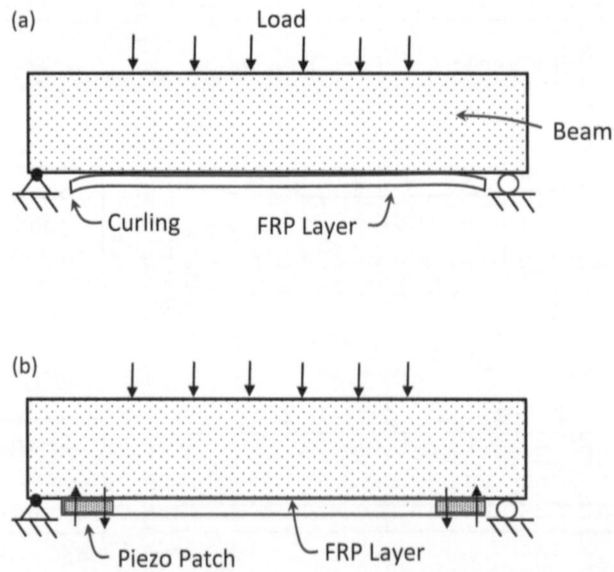

FIGURE 9.4 Piezoelectric curling prevents delamination of composite strengthening layer on concrete beam. (a) Concrete beam reinforced with FRP layer that curls at ends. (b) Piezo patches counter-curling of ends of FRP layer. (Adapted from [Rabinovitch 2007].)

yields and concrete cracks before collapse, and avoids dangerous shear failures that lead to rapid collapse. Another example is a control system with redundant sensors. Fuzzy-set voting systems are robust against individual sensor failures and provide graceful degradations of control system performance as individual sensors fail [Oosterom 2002].

9.2.6 Passive Damage Mitigation

Passive methods use prepositioned elements to mitigate damage-causing loads. An example is fuel tanks penetrated by projectiles. Fluid shock wave loading, known as hydrodynamic ram, caused by the high-speed projectile passing through the fuel is a major design challenge for self-sealing fuel tanks. Failure to mitigate fluid shock loads causes the tank to burst. An early technique used a gelatin-coated rubber lining as a flexible shock-resistant and leaking-sealing structural component of gasoline fuel tank for airplanes [Kraft 1927]. The gelatin prevents the gasoline from contacting the rubber, unless pierced by a projectile, which then leads to gasoline-solvent-induced swelling of the rubber for sealing. Flexible rubber walls resist bursting due to hydrodynamic ram. Compressible foam walls that crush with hydrodynamic ram provide a further degree of mitigation (Figure 9.5) [Dasher 1948] [Winchester 1974]. A similar issue arises in space-based vehicles with liquid fuel tanks. Small hypervelocity particles penetrate the tank walls to produce hydrodynamic ram. An outer layer of metal breaks up the projectile into smaller pieces and minimizes the ram [Fry 1976].

The folding back of metal tank walls into sharp petals or flowers surrounding the hole is a primary form of damage following projectile penetration. Petals aggravate leaking and cause difficulties with self-sealing. Fiberglass-reinforced hardboard layers resist petal formation [Hoover 1947]. Composite elastomer and fabric layer walls reduce the size of the bullet hole and leave the hole filled with wall-attached fibers [Howard 1952]. In a manner that mimics fibrin in a blood clot, the fibers in the hole form scaffolding for self-sealing plugs.

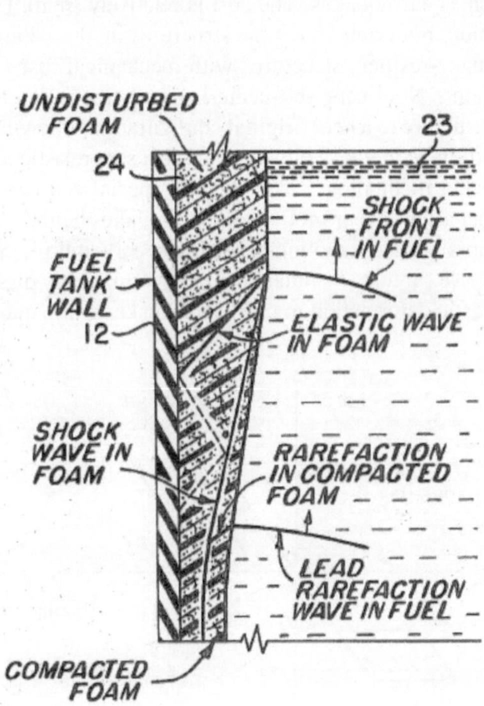

FIGURE 9.5 Crushing foam attenuates hydrodynamic ram in fuel tank subjected to projectile penetration and impact. (From [Winchester 1974].)

9.2.7 Active Mitigation

Most cases of active damage mitigation are application-specific. Nonetheless, most of these cases have common themes of decoupling from damaging forces, dissipating the energy of damaging forces that manage to couple into the structure, and arresting damage as it forms. Active mitigation often only requires relatively minor modifications to the state of the system, but does require coordinating the timing and location of the efforts. Important steps are as follows:

1. *Preparation:* The mitigation system must respond at a moment's notice. This requires prepositioning the energy, materials, and switching capabilities.
2. *Watchdog Sensing:* The system must be alert and able to sense – even anticipate – damage-causing events that may occur randomly and unexpectedly. Minimizing false alarms is important.
3. *Decision-making upon Receipt of Signal:* The need is for rapid and near real-time decision-making. Analog systems, including preloaded mechanical elements, can be decisive.
4. *Timely Response:* Early-stage mitigation prevents damage from growing at low cost.

The following are some examples of arresting damage:

Freeze-out Plugs: Used on engine blocks, these plugs release and leak cooling water if it starts to freeze, as is common when antifreeze solutions fail. The release of coolant fluid through freeze-out plugs prevents catastrophic engine block breakage by the expansion of freezing water.

Decoupling: Decoupling from a large source of propagating mechanical energy prevents damage to the structure. An example from biology is conifer trees with specific structural details at the roots of their branches that allow for the branches to detach under heavy loading, thereby decoupling the load from the rest of the tree, without causing much secondary damage [Müller 2015]. Branch root tracheid structures have a zigzag form that cracks with energy dissipation and heal rapidly when the limb is not fully detached.

Joints, connection details and isolated structural members, and similar mechanisms in structures provide excellent opportunities for decoupling and altering the paths of potentially damaging loads in highly energetic events, such as earthquakes. The cost is relatively small. Fusing saves structures from severe damage and destruction, but often leaves the structures in altered and unusable states following the energetic event. Earthquake-resilient structures with mechanical fuses use recentering to stabilize post-earthquake configurations. Nonlinear self-centering isolation systems are a more sophisticated alternative that returns the structure to a near original configuration following energetic shaking [Berton 2007]. Joints with embedded shape memory alloys acting in a superelastic mode control the recentering [Shrestha 2015]. Hybrid design methods are available for specialized beam–column joint assemblies with machined fuse bolts that allow linear motions in earthquakes below a specific level and nonlinear dissipative hysteretic behaviors when the motion exceeds a specified limit [Solberg 2008].

A flag-shaped hysteresis curve captures the intent behind hybrid energy dissipation devices (Figure 9.6). This behavior minimizes plastic deformation in mild earthquakes while maintaining structural stability

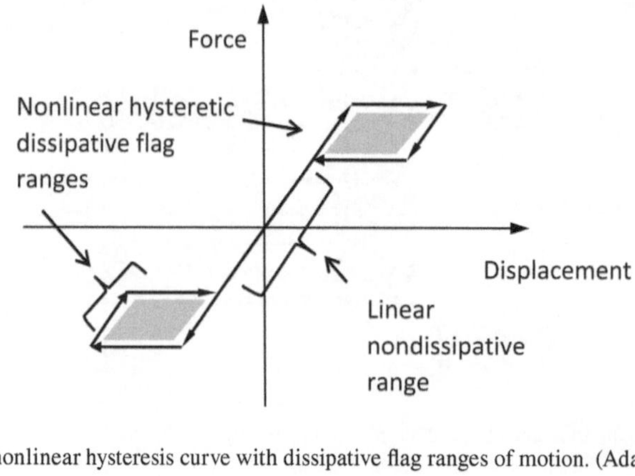

FIGURE 9.6 Linear to nonlinear hysteresis curve with dissipative flag ranges of motion. (Adapted from [Solberg 2008].)

in large earthquakes [Fortney 2007]. For small displacements, the mechanics are linear and nondissipative. Large displacements induce flag-shaped nonlinear hysteresis. The area inside the flags is the amount of energy dissipated each cycle.

More general material configurations are possible. Composites containing internal skeletons of stiff links held in place with rigid joints preferentially fail under large loads and then heal in a subsequent step, such as a heat cycle [Boncheva 2003].

Sometimes altering the stiffness detunes the dynamics of a system and prevents damage, such as due to resonance-type vibrations [Dry 1996e].

Feathering and/or furling windmill: Windmills have a severe design constraint of being light enough to turn in light winds, yet strong enough to withstand extremely strong winds. Feathering or furling the blades to disengage from strong winds is a common damage mitigation maneuver. The Ogallala windmill furls in strong winds to rotate about the vertical yaw axis to reduce exposure to the wind (Figure 9.7). A key feature is a hinged tail vane that provides a controlling yaw moment, typically with a gravity assist from a tilted hinge axis [Muljadi 1998].

9.2.8 Damage Isolation and Recovery

For damage modes that are well understood, it is often possible to fabricate the structure so that damage remains isolated to a portion of the structure that can be repaired with ease. An example is the self-sealing fuel tank which separates the sealing and resealing functions from the structural supports [Davies 1948].

9.2.8.1 Arresting Active Damage

Issues that favor arresting active damage are as follows:

1. *Self-propagation and Growth:* Damage self-propagates and/or is growing.
2. *Signaling:* Damage produces signals that can trigger a response.
3. *Competent Response:* Response arrests the damage.

Damage processes that are good candidates for active arresting include fires, fractures, corrosion, impacts, progressive collapse, and complicated damage scenarios that comingle these processes.

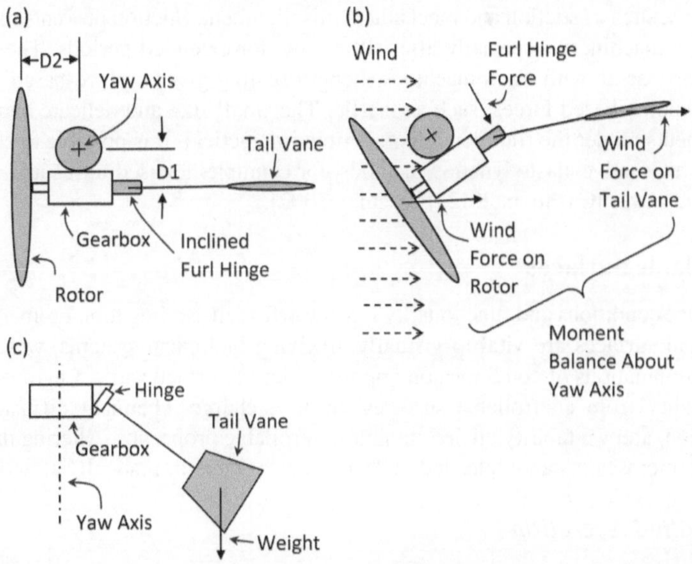

FIGURE 9.7 Furling of an Ogallala windmill by rotation and moment balance about yaw axis rotates rotor into weak winds for maximum power harvesting while also rotating rotor out of strong winds to prevent damage. (a) Unfurled, no wind (top view). (b) Furled in wind (top view). (c) Inclined furl hinge (side view). (Adapted from [Muljadi 1998].)

9.2.8.2 Arresting at Damage Source

Progressive damage occurs when a small amount of damage alters conditions with positive feedback that causes the damage to grow more quickly. In brittle fracture, a growing crack alters stress in a way that promotes more cracking. Altering stress patterns can arrest growing cracks, such as by drilling a hole at the crack tip, or using shape memory tendons and actuators [Garcia 2010]. Superelastic SMA tendons are useful retrofits for mitigating earthquake damage in historic masonry structures [Indirli 2008].

9.2.8.3 Arresting by Energy Dissipation

Sometimes, it is possible to dissipate the energy in forces that cause damage. Passive energy dissipation devices include tuned mass dampers, damping layers, controlled plastic deformation of metal and concrete in structures, and SMA superelastic tendons. Active energy dissipation systems are less common, including controlled servohydraulic dampers [Dyke 1996] and piezoelectric dampers.

Tuned mass dampers dissipate vibration energy with a vibrating auxiliary mass. Tuned mass dampers operate in either an elastic mode, where the tuned mass and spring combine to split and shift troublesome resonance frequencies away from that of a periodic input force, or in an inelastic mode where the auxiliary mass dissipates vibratory energy with damping in the support [Frahm 1911] [Ormondroyd 1928].

Shifts in the properties of the applied force or that of the primary structure detune a mass-damper and render it ineffective. Active vibration control with adaptive self-tuning compensates for frequency-dependent changes that result from damage and degradation [Casado 2007]. An example is a self-healing servo-control system for a computer hard disk memory read/write head. Wear to the bearings and mechanical components causes mechanical defects, such as stiction, leading to dynamic feedback errors. An adaptive system detects incipient problems and extends the life of the hard drive by adjusting feedback gains and notch filtering specific frequency bands [Dang 2005].

9.3 Mechanical Action

9.3.1 Coordinated Tapping and Elastic Waves

Judiciously applying mechanical forces and impulses, that is, hammer taps, resolves many mechanical faults in machinery, such as stiction and mechanical misalignment. Stiction prevents relative motion and leads to seizing in machinery, especially after sitting idle for extended periods. The issues of stiction become particularly acute with micromachines. Length scaling gives surface-based stiction forces an advantage over volume-based forces, such as inertia. The small size and delicate nature of the micromachine components render the traditional hammer tap impractical. It is possible to create hammer tap equivalents with alternative elastodynamic methods, for example, focused ultrasound with laser-generated thermoelastic waves [Cushman 2012] [Leseman 2007].

9.3.2 Controllable Surfaces

Controlling surface conditions and functionality is a powerful self-healing tool. From a biomimetic point of view, functional surfaces are vital to virtually all living biological systems, which thrive through differentiated manipulations of conditions on opposite sides of surface walls. Engineered surfaces gain similar functionality from controllable surfaces. Surface charge, chemical diffusion, permeability, geometry, adhesion, and wettability all are possible controllable properties. Altering the surface texture and properties at microscale, nanoscale, and molecular scale often dramatically affects wetting.

9.3.2.1 Controlled Accretion

Switchable molecular recognition creates surfaces with switchable morphologies. One possibility is networks of molecules and nanoparticles switched into different configurations by the presence of different chemical compositions. Dangling polymer chains with reactive hydrogen bonding between regions of the chain provide a wide range of reversible and controllable interactions. Elongated molecules assembled in a brush-like structure where molecular bristles change properties under different conditions produce a surface with switchable properties [Minko 2003] [Sidorenko 1999]. Switching the hydrogen bonding on or off changes the shape of the dangling polymers, such as from helical coils to elongated blobs [Sun 2011]. Changes in temperature, pH, or specific chemical concentration reversibly changes the shape of the dangling surface molecules and the macroscopic physical properties such as wettability. Poly(*N*-isopropylacrylamide) and its copolymer films are an effective material for this smart behavior.

The appearance of water or changes in relative humidity changes surface wettability of thin films. In one example, poly(ethylene glycol) (PEG) with fluoroalkyl buries end groups to increase the contact angle in high humidity [Mackel 2007]. In a second example, hybrid polymer brushes made of poly(dimethylsiloxane) (PDMS) and a highly branched ethoxylated polyethylenimine exhibited similar macroscopic behavior with a microscopic behavior being the absorption of water and subsequent swelling of the layer [Motornov 2007]. Water droplets also produce this change and induce a large contact line hysteresis that restrains receding droplets – a process that would tend to pin droplets to the surface, rather than shed them.

Applying electric voltages changes the wettability of conductive liquids over ranges from superhydrophobic to superhydrophilic [Krupenkin 2004]. Electrostatic fields alter the geometry of surface-attached molecules from a stand-up to a bent-over configuration (Figure 9.8) [Lahann 2003]. The different molecular geometries produce different wettabilities. Electrostatic forces drive nonwetting liquids into yarns [Konstantin 2007].

Switching fluorinated polymers from an ordered smectic semicrystalline state to a disordered amorphous state switches a surface from smooth and hydrophobic to tacky over a small temperature range, for example, a 2°C range centered at 35°C [de Crevoisier 1999].

UV light alters the wettability of photoresponsive materials, such as pyrimidine-terminated molecules with charged ends attached to a substrate [Abbott 1999]. A 280-nm wavelength UV light induces dimerization at the ends of the dangling polymers, causing the ends to join, eliminating the charge, and increasing wettability. UV light at a wavelength of 240 nm cleaves the dimers to reverse the process, including the change in wettability.

An alternative approach to controlling surface behavior is to affect molecular accretion with environmental conditions that alter the structure and behavior of chemical species. In a finely tuned example of recognition, DNA-functionalized nanoparticles perform switchable recognition where they either repel or latch on to one another depending on the NaCl concentration. This affects the form and extent of nanoparticle assemblies that deposit on a surface (Figure 9.9) [Srivastava 2014].

9.3.2.2 Controlled Diffusive Permeability

Regulating the diffusivity of membranes controls the through-surface flow of material. Regulating the temperature of materials with temperature-dependent pore sizes controls the permeability. Controlled

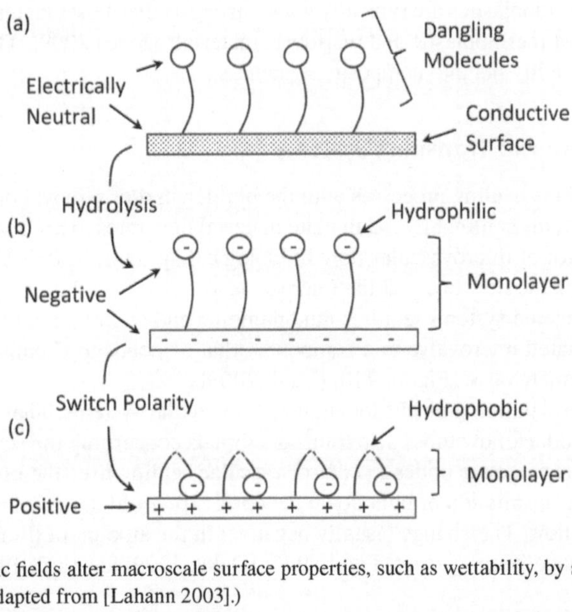

FIGURE 9.8 Electrostatic fields alter macroscale surface properties, such as wettability, by switching the configuration of dangling molecules. (Adapted from [Lahann 2003].)

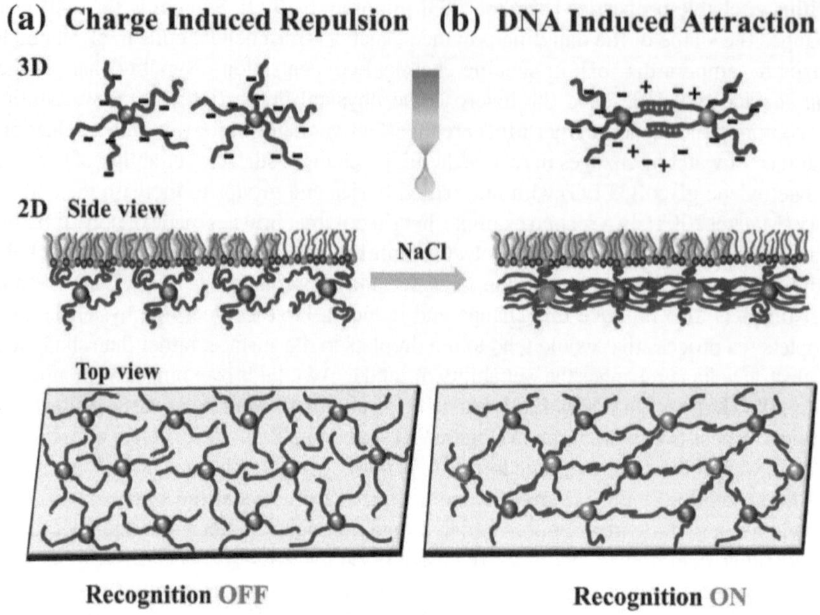

FIGURE 9.9 DNA molecules perform the switchable recognition and alter surface properties. (a) Charge-induced repulsion among dangling molecules. (b) Addition of NaCl enables molecular recognition and attraction between molecules and affect nanoparticle deposition at liquid interfaces. (Reprinted with permission from [Srivastava 2014]. Copyright 2014, *American Chemical Society.*)

drug delivery using nanogels increases the porosity when the temperature rises above a threshold value [Hoare 2009]. Embedding superparamagnetic particles subjected to oscillating magnetic fields provided the heat source. Reversing the permeability requires supplemental cooling, such as the body of a warm-blooded animal that acts as a temperature-regulated thermal reservoir.

9.3.3 Controllable Mending and Material Properties

Materials that mend under control enable a variety of controllable healing processes. Heat treatment controls thermoplastics. Thermoplastics are typically not as strong as thermoset materials. One possibility is to form a solid solution of thermoplastic and thermoset materials [Jones 2009]. These blended materials are a compromise that readily fits into composite structures.

9.3.4 Controlled Material Transport Systems

Controlled transport affects healing processes with the buildup and/or removal of material from remote locations. Biological systems continually manipulate material flow rates to promote healing and growth. An example is the control of microvascular flow in biological systems. Tissue and vascular dilation to increase blood and lymphatic flow occur at the microscale as a part of inflammatory response to injury. Doing the same in engineered systems requires fluid pumping and control of delivery. Somewhat biomimetic is a hydrogel-actuated microvalve in a component that responds to chemical stimuli by swelling that opens and closes a microvalve (Figure 9.10) [Baldi 2003].

Hollow tubes are natural building blocks for engineered vascular systems. They contain repair materials for transport as needed. Hollow tubes also transport signals concerning the state of damage and self-repair. An approach that integrates optics to control vascular healing uses the presence of liquid in the waveguide to guide the transmission of light down the core. Structural damage causes the fiber to crack and healing material to flow. The change (usually negative) in the amount of transmitted light indicates the occurrence of a damage/healing event (Figure 9.11) [Dakai 1999] [de Vries 1994] [Kim 2004b].

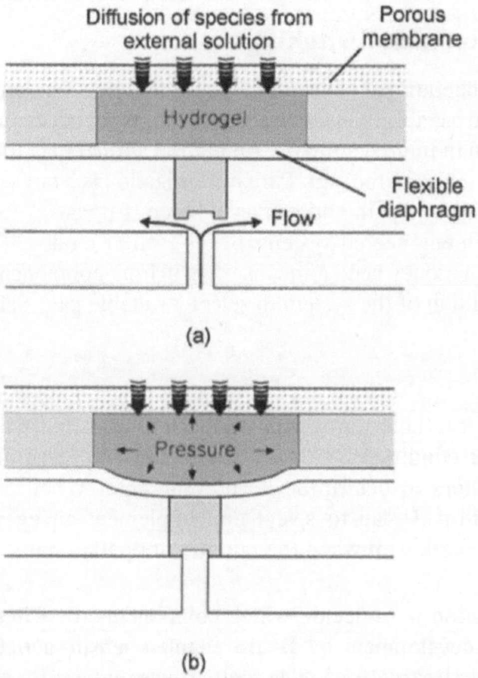

FIGURE 9.10 Hydrogel-actuated environment-sensitive microvalve for active flow control – open (a) and actuated (b) by diffusion of selected species and closed. (From [Baldi 2003].)

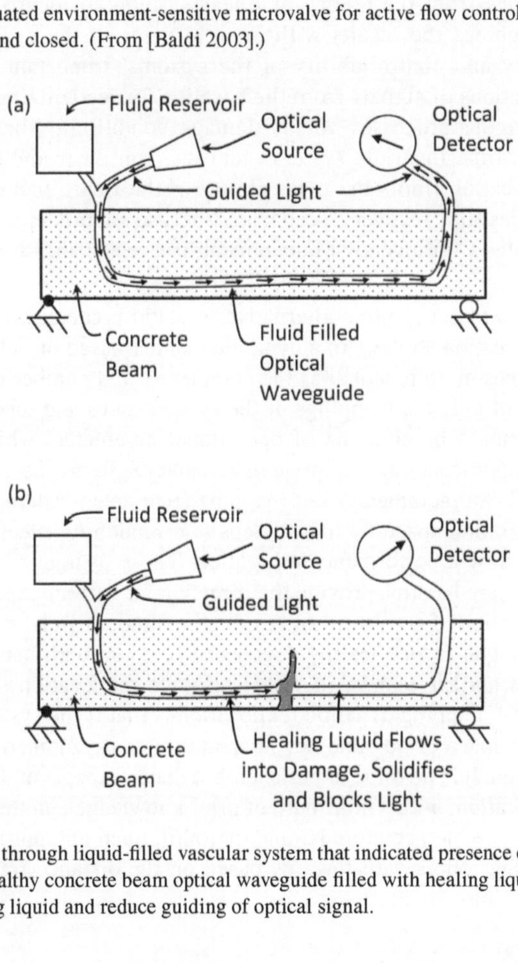

FIGURE 9.11 Light sensing through liquid-filled vascular system that indicated presence of damage and prompts supply of more healing liquid. (a) Healthy concrete beam optical waveguide filled with healing liquid. (b) Damaged beam cracks waveguide to dispense healing liquid and reduce guiding of optical signal.

9.4 Limping and Component Switching

Limping is a change in the operational behavior of a system following damage. The system still functions, but often with reduced capability in a move that attempts to reduce further damage and facilitate repairs. Limping is common in injured animals. An injured animal may find it difficult to run but may be able to walk slowly with an impaired gait. Pain minimization is a key parameter. In dire situations, the same animal can suppress the pain and run away, even if it causes further injury. Implementing similar limping behaviors in engineered systems offers similar modes of damaged-but-still-running operation. Most limping techniques need a means of switching components, and some sort of cognitive map depicting the condition of the system to select a suitable gait. Selection methods include the following:

1. *Predetermined Map:* The system has a map of how components interact and operate in degraded modes. This top-down approach is possible when the states comprise a known finite set of conditions with selectable and predetermined configurations. The system the self-monitors to determine its present state. When damage occurs, the system uses a predetermined map to select and implement an operational limping mode. Predetermined maps work well when the number of operational states and damage modes is small.

   An important question is to decide which components to switch on and off and when. An early (ca. 1971) development by Beard enabled sensor–actuator linear systems that recover from damage [Beard 1971]. The method uses internal diagnostics based on state estimation techniques from the theory of linear systems to identify the damage-inducing events and combines the results with internal maps of the dynamic behavior based on the observability and controllability of the systems. Important issues are setting of thresholds for deviations of signals from the baseline for recognizing damage and accommodating normal aging processes versus damage conditions. Since the basis of these techniques is linear time-invariant systems analysis, nonlinear and time-varying damage and resource reallocation limits the extensibility of the Beard technique. For cases with known failure modes and recovery scenarios, the use of predetermined switching patterns, as in Automated Contingency Management, is a reasonable approach [Saxena 2007] [Pecht 2010].
2. *Dynamic Reallocation:* A cognitive alternative to static predetermined limping strategies is to develop a reallocation strategy in a dynamic manner based on self-observed behaviors. Self-mapping becomes more practical as the complexity and number of degrees of freedom increases the range of possible mappings of the system states and topologies. Self-mapping begins with measuring a baseline set of operational parameters while the system is in a healthy state. Such approaches are common in computer systems that self-diagnose and self-repair [Ghosh 2007]. An incremental self-mapping technique for damaged systems explores the possible configuration space by initial steps that randomly select and activate degrees of freedom, followed by a performance evaluation. Those degrees of freedom giving good performance, that is, are healthy, provide the direction for further exploration in subsequent steps of random selection of activating and evaluating randomly chosen degrees of freedom. A greedy bias technique favors healthy degrees of freedom with biased random selection. The process repeats until it achieves a viable resource reallocation strategy (Figure 9.12). The advantage of self-mapping dynamic reallocation is that if the system is damaged, it can begin to explore the space of possible new maps through a combined process of modeling, testing, and prediction. It is possible to build a robot that is capable of developing a map of its own system configuration, with a minimum of prior knowledge and then walk, even in cases of damaged and inoperable actuators [Bongard 2006]. Such an approach develops limping strategies for walking robots with rewards based on the distance traveled in a fixed time [Christensen 2013] [Cully 2015].

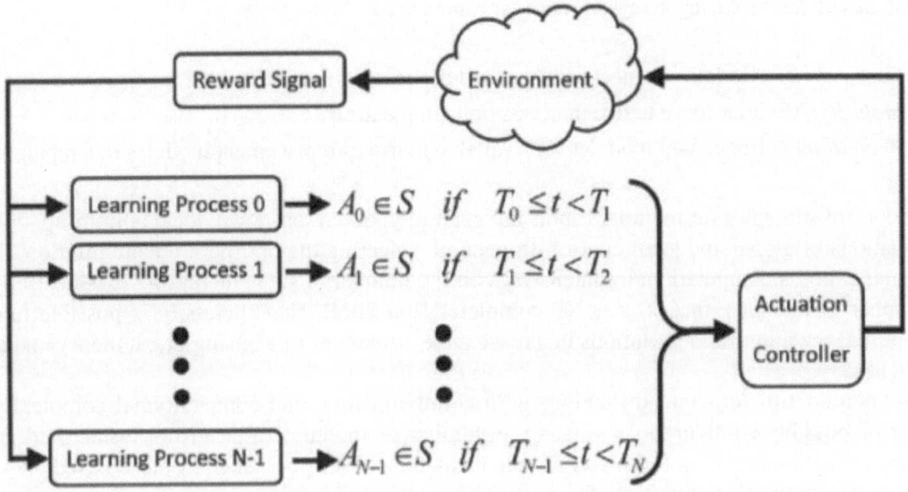

FIGURE 9.12 Self-healing gait through learning-based resource reallocation. (From [Christensen 2013].)

9.4.1 Limping of Mobile Robots and Autonomous Vehicles

Many mobile robots and autonomous vehicles use redundant components to retain functionality in locomotion and maneuver. A multi-legged robot that injures a leg needs to alter its gait. In general, more legs imply more stability and ease of control. Walkers with four or more legs are inherently stable as they can always keep a tripod of legs in contact with the ground. For example, a symmetric six-legged robot can lose up to two legs and still execute a statically stable walking gait [Yang 1998]. One-, two-, and three-legged gaits are possible, but typically require sophisticated dynamic control, similar to that needed for running or galloping.

Autonomous vehicles are specialized robots for transport and related activities. Mass production and mass usage affords the luxury of identifying many of the common failure modes and use of preprogrammed tree-like logic architectures to detect, diagnose, and recover from faults [Duan 2005].

Modular robots are assemblies of modules that can attach and reattach to each other in different locations. Force, material, power, and signals pass through the attachment points. These robots can be highly redundant and are capable of self-repair through component switching and topological reconfigurations [Castano 2002] [Goldstein 2009]. The primary architectures of modular robots are as follows:

1. *Lattice:* Robots aggregate into 3D lattices.
2. *Hard and Soft:* Most robots are chains and trees of interacting hard (rigid) links. Movement comes from rotation and sliding between links. Soft robots do not contain rigid links. Movement comes from large deformation of soft elements, much like that of an octopus. Modular self-repairing hard robots are more common than soft versions.
3. *Mobile:* Robots can maneuver in a 3D environment to assemble into a structure [Yim 2007] [Jorgensen 2004] [Spröwitz 2014].

Primary forms of redundancy are as follows:

1. *Component Redundancy:* Replace a failed component with a healthy one. Arrays of homogeneous modules facilitate the replacement.
2. *Functional Redundancy:* Perform the same task with more than one configuration of robot components [Murata 2001].

Attributes of self-repairing arrays of modular robots are as follows:

1. *Connectivity:* Modules connect to other modules in specified locations.
2. *Mobility:* Modules move to different positions in the array.
3. *Redundancy:* Redundant modules are available to use as replacement modules in a repair.

The control strategies for modular robots are generally global top-down, local bottom-up, or hybrid combinations of global and local control. In general, selecting the optimum reconfiguration plan for a modular robot is computationally intensive with the number of steps increasing exponentially with the number of modules, that is, it is NP-complete [Hou 2014]. Nonetheless, it is possible to obtain near optimal reconfiguration solutions in a reasonable amount of time, using graph theory and related approaches.

Top-down control runs into difficulties with communication and computational complexity. The number of possible configurations versus the number of modules often grows faster than polynomial rates [Chirikjian 1996]. Relatively simple local control schemes, such as those based on preprogrammed connectivity requirements for individual modules, along with dynamic stirring methods that prevent deadlocks, create self-healing and self-assembling structural arrangements [Yoshida 1999] [White 2009].

The following are the strategies and characteristics by which a team of robots act to maintain their capability through a combination of team-based diagnosis and repair [Kutzer 2008]:

1. *Robustness:* The failure of a single robot does not disable the entire team.
2. *Diagnosis:* Robots self-diagnose their own health and diagnose the health of other team members. Diagnosis checks the ability to perform test tasks and infers health from the results.
3. *Plan:* Form a plan of repair, including a triage scheme for removal, repair, and leave as is the robot modules.
4. *Removal:* Remove unrepairable robots from duty.
5. *Repair:* Repairable robots undergo a repair in which other team members fetch replacement parts from a repository. The replacement parts are inserted into the damaged robots.

9.4.2 Passive Modes and Component Switching

Passive mode switching uses damage or deformation to drive the switching of the system behavior without direct cognitive intervention. Passive mode switching typically uses a direct cascading of load paths as damage occurs. A biological example is the byssus threads that latch mussels onto rocks in harsh surf conditions and use networks of stiff and compliant components. The stiff components require energy to break, and the compliant components hold the stiff components together after breakage [Qin 2013].

9.4.3 Active Operational Mode and Component Switching

Active operational mode switching senses damage, decides what to switch, implements the switch, and evaluates performance of the new configuration to see if the component swap is adequate. Alteration of operational behavior, that is, limping, requires more sophistication than passive techniques. Airplanes, having intense safety requirements, routinely use active switching techniques. If a hydraulic pump for aileron actuation fails, active switching swaps the failed pump with a redundant preconnected standby unit, and the flight may continue with minimal change in operation. However, the use of the redundant standby pump further reduces redundancy. Prudence may weigh in favor of quickly landing and replacing the broken pump before proceeding further.

9.4.3.1 Contingency Planning

Contingency planning identifies possible future faults and puts in place a plan for correcting and recovering from the fault. The main steps are as follows:

1. Identify possible future faults.
2. Create a plan for recovering from a fault.
3. Implement fault recovery capability into system.
4. Operate the system.
5. Check for and identify faults as they occur in operation.
6. The system selects fault recovery based on an original map of fault onto contingency plans.
7. System implements recovery plan.
8. System operates in a recovered mode, possibly in a degraded limping mode.

Contingency planning for redundant component switching works well for systems with well-understood failure modes. Some examples:

- Figure 9.13 shows an automated contingency planning system for aircraft propulsion systems through the integration of adaptive intelligent controllers (AICs) into an automated contingency and life management (ACLM) system [Kallappa 2005].
- Spacecraft use arrays of gyroscopic rotors for inertial navigation and orientation control. Determination of orientation and actuation used to point spacecraft in arbitrary 3D directions requires a minimum of three rotors with non-coplanar axes. Adding more rotors increases redundancy. An optimal configuration for six sensors and six gyroscopes aligns the axes as normal to the faces of a dodecahedron (Figure 9.14) [Gilmore 1972]. In the event of failure of one, two, or three rotors, the remaining components have an appropriate non-coplanar configuration to provide appropriate sensing and control. It is possible to regenerate levels of redundancy by simply adding a rotor back into the control system as it heals itself.
- Cables with bundles of parallel electrically conductive wires support redundant switching for connectivity fault recovery. Damage to an individual wire may disable an entire system. Being

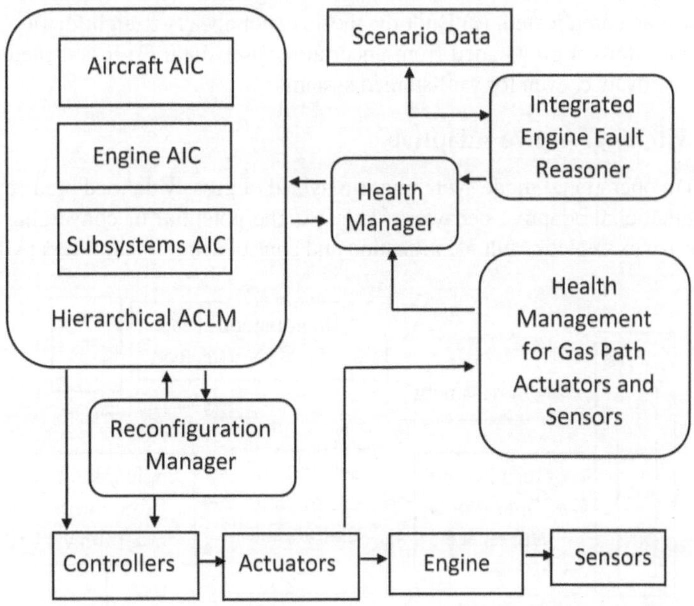

FIGURE 9.13 Aircraft propulsion system contingency planning architecture. (Adapted from [Kallappa 2005].)

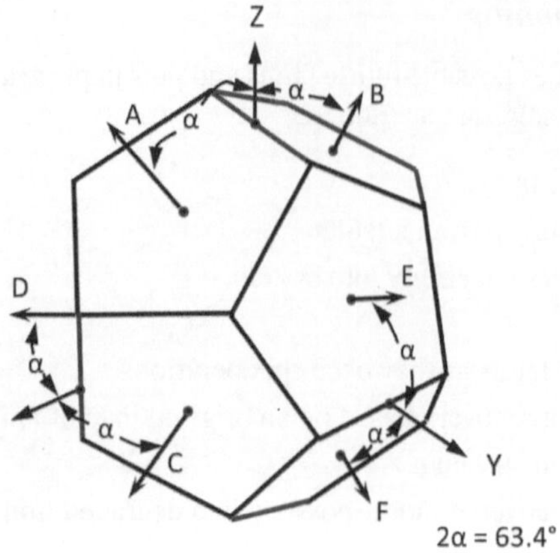

FIGURE 9.14 Optimal axis configuration for redundant rotation sensor and six-gyroscope inertial orientation control system. The geometry is of a dodecahedron with axes aligned normal to the faces. (Adapted from [Gilmore 1972].)

buried inside a cable bundle often makes it difficult to diagnose and repair the faulty cable. Removing the system from service and replacing the entire cable bundle and harness may be more economical than debugging and fixing an individual wire. An alternative is a fault-detecting and route-switching cable system that relies on redundant wires in the bundle to switch connectivity to carry the electric signals and power as needed [Medelius 2013].

The control and switching of multiengine airplanes with failed engines are complicated due to the number of possible scenarios. A map derived from graph theory provides guidance (Figure 9.15). An example is a multiengine airplane with three engines. A simplified operational representation is that each engine is either ok (1) or not ok (0). There are $2^3 = 8$ possible engine states and at least 4 different flight states – taxiing, takeoff, cruising, and landing, leading to a maximum of $8 \times 4 = 32$ operational states. Not all configurations are viable, which reduces the solution space. An airplane cannot taxi or takeoff with all three engines in the 0-state, that is, not working.

Implementing contingency planning has challenges: (1) Systems with even modest levels of sophistication quickly become complicated. (2) Fault diagnosis is not always straightforward, and instead uses reasoning based on information gleaned from uncertain sensor data. (3) It is difficult to anticipate all failure scenarios in advance, even for well-studied systems.

9.4.4 Mode Switching: Active Adaptive

A third approach to operational mode switching is a hybrid of grossly defined predetermined behaviors combined with situational adaptive behaviors. This has the potential of converging more quickly on optimal behavior. An example is fault identification and control using linear filters [Mladenov 2008].

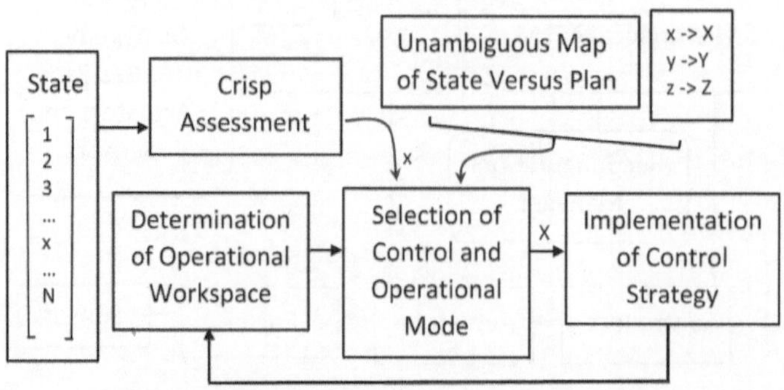

FIGURE 9.15 Predetermined self-referencing system operation that enables limping.

A more complicated problem for aircraft is coping with damage to flight control systems. Redundancy in the control system permits flying in severely damaged conditions by rapidly altering control algorithms. Motivated by cases of pilots successfully flying severely damaged aircraft in unanticipated manners, automated self-repairing flight control systems emerged with some efforts dating back to the early 1960s [Tomayko 2003] [Amin 2005] [Burken 2006] [Deng 2006]. This led to an effort to replicate the pilot's skill with a neural network approach known as the damage-adaptive intelligent flight control system (IFCS).

Shipboard electric power distribution systems use active mode switching to deal with power disruptions and recovery. Disruptions to available power sources and/or the distribution network mandate selectively removing loads and/or sectors of the power grid. Similarly, power restoration selects which loads to add back to and the timing of the additions. Competing considerations complicate the choice. through Expert system, nonlinear optimization, and machine learning techniques can automate the priority ranking of loads [Ding 2009] [Bose 2013].

9.4.5 Hierarchical Actuators

Recruiting redundant subcomponents to function as needed coordinates repair. One example is the bio-inspired hierarchical actuator (HA) [Huston 2005]. An interesting feature of biological muscles is the coordination of individual molecular motors with triggered non-smooth pulsatile forces into an assembly with smooth, variable speed, and variable force motion. Enabling this force transformation is a multi-scale hierarchical physical and functional organization that runs from the molecular scale to the macroscopic with myosin molecular motors, sarcomeres, myofibrils, fibers, fascicles, and whole muscles. In addition to being actuators with unmatched capability, muscles self-heal through biological reconstruction, growth, and limping with load reallocation among the muscular subunits. HAs mimic the healing by limping and other aspects of control of biological muscles.

HAs come in a wide variety of possible actuator and subactuator configurations. One form is an actuator created from subactuators connected either in series or in parallel by whiffletree style links. Figure 9.16 shows 1-, 2-, and 3-level whiffletree actuators, along with a 3-level whiffletree in an asymmetric arrangement.

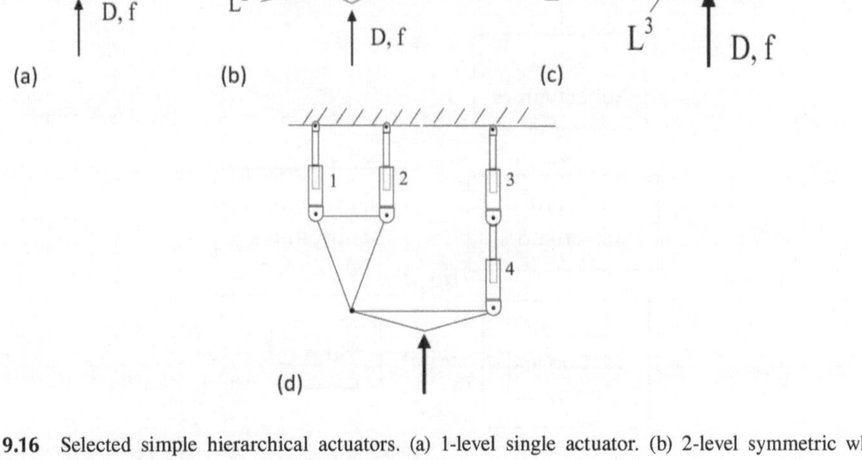

FIGURE 9.16 Selected simple hierarchical actuators. (a) 1-level single actuator. (b) 2-level symmetric whiffletree. (c) 3-level symmetric whiffletree. (d) Asymmetric whiffletree.

A simplified version of HA mechanics considers the lowest-level actuator to operate as either off and extended with the normalized displacement being 0 or on and retracted with the normalized displacement being 1. Using the notation s^j_i to indicate the state (displacement) of actuator i at level j, the displacement D and state matrix [D] of *1-level single actuator* is as follows:

$$D = s_1^1 = \left\{ \begin{array}{c} 0 = \text{OFF} \\ 1 = \text{ON} \end{array} \right\}; \quad [D] = \begin{bmatrix} s_1^1 \\ 0 \\ 1 \end{bmatrix} \tag{9.1}$$

A symmetric *2-level whiffletree* has two subactuators and a displacement averaging lever (Figure 9.16c). The actuator has four possible states and three possible displacements with two states being redundant and producing the same displacement:

$$D = s_1^2 = \frac{1}{2}\left(s_1^1 + s_2^1\right); \quad [D] = \begin{bmatrix} s_1^2 & s_1^1 & s_2^1 \\ 0 & 0 & 0 \\ 1/2 & 1 & 0 \\ 1/2 & 0 & 1 \\ 1 & 1 & 1 \end{bmatrix} \tag{9.2}$$

$$D = 0\,(1\text{ state}) \tag{9.3}$$

$$D = 1/2\,(2\text{ states}) \tag{9.4}$$

$$D = 1\,(1\text{ state}) \tag{9.5}$$

The *3-level whiffletree* actuator has two layers of subactuators. There are four 1-level subactuators, 16 possible states, and five different possible displacements. For an *asymmetric 3-level* actuator, as in Figure 9.16d, there are three 1-level actuators, eight possible states, and five possible displacements. The redundancy of a hierarchical actuator enables recovery from a subactuator failure by recruiting additional subactuators in an on-demand fashion that mimics the control of muscles in animals (Figure 9.17). Experiments demonstrate recovery from selectively damaged subactuators (Figure 9.18).

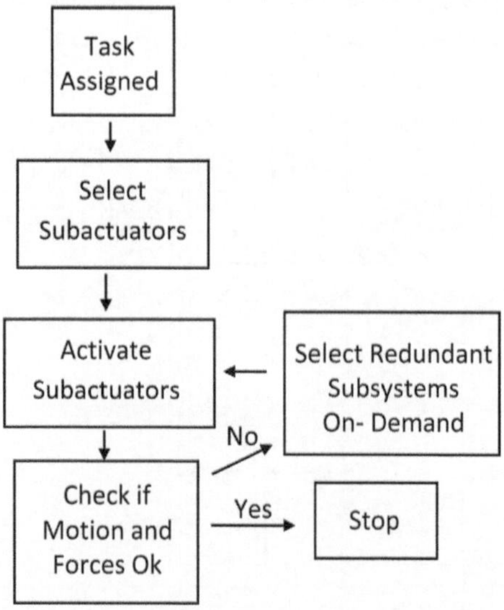

FIGURE 9.17 Algorithm for on-demand recruitment of subactuators in hierarchical actuator, as a muscle bio mimic.

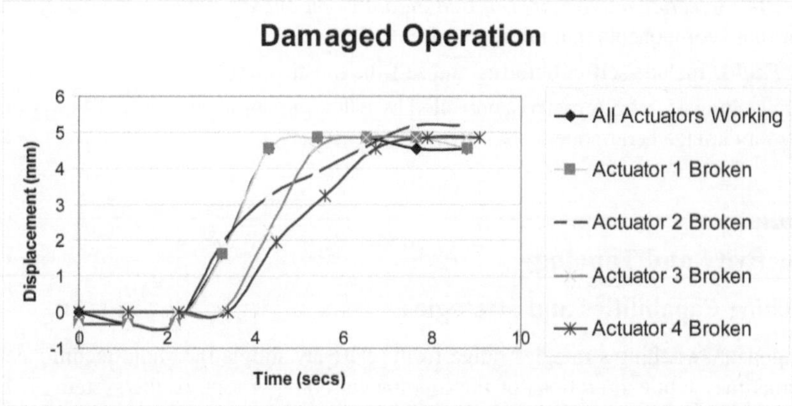

FIGURE 9.18 Automated damage recovery by subactuator recruitment on demand for 2-level symmetric hierarchical actuator.

Electronic systems implement similar strategies. Systems that switch at the component level provide a low-cost alternative to the high cost of fully replaceable redundant systems. Figure 9.19 is an example of a potentially viable switching strategy that considers known failure rates of components in a data acquisition system [Medelius 2002]:

1. *Determine Failure Rates of Individual Components:* The failure rates may vary between components. For example, components that comprise the analog inputs in a data acquisition system may experience unexpected transients and fail more often than internal digital signal processing components.

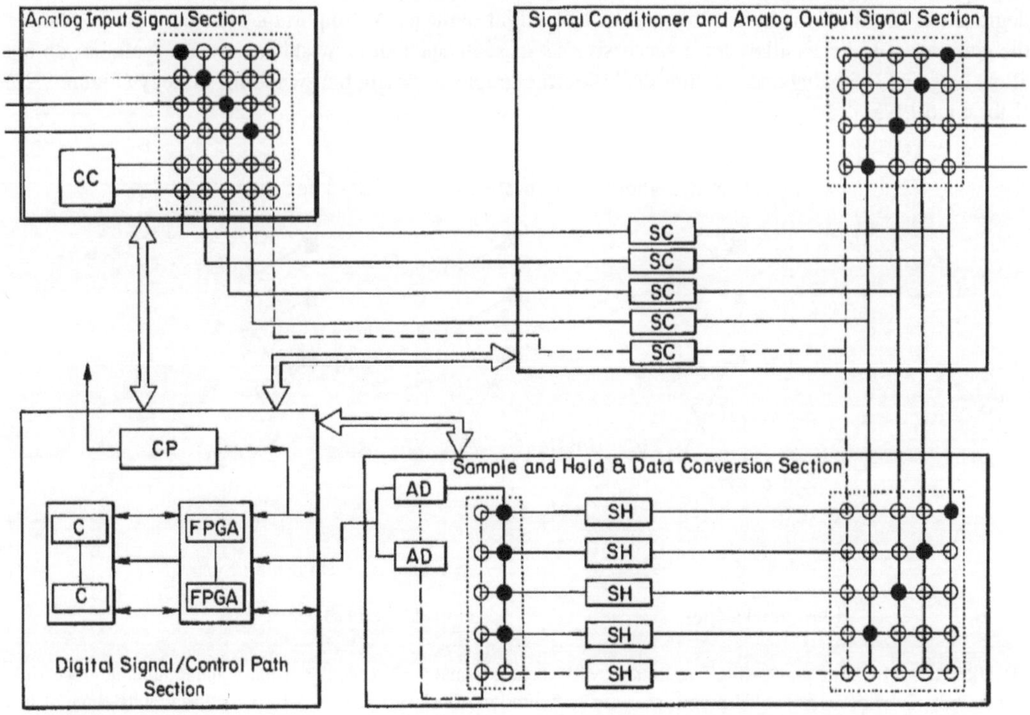

FIGURE 9.19 Electronic component switching matrix controlled by FPGA. (From [Medelius 2002].)

2. *Balance the Selection and Number of Redundant Components:* Select redundancy to compensate for those components that fail more often.
3. *Detect Faults:* Include self-calibrating and self-diagnostic circuitry.
4. *Use a Switching Matrix:* A matrix controlled by a floating point gate array (FPGA) switches in healthy for damaged components [Akoglu 2009].

9.5 Connectivity and Topology

9.5.1 Switching Capabilities and Strategies

Possible component switching strategies range from relatively simple threshold techniques to sophisticated and sometimes subtle alterations of the topological configuration of the system. The damage or intelligent controller drives the switching options and actuation. Component switching is hysteretic and history dependent.

Aircraft and spacecraft employ redundant switching self-healing techniques to improve system reliability [Eckhoff 2001]. The control systems are inherently complex, but open versions promote rapid development and programming [Wills 2001].

9.5.2 Component Switching, Connectivity Control, and Limping

Switching loads from damaged to undamaged components restore health and functionality. The switch may be a physical replacement of the damaged component, may redirect the effort onto prepositioned undamaged components, or may be a change in task. Component switching may produce a system with fully restored functionality or may leave the system operational at a reduced level of functionality. In either case, if the system continues to operate without replenishment, it has reduced redundancy for future damage recovery efforts.

In general, switching needs a plan. Otherwise, switching runs the risk of being ineffective, and even degrade the health of the system beyond that of initial damage. An important consideration is whether the component switches allow for a successive set of switches, that is, whether they are "safe" or do not allow for more switching and are "unsafe." Bipartite graphs represent the switching strategy (Figure 9.20) [Libeskind 1995].

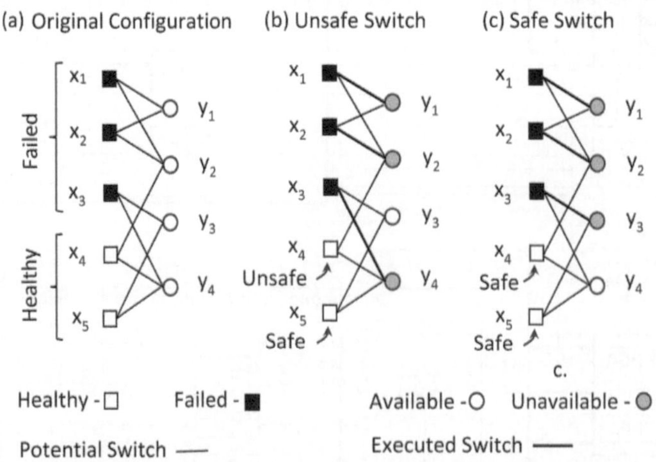

FIGURE 9.20 Bipartite graph switching strategy with vertices x_i and y_i representing primary and redundant processors, respectively. Solid boxes are failed processors. (a) Configuration of original system as bipartite graph with three of five primary processors failed. (b) Replacement strategy that leaves x_4 in an unsafe condition. (c) Safe replacement strategy that leaves both x_4 and x_5 in a safe condition. (Adapted from [Libeskind 1995].)

If switching is easy and low cost, then it is reasonable to switch quickly upon detecting an anomaly [Takahashi 2010a]. A concern is that sensor failure may cause a controller to misinterpret the state of the system and lead to a cascade of actions or inactions that disable the system. Rapidly detecting sensor anomalies and switching a redundant sensor into the control loop alleviates this vulnerability. If the sensor detects an anomaly in the system, then switching in a new sensor provides the same information and the controller can act with confidence on correcting the situation. If the anomaly is in the sensor, then switching sensors gives a different signal and the controller ignores the signal from the first sensor. This assumes that the second sensor is healthy. More complicated multi-sensor voting schemes and the use of heterogeneous sensor types with different operating principles and failure modes increases confidence in the results.

Multi-legged walking robots with defective legs compensate to walk by gait alteration. Four-legged robots walk and run with three legs with a suitable change in gait. The kick-and-swing three-legged gait is in many aspects like running or galloping. The gait first elevates the body with a kick to take the weight off the isolated leg and then swings it forward while airborne. Such a gait can be a preprogrammed contingency limping strategy [Lee 2002].

Systems that rely on the flow of material, forces, information, and/or energy between components are particularly sensitive to changes in topological and connectivity. Electrical systems rely on the flow of electrons or voltage between components for operation. Changing the connectivity of electrical components may cause short circuits with rapid flows of current or create an open circuit where no electricity flows. In structural systems, the addition of a single connection may convert a statically determinant structure into a redundant statically indeterminant system. Conversely, breaking a single connection in a statically determinate structure forms a kinematic mechanism and may lead to collapse.

Topological control that restores the connectivity to damaged components is a potentially powerful facilitator of self-healing. Sensing loss of connectivity feeds into a topological self-repair control loop. Connections that retain some degree of connectivity following damage may use the remaining connectivity as a scaffolding for operations that restore full connectivity. Similarly, recovering from situations with complete loss of connectivity often benefits from first establishing connectivity with a light pilot line as an initial scaffolding.

Representations with graph theory describe and quantify the connectivity between components in a system. Bond graphs contain more information such as quantifying the flow of material, energy, and/or information between components.

Possible autonomous changes to topological configuration include the following:

1. *Decoupling:* Decouple components in a mitigation maneuver to prevent further damage. An example is breakaway lines on motor vehicle fuel dispensers, that is, gasoline pumps at filling stations. Excessive force causes the line to separate and a valve to trip and stop the fuel flow. These devices prevent large fuel spills that happen when a car pulls away from a pump with the fuel nozzle and line accidentally left inserted into the vehicle (Figure 9.21) [Healy 1994].

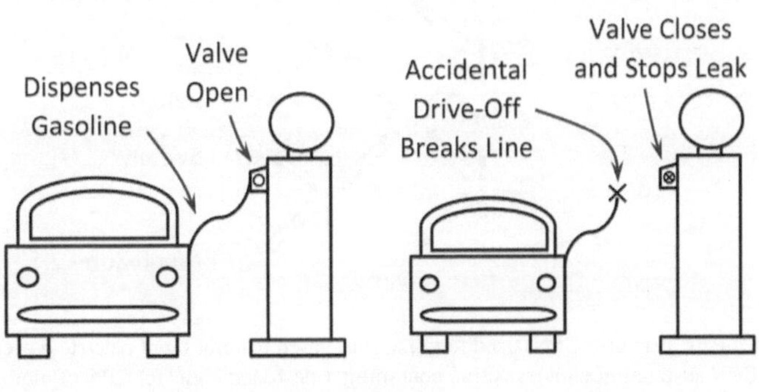

FIGURE 9.21 Breakaway valve on gasoline pump prevents major leak and possible fire.

2. *Partial Decoupling:* Decouple, or partially decouple, components, yet leave the system in a functional state, but with degraded performance. This is limping. An example is a two-engine airplane that loses power in one engine. The remaining operational engine has enough thrust to fly the airplane, but usually has degraded fuel efficiency and flight characteristics.
3. *New Coupling:* Heal by creating new connections.
4. *Altered Coupling:* Heal by altering the strength of a connection.

Bolted connections are an example. Bolts require preloading for solid mechanical contact and tolerance of fluctuating loads. Loss of preload causes rattling, fatigue, loss of friction in the joint, increased load on the bolt, and loosening of nuts. While there are a variety of passive techniques for tightening the bolt, these are almost all preventative and do not provide the means of autonomically recovering from lost tension. Automated tightening systems recover bolt tightness, such as a combined system with piezoelectric sensors that detect looseness and an SMA-laden washer to swell and tighten the connection [Peairs 2003] [Park 2003a]. Similarly, SMA bolt tightening suppresses transverse cracks in FRP airplane fuselage structures [Ogisu 2003].

Networked systems being intimately dependent on component connections routinely alter connectivity for damage recovery. How to choose an optimal solution for recovery? An example is electric power distribution networks. Damage to power sources and the interconnections causes power outages. Recovery coordinates combinations of topological reconfiguration and fault-repair. Reconfiguration with on/off switching operations isolates faults, restores power to nondamaged areas, and removes power from low-priority areas. Deciding how to switch is not always obvious. The controller can use predetermined criteria or can sense conditions and choose a more optimal configuration, perhaps by solving a nonlinear mixed-integer problem. An omniscient controller that knows the health and operation of all components of the network is effective, but difficult to implement, especially with large-scale networks. Such optimization problems become intractable for large networks, but may be manageable for smaller networks, such as appear on a ship [Khushalani 2008]. Petri net models develop rules for recovery in electric power distribution systems based on sequentially connecting isolated, healthy, out-of-service lines (Figure 9.22) [Chen 2002a].

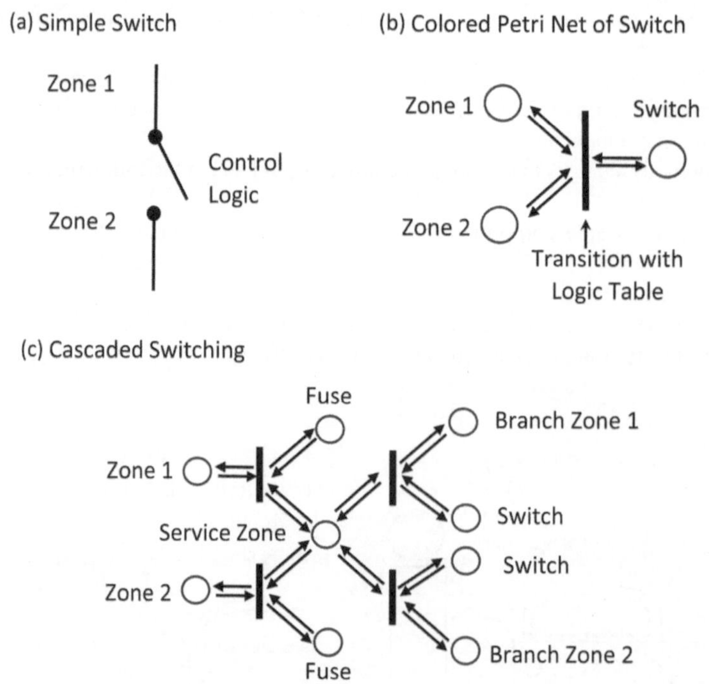

FIGURE 9.22 Colored Petri network (CPN) used to guide rule-based restoration of electric power distribution network. (a) Simple switch. (b) CPN abstraction with transition containing rule-based logic. (c) CPN of more complicated cascaded switch. (Adapted from [Chen 2002a].)

9.6 Networks

Networks are distributed systems of interconnected components and subsystems. A feature of networks is that the interconnections between the various components govern large parts of the overall behavior. Small changes to the interconnections sometimes radically alter the behavior of the network. Such behavior enables switching to redundant transmission paths for self-healing. The following are some of the common issues:

1. *Redundant Capacity:* Networks need redundant capacity for switching healing.
2. *Independence and Dependence:* Network components are often independent in failure behavior, but in many situations the unmanaged load switching causes a local failure of one component to promote a cascading failure to other components.
3. *Self-diagnosis:* The network detects failure of transmission in components.
4. *Plasticity:* The network has the ability to reconfigure itself and direct flow in new directions.
5. *Timely Switching:* The network switches in a timely manner to a healed state.
6. *Stability:* The network has a stable and effective algorithm for assigning the new configuration when damaged.
7. *Hysteresis:* Switching processes have sufficient hysteresis to prevent excess switching.

9.6.1 Examples of Self-healing Network Systems

1. *Hybrid Fiber-Optic Coaxial Information Transmission:* These networks use multiband transmission and side leg transmission paths for redundant capacity [Bhagavath 1998].
2. *Tensegrity Structures:* These are active truss-like structural systems comprised of strut compression members and actuatable cable tension members. The structural form is a complicated 3D network that reconfigures itself by changing the lengths of the individual cables [Snelson 1960] [Skelton 2001] [Sultan 2001]. Most tensegrity structures are statically indeterminate with redundant load paths (Figure 9.23). Following damage, reconfiguration of the structural form and associated load paths sustains the load in stable configurations. Selecting a control strategy is a challenge. A successful implementation is tube-shaped tensegrity structures used in foot bridges. The analysis considers damage scenarios and possible control strategies with multiobjective optimization methods [Korkmaz 2012]. The footbridge structure recovers to a safe and serviceable state for many, but not all, of the damage states considered.
3. *Wireless Sensor Networks:* Such networks can change the connection configurations with ease. Figure 9.24 shows a multi-hop mesh network that autonomously reconfigures itself following network damage to remove sensor nodes from operation based on a periodic, event-driven, and query-based protocol [Boukerche 2006].

The choice of switching strategy depends on the failure modes and task requirements. A versatile approach fits a system with a redundant array of identical programmable components, each of which performs a variety of tasks. In a nonlimping mode, the system allocates separate tasks to the components.

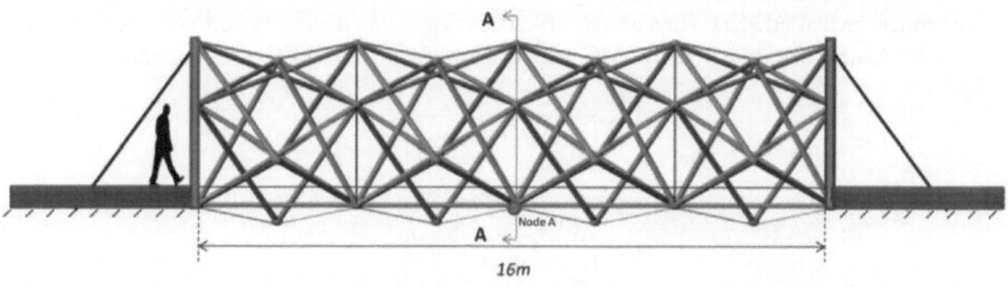

FIGURE 9.23 Tensegrity footbridge structure. (From [Korkmaz 2012].)

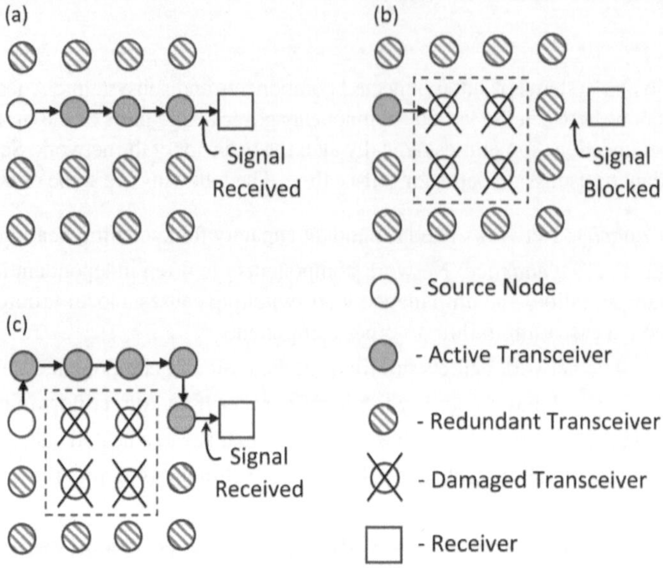

FIGURE 9.24 A multi-hop mesh network that autonomously reconfigures itself following network damage that removes sensor nodes from operation based on a periodic, event-driven, and query-based protocol. (a) Healthy signal path. (b) Damage blocks signal. (c) Rerouting heals path. (Adapted from [Boukerche 2006].)

Some components may sit idle. When damage disables individual components, the system switches to a limping mode, discards or removes from duty failed components, and allocates tasks to the remaining components, some of which may multitask at reduced performance. Voting techniques help to identify failed components from faulty outputs [Shuler 2010].

9.7 Sense, Repair, Monitor

9.7.1 Triggered Healing

Simple controls trigger self-healing (Figure 9.25).

When signals from damage exceed specified threshold values, the sensing identifies and determines the location of the damage. The actuators release and direct healing action at the damaged location. Some example applications are as follows:

1. *Thermal Actuation of Healing:* Thermal activation can be a simple healing technique. Upon sensing damage, the system triggers a heater that raises the temperature of a healing material to melt and flow or otherwise change state to heal the structure. Localized heating is favorable and must be at a level that does not damage the structure. Healing with heat-controlled reflowable polymer, such as ethylene vinyl acetate (EVA), is viable for concrete beams and structural panels. EVA melts at approximately 93°C, which is lower than the 100°C that boils entrained water

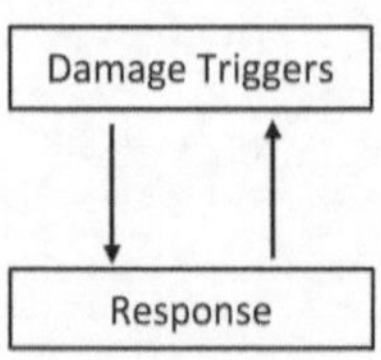

FIGURE 9.25 Simple trigger and response coordination of health.

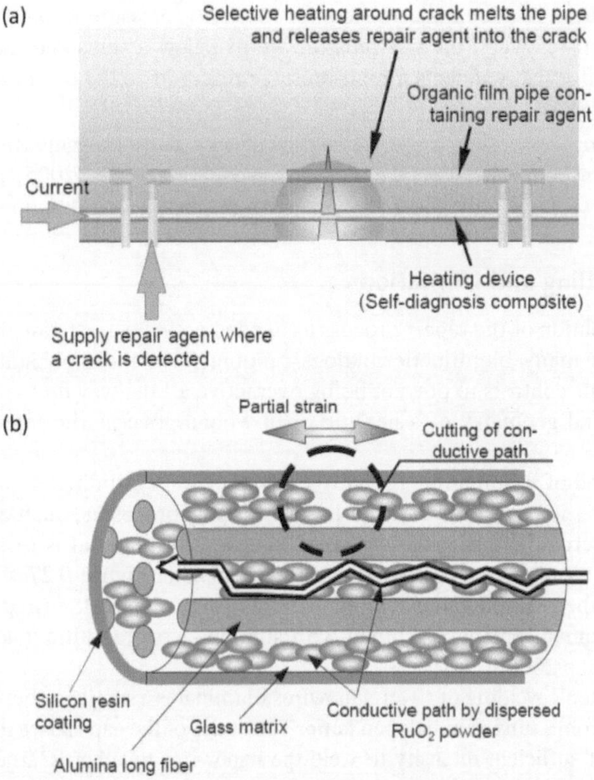

FIGURE 9.26 Localized heating due to damage of multipath conductor prompts vascular concrete repair. (a) Healing system. (b) Multipath conductor based on percolating conductive RuO_2. (From [Nishiwaki 2006].)

and damages concrete [Nishiwaki 2006]. Using EVA for healing runs into practical issues of sensitivity and control, that is, how, when, and where to heal? One solution uses logic inherent in the network with specialized multipath conductors, which lose some of the paths, but not all of the continuity, when damaged. The remaining conductive paths carry more current and produce Joule heating at locations corresponding to localized damage. This initiates a localized reflowable repair (Figure 9.26). Similar systems use arrays of point sensors to locate the damage and heating loops to provide localized heating and healing response through combinations of effects, such as annealing, shape memory materials, polymer reflow, and increased reptation rates [Takahashi 2010b] [Hemmelgarn 2014].

2. *Internal Forces:* Placing piezoelectric patches on the outer faces of a composite beam with internal delamination provides an active bending moment to counteract the loss of stiffness [Wang 2004b] [Duan 2008].
3. *Microencapsulated Delivery of Healing Liquids:* Damage or damage precursors trigger the microcapsules to rupture and release the healing liquid. An example is capsules embedded in reinforced concrete that are functionalized to rupture in the presence of corrosion-inducing chloride ions [Xiong 2015].
4. *Controlled Vascular Flow:* Effective application of healing liquids requires methods of controlling vascular flow, such as with localized or district pumping and/or damage-induced leakage. Fiber-reinforced composite panels often suffer from delamination. Integrated fiber-optic sensors or a network of electrically percolating carbon nanotubes track such damage and provide information for directing healing liquids through a hierarchical vascular system [Minakuchi 2014] [Wu 2012a].

5. *Self-inflating Tires:* A pneumatic tire equipped with a pressure sensor and high-pressure reservoir self-inflates when the tire pressure drops below a threshold. Sequential opening and closing a valve that connects a high-pressure reservoir to the tire replenishes the air as needed [Olney 1994].
6. *Self-cleaning Surfaces:* Issues of sensing and feedback control arise in self-cleaning surfaces. An example is traveling wave electrostatic dust removal [Mazumder 2005]. Sensing dusty conditions and then cleaning only when necessary reduces energy consumption.

9.7.2 Repair Throttling and Limitation

Throttling and termination of the repair process as needed prevents waste and harmful over repairs. Repair throttling has many biomimetic analogs ranging from the molecular scale where DNA repair machinery needs controls to prevent being overactive all the way up to the macroscale where the changing stress and geometry of a healing open wound governs the rate and type of healing [Ohouo 2013].

A convenient method of terminating a repair is to supply only a limited amount of repair materials and then stop. This is typical of damage-initiated repair processes, such as microencapsulated liquid healing. An alternative that makes more efficient use of material is to monitor the repair as it proceeds and then terminates the repair once it is complete. Figure 9.27 shows a thermoplastic panel with controlled thermal reflowing healing using subsurface optical sensing to detect the damage [Hurley 2011a]. Capacitive sensing combined with sunlight-activated ultraviolet curable epoxy is a viable approach [Carlson 2006b].

Light-induced plasmonic welding of silver nanowires illuminates a small gap between silver nanowires with broadband light from a tungsten–halogen lamp. The optics of the gap causes the light to concentrate and create a plasma of sufficient intensity to weld the nanowires together [Garnett 2012]. Completion of the welding process changes the optics of the gap so that it no longer concentrates the light and stops feeding the plasma, which then extinguishes.

The living polymer technique provides a method of controlling and limiting the extent of healing reactions. Quenching living polymer reactions on cue is possible. ROMP is one of the possible

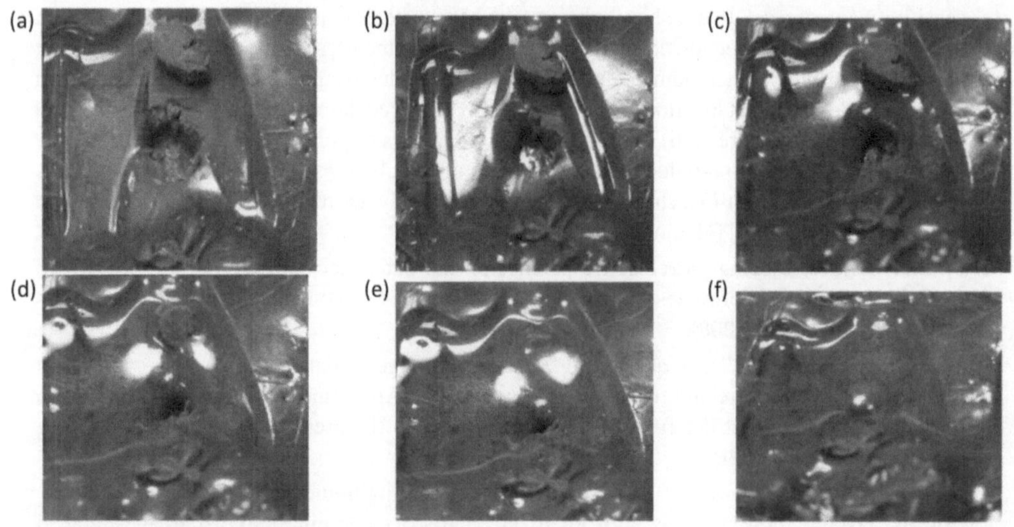

FIGURE 9.27 Coordinated healing of a panel made of reflowable thermoplastic. Subsurface optical sensors detect the damage, initiate a heating-controlled healing response, and then terminate once the sensors detect a healed condition. Sequence starts in top-left panel with (a) perforating gouge wound, and (b–e) proceeds sequentially to (f) a healed state. (Adapted from [Hurley 2011a].)

controllable reactions using the timed addition of a reagent that specifically removes the metal catalyst from the end of the polymer molecule and replaces it with a molecule that arrests the addition of more mers [Bielawski 2007].

9.7.3 Circadian Healing and Going Down for Repairs

Healing a system while remaining operational is often impractical. Shutting down for repairs is an alternative. Shutdown may occur as needed or on a periodic basis, as with circadian sleep rhythms in biological systems. In particular, it appears that brains and neural systems need to sleep to function normally. The cross-layer accelerated self-healing (CLASH) system is a system that heals computer chips from bias temperature instability (BTI). CLASH uses an extended period of operation (31 h) followed by a short repair cycle (1 h) with high temperatures and reverse voltage to heal BTI [Guo 2017]. Another example is copper-based SMA structural members that cycle through superelastic deformations and build-up damage with each load cycle. Thermal cycling heals the SMA. Built-in thermal cycling that activates after a present number of load cycles heals the system on a time of usage basis [Casciati 2007].

9.8 Wound-closing Control

Wound closing uses mechanical forces, material deformations, and material movements to close cuts, perforations, and other forms of localized surface breaches. The intent is to accelerate repairs by minimizing the volume of material repair required. Wound closing combines sensing and control [Duenas 2007]. Subtleties include working with the statically indeterminate nature of autonomically generated internal stresses and deformation.

Wound closing is bioinspired. Virtually all forms of life, unicellular and multicellular, use wound healing to survive from injuries that breach outer or internal tissue barriers. Wound healing in mammals migrates epithelial cells from the edge of an open wound inward to cover and heal the wound. Experiments supported by agent-based numerical calculations indicate that much of the movement has the epithelial cells crawl on subcutaneous tissue into the unconstrained direction of the open wound as a mechanical wave [Bindschadler 2007] [Serra 2012].

Autonomic repair of delamination benefits from wound closing. Delamination between fiber layers is a principal failure mode of fiber-reinforced polymer panels. Closing a delamination requires applying forces with significant components directed perpendicular to the delamination. Closing the gap with intrinsic actions is challenging due to the typical geometry of elongated structural elements with delaminations running lengthwise (Figure 9.28). Weaving SMA wire into the fiber layup that causes the wires to contract when heated and pull to close the delamination. This technique combines with reptation healing of thermoplastics for a synergistic healing effect (Figure 9.29) [Bor 2010].

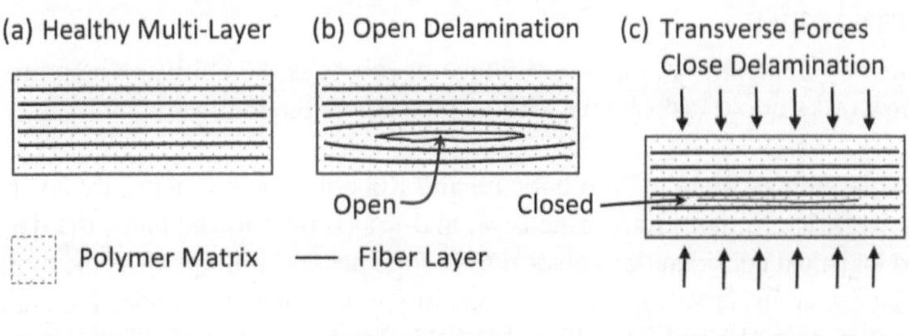

FIGURE 9.28 (a) Fiber-reinforced composite element in healthy state. (b) Delamination forms between fiber layers. (c) Closing of delamination by forces transverse to axis of element.

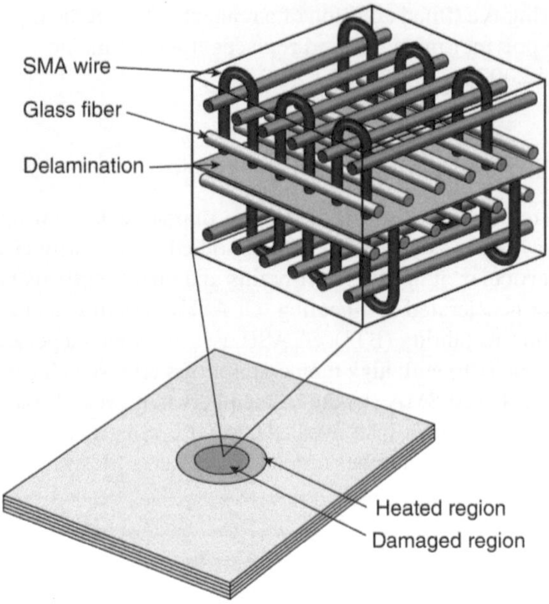

FIGURE 9.29 Shape memory alloy synergistically closes and reflows polymer with thermal effects to heal delamination in composite. (From [Bor 2010].)

9.9 Robotic and Robot Repair

Modern trends point to ever-increasing sophistication and ubiquity of robots – including the use of robots for repair [Kurzweil 2006]. Robots embedded as subsystems in larger systems can be agents for autonomic system repair, even forming self-repairing robots. Most of the reported repair robots for applications, such as bridge and spaceborne solar panels, are not yet fully autonomous, but instead rely on human intervention for operation [Lim 2008] [Oghenekevwe 2009].

Some principles for robot repair and maintenance:

1. *Competence:* Robot must be able to execute the desired maintenance and repair tasks.
2. *Mobility:* Robots must be able to move to damaged regions, and then vacate the premises, as needed.
3. *Safe:* Robots should be inherently safe and pose minimal risk of collateral damage.
4. *Damage Assessment and Repair Planning:* Autonomous robots must be able to assess the damage and formulate a plan for repair. Semiautonomous robots that include humans as part of the control loop involve consideration of human factors and organizational factors, such as technical standards and allowing for redundant human controllers [Rembala 2009].
5. *Quality Control:* Robots need to conduct quality control and inspection procedures on their repairs [Höfener 2014].
6. *Improvement by Learning:* A major advantage of robots is the ability to be reprogrammed to perform new tasks or old tasks better without changes to the hardware.

This raises the question as when to do robotic repair? Robotic repair systems are favored in those situations where the work is too dangerous, expensive, and impractical for humans, the damage modes are well-defined, and logistical considerations favor robots. Pipe networks are perhaps the most common users of robotic self-repair systems, largely due to inaccessibility to humans, well-understood failure modes, and relatively simple geometries [Ehsani 2010] [Goodell 2012]. Inserting inflatable bladders and balloons into a damaged pipe, which can press repair patches against the interior walls and cure them in spot by heat or other agents, is an effective repair technique [Blackmore 2006]. Pavement repair is dangerous and potentially well suited for robots. Automated pavement crack repair robots use machine vision and pattern recognition to detect cracks, and then use an articulated multi-nozzle spraying system to seal the cracks [Holmes 2010]. The robotic repair of in-orbit satellites is a challenging, but potentially valuable, use of autonomous systems.

9.10 Universal Principles

Coordination raises interesting, and perhaps fundamental, questions.

9.10.1 Adaptability and Survival

Systems that survive for long times in the face of unpredictable and changing environments often share multiple characteristics:

1. *Vigilance and Autonomy:* Self-preservation requires vigilance and existing and emerging threats. Autonomy enables vigilance.
2. *Perception and Quantification of Safe versus Dangerous Conditions:* This allows for the appropriate level of response.
3. *Multilayered Multi-agent Defense and Repair System:* In humans, this consists of physical barriers, and the immune system with B-cells, T-cells, phagocytes, and so on.
4. *Adaptive and Innate Capabilities:* A suitable mix of adaptive and innate responses creates a highly resilient system. The continuous debate about the importance of nature versus nurture in the training of humans and animals confirms the strength of a combined approach.
5. *Self-recognition:* The recognition of self versus non-self guides the planning and execution of self-repair.
6. *Reversibility:* Perfect repair represents reversibility, even though intermediate steps are irreversible [Eppel 2014]
7. *Homeostasis:* Maintaining a healthy internal state by exchanging material, energy, and entropy with the environment.

9.10.2 Systems and Biomimetic Considerations

Coordinating self-healing benefits from considerations of biomimetic analogs, self-awareness, and perceived senses of healthy versus unhealthy states [Takeda 2003].

Idealized physical models consider time-invariant systems of infinite sources and sinks. Most real systems are nonideal and operate in environments with input and output actions that alter the surrounding environment and affect the internal state and performance of the system. Self-healing actions address the internal states. An important complication arises with uncertainty in the boundary between a system and the environment. A related important case examines the state and actions of healing subsystems within a system (Figure 9.30). Judiciously selecting the system boundaries is advantageous for analysis and design.

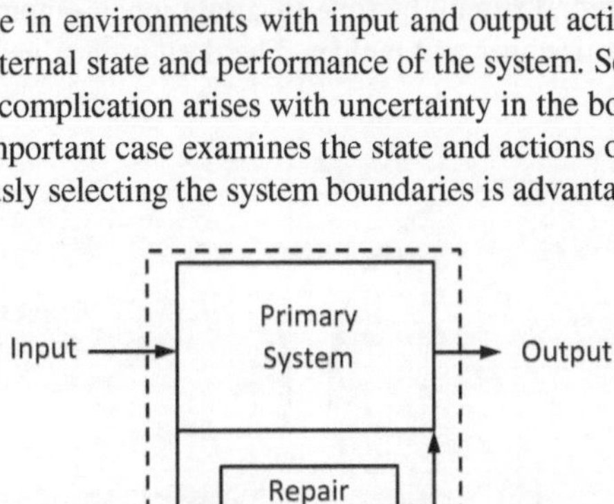

FIGURE 9.30 Primary system with repair subsystem.

An example is a ship at sea. Virtually all ships leak. Normally, the leaks are slow and manageable with automatic bilge pumps. Larger leaks may require more aggressive actions, such as sealing internal compartments and rearranging fuel or ballast water to stabilize and minimize list. If the ship sinks, survivors may take to lifeboats.

9.10.3 Sense of Self

Distinguishing self from non-self is crucial to self-healing and self-maintenance. There may be subtle issues:

1. If a system behaves as a set of agents acting together as a team, the team members must distinguish themselves in a threefold manner: (a) Distinguish themselves from members of other teams. (b) Distinguish themselves from members of their own team. (c) Recognize components and subsystems of themselves.
2. A system that autonomously constructs a network by linking individual components should link components preferentially to other components in a pattern that matches the needs of the network. A biological example is the formation of a network of neurons by neurite interconnections. A neuron must be able to distinguish between a neurite from itself and a neighboring neuron, even though both neurons come from the same cell line and have the same DNA. This is accomplished by the individual neurons identifying themselves as distinct with neuronal cell recognition proteins that differ from one another. The creation of these distinct proteins is a randomized combinational process that assembles proteins from thousands of isoforms [Hattori 2009]. A biomimetic variant is autonomic computer systems that use self-recognition as a basis for self-healing [Ganek 2003].
3. In an application aimed at protecting computer systems and networks, it is possible to use the human immune system to establish the framework of self-maintenance and resilience [Elsadig 2009].
4. A system for autonomous fighter aircraft uses an artificial immune system, with different settings corresponding to different levels of aggression [Kaneshige 2007].

The recognition of self from non-self is an inherent generator of complexity, [Zak 2007] [Hofstadter 1999]. The use of a map of self and non-self induces a cascade of reflective maps (Figure 9.31). Self-reflection produces complications and sometimes paradoxes (Figure 9.32).

9.10.4 Self-Diagnosis and Homeostasis

Systems that recognize their own state can interact with the environment to promote health. Relatively simple systems respond to external stimuli by time-invariant input–output relations (Figure 9.33). Such systems require an intelligent designer and builder. The design must perceive future conditions to be robust and resilient.

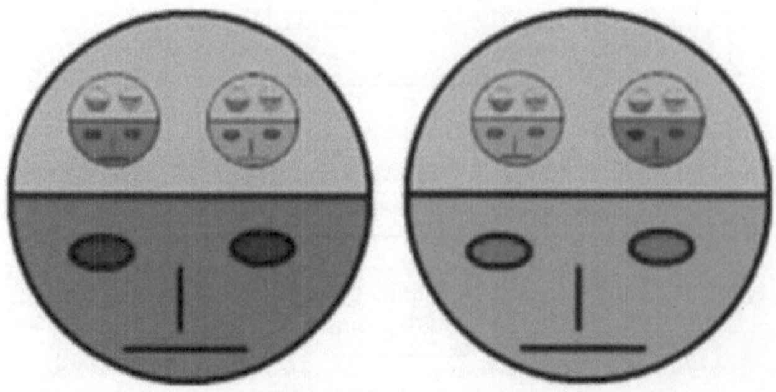

FIGURE 9.31 Chain of self-reflections. (From [Zak 2007].)

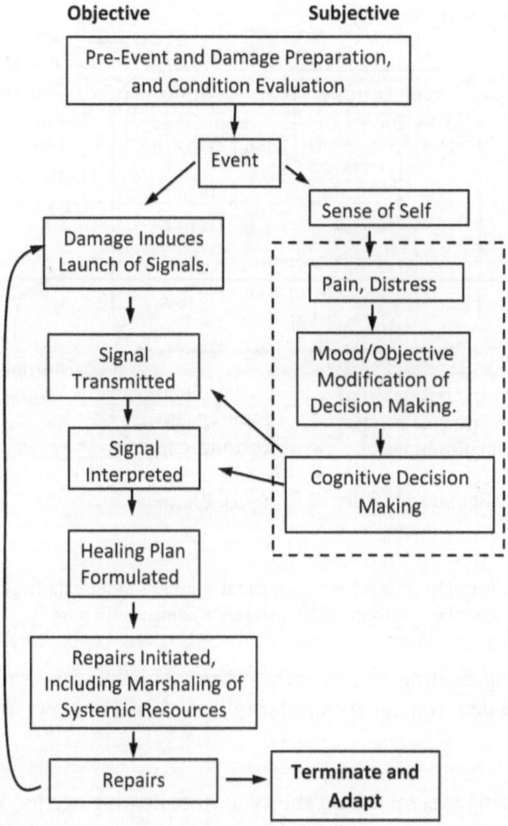

FIGURE 9.32 Chain of self-reflections increases complexity of system.

More sophisticated systems maintain homeostasis. Key steps include ingesting healing agents, storing healing agents, and exuding high-entropy waste material (Figure 9.34). Power management strategies for energy harvesting systems provide some guidance for designing and describing the behavior of these systems [Kansal 2007]. The available external supply of healing agents has the following forms:

1. *Uncontrollable but Predictable:* The supply of external healing agents is uncontrollable, but predictable and usually of sufficient quality and quantity to be useful.
2. *Uncontrollable and Unpredictable:* The supply of external healing agents is uncontrollable and unpredictable. Harvesting of healing agents is an adaptive and reactive process.
3. *Fully Controllable:* The supply of external healing agents is fully controllable and usually of sufficient quality and quantity to be useful.
4. *Partially Controllable:* The supply of external healing agents is partially controllable and usually of sufficient quality and quantity to be useful.

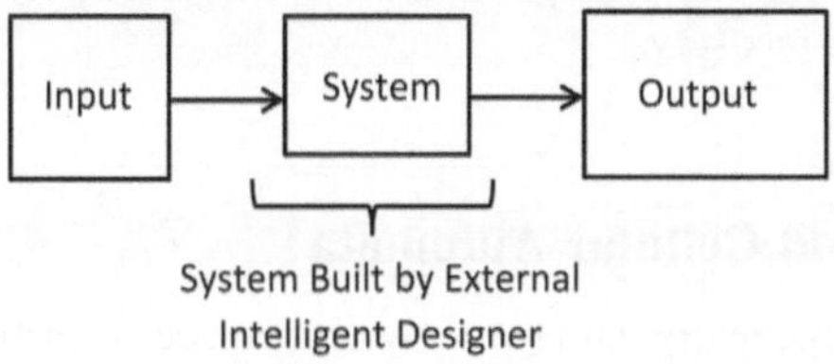

FIGURE 9.33 System built by external intelligent designer.

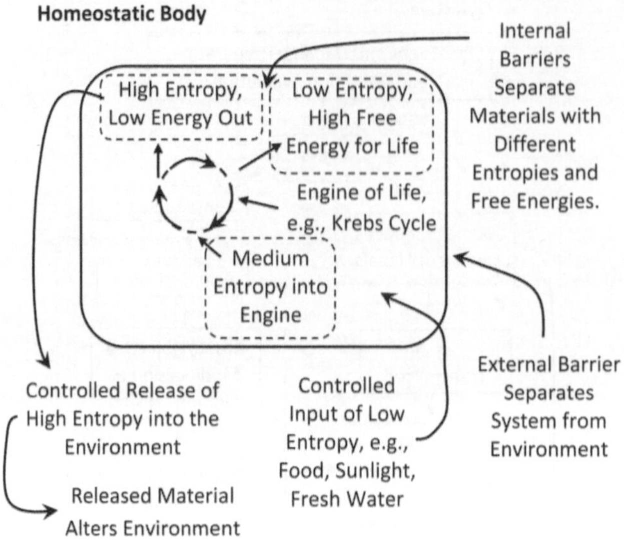

FIGURE 9.34 Living system uses low-entropy environmental sources to maintain healthy state. The process increases entropy, which must be ejected into the environment to maintain a homeostatic state.

Harvesting and ingesting healing agents makes them available for storage and then subsequently available for repairs as needed in manners similar to that of energy harvesting systems [Kansal 2007]. Possibilities for storage:

1. *No Storage:* Harvesting system has no ability to store healing agents. The agents must be used immediately.
2. *Ideal Buffer:* The harvesting system stores healing agents in a buffer that can store any amount of healing agent, release the agents for use whenever needed, and does not lose any healing capability in storage and release.
3. *Nonideal Buffer:* Harvesting system with nonideal buffer, with potential limitations on storage capacity, limitations on healing release rates, and degradation of healing capabilities in storage.

9.11 Templated Repair

Templates are preformed patterns that provide a plan for repair. The template may be a physical structure with the plan directly encoded in the structure, or a plan stored in the memory of an active system. Templates may be isolated entities or may be diffused and stored collectively throughout the structure. An example is material systems that retain the pattern for the healed system within the molecular and crystal structures. Diblock copolymers, such as poly(styrene-*b*-methylmethacrylate), exhibit this behavior with the formation of nanoscale crystalline patterns and structures form [Yufa 2009]. These materials recover completely from nanoscale damage, that is, a dent from a nanoindenter. A thermal cycle activates material degrees of freedom so that crystal patterns in the deformed polymer surrounding the dent provide the template to guide the recovery.

9.12 Self-replication and Cellular Automata

Self-replication of subsystems repairs damage to systems made of distinct subsystems. Multicellular biological systems routinely perform subsystem self-replication, but this remains a difficult challenge for human-built systems. von Neumann espoused much of the early theoretical development of self-replicating machines during the 1940s and 1960s [Neumann 1966]. The process has both physical and information components, many of which are inseparable. The machine must contain the information required to build a copy of itself and must replicate the same (or improved) information in the new machine. Improvement in terms of fitness for survival and breeding a new generation often comes with increased complexity, as evidenced by the higher organisms. An important aspect of biological self-replication is that living systems do not replicate in isolation, but instead thrive in an environment with many complex molecules already available for absorption and incorporation into the system. It is possible for robots to assemble copies of themselves from prebuilt incomplete components placed in the surrounding environment [Freitas 1982] [Suthakorn 2003] [Park 2004b]. Both living and human-built self-replicating systems must follow the laws of physics. Statistical mechanics being closely related to information theory provides a framework for describing the kinetics of self-assembly and can put thermodynamic bounds on self-replication [Napp 2006] [England 2013].

Self-replicating modular robots have received considerable interest as possible self-replicating machines. The concept is for the robots to have enough information to self-replicate with prefabricated parts available in the environment [Zykov 2007]. These efforts have been successful at relatively simple replications, usually in the form of snapping parts together. Measures of complexity, including Sanderson's parts entropy, quantify the level of complexity in the assembled final robot [Lee 2008].

Embryonics and unitronics are electronic extensions of the concepts of von Neumann's self-replicating machines to electronic systems [Mange 2000]. The approaches are biomimetic where the electronic systems act as multicellular organisms. These systems produce self-healing and fault-tolerant electronic systems based on mimicry of evolving biological organisms. Unitronic systems mimic the evolution and development of prokaryotic bacterial colonies, whereas embryonics mimic eukaryotic multicellular organisms [Samie 2013]. The complexity of unitronic systems is less than that of embryonic systems, yet can produce similar self-healing and fault-tolerant behaviors. Individual cells in embryonics take on many of the traits of von Neumann's self-replicating automata, both of which lie in a biomimetic hierarchy of four basic levels:

1. *Molecules:* These are electronic building block elements of programmable circuits.
2. *Cells:* Assemblies of molecules that make up a processor and memory.
3. *Organisms:* Multicellular systems that can do specific tasks.
4. *Populations:* Collections of organisms.

9.13 Chemical Control System

Biological systems routinely exert nonlinear control through chemistry. An example is using thresholds to control coagulation through competing reactions (Figures 9.35 and 9.36). This approach satisfies the control requirements where coagulation should only occur when a particular damage signal exceeds a threshold value and is otherwise stable against perturbations.

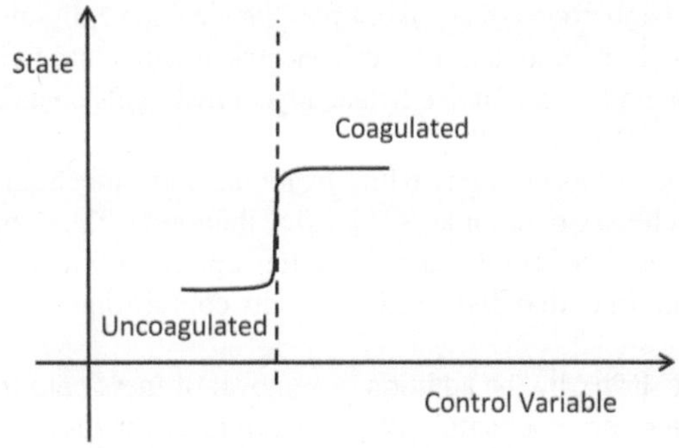

FIGURE 9.35 Crisp nonlinear coagulation response to change in conditions.

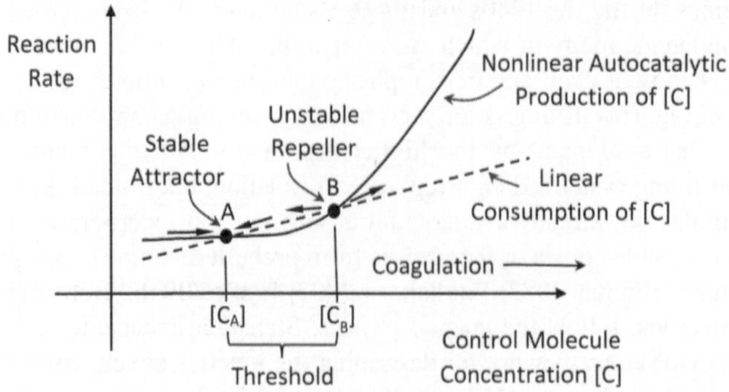

FIGURE 9.36 Competition of linear consumption with nonlinear production creates threshold trigger that works with chemical reactions, such as coagulation. Point A is a stable attractor. Point B is an unstable repeller. If [C] > [C_B] triggers coagulation, [C] < [C_B] moves to attractor point A. (Adapted from [Runyon 2004].)

9.14 Apoptosis for Healing

Apoptotic systems and subsystems self-degrade in controlled manners that improve overall health. Apoptosis is a ubiquitous healing technique in multicellular biological systems that removes infected, aged, excess, or otherwise undesirable tissue [Meyer 2014]. Apoptosis generally requires the following:

1. *Decision:* Deciding when subsystems should self-destruct. Sometimes an entire system should self-destruct. This occurs in cases where the system can cause damage to the surrounding environment.
2. *Destructive Capability:* A means of destroying unwanted subsystems.
3. *Removal Capability:* A means of removing the defunct subsystems and detritus.

9.14.1 Apoptotic Material Systems

Heterogeneous apoptotic material systems typically need only a small fraction of the material to change state and cause disintegration. For example, introducing sucrose into the matrix of a PVA polymer results in a material with modest changes of properties, yet is long-term stable. Exposing sucrose-modified PVA to water dissolves the sucrose and the entire material matrix [Acar 2014]. Varying the mixture proportions provides different dissolution rates. Similarly, polystyrene self-folding origami robots self-degrade when placed in acetone [Miyashita 2015]. Another example formulates epoxies that solidify with DA reactions. The DA bonds break down by heating and application of a solvent to make thermally removable epoxies [Loy 2002]. Additional examples include the controlled dissolution of material into a host fluid that lead to a variety of apoptotic effects. Examples also include a dissolvable silk fibroin substrate that enables the mounting and integration of ultrathin electrical wires into living biological tissue [Kim 2010a] and the hydrolytic degradation of polyethylene glycol hydrogels containing encapsulated biomaterials for healing and growth [Hoffman 2014].

Thermally functional epoxy-based foams with DA adducts are mechanically resilient, yet easily removed with solvents, such as 1-butanol at 90°C [McElhanon 2002]. Another apoptotic approach is chemically inert and long-term stable thermosetting epoxy polymers made with cyclic anhydrides with cleavable acetal links that dissolve in solvents containing acids, such as carboxylic acid [Buchwalter 1996].

Living polymers grow or shrink by the addition or removal of mers onto the end of polymer chains. In certain temperature ranges, thermodynamic behavior can favor the dissociation of mers and a length reduction. Attaching specialized end cap mers stabilizes these unstable molecules. A trigger stimulus that removes or disables the protective end cap mers can disintegrate the molecules [Kaitz 2015].

Biodegradable and biocompatible metal foils made of magnesium, iron, or tungsten packaged inside of biodegradable polymers, such as polyanhydrides, are suitable building blocks for biodegradable electronics, such as batteries for transient implanted electronic devices [Yin 2014a].

9.14.2 Apoptotic Systems

Complex hierarchical systems and cyberphysical systems with embedded controllability provide multiple opportunities for self-improvement by apoptosis Enabling and then programming modular robotic assemblies to disassemble can help with more rapid reassembly into new configurations [Gilpin 2008]. A cyberphysical example is a Shannon box that autonomically turns itself off when turned on (Figure 9.37).

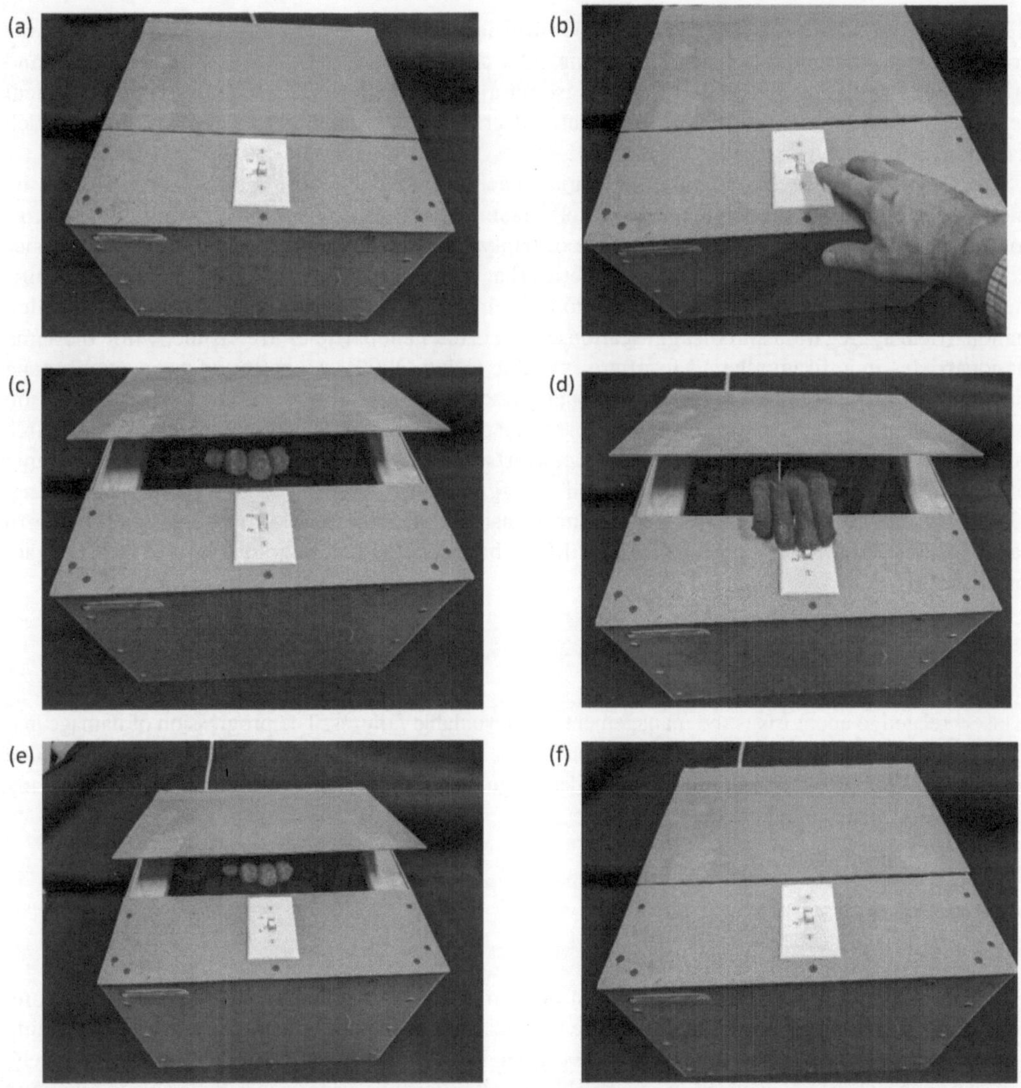

FIGURE 9.37 Shannon box exhibits apoptotic behavior by switching itself off when turned on. (a) Initially off. (b) External hand switches on. (c) Internal hand emerges to begin apoptotic maneuver. (d) Internal hand switches off. (e) Internal hand retreats. (f) Box reverts to off state.

Many missiles and rockets are apoptotic with internal self-destruct systems. A decision must be made if and when to self-destruct. A simple criterion is to trigger a self-destruct sequence when the flight path or flight characteristics deviate outside of a desired range. An autonomous method for pyrotechnic missiles, that is, fireworks, is to have the missile self-destruct after flying a desired distance by causing a shape alteration that leads to an aerodynamic instability and tumbling [Kim 2015a]. Reactive burning of the projectile body initiated at launch occurs at a sufficiently controllable rate to cause tumbling after a preset distance.

Selective removal of system components is an apoptotic action that aids healing actions. Softening and removing a tough protective layer enables the repair of sensitive underlying components, such as the components of an electric circuit board that have been encapsulated to protect them from a harsh operating environment [Small 2001].

9.14.3 Self-erasing Inks and Written Patterns

Self-erasing inks are a specialized apoptotic material technology that eliminates printed features from the surface of documents. Self-erasing inks come in two main varieties – autonomic and stimulated. The autonomic varieties self-erase on a fixed schedule, usually due to slow chemical reactions. The externally stimulated varieties self-erase upon receiving an external stimulus, such as the UV radiation in sunlight.

Much of the molecular and supramolecular machinery used in self-healing materials can be modified and harnessed for apoptotic and self-erasing ink applications. Some examples are as follows: (1) Supramolecular polymer gel made of triptycene-based bis(crown ether) and a copolymer containing dibenzylammonium (DBA) moieties that convert a material with self-healing capabilities into a self-erasing material by the addition of spiropyran molecules [Zeng 2013]. (2) Controlled nanoparticle aggregation and disaggregation. Aggregated nanoparticles are visible, while the same nanoparticles in a disaggregated form are invisible [Klajn 2009]. A coating of *trans*-azobenzene on the nanoparticles changes state from nonpolarized to polarized with an isomer transformation upon UV light irradiation. The polarized form causes the particles to aggregate and form a visible ink. The polarized form is metastable and converts back to the nonpolarized form over a couple of hours. The nanoparticles follow suit and disaggregate to erase the visible pattern of writing. (3) Thin gold films on carbon-based pressure-sensitive adhesives heal damage corresponding to laser-written patterns of text [Alessandri 2010]. The thickness of the gold layer controls the rate of healing.

9.14.4 Graceful Failure

Closely related to apoptosis is the management of unavoidable failures. The progression of damage and failure is often complicated. Unexpected second-order effects exert confounding and counterproductive actions. An example is the leak before burst design guideline for pressure vessels. Leaks are annoying but generally preferred to bursting.

9.15 Reactive Armor

Reactive armor actively arrests the motion of projectiles before they impact and damage a structure. Reactive armor is effective in situations where passive armor is not practical or capable of withstanding the energy of extremely energetic impacts. Many details of this technology are not disclosed publicly, but some concepts can be gleaned from available sources. Reactive armor needs to be able to sense and react to incoming high-speed projectiles in a very short time frame. A space-borne application for countering micrometeorites uses a high-potential capacitor between two layers in an external skin. When a

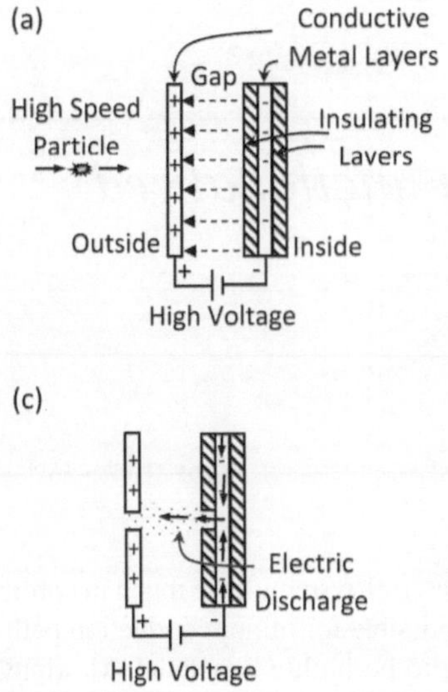

FIGURE 9.38 Capacitive-type spacecraft micrometeor hypervelocity impacts protection system. (a) Healthy skin acts as charged capacitor. (b) Particle penetrates and breaks into ionized debris. (c) Electric discharge vaporizes debris. (d) Skin heals by reverting to capacitive state. (Adapted from [Edwards 2002].)

micrometeorite strikes the layered capacitive skin, it disintegrates into a plasmonic conductive cloud. The conductive cloud connects the charged layers, causes electrical current to rush in, and vaporizes the projectile before it does any more damage (Figure 9.38) [Edwards 2002]. The action is similar to that of self-clearing electrodes.

Shock waves due to nearby explosions also pose a high-speed mechanical threat to structures. It is possible to attenuate and deflect shock waves with an electromagnetic arc launched on laser-ionized air or a rapidly inflating protective airbag [Tillotson 2014] [Tillotson 2015].

10

Electronics, Optics, and Functionalized Material Systems

10.1 Introduction

Electrical, optical, and related functionalized material systems have much in common. The confluence of concentrated energy and signal paths makes it possible for minor conduction path disruptions to disable an entire system, yet small amounts of healing to be highly beneficial. Multiphysics functionalization expands the activation of the material through combinations of thermal, electrical, optical, and magnetic effects that interact at multiple length scales to enable self-healing [Zhu 2014].

10.2 Damage Mitigation Devices and Systems

Unmitigated faults in electrical and electronic systems cause serious damage. The actions of stray currents and voltages disable subsystems and components, and lead to overall system failures. High-performance systems, such as aircraft, are prime examples. Minor disruptions to electrical wiring can cause severe problems. Autonomic recovery can prevent a catastrophe [Johnson 2000] [Stefani 2000].

Many simple and low-cost methods mitigate damage due to electrical faults. These include fuses, metal oxide varistors, surge suppressors, and supplemental conductive paths to ground. In a fuse (Figure 10.1), high current levels overheat a metallic element that burns out. The operation opens a circuit in the event of an electrical overload and prevents the overload from propagating elsewhere and causing damage. The metal oxide varistor (MOV) is an inverse where large electrical transients reduce resistance and cause a circuit to close and short the transient to ground. A ground fault current interrupter (GFCI) acts as a smart fuse. The operational basis is that the current flowing through a closed circuit is conserved and equal at both the live and ground ends. When current strays from the closed circuit and returns to ground by an undesired path, such as through shorts, insulation breakdowns, and other dangerous conditions, a ground fault occurs. A GFCI device detects the missing current and

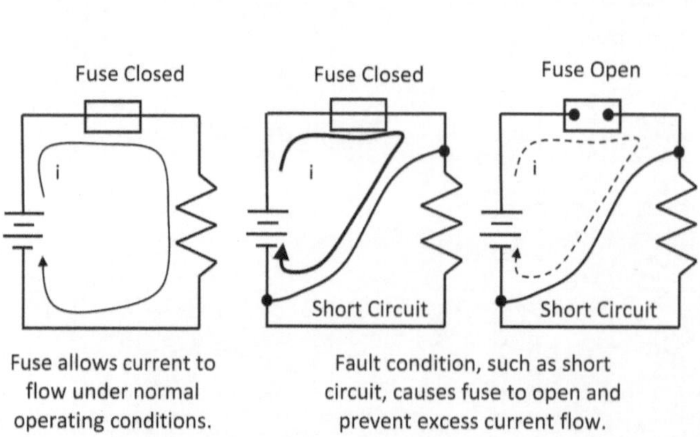

FIGURE 10.1 Fuse acts to protect circuit and mitigate damage by opening when current exceeds a threshold.

DOI: 10.1201/9781003396048-10

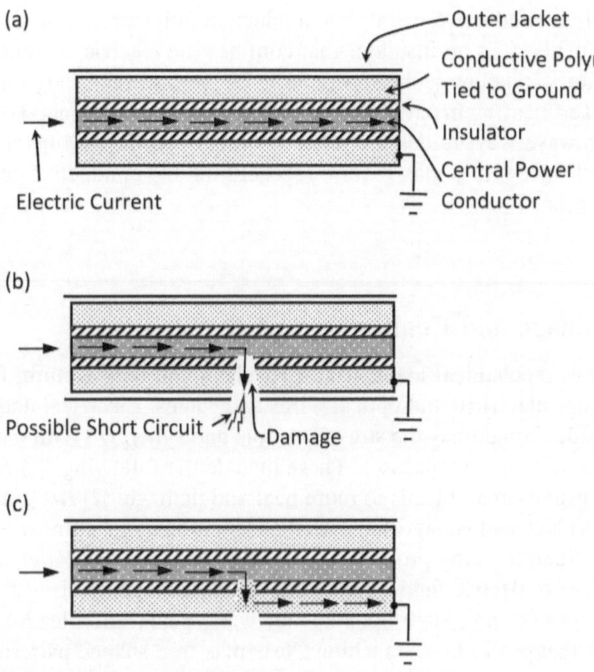

FIGURE 10.2 MOV type distributed self-clamping wire insulation. (a) Healthy power cable. (b) Damage breaches insulation. (c) Polymer conducts current to ground. (Adapted from [Fisher 2006].)

trips a breaker to open the circuit. Effective usage requires setting the threshold to account for normal differences that result from reactive loads, such as from an electric motor. The fuse, MOV, and GFCI mitigate against serious damage by circuit topology changes, but typically leave a system in a disabled state. More sophisticated systems restore the functionality of the original circuit by restoring the original topology, if it can operate fault-free.

Fuses, MOVs, and GFCIs are usually discrete components. Distributed versions run continuously along the wires. When damaged, wire insulation causes electrical faults, a conductive polymer layer shorts faults to ground and mitigates against damage that may occur anywhere along the wire [Fisher 2006]. A concentric vascular arrangement with a low-cost flowable conductive polymer ensures good contact with the wire (Figure 10.2).

10.3 Electric Wire and Cable

Loosely defined, *wires* and *cables* are often synonyms. A more accurate definition provides distinction with wires being single strands and cables being bundles of strands. Braided-strand, parallel-strand, and combinations of material layers with wires all are cables. Waveguides are linear conductive elements, and/or cavities, that confine, guide, and propagate waves.

The importance of signal and power transmission through wires and cables to modern engineered systems should not be underestimated. Relatively minor wiring faults receive blame in multiple aircraft failures, such as the Swiss Air Flight 111 and TWA Flight 800 calamities that resulted in hundreds of fatalities [Slenski 2004]. Faults in wiring are difficult to repair. The replacement of entire bundles of cable, wiring, and wiring harnesses is often easier and less expensive than debugging and repairing a minor insulation breach, especially those that cause intermittent faults. Similarly, the failure of wires in implanted heart defibrillators has caused multiple deaths [Hauser 2012].

Typical electrical wires, optical fibers, and microwave waveguides are composite assemblies with a central conductor surrounded by guiding material on the perimeter. The central conductor in an electrical wire is usually a conductive metal, but conductive polymers, ionic gels, and neural axons also appear. The outer material is an insulator that confines the electric current wave to the central core. Since many insulators and outer jackets in electrical wires and cables are made of polymers, it is usually easier to imbue self-healing in outer polymer insulators than inner metal or optical conductors. Optical fibers and microwave waveguides use reflective media on the perimeter to guide the waves. Self-healing of damaged waveguides may require regenerating the conductive core and/or the reflective guiding perimeter layer.

10.4 Insulation Damage and Healing

External stressors such as mechanical loads, heat, fire, chemical, temperature, fungus, water, moisture, and UV light damage electrical and optical wires and cables. Electrical and optical fields stress conductors from the inside. Sometimes the stresses couple and synergize with material breakdown to create aggressive localized damage processes. These include the following: (1) *Hot spots* where heat damages the wire and insulation and leads to more heat and damage. (2) *Arc tracking* where comingling of saline water and electrical voltage damages the insulation along a linear track, with polyimide insulators, for example Kapton, being prone to this type of damage. (3) *Water-tree* (aka *Christmas-tree*) damage arising due to electric field interactions with water penetrating into microcracks and openings causing crack growth and more openings for water penetration leading to a repeating and often branching process that produces a branching Christmas-tree-shaped pattern of damage [Bahder 1976] [Bulinski 1998].

Most electric current-carrying cables have dielectric insulating layers. Damage to the dielectric insulation breaks down the ability to confine the conductive paths for the current, with short circuits, loss of functionality, and the ignition of fires being common outcomes. Due to the industrial importance of these techniques, much of the research and development occurred outside of academia, with many of the key developments appearing in patent literature. Appendix B lists selected patents used in self-healing wires and cables.

10.5 Flooding and Vascular Methods

Vascular approaches take advantage of the elongated geometry of wires and cables to run healing fluids in channels alongside the electrical and optical conductors often in a concentric arrangement. *Flooding cables* are a prepackaged subset of these techniques.

A long-standing self-healing application is maintenance of the dielectric insulation surrounding conductors in high-voltage electric power lines. A typical vascular configuration is a coaxial arrangement with replenishing dielectric fluid flowing around a center conductor [Whitehead 1968]. The motivation for a vascular system is that the transmission of high power electricity favors using high voltages to reduce the currents and losses due to current flow and Joule heating. High voltages break down most solid dielectric insulators. Recirculation of liquid or gas dielectric insulation heals by replenishment.

Moderate-voltage electric power delivery systems place less demand on the insulation than high-voltage systems. Nonetheless, environmental loads and cost constraints lead to persistent performance challenges. Water penetration through polymer dielectric barriers is problematic, especially for buried cables carrying greater than 600 V. The water sets up a corrosive electrolytic cell driven by a rectified portion of the alternating current in the wire that breaks down aluminum sheaths and conductors. Aluminum converts into aluminum hydroxide, a whitish chalky powder [Spruell 2013]. The chemical reaction also releases hydrogen gas – a potential explosion hazard, especially for wires contained in underground structures. For these medium- and light-duty applications, a viscous fluid or gel-like material in one of the layers acts as a self-healing agent. Unlike high-power cables that circulate the flow, the gels in medium power cables remain relatively stationary. In the event of damage to the outer jacket, the gel

FIGURE 10.3 Self-healing flooding cable for medium-power underground electric power distributions using viscous PIB to seal insulation breaches undergoing puncture test.

flows into the gap, provides a dielectric barrier, and perhaps more importantly blocks the penetration of water. The requirements of the flooding fluid or gel are as follows:

1. *Insulating:* Electrically insulating.
2. *Insoluble:* Insoluble and impermeable with water-blocking capability.
3. *Adherence:* Must adhere to inner and outer walls to prevent flow of water.
4. *Environmental Tolerance:* Must exhibit appropriate viscosity under a moderately large range of conditions.
5. *Stability:* Must retain water-blocking capabilities long term.

Polyisobutylene (PIB) is a good candidate material for flooding cables [Reece 2002] [Ware 2005]. PIB is a versatile sticky polymer with many applications, including being a base ingredient in chewing gum (Figure 10.3).

As with all vascular systems, actuating and controlling the flow of the liquid is key to success. One method is simply a fluid-filled concentric layer. Potential concerns are gaps and nonuniformity. Wicking of the fluid along braded fibers in an outer layer can promote uniformity. A finned internal vascular structure enhances the flow of viscous fluid, while also armoring the cable (Figure 10.4) [Spruell 2003].

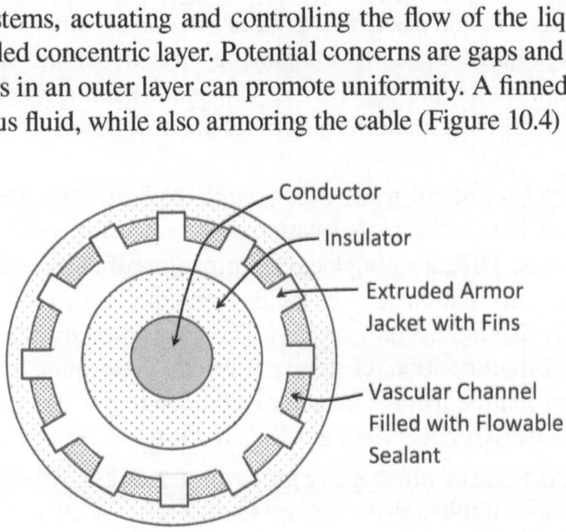

FIGURE 10.4 Cross section of electric power cable with extruded finned layer providing both armor and vascular channels for flooding compounds. (Adapted from [Spruell 2003].)

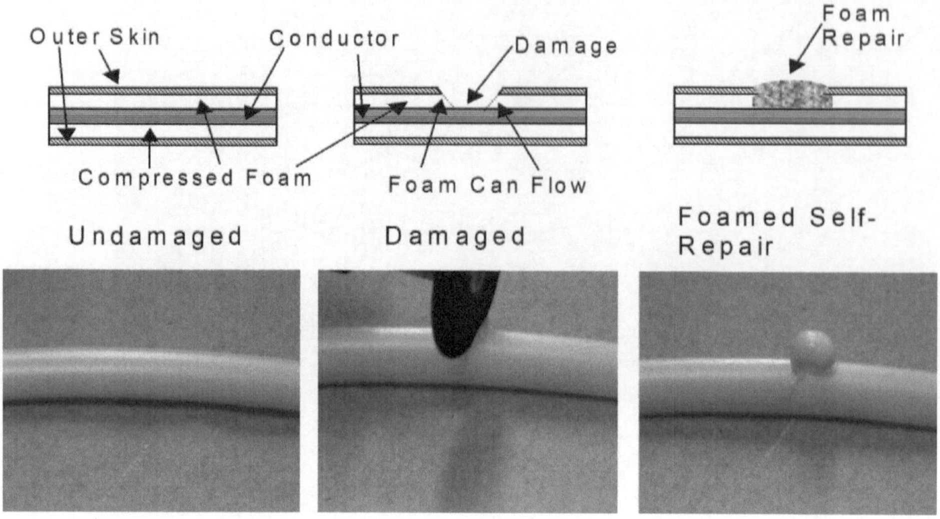

FIGURE 10.5 Foam expands and fills cut in tube. (From [Esser 2005].)

Expanding foams that fill damage are an alternative to flooding liquids or gels [Lucas 2001]. One method uses liquid foams to expand and then solidify in place. Laboratory experiments have demonstrated the potential viability for use in wire and cable applications (Figure 10.5) [Esser 2005]. Joule heating of conductors provides a potential trigger and energy source to activate thermally reflowable damaged insulation [Castellucci 2009].

10.6 Conductors

The repair of conductive elements in electrical wiring requires reestablishing continuity of a desired conductive path and avoiding creation of unwanted paths. Most conductors are metals with melting temperatures that are much higher than can be sustained by polymer insulators.

Electrical conductors made of solid metals do not tolerate large deformations very well. Instead, metallic conductors require a variety of mechanical strain relief methods to prevent damage during normal operations. Stretchable conductors and electronic devices offer an alternative in that they undergo much larger deformations without damage. In addition to playing a role in damage mitigation, stretchable electronics also enable large deformation damage recovery in smart healing systems. Some of the primary methods of achieving stretchability in modern electronics are as follows:

1. *Serpentine Patterns of Thin Metal Wires in a Compliant Elastomer Matrix:* Thin metal wires placed in serpentine buckled and curled patterns can uncurl when loaded to produce large macroscale deformations, while keeping local deformations of the metal small. Placing an elastomer capable of large deformations guides the deformation of the metal so that it is capable of repeated macroscale large motion deformations. The once ubiquitous helical telephone handset cord is an example. Miniaturization, customized fractal patterning, and advanced packaging methods extend the serpentine metal conductor technique to a wide range of flexible electrical interconnect applications [Xu 2012] [Fan 2014] [Lin 2014].
2. *Adaptive Stiffness:* Electrodes with adaptive properties, such as stiffness, can improve durability. An example is a neural implant that connects to the brain. Electrodes that are stiff enough to penetrate a brain for insertion are too stiff for long-term compatibility with soft brain tissue and produce scars. There is a need for electrodes that soften after insertion, such as with desiccated hydrogels as stiffening elements that soften once inserted into a wet brain [Capadona 2012].

Restoration of conductivity by depositing conductive material near to a break requires information regarding the desired configuration, a material system that places material at the correct location, reestablishes conductivity, and does not allow conductivities to stray into unwanted areas. Some possibilities are as follows:

1. *Electrodeposition:* Charge transfer salts, such as TTF-TCNQ (tetrathiafulvalene–tetracyanoquinodimethane), are potentially useful materials. They are nonconductive in solution, but become conductive upon evaporation of a solvent, often in a one-dimensional pattern that follows crystal orientation, and may exhibit superconductivity at cold temperatures [Coleman 1973] [Ferraris 1973], for example, Electrodeposition of metals such as copper from a $CuSO_4$ solution. It may be possible for the activation of electric potentials across the open circuit to drive the deposition to restore conductivity. Dielectrophoresis uses the large electric field gradients formed on the tip of a charged nanowire to drive the deposition of conductive nanoparticles that lengthen the nanowire [Hermanson 2001].
2. *Vascular Systems:* Liquid metal conductors flow in vascular self-healing systems, such as antennas or specialized sensors. Liquid electrical conductors can restore conductivity by flowing into the gap of an open circuit but have potential difficulties with staying in place once deposited. Mercury is liquid at room temperature but is toxic. Eutectic indium-gallium (EIG) is liquid at room temperature, is much less toxic, and forms an oxide skin when exposed to air that prevents leakage and aids with combined systems using self-healing polymers to heal electric circuits following a complete mechanical cut of the circuit [Palleau 2013b]. Ionic liquids, such as those based on imidazolium salts, are conductive, but not as conductive as liquid metals [He 2015]. EIG forms a thin oxide skin that makes it compatible with elastomers for large deformation antennas and clots small cuts through circulatory vessel walls [So 2009b]. The oxide layer also prevents the liquid from flowing through capillaries. Inserting supplementary electrolytes into the capillaries with an electrowetting technique creates a slip condition at the wall. The slipping allows the EIG to flow freely and provides a means of fine-tuning antenna resonance frequencies as part of a healing process [Wang 2015a].
3. *Microencapsulation:* Liquid materials that restore conduction for release at damaged sites is a potentially viable delivery method. Tests with systems using charge-transfer salts and EIG demonstrate viability [Odom 2010] [Blaiszik 2012].
4. *Shape Shifting and Healing Conductive Gels and Meshes:* Composite materials and inks made of a matrix with embedded nanowires and particles are electrical conductors when the particles achieve density levels to sustain percolation-type continuity. Polyelectrolyte matrices with embedded Ag nanowires restore conductivity through cuts with a water-initiated reformation of the matrix [Li 2012d]. The integration of supramolecular Zn-polypyrrole hydrogels introduces conductivity at the molecular level. These gels restore electrical conductivity, even after multiple cut-and-heal cycles [Shi 2015].

10.7 Connectors, Contacts, Electrodes, and Switches

Connectors, contacts, electrodes, and switches provide crucial links in continuity between components in electrical circuits. While failure of one of these components can disable an entire system, the amount of material movement and energy needed for repair is minimal. Connectors are integrated mechanical, electrical, optical, and/or fluidic devices that support and protect the contacts. Contacts are mating pairs of conductive components that provide for continuous flow of current when connected and prevent the flow of current when open. Electrodes are specialized contacts that provide electrical continuity between a conductive wire (usually metallic) and a semiconductive or nonconductive material, as in electrical batteries, other electrochemical devices, and capacitors. Switches open and close contacts in response to external commands or actuations.

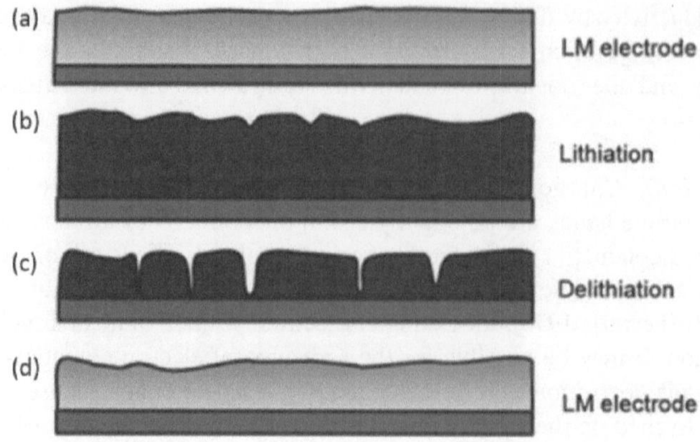

FIGURE 10.6 Schematic of low melting point metal healing cracking in lithium battery negative electrode caused by charge–discharge lithiation–delithiation cycling. (a) Liquid metal electrode before electric cycling. (b) Electrode solidifies and swells during lithiation. (c) Delithiation causes solid electrode to shrink and crack. (d) Electrode returns to liquid state and heals cracks. (From [Deshpande 2011].)

Electrodes lie at the interface of electrical conductors and other electroactive materials. The mismatch of material properties, electrochemical transport, and large deformations combine to produce severe loading requirements for the electrode. Cracking, shorting, and the combined degradation of electrical and mechanical performance are common failure modes.

Materials that self-weld may act as self-healing electrodes. An example for harsh environments is to use lithium to promote electrochemical self-welding in silicon-to-silicon nanowire contacts in lithium-ion batteries [Karki 2012]. Sometimes it is possible to find an electroactive effect to counter the damaging effects and heal the electrode. Low melting point metals heal cracked electrodes by liquid migration and deposition. An application is lithium batteries that undergo charge–discharge lithiation–delithiation cycling to crack electrodes (Figure 10.6) [Deshpande 2011]. Oxide scales forming on the surface of liquid electrodes have large electrical resistivities that impair self-healing performance. Alloying the base metal often resolves these difficulties [Aindow 2010].

Electrodes attach to materials that undergo large deformations, which raise issues of strain compatibility and material durability. A mitigating technique to accommodate the large deformations in electrodes is to place a metal conductor as a thin film on top of a compliant elastomeric foam. Upon undergoing large deformation, the metal film cracks, while the underlying foam layer remains intact. Properly sizing the layers forces the metal film to crack in controlled patterns that retain electrical conductivity [Vandeparre 2013].

Electrodes for dielectric electrically actuated polymer (DEAP) actuators face severe loading requirements due to the combination of large deformations in the elastomers and large voltages required to produce the deformation. Cracking is a common failure mode. The electrochemical deposition of copper from platinum–copper in electrodes is a healing mechanism for ionomeric polymer–metal composite actuators. Copper ions migrate into cracks in the electrodes [Johanson 2008].

High-performance electrical batteries pose another set of demanding requirements for electrodes. In addition to electrical, chemical, and thermal loads during normal operation, some batteries, including lithium-ion batteries, undergo large mechanical deformation during charge–discharge cycling. Such deformations can quickly crack the electrodes. Wrapping the microparticles in a self-healing dynamer solves many of these durability issues in silicon microparticle anodes [Wang 2013a].

10.7.1 Self-Clearing Electrodes

The spurious formation of shorts is a common failure mode of electrodes and capacitors. Self-clearing electrodes heal electrical shorts with short-duration high-current surges that oxidize and/or melt the short with Joule heating. Self-clearing electrodes are particularly useful when the electrodes span thin films,

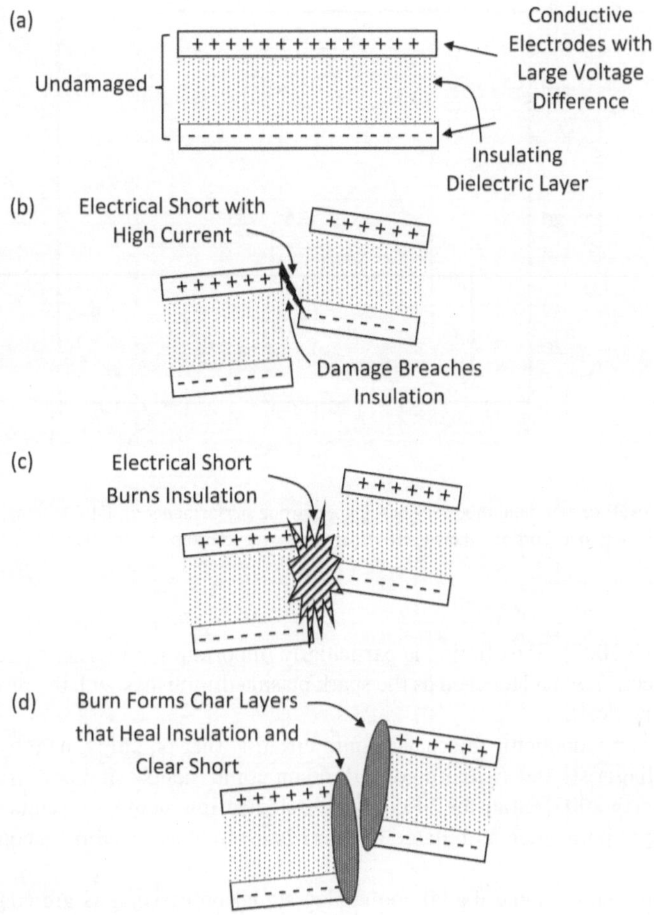

FIGURE 10.7 Self-clearing electrodes correct shorts in high-voltage devices.

which have large in-service deformations, that combine large voltages and large deformations, such as in capacitors or DEAP actuators. A small, even pinhole size, breakdown of the dielectric causes large current shorts. Such breakdowns normally disable a device. Following self-clearing, the electrode works but has a reduced area for operation, often only a negligible reduction (Figure 10.7) [Martin 2013a].

The dielectric material between the electrodes plays a large role in self-clearing. Self-clearing favors materials that form char layers, as in polymer/foil capacitors, but not in metalized foil capacitors [Reed 1994]. Charcoal powder and silver layers create self-clearing electrodes for DEAPs with breakdown voltages of 10 kV and 1.9 kV, respectively [Lau 2012] [Low 2012]. A two-part electrode made of a silicon sponge filled with silicon oil self-clears with oil flowing out of the sponge into the gap [Hunt 2014]. Electrodes made of single-wall carbon nanotubes self-clear in DEAP actuators (Figure 10.8) [Yuan 2008a]. Laser-assisted self-clearing heals DEAP actuators in flying microrobots [Kim 2023].

Organic light-emitting diodes have short-circuit failure modes that occur in transparent indium tin oxide (ITO) electrodes. Transparent metal nanomesh electrodes heal by ramping up the voltage to clear the short by burning the nanowires [Belenkova 2012].

Plasma arcing due to the opening of circuits damages the electrodes in switches and circuit breakers for high-power electricity transmission circuits. The self-blast interrupter mitigates spark plasma damage by forcing a dielectric gas, typically sulfur hexafluoride (SF_6), into the opening gap to limit and extinguish the electric arc. Active methods of forcing the gas to flow use pumps and represent a reliability concern. An alternative is the self-blasting circuit breaker. Energy of the electric arc heats the dielectric gas in a chamber in a rapid manner so that it blasts out of a nozzle into the arc for quick extinguishment

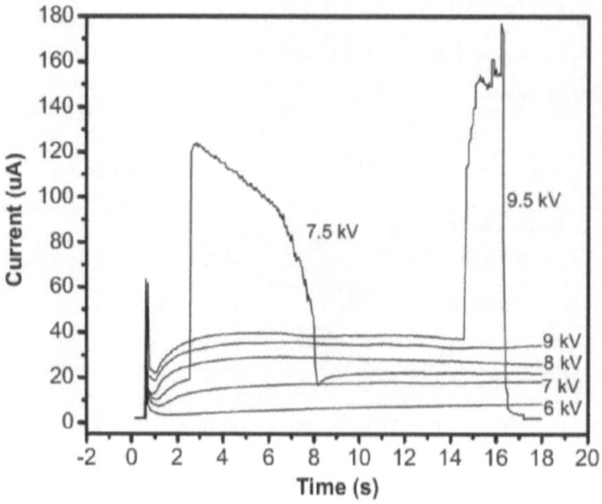

FIGURE 10.8 Single-wall carbon nanotube self-clearing electrode performance in DEAP actuators. The 7.5 kV curve shows a spike corresponding to a short and then a self-clearing recovery. (From [Yuan 2008a].)

[Sedlacek 2003] [Lee 2005]. Blast timing is particularly important for alternating current circuits. The blasting is more effective when launched as the spark plasma diminishes with the reversal of the current during the alternating cycle.

Most integrated semiconductor microelectronic circuits, that is, chips, heal by combining internally driven fault diagnosis and switching of redundant components. Memory circuits are an important application [Barth 2002] [Stauffer 2009]. Some self-healing memory circuits use programmable fuses to open a circuit path when stimulated and anti-fuses to close circuits on command [Lee 2001] [Geppert 2004].

Integrated circuits that combine digital and analog signal processing, as are common in radio and microwave electronics, introduce additional complications and failure modes. Damage may be catastrophic where the device fails to operate, or may be parametric where the device still operates, but degradation of the analog components inserts unwanted noise and distorts the signals. In the case of parametric damage, the introduction of onboard digital components allows for healing the performance through the adjustment of analog-operating parameters, along with the inclusion of digital compensation [Goyal 2012]. The process begins with a self-assessment of performance, followed by optimization algorithms that select corrective actions, implementation of corrective actions, and then reassessment of performance. Figure 10.9 shows a transmission line stub that shorts to ground in a stepwise manner that changes the length of the stub and enables self-healing by tuning [Bowers 2013]. Self-healing with redundant component swapping improves the yield of viable chips during manufacture at rates that may justify the parasitic costs (Figure 10.10) [Maxey 2012] [Chien 2012].

Electromigration is a primary failure mode of microelectronic circuits. The flow of electrons in DC currents exerts forces by momentum transfer on the atoms of the conductor. Over time, electron forces combine with thermal stresses and diffusive motions to change the shape of the conductor resulting in whiskers, hillocks, and voids [D'Heurle 1971] [Ren 2006] [Ho 2022]. Discontinuities in the conductive paths, such as crystal grain boundaries and thin metal films, are particularly prone to electromigration damage [Ogurtani 2005]. Occasionally, electromigration-based material movement improves the condition of an interconnect, that is, the interconnect self-heals. An example is the favorable electromigration-driven diffusion of Cu through an Ag passivation layer into Cu in Cu–Cu hybrid bonding used in 3D heterogeneous packaging of computer chips [Chou 2021] [Hong 2023]. The diffusion improves the interconnect and allows for bonding with little or no need for annealing. Similarly, a self-healing scheme for through-silicon vias rendered defective by electromigration combines diagnosis, isolation, and reversal of DC current [Cheng 2016].

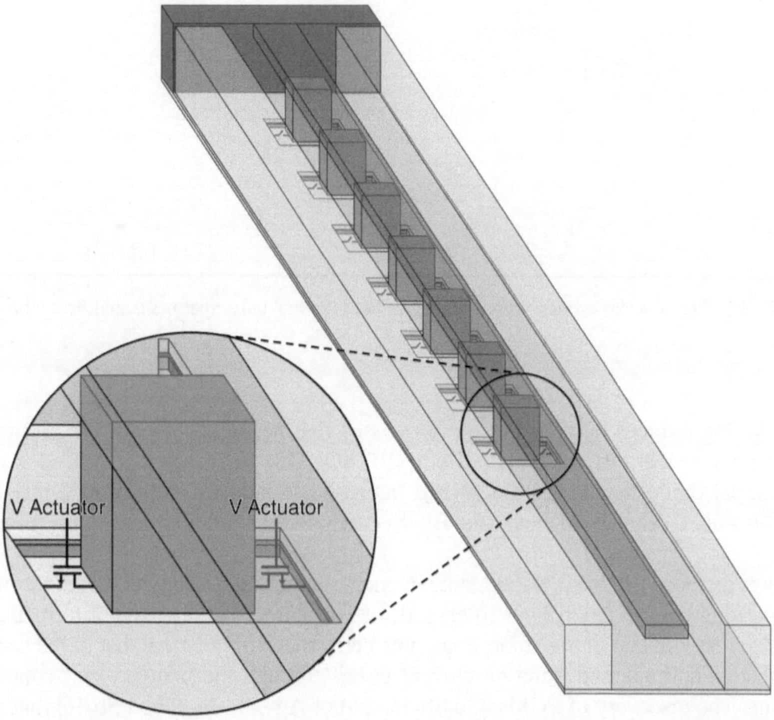

FIGURE 10.9 A transmission line stub that shorts to ground in a stepwise manner to change the length of the stub and enable self-healing tuning. (From [Bowers 2013].)

Additional cases arise where nonequilibrium thermodynamic processes drive a material system into a conductive state. An example is electrorheological fluids containing wiggling carbon nanotubes [Belkin 2015]. The application of an electric field to the fluid causes the movement of the nanotubes to configure in patterns that reduce conductivity, increase Joule heating, and maximize entropy production.

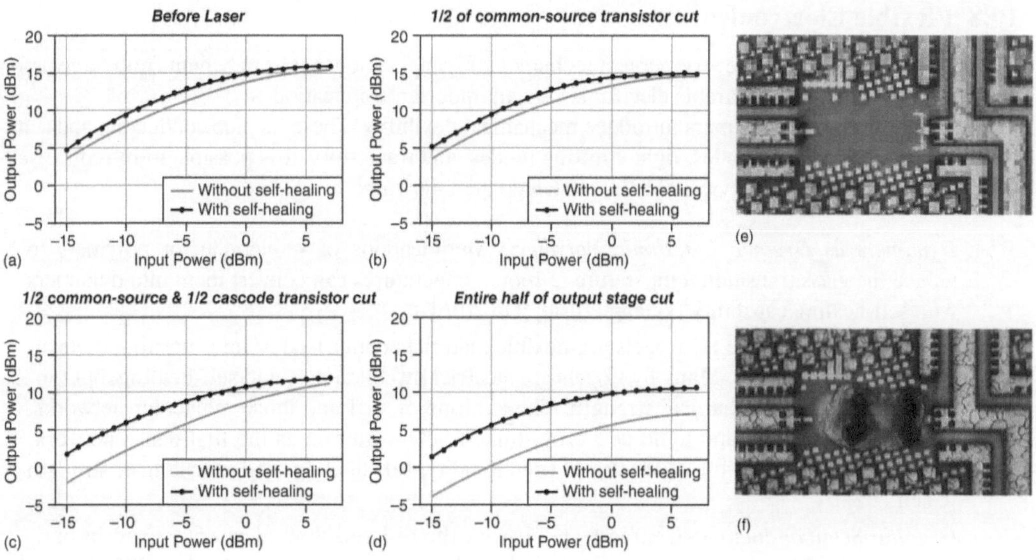

FIGURE 10.10 Recovery of power amplifier performance on microwave SoC chip following laser damage. (From [Maxey 2012].)

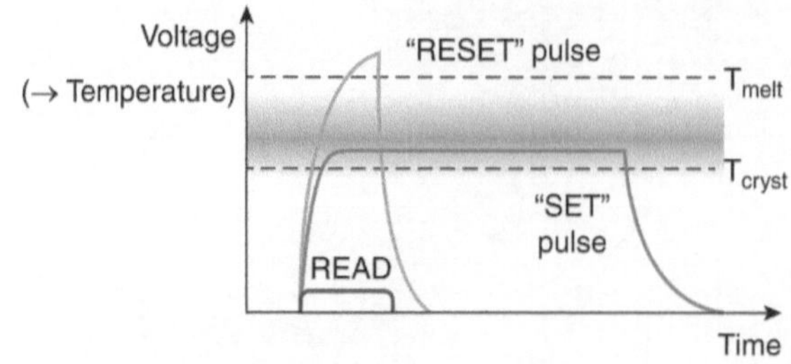

FIGURE 10.11 PCM read, write cycling with a higher voltage RESET pulse that melts and heals the material. (From [Burr 2010].)

Thermal cycling induces annealing-type processes that heal microelectronic circuits. Bandgap-engineered silicon oxide–nitride oxide–silicon (BE SONOS) flash memory degrades with repeated cycling by tunnel oxide formation. Localized micro-Joule heaters with short (~ms) on/off cycles heating above 800°C anneal to heal the BE SONOS, while not damaging the rest of the chip [Lue 2012].

Phase change memory (PCM) is a solid-state memory storage technology that uses thermally induced phase changes to transform amorphous to crystalline structures. The memory information encodes as electrical resistivity with that of the amorphous state being much higher than that of the crystalline state. Sending suitable amplitude and duration current pulses through the memory cell controls the phase state switching. The discovery of PCMs ($Ge_2Sb_2Te_5$ and/or Ag- and In-doped Sb_2Te) that switch phases in less than 100 ns makes them suitable candidates for high-speed nonvolatile memory applications (Figure 10.11) [Burr 2010]. The technique requires switching endurances that exceed competing methods. Electromigration and segregation are principal failure modes that leave a memory cell stuck in a state of high resistance. The sensing of this failed condition followed by the cycling of currents sufficiently large to melt the PCM heals damaged states [Du 2012].

10.8 Flexible Electronics

Flexible electronic circuits are an emergent technology for applications that can benefit from large deformations without damage. Wearable electronics are an important application.

Electronics built from polymers introduce mechanical flexibility. These include conductors and semiconductor devices, that is, diodes, light-emitting diodes, and transistors. It is possible to introduce self-healing into these polymer devices. Methods include the following:

1. *Dynamers as Organic Electronic Polymers:* Modifications of semiconductor polymers to reduce the glass transition temperature to room temperatures can convert them into dynamers with self-healing capability [Huang 2015b] [Oh 2016].
2. *Conductive Hydrogels:* Hydrogels are flexible materials comprised of intermeshing molecular networks and fluids. Many hydrogels are electrically conductive and self-healing, but typically lack robust mechanical strength. The addition of stiff and loose molecular networks, such as Ca^{2+} alginate and ionic pair cross-linked polyzwitterion as the first dense network, provides a mix of flexibility, electrical functionality, self-healing. and mechanical strength [Chen 2022].
3. *Organic Semiconductors:* Charge traps disrupt the ordered flow of charge, even in optimized single-crystal devices. Inert dipolar liquids, such as perfluoropolyether (PFPE), that is, Fomblin, neutralize the charge traps to produce better performing semiconductor crystals [Lee 2013b].

10.9 Optics

Optical beams are collections of electromagnetic waves that travel in the same direction, often defined by a central axis. Optical beams that propagate through free space can self-reconstruct much of the shape and intensity pattern following a partial obstruction through clever exploitation of the geometry of scalar diffraction.

10.9.1 Self-reconstructing Light Beams

The beam shape favors self- repair by scattering from small objects in a manner that reconstructs the waveform. Diffraction-free, that is, propagation-invariant, and related beams form the basis of these methods. The propagation-invariant effect produces an intensity pattern in a propagating beam that does not change, even though diffraction processes are active in the propagation. Diffraction mixes and spreads the waves as they propagate, but the kinematics leads to beams with specific patterns that do not change with propagation. These include Bessel and Airy beams. True propagation-invariant beams have infinite lateral spreads. The amplitudes decline with lateral distance slowly enough to prevent convergence of a total energy integral and lead to physical unrealizability. Accelerating the decline of amplitude with lateral distance using Gauss–Bessel and Gauss–Airy patterns produces physically realizable truncated beams that are approximately propagation-invariant with an initial pattern that heals and persists for significant distances. (More details in Appendix C.)

The propagation-invariant beams and their truncated physically realizable counterparts reconstruct the intensity pattern following partial obstruction of the beam by an opaque object. The qualitative explanation for this self-reconstruction effect is that the light at each point in a beam is the superposition of light diffracted from points upstream. The farther the point moves away from an obstruction, the larger the contribution from transverse points that are independent of the shadow produced by the obstruction. Babinet's principle uses linearity and superposition to provide a quantitative explanation [Bouchal 1998]. Figure 10.12 shows a nondiffracting beam propagating in the z-direction and interacting with an opaque obstruction at $z = 0$. A broad class of z-direction propagation-invariant beams, including plane waves and Bessel beams, have a complex scalar wave field amplitude, $U(x, y, z)$, of the form

$$U(x,y,z) = u(x,y)\exp(i\beta z) \tag{10.1}$$

For these beams, the intensity field, $I(x,y,z) = UU^* = |U^2|$, is independent of the propagation coordinate z, that is,

$$I(x,y,z) = \left|U^2\right| = u(x,y)u^*(x,y)\exp(i\beta z)\exp(-i\beta z) = I(x,y) \tag{10.2}$$

Here, β is a propagation constant. When a beam with wave amplitude $U_I(x,y,z)$ propagates in a positive direction from $z < 0$ and impinges on an absorbing obstruction at $z = 0$, the result is a beam with an altered pattern for $z > 0$. Corresponding to the obstruction is the complementary wave amplitude

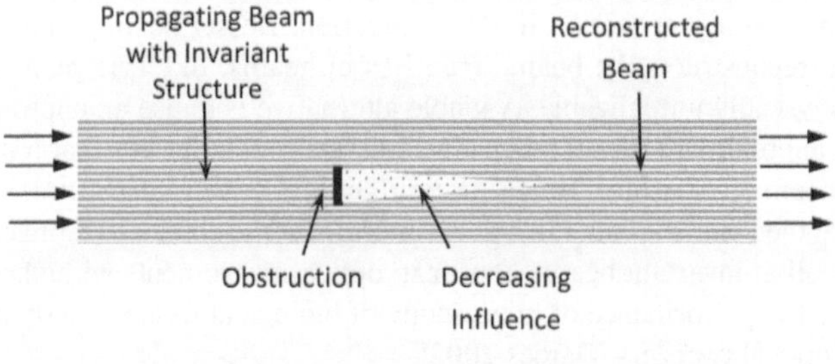

FIGURE 10.12 Babinet's principle explains reconstruction of propagation-invariant beam.

$U_C(x, y, z > 0)$ defined as the field produced by a uniform illumination source from $z < 0$ impinging upon a perfectly absorbing wall at $z = 0$ with an aperture matching the geometric outline of the obstruction. Babinet's principle states that the disturbed field, U_D, due to the obstruction is

$$U_D = U_I - U_C \tag{10.3}$$

The intensity of the disturbed field, I_D, is

$$I_D = U_D U_D^* = I_I + I_C - U_I U_C^* - U_C U_I^* \tag{10.4}$$

As z increases, U_C decreases as 1/z. Hence, in the limit as $z \to \infty$, $U_C \to 0$ and $I_D \approx I_I$. This allows for the propagation-invariant beam to reconstruct after an obstruction.

The reconstruction of a beam beyond an obstruction is not unique to the propagation-invariant beams. In fact, Equation (10.4) applies to all propagating beams, but it is the propagation-invariant beams that create a framework for easily visualizing the Babinet reconstruction. The supporting structures for many optical elements create unavoidable obstructions. Babinet's principle guides the placement of these obstructions so that the shadows defocus and do not appear in the image. For example, a Newtonian reflective telescope places internal optics in the beam line, but these defocus from the image plane and do not appear. Similarly, optical semiconductor lithography uses thin transparent pellicles to protect mask structures by picking up dust particles that do not image in the projected optics.

An example is the diffraction pattern behind a circular disk lying in the x–y plane with center at the origin produced due to a plane wave propagating in the z-direction. Babinet's principle first calculates the complementary wave amplitude as the diffraction pattern behind a circular hole in an opaque plane.

A handful of beam shapes have been found that are self-reconstructing. Perhaps the most common are the Bessel beams [Durnin 1987a] [Durnin 1987b]. The form of Bessel beams is axisymmetric:

$$E_B(x, y, z \geq 0, t) = \exp\left[i(\beta z - \omega t)\right] J_0(\alpha R)$$

Here, R is the radial distance in the x–y plane. α and β are wave parameters with units of length^{-1}. J_0 is a Bessel function of the first kind. ω is the circular natural frequency. The wave speed is c, leading to the constraint:

$$\beta^2 + \alpha^2 = \left(\frac{\omega}{c}\right)^2 \tag{10.5}$$

The intensity I_B of the Bessel beam depends on the radial distance R, but is independent of the propagation distance z, that is, it is propagation-invariant:

$$I_B = \frac{1}{2}\left|J_0(\alpha R)\right|^2 \tag{10.6}$$

The Bessel beams have an intensity that decreases with radial distance from the beam center. Like plane waves, the beam spreads, that is, diffracts, in a perfectly balanced manner so that the intensity pattern does not change with propagation in the z-direction. Bessel beams scatter off nonhomogeneities in a manner that reconstructs the beam. True Bessel beams, like true plane waves, require infinite energy and are physically unrealizable. A viable alternative is to use an approximate Bessel beam. Methods for creating approximate Bessel beams include fiber-optic devices that initially collimate light into phase-coherent beams into a fiber Bragg grating and selectively form an azimuthally symmetric higher order mode that then launches out through an aperture at the end face [Ramachandran 2013]. Self-reconstructing propagation-invariant beams appear in optical instruments with challenging conditions. Bessel beams increase the performance of microscopy of biological tissue heavily laden with scatterers [Fahrbach 2010] and optical tweezers [Garcés 2002].

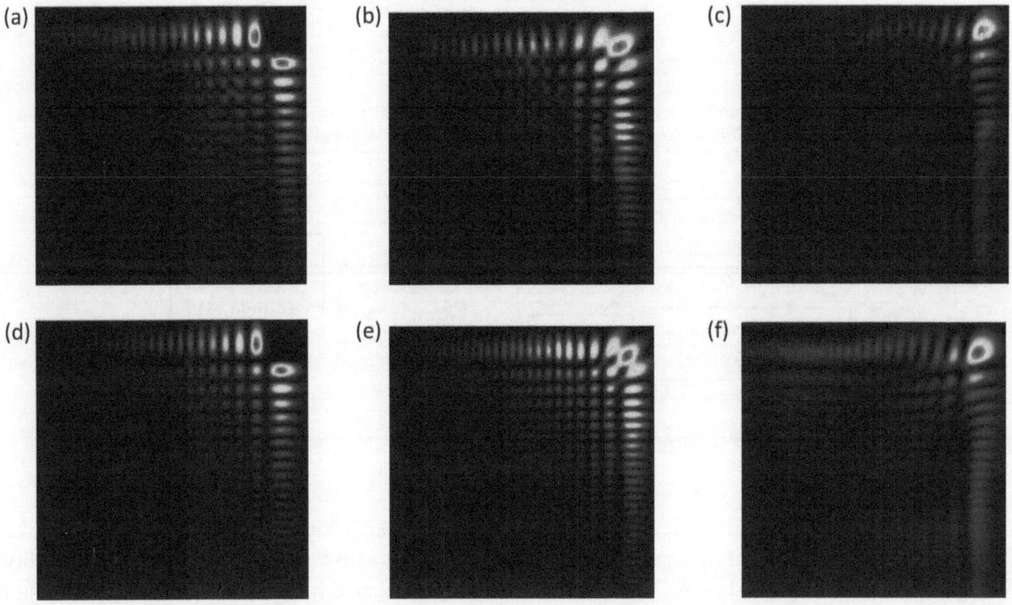

FIGURE 10.13 Experimental confirmation of the self-healing of the main lobe of the transverse pattern of an Airy beam. (From [Broky 2008].)

Self-healing capabilities of Bessel beams extend to the quantum domain with the healing of the entanglement orbital angular momentum following an obstruction [McLaren 2014]. The healing is more robust when measured in a Bessel versus a Gauss–LaGuerre basis. Diffraction-free self-healing beams appear in other domains, such as cosine-Gauss beams in the 2D domain of propagating surface plasmons on metallic surfaces [Lin 2012].

In contrast to the circularly symmetric Bessel, the Airy beams represent a class of propagation-invariant diffraction beams without circular symmetry [Broky 2008] [Bongiovanni 2015]. The principal beam pattern is the product of two Airy functions. This beam pattern regenerates and preserves itself approximately by linear passive diffraction with propagation, but the position moves transversely, that is, it accelerates (Figure 10.13).

Airy beams transfer photons from the side lobes to the main lobe in the corner, while the whole pattern, including the main lobe, moves laterally in a parabolic path. Momentum transfer from the photons can move small particles from the side lobes to the main lobe and then laterally out of the field as a self-cleaning effect [Baumgartl 2008].

In addition to the Bessel and Airy beams, several other waveforms exhibit self-focusing and self-healing. General forms include superpositions of Bessel and Airy beams that move and spin as they propagate, as well as caustics associated with Pearcey beams and axicons [Piestun 1998] [Ring 2012] [Anguiano 2007]. Self-healing light bullets are short pulses of light that do not spread and recover after passing through small obstructions. A viable form of light bullet operating in a linear medium uses a Bessel pattern in the transverse direction and an Airy pattern in the direction of propagation [Chong 2010].

10.9.2 Waveguide Repair

Light propagates through a vacuum and through transparent media. A vacuum exerts little influence on propagating light. Non-vacuum transparent media, such as air, water, or glass exert a large effect on the propagation. Spatial gradients of the refractive index focus, guide, and reflect light. Damage to the

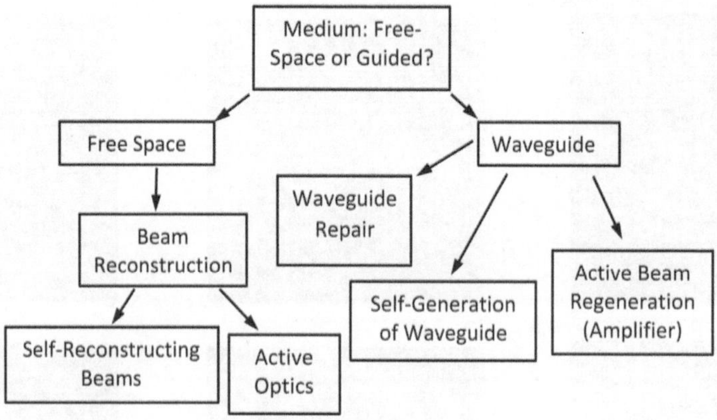

FIGURE 10.14 Taxonomy of self-reconstructing beams.

transparent medium alters the refractive properties and may make it lossy so that it absorbs the waves. The net effect is an undesired distortion and diminution of the beam leading to a loss of functionality. Waveguides transmit optical beams, with spatially varying refractive indices that guide waves along a particular direction. In general, waveguiding occurs when the index of refraction of the central core is higher than that of the periphery.

Optical beams propagating in free space and in waveguides are open to self-healing with a handful of different techniques. These techniques segregate by whether they reconstruct the optical beam, transmission medium, and/or waveguide (Figure 10.14). Optical waveguides use a central core to carry the bulk of the light and an outer region that confines and guides the wave back to a central core for propagation. Low loss transmission over appreciable distances requires precisely fabricated components and tight control over material properties. Damage to the waveguide affects beams that propagate through confined waveguides, such as optical fibers. Damage to either the central core, or to the outer guiding region results in the loss of transmission. Scattering off particles, refractive index inhomogeneities, and absorption all are effects that cause a cleanly launched beam to fail to reach the intended target in a suitable form.

10.9.2.1 Self-generation of Waveguides

Self-guiding waves use the energy in a light beam to alter the transmission medium in a way that produces a waveguide. The typical approach alters the refractive index so that it is higher inside the waveguide than outside. The light channels inside along the waveguide axis, often by reinforcing the change in refractive index properties.

In air, self-guiding waves occur when the intensity of the light is sufficiently high to cause weak ionization. The threshold is $\sim 10^{13}$ W/cm^2. Under normal nonionizing conditions, diffraction causes a collimated light beam to spread with distance and the intensity to decrease. Focusing and high power increase the intensity. The associated large electric field activates the optical Kerr effect to increase the index of refraction of the air. The nonlinear effect of altering refractive index acts as a lens to focus the beam. Further increases in the intensity ionize the air which competes with the Kerr effect by reducing the index of refraction and defocusing the beam. When properly tuned, the combination of Kerr and ionization effects produces a confined self-generating channel through which laser beams propagate unimpeded and without spreading. This effect is known as "optical filamentation," with "optical bullets" being suggested as a more descriptive term of the light pulses (Figure 10.15) [Couairona 2007] [Lim 2014]. Supplemental dressing beams further extend the range [Scheller 2014]. Such lasers have the potential to transmit light at high intensities over extended distances. The resulting ionized air path forms a channel capable of carrying high-current electrical beams, including high-voltage discharges and lightning strikes (Figure 10.16) [Rodriguez 2002] [Rubenchik 2014].

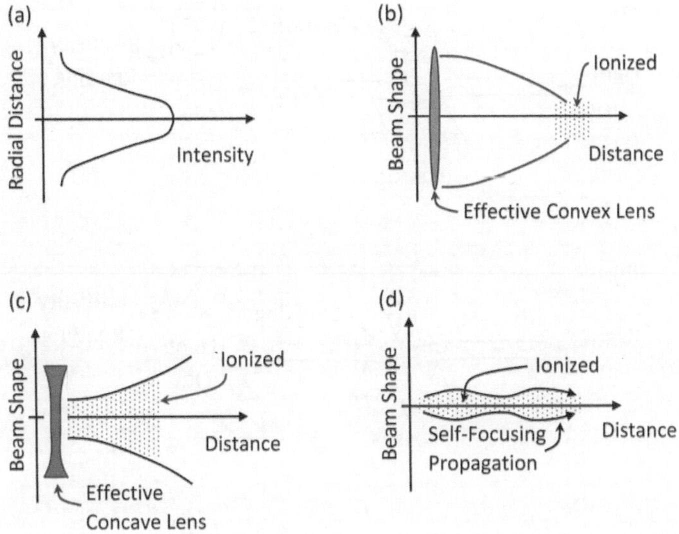

FIGURE 10.15 Competing focusing Kerr and defocusing ionization effects create self-focusing beam. (a) Radial intensity of laser beam at launch. (b) Kerr effect focuses laser beam to ionize air. (c) Ionized air defocuses laser beam. (d) Combined effect is self-focusing beam. (Adapted from [Couairona 2007].)

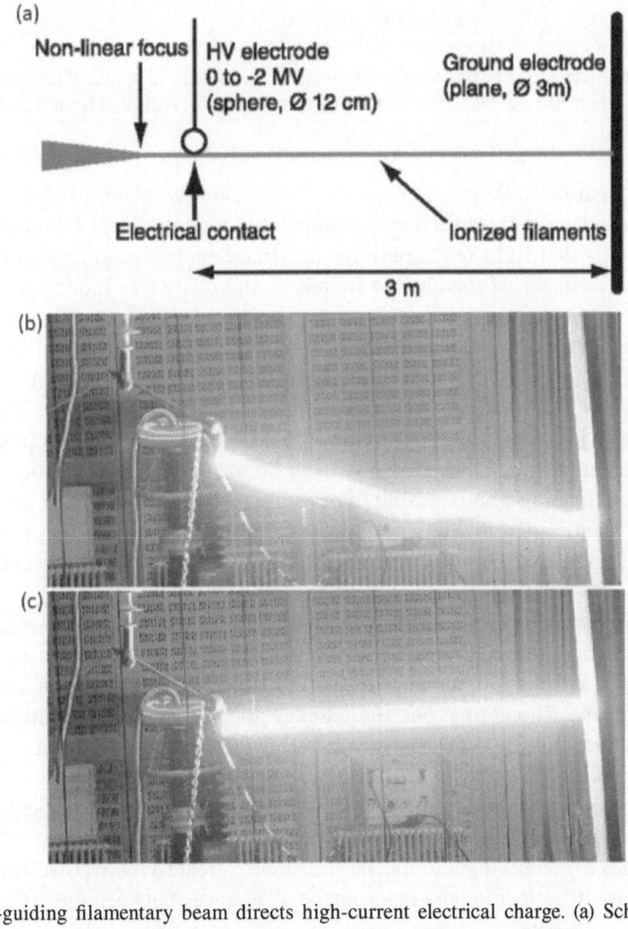

FIGURE 10.16 Self-guiding filamentary beam directs high-current electrical charge. (a) Schematic of experimental setup. (b) Discharge without guiding. (c) Discharge with guiding. (From [Rodriguez 2002].)

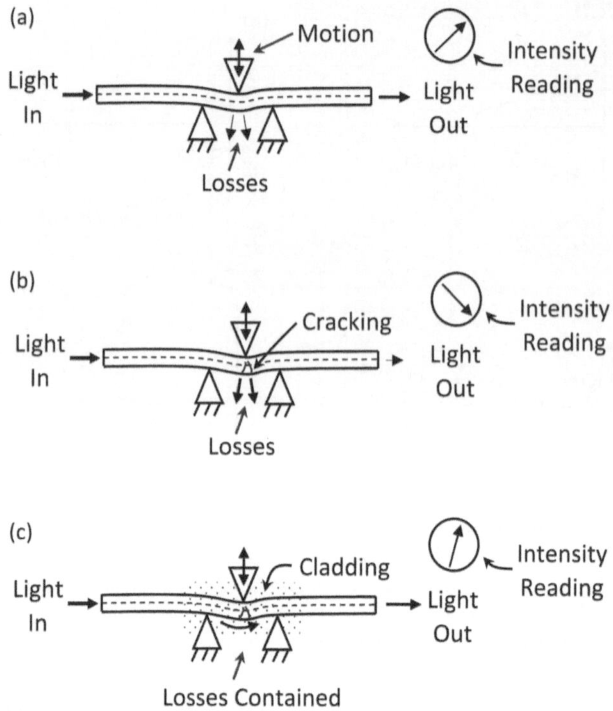

FIGURE 10.17 Photoactive cladding heals break in fiber-optic waveguide used in microbend sensor. (a) Microbend sensor. (b) Cracking causes sensor loss. (c) Photoactive cladding restores sensor. (Adapted from [Li 2008d].)

Self-guiding also occurs in photoactive solids. The technique generally uses the energy present in a light beam to produce refractive index changes to steer the light beam in a self-reinforcing process. A key effect is that light with spatially localized energy produces a corresponding localization of the refractive index. If the change increases the refractive index, then an increase in the gradient focuses the light at the center of the spot, similar to the Kerr effect in Figure 10.15. This in turn localizes the refractive index increase at the center of the spot, leading to self-reinforcing propagation of up taper and a filamentous waveguide. Among the possibilities are UV-curable epoxies stimulated by UV light, silica glasses subjected to femtosecond laser pulses, and photosensitive polyimide resins [Frisken 1993] [Yamada 2001] [Kagami 2001]. It is possible to align fiber cores that transmit light with UV-curable epoxies to create self-aligning optical waveguides [Song 2011] [Jradi 2008]. Preplacing photocurable material on the periphery of an optical fiber enables a more aggressive self-healing approach where damage to the fiber, including cracking, causes light to leak, which initiates a localized healing reaction to reconstruct the waveguide (Figure 10.17) [Li 2008d].

Lasers use specialized waveguides made of matter which receives and releases electromagnetic energy at specific wavelengths and phases to produce coherent beams of light. Lasing molecules often suffer damage from heating and radiation, leading to spectral broadening and loss of coherence of the laser beam. Sometimes, as with Rhodamine 6G, the damage appears as a reconfiguration of the molecules into a higher energy state. Turning off the laser and letting the system rest permits the laser molecules to revert to a desired low-energy state, which heals the laser performance [Anderson 2015]. More aggressive healing uses the lasing medium in a metal organic framework hydrogel that self-heals (Figure 10.18) [Zhu 2023].

There are advantages to combining multiple laser beams to create a more powerful laser. This is a nontrivial task due to phase, wavelength, and mode shape differences between the different beams. Lack of

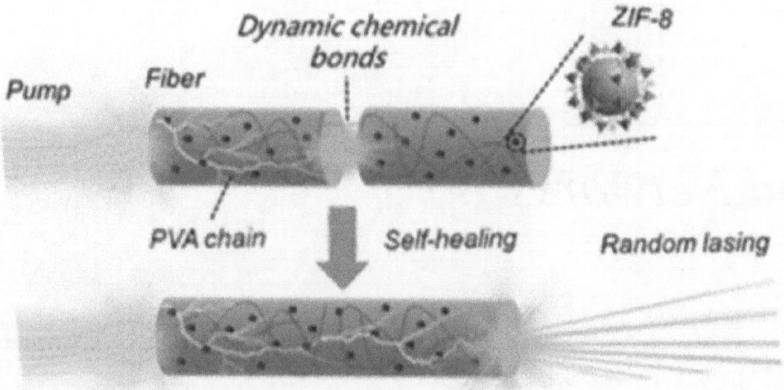

FIGURE 10.18 Self-healing hydrogel laser. (From [Zhu 2023].)

coherence, cross talk, and feedback among the laser beams limits performance. Coupling the beams with evanescent waves introduces a healing effect that separates the combined beam into a coherent multi-laser strong beam and a separate multi-laser beam that dissipates the noncoherent parts. Supersymmetry provides a framework for explaining this beam-healing effect as a stable dissipative structure far from equilibrium [El-Ganainy 2018] [Qiao 2021].

11

Systems and Networks

11.1 Introduction and General Principles

Systems and networks are collections of entities that interact in largely predictable and controlled manners. The collective behavior differs and may be superior to the sum of the individual component behaviors. The distinction between a system and a network is not crisp. Systems tend to have distinct boundaries that separate themselves from an environment. Networks have less distinct boundaries. The interconnections between components in a network often dominate behavior and performance.

Autonomous systems operate and maintain themselves without human intervention. Features that aid in the long-term endurance of autonomous systems are as follows [Truszkowski 2009]:

1. *Self-configuring:* Configure operational modes, geometries, and topologies as needed by the circumstances.
2. *Self-optimizing:* Learn and improve from experience.
3. *Self-healing:* Recover from damage.
4. *Self-protecting:* Protect themselves from potential damage-causing threats.

Networks are loosely defined as systems comprised of interconnecting subsystems. Examples include public utilities, communication networks, social interactions, sensor grids, and protein networks in biological systems. Damage to the subsystems, interconnections, or organization degrades the performance, operation, and resilience of a network. A variety of techniques enable networks to self-heal. Many healing techniques reconfigure the interconnection topology.

Desirable features in a resilient and self-healing network are as follows:

1. *Lightweight:* The overall network is lightweight in terms of resources needed to construct. This reduces the cost of original construction and repair.
2. *Robust:* Local nodes and links are inherently robust.
3. *Graceful Failure:* Failure of local nodes leads to a graceful redistribution of function, while not causing propagating failure modes.
4. *Self-preserving Large Load Responses:* Macroscopic responses to large-scale loads act to mitigate the possibility of macroscopic damage while preserving functionality.

An example from nature is the spider web [Cranford 2012]. Trapping insects with minimal biological cost is a primary function. A web is an exceptionally light heterogeneous network of silk threads. Spider silk is a lightweight, highly resilient material that resists axial tension loads with a steep nonlinear stiffening behavior (Figure 11.1). The properties of the radial and spiral threads differ. Application of a point load to a strand induces nonlinear geometric and material behaviors. The compliance of the silk at low strains combined with geometric rotation enables large movement of the applied load before stiffening occurs – a behavior that helps to trap insects, while dissipating and/or redirecting severe loads to localize onto a small part of the web.

DOI: 10.1201/9781003396048-11

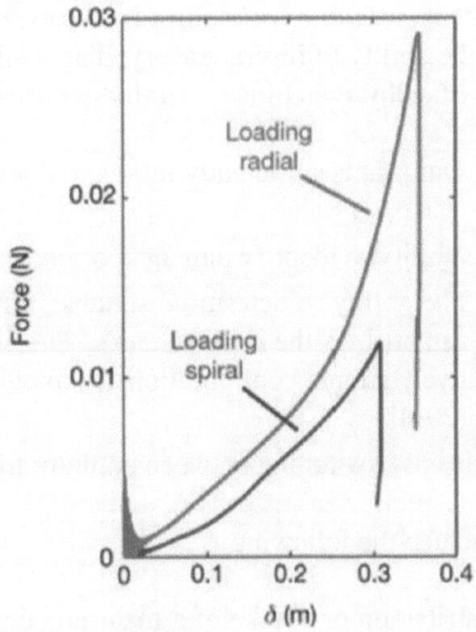

FIGURE 11.1 Nonlinear stiffening behavior of spider silk thread with the radial threads being stronger than the spirals. (From [Cranford 2012].)

If the load is too severe, the silk breaks and confines the point load damage to a small region. The web usually retains span-wise continuity, which allows for easy reconstruction of damage (Figure 11.2).

Stable networks have interconnection topologies that tend to not change when stressed. Dynamic networks add, delete, or otherwise alter the characteristics of the interconnections as needed. Examples include transportation or other distribution networks. Sometimes small changes dramatically alter the behavior of networks. Potential network failure modes include (1) inability to deliver services and goods, (2) collapse, (3) partial degradation, and (4) overgrowth. Networks with nodes that fail randomly, and fail when neighboring nodes fail, may be fragile and undergo catastrophic collapse. Networks that are close to tipping points of collapse exhibit precursor behaviors such as slow recovery for minor perturbations [Veraart 2012]. Conversely, networks with nodes that both fail and heal randomly, and fail when neighboring nodes fail, can undergo phase-change-type swings between collapse and recovery. Financial markets and brains seizures are examples [Majdandzic 2014]. Anti-fragile systems thrive on environmental variabilities [Taleb 2013]. Some even get stronger.

A wide variety of systems and networks self-heal. Each differs in detail. Some general principles are as follows [Xiao 2007]:

1. *Detection:* Sensors measure quantities generated by fault conditions.
2. *Assessment and Prognosis:* Intelligent condition and prognosis assessments based on sensor data and system information.
3. *Recovery Capabilities:* Method of recovering from damage, including switching loads onto redundant standby subsystems and operations continuing, but at reduced levels of performance.

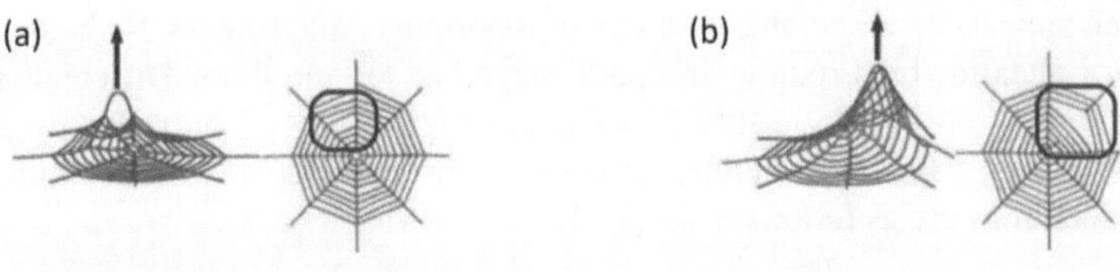

FIGURE 11.2 Geometric and material nonlinearity focuses severe point loads on radial arms of spider wed, which leads to localized failures that preserve continuity of the web. (From Cranford 2012].)

Factors favoring standby subsystem switching for healing are the availability of relatively low-cost and lightweight standby systems and the ability to insert sensory, diagnostics, and control methods, also at low cost, into the standby system. Standby switching generally requires the following:

1. *Mild to Moderate Damage:* Damage is sufficiently mild and does not disable supervisory control of the system.
2. *Damage Identification:* The ability to identify damaged or underperforming subsystems.
3. *Reconfiguration Planning:* The ability to determine suitable, alternative configurations using redundant subsystems that can still do the desired tasks. The strategy may be to use a preplanned approaches for relatively simple configurations or may be adaptive based on local or global rules [Quattrociocchi 2014].
4. *Ease of Switching:* Relatively easy switching between configuration states.

Network self-healing metrics include the following:

1. *Performance of a Task:* Distribution or uptake of a material commodity. Is it being delivered as desired?
2. *Use Redundant Connectivity to Reroute Transport in the Event of an Outage:* Statically indeterminate structures that gracefully shift loads are an example.
3. *Strength of Connectivity:* Measures of the ability to modify strength of connectivity.
4. *Node and Connection Addition:* Measure of the ability to add new nodes and connections.

Some approaches to dynamic reallocation are as follows:

1. *Peer-to-peer Network:* Graph measures, such as path length between two nodes, width and nodal degree, quantify the health of the network. One mode of damage to peer-to-peer networks is the removal of nodes and the associated edges. If a network has the ability to respond by adding edges, but not new nodes, then it is still possible to recover health metric. The Forgiving Graph algorithm assigns new edges based on need [Hayes 2012].
2. *eDNA:* A biomimetic architecture for fault-tolerant switching between subsystems. *eDNA* is viable in applications that include the control of a liquid-crystal interferometer for space-borne sensing applications, but exhibits a high parasitic computational load [Boesen 2011].
3. *Digital Twin:* A system can run an external simulation of its operating behavior and configuration known as a digital twin. If deviations of the simulated versus observed behaviors occur, twins help to diagnose the fault and recover [Garlan 2002].

11.2 Communication Networks

A communication network graph consists of nodes with attributes of being sender, receiver, transceiver, data buffer, data processor, and edges as communication channels that connect two nodes allowing one-way or two-way communication through a variety of physical layers. Damage to digital data communication networks often appears as inoperable data transmission links and routers. Redundant transmission paths enable network healing by rerouting around damaged nodes and links. Different network topologies facilitate healing. A self-healing metric is the ability to switch connections between nodes within the context of constraints, such as the number of connections per node and the total number of nodes. Some switching modalities are as follows:

1. *Nonlocal Free Space:* Any node can connect to any other node.
2. *Local Free Space:* Any node can connect to any other node within a local neighborhood.
3. *Wired:* Physical constraints of wiring fix the connections in place. Rerouting is the option.

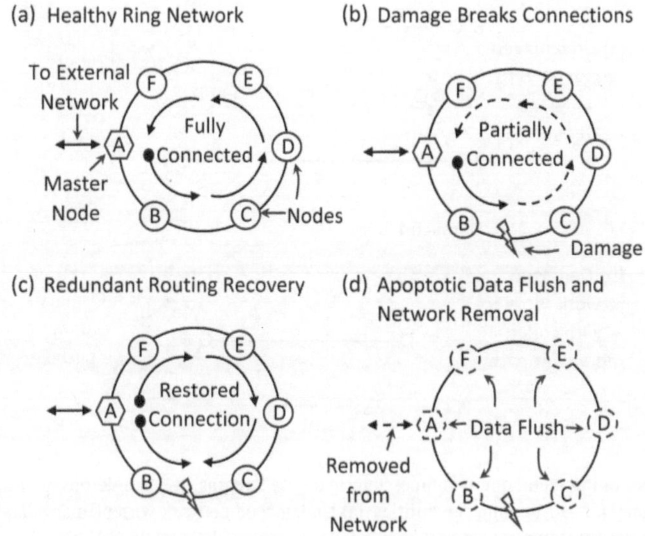

FIGURE 11.3 Ring network for digital communications. (a) Healthy network provides connectivity to all nodes and external network. (b) Damage restricts connectivity. (c) Redundant rerouting restores connectivity. (d) Apoptotic actions flush data from nodes and remove ring from external network. (Adapted from [Lee 2012b].)

The ring topology links nodes into a ring arrangement (Figure 11.3). When fully connected, the ring offers two different communication paths between nodes. Damage to a single link or node leaves a redundant path, around the other way, for connectivity, but in a degraded performance mode. Damage to more than one link leaves functional subnetworks. Intersecting links improves redundancy. When needed, apoptotic actions flush data from the nodes and remove the ring from communications with external network. Modified priority protocols improve communication performance [Lee 2012b]. Placing a mesh network into a 3D surface with topology equivalent to that of a sphere or torus opens modalities for additional reconfiguration self-healing [Wu 2010a] [Perez 2012].

Ad hoc wireless communication networks form from spatially distributed transmitter/receiver nodes. These nodes generate information for their own signals and then transmit the signals to neighboring nodes. Alternatively, the nodes receive information in signals from neighboring nodes and retransmit the signal to neighboring nodes on the other side. Flexibility as to which nodes talk to other nodes creates possibilities for autonomously reconfiguring transmission paths. It is possible to accommodate changing node configurations, alleviate information traffic jams, and heal to recover from damaged nodes. Technical challenges facing the formation of ad hoc networks are channel access arbitration and routing determination [Elliott 2000]. Channel access arises as an issue when more than one node tries to transmit at the same time over the same node. Protocols for arbitrating between nodes mostly have one of the nodes wait for a random period and then try to resubmit to find an open channel. Waiting methods sacrifice overall transmission rate performance for the ability of nodes to transmit in arbitrary asynchronous manners. These network-healing protocols accommodate the insertion and removal of nodes and devices from the network [Pandey 2014].

Network coding uses redundant information streams to combine multiple data streams on a single physical transmission path as an alternative to redundant path routing. Figure 11.4 shows the case where a network needs to transmit data "a" from A to T1 and data "b" from B to T2. Direct paths from A to T1 and from B to T2 are available. The intermediate paths A-C-D-T1 and B-C-D-T2 are also available. The network coding approach for node C combines a and b and sends it as a + b to D, and then as a + b to both T1 and T2. The receivers decode the streams to select and choose only the data that are needed, that is, T1 selects a from a + b, and T2 selects b from a + b. Sending the same combined data over different data paths produces a communication network that self-heals following damage. The network recovers from damage to lines B-T2, C-D, and A-T1 [Aly 2012].

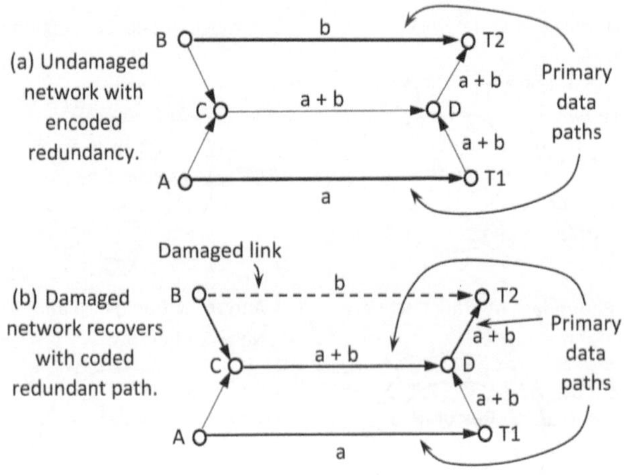

FIGURE 11.4 Network coding technique combines multiple data streams over single physical connection to increase overall network redundancy for self-healing capabilities. (a) Undamaged network with primary data paths B-T1 and B-T2 and encoded redundant path. (b) Damaged network recovers by using encoded path B-C-D-T2. (Adapted from [Aly 2012].)

11.3 Information Systems and Networks

Modern information systems and networks are dynamic and self-heal through internal state sensing and reconfiguration. Most of the techniques activate redundant network components as needed.

Sensor networks collect and transmit distributed measures of physical quantities. When the physical placement of the sensors is dense enough for the spatial domains of sensitivity to overlap, the sensor network acquires a degree of redundancy. Healthy redundant sensors provide similar readings. Defective sensors disagree with others. This creates the opportunity to remove defective sensors from the network while still maintaining adequate sensor coverage. The decision to remove sensors is easy if the sensors fail completely and give no reading. If the sensors do not fail, but instead give possibly inaccurate readings, then the decision to remove or perhaps deemphasize by weighting factors is not so simple. Sensor voting and fuzzy information combination techniques help to provide arbitration [Yen 2001].

Control systems with networks of sensors and actuators benefit from self-healing behavior. Many systems, such as airplane flight control systems, are too complicated for healing strategies based on consideration of every possible contingency. Neural network methods of reconfiguring the control system network may be practical [Urnes 2001] [Shin 2005].

Digital computers represent some of the most sophisticated human-built systems with large number of components, subsystems, and complicated operating systems. Digital computers capable of self-diagnosis and dynamic switching of failed subsystems appeared as early as 1971 with the STAR (self-testing and repairing) architecture [Avizienis 1971]. Operating systems are complex and present many possible failure and recovery scenarios. Some possible self-healing techniques are as follows [David 2007] [Saha 2007] [Appavoo 2003]:

1. *Exception Handling:* The system identifies fault events and throws the result to a handler without aborting the program. This requires prior knowledge of potential fault events and behaviors.
2. *Component Isolation:* The system encapsulates individual components and processes to isolate them from the rest of the system to prevent propagating faults.
3. *Code Reloading:* Random faults contaminate codes. Reloading and rerunning the code afresh, that is, rebooting, leads to recovery.
4. *Automatic Restarts:* Restarting the operating system on scheduled or random intervals recovers from faults that hang up the system.
5. *Watchdogs:* Internal monitors with autonomy and self-awareness that watch for fault or hang-up conditions and restart or initiate a recovery process.

6. *Adaptability:* Modification of system structure and topology to cope with faults.
7. *Observability and Traceability:* Tracking of faults and recovery techniques to enable continued improvement of system resilience.

Recovery of information systems, from worms, viruses, and cyberattacks, requires similar automated healing methods. It is essential to learn from attacks to adapt and repel similar attacks in the future [Brumley 2007].

11.4 Electrical Networks

The following are items that favor self-healing techniques in electric power grids:

1. *Small Faults Cause Large Failures:* Repairing the faults may not require much capital effort.
2. *Switching Techniques Isolate Faults:* Aggressive, agile, and intelligent methods localize the isolation to only small parts of the network surrounding the fault.
3. *Distributed Intelligence, Control, and Power Supply:* Isolated systems and network sectors recover independent of centralized control.
4. *Multipath Redundancy:* The grid topology of electric power distribution sets up redundant paths for power delivery and recovery.

Things that complicate using self-healing in electrical grids:

1. *Cascading Failures:* Local failures unbalance the grid, often with excess power that leads to propagating and cascading global failures.
2. *Legacy Technology:* Electric power grids are built over decades with equipment deriving from multiple technology nodes. Implementing new technologies, such as self-healing, is a challenge.
3. *Minimal Redundancy:* Electric power grids often run at near capacity.
4. *Decentralized Control:* Networks not under unified control, which confounds centralized self-healing techniques.
5. *Outages Disable Self-healing:* Damage to the electric power grid may disable self-healing systems that rely on electric line power.
6. *Zero Tolerance:* Customers have little tolerance for outages, even those of short duration.

The topology of electric circuits has a large influence on resilience. A simple version arises in serial versus parallel networks. In a serial network, the failure of a single component or interconnect disables an entire system (Figure 11.5). In a parallel network, failure of a single component causes only a local failure (Figure 11.6).

The operation, failure, and recovery of an electric power grid is complicated. Figure 11.7 shows a high-level view of four main states and transitions [Ćurčić 1996]. Power restoration efforts require selecting and implementing a combination of power supply, intact power lines, and choice of switching sequence. The restoration sequence is not unique. Optimization rules and algorithms guide selection of the best sequence [Hou 2011] [Miu 1997]. The following are the items to consider:

1. *Isolation:* Progressive detection and isolation of faults into smaller spatial domains that surround the fault
2. *Good Neighbors:* Finding neighboring parts of the grid that are energized with viable transmission paths
3. *Limited Power:* Determining acceptable operating points for the network during transient levels of recovery and restoration
4. *Limited Transients:* Determining how to manage transients during load restoration
5. *Priorities:* Restoring service to priority customers on a timely basis

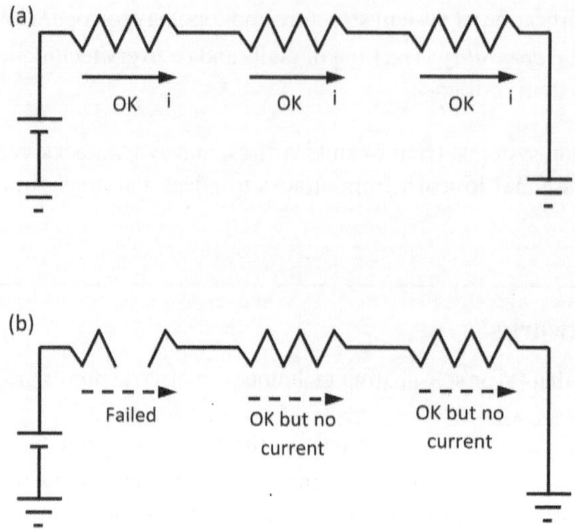

FIGURE 11.5 Single-component failure disables entire system in a serial network. (a) All components healthy and working. (b) Single-component failure disables entire system.

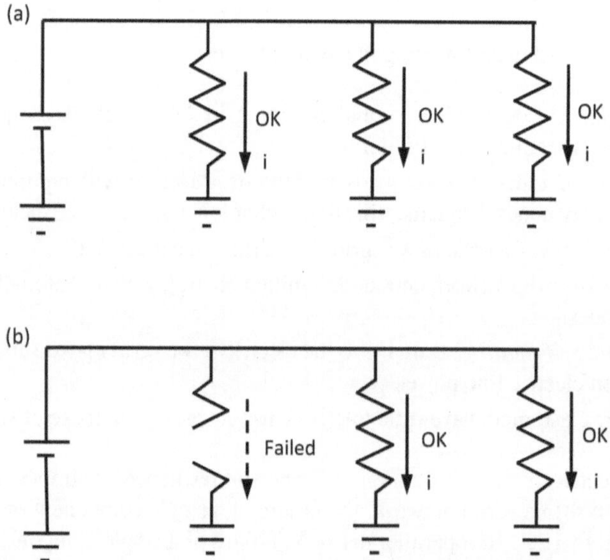

FIGURE 11.6 Parallel circuit is resilient and operates even if individual components fail. (a) All components healthy and working. (b) Single component failure does not disable entire system.

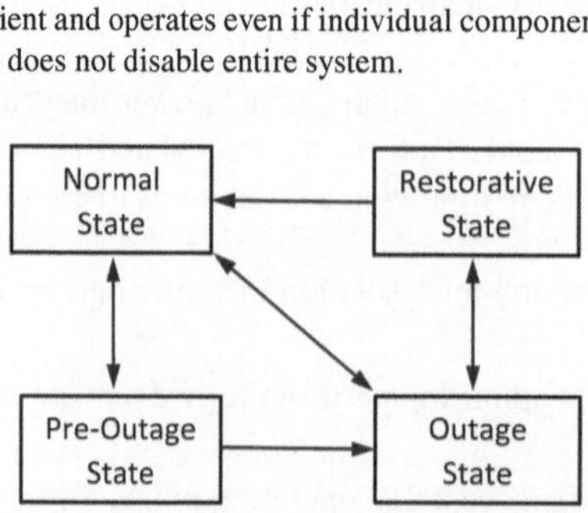

FIGURE 11.7 Electric power grid system operating states, with failure and recovery transitions. (Adapted from [Ćurčić 1996].)

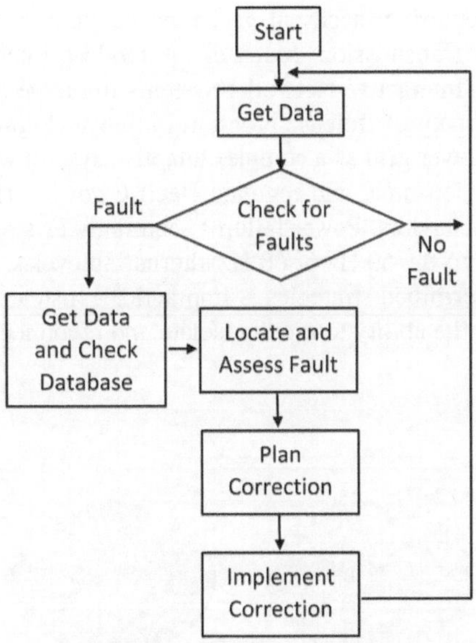

FIGURE 11.8 Flowchart for a shipboard self-healing power system using combination of automated failure assessment and graph-based failure recovery strategies to restore power to critical systems in a prioritized sequence. (Adapted from [Butler 2004].)

Use of fault detection, isolation, and restoration schemes reduces average power restoration times for undamaged parts of the electric grid from hours to minutes [Angelo 2013].

The electric power systems that run ships at sea use self-healing as a built-in feature. Ships use three-phase electric power grids with multiple redundant power generators. Parallel configuration of the circuits combined with intelligent switching enables recovery from faults. While humans normally sense and assess damage, develop reconfiguration plans, and implement reconfiguration, many of these steps can be automated (Figure 11.8) [Butler 2004]. One is formulating an optimal reconfiguration plan that both satisfies the physical constraints of power flow restoration and prioritizes the limited power distribution. Due to the complexity of the interactions, it is often not practical to preplan for every contingency. An effective alternative solves the optimization problem quickly using modest computing resources, such as with mixed integer and graph-based methods.

The following are some of the considerations for the healing of electric power transmission networks:

1. *Continuity:* Electrical systems need continuous paths for the current to flow.
2. *Danger:* The amount of power transmitted is large enough to do serious damage to equipment or cause personal injury.
3. *Synchrony:* The flow of electricity through the various components on the network are synchronized and tightly coupled. For example, the North American electric power grid operates at 60 Hz, with large sectors operating in phase synchronization. Tight coupling means that connected components must stay connected with synchronized frequency and phase for everything to work. The drop of a single connection or component may cause unfavorable buildups of electricity or loss of synchronization elsewhere in the network, sometimes leading to cascading failures, that is, blackouts.
4. *Transient:* Electricity is transient and difficult to store in bulk.
5. *Uncoordinated Autonomy:* Networks have multiple users and generators that sometimes act independently, and other times act cooperatively and coherently, in often uncontrollable and marginally predictable manners.
6. *Difficult Geometry:* Geometric constraints make it difficult to repair wires.
7. *Live Switching:* Dynamic load allocation systems and live network configuration are possible forms of self-healing.

The importance of a self-repairing electrical grid was recognized in some of the earliest days of electric power generation and transmission. Much of the modern interest initiated at the turn of the millennium with the Complex Interactive Networks/Systems Initiative (CIN/SI), a joint Electric Power Research Institute–US Department of Defense program initiative (Figure 11.9) [Amin 2000]. This initiative examined the electric power grid as a complex adaptive system with a mix of tightly and loosely coupled agents that generate, consume, and regulate electric power. The grid exhibits collective and difficult to control emergent behaviors. Power failures seemingly arise with little warning from minor events, often at speeds related to the 50 Hz or 60 Hz alternating cycles. Using global top-down control of these networks with predetermined strategies is impractical. Instead, networks use hybrids of local and global control, along with the ability to rapidly isolate and reconnect parts of the system for failure mitigation and recovery.

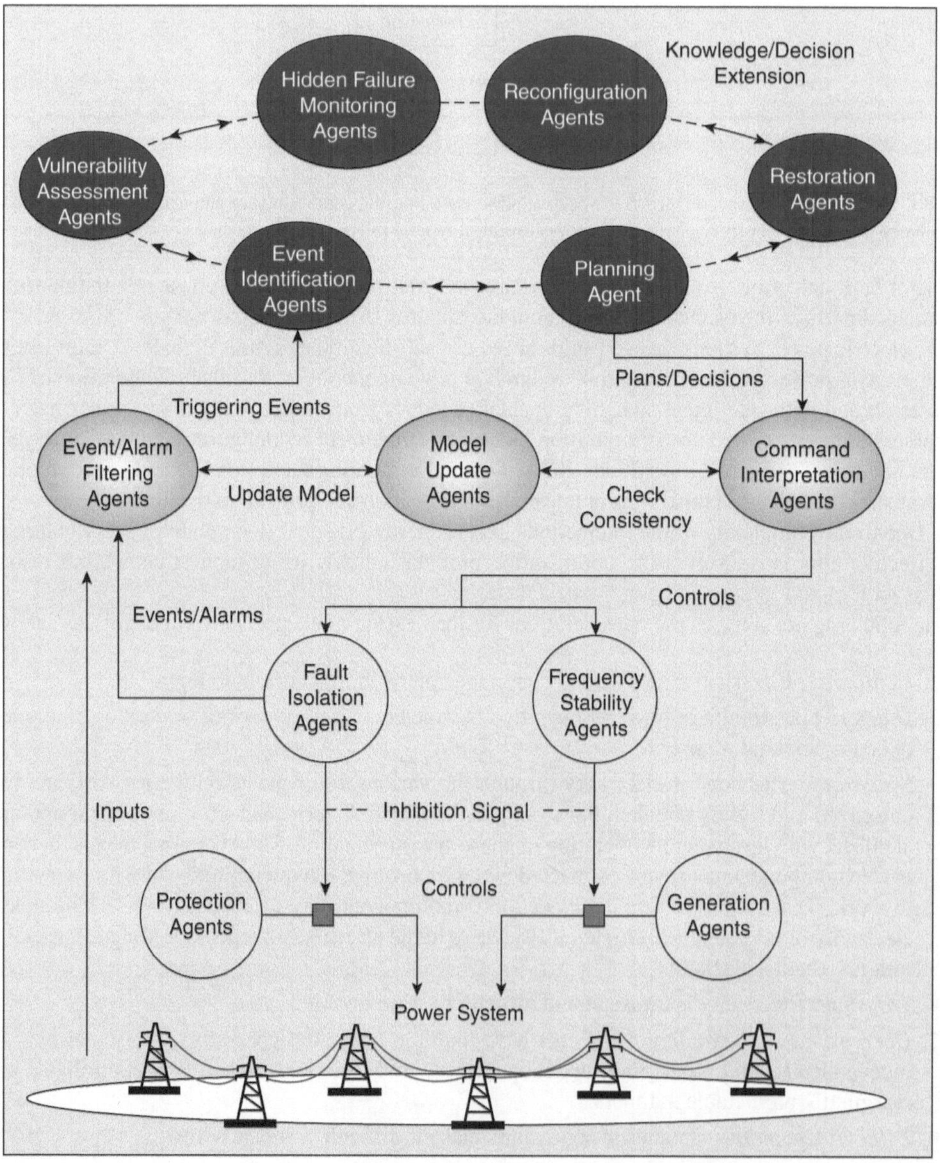

FIGURE 11.9 Self-healing grid as a set of interacting semiautonomous agents. (From [Amin 2001].)

11.5 Pipeline Utilities

Important commodities, that is, drinking water, wastewater, storm water, steam, gas, and petroleum products, move through pipeline networks as liquids, slurries, or gases. Damage to fluid-carrying pipeline networks causes leaks, failure to deliver material as needed, and contamination by leakage into and out of the pipes. At a network level, the primary means of autonomic repair is to shut off leaking sections and to switch into service preexisting redundant network components.

11.6 Swarms and Flocks

Flocks of birds, schools of fish, ant colonies, and fleets of autonomous vehicles are cooperative groups with local agents acting both autonomously and collectively. An interesting feature of many collective systems is that the sensing, intelligence, and control are distributed among the various individual agents, with no strong leader. In both the animal and plant kingdoms, collective groups are very effective in accomplishing and/or adapting to difficult tasks. In many respects, multicellular organisms are also collective groups. The notions of self-repair and self-identification apply to both the group and the individual cells. Engineered swarms of cooperating and autonomous agents appear in a variety of applications. An interesting biomimetic example is fire ants that form rafts, bridges, and bivouacs by linking onto one another [Foster 2014]. The heterogeneous size distribution of fire ants allows for smaller ants to fit in between bigger ones and to increase the average number of links per ant and the overall strength of the structure.

There are several strategies available for organizing kinematics and control of swarms of cooperating autonomous vehicles. These range from a global supervisor and controller that coordinates all of motions of the vehicles to local control where each vehicle has a control law that causes it to form into a flock. An intermediate level of control has the vehicles coordinate by communicating among themselves.

Possible uses of cooperative agent collectives in self-healing are as follows:

1. *Self-healing of the Collective:* A desirable characteristic of robot swarms is that the individual robots have simple localized systems for control, communication, and movement, while the swarm executes robust programmable macroscopic behavior. For example, experiments with a 1,000-robot swarm indicated the ability to assemble into preprogrammed shapes using robots with the three simple behaviors of edge-following, gradient formation, and localization along with a bitmap of the desired shape [Rubenstein 2014]. Key to success is the ability to identify and move around defective immobile and sluggish robots, and to recover from damage to the collective shape. Figure 11.10 shows some of the robust shape forming capabilities of the 1,000-robot swarm.

   Load-responsive control is an alternative control scheme for modular robots. Rather than striving to achieve or maintain a particular geometric configuration, load-responsive control moves robot modules in a manner that alters the internal loads. A typical maneuver adds modules on demand at stress concentrations, in a manner that mimics Wolff's law for bone remodeling (Figure 11.11) [Suzuki 2011].
2. *Collective Performance of a Repair:* Such situations appear in biology, especially with social insects where simple local rules enable the seemingly uncoordinated construction of structures significantly larger than the builders, that is, ant colonies, beehives, and termite mounds. Teams of robots mimic preprogrammed behavior that follows local rules and respond to the present configuration to build large structures [Werfel 2014]. Embedded in each robot is information, perhaps only localized, regarding the final desired structural configuration.

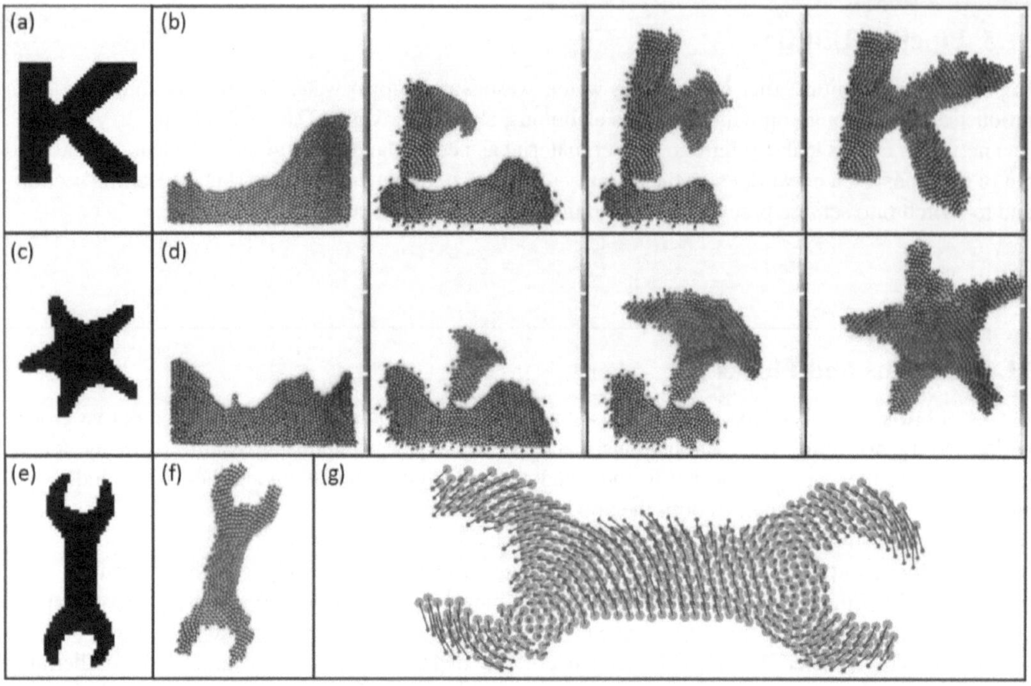

FIGURE 11.10 Robust assembly with embedded self-repair of 1,000-robot swarm, with each dot representing a single autonomous robot. (From [Rubenstein 2014].)

3. *Hybrid Performance of a Bigger System:* Structural systems with arrays of active struts collectively move to a position of homeostasis by relatively simple rules and local communication between struts. An example is a structure that maintains a specific shape even though certain elements have externally varying loads or displacements (Figure 11.12) [Huston 1998] [Yu 2006]. How to control? One method is an abstraction based on the concepts of agents and homeostasis. The agents a_i are the active struts that can change length or angle. The agents also sense their position relative to the neighboring connecting agents (x_i–x_j),

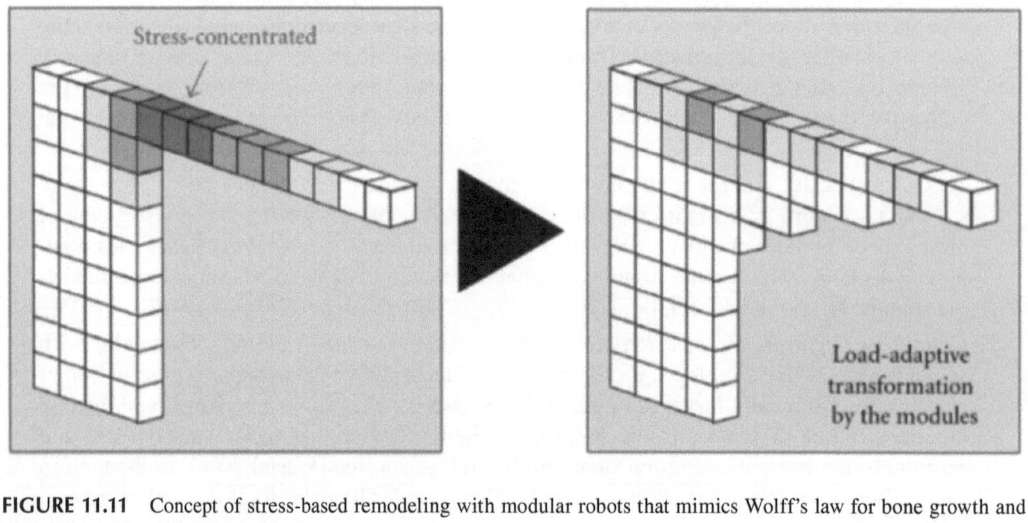

FIGURE 11.11 Concept of stress-based remodeling with modular robots that mimics Wolff's law for bone growth and remodeling. (From [Suzuki 2011].)

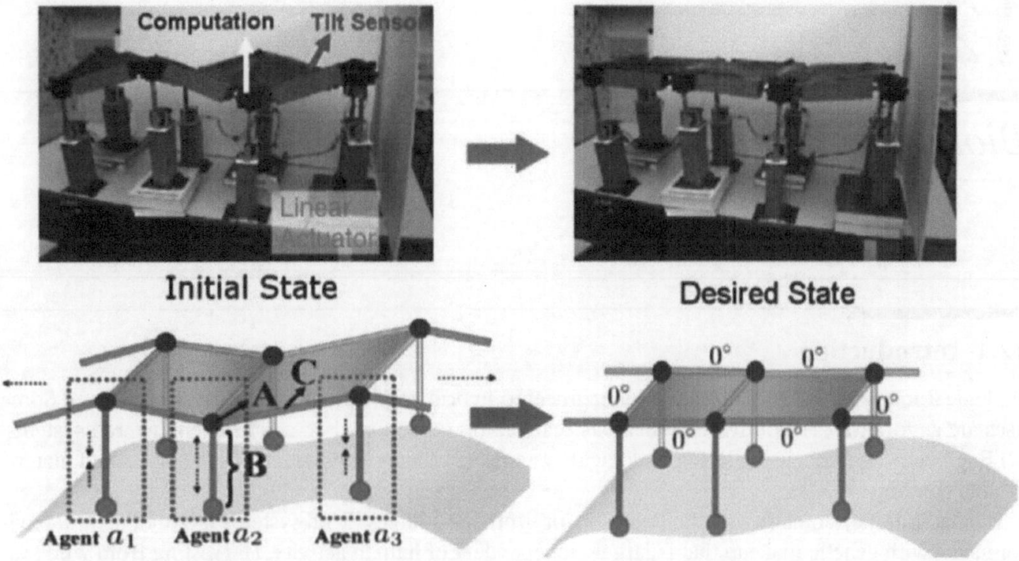

FIGURE 11.12 Structural system with active struts acting as agents preserving homeostasis, with a self-leveling bridge that compensates for movement of the legs. (From [Yu 2006].)

and then make simple calculations for adjusting the lengths or angles. The structural system is given a particular configuration that it attempts to maintain, that is, preserve homeostasis by attempting to maintain Δ_{ij}. External disturbances, such as base motions or applied forces, deform the structure into a configuration that differs from homeostasis. The agents detect the disturbance as motion relative to their neighbors and correct the disturbance by local behaviors. A simple control law produces an overall system that is robust and reconverges to homeostasis under a wide range of conditions:

$$x_i(t+1) = x_i(t) + \alpha \sum_{a_j \in N_i} \left[x_j(t) - x_i(t) - \Delta_{ij}^* \right] \quad (11.1)$$

Here, N_i is the set of neighboring agents that connect to a_i, with total number of members $|N_i|$. The parameter α is a constant with value $0 < \alpha < 1/|N_i|$.

Apoptotic behavior heals swarms by removing defective members from the swarm, leaving the swarm with fewer members, but with an overall better state of health [Truszkowski 2009].

12

Biological Hybrids

12.1 Introduction

Biological material forms the basis of bioengineered hybrid self-healing systems [Brochu 2011]. Some uses are as follows: (1) Nonliving biomass as scaffolding for self-healing action in engineered systems. (2) Engineered systems using living biological systems. (3) Biological systems using engineered materials and components.

Biomaterial may come from wild, domestic, or synthetic sources. Wild systems thrive in natural environments with genetic makeup that is largely independent of human activity. Harvesting from wild systems is often a low-cost source of robust material but can be difficult to control. Wild biological systems have evolved materials and functional structural behaviors that promote individual survival and species propagation. Serving the needs of humanity is a matter of happenstance. For example, trees use wood to stand up and thrive. These requirements make wood an excellent material for buildings, furniture, and baseball bats. Domestication by selective breeding, genetic engineering, or behavior modification alters the characteristics of the wild system to be more compatible with human needs. Successful domestication generally follows the "Anna Karina" principle. The domesticated life-form must satisfy simultaneously numerous functional criteria to synergize with human society [Diamond 2005]. It is only possible to domesticate those wild systems that are sufficiently close to meeting multiple criteria of compatibility. For example, wild horses, canids, apple trees, and Indian elephants are readily domesticated, but zebras, raccoons, oak trees, and African elephants are not.

12.2 Homeostasis Driving Healing

The exceptional functionality of living biomaterials, organisms, and ecosystems makes for useful self-healing materials. Figure 12.1 shows some possible combinations. Maintenance of homeostasis, that is, keeping internal chemical, mechanical, and thermal states within acceptable ranges, is the universal feature of living biological systems. From a thermodynamic point of view, the difference between homeostatic internal states and the external environment represents a state of lower entropy (higher negentropy) [Schrödinger 1944]. Homeostasis transports entropy produced by internal engines of life out into the environment. Some of the by-products of these processes heal inanimate systems.

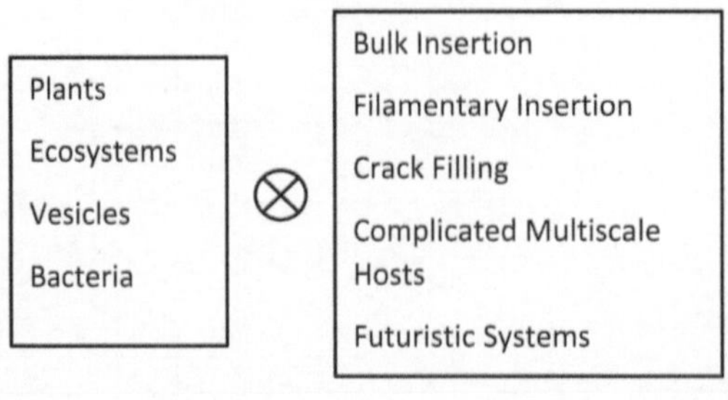

FIGURE 12.1 Some possible combinations of living biological systems used in self-healing.

DOI: 10.1201/9781003396048-12

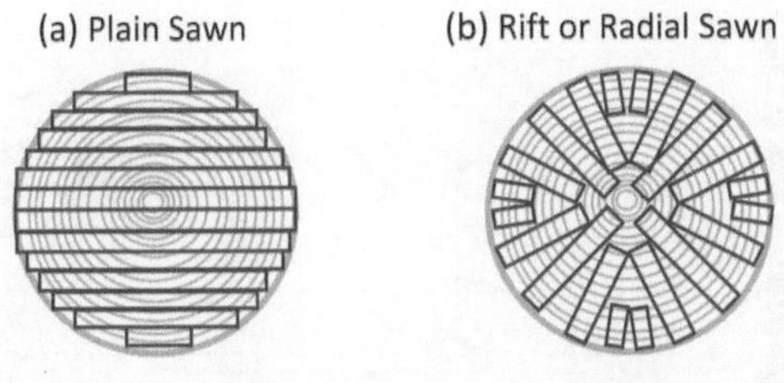

FIGURE 12.2 Saw cut geometry in wood log affects warping and swelling anisotropy. (a) Plain sawn boards swell nonuniformly when wet and warp. (b) Rift or radial sawn boards swell uniformly when wet and tend to not warp, but waste wood.

12.3 Nonliving Biomass Materials

Nonliving materials derived from the formerly living often retains functionality with potential for adaptation to self-healing. An example is water-induced swelling of wood that self-seals containment vessels, for example, ships, barrels, and pipelines. The amount of swelling is anisotropic and depends on the direction of the grain. Most logs have circular symmetry. Radial cuts form boards with homogeneous anisotropic properties to provide uniform and controlled swelling, with minimal warping, at the expense of wasted material (Figure 12.2). The more common plain sawn or bread loaf cut is less wasteful, but the boards warp when soaked in water.

Plant-derived fibers are strong and pliable. Cellulose fibers reinforce cement mortar and adobe bricks. Unreinforced cement mortar is strong in compression, but weak in tension, which leads to cracking, especially drying shrinkage cracking. Inserted plant fibers minimize cracking rates and crack widths, and promote autogenous crack healing, especially in moist atmospheric conditions [Toledo 2005].

Collagen is a matrix of animal-derived connective tissue protein. It is a coating in a variety of applications requiring biocompatibility. Abrasion wears collagen coatings, which then self-heal through a water-induced swelling-driven regeneration of subsurface material [Kim 2016a].

12.3.1 Engineered Systems Using Living Biological Systems

Nonliving biomaterials self-heal at the molecular scale. One set of techniques uses the orthogonal binding nature of DNA molecules to build useful molecular structures. An example is a tensegrity structure built out of DNA molecules (Figure 12.3) [Liedl 2010]. These molecular structures have many of the properties of macroscale tensegrity structures with compression struts and tension members and structural-kinematic self-healing properties. Molecular functionality extends to other DNA-based applications, such as self-assembling and self-directing conductive nanowires [Chen 2010]. DNA also forms dynamers with reversible bonds in the backbone molecules of linear polymers [Xu 2004].

12.4 Living Biohybrids

These are self-healing biohybrids using living organisms.

12.4.1 Bacterial Hybrids

At the microscale, sulfur-oxidizing bacteria, such as *Thiobacillus*, reduce and remove black organic lichen buildups from concrete [DeGraef 2005]. The bacterium *Bacillus subtilis* binds to metals and nucleates reactions that form metal phosphates, metal sulfides, and organometallic polymers, producing diagenetic solidification of soils [Beveridge 1983].

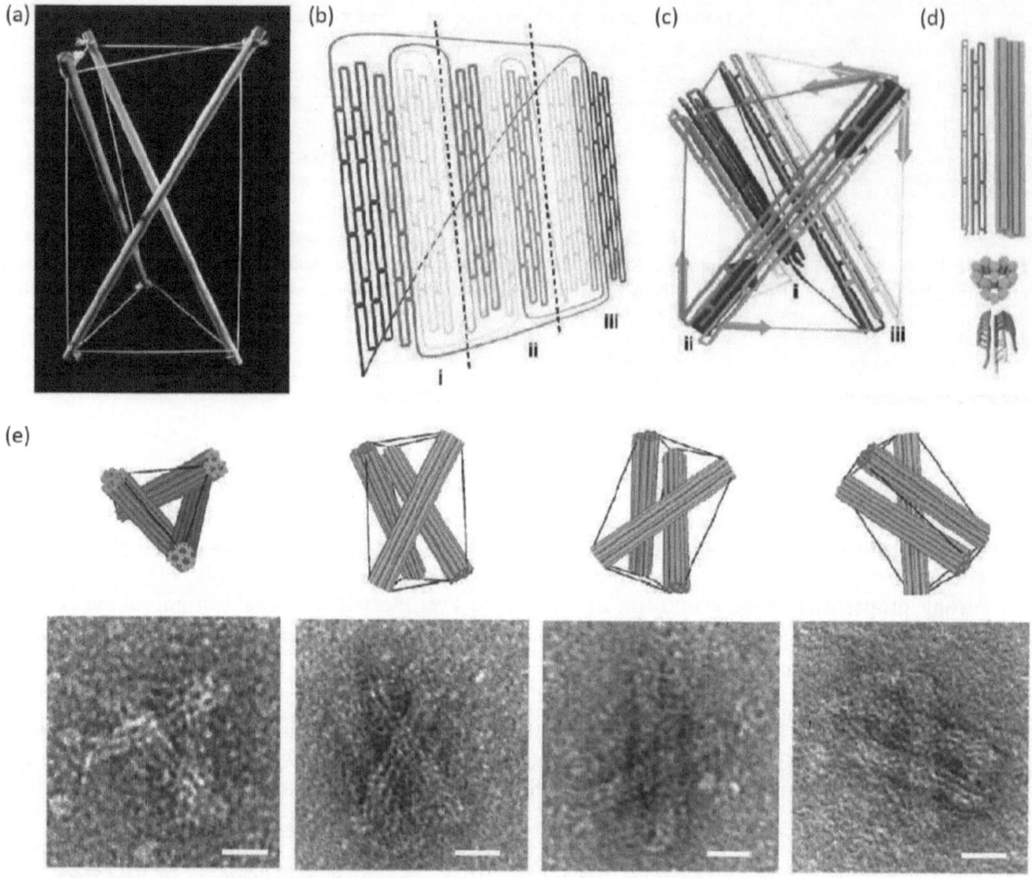

FIGURE 12.3 Molecular scale DNA-based tensegrity structures. (From [Liedl 2010].)

12.4.2 Entire Organisms

Entire multicellular organisms may be self-healing or self-cleaning components of larger systems. Living snow fences (LSFs), that is, trees, shrubs, and standing corn rows, prevent the buildup of snow on adjacent roads during winter conditions. The living fence is less costly than snow removal [Wyatt 2012]. The primary competition with LSFs are wood slat fences. LSFs have some advantages. The service life of an LSF is 50–75 years versus 20–25 years for a slat fence. The overall lifetime cost of ownership is less than that of a slat fence. Living mature trees provide up to 12 times more snow capture but require an offset from the road into the adjacent field of 35–70 m [Sundstrom 2015]. A different cleaning application is air inside buildings. Plants remove volatile organic compounds from the circulating air [Darlington 2001].

12.4.3 Ecosystems

At a high level of organization, ecosystems of organisms may collectively affect the health of structures. Biofilms with multiple species of bacteria and other small organisms are particularly active in corrosion acceleration and/or prevention. Harnessing the activities of a biofilm can be advantageous. Steps to help form a bacterial biofilm on a surface are as follows: (1) Reversible attachment of cells to the surface. (2) Use of extracellular substances to create an irreversible attachment of the cells. (3) Cells proliferate and form encapsulating layers. (4) The cell community grows into a 3D structure. (5) The community releases cells into the surrounding fluid so as to populate and create new biofilm communities [Renner 2011].

12.5 Anti-biofouling

Anti-biofouling is a self-cleaning process that removes and/or prevents the buildup of unwanted organisms, both single and multicell. Environmental compatibility is a concern. Anti-biofouling activity should be limited to the local material and structure to prevent collateral injury to non-fouling organisms [Almedia 2007] [Yebra 2004]. The variety of organisms often precludes a single universal effect from limiting adhesion. Some species adhere strongly to surfaces that prevent the buildup of other species.

12.5.1 Mechanical Anti-biofouling

Mechanical action prevents biofouling. Most organisms find it difficult to attach and thrive on flexing elastomers [Shivapooja 2013]. Being slippery prevents organisms from getting a good grip onto surfaces. Oil-impregnated silicone is a possibility for marine structures [Stein 2003].

12.5.2 Microtextured Surface Anti-biofouling

Most organisms thrive within specific ranges of geometric and mechanical milieu. Specialized surface textures are anti-biofouling. Examples include biomimetic surfaces with microtextures similar to sharkskin or insect wings [Diu 2014]. This sharklet texture on an elastomeric surface is highly resistant to biofilm formation [Chung 2007] [Schumacher 2007]. Black silicon surfaces with a microtexture that absorbs most visible light are also bactericidal [Ivanova 2013]. Vertical arrays of silicon nanowires prevent biofouling while also providing a means of slow drug delivery [Brammer 2009]. Hydrophobic and omniphobic surfaces prevent bacterial buildup on medical devices, such as catheters, without resorting to toxic chemicals [Ding 2012]. Nano- and microtextured perfluorocarbon surfaces combine anti-biofouling geometry with a relatively bio-inert material [Leslie 2014] [Gudipati 2005].

The design of anti-biofouling microtextured surfaces is complicated. The design must accommodate multiple competing interactions. The use of artificial intelligence and neural network techniques shows promise. One approach yields a microtexture pattern for urinary catheters with texture that confounded the movement of *Escherichia coli* bacteria that swim upstream. The texture creates vortices that redirect the bacteria downstream, unlike a smooth surface that encourages the bacteria to swim upstream and cause infections [Zhou 2024].

12.5.3 Energetically Active Surfaces Incompatible with Biofilm Growth

Materials with energetic and active surfaces resist biofouling. Zwitterionic molecules have separate negative and positively charged regions that form hydrated anti-biofouling [Chen 2007] [Cheng 2007b]. Similarly, attaching PDMS to amphiphilic fluorinated block copolymers creates a material that prevents the adhesion of both zoospores and diatoms in marine environments [Sohn 2006].

Light-sensitive coatings exert bactericidal effects. While UV light is often used, it is possible to create surfaces that activate under a broad spectrum of light using gold nanoparticles and multiple wavelengths of dye sensitizers [Noimark 2014].

Filters made of electrically activated silver nanowires combined with carbon nanotubes and cotton kill and/or filter waterborne bacteria with >98% removal rates [Schoen 2010]. Janus particle microrobots with bactericidal silver nanoparticles swim and kill 80% of *E. coli* bacteria versus 35% for nonswimmers [Vilela 2017].

12.5.4 Slow Exposure and Controlled Release of Toxic Reagents for Anti-biofouling

Loading small amounts of toxic material into solid matrices for slow release to the environment creates durable antibacterial material systems. Weaving antibacterial and insoluble silver nanowires into yarns creates antibacterial fabrics [Davoudi 2014]. Single-walled CNTs form a protective scaffolding for the slow release of the ova-antimicrobial enzyme lysozyme [Nepal 2008]. Rosin-based coatings impregnated

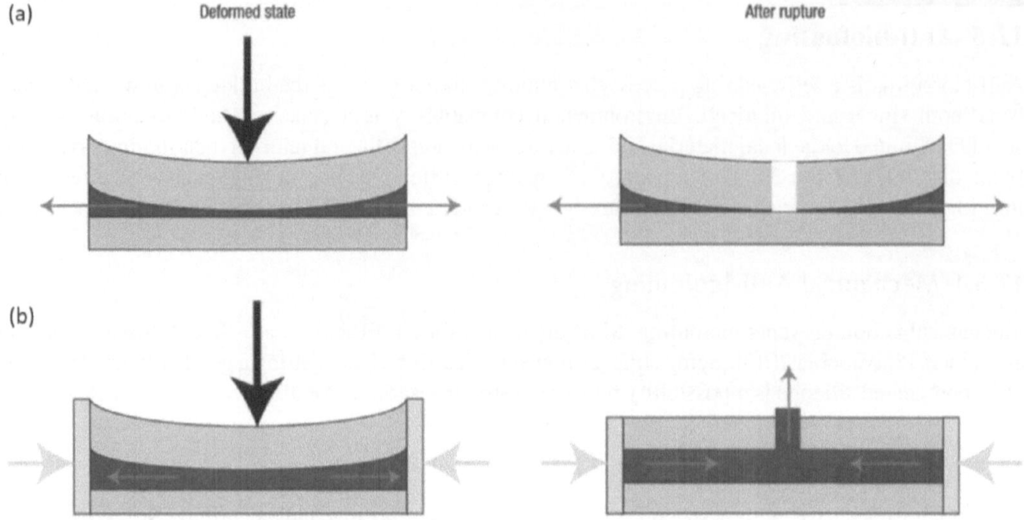

FIGURE 12.4 Active elastomeric liquid trilayer system delivers biocide to damage sites that breach an outer skin layer. (a) Unconstrained system does not deliver biocide. (b) Elastic constraints deliver biocide in the proper direction. (From [Sonntag 2004].)

with ivermectin inhibit the growth of barnacles on ship bottoms by both mechanisms as do selenium nanoparticles [Pinori 2011] [Tran 2013]. At a larger length scale, microcapsules release liquid by rupture with mechanical damage or may slowly release the chemicals through diffusion – a trait that is useful in environmentally safe anti-biofouling coatings [Szabó 2011]. Vascular systems and liquid-filled layers deliver biocidal liquids upon breaching of an outer skin layer (Figure 12.4) [Sonntag 2004].

12.5.5 Biological Anti-biofouling

Extracorporeal membrane oxygenators (ECMOs) are synthetic externally attached surrogates for lungs. Gas exchange through thin membranes oxygenates blood. Biofouling is a very dangerous condition. The addition of titania (TiO_2) to the membrane surface promotes the buildup of a single cell endothelial layer. This layer prevents biofouling at a cost of a reduction in oxygen transmissivity of 22%. The layer is self-healing and reconstructs by growing new endothelial cells [Pflaum 2017].

12.6 Prosthesis Structures

A prosthesis is an artificial component that replaces or augments the functional action of defective or missing tissue in living systems. The loading on many prostheses is severe. Imbuing prosthetic structures with self-healing capability extends operational lifetimes.

12.6.1 Prosthesis and Tissue Ingrowth

Prosthetic implants must be compatible with living biological tissues. Complex interactions can be a challenge [Anderson 2008]. Compatibility generally takes on one of the following forms:

1. *Prosthetic Material Is Inert and Is Largely Ignored by the Tissue:* Inert materials are particularly useful for implants that are intended for long-term usage.
2. *Tissue Grows into and Adapts to the Implant:* The implant serves as scaffolding and reinforcement of the tissue as it heals. Implant geometry, both macroscopic and microscopic, along with mechanical stiffness has a big influence on success.

3. *Implant Dissolves and Absorbs Prosthesis:* The implant serves a temporary purpose and slowly dissolves or resorbs as the tissue heals. Controlling the rate of dissolution is important. Hydrogels, for example, provide a long-term slow release of drugs in a particular location, and then resorb slowly as their effectiveness wanes [Rahaman 2014]. Tissue growth matrices based on natural polymers, such as alginate, chitin, collagen, glycerol, sebacic acid, and silk have an inherent advantage over many artificial polymers in that they degrade slowly in a biocompatible manner.

12.6.1.1 Orthopedic Implants

Resorbable bone implants combine fast and slow resorption rates, such as a fast-resorbing bioactive glass that promotes pre-osteoblast infiltration combined and slow-resorbing calcium phosphate core for bone ingrowth. Laser sintering and 3D printing fabricate custom shapes for the implants [Comesaña 2015]. Severe shot peening of stainless steel bone implants produces a surface texture that promotes the ingrowth of bone-forming osteoblast cells while also reducing the growth of several bacterial species [Bagherifard 2015]. Microencapsulated healing liquids enhance the durability of bone cements [Wilson 2010].

12.6.1.2 Electronic and Neural Implants

Interfacing electronic devices with living tissue, including neural tissue, requires mechanically compliant electrodes. Conventional metal and semiconducting electronics are too stiff. Soft electrically conductive polymers provide a sufficiently compliant interface [Martin 2015].

12.6.1.3 Vascular Implants

Replacement vasculature in animals with implants requires biocompatibility, including minimizing clot formation, matched mechanical compliance, and potential for resorption as the vessel heals. Thermoplastic polyurethane grafts manufactured by electrospinning leaves sufficient porosity for cell migration to fill most of these requirements [Bergmeister 2015]. Magnesium is a nontoxic metal that resorbs and is suitable for coronary artery stents [Erbel 2007].

A combination of silk-fibroin and chitosan shows promise as ingrowth scaffolding for vasculature [She 2008]. Poly(glyceral sebacate acrylate) combined with a photoinitiator makes a hydrophobic elastomer-like resorbable adhesive that joins cardiac and vascular tissue without stitches [Lang 2014].

12.7 Applications

12.7.1 Concrete, Stones, and Soil

Bacteria modify local environments to promote homeostasis. A by-product of these processes for bacteria living in concrete, stones, and soil is to heal the material by plugging pores and microcracks. Some bacterial species precipitate carbon dioxide as solid crystals, usually calcite, and sometimes vaterite [Stocks 1999] [Rivadeneyra 1991]. Calcite deposition follows a urease-based pathway with the bacteria exuding both urea and the enzyme urease that breaks down the urea into $CaCO_3$ and ammonia. The ammonia raises the pH of the extracellular medium, which drives the calcite to precipitate and form extracellular crystals [Bang 2010]. Wild bacteria, including *Bacillus pasteurii* and *Sporosarcina ureae,* follow the urease pathway and deposit calcite that plugs holes in concrete and geologic structures [Gollapudi 1995] [DeJong 2010].

It is possible to domesticate organisms to deposit calcite for self-healing. *Bacillus tequilensis* isolated from concrete kiln dust promotes the self-healing of foamed bricks, mainly through $CaCO_3$ deposition in pores [Alshalif 2022]. *E. coli* bacteria recombinant-modified from *B. pasteurii* exude urease deposit calcite, as will urease extracted from the bacteria and embedded in polyurethane foam [Bachmeier 2002]. Industrial by-products, such as lactose mother liquid, can feed the process [Achal 2009].

A disadvantage of urease-based pathways for $CaCO_3$ deposition is the release of ammonia. An alternative pathway metabolizes calcium lactate with the reaction using strains of *Bacillus pseudofirmus* and *Bacillus cohnii* isolated from alkaline soils [Jonkers 2010] [Wiktor 2011].

$$CaC_6H_{10}O_6 + 6O_2 \rightarrow CaCO_3 + 5CO_2 + 5H_2O$$

Bacteria can provide seed crystals for an accelerated plugging process in liquids saturated with ions. Bacterial plugging of oil wells is an example [Ferris 1992].

The plugging of holes with $CaCO_3$ may not always be desirable. Sometimes stones and masonry structural materials do better with open surface pores that allow passage of fluid and vapors. Strengthening the pore walls while leaving them open strengthens overall material. *Myxococcus xanthus* fulfills this need in ornamental limestone with carbonate depositions that cement existing calcite crystals [Rodriguez 2003]. The mechanism appears to be an intermixing of calcite and vaterite polymers to form nano- and microstructures that follow the geometry of bacterial skeletons.

The delivery of bacteria into concrete for self-healing requires predispersal of seed bacteria, or spores, and survival through the harsh environments of mixing and curing. Packaging the bacteria in suitable water-soluble containers provides a viable solution. The packages disintegrate and release the bacteria at appropriate times and locations [Wang 2012c]. Water-soluble polyurethane packaging produced healing responses, with efficiencies approaching 60%.

12.7.2 Corrosion Inhibition

Many corrosion processes occur on surfaces. Surface-attached microorganisms may either promote or inhibit corrosion. Typical interactions are complicated metabolic and electrochemical interactions across a multi-species community [Little 2006] [Herrera 2009] [Shi 2011]. Neutralizing corrosion chemistry or creating a protective impermeable barrier are possible protective mechanisms.

12.7.2.1 Neutralizing Corrosive Chemistry

Aerobic bacteria such as *Pseudomonas fragi* form biofilms that scavenge oxygen from the environment and prevent corrosion of mild steel [Ismail 2002]. Engineered and wild versions of *Bacillus licheniformis* secrete γ-polyglutamate, which inhibits anionic pitting corrosion aluminum [Örnek 2002].

12.7.2.2 Impermeable Biofilms Isolate Surfaces from a Corrosive Environment

Bacteria that metabolize iron inhibit the corrosion of steel through combinations of metabolic activity and the formation of biofilms with favorable properties [Videla 2009]. A principal metabolic reaction is the reduction of ferric iron (Fe^{3+}) to ferrous (Fe^{2+}) with anaerobic respiration. Ferric iron ions form insoluble corrosion-inhibiting iron oxide compounds, such as magnetite (Fe_3O_4). Conversely, ferrous iron forms iron oxide compounds, such as hydrated ferrous oxide ($Fe_2O_3{\cdot}xH_2O$), which are soluble in water and provide minimal corrosion protection. Bacteria such as *Shewanella oneidensis* reduce ferric to ferrous iron, and accelerate corrosion, especially steels with passivating layers of magnetite. *S. oneidensis* biofilm communities break down H_2 passivating layers that form on iron surfaces in oxygen-deprived environments to promote corrosion [De Windt 2003]. Sometimes these biofilms collectively form passivating layers and protect metals against corrosion (Figure 12.5) [Dubiel 2002]. Related are the iron and manganese-depositing bacteria that create pipe-clogging extracellular iron hydroxide and manganese oxide structures [Ghiorse 1984]. Similarly, iron-reduction processes in bacteria stabilize the remediation of soils and sediment contaminated with metals, radionuclides, and noxious organic chemicals [Fredrickson 1996].

12.7.3 Digestion for Self-cleaning

Living organisms ingest material from their surroundings. Digestion breaks down the material into simpler chemical compounds. The organism then uses the simpler compounds as building blocks for the functions of life, including energy harvesting and building molecular structures and tissues. These chemical processes are notable for their efficiency, specificity, and locality of action. There is a wide range of self-cleaning and similar biohybrid structural and machine applications based on digestion.

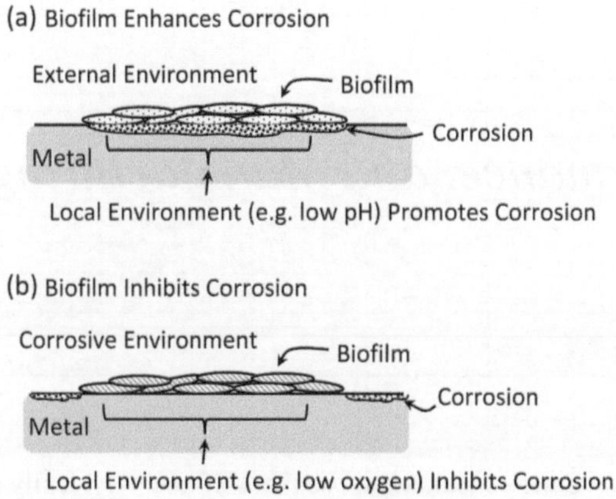

FIGURE 12.5 Biofilms form local surface environments that affect corrosion growth rates. (a) Enhanced corrosion. (b) Inhibited corrosion.

Digesting particulates to alleviate clogging of plumbing and filters with conglomerations of partially digested and decomposed biological structures and molecules, that is, sludge, is an application. Bacteria digest sludge into simpler chemical compounds that sustain growth and reproduce bacteria, all while preventing and alleviating clogging [NASA 2007]. Figure 12.6 shows a bacteria-based self-regenerative system for sludge removal from sewage wastewater.

Microbes, primarily bacteria, digest metals by lowering the redox state [Lovley 1997]. Amassing and removing the microbes through flocculation followed by mechanical removal remediates metal-laden water.

12.7.4 Fire Prevention

Alkaline deacetylated mycelium fungus is a fire-resistant insulating and filling material that replaces flammable polyurethane foams. The fire resistance largely comes from the formation of char layers [Chulikavit 2023].

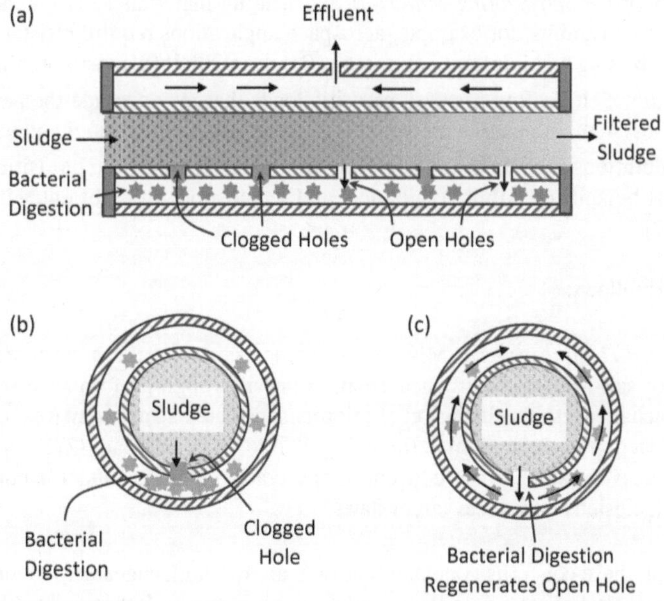

FIGURE 12.6 Bacteria digestion of sludge alleviates clogging to create biohybrid self-regenerating filter. (a) Side view with self-regenerating filter holes. (b) Axial view with clogged hole. (c) Axial view with unclogged hole. (Adapted from [NASA 2007].)

13

Design, Performance, and Manufacturing

13.1 Design

The design of engineering systems – including those that self-heal – typically starts with high-level conceptual designs and moves through subsystems and the detailed design of individual components [Pahl 1988]. Successful designs reduce cost, ease manufacture, and increase performance.

13.1.1 Overall Design Issues

The following are the high-level concepts to consider in the design of self-healing systems:

1. *Importance and Need for Self-healing:* Important systems that have difficulty enduring without self-healing and self-healing does not impose large costs.
2. *Role of Mitigation:* Should mitigation methods and/or robust components be used instead of, or in conjunction with self-healing?
3. *Choice of Self-healing Technique:* Polymers provide a variety of self-healing methods [Yuan 2008b]: (i) Crack healing effects based on molecular chain interdiffusion and entanglements, catalyzed polymerization, post-curing of residual functional groups, and Diels–Alder (DA) reactions. (ii) Crack-induced damage of 3D microvascular networks, releasing healing liquid that solidifies by ring-opening metathesis polymerization. (iii) Crack-induced rupture of microencapsulated healing liquids.
4. *Compatibility of the Self-healing Material:* Self-healing materials satisfy a variety of application-specific constraints, for example, aerospace applications require resistance to radiation and thermal stressing and minimal outgassing [Bererton 2002] [Semprimosching 2006].
5. *Parasitic Loads:* Self-healing imposes parasitic loads that may degrade the performance and reliability of a system. In general, the benefits of using self-healing should outweigh the costs.
6. *Cost:* Considerations include the fabrication and material costs, the parasitic loads on performance, the benefits of failure avoidance, and the cost of timing of entry into the market [Huang 2007].

Some cases are as follows:

Microencapsulated Healing Agents: Microencapsulation is one of the more thoroughly studied techniques for self-healing. Microencapsulation provides high-performance self-healing, but also presents challenges. The inclusion of microencapsulated agents may weaken a bulk material by (1) reduction in the bulk material being able to carry the load, (2) inclusion of a small radius introduces possible stress concentrations crack initiation, and (3) nonhomogeneous bunching of inclusions introduces larger flaws.

Studies addressing these concerns conclude that they are often manageable. A biomimetic approach provides mitigation of weakening due to inclusion of microstructures. Bones and teeth have many holes but they do not suffer from stress-riser cracking. In bones, millimeter-scale holes, known as nutrient

DOI: 10.1201/9781003396048-13

foramina, allow for fluid to flow through the bone wall and sustain life. Observations indicate that nutrient foramina holes are rarely the site of bone fractures and do not produce a combination of weakening and stress concentration commonly found with holes in homogeneous materials. Reasons for the mechanical resilience of bone tissue around a foramen include the following: (1) Stress concentration reduction due to an increased compliance (reduced stiffness) of the bone near the foramen, which blunts the sharp stress discontinuity due to the hole. (2) A reinforcing ring of increased stiffness placed at some distance from the foramen. (3) A ring of lamellar bone placed along the foramen inside edge, which may prevent crack formation [Götzen 2003]. This contrasts with the introduction of holes in FRP composites which for the inclusion of fiber-optic sensors typically reduces the strength of structural elements by values running from 10% to 40% [Shivakumar 2004].

Introducing micro- and nanoscale solid inclusions into polymer matrices may strengthen or weaken the material. Stiff particulates in polymer composites increase the tensile strength by values up to 50% when volume packing ratios reach over 20% and there is a good transfer efficiency of stress between the particles and polymer matrix [Sudduth 2006]. Numerical simulations of the effect of breaking of capsules indicate that nonlocal effects can lead to an amplification of the healing [Privman 2007].

Vascular self-healing systems may introduce large quantities of parasitic material. The potential loss of strength is correspondingly high. For example, losses of compressive strength in fiber-reinforced polymer composites range up to 30% for 0.25 mm and 0.5 mm vessels in panels with nominal thicknesses of 1 mm [Norris 2011]. Conversely, the judicious removal of material creates high-performance cellular and metamaterial systems out of solids [Gibson 1997].

13.1.2 Taxonomy Approaches

Taxonomy approaches aid conceptual design by organizing concepts in a hierarchical framework. A self-sealing spacesuit may make use of multiple technologies to manage leaks. Possibilities are microencapsulation, foaming, moving gels, and actuated wound closing, either individually or as a combination. A taxonomy table presents the information in a convenient format (Figure 13.1) [Ferl 2007].

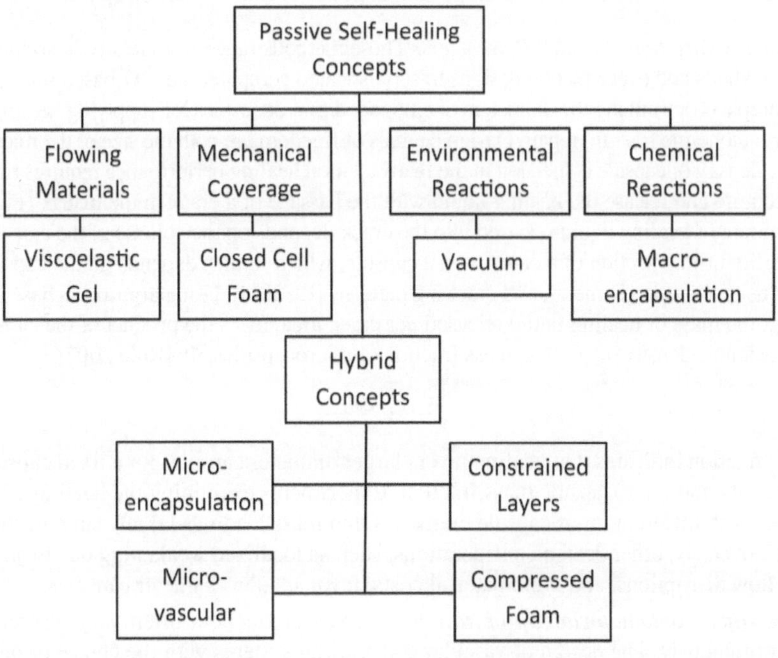

FIGURE 13.1 Taxonomy of self-healing concepts for space suits. (Adapted from [Ferl 2007].)

13.1.3 Multiscale and Multifunctional Methods

Controlling Wall Thickness and Strength: Customized configurations of microencapsulation tailor the capsules to react to specific types of damage. An example is impact damage that causes delamination in an FRP composite. Repairs require bridging the delamination or crack. A semiautonomic technique combines microencapsulated healing with post-damage compression forces transverse to the delamination and fiber ply layers [Yin 2008]. The addition of microencapsulation bruising agents produce recognizable optical signatures, for example, fluorescence, upon impact, provides further functionalization.

Supplemental Effects Causing Rupture: Chemical effects, including chemically amplified reactions and signaling, augment mechanical rupture.

13.1.4 Optimization

Optimization follows concept selection. Typical steps are as follows:

1. *Set a Goal:* Optimal designs achieve goals within constraints. Conventional optimization begins with the selection of a scalar measure, that is, a cost function, for comparing various design alternatives. The typical cost function combines multiple requirements, including specific measurable quantities, such as weight, and more holistic measures, such as robustness, durability, serviceability, and lifetime cost of ownership [Dry 2000]. Constraints restrict the optimum design space.
2. *Performance Assessment:* Numerical simulations assess performance. Particularly challenging are large-scale or long-lived systems subject to various and unpredictable operating conditions and stressors.
3. *Optimization Strategy or Algorithm:* Many conventional techniques for the optimization of engineering systems extend directly to the design of self-healing systems.
4. *Available and Controllable Degrees of Freedom:* Optimization requires a sufficient number of adjustable degrees of freedom to produce the desired performance.

A variety of tools can find optimum configurations in the available solution space:

1. *Continuous Variation of Scalar Parameters:* These methods range from relatively simple calculus-based methods and experimental designs to sophisticated computer- and AI-based search methods. An example is optimizing the diameters of embedded microcapsules for supplying healing agents to support autonomic fracture repair. Design degrees of freedom are both the size of the microcapsules and the density of capsule dispersion in the matrix. Good healing performance requires that the capsules rupture and release the healing liquid with the passage of a crack in the matrix (Figure 13.2). The amount of healing liquid released into the crack depends on the volume of the capsule and the probability of intersection of a crack with a capsule, which in turn depends on the capsule dispersion density, capsule geometry, and cracking patterns [Lv 2014]. Using arguments based on length scaling, the mass of healing liquid released per crack area, $\bar{m}$, is the product of the capsule diameter d_c, sample density ρ_s, and the mass fraction of microcapsules, Φ: [Rule 2007].

$$\bar{m} = \rho_s \Phi d_c \tag{13.1}$$

This relation indicates that healing favors larger diameter capsules for a fixed capsule dispersion density and microcapsule mass fraction. Experiments examining the healing efficiency of samples with different microcapsule diameters and mass loadings largely support this conclusion. Conversely, other design considerations, such as localized weakening due to stress risers, stress flow disruptions, and bulk material costs, favor smaller diameter capsules.

2. *Number and Combinatorial Optimization:* Degrees of freedom often vary discretely rather than continuously. The design of vascular self-healing systems with the choice of number and topology of vessels benefits from discrete approaches. The branching of vasculature produces a

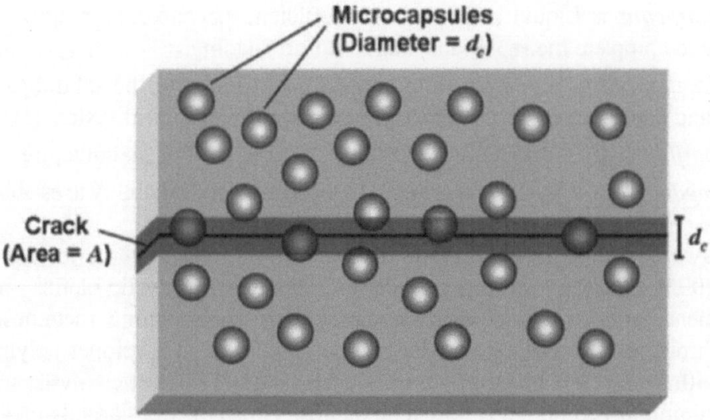

FIGURE 13.2 Intersection of crack surface with randomly distributed capsules. (From [Rule 2007].)

tree-like structure. Recirculation requires return paths. Connecting tree canopies for outbound and return flows offers additional design degrees of freedom in terms of the number and combination of the links [Kim 2006b].

3. *Combined and Mixed Optimization:* A consideration is to reduce the parasitic mass of a vascular self-healing system [Williams 2008a]. It is also important to maintain connectivity in the event of damage through the loss of vascular links. If the flow and network configurations are stable, then a hierarchical architecture with larger pipes branching into smaller pipes in tree-like fashion may be optimal. Distribution networks with tree-like architectures are susceptible to widespread flow loss in the event of damage to lower level primary pipes. Biological systems can provide some clues as to how to proceed – they use loops when needed to provide redundancy and/or respond to uneven load demands, as in the leaves of deciduous trees [Katifori 2010]. Since looping comes at a cost, many biological systems use a mix of branching and looped architectures. The loops serve only in critical regions. In humans, loops, also known as circulatory anastomoses, appear in the brain as the circle of Willis, around the knee and in the hand [Williams 2008a]. Numerical studies of loop architectures indicate that they impose an overall penalty of flow resistance, but readily maintain flow following the removal of conductive links [Wechsatol 2005].

Optimal designs of self-healing systems for multifactor compatibility consider combined effects. An example is selecting a polymer for use in medical and rehabilitation applications that is compatible with living tissue, such as human skin, but resists the buildup of bacteria and biofilms. A design challenge is that the wide choice of polymeric materials is daunting and prohibitively expensive, especially from a testing point of view. A workaround is automated machines that perform empirical testing at high-throughput rates and lower costs. The selection of materials for antibacterial coatings based on a combination of esters and cyclic hydrocarbon moieties in silicone automated combinatorial methods arrived at a material solution in a time frame not possible with conventional screening methods with reports of 30-fold reductions in pathogenic bacteria (*Pseudomonas aeruginosa*, *Staphylococcus aureus*, and *Escherichia coli*) versus commercial silver-hydrogel coatings [Hook 2012].

13.1.5 Liquid-to-solid Healing Methods

Converting a liquid into a solid is the basis of multiple material healing methods. Success with liquid-to-solid healing requires the following:

1. *Delivery:* Liquid delivery to damaged region, perhaps flowing into a tight crack and similar confined regions in an on-demand basis.
2. *Stickiness:* Liquid wets and sticks to the solid at the damaged site.

3. *Timely Solidification:* Liquid solidifies with sufficient mechanical strength and mechanical compliance to complete the repair and survive future loading.
4. *Stiffness Compatibility:* Healing system matches the stiffness of the solid repair to that of the base solid and does not excessively weaken or burden the structural system [Dry 2001a].
5. *Cost Compatibility:* Cost-efficient and cost-effective methods of manufacture.
6. *Design Compatibility:* Sufficient knowledge of performance and reliability to establish design rules.

Catalytic and two-part mixing are primary variants of polymer systems for use in liquid-to-solid healing [White 2005]. Catalytic variants include the following: (1) Cyclic olefins – dicyclopentadiene (DCPD), norbornene, and cyclooctadiene polymerized by ring-opening metathesis polymerization (ROMP) with Schrock and Grubbs catalysts or two-part catalysts. (2) Lactones polymerize into polyesters and nylons with cyclic ester and cyclic amide catalysts. (3) Two-part catalysts, such as those based on tungsten and aluminum compounds for cyclic olefins. Two-part mix variants include the following: (1) Unsaturated compounds, such as acrylics, styrene, isoprene, and butadiene polymerized by two-part atom transfer radical polymerization with organohalides and peroxide compounds. (2) Isocyanates and compounds with hydroxyl groups.

Several norbornene-based healing liquids solidify by ROMP in the presence of Grubbs catalyst, including 5-ethylidene-2-norbornene (ENB), *endo*-DCPD, *exo*-DCPD, and mixtures of these three (Figure 13.3). Different mix ratios take advantage of the different properties of the individual components. ENB tends to have the fastest polymerization kinetics, offers superior cold temperature performance, and requires less catalyst [Lee 2006] [Sheng 2009]. *endo*-DCPD solidifies at 30°C. Self-healing composites made with ENB have high dynamic elastic moduli in the temperature range between −50°C and 70°C and glass transition temperatures of 110°C [Raimondo 2013]. Aerospace applications require operation in wider temperature ranges of −150°C to +150°C and in a vacuum. Embedding the catalyst in a mixture of oil, perfluorinated solvents, and hydrophobic ionic liquids enables healing in these demanding conditions [Merle 2014].

13.1.5.1 Design of Microencapsulated Liquid Healing Systems

Success with microencapsulated healing systems accommodates multiple issues of chemical and mechanical compatibility.

Microcapsule Size and Dispersion Distributions: Measuring size distributions is often a matter of relatively straightforward micro-optical methods that image sets of capsules and then determines the diameters of individual capsules. Similar optical methods determine statistical distributions of the dispersion of capsules throughout a matrix. Uniform dispersions may be favored over purely random or aggregating distributions.

Induced Rupture: A strong bond between the capsule wall and surrounding matrix serves two purposes: (1) Transfer the propagation of cracks from the matrix to the capsule wall to induce capsule rupture and healing liquid release at the proper time. (2) Produce a stronger composite capsule–matrix structure. Typical capsule walls are urea formaldehyde, which do not readily bind to many composite matrix materials, such as epoxy. Functionalizing the capsule walls, such as with silane-based coupling agents, increases the bond strength [Wang 2009] [Li 2008b].

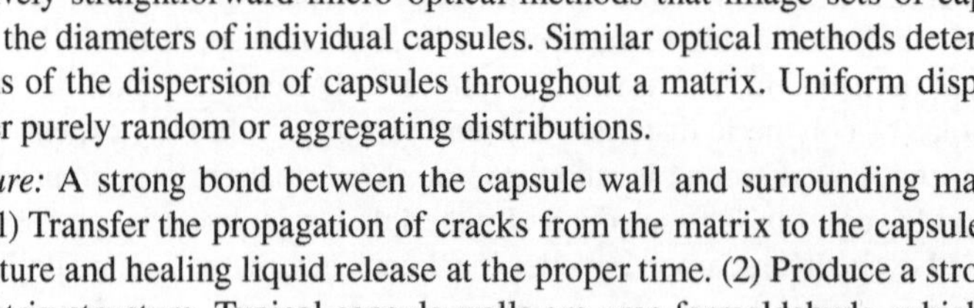

FIGURE 13.3 Monomer structures of norbornene-based healing liquids. (a) ENB. (b) *endo*-DCPD. (c) *exo*-DCPD. (Adapted from [Larin 2006].)

Long-term Stability of Encapsulated Healing Fluid and Polymer Matrix: An encapsulated healing liquid typically has high available energy so that it can heal when needed. These liquids may not remain stable in the capsules. The use of solvent-based healing liquids suffers from diffusion of the solvent through the capsule wall [Neuser 2013].

Thermal Stability: Temperature extremes adversely affect microcapsules and healing liquids. DCPD has a melting point of 19.5–33.0°C and breaks down at 170°C. Some epoxy healing liquids are stable at temperatures over 200°C. PUF wall thickness, size, and surface morphology, along with choice of healing liquid, affect the resiliency. Smooth PUF walls with a thickness of 30.5 ± 5 μm may be optimal for thermal stability [Yuan 2006].

Chemical Stability: Liquids that solidify primarily in the presence of a catalyst have enhanced long-term chemical stabilities. An example is a healing liquid comprised of hydroxyl-end functionalized PDMS and poly(diethylsiloxane) and catalyzed by dibutyltin dilaurate. Placing the healing liquid mixture and catalyst in separate, but intermingled, capsules that rupture with damage to the surrounding matrix creates a healing system that is stable at temperatures over 100°C [Cho 2006]. Similarly, the capsule wall must be stable with respect to both internal chemistry and external chemistry. Some curing agents in epoxy–amide matrices attack capsule walls leading to premature release in microcapsule-based anticorrosion coatings [Kashi 2009]. A compromise is a matrix with a less aggressive cross-linker that spares the capsule coating, at the expense of a weaker cross-linked epoxy matrix.

13.1.5.2 Vascular System Design

The ability to fill cracks is a primary performance target for vascular healing systems. The optimization of a vascular network combines the configuration topology with the variation of a parameter. This is a complicated problem with few closed-form solutions. A dimensionless parameter for the global flow resistance of a network is f.

$$f = \frac{\Delta P}{\dot{m}} \frac{V^2}{8\pi \upsilon R^3} \tag{13.2}$$

Here ΔP is the pressure drop across the network, V is the volume of the network, $\dot{m}$ is the mass flow rate, and R is the tube radius [Wechsatol 2005]. Figure 13.4 shows a vascular grid with independent damage

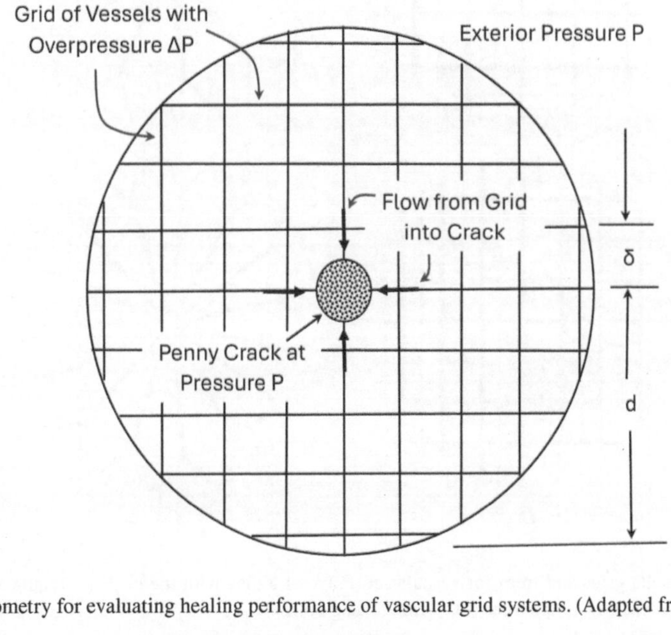

FIGURE 13.4 Geometry for evaluating healing performance of vascular grid systems. (Adapted from [Bejan 2006].)

sites, separated by a minimum radius of influence, d. An important parameter is the grid pitch δ. The healing liquid flows out when the crack traverses a vessel. The speed of fluid dispensing depends on the effective permeability K of the grid when damage extends across vessels. Higher values of K correspond to more rapid dispensing. Based on flow calculations at the loop level, triangular loop meshes have faster dispensing rates than either square or hexagonal meshes.

Using meshes with a heterogeneous set of tube diameters, including those with tree-like architectures, increases performance (Figure 13.5) [Bejan 2006] [Wang 2007a] [Lorente 2009]. Some configurations have flow rates $\dot{m}$ that exceed that of equivalent parallel grids $\dot{m}_{ref}$, with the performance ratio r:

$$r = \frac{\dot{m}}{\dot{m}_{ref}} \tag{13.3}$$

Favorable designs are those with r >1 [Zhang 2007a].

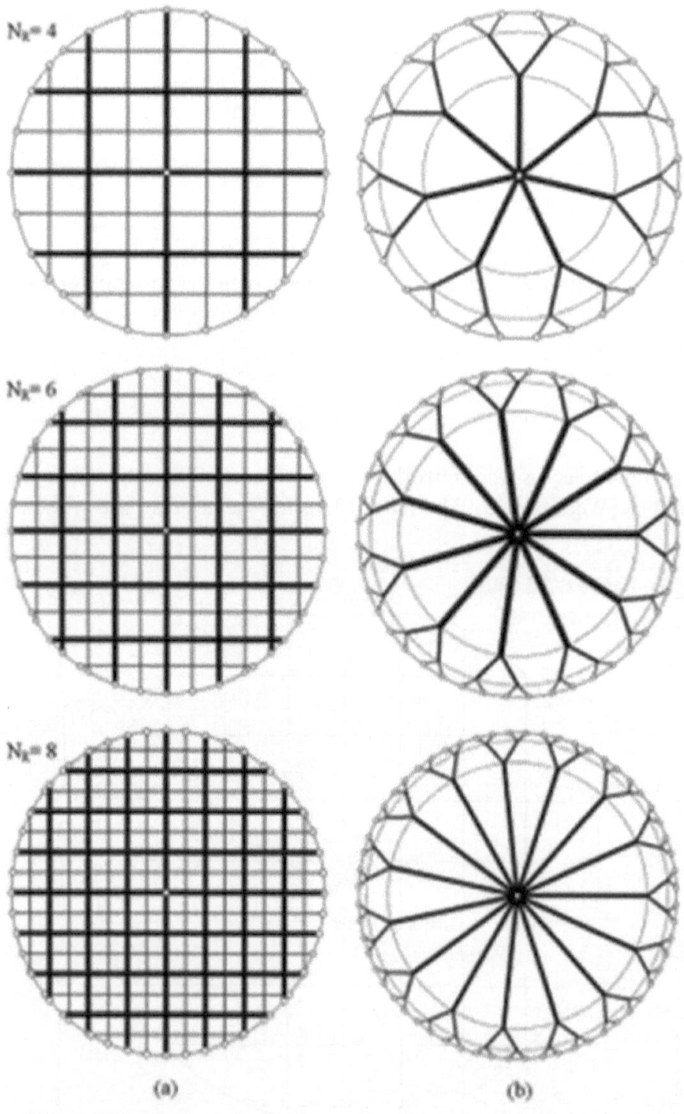

FIGURE 13.5 Multiscale grids and trees for vascular coverage of a disk with the grids (a) requiring 3.7–6.8 times the pressure drop for equivalent flow to the trees (b). (From [Wang 2007].)

In a seminal paper, Murray determined the optimum configuration of a network of capillaries for blood circulating in animals [Murray 1926]. The following is a synopsis of Murray's development:

The total work per unit time, E, performed by blood in a circulatory system is the sum of the mechanical power, P, that moves the blood and the work to support living metabolism. The mechanical work operates on a capillary with surface area s and volume V:

$$s = 2\pi r l \tag{13.4}$$

$$V = \pi r^2 l = \frac{sr}{2} \tag{13.5}$$

Here r is the radius. l is the length, and Q is the volumetric flow rate:

$$Q = u\pi r^2 \tag{13.6}$$

Here u is the flow velocity. n_f is the total rate of blood flow in a body.

Poiseuille's law mechanical power is

$$P = \frac{8Q^2 l\eta}{\pi r^4} \tag{13.7}$$

Here η is the viscosity. When the metabolic power and other energy expenditures are proportional to the volume of blood with b as constant of proportionality, the total power is as follows:

$$E = \frac{8Q^2 l\eta}{\pi r^4} + bl\pi r^2 \tag{13.8}$$

Minimizing E by differentiating with respect to the radius leads to the optimal criterion:

$$b = \frac{16Q^2\eta}{\pi^2 r^6} \tag{13.9}$$

Alternatively, optimizing with constant surface area s begins with rewriting the total work in terms of s:

$$E = \frac{Q^2 s 4\eta}{\pi^2 r^5} + \frac{brs}{2} \tag{13.10}$$

Differentiating to obtain the optimal criterion for b:

$$b = \frac{40Q^2\eta}{\pi^2 r^6} \tag{13.11}$$

Using nominal values of physiologic parameters, Murray concluded that the constant surface area criterion (13.10) provides a better match to existing data. Supporting this result is the notion that capillaries do much of their important work via chemical diffusion through the walls. An examination of the optimum conditions (13.9) and (13.11) indicates that the flow through a vessel is proportional to the cube of a radius. Combining with a continuity assumption appearing in a branch from a parent vessel with radius r_p to a set of child vessels with radii r_{ci} leads to

$$Q = kr_p^3 = k\sum_{i=1}^{n} r_{ci}^3 \tag{13.12}$$

The ZKM model describes branching vasculature with consideration of stems and crowns [Zhou 1999]. A stem is the top tube in a branched hierarchy (Figure 13.6). The crown is the collection of smaller branching tubes that attach to a stem. A branching vasculature is a cascade of stems and crowns with each crown branch forming a smaller new stem and crown.

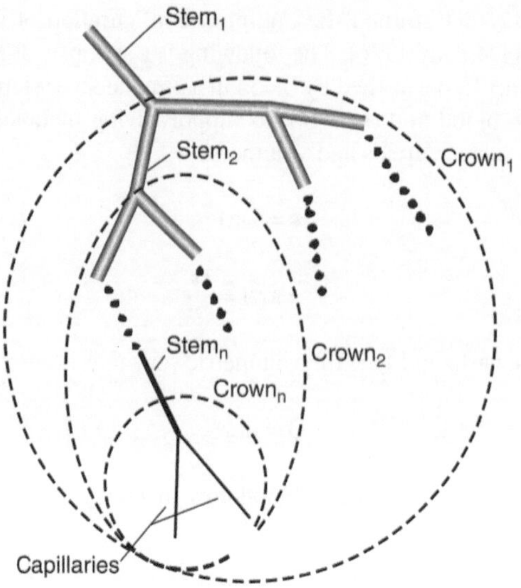

FIGURE 13.6 Stem and crown hierarchical structure of vasculature. (From [Kassab 2006].)

The ZKM model predicts the following relations [Kassab 2006]:

$$\frac{V}{V_{max}} = \left(\frac{L}{L_{max}}\right)^{\frac{5}{\epsilon'+1}} \tag{13.13}$$

$$\frac{D}{D_{max}} = \left(\frac{L}{L_{max}}\right)^{\frac{3\epsilon'-2}{4(\epsilon'+1)}} \tag{13.14}$$

$$\frac{Q}{Q_{max}} = \left(\frac{r}{r_{max}}\right)^{\frac{4(\epsilon'+1)}{3\epsilon'-2}} \tag{13.15}$$

The subscript max denotes the maximum value. ϵ' is the ratio of metabolic to viscous power dissipation in the crown with typical values running from 2.5 to 3.5. A separable convex approximation yields optimal results – including a confirmation of Murray's law [Klarbring 2003] [Sherman 1981]. Genetic algorithms are effective at solving such mixed optimization problems [Aragon 2007

13.1.5.3 Solid and Surface Design

Resilience-based Design: Failure mode, effects, and criticality analysis (FMECA) (aka failure mode effects analysis [FMEA]) assesses resilience. The intent is to consider the full spectrum of possible failure modes followed by the possible cascading consequences. A FMECA for a vascular self-healing system examines the following: (1) Pump/reservoir underperformance or complete failure. (2) Fluid leakage. (3) Vessel blockage or constriction by cured resin or debris. (4) Failure to rupture and bleed from a vessel on cue from damage. (5) Healing resin cured properties degraded. (6) Healing resin performance degraded [Williams 2008b].

Brittle Material Healing Design: Brittle materials are prone to cracking. Successful crack repairs need to close the cracks while accommodating post-repair loading and movement. If the repair is too stiff, it breaks away leading to more cracking [ACI 2010].

Ceramics are fracture-prone. Most self-healing techniques for ceramics use oxygen as a promoter of in-filling. Surface cracks connected with ambient oxygen can heal, but internal oxygen-starved cracks cannot. A recommendation is to adopt a combined healing + proof testing + in situ healing methodology to ensure better endurance of ceramic parts [Ando 2005]. The initial healing process corrects cracks while the part is out of service or being prepared for service. The proof testing identifies internal cracks that cannot be corrected by oxygen-based self-healing. In situ healing allows for cracks to heal while the part is being used, even under cyclic loading. Some Mullite/SiC and Si_3N_4/SiC ceramics exhibit excellent in situ healing properties.

Artificial Intelligence and Machine Learning (AI and ML): Concrete is a multicomponent material that self-heals through multiple mechanisms, including hydration, carbonation, swelling, precipitation of impurities, and biochemical activity. The interactions of healing with damage are sufficiently complex that AI/ML methods of modeling are effective [Suleiman 2017].

Wound-closing Design: Mechanically actuated closing of wounds for healing with discrete actuators requires consideration of load paths and actuator sparsity. For example, if shape memory alloy (SMA) actuators close wounds, then the entire load path, including bonding anchorages, must be strong enough, and have enough discrete actuators to actuate and then carry the loads [Wang 2002].

Superhydrophobic Surface Design: The design of superhydrophobic surfaces is often an issue of mixed discrete and continuous optimization that considers the size, scale, and arrangement of micro- and nano-features as variables. Such approaches achieve multifunctionality from a textured surface, such as a combination of self-cleaning super-hydrophobicity and omnidirectional broadband optical super transmissivity [Park 2012]. Synergistic designs with multifunctional micro- and nanotextured surface effects create high-performance systems, such as micro- and nanotextured surfaces that combine "Lotus leaf" self-cleaning superhydrophobic behavior with "shark skin" low fluid dynamic drag characteristics [Bixler 2012].

The mechanics of superhydrophobic surfaces is sufficiently well understood to design droplet-shedding surfaces from fundamental principles [Patankar 2004] [Tuteja 2008] [Bartolo 2006] [Extrand 2006] [Liu 2014] [Yang 2006]. Materials with smooth surfaces may be either hydrophilic or hydrophobic. Micro- and nanoscale textures combined with hydrophilic materials create superhydrophobic materials. The principle is to trap air under a droplet with microtexture, in a combined concave and convex, that is, reentrant, microtexture (Figure 13.7). Hydrophilic forces on reentrant surfaces pull droplets away from the surface and trap air underneath in a Cassie state.

Eluting Self-cleaning Surface Design: Eluting surfaces with designs inspired by nepenthes pitcher plants combine surface texture and specialized surface-embedded lubricating liquid to repel impinging liquids, as with the slippery liquid-infused porous surface (SLIPS) (Figure 13.8) [Wong 2011]. Design considerations include the following: (1) The surface must have microtexture and material properties that enable the liquid to flow, infuse and lubricate the surface, and then stay in place. (2) The lubricating liquid must wet the surface. (3) The lubricating liquid and impinging fluids need to be immiscible.

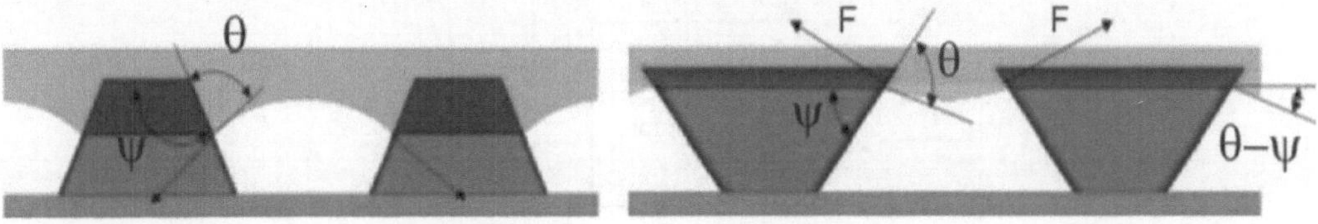

FIGURE 13.7 Cassie state resulting from hydrophobic and hydrophilic wetting with different values of the contact angle θ and pillar texture angle ψ. (From [Tuteja 2008].)

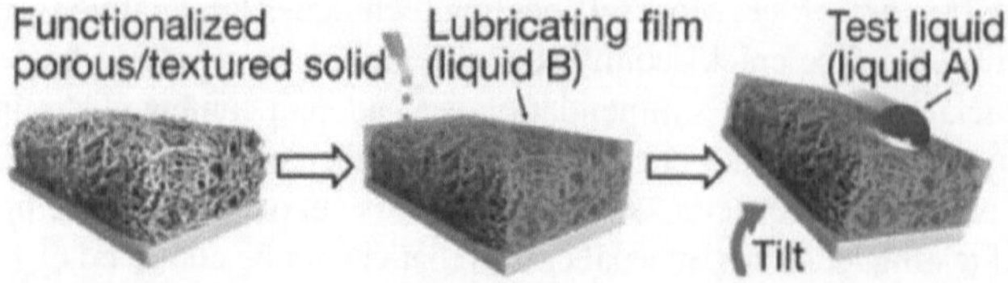

FIGURE 13.8 Slips surface design with functionalized porous surface filled with inert lubricating liquid creates a solid that easily sheds drops of immiscible liquid. (From [Wong 2011].)

Design criteria based on surface energy differences are as follows:

$$\Delta E_1 = E_A - E_1 = R(\gamma_B \cos\theta_B - \gamma_A \cos\theta_A) - \gamma_{AB} > 0 \quad (13.16)$$

$$\Delta E_2 = E_A - E_2 = R(\gamma_B \cos\theta_B - \gamma_A \cos\theta_A) + \gamma_A - \gamma_B > 0 \quad (13.17)$$

Here E_A, E_1, and E_2 are the surface energies for the cases of being wetted with an arbitrary impinging liquid, lubricating liquid, and combined lubricating and impinging liquid, respectively. R is a surface roughness parameter. γ_A, γ_B, and γ_{AB} are the surface tensions for the impinging liquid, lubricating liquid, and interface between impinging liquid and lubricating liquid, respectively.

13.1.6 Redundant System Design

A cognitive approach to redundancy assesses the present situation and uses onboard intelligence to select a workable, optimized, and new operational configuration. An alternative preassembles a library of configuration graphs of possible operational configurations and uses sequential configuration switching and checking to arrive upon an acceptable configuration. Sequential configuration state switching methods can be slower than cognitive approaches but are simpler (Figure 13.9) [Peng 2006]. More complicated approaches combine offline diagnostics and tuning of analog components and operating parameters with component switching [Sylvester 2006].

Microelectronic ICs self-repair by swapping the loads carried by damaged components onto healthy equivalents. Successful component swapping and healing requires (1) an IC with redundant components, (2) switching circuits that alter the topological configuration of components, (3) fault detection system, and (4) decision-making process that selects new healthy configurations.

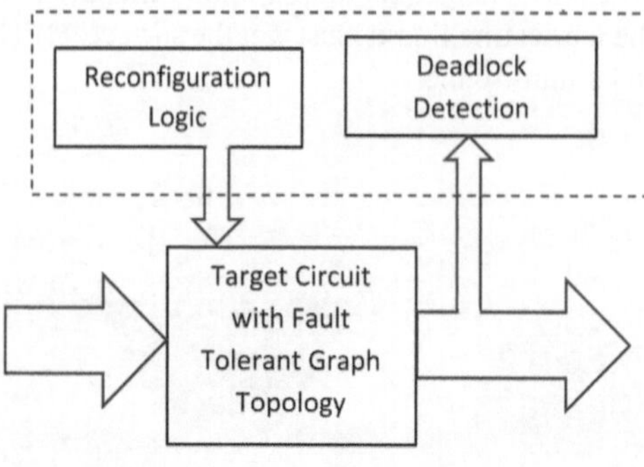

FIGURE 13.9 Scheme for component swapping to self-heal integrated circuits. (Adapted from [Peng 2006].)

13.2 Self-healing Performance Evaluation

Many self-healing performance evaluations are modified conventional performance assessment tests [Takahashi 2003]. Some are specialized.

13.2.1 Metrics for Self-healing Performance Evaluation

Performance metrics quantify the behavior and properties of a system. Ideal metrics evaluate and enable comparing a wide range of technologies. Simplicity and ease of use favor condensing multifaceted effects into scalar metrics. *Healing efficiency* η is the ratio of a post-damage recovered property, θ_{Healed}, to that of an original pre-damaged property, $\theta_{Predamage}$:

$$\eta = \frac{\theta_{Healed}}{\theta_{Predamage}} \% \tag{13.18}$$

Systems with 100% efficiency completely recover from damage. A variant definition is the ratio of the change in structural property due to damage, θ_{Damage}, and healing:

$$\eta = \frac{\theta_{Healed} - \theta_{Damage}}{\theta_{Predamage} - \theta_{Damage}} \% \tag{13.19}$$

The choice of parameter θ depends on context. Crack-healing efficiency is the percent healing ratio based on the fracture toughness parameter K_{IC} [Brown 2011a] [Brown 2002] [Boresi 2002]. Other variants use measures of structural fitness, such as the resonant frequency of concrete test bars, the elastic energy stored in a structural member under load, and the tear strength [Lepech 2006] [Park 2009] [Keller 2007]. Two viable healing indices are HI^1 and HI^2 [Tan 2012b]. HI^1 is essentially the η efficiency of (13.1) using the dynamic shear modulus G^* for θ:

$$HI^1 = \frac{|G^*|_{after}}{|G^*|_{initial}} \times 100 \tag{13.20}$$

HI^2 accounts for the terminal shear modulus following rest cycle with inclusion of the number of cycles before, N_{before}, and after, N_{after}, the rest period:

$$HI^2 = \left(\frac{|G^*|_{terminal}}{|G^*|_{initial}} \right) \frac{(N_{after} - N_{before})}{N_{before}} \times 100 \tag{13.21}$$

It is possible for healing efficiencies to exceed 100%. An example is a report of a healing efficiency of 800% in concrete beams. The healing process (encapsulated three-part methyl methacrylate) converts material with inherently weak tensile strength into one that better tolerates tension [Dry 1996b].

Combining several parameters into more holistic measures often has value [Laver 2009]. Fidelity is the ratio of the number of desired features realized in the structure to total desired features. Fidelity describes supramolecular assembly performance based on the ratio of the concentration of desired complexes to all complexes [Todd 2005]. Fidelity extends to graphs of topological configuration, where fidelity is the ratio of number of desired links to total number of links. Similar are control strategies in the self-repair through tendon actuation of tensegrity bridges [Korkmaz 2012]. Healing in asphalt, being a complicated interaction of material and mechanical effects acting over multiple timescales, is natural for holistic measures of healing performance. S is a healing efficiency for asphalt, η_A is

$$\eta_A = \frac{S_f - S_i}{S_i} \times 100 \tag{13.22}$$

where S_f and S_i are the final and initial damage indices for a given rest period. The index S depends upon pseudo compliance derived from curve fits to mechanical compliance test data [Palvadi 2012].

The width heal ratio (WHR), defined as average width of damage before, w_b, and after, w_a, healing quantifies the materials with cracking or similar line-shaped damage [Sundaresan 2013]:

$$WHR = \frac{w_b - w_a}{w_b} \tag{13.23}$$

Caution should be exercised when interpreting performance based on healing efficiency, as defined in (13.18). The deleterious effects of parasitic weakening of the undamaged material by the addition of self-healing agents may bias the healing efficiency metric by reducing the overall strength term appearing in the denominator of the healing efficiency ratio [Sanada 2006]. Certain systems with large parasitic loads appear favorably, such as a mercaptan–epoxy microencapsulated liquid healing system which achieves 105% healing efficiency with only a 5% by weight parasitic load of healing [Yuan 2008b].

Functional efficiency metrics provide the framework for evaluating reconfigurable versus dedicated nonreconfigurable systems [Siddiqi 2009]. The functional efficiency η_f of a generic system is the ratio of functional capability to resources, such as cost, mass, or energy:

$$\eta_f = \frac{\text{Total functional capability}}{\text{Resources}} \tag{13.24}$$

The functional efficiency of a reconfigurable system, η_{fR}, is

$$\eta_{fR} = \frac{\sum_{i=1}^{m} \upsilon_i}{\rho} \tag{13.25}$$

Here m is the number of configuration states; υ_i is the capability of configuration i; and ρ is the resources consumed. The functional efficiency of a dedicated system is

$$\eta_{fD} = \frac{\sum_{i=1}^{k} \upsilon_i}{\sum_{i=1}^{k} \rho_i} \tag{13.26}$$

Here k is the number of dedicated systems; υ_i is the capability of configuration i; and ρ_i is the resources consumed by the ith system. The ratio of the functional efficiency of the reconfigurable system to that of the dedicated systems is the relative functional efficiency ratio Ξ_f:

$$\Xi_f = \frac{\eta_{fR}}{\eta_{fD}} \tag{13.27}$$

The ratios of $\Xi_f > 1$ favor reconfigurable systems.

13.2.2 Methods of Assessing Self-healing Performance

13.2.2.1 Long-term Stability

Establishing long-term durability of engineered materials with long-term performance testing is often impractical. An alternative is to conduct short-term accelerated tests under aggressive stress conditions that mimic the effects of long-term tests. Highly accelerated life testing (HALT) and highly accelerated stress screening (HASS) are standardized approaches [Hobbs 2000].

13.2.2.2 Methods of Determining the Self-healing Performance of Polymers – Diffusion and Dynamers

Quantitative measures of the self-healing properties of polymers include dynamic modulus, fracture toughness, fracture energy, and rehealing velocity. Many of these start with conventional damage assessment tests and then allow for healing in a rest period. Specifying details on the controlled production of damage is important, such as using knife edges or lasers [Shen 2002].

Tests to measure dynamic modulus apply dynamic loads to materials or structures and measure the response. Sinusoidal shaking measures the elastic modulus as a complex frequency-dependent quantity. The real component represents conservative forces, that is, stiffness. The imaginary component represents losses. The dynamic shear modulus G^* is

$$G^* = G' + iG'' \quad (13.28)$$

The loss factor δ is

$$\delta = \tan^{-1}\left(\frac{G''}{G'}\right) \quad (13.29)$$

The loss tangent is tan(δ).

Measurements of the frequency dependence of the complex elastic modulus of dynamers and polymers with reversible bonds and gelation quantify the healing performance [Liu 2012c]. Increases in the real part and decreases in the imaginary part of the modulus correspond to healing. Conversely, decreases in the real part and increases in the imaginary part correspond to damage.

Crack-healing tests first initiate and open cracks with applied loads and then relax the load to allow the cracks to close tightly, while providing an easy measurement of the crack size versus loading history. Many fracture-healing test instruments reduce the variability of crack formation with specialized geometries and load paths.

The *four-point torsion test* measures the fracture energy of undamaged and rehealed polymers following cracking (Figure 13.10). Advantages of this test include the ease of setup, control of loading rate, ease of data interpretation, the formation of cracks that do not run entirely through the specimen, and the tight closing of the crack upon unloading. Early (ca. 1969) measurements on epoxy resin polymers examined the rate at which cracks heal, that is, the healing velocity. Observed healing velocities range from 5×10^4 cm²/s to 32×10^4 cm²/s, with little temperature dependence once above the glass transition temperature [Outwater 1969].

Double cleavage drilled compression (DCDC) is a reversible fracture test. The test specimen geometry is a short column with a transverse hole passing through at midspan (Figure 13.11). Loading the column in compression produces hoop stresses on the circumference of the hole. The maximum hoop tension is at the crowns, that is, top and bottom of the hole. Notches machined into the crowns are stress risers that combine with the hoop stresses to initiate cracks to propagate along the axis of the specimen. Internal gradients reduce tensile stresses with distance from the hole and limit the lengths of the cracks.

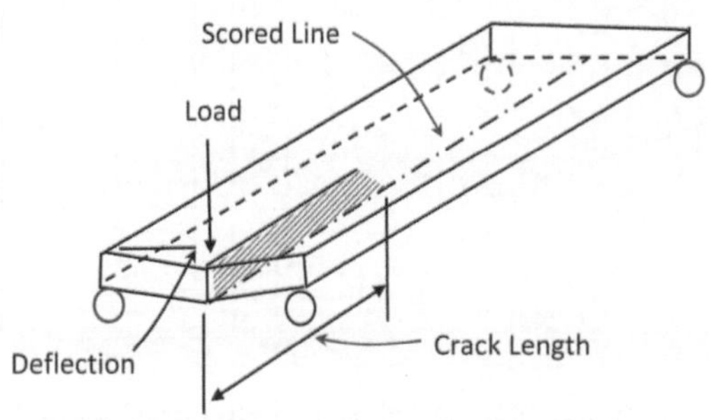

FIGURE 13.10 Four-point panel torsion test to measure fracture energy. (Adapted from [Outwater 1969].)

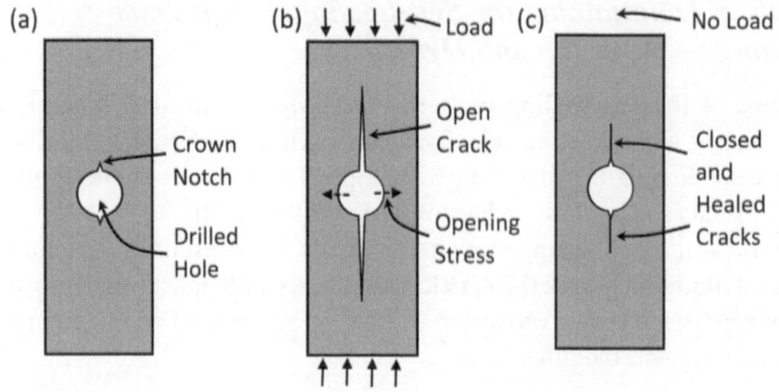

FIGURE 13.11 DCDC crack test specimen. (a) Unloaded pristine condition. (b) Axial loading opens cracks that initiate at crown notches. (c) Unloading closed cracks for healing.

Releasing the compression load releases residual elastic energy and closes the cracks. The DCDC method relies on geometric nonlinearities to open and close the cracks. Sensitivity to these nonlinearities poses potential difficulties with control.

The *double cantilever beam* (DCB) provides improved reversible control of the gap-opening geometry and stress with tapered cantilever beams and polyimide tape inserts as crack initiators (Figure 13.12) [Meiller 1999] [Raghavan 1999]. The overall concept is that the moment arms of the cantilever increase as the crack grows. Constant cross-sectional beams increase the stress intensity at the crack tip. Tapered cantilevers compensate by increasing the cross section and reducing the intensity [Marcus 1971] [Mostovoy 1976] [Tsangouri 2015] [Blackman 2002]. Different shapes fulfill this requirement, including specialized short beam variants for testing inside of climate-controlled chambers [Beres 1997].

A combination of a *nanoscratch tester and an atomic force microscope* (AFM) assesses nanoscale and molecular scale healing behaviors of dynamers and similar noncovalent and thermally reversible healing systems. The AFM measures the rate of healing by observing a scratch as it spontaneously disappears [Brancart 2014].

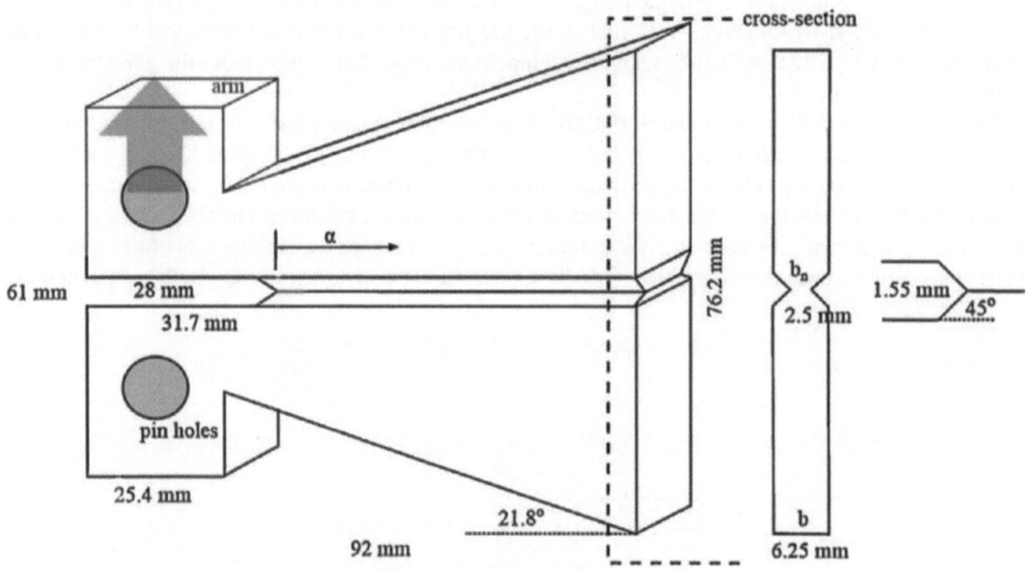

FIGURE 13.12 Double cantilever beam test specimen. Pulling vertically through the pin holes introduces cantilever style loading on the material in the joining strip between the beams. The varying beam geometry aims to produce uniformly controlled loading of the crack as it grows or heals. (From [Tsangouri 2015].)

Dark field microscopy and light-transmission photometry measure micrometer-scale voids that form and heal in transparent materials [O'Connor 1980]. Small internal damage, that is, voids and crazes, clouds the transmission of light through transparent polymers. Polymer-based healing, such as reptation, repairs materials and recovers transparency.

Polymers with reversible-bonding molecular healing often display complicated and subtle material-dependent effects that are difficult to predict a priori. *Differential scanning calorimetry* (DSC) pre-screens candidate polymer materials by examining the specific heat versus temperature to identify glass transition temperatures [Nielsen 2010]. DSC is not always a reliable predictor of self-healing capability. Favorable DSC results require confirmation with subsequent mechanical testing.

Dynamic tests assess ionomers that heal by dynamic thermomechanical effects. Impacting stationary plates with projectiles of various speeds and diameters provides dynamic loading. Outcomes are visible appearances of hole closure and pressurized leak testing across the hole. Leak testing is the more demanding test. Experiments on the ionomer Surlyn™ confirm that both low-velocity (180 m/s) and high-velocity (400 m/s) projectiles produce healing in specific ranges of plate thickness to projectile diameter ratios. A ratio of 0.2 or greater heals in the low-velocity case [Grande 2012].

Polymers with noncovalent bonds that detach and reattach without breaking other molecular bonds are a class of materials capable of multicycle healing. Measuring the strength of attachment over a number of healing cycles is a healing efficiency performance metric. Microscopic optical imaging of the attachment surface texture reveals patterns of texture. Regular fringes of equal chromatic order correspond to elastic materials. Blotchy patterns indicate materials with more viscous behaviors (Figure 13.13) [Faghihnejad 2014]. High-speed imaging clarifies the impact and recovery dynamics. Infrared imaging quantifies

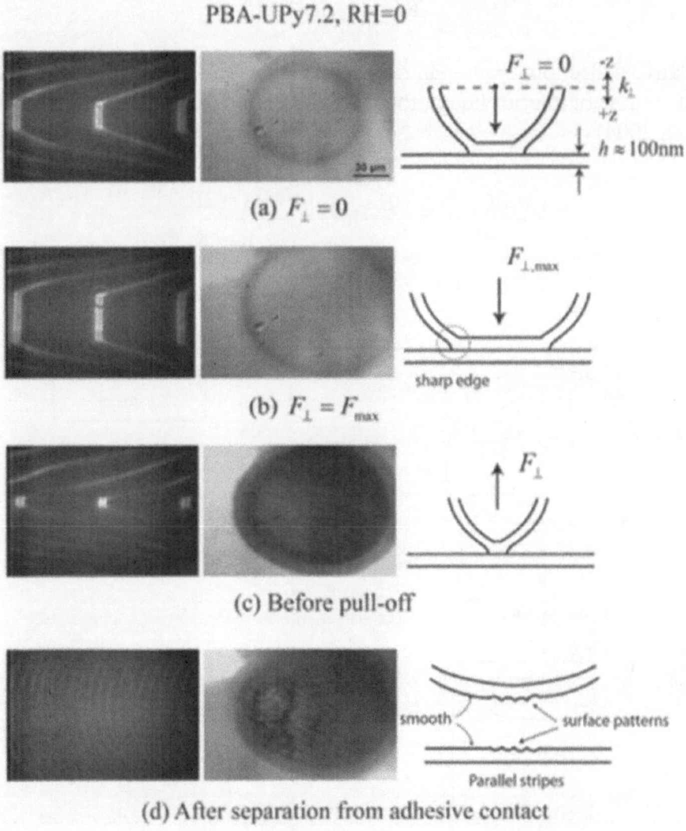

FIGURE 13.13 Measurement of strength of interpenetration and reversible attachment of noncovalent bonds with load test apparatus combined with interferometer. Results shown from testing 2-ureido-4[1*H*]-pyrimidinone quadruple hydrogen bonding groups. (From [Faghihnejad 2014].)

thermodynamics and helps to assess the material ejected by the projectile in polymer-based structural elements that recover from ballistic impact [Zavada 2015].

13.2.2.3 Methods of Determining the Self-healing Performance of Materials with Microencapsulated Healing Liquids

Microencapsulated liquid self-healing is a multistep process with multiple demanding performance requirements. Success requires addressing a host of practical issues, including overall durability of the healing system, damage to the matrix properly triggering the rupture of the capsules, and the healing liquid properly flowing into the damaged region.

Microcapsule strength is an important performance parameter. The capsule should be strong enough to survive manufacturing and routine operational use of the structure, yet frangible enough to release with loads that damage the structural element. Specialized apparatuses determine the strength of isolated capsules [Liu 1996]. The load force f versus deformation Δ, in terms of Young's modulus E, Poisson's ratio ν, wall thickness h, and sphere radius R, for a permeable membrane sphere with, and stretching only deformation, is as follows (Figure 13.14) [Dubreuil 2003]:

$$f = \frac{\Delta E h^2}{R(1-\nu)\sqrt{6(1+\upsilon)}} \tag{13.30}$$

For shear only,

$$f = \frac{\Delta E h^2}{R(1+\nu)\sqrt{6(1-\upsilon)}} \tag{13.31}$$

Based on the relative deflection, $\varepsilon = 1 - H/(2R_0)$, defined in terms of the geometry of Figure 13.15, and assuming constant volume of interior liquid, the force deflection relation for a spherical microcapsule is as follows [Lulevich 2004]:

$$\mathcal{F}_s = -\frac{\partial G_s}{\partial H} = 2\pi \frac{E}{1-v} hR_0 \frac{\left[1+R_0/(2R_s)\right]^2}{(1+R_0/R_s)^4} \in^3 \tag{13.32}$$

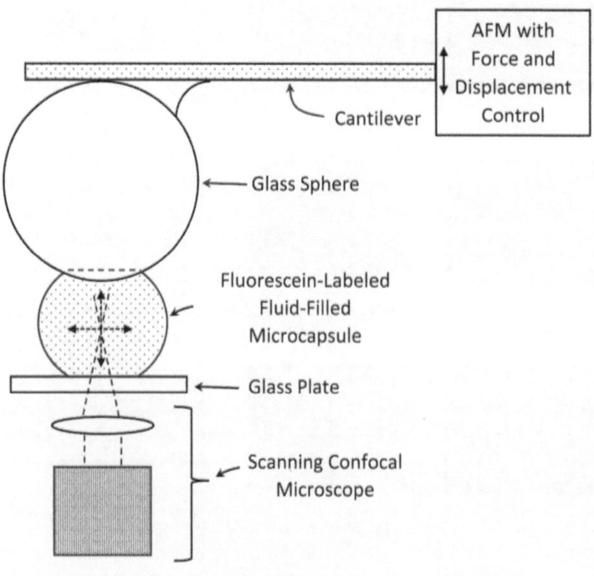

FIGURE 13.14 AFM with glass sphere probe combined with scanning confocal microscope to determine stiffness of fluorescein-labeled microcapsule. (Adapted from [Lulevich 2004].)

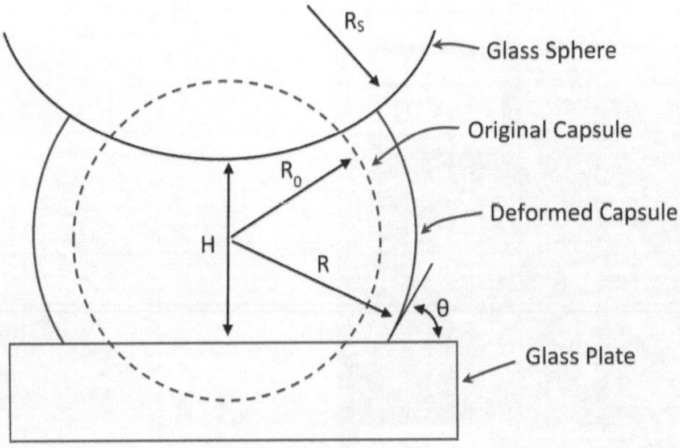

FIGURE 13.15 Geometry of load and deflection of a microencapsulated sphere. (Adapted from [Lulevich 2004].)

Capsule rupture and release of healing liquid are central to microencapsulated liquid healing. It is difficult to ascertain whether the capsules rupture on cue. One option for assessing rupture timing and extent is to supplement the healing liquid with chemical agents that signal the presence of rupture with easily observed techniques, such as fluorescence under exposure to UV [van den Dungen 2010].

Using microcapsule loading and a membrane model of the elastic mechanics (ignores wall bending stiffness) leads to typical values of the mean elastic modulus of urea-formaldehdye capsules running from 3.6 to 3.9 GPa, and rupture strengths ranging from 0.14 MPa for 213 μm diameter wet capsules to 0.8 MPa for 65 μm dry capsules [Keller 2006b].

Elastic fluid-filled spheres buckle according to a linear theory when the exterior pressure exceeds the interior pressure by the critical difference p_{cr} [Timoshenko 1961]:

$$p_{cr} = \frac{2E}{\left[3\left(1-\vartheta^2\right)\right]^{1/2}}\left(\frac{h}{R}\right)^2 \tag{13.33}$$

Here E is the elastic modulus; t is the thickness; and R is the radius. Buckling measurements help to determine the elastic properties of spheres, such as polyelectrolyte multilayer microcapsules [Fery 2004].

13.2.2.4 Methods of Determining the Self-healing Performance of Superhydrophobic Surfaces

High-speed optical imaging reveals the dynamics of droplet interactions with superhydrophobic surfaces [Brown 2011b].

13.2.2.5 Methods of Determining the Self-healing Performance of Asphalts

The dynamic modulus and number of load cycles to failure are principal parameters that help to assess the fatigue performance of asphalt [Daniel 2001]. Cyclic testing readily measures the maximum shear stress load τ_0 and maximum shear strain γ_0. The ratio of the two is the amplitude of the complex-valued shear modulus [Kim 2003a]:

$$\left|G^*\right| = \left[G'^2 + G''^2\right]^{1/2} = \frac{\tau_0}{\gamma_0} \tag{13.34}$$

Dynamic tensile testing quantifies microdamage healing [Chen 2013b]. Four-point bending fixtures apply uniform bending moments across the center of a beam for both dynamic and fatigue testing. Four-point bending tests also provide access to the asphalt specimen during healing, as for the case of induction heating healing [Liu 2012b].

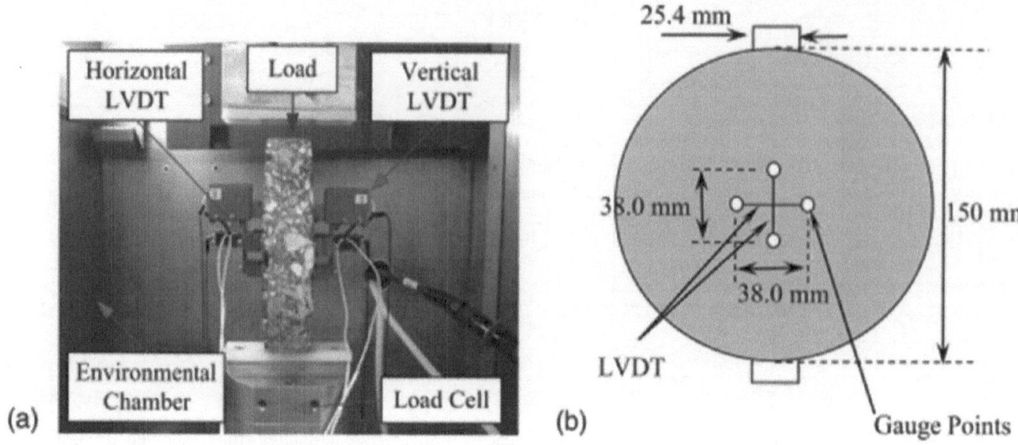

FIGURE 13.16 IDT system for testing fatigue and self-healing of asphalt. (a) Test setup. (b) Specimen geometry. (From [Lee 2015.)

Direct tensile testing is impractical with composite granular materials, such as concrete and asphalt. Indirect tensile testing (IDT) methods with transversely loaded cylinders are effective in such cases. IDT applies compression to the edges of a cylindrical disk. Stress sets up in the disk to induce a corresponding tension stress at the center of the cylinder (Figure 13.16). It is necessary to avoid creating excessively open cracks, as they will not heal. A haversine load of 0.1 s followed by a 0.9 s rest period works well for split cylinder fatigue testing of asphalt [Li 2009b].

The dynamic shear rheometer provides controlled loading for the determination of the intrinsic healing properties of asphalt binders [Bommavaram 2009].

13.2.2.6 Methods of Determining the Self-healing Performance of Concrete

Autogenous self-healing of concrete uses crack in-filling and closing. Three-point bending combined with crack opening displacement (COD) measurements is compatible with environmental test chambers [Ferrara 2014].

Water permeability can measure crack healing in concrete and mortars [Roig-Flores 2016]. Healing and closing of the cracks decrease permeability. The test is to first introduce controlled cracks in a concrete specimen by the split cylinder method or a beam in three-point bending. The cracks should be kept small, preferably with a width of less than 0.3 mm. Next, water permeability testing measures the rate of water flow using gravity and capillary action as the motive force. The healing rate quantifies the amount of healing.

$$\text{Healing rate} = 1 - \frac{\text{Final flow}}{\text{Initial flow}} \tag{13.35}$$

An alternative measure is the increase in mass of the specimen by the amount of water absorbed into the beam [Park 2016a].

Static and dynamic assessments of autogenous concrete crack-healing strength require that the cracks remain tight for the healing process to occur. Studying a specified crack under controlled static load conditions requires test apparatus that isolates and controls the loading on the crack, such as with an apparatus shown in Figure 13.17, or with four-point bending measuring the healing recovery of more diffuse and uncontrolled cracking [Schiessl 1996] [Qian 2010]. Dynamic measurements of the natural frequencies of concrete beams determine strength and readily extend to measuring the performance of self-healing in concrete and cementitious materials [Herbert 2013]. Ultrasonic pulse velocity (UPV) is a

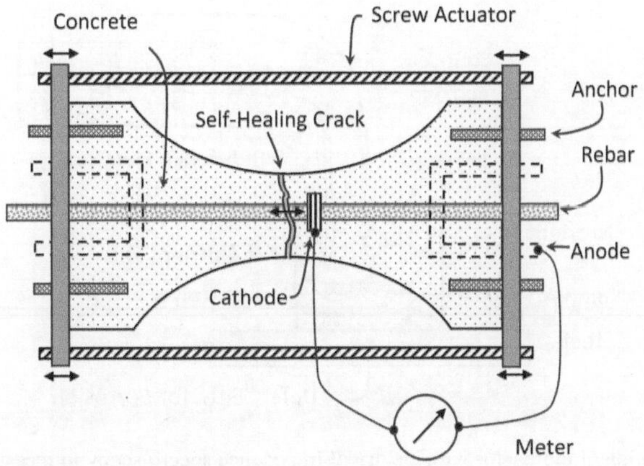

FIGURE 13.17 Test setup for assessing reinforcement corrosion during autogenous healing of concrete. (Adapted from [Schiessl 1996].)

similar dynamic test. Damage reduces the elastic wave velocity. Higher velocities correspond to higher strength and healing. The damage index D, in terms of the original undamaged velocity v_0 and the present damaged v, is as follows:

$$D = 1 - \frac{v}{v_0} \tag{13.36}$$

13.2.2.7 Methods of Determining the Performance of Corrosion Self-healing

Electrochemical impedance spectroscopy measures the frequency response of electrochemical reactions with a three-probe system. Fitting curves to frequency domain data identifies the action, performance, and synergistic self-healing of separate corrosion inducing and inhibiting processes [Tan 1996] [Montemor 2012] [Macdonald 2006]. Figure 13.18 shows a setup that bathes the material under test in a saline solution, while three electrodes probe and measure electrochemical behavior. The working electrode attaches to the material, by a coated metal sample with scratch damage. An inert platinum plate acts as a counter electrode to complete the circuit with the working electrode. An Ag/Ag-Cl reference electrode provides a stable reference voltage. Both static and dynamic measurements produce useful data. Tafel diagrams represent the static data, normally in a semi-log plot of applied voltage versus current through the electrodes. Bode plots present the dynamic data. The area under the Bode plot

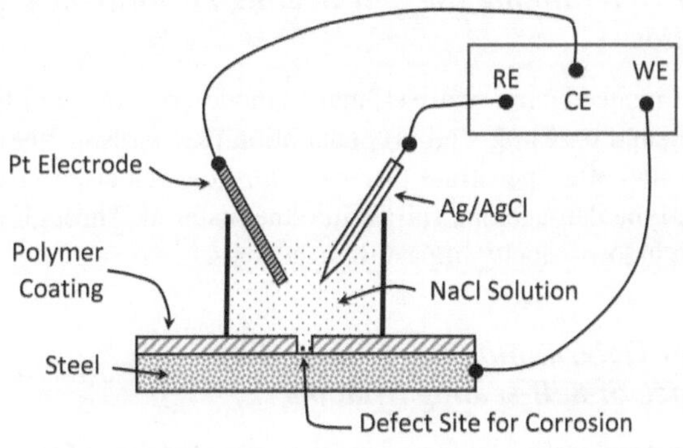

FIGURE 13.18 Electrochemical impedance spectroscopy unit used to assess corrosion healing. RE is reference electrode. CE is counter electrode. WE is working electrode. (Adapted from [Kouhi 2012].)

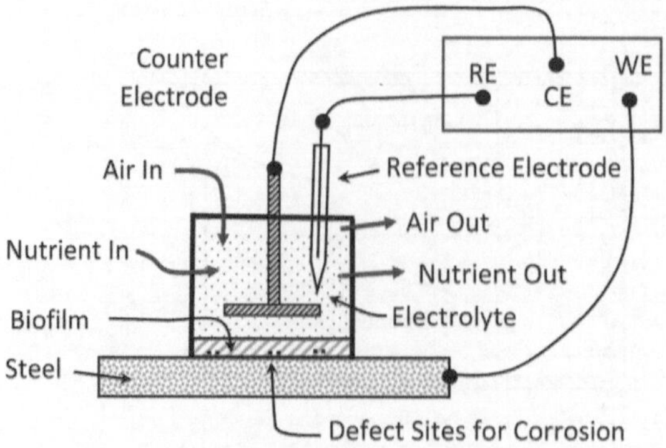

FIGURE 13.19 Combination of bioreactor with electrical impedance spectroscopy to measure corrosion inhibition and promotion by bacterial biofilms. (Adapted from [Ismail 2002].)

quantifies the performance of active anticorrosion coatings, such as those with microencapsulated linseed oil [Kouhi 2012]. Combining a bioreactor with electrical impedance spectroscopy creates a system for measuring the ability of a biofilm to suppress, or accelerate, corrosion (Figure 13.19).

13.2.2.8 Methods of Determining the Self-healing Performance of Vascular Healing

The following are the performance tests for flooding self-sealing electrical power wire [Sasse 2005]:

1. *DC Impulse Overvoltage Recovery:* Insert high-voltage spikes (>70 kV) into the conductor to break down the insulator. Allow the insulator to heal by sealant flow. Retest the breakdown voltage of the insulator.
2. *Fish Tank Wedge Cut Test:* Cut a wedge out of the insulator down to the central conductor. Bend the wire to open the wedge cut. Place the cut wire in a tank, for example, fish tank, of salt water. Run current through wire and measure the amount that leaks into fish tank versus time as flooding proceeds.
3. *Sealant Downhill Runout:* Expose a section of the insulation at one end of a wire. Hang the wire so that gravity drives the sealant downhill and out of the exposed section. Cycle through high temperatures, for example, 90°C. Measure the amount of material that runs out the bottom end.

13.2.2.9 Methods of Determining the Self-healing Performance of Composites

Principal types of impact-induced damage are (1) matrix mode-cracking parallel to the fibers, (2) delamination, (3) fiber breaking and buckling, and (4) penetration [Richardson 1996]. High-velocity impacts, perhaps faster than 100 m/s, often penetrate the material. Low-velocity impacts do not penetrate and often leave delamination-type damage that spirals into the material. Through-thickness stress waves do not play a significant role in low-velocity impact damage tests.

13.2.2.10 Methods of Determining the Self-healing Performance of Self-sealing Systems

Self-sealing material and structural systems prevent the unwanted flow of fluids by (1) sealing leaks and breaches that allow the fluid to flow, (2) restricting the flow of fluids across defined surfaces, and (3) restricting the flow of fluids through porous, semiporous, cracked, and seamed solids.

Performance tests include puncturing or cutting a small hole in a self-sealing barrier wall, followed by allowing it to heal under no pressure and then measuring the ability to hold differential pneumatic pressure across the wall [Beiermann 2009]. The nail test measures self-sealing in tires [Schutz 1959].

13.2.2.11 Methods of Determining the Performance of Self-cleaning Surfaces

Durability tests of self-cleaning surfaces include finger wipe, sandpaper abrasion, scratch, and oil contamination. Anti-biofouling surface tests on ships include applying pressurized water jets to see how much scours from the surface and conducting shear strength measurements of scrape off strengths of macroscale infestations, such as barnacles [Lu 2015] [Swain 1996]. Instrumented analysis, such as with scanning electron microscopes and energy-dispersive X-rays, assesses surface cleaning and anti-biofouling at microscopic and submicroscopic length scales [Fay 2005].

The contact angles between a liquid drop and solid surface assess the strength of hydrophobicity or hydrophilicity. The contact angles for advancing and receding droplets assess movement hysteresis [Chen 1999].

The following is the procedure to measure the performance of superhydrophobic droplet roll-off-based self-cleaning surfaces: (1) Place fluorescent and luminescent particles on the surface under test in a controlled manner with an air-fan-driven deposition. (2) Apply fog and/or rain treatment to induce water droplet formation and roll-off cleaning. (3) Sputter coat the surfaces to fix the remaining particles to the surface. (4) Use a scanning electron microscope with custom cathodoluminescence sensing capability to examine the surface and distinguish the contaminant particles from the surface. (5) Measure the tilt angle of a flat plate that causes a premeasured drop to roll off the surface due to gravity forces [Fürstner 2005]. More detailed measurements account for texture geometry on area calculations, and electron microscope imaging of the pinning and depinning interactions of droplet edges with microtexture as the droplet moves [Lee 2010a] [Paxson 2013]. Magnetic actuation of resonant and decaying oscillatory behaviors of droplets laced with superparamagnetic nanoparticles provides additional insight into the mechanics of roll-off [Timonen 2013].

Rough or soft surfaces often make it difficult to measure contact angles. In such cases, the dimensionless wetted area A_w is a useful measure of superhydrophobic performance:

$$A_w = \frac{a_w}{V^{2/3}} \tag{13.37}$$

Here a_w is the wet contact area and V is the volume of the droplet [Marmur 2004]. Values of A_w less than about 0.5 correspond to superhydrophobic surfaces with contact angles in excess of 150°.

Scratch, indent, and similar mechanical tests quantify the durability of superhydrophobic surface textures. Droplet loadings in the quasi-static regime with relatively slow impacts leave large opportunities for air to remain entrapped in the micro- and nanotextures to produce a Cassie–Baxter configuration. High-speed droplets may have sufficient kinetic energy to drive the liquid through surface tension and residual air pressures to wet the surface and form a Wenzel state. The ratio of kinetic to surface energy in a liquid droplet, that is, the Weber number, characterizes the influence of dynamic effects on superhydrophobicity:

$$We = \frac{\rho R V_i^2}{\sigma} \tag{13.38}$$

Here ρ is the mass density of the liquid; R is the radius of the droplet; V_i is the droplet impact velocity; and σ is the surface tension. Experiments on surfaces with polymeric micropatterns and carbon nanofiber features find that droplets behave quasi-statically for $We < 120$ and dynamically for $We > 120$ [Tsai 2009].

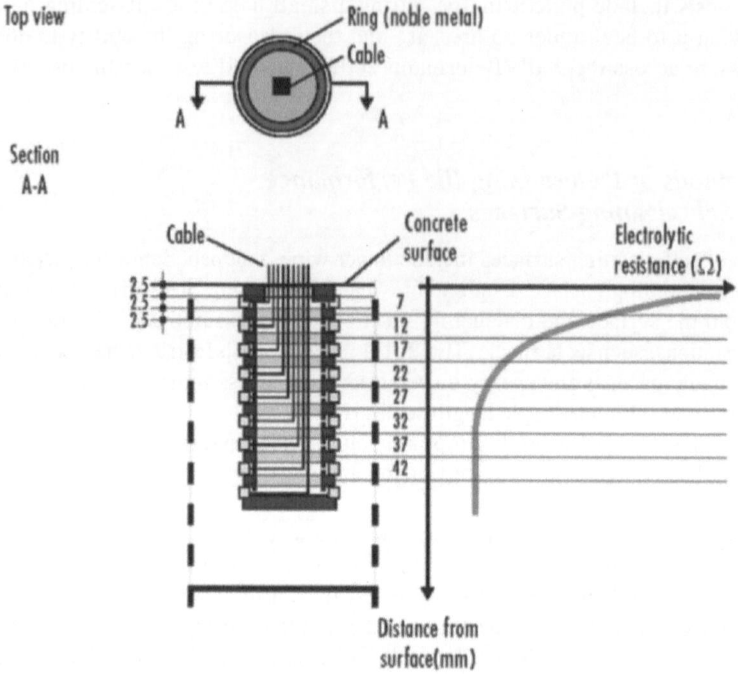

FIGURE 13.20 Multiring electrode detects electrolyte penetration into concrete with hydrophobic surface. Each ring reports conductivity at a depth. (From [Raupach 2005].)

Hydrophobic surface treatments prevent liquid penetration into naturally porous materials. An embedded multiring probe assesses the performance of electrolytic fluid-blocking surface treatments with depth-dependent electrical resistance measurements (Figure 13.20).

13.2.2.12 Methods of Determining the Self-healing Performance of Electrical Systems

Scintillation measurements monitor the performance of self-healing in capacitors. The tests apply a constant current load to the capacitor and measure the resulting voltage. A healthy capacitor charges with the current load and increases the terminal voltage up to a preset limit. An unhealthy capacitor may suffer from incipient dielectric layer insulator breakdowns in the form of rapid voltage drops and scintillating flashes at the location of a short. The capacitor recharges if self-clearing healing occurs at the hot spots (Figure 13.21) [Teverovsky 2021] [Prymak 2006].

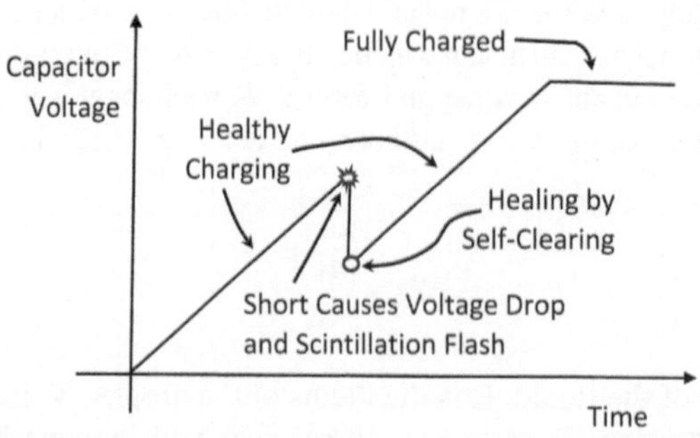

FIGURE 13.21 Testing of capacitor with short leading to voltage drop, scintillation ramp, and hot spot self-clearing healing. (Adapted from [Prymak 2006].)

13.2.2.13 Methods of Determining the Self-healing Performance of Systems

Complicated and unpredictable situations prevent comprehensive testing self-healing performance in all possible damage and healing scenarios. Complicated cases benefit from the use of alternative performance evaluation strategies. Intelligent airplane systems, including those that reconfigure from unexpected damage, recommended the following considerations [Crum 2004]: (1) *Mixed Criticality*: Since not all components of the system are critical for survival, it may be advantageous to use a mixture of testing protocols. (2) *Multi-entity*: Consider multiple interacting systems in the context of mitigation, healing, and limping. (3) *Mixed Initiative/Authority Management:* Components and subsystems operate under a mixture of local autonomy and centralized supervision. When and under what circumstances should the exertion of control be local or central depends on considerations of situational/environmental awareness, present state of health, and response time for corrective action. (4) *Sensor Fusion*: Integrate multiple sensors, sensor types, and sensor health measurements to produce information in a condensed form for decision-making. (5) *Adaptive Learning, and Multimodal Functions*: The system should adapt and learn from experiences in the environment. Ideally increased knowledge can be passed from system to system.

13.2.2.14 Methods of Determining the Self-healing Performance of Biohybrids

Evaluating the performance of biohybrid healing systems may adapt conventional healing performance tests to the specialized conditions of engineered biosystems. Examples include measuring crack healing by *Bacillus subtilis* bacteria in mortars and concrete and multivariable lifetime assessments of energy and resources consumed by green façades on buildings [Yamasamit 2023] [Ottelé 2011]. For living and green facades on buildings, the living skin acts as an insulating layer and windbreak keeping excess heat out in the summer and heat in during the winter. The live biomass purifies the air through metabolic processes and helps to reduce heat island effects, especially in urban environments.

13.2.2.15 Material Property Testing with Healing

Observations of healing processes provide information on the underlying properties of materials and provide tools for studying materials. An example is the measurement of reptation healing with molecular chain scission corresponding to the molecular basis of tensile strength and fracture energy in polystyrene [Mohammadi 1993]. Similarly, rates of annealing-type healing of fission track damage in rock crystals containing trace amounts of uranium provide information about the age of the rocks [Dumitru 2013].

13.2.3 Selected Self-healing Performance Results

13.2.3.1 Self-healing Performance of Polymers with Diffusion and Dynamers

The dynamic nature of dynameric versus diffusion healing affects the results of rate-dependent tests with dynameric healing being rate-independent and diffusion healing being rate-dependent. For example, when split-anvil testing of solid polymers formed by Diels–Alder reactions is able to recover virtually all of the pre-damaged strength independent of time, the conclusion is that this polymer exhibits dynameric behavior as the cracking was largely a breakdown of DA bonds that reversibly rejoin upon reclosing the crack [Plaisted 2007]. Cases with time-dependent healing may heal by reptation-type diffusion processes.

13.2.3.2 Performance of Self-healing Shape Memory Materials

The healing performance of shape memory materials depends on the temperature at which switching of geometric and mechanical states occurs. The amorphous molecular structure of most polymers leads to gradual temperature-dependence of behaviors. Many applications require a crisper switching temperature. Isolating the molecules that switch into crystalline domains and attaching the domains to the rest of polymer matrix with flexible elements provides the order necessary for a well-defined switching temperature [D'Hollander 2009]. An increase in the enthalpy required to switch states indicates a crisper phase change, with values on the order of 10 J/g for co-block polyurethanes.

13.2.3.3 Microencapsulation Self-healing Performance Results

Microcapsules placed in polymer matrices alter the strength and fracture toughness. Liquid droplet inclusions affect the stiffness of elastic solids. Based on energy considerations that extend Eshelby's inclusion model, the surface tension Υ, of a dilute distribution of small liquid inclusions of radius R and solid with elastic modulus E, stiffens the material when $R < 1.5\ \Upsilon/E$ and softens otherwise [Style 2015]. With regard to fracture, a series of experiments dating back to the 1970s indicates that the insertion of sub-100 μm particles with volume ratio loadings of 15–20% increases fracture and fatigue resistance. Adding rubber or hollow microparticles increases the fracture toughness and fatigue resistance of epoxies. Volume ratio loads ranging from 5% to 15% seem optimal for fracture toughness with increases of up to 300% observed with 40 μm diameter particles [Bagheri 1996] [Azimi 1996]. The fatigue resistance of highly cross-linked epoxy composites increases with the Paris law exponents reducing from 9.7 to 4.5 in 10% by weight loading of microcapsules [Brown 2006]. A plausible explanation is that microcapsules arrest, pin, and blunt the formation of early-stage fatigue microcracks. However, measurements on composites with embedded, hollow microspheres did not find any fracture pinning toughening mechanisms [Kim 2001]. Microencapsulated DCPD increases the resistance to fatigue with boosts in the number of load cycles to failure by factors of 5 or more [Jones 2007]. Loose bonding between the particle and the matrix favors fracture toughness [Broutman 1971]. Tight adhesion increases compression strength [Cardoso 2002].

The toughness and strength of microcapsules controls the overall performance of a microencapsulated liquid-healing system. Mechanical testing indicates that poly(urea–formaldehyde) (PUF) capsules rupture by brittle fracture with a strength that decreases with diameter following a power law with exponent $n = -0.77$ [Yang 2008]. The insertion of nanoparticles, such as carbon nanotubes and nano alumina, increases the stiffness of PUF microcapsules by up to 46% and the plasticity ratio from 0.23 to 0.54, with 0 corresponding to perfectly elastic and 1 corresponding to perfectly plastic [Ahangari 2014].

Heat damages microcapsules. Thermal breakdown mechanisms for epoxy-filled PUF capsules below 251°C are thermal expansion mismatch between the wall and encapsulated material and diffusion through the walls; above 251°C are largely the breakdown of the wall material [Li 2006]. Thickening the walls increases endurance. The addition of melamine to PUF capsule walls increases resistance to water, heat, acids, and alkalis through the action of a triazine ring in melamine [Tong 2010].

The ability of microcapsules to fracture and release liquid on cue with crack formation affects healing. A study using high-resolution X-ray imaging observed the release and flow of microencapsulated liquid with crack formation, 60 μm solvent-laden capsules ruptured preferentially within a thin 75 μm layer next to the crack [Mookhoek 2010]. The solvent flows into the crack and provides partial, but not complete, coverage of the crack surface. Related studies using microencapsulated DCPD and matrix-embedded Grubbs catalyst as the healing method and optical microscopy to observe capsule rupture and healing agent activation also found that the capsules ruptured in a thin layer near to the crack surface [Thoma 2003]. Sometimes the healing liquid flows properly into the crack for healing, other times the liquid does not fill and bridge across the crack, causing the repair to fail.

DCPD is a principal liquid in many microencapsulated healing systems. DCPD is a low-viscosity liquid at room temperature formed from a pair of small organic ring molecules, that is, norbornene, a six-sided ring, and cyclopentene, a five-sided organic ring. Upon encountering ruthenium-based Grubbs catalyst, the norbornene rings open to reveal active binding sites that bind to similar open rings. The process, known as ring-opening metathesis polymerization (ROMP), creates polymer chains and networks of molecules that solidify the liquid [Grubbs 2004]. Multiple variants of ruthenium catalysts for ROMP curing of DCPD are available, including first-generation Grubbs, second-generation Grubbs, and second-generation Hoveyda–Grubbs. The different variants have different reaction rates and temperature sensitivity. Second-generation catalysts have faster reaction rates and provide better high-temperature performance [Wilson 2008]. The increased reaction rates of the second-generation catalysts do not provide an improvement in healing performance.

Since ruthenium-based catalysts are expensive, an important question is the sensitivity of the ROMP reaction and bulk solidification to the amount of catalyst. Experiments involving Raman and Brillouin scattering indicate that the reaction and reaction rate are sensitive to the amount of catalyst, with the degree of cure and elastic modulus varying by amounts of 10% and 15%, for catalyst concentrations varying from 0.35% to 0.50% by weight, respectively [Aldridge 2010]. The rate of DCPD polymerization induced by ruthenium indenylidene catalysts, such as the first Grubbs catalyst, is highly sensitive to isomer variation. The *endo*-isomer of DCPD heals 20 times faster than the *exo*-stereoisomer, but with a lower healing efficiency [Mauldin 2007]. *N*-Heterocyclic carbine and Schiff base–substituted analogs show superior chemical compatibility and healing efficiency for epoxy-based self-healing systems [Öztürk 2015]. Including wax particles in self-healing composite materials protects the Grubbs catalyst [Moore 2009].

Microencapsulated epoxies in cementitious composites have healing efficiencies of up to 60% [Li 2013a]. Acrylics, such as those used in dental restorative materials, do not fare as well as epoxies with standard urea-encapsulated DCPD healing systems with 50% versus 100% efficiencies for equivalent epoxy-based systems [Wertzberger 2010].

Studies on the long-term performance of microencapsulated liquid healing reveal the following: (1) *Aging Has a Negative Effect*: Long-term ambient condition storage caused the healing efficiency to drop from an initial 85% to 68% over the first month, and then plateau at that value for the next five months of tests. (2) *Elevated Temperatures Cause Rapid Aging:* Short-term elevated temperature stress at 121°C causes a reduction in healing efficiency from an initial level of 85% to 45% after 1 h, and with smaller, but significant, reductions appearing in subsequent hours [Jin 2012]. Microscopic examination reveals that one mechanism of thermal breakdown is the high-temperature disintegration of amine capsules containing the hardener component.

13.2.3.4 Asphalt Self-healing Performance Results

Two complementary mechanisms for damage healing in asphalt are viscoelastic recovery and fracture healing. Reports dating back to as early as 1967 find that adding rest periods between loading increases the fatigue life of bituminous asphalt pavement. Values range from 8.7% to 23.9% [Kim 2003a].

Asphalt binders with minimal self-healing capabilities perform poorly under field conditions. Cracks appear in nonhealing pavements after one or two years, especially in cold climates. Binders that heal and reverse aging perform much better. Measurements of the non-Newtonian thixotropic properties of the binder help with identifying molecular healing proclivities [Shan 2010].

The following are the effects of rest periods on the fatigue life of bituminous mixes [Bonnaure 1982] [Hesp 2007]:

1. *Overall Performance:* The insertion of rest periods between loading cycles improves the fatigue life of asphalt pavements.
2. *Optimum Rest Duration:* A rest period equal to about 25 times loading cycle or longer produces the maximum healing benefit.
3. *Temperature Dependence:* Healing is more active at higher temperatures.
4. *Stiffness Dependence:* Healing is more active with soft, rather than stiff, binders.
5. *Molecular Structure:* High levels of waxy and linear molecules exert a negative effect on performance.

Miner's rule quantifies the fatigue performance for cases of loading with cycles of different amplitudes with fatigue index M:

$$M = \sum_{i=1}^{q} \frac{n_i}{N_i} \tag{13.39}$$

Here q is the total number of different amplitudes in the load regime; N_i is the total number of cycles needed to produce fatigue failure for repeated loads at level i; n_i is the number of cycles experienced at load level i; fatigue failure occurs when M = 1. A generalized fatigue index M′ includes the effect of rest periods [Francken 1979].

$$M' = N_T \sum_{i=1}^{I}\sum_{j=1}^{J} \frac{f(i,j)}{N_i M(j)} \tag{13.40}$$

A simplification is to use load levels that vary by equal increment ε_1 (i = 1, 2, …, I), and with rest periods that vary by an equal increment T_L (j = 1, 2, …, J). N_T is the total number of cycles required to produce fatigue failure for an initial strain level ε_1. Experiments determine values of f(i, j) and M(j) for rest periods T_R. A curve fit for M(j) is as follows:

$$M(j) = M\left({}^{T_R}\!/\!{}_{T_L}\right) = 1 + 2.8\left({}^{T_R}\!/\!{}_{T_L}\right)^{0.44} \tag{13.41}$$

Aging appears as a slowdown of healing recovery rates with repeated loading and healing cycles [Roque 2012]. Eventually, the healing cannot keep up with the damage and the material fails.

A modified Avrami equation includes measurable parameters that describe healing during rest periods:

$$R_h(t) = R_0 + (1 - R_0)\left[1 - \exp\left(-qt^r\right)\right] \tag{13.42}$$

Here R_h is the rate of healing; R_0 is the instantaneous state of healing; and q and r are empirical parameters. Measurements on typical binders give values of R_0 in the range of 53–95%, q in the range of 0.002 to 0.38, and r running from 0.2 to 1.02, with t in minutes [Bhasin 2011]. These experiments support the hypothesis that healing rates increase with temperature and decrease with age.

It is possible to separate the effects of viscoelastic recovery from fracture healing by assuming an additive model of viscoelastic recovery and fracture healing effects, and that creep measurements provide an equivalent measure of viscoelastic recovery [Shen 2014]. Steric hardening stiffens the binder matrix with molecular interactions caused by repeated loading and may confound healing matrix performance measurements [Santagata 2013].

13.2.3.5 Concrete Self-healing Performance Results

Concrete and cementitious materials are composites of sand and gravel aggregate held together by cement. Unreinforced, these materials tend to be strong in compression, but are prone to tension cracking. The immersion of concrete with unreacted calcium hydroxide into water containing calcium carbonate and other dissolved mineral electrolytes infills cracks and grows protuberances [Haegermann 1930]. Early tests (ca. 1956) achieved healing efficiencies in tensile tests of up to 25% [Kenneth 1956].

UPV experiments indicate that autogenous healing works best when the damage is below a threshold value. For normal concrete, the threshold value is D ≈ 0.6–0.7, and for high-strength concrete D ≈ 0.4 [Zhong 2008] [Aldea 2000].

Expansive and swelling agents, such as hauyne, anhydrite, free lime, and montmorillite, increase the effectiveness of autogenous healing [Ahn 2010] [Jiang 2014]. Actuation to close cracks also enhances healing. Cement mortar beams reinforced with 0.5% by volume of nitinol SMA fibers exhibited superelastic recentering recovery of 60% following large deformations versus minimal recovery in similar beams reinforced with conventional steel fibers [Shajil 2012].

When the source of corrosion is penetration of water through a covering layer to the rebar, the sealing of cracks with carbonate reduces the rate of corrosion. Tests of mortars with water, alcohol, and gas indicate that the permeability declines with water, less so with alcohol and not with gases. These results confirm the hypothesis that water-based carbonation reduces permeability. Resting periods between freeze-thaw cycles increase durability up to 90%, an effect attributed to autogenous crack healing preventing the ingress of liquid water [Sukhotskaya 1983]. Nonetheless, carbonate sealing of cracks does not provide much mitigation against corrosion when the water contains chlorides [Schiessl 1996]. Customized encapsulated healing systems are more effective. Cylindrical borosilicate glass capsules (length: 35 mm, internal diameter: 3 ± 0.05 mm, wall thickness: 0.175 ± 0.03 mm) filled with one component polyurethane-based healing agent reduces chloride concentration near cracks by 75%, leading to a predicted tenfold extension of the life of concrete structures in aggressive marine environments [Van Belleghem 2017].

Engineered cementitious composites (ECCs) are specialized cementitious materials with customized functionalities, including self-healing. Hydration filling of tension cracks produces resonant frequency-healing efficiencies in ECCs in excess of 100% [Kan 2010] [Yang 2009a]. Hydration filling of micro-cracks reduces chloride diffusivity by 4× versus mortar [Sahmaran 2007].

13.2.3.6 Ceramic Self-healing Performance Results

Multiple cycles of cracking and healing are possible with ceramics that oxidize and fill cracks. Ti_2AlC ceramics heal up to seven cracking cycles by the formation of α-Al_2O_3 and rutile TiO_2 through oxidation at elevated temperatures [Li 2012c].

13.2.3.7 Corrosion Self-healing Performance Results

Microencapsulation supplies healing liquids to damaged surface coatings. One example is a sealant overcoat that prevents the release of lead dust pollution from lead-based paints. Damage to the overcoat causes dust to shed from the undercoat. Microencapsulated sealing liquids, such as polybutene, calcium hydroxide, tung oil, and spar varnish, suppress dust. In general, the larger the microcapsule, the more the sealing liquid available upon rupture. However, the thickness of the overcoat layer constrains the maximum size of the capsules. Diameters of 63–150 μm may be optimal [Kumar 2008].

13.2.3.8 Vascular Self-healing Performance Results

Performance testing of a two-part vascular healing system reveals crack healing, with efficiencies starting at 85% for an initial loading-healing cycle and up to 60% over 16 healing and damage cycles (Figure 13.22) [Hamilton 2009].

Vascular systems need to dispense healing liquids, when and where needed. Short tubes with sealed ends create back pressure that restricts the flow. Low-viscosity single-agent materials, such as cyanoacrylate adhesive, are good for capillary-driven flows in borosilicate glass tubes in concrete [Joseph 2008]. Murray's law sets a standard for optimal flow resistance minimization of a vascular network [Wu 2010b].

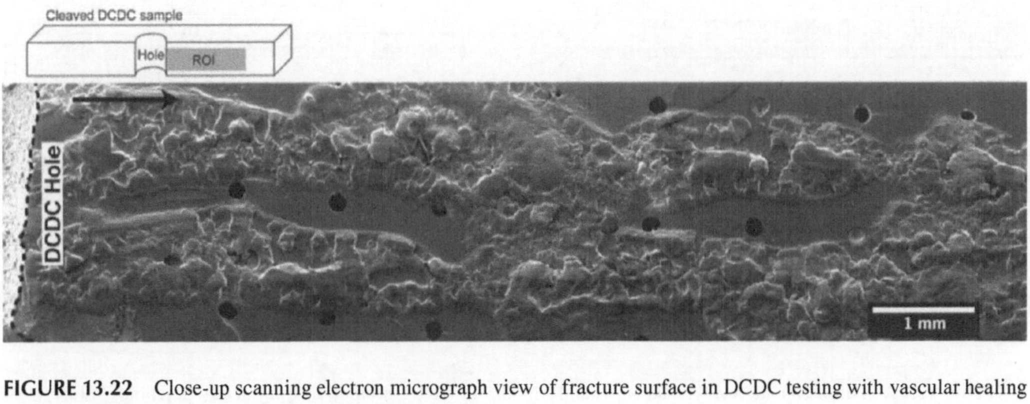

FIGURE 13.22 Close-up scanning electron micrograph view of fracture surface in DCDC testing with vascular healing after ten damage healing cycles. (From [Hamilton 2009].)

Vascular hollow glass fibers embedded into CFRP composite panels deliver premixed two-part epoxy-healing liquid in a technique that repairs delamination and shear cracking due to impact damage [Williams 2007a]. A layup with a tight pitch of vessels (70 μm) heals up to 98% flexural stiffness and 89% of baseline interlaminate performance. Increased levels of wetting between the epoxy and glass increases the healing performance [Motoku 1999].

The post-cure stiffness of the healing liquid has a modest effect on the performance of vascular crack repair. The effects are modest for concrete repair [Dry 2003a]. Liquids that turn into stiff materials, such as cyanoacrylates, provide a repair strong enough for new damage to appear in new locations upon reloading. More flexible materials, such as two-part epoxy, also perform well. A hypothesis explaining the good performance of flexible materials is that the crack surface constrains Poisson-type deformations and effectively stiffens the more flexible polymer repair material. Penetration into tight cracks and the ability to tolerate opening and closing of crack widths as the system continues to operate are also important factors. Carbon nanotubes mixed into healing liquids increase the performance of repairs [Aissa 2012]. Expanding polyurethane fills, bridges, and adheres across relatively large crack gaps to produce healing efficiencies in excess of 50% in three-point bending and split-cylinder laboratory tests [Van Tittelboom 2011].

An embedded vascular system carries with it many of the same issues of parasitic weakening as embedded microencapsulation healing. In general, the strength and stiffness of composite members decrease with increased vascular material due to combinations of fiber ply waviness, reduced load bearing area, and reduced fiber volume [Kousourakis 2008]. Experiments confirm that increasing the vessel diameter weakens the composite material, with compressive strength reductions of 13–70% [Huang 2010]. Vessels with axial directions nonparallel to the FRP fiber grains create points of structural weakness, often as resin pockets (Figure 13.23). Table 13.1 lists selected performance results for FRP composite repair techniques.

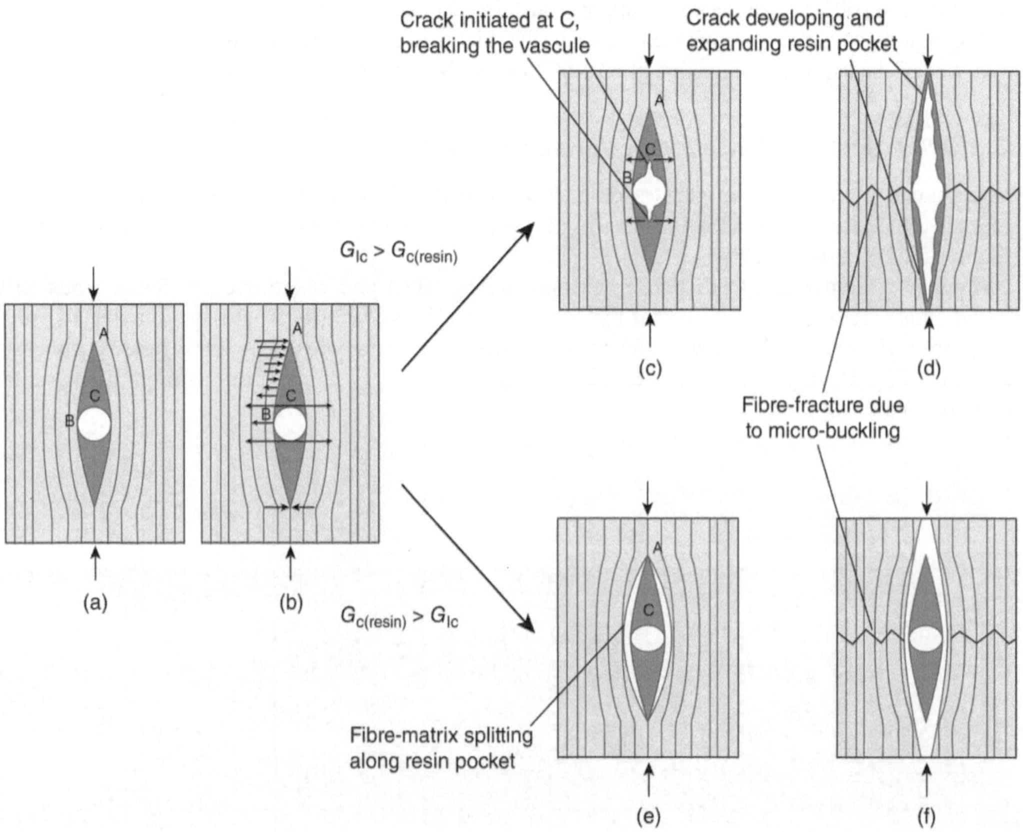

FIGURE 13.23 Formation of resin pockets around vessels in FRP vascular healing seeds potential failure sites. (From [Huang 2010].)

TABLE 13.1
Reports of the Performance of Self-Repair of FRP Composites

Type of Material	Damage Mode	Healing Type	Healing Measure	Performance	References
GFRP	Matrix cracking by impact	Thermoplastic diffusion in thermoset	Impact efficiency	65% first healing, 30% second healing	[Hayes 2007a] [Hayes 2007b]
GFRP	Matrix cracking by impact	Microencapsulated DCPD with fluorescent bruise indicator	Crack width reduction	51% reduction	[Patel 2010]
GFRP with foam core	Cylinder drop impact damage	Two-part premixed epoxy vascular	Load-displacement efficiency	100%	[Williams 2007b]
GFRP with woven fabric	Double cantilever cracking	Microencapsulated solvent	Load-displacement efficiency	1–10%	[Manfredi 2014]

13.2.3.9 Biohybrid Self-healing Performance Results

- Bacteria fill cracks and plug holes in concrete by excreting urea and urease that react extracellularly to produce $CaCO_3$. An advantage of bacterial versus autogenous $CaCO_3$ in-filling is the potential to provide a more uniform coverage of cracks (Figure 13.24). The high pH of ordinary Portland cement concrete and modern low carbon footprint concrete mixes hinders most bacterial activity. Encasing spores of spore-forming bacteria in compatible materials, such as silica or polyurethane, both immobilizes the bacteria and provides an environment in which they thrive. *Bacillus sphaericus* encased in a silica sol–gel with calcium-based additives, such as $Ca(CH_3COO)$, performs better than neat epoxy or mortar [Van Tittelboom 2010]. *B. sphaericus* decreases carbonation rate and chloride migration by about 25–30% and 10–40%, respectively, versus untreated Portland cement concrete [De Muynck 2008]. An alternative technique uses porous clay particles laced with calcium lactate and activated by soaking in water [Jonkers 2011].

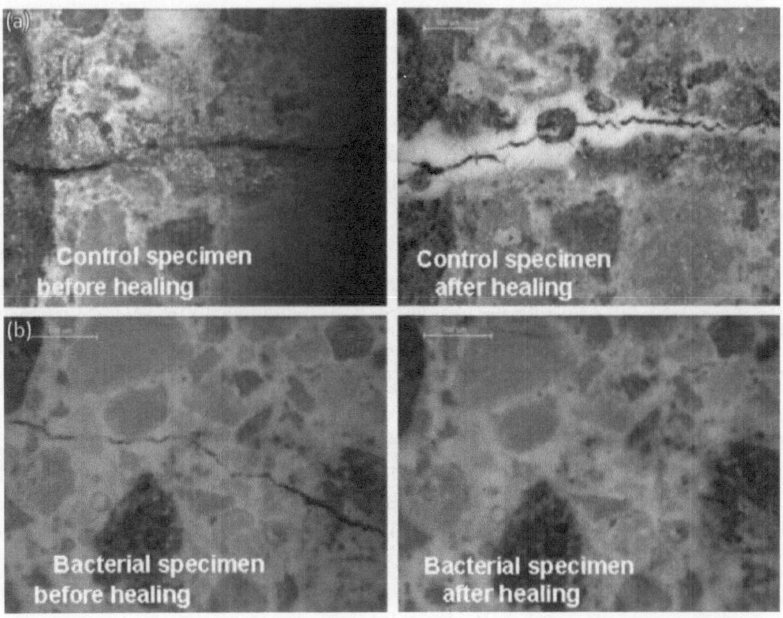

FIGURE 13.24 Bacteria versus autogenous $CaCO_3$ in-filling of cracks where the bacteria provide more uniform coverage of crack. (From Jonkers 2011.)

Bacteria films may promote or impede corrosion. Thin films of *Pseudomonas fragi* attach to mild steel along with planktonic suspensions of bacteria to decrease corrosion activity by 10× [Campaignolle 1997]. Conversely, marine sulfate-reducing bacteria increase the electrical potentials that drive corrosion [Sahrani 2008].

At the higher organism level, using plants as living fences limits wind-blown material transfer, while providing carbon sequestration and biomass. Living fences require maintenance, proper siting, and proper species. The regenerative capability of shrub willows makes them economically competitive for controlling snow drifts, particularly on roadways [Zamora 2015].

13.2.3.10 Surfaces and Sealing Self-healing Performance Results

Tests confirm that self-sealing surfaces can hold differential pressures. FRP composite panels with embedded encapsulated DCPD monomer liquid and wax-protected Grubbs catalyst seal fully following indentation-induced microcracking, as does the use of reflowable thermally activated polymers [Moll 2010] [Hurley 2011b].

Low earth orbit (LEO) is a very aggressive environment that works against surface durability, largely due to the presence of atomic oxygen. Polyimide-block-PDMS copolymers provide effective self-healing coatings for LEO applications. The atomic oxygen interacts with the PDMS during erosion to produce a smooth toughened silica layer in a process analogous to forming a char layer on TPS materials [Fischer 2010b].

13.2.3.11 Self-cleaning Surface Systems Performance Results

Superhydrophobic surfaces increase the durability of structures by preventing damage due to water ingress, such as freeze-thaw. Cementitious composites with superhydrophobic surfaces show an increase in freeze-thaw durability of over 100% [Muzenski 2015]. Concrete with hydrophobic surface treatments reduce chloride ingress to values on the order of 20% of that of untreated concrete [De Vries 1997].

The following are some of the techniques for increasing the durability of superhydrophobic surfaces, especially those with functionalized and easily abraded micro- and nanoscale textures:

1. *Alteration of Manufacturing Process Parameters:* Tuning of process parameters when using microwave-assisted chemical vapor deposition optimizes both mechanical durability and water repellency [Wu 2004].
2. *Sacrificial Surface Elements:* Large-scale sacrificial surface elements, such as bumps and pillars, protect fragile fine-scale surface textures [Huovinen 2014].
3. *Dual Length Scale Texture:* Superhydrophobic surfaces with texture at the micro- and nanoscales have mechanical stability superior to that of stand-alone micro- or nanotextures [Groten 2013].
4. *Recovery from and Avoidance of Water and Ice Condensation:* The condensation of water and ice onto a superhydrophobic surface negates the effects of surface texture. Recovery is possible through evaporation, but preventative mitigation has advantages [Yin 2011]. Superhydrophobic surfaces delay the condensation of ice, that is, frosting. Preventing the initial ice formation at seed points, such as due to dirt and debris, further delays frosting [Hao 2014].
5. *Solvent-based Treatments:* Laboratory tests of the durability of hydrophobic treatments on concrete surfaces in aggressive weathering environments find that solvent-based and heavily concentrated surface treatments perform better than the water-based treatments. The effect seems to be confined to the surface [Raupach 2005]. Increasing the binding of microtexture with solvents, such as tetrahydrofuran (THF), increases the durability of polycarbonate surfaces [Rios 2008].

Temperature affects the ability of superhydrophobic surfaces to repel water penetration by roll-off and droplet impact resistance. The increased viscosity of supercooled water causes droplets to stick to superhydrophobic surfaces more readily than those at room temperature [Maitra 2014].

For photocatalytic self-cleaning surfaces, doping with low valence (Fe^{3+}; Co^{2+}; and Ni^{2+}) and high valence (Mo^{5+}; Nb^{5+}; and W^{6+}) cations improves the performance of photocatalytic titania-based surfaces that break down unwanted organic chemicals [Park 2004a]. $LiNbO_3$ improves air quality better than TiO_2 [Nath 2014]. Al + W additives increase hydrophilicity while maintaining photocatalytic properties for applications, such as antifogging optical surfaces [Lee 2003].

13.2.3.12 Active Matter Self-healing Performance

Origami usually folds materials along predetermined lines. The material must be thin enough to fold as a planar sheet, but thick enough to provide structural rigidity, along with carrying sensing and actuation mechanisms. Thickening with compound folding, that is, the folding of a fold, is a significant complication [Hawkes 2010].

Active origami material micropatterned with an internal anisotropic swelling actuation layer has a figure of merit A/W^2. A is the area of the flat sheet. W is the width of the folds. Values on the order of 5,000 enable complex bending patterns, including the Miura-ori tessellation and Randlett's bird [Na 2014]. Shear and torsion are often superior methods of actuation, due in part to the homogeneous loading of the material [Liu 2013].

Many SMAs suffer from both functional and material fatigue. The introduction of Ti_2Cu into precipitates increases the fatigue life of nitinol SMA up to 10^7 cycles by acting as sentinels that guide the phase transitions [Chluba 2015]. Fe-Mn-Si-based alloys have potential advantages over nitinol due to the increased temperature ranges and lower costs [Lee 2013d]. Ionic polymers recover shapes with hole closure efficiencies ranging up to 80–95% over several hours [Huber 2005].

SMPs are sensitive to temperature extremes. Polyimide materials are durable high-temperature polyimides that are stable over temperature ranges of +150°C to −150°C [Xiao 2015].

13.3 Manufacturing of Self-healing Systems

A complication with self-healing material systems is that many of the material transformations of manufacturing mimic damage processes. For example, high-performance FRP structural composites undergo heat-cure cycles during manufacture that may damage many liquid-healing agents. A workaround is vascular systems with two-part liquid-healing agents that withstand heat cycles up to 300°C [Dry 2008].

13.3.1 Self-healing Bulk Polymer Manufacturing Techniques

The wide range of polymer types with potential for self-healing include crystals, liquid crystals, linear molecules, branching molecules with star and more complicated dendrimer shapes, supramolecular networks, and block copolymers. Controlling polymer variations generally comes from control of process conditions, especially those that affect hydrogen bonds with potential for self-healing [Montarnal 2009].

13.3.2 Microencapsulated and Heterogeneous Material Systems Manufacturing Techniques

The mass production of microencapsulated material started in earnest in 1954 at the National Cash Register Corp. following the invention of formaldehyde shell strengthening by Barrett Green, originally for carbonless carbon paper applications [Gardner 1966].

The manufacture of microencapsulated liquid healing agent capsules requires the following:

1. *Emulsifying the Healing Liquid into Microdroplets:* Rapid stirring is simple but results in a large variance of the droplet diameters. The Shirasu porous glass method forms droplets by pushing the healing liquid through a porous glass material producing droplets with a more uniform size distribution [Liu 1987].

2. *Encapsulating the Microdroplets:* Placing a capsule around the microcapsule.
3. *Hardening and Strengthening of the Capsule Walls:* Process the capsule to strengthen the walls so as to survive until needed for dissolution.
4. *Incorporating Functionality into Capsule Wall:* Functionality increases the performance of the capsule, such as with temperature-dependent poly(*N*-isopropylacrylamide) that controls shape [Cheng 2007a].

Concrete is a composite material with properties that favor microencapsulated liquid healing to solid healing. The fabrication conditions are reasonably mild. The composition includes stones, sand, reinforcing, cement, multiple chemical admixtures, and microfoaming air-entrainment components. The chemical composition is an alkaline mixture of several constituents that hydrate and cure into a solid paste after mixing with water. Temperatures range between 0 and 50°C. An expedient formulation for concrete applications uses two-part microencapsulated system based on an oil core/silica gel process to encapsulate methylmethacrylate monomer as the healing agent and triethylborane as the catalyst, with an average capsule size of 4.15 μm (Figure 13.25) [Yang 2009b]. Encapsulation uses sulfonated latex particles to self-assemble into a formwork to create an encapsulation layer. A subsequent step releases the encapsulating formwork leaving a thin encapsulation layer of silica gel.

The manufacture of microencapsulated liquid healing systems presents several practical issues:

1. *Dimension Control of Encapsulation:* Key dimensions are wall diameter, wall thickness, and amount of encapsulated material. Capsule diameter and wall properties ranging from stiff to supple and smooth to spongy depend on controllable process parameters, such as temperature and surfactant concentration [Frère 2008].
2. *Wall Toughness:* Microencapsulation forms thin walls between two distinct liquid phases. Urea-formaldehyde capsules precipitate from an aqueous phase to the interface surrounding microdroplets of dispersed healing liquid. Success depends on proper interactions of reagent dispersal, stirring speed, and temperature [Cosco 2006].

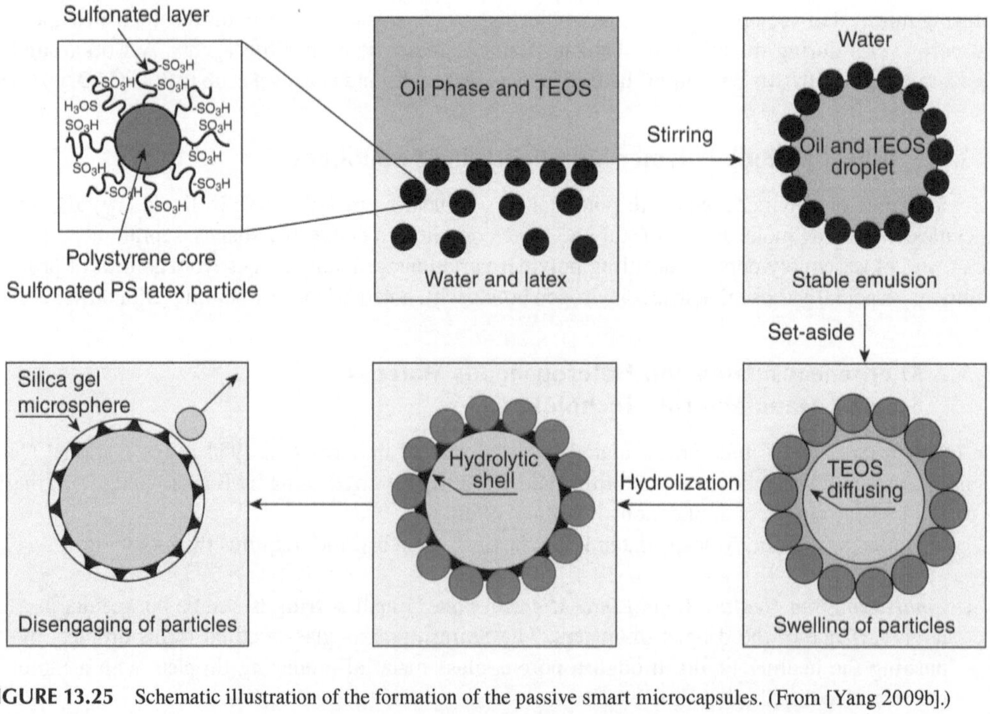

FIGURE 13.25 Schematic illustration of the formation of the passive smart microcapsules. (From [Yang 2009b].)

3. *Functionalized Wall Properties:* Capsules with heterogeneous wall structures provide specialized functionality but require additional manufacturing steps. Microcapsules made with a layer-by-layer polyelectrolyte process provide highly selective triggers for release of the internal contents [De Geest 2007]. An alternative is electroformation to produce capsules with heterogeneous wall properties, including Janus particles [Rozynek 2014]. Electroformation uses capsules with mobile nanoparticle and/or colloidosome wall structures made of two or more different varieties. In a first step, a DC electric field rearranges the surface structure into an equatorial configuration. A second step merges the different types of microcapsules with electrohydrodynamic flows. Merging particles leaves the spatial structures formed in the first step largely intact but merged into a larger capsule.
4. *Dispersion Control:* Proper dispersion in the material matrix controls overall healing system performance, with the usual preference being a widespread uniform distribution, not aggregated. Silica coatings on microcapsules minimize aggregations [Jackson 2010]. Molecular scale dispersion of Grubbs catalysts improves the efficiency of DCPD polymerization healing [Guadagno 2013].

13.3.3 Vascular Systems Manufacturing Techniques

Vessel wall material is an important consideration. Success requires placing tubes in suitable locations, geometries, interior smoothness, wettability, fluidic conductivity, and leak integrity [Olugebefola 2010]. The walls may be neat solid matrix or may be lined with a different and potentially functionally active material.

The following are some of the methods of manufacture:

1. *Insertion of Hollow Tubing:* Place hollow glass fibers into the solid matrix concurrent with a liquid-to-solid matrix manufacturing step [Motuku 1999] [Trask 2007a]. Tube insertion helps with forming vascular channels, such as for cooling, in microelectronic or microelectromechanical devices. An issue is that many manufacturing techniques derive from planar geometries, while vascular systems benefit from 3D architectures. Preplacing through-plane microfluidic channels in planar micro-manufactured systems is an alternative [King 2008].
2. *Subtractive:* Subtractive processes, such as laser ablation, cut vascular channels into an initially homogeneous solid. The geometry of subtraction favors 2-D over 3-D network structures [Qi 2009]. A semiautonomous 3D subtractive process forms Lichtenberg trees inside solids with the discharge of high-electric-voltage potentials. The discharge burns a branching pattern that forms a viable vascular network [Huang 2009].
3. *Additive then Subtractive:* An additive process initially assembles the solid two-part mix of an insoluble solid matrix and a soluble vascular-forming material, such as wax or sugar. Dissolution removal of the wax or sugar leaves the solid matrix with hollow vascular network [Therriault 2005] [Lewis 2006] [Toohey 2007] [Bellan 2009]. Additive then subtractive methods allow for creating a material with two interpenetrating networks to supply two parts of an epoxy-based healing liquid system [Hansen 2009]. Tests with dual-component healing liquid vasculature show an ability to heal repetitively up through 30 damage cycles.
4. *Molding and Imprint:* Templated molds allow for the mass production casting, embossing, and imprinting of features that form microchannels in solids. Patterning photoresist on flat surfaces produces molds with micrometer-scale features that readily print onto PDMS, which effectively seals with joining top and bottom pieces to form an elastomeric structure with microchannels [Duffy 1998].
5. *Multicomponent Extrusion and Electrospinning:* Flooding cables use self-healing liquids in a parallel-channel vascular arrangement. Multicomponent and multivoid extrusion produces cables with these configurations in large quantities [Maunder 2007]. Related is electrospinning which uses high electric voltages to drive a thin filament of liquid out through a nozzle that then solidifies quickly. Electrospinning in a two-component core–shell arrangement makes hollow tube fibers with inner and outer layers and 5–10 μm diameters with PMMA/PAN solutions in dimethylformamide to precipitate tube-forming PMMA droplets [Bazilevsky 2007].

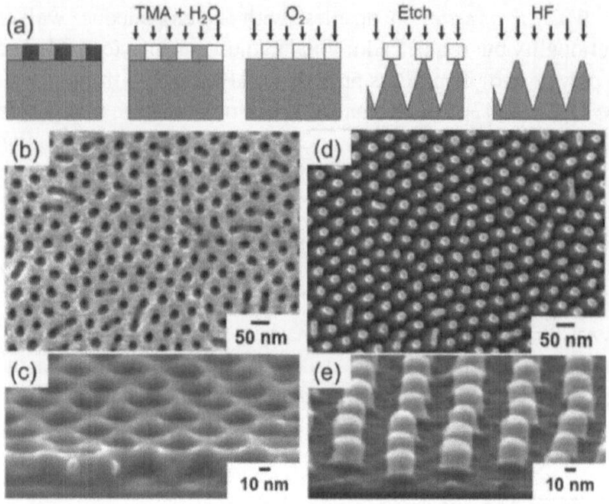

FIGURE 13.26 Lithographic manufacture of superhydrophobic surface using PS-PMMA base. (From [Checco 2013].)

13.3.4 Surface Manufacturing Techniques

Manufacturing functionalized self-cleaning surfaces places micro- and nanostructures with embossing, imprinting, paraffin-coated thin wire arrays, and electrospinning of fluorinated polymers [Agarwal 2006] [Li 2007b] [Ma 2006]. Figure 13.26 shows lithographic techniques. Table 13.2 lists manufacturing methods for self-cleaning micro- and nanoscale patterned superhydrophobic surfaces.

13.3.4.1 Manufacture of Photocatalytic Self-cleaning Surfaces

Most photocatalytic self-cleaning surfaces use particles of titania or similar semiconductors for activation. Standard manufacturing processes fix the titania to the outer layer of the surface. A method for antireflective coatings combined with photocatalytic self-cleaning inserts titania in micelles that self-assemble into a coherent matrix layer held together by copolymers [Guldin 2013]. Steps must be taken when working with soda-lime glass to reduce the influence of sodium oxide which lowers the photoefficiency of the titania [Paz 1997].

Process flexibility combines self-cleaning with other properties, such as optical transmittance. Most antifogging surfaces use superhydrophilic techniques, typically with titania. Depositing a mixture of titania and a fluidic ester on glass at temperatures over 600°C produces an antireflective self-cleaning surface [Hurst 2005]. Spin-coating SiO_2 particles in an SiO_2 sol combines antifogging and antireflective optical coatings [Zhang 2012a]. Another example is antireflective surface coatings with hierarchical layers of silica particles. Such coatings may be both optically transmissive and either superhydrophobic or superhydrophilic [Li 2010b].

13.3.4.2 Manufacture of Self-cleaning Fabrics

Photocatalytic self-cleaning fabrics remove stains, deactivate chemical toxins, and exert bactericidal effects. The methods of manufacture are as follows:

1. *Binding:* Bind particles to the surface, such as anatase nanocrystalline titania particles attached with a low-temperature sol–gel process followed by heat treatment with boiling water [Daoud 2004].
2. *Additive:* Nanoscale layer-by-layer additive technique with the possibility of functionalization and inclusion of nanoparticles. Radiation, such as UV, cross-links and toughens the interfaces between layers. Adding a durable superhydrophobic multilayer coating of photoreactive azido-containing silica nanoparticle/polycation onto cotton fabrics [Zhao 2012].
3. *Functionalizing:* Functionalize polymer layers with covalent binding and stabilization of reactive metal oxides, such as TiO_2 [Drechsler 2009].

TABLE 13.2
Methods of Manufacturing Superhydrophobic Surfaces

Approach	Technique	References
Surface deformation	Embossing	[Thieme 2001] [Vorobyev 2015] [Guo 2004] [Telecka 2016]
	Wrinkling	[Zang 2013]
Removal of materials	Lithography	[Checco 2013]
	Plasma etching	[Vourdas 2007]
	Laser etching	[Nayak 2013]
	Electrolytic etching and machining	[Ohkubo 2010] [Yang 2016] [Ruan 2013]
	Solvent etching	[Erbil 2003]
Addition of materials	Electrochemical growth of porous layers	[Vorobyev 2015]
	Dipping glass to coat with mix of polymer and silica particles	[Sasaki 1998]
	Plasma deposition and roughening	[Woodward 2003]
	Multilayer self-assembly with clay, Janus, and latex layers	[Jisr 2005] [Zhang 2014] [López 2016]
	Molding	[Zhang 2006]
	Functional nanoparticle layer, including Janus particles, silicone nanofilaments, etched silica, silica-fluoropolymer hybrids	[Gao 2010b] [Yang 2015a] [Artus 2012] [Du 2011] [Lee 2013c]
	Covalently attached slippery liquids	[Wang 2016]
Combined methods with deposition of material on micro- and nanopatterns	Chemical vapor deposition on micropattern with hydrophobic fluoropolymers, PTFE on carbon nanotube forests, wax on fractal surface patterns, pillar on pore anodizing and stearic acid on paper sludge ash, and nanoporous polymer chalk	[Ma 2005] [Lau 2003] [Shibuichi 1996] [Kemell 2006] [Jeong 2012a] [Spathi 2015] [Wong 2015] [Zhang 2011]
Combined methods with removal of material in micro- and nanopatterns	Erodible colloids in PTFE, etching of steel followed by silane surface layer addition	[van der Wal 2007] [Wang 2015b]

13.3.4.3 Self-cleaning during Manufacture

Self-cleaning of components during manufacture enables high-performance assembly and minimizes the need for subsequent cleaning process steps. Self-cleaning benefits the epitaxial assembly of 2D materials for micro/nanodevices [Kretinin 2014]. Graphene self-cleans surfaces by attaching to clean materials with high carrier mobilities (60,000 $cm^2/(V\ s)$), such as hexagonal boron nitride and tungsten and molybdenum disulfides. The attachment kinetics are strong enough to squeeze contaminants into island pockets that minimize deleterious effect on device performance. Graphene does not self-clean materials with low carrier mobilities (1,000 $cm^2/(V\ s)$), such as the flat-layered oxides mica, bismuth strontium calcium copper oxide, and vanadium pentoxide.

13.3.5 Tank, Tire, and Vessel Manufacturing Techniques

The manufacture of self-sealing fluid containment vessels, for example, fuel tanks and pneumatic tires, generally adds the capability as integrated manufacturing steps. Many processes use layup techniques that add materials to an inner mandrel, followed by a curing or fixation step, with the layup process inserting the functionality. Appendix B lists selected patents for the manufacture of self-sealing fuel tanks and tires.

13.3.6 Self-healing Wiring and Cabling

Elongated cylindrical geometries drive much of the technology underlying self-healing wires and cables. Many techniques add active components around central conducting cores with tape wrapping, exterior braiding, or extrusion. A primary advantage of tape wrapping and braiding is the fabrication of high-performance, possibly functionalized, materials as a separate preliminary step. Extrusion has the advantage of supporting long continuous production runs but requires the simultaneous addition of high-performance and/or functional components as part of the process. Expanded polymer foams are extrudable materials with versatile properties that can be used in self-healing cables. Water-induced swelling heals medium-power underground electric power transmission cables [Belli 2006].

13.3.7 Biohybrid Manufacturing Techniques

Most biohybrid self-healing systems and materials are heterogeneous arrangements of materials. As an example, spinal implants benefit from the properties of high-performance polymers, such as PEEK, while also benefiting from bone in-growth properties of ceramic hydroxyapatite. A manufacturing challenge is that ceramics require higher processing temperatures, motivating the use of bonding and thermal protection layers, such as yttria-stabilized zirconia [Rabiei 2013]. Similarly, materials that promote tendon ingrowth have a geometry similar to tendon, that is, an aligned fibrous form. Electrospinning biocompatible and degradable polymers forms tendon scaffolding [Hakimi 2013].

13.3.8 Self-healing during Manufacture

Manufacturing processes modify and assemble raw materials and components into finished products. Many of these processes cause dramatic changes to materials and run the risk of introducing unwanted damage. Self-healing during manufacture can be very useful. Some examples are as follows:

1. *The in-place manufacture of concrete structures* uses hydration processes in cementitious materials to solidify, that is, cure. Uneven distributions of water, especially at exposed surfaces, causes uneven swelling, which leads to cracking. Concrete with specialized additives, such as glycol and alcohol, absorbs moisture from the surrounding air and releases the moisture during cure to prevent cracking [Jau 2011].
2. *Exfoliated hectorite clay* heals defects in the manufacture of multilayer nanostructured materials with new layers that expand and heal incompletely formed underlying layers using UV-based cross-linking [Vuillaume 2002].
3. *The joining of polymers* by reptation and similar diffusion processes is a healing tool for manufacturing. One application uses polymer diffusion to heal a skin of discrete latex-PMMA particles with a thermal annealing process [Canpolat 1996].
4. *Directed self-assembly* processes, including 3D printing, rapidly create a wide variety of useful shapes. Self-healing melds separate parts together at room temperature, such as ionic polyamides [O'Harra 2020a]. Self-healing can also aid self-assembly based on minimizing free energy that produces localized defects with trapped free energy. Relieving these defects is possible with process-specific solvent and thermal annealing [Hur 2015]. Small, even molecular scale, material bridges across the barrier reduce the required activation energy and promote defect healing.
5. *The manufacture of FRP composites* is often a layup process that inserts components in heterogeneous arrangements to create parts with enhanced capabilities. It is possible to fabricate a composite with islands of thermoplastic and thermosetting plastic using a heat cure cycle that hardens the thermoset while the thermoplastic island remains largely intact. Such an arrangement heals damage, including delaminations, by thermoplastic reflow [Hodzic 2016].

6. *Electronic components* provide multiple opportunities for self-repair during manufacture. Tantalum capacitors with magnesium dioxide (MnO_2) cathodes suffer from a failure mode where large surging currents form a localized conductive path that heats the dielectric matrix to over 480°C and converts Ta_2O_5 from an insulating amorphous state into a crystalline conductive state. The phase change leads to positive feedback with increased conductivity that increases the heat load and causes ignition. It is possible for the capacitor to self-heal with a cathode that seals off the conductive path through the formation of an insulating Mn_2O_3 cap when temperatures go over 380°C. Scintillation conditioning during manufacture accentuates the healing by applying microampere-level constant current loading to the capacitor to induce insulator breakdown that then self-heal forming a scar that resists future breakdowns [Fritzler 2014].

14

Future Directions

This chapter contains a selection of present-day trends, potential tools, and opportunities for future developments in self-healing technologies.

14.1 Materials with Agency for Wound Healing

Materials with agency, that is, programmable and active matter, change shape, properties, and overall state in response to a stimulus by acting with a purpose [Ball 2023]. Imagination and the laws of physics are perhaps the only limits on the possible uses of these versatile material systems [Meng 2010] [Popkin 2016].

Living biological tissue is an extreme example of active matter with agency. Living tissue transduces available chemical energy into actuation, growth, and healing [Ramaswamy 2010]. In some animals, such as salamanders, the loss of a limb produces biochemical and mechanical signals to induce cells to transform from a differentiated state to a germ-like state, which then initiates regenerative healing growth and future differentiation [Zhu 2012]. Plants have stronger abilities to change and adapt body form to heal injuries and adapt to local circumstances.

Softening, swelling, flowability, and solubility are possible state changes that promote healing on the scale of large wounds. Change in shape and stiffness combined with growth closes wounds and regenerates damaged components. The statistical mechanics of active matter extends simple collision and interaction behaviors between particles to describe emergent macroscopic behaviors. For example, angle-dependent interactions between filamentary particles that align particles at the molecular scale produce polar vortex-type patterns at the macroscale [Suzuki 2015]. Antagonism and prestress enhance the performance of active material systems [Baumann 2018].

14.1.1 Programmable and Active Matter: Enabler of Wound Closing

14.1.1.1 Stiffness Change

Solids with strength and stiffness that change when needed enable large material movements needed to support wound closing. Molecular scale, nanoscale, and microscale mechanisms lead to stimulus-response stiffness change [Flores 2013]. Controlling the percolation of mechanical contacts between embedded stiff particles switches a material matrix between soft and stiff states [Shanmuganathan 2010]. The biomimetic approach of adding and removing materials, as in Wolff's law description of load-driven remodeling of bones, is conceptually simple but challenging to execute with most materials. Anisotropic material property change may be easier to implement. Table 14.1 lists additional possibilities.

14.1.1.2 Shape Changing Solids that Undergo Large Controlled Deformations

Wound closing, load shifting, topological change, and movement of healing materials all are possibilities:

1. *Volume Changes with Minimal Shape Distortion:* These are isotropic shape changes in a minimally constrained homogeneous body. Possible mechanisms are as follows: (a) Changes in temperature that cause thermoelastic swelling or contraction. (b) Swelling by imbibing liquids. (c) Molecular cleavage and/or cross-linking of isotropically oriented molecules [Shi 2008].
2. *Shape Distortion with a Relatively Small Volume Change*: Many elastomers and gels deform by distortion with minimal volume change. Liquid crystal elastomers change shape with anisotropic and spatially varying patterns due to micro- and nanoscale features with macroscopic orientations [Pei 2014].

 DOI: 10.1201/9781003396048-14

TABLE 14.1

Selected Stimulus-Response Stiffness Changes in Solids

Material	Stimulus	Response	References
Mix of thermoplastic with elastomer	Thermal cycles	Reversible softening and stiffening of thermoplastic component combined with elastomer	[Koerner 2004]
Cross-linked shape memory polymer (SMP)	Thermal cycles	Reversible alteration of cross-links softens and stiffens	[Gheneim 2002] [Leng 2009]
SMP with magnetic nanoparticles	Cyclic magnetic fields	Heating of nanoparticles induces thermal shape memory alteration of stiffness	[Kumar 2010]
Liquid crystal elastomers with a nematic phase	Repetitive low-frequency compressive loads	Stiffens by up to 90% with alignment of the nanocrystal structure	[Agrawal 2013]
Composite metal mesh with electrolytic phase	Electrical stimulation	Strength changes up to 45% and localized self-healing	[Jin 2011]
Elastomers with embedded fibers that reversibly cross-link	External neural mimic	Reversible softening and stiffening that mimic echinoderms	[Trotter 2000]
Deep subsurface printing-patterned solidification	Acoustic and ultrasonic	Self-enhanced thermal acoustic absorber with thermal initiator for thermosetting hydrogels and nanocomposites	[Kuang 2023]
Rat and human fibroblasts	Cyclic stretching	Anisotropic frequency-dependent reorientation	[Jungbauer 2008]

3. *Volume and Shape Distortion with Constraining Layers:* Materials with embedded anisotropic molecular structures, nanostructures, and/or microstructures change macroscopic shape in response to simple bulk deformations, such as swelling. Specific geometric arrangements of layer plies produce specific movements, with dimensions that are often smaller than that of the overall macroscopic motion (Table 14.2). Flat rectangular members with swelling layers running as stripes at 45° cause a flat plate to curl into a helix (Figure 14.1).

TABLE 14.2

Controlled and Anisotropic Deformation of Materials and Structures with Homogeneous Swelling and Heterogeneous Stiffness Constraints

Material or Structure	Constraint Geometry	Motion	References
Flat plate	Bands with 45° alignment	Plate curls into helix	[Wu 2012b]
Swelling needle tip to produce strong mechanical interlock	Stiff inner layer with compliant outer layer that guides swelling intermediate layer	Bulbous swelling of tip	[Yang 2013]
Elastomeric pneumatic actuator, pneumatic springs, McKibben actuators, soft robots, and bimorphs	Elastomer chamber walls constrained with stiff material	Anisotropic motions	[Liu 2012a] [Shepherd 2011]
Thermoresponsive actuation	Hydrogel with layered nanosheets	Electrostatic actions cause gels to resist compression, but deform easily in shear	[Kim 2015c]
SMP	Internal anisotropic patterns of rod and fiber-shaped nanoparticles, and crystal structure	Enhanced anisotropic motion, actuation	[Xu 2010a] [Behl 2013]
Dielectric polymer actuator	Internal fibers and molecular structures	Controlled deformation and improved performance of actuator motion	[Koh 2011]
Crystalline solid	Nanolattice fence	Collosal anisotropic negative and positive thermal expansions	[Goodwin 2008]
Soft solid	Tying with loops, knots, and stitches	Constrained movement of soft material	[Wisner 2002]

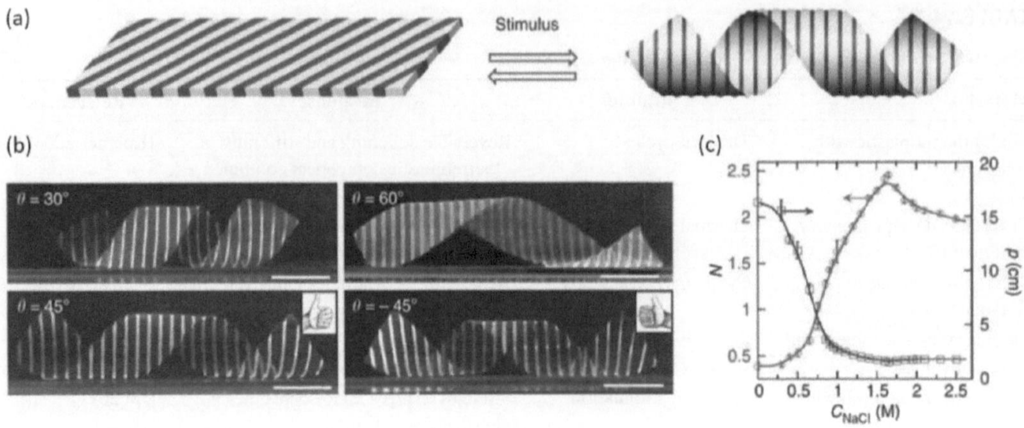

FIGURE 14.1 Striped layers with different swelling convert flat plate into helix. (From [Wu 2012b].)

An example application of state change for healing is the binding of soft materials together with an active barb needle that initially penetrates as a sharp point and then swells to introduce a strong interlock (Figure 14.2) [Yang 2013].

4. *Stimulated Shape Changes:* External stimuli, such as electric fields, magnetic fields, and illumination, lead to large shape changes (Table 14.3).
5. *Prestressed Solids with Elastic Shape Change:* Internal prestress in solids alters mechanical behavior and failure modes. Prestressing eliminates tension stresses that may cause cracking and accentuates self-healing through both antagonistic and material bias effects. The release of prestress activates large healing actions, such as wound closing or shape recovery.

Prestress spans molecular to macro length scales. Polybutadiene treated with difluorocarbene is a polymer with antagonistic tension-induced contraction properties [Lenhardt 2010]. Stretching the polymer triggers a transition to a shorter isomer, which promotes macroscopic shortening. Inserting cuts with coherent aligned patterns during specific thermal cycle phases introduces large prestresses into SMPs that improve and dramatically alter performance [Viry 2010].

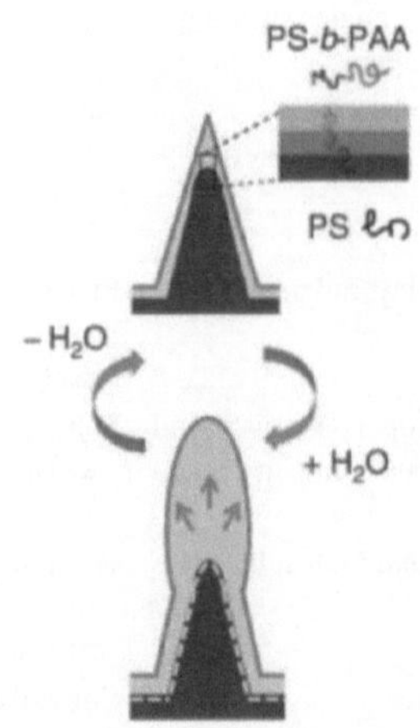

FIGURE 14.2 An active barb swelling needle tip produces a strong interlock into a soft material. ([From Yang 2013].)

TABLE 14.3

Electric Field, Magnetic Field, and Illumination Stimulated Large Deformations of Solids

Material or Structure	Stimulus	Motion	References
Elastomer	Large electric fields inducing Maxwell stresses	Large deformations in directions normal to electric field vectors	[Anderson 1986] [Rasmussen 2012]
Perovskite multiferroic crystal, bismuth ferrite	Electric field	Shape memory martensitic transformations with strains up to 14%	[Zhang 2013]
Carbon nanotubes (CNTs) in aerogel matrices	Thermal or electrothermal	Shape memory transitions into torsion and axial motions of coiled spirals	[Aliev 2009]
Stimulus-responsive polymers	Electrochemical	Molecular folding and unfolding of polymer chains	[Ionov 2010]
Protonated oxides or solid acids with nanolayered structure	Infusion of water between layers and long-range structuring of H_2O	Anisotropic swelling of 100×	[Geng 2013]
Graphene-elastic composite hydrogel	Light with spatial patterns	Finger-like flexing and crawling	[Wang 2013b]
Photosensitive reticulating polymers	Light with spatial patterns	Spatial control of actuation of polymer gels	[Li 2015]

14.1.2 Intermediate 2D and 3D Foams and Cellular Materials

Cellular macroscopic materials are made of small cellular components. Cellular materials may undergo large, rapid, adaptive, and tunable material property changes while retaining all the underlying cellular structure. Internal mechanical structures in cellular solids, such as latches and shape memory effects, control deformations that may be relatively small at the cellular level yet manifest as large deformations at the macroscopic scale (Figure 14.3) [Mehta 2008].

SMPs provide thermomechanical functionalization for active foam materials [Di Prima 2007]. This includes foams that can undergo 90% volume deformation while retaining original foam structure and containing viable with living cells [Bencherif 2012]. It is challenging but possible to make cellular materials out of tough and durable materials such as polyimides (Kapton™), PTFE (Teflon™), and metals [Vazquez 2005] [Weiser 2000].

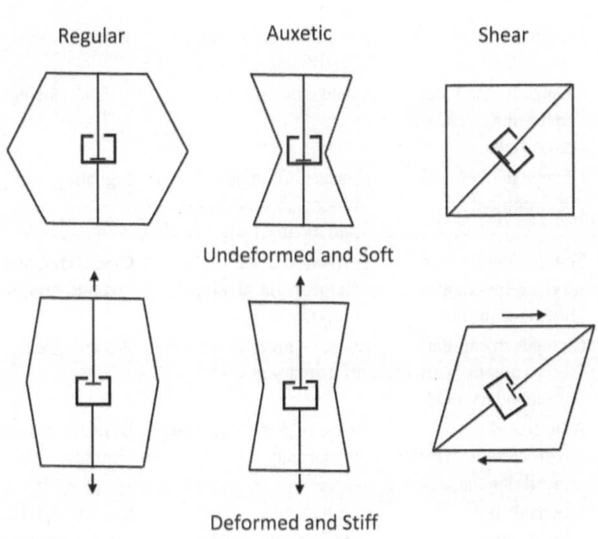

FIGURE 14.3 Cellular material with large deformation constrained by contact mechanism to produce combined compliant and stiff material. (Adapted from [Mehta 2008].)

14.1.3 State and Property Changing Gels, Blobs, and Liquids

Gels have a microstructure consisting of a filamentary cross-linked network supported by interstitial molecules. This heterogeneous structure provides a platform for a variety of functional behaviors [Oh 2008]. Most gels, liquids, and blobs of soft material deform easily under load, with small expenditures of energy. As such, these materials have a large capacity for advanced self-healing (Table 14.4). Toughening the materials and controlling the behaviors remains a challenge.

Stiffness, Viscosity, and Flowability: The stiffness and flowability properties of liquids, gels, and blobs derive from interactions at the molecular, nano, and micro levels. Altering these small-length-scale interactions alters the macroscopic stiffness and flowability properties.

Conglomeration and Growth: Controlled conglomeration at point of damage heals the damage. Liquids that are normally free-flowing repair damage by conglomerating to fill a wound with a solid mass of healing material. Control requires particles that conglomerate only near to an unhealed wound. Figure 14.4 shows muscle-like self-healing and growth of a complex gel following mechanical loading. Table 14.5 lists active conglomeration systems with potential for self-healing.

TABLE 14.4

Selected Gels and Blobs with Controllable Shape and Material Property Changes

Material or Structure	Stimulus	Motion	Possible Healing Action	References
Soft material inside another soft material	Electric field	Shape change driven by Maxwell stress	Wound closing	[Trau 1995]
Blob with printed ionic patterns	Electric field	Controllable shape change driven by ionic patterns and field	Wound closing	[Palleau 2013a]
Active skin surrounding blob	Skin movement and shape change	Controllable area and volume preserving shape changes, as in euglenoids	Wound closing	[Arroyoa 2014]
Myosin molecular motors in cytoskeletal meshes	Activation of myosin motors	Shape change and even damage to blob	Wound closing	[Alvarado 2013]
Metallo-supramolecular Polymers	Thermal cycling	Reversible large changes in stiffness	Wound closing	[Beck 2003]
Cross-linked molecular networks	Change in chemical environment cleaves cross-links	Change in stiffness	Wound closing	[Scott 2005]
Cross-linked molecular networks	Change in chemical environment	Release of trapped functional molecules into environment	Signaling	[Oh 2008]
Lipid vessels with nematic films	Topological defects formed by aligning fibers on surface	Formation of dendritic elements mimicking filopodia protrusions	Crack, tear, and wound bridging	[Keber 2014]
Gel with secondary immiscible fluid	Changes in capillary forces among particles of secondary fluid	Large changes in stiffness of primary gel	Wound closing	[Koos 2011]
Gel with micelles that change shape from worm-like to spherical	Addition of hydrocarbon switches micellular shape	Factor of 5,000 decrease in viscosity	Delivery of healing liquids	[Molchanov 2007]
Generic simulated active gel	Mechanical deformation	Oscillatory mechanochemical Belousov–Zhabotinsky waves	Mechanical damage stimulates healing chemical reaction	[Kuksenok 2007] [Yashin 2007]

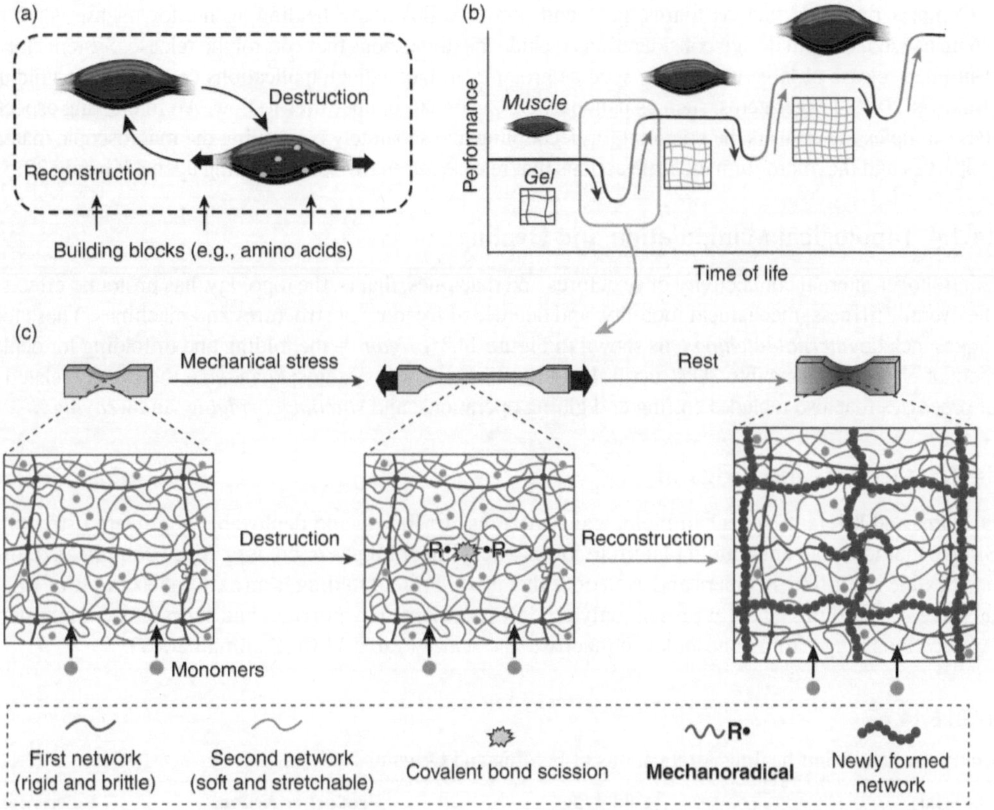

FIGURE 14.4 Self-growth of complex gel material by mechanical loading that introduces local damage and then stimulates conglomeration growth. (a) Mechanical training of a muscle. (b) Self-growing materials. (c) Mechanical training of a double-network gel. (From [Matsuda 2019].)

TABLE 14.5

Active Conglomeration Systems with Potential for Healing

Material or Structure	Stimulus	Action	Possible Healing Action	References
Gel that dissolves and reforms as a gel	Star-shaped polymer molecules with functionalized arms	Switching stickiness of arms switches gel to solution and back to gel	Release of gel to flow into wound and re-gel	[Kamada 2010]
Transmutable nanoparticles	Programmable DNA response	Reconfigurable surface ligands	Damage-specific conglomeration	[Kim 2016b]
Linear co-block polymers with differing solubility in blocks	Temperature change	Thermoresponsive aggregation and flocculation	Clotting	[Skrabania 2007] [Wang 2012f]
Living supramolecular polymer	Artificial prion infection	Porphyrin-based monomers form nanoparticles and nanofibers	Clotting	[Ogi 2014]
Gel with electrically sensitive ellipsoidal Janus particles	Electric field	Stiffness and shape change	Electro-coagulation and wound healing	[Shah 2015]
Complex fluid with active nanoparticles	High shear rate with leak	Jamming and gripping of particles	Mimic clotting to stop leaks	[Chen 2013]
Complex gel	Mechanical loading	Conglomeration after local damage due to mechanical loading	Muscle-like repair and growth	[Matsuda 2019]

Complex materials such as foams, gels, and porous solids house healing agents for release when and where needed. Critical design considerations include the dimensions that control the release. Systems for the controlled release of large molecules, such as proteins, in biomedical applications could provide guidance [Baldwin 2012]. Larger items, such as nanoparticles, need larger apertures to flow. An interesting aspect of these complex materials is the potential for decoupling and separately controlling the macroscopic material properties and the micro- or nanoscale actions that release internally stored healing agents [Grindy 2015].

14.1.4 Topological Manipulation and Healing

Control over internal connectivity of structures and machines, that is, the topology, has profound effects on the overall stiffness, mechanical mobility, and degrees of freedom of structures and machines. These techniques include *internal latching* – as shown in Figure 14.3; *origami* – the folding and unfolding for deployment of planar elements into 3D structural and machine element systems; *kirigami* – the closely related set of processes that also includes cutting and gluing operations; and *stitching*, *bridging*, and *weaving*.

14.1.5 Origami and Kirigami

Origami and kirigami appear in biological structures, art forms, and deployable engineered structures. Simple methods fold and unfold plate-like forms along linear hinges or creases. More complicated forms use flexible cloth-like members and even ones that grow. Protein folding is an extreme example of what is possible – something that is at present only crudely represented by polymer and supramolecular systems. Applications range from nanoscales to macroscopic scales (Table 14.6) [Stellman 2007].

TABLE 14.6

Engineering and Self-healing Applications of Folding and Origami

Material	Fold Type	Product or Action	References
Submillimeter scale self-folding polymer gel	Hidden DOF with geometrically restricted square Miura-ori folds	Mechanical metamaterials	[Mahadevan 2005]
Thermoresponsive polymer (PDMS) constrained by insensitive material layer	Reversible curling of sheets	Thermal actuator that causes materials to bend and curl	[Simpson 2010]
Silicon nitride plates with flexure hinges	Elastocapillary folding	Actuated reversible folding of plates with addition and removal of water	[Legrain 2014]
Submillimeter scale planar devices with lithographic pre-patterning of fold geometry	Enzyme-specific triggered shape changing material layers	Mesoscale mechanical actuation with chemical stimulus	[Bassik 2010]
Composite panels with embedded electronics	Origami for functional devices	Actuated folding of plates with shape memory composites	[Felton 2014]
Microscale silicon ribbons	Multi-wave buckling	Self-assembling micro spirals, waves, and structures	[Xu 2015]
Elastic structure with in-plane stress	Hinged joint and buckling with snap-through elastic instability	Rapid out-of-plane motions, for example, Venus fly trap and lids on baby food jars at macroscale, and active materials at microscale	[Holmes 2007]
Buckliball spherical structures	Snap through folding buckling	Reversible collapse to smaller volume	[Shim 2012]
Planar structures with in-plane deformations and compliant free edges	Curling and warping of edges, often with fractal geometries	Spontaneous regenerative formation of favorable edge geometries	[Shenoy 2012] [Sharon 2004] [Kim 2012a]
Living animal tissue	Tissue folding	Apoptotic epithelial cells pull and induce cellular remodeling at creases	[Monier 2015]

14.1.5.1 Kinematic Synthesis

The sequence of folds, cuts, and gluing of joints to produce desirable outcomes with origami is relatively easy to reproduce. It is generally much more difficult to do the inverse, that is, synthesize a sequence with desired properties [Hawkes 2010] [Castle 2014]. While methods of determining favorable folding sequences largely remains more of an art than matters of scientific investigation and engineering analysis, recent advances are moving toward more systematic approaches [Silverberg 2015] [Rafsanjani 2017] [Walker 2022] [Liu 2024].

14.1.5.2 Activation and Regulation of Origami Folding

Autonomous origami systems require the means to actuate folding. Folding and latching into place results in an equilibrium state that minimizes potential energy. The changed geometry creates new vistas of energy minimizing configurations for subsequent folding actions [Hawkes 2010]. Random agitation with selective click binding, often with DNA molecules, implements origami at the molecular and nanoscale (Figure 14.5) [Rothemund 2006] [Wei 2012] [Douglas 2009] [Zhang 2015a].

Similar to folding, unfolding produces large shape changes. Prestress and antagonism, largely implemented through elastic and plastic deformations, promotes unfolding [Arora 2007]. Looping of polymers, such as with quadruple hydrogen bonding of 2-ureido-4-pyrimidone (UPy), creates a designer molecule that unloops when loaded, but remains intact, allowing for large and potentially recoverable deformations [Guan 2004]. Likewise, enzyme activation of spatially selective material swelling induces folding for submillimeter scale gripper actuation [Bassik 2010].

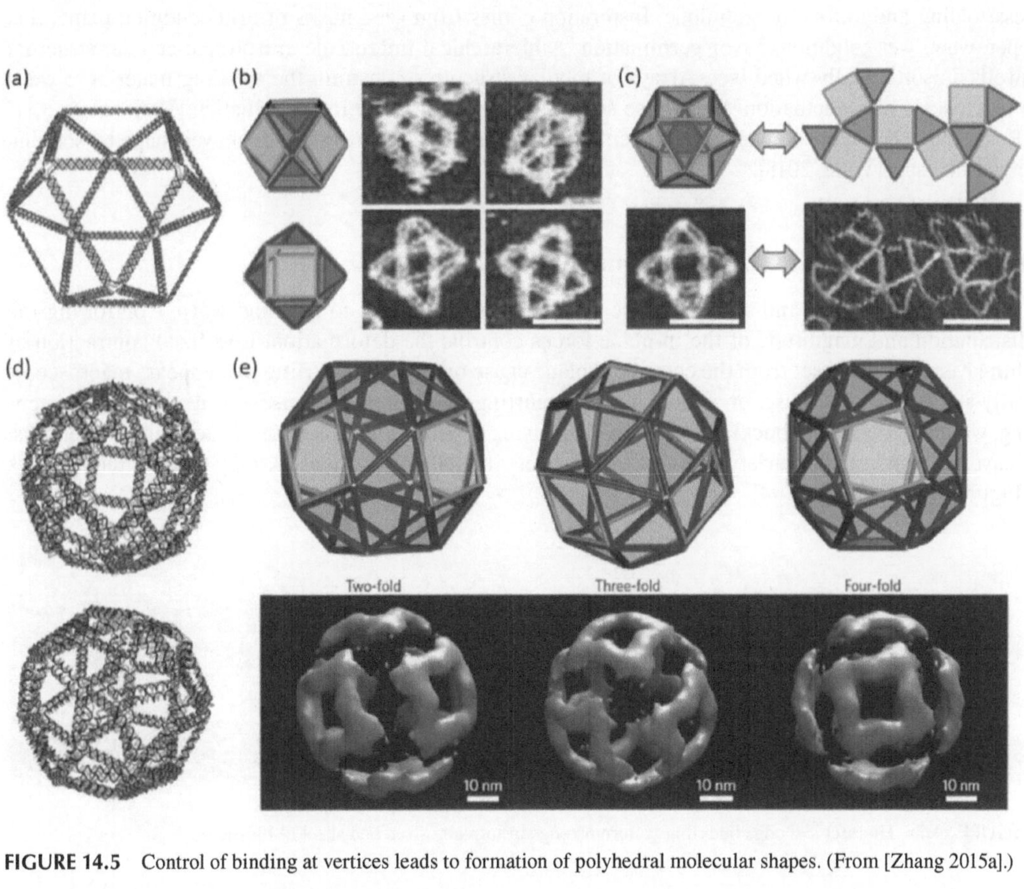

FIGURE 14.5 Control of binding at vertices leads to formation of polyhedral molecular shapes. (From [Zhang 2015a].)

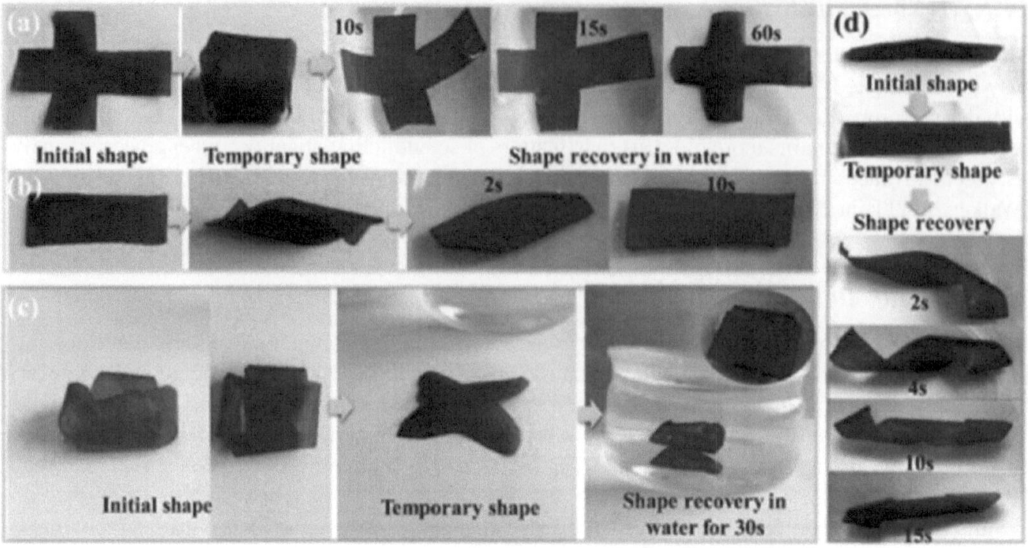

FIGURE 14.6 Coiling, unfurling, and recovery of planar SMP shapes actuated by soaking in warm water. (From [Wang 2018].)

Chiral wrapping and unwrapping of structural elements from arrays of central spools is a hinge less folding and unfolding technique. Inspiration comes from seed husks of arid condition plants that open when wet conditions favor germination. A hierarchical microscale anisotropic cellular structure swells anisotropically when wet. Arrays of tubular structures constrain the swelling material to exert a macroscale hinge actuation. Water also softens the material to aid in the actuation [Harrington 2011] [Rossiter 2012]. Figure 14.6 shows a system using SMPs that coil, unfurl, and recover shape by soaking in warm water [Wang 2018].

14.1.6 Distributed in-plane Deformations

In-plane deformations and stresses cause compliant planar sheets to curl and warp. Controlling the distribution and magnitude of the in-plane forces controls the deformation. Localized contraction of thin layers placed offset from the centroidal plane cause out-of-plane curling that appears macroscopically as folding at a crease, or as bucking. This curling–folding effect causes local geometric stiffening, which increases the buckling wavelength, causing a hierarchical cascade of buckling wavelengths. Leaves, flowers, and a variety of engineered sheets, including torn plastic, may take on these forms (Figure 14.7) [Sharon 2004].

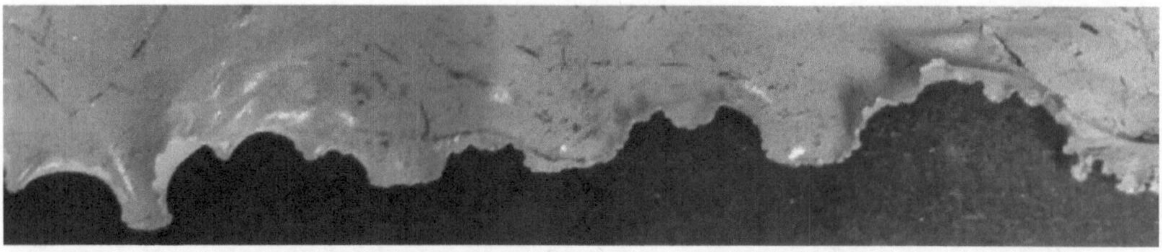

FIGURE 14.7 Hierarchical edge buckling pattern on edge of torn and stretched sheet of plastic.

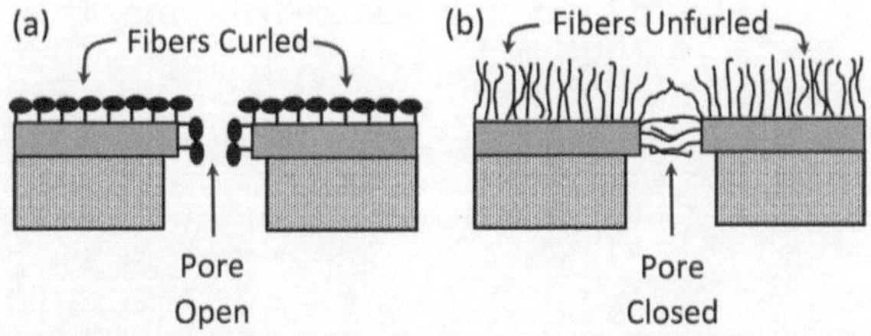

FIGURE 14.8 Temperature-controlled infilling and plugging of pores with nanoscale polymer brushes: (a) $T > T_c$ fibers curl and pores open. (b) $T < T_c$ fibers unfurl and pores close. (From [Lokuge 2007].)

14.1.7 Controlled Porous Solids

Solids that switch between different states of fluid permeability are useful in a variety of self-healing, self-sealing, and self-cleaning applications. Pores vary in size from the molecular to macroscopic scale.

Molecular scale opening and closing of pores through torsional motions of peptides arranged in a crystalline structure is a viable gating technique. Pressure and the presence of chemical species controls the torsional configuration and porosity of the solid [Rabone 2010].

Combinations of active gels and mesoporous networks switch permeability. Hybrids made of ordered nanoparticles (silica) with interstitial active polymers (poly(*N*-isopropylacrylamide)) form a porous network that regulates by thermally activated swelling and shrinking polymers [Fu 2007]. Nanoparticles act as keys to open and close pores while providing specificity as to which molecules traverse the pores [Astier 2009]. Thermally switchable nanopore plugs made of nanoscale polymer brushes seal and open surfaces to material transport (Figure 14.8). Magnetic fields control the porosity of gels made of ferrous material [Zhao 2010].

Chemically selective switchable surface permeability enables chemically stimulated delivery of healing agents. For example, the presence of sugar on colloidal particles makes them sticky so that they form surface films that differentiate protein adsorption [Bae 2005]. Stimuli responsive behavior with respect to pH, ionic strength of the colloidal solution, and temperature governs the formation of the sugar-colloidal protein selective films.

14.1.8 Yarns, Textiles, Basket Weaving, and Stapling

Yarns are twisted strands of fibers. Yarns comprise hierarchical layers in fabrics, ropes, and cables. Some functional materials twist and braid into yarns. Others require assistance, such as rolling sheets of thin material into spirals with layers of guest particles much like a stromboli [Lima 2011]. Figure 14.9 shows the 3D printing of helical multi-material filaments using a rotating print head.

The geometry of a helix converts torsional deformations and motions into length changes, and vice versa. Axial stretching or compression of a coil spring corresponds to torsional deformation of the spring wire. Twisting active materials into yarns and then into hierarchical helical arrangements creates length-changing string actuators with high-performance long-length strokes driven by a wide variety of physical mechanisms, that is, light, electricity, chemical reagents, solvents, and vapors [Lima 2012] [Chen 2015c].

DNA assembly of particles with unique programmable click binding between particles promotes the molecular scale assembly of a variety of sophisticated molecular structures, including those built with molecular scale basket-weaving techniques [Ke 2011].

14.1.9 Functionalized Surfaces

Surfaces between structural components and the environment and are the site for many self-healing applications. Surfaces selectively transport materials, stress, and heat. Functionalized surfaces enhance

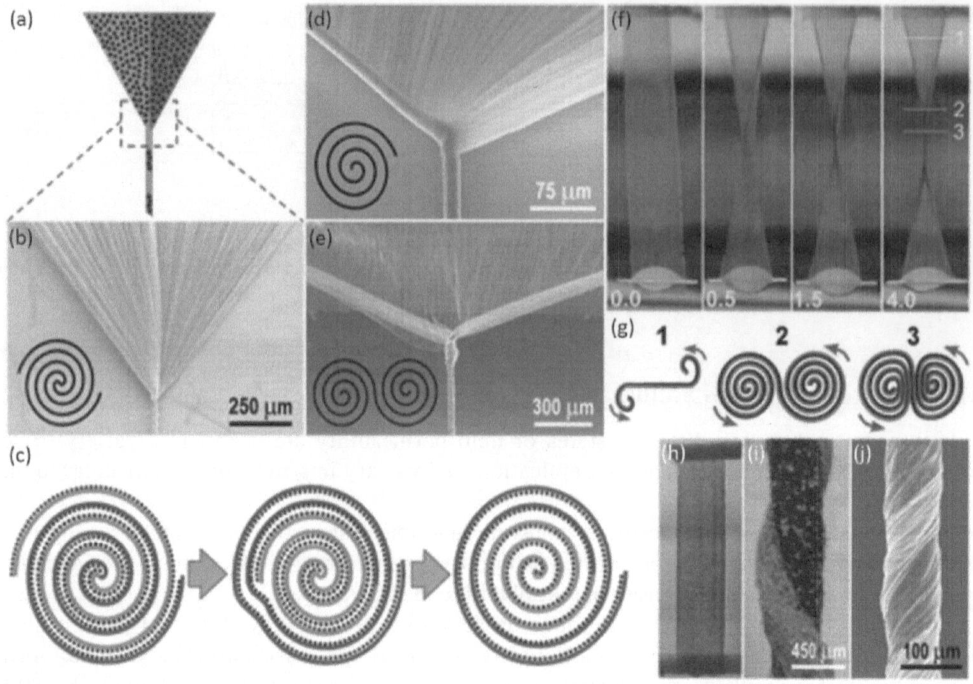

FIGURE 14.9 Three-dimensional printed helical structure using rotating multimaterial print head. Scale bar is 1 mm. (From [Larson 2023].)

self-healing by changing properties under control, such as adherence, porosity, stiffness, shape, and color, as needed (Table 14.7).

14.1.10 Highly Integrated Synthetic Skins

Material systems with many of the properties of animal skins, for example, mechanical compliance, controlled permeability, sensation, regeneration, and self-healing, have many advantages. Technical

TABLE 14.7

Selected Mechanically Active Surfaces

Surface Type	Product or Action	Potential for Self-healing	Reference
Self-assembled molecular layers that contain functional supramolecular structures	Molecular motor that switches from capturing specific molecules to releasing them	Capture or release of specific self-healing reagents, possibly as chemical signals	[Bunker 2007]
Photochromic diarylethene single crystal	UV light reversibly switches from smooth to nanotextured surface	Nanotexture change affects hydrophobicity and self-cleaning of surface	[Irie 2001]
Hydrogel with switchable surface friction	Thermally switchable hydrophobic phase collapse switches surface friction	Self-cleaning surface	[Chang 2007]
Hydrogel with switchable adhesion	Functionalize the surface with nanoparticles	Active patches for hydrogels	[Rose 2014]
Dry adhesive combined with SMP patch	Thermally reversible peeling	Active patches for solid elements	[Xie 2008]
Graphene oxide membrane with switchable liquid permeability	Electric field creates ionized nanochannels that allow liquid to flow	Active delivery of liquids through membranes	[Zhou 2018]

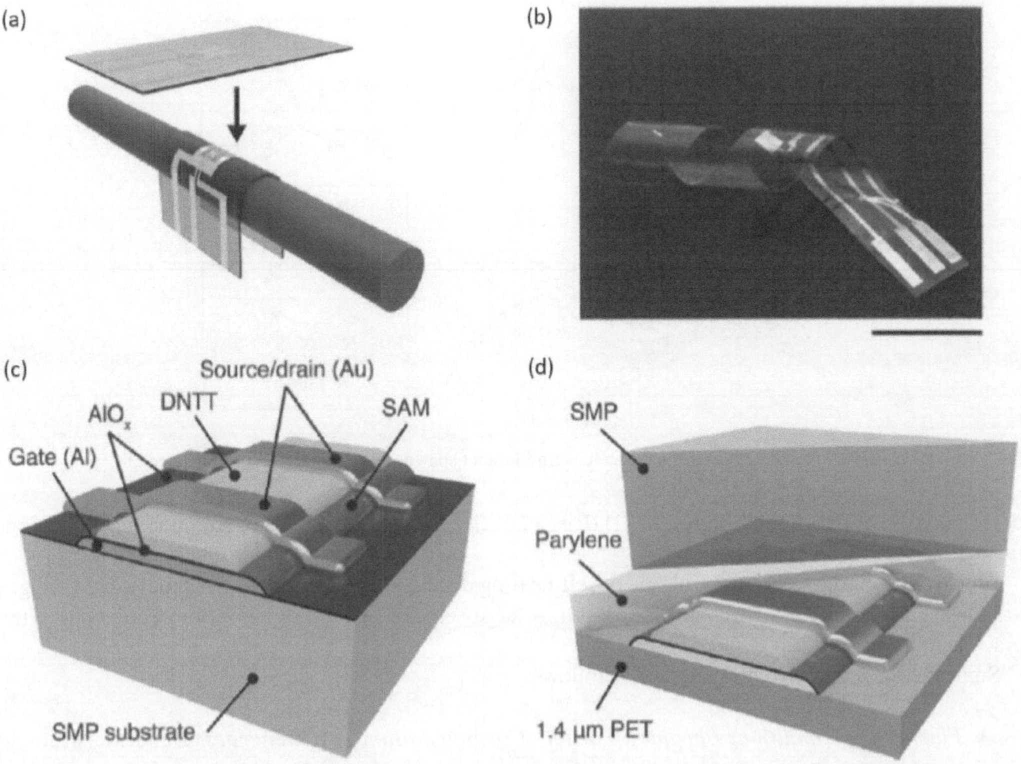

FIGURE 14.10 Circuit boards with capability of transitioning from stiff to soft and compliant using combination of organic thin film transistors and SMP substrates. (From [Reeder 2014].)

advances toward this end include (1) a nanowire active-matrix circuitry for low-voltage macroscale artificial skin based on a grid of addressable crossing arrays of parallel strand nanowires [Takei 2010] and (2) electronic circuit boards made with organic thin film transistors and SMP substrates that transition from stiff to soft and compliant electronic systems (Figure 14.10).

14.2 Active Agents

14.2.1 Active Dendrite Growth

Dendrites are elongated micro- and nanoscale features that grow from solid surfaces, sometimes with branching. Dendrites usually grow by adding material at the free end, but sometimes grow by adding to the base. Dendrites that grow long enough to span from one structure surface onto another change structural topologies and exert self-healing action with minimal material transport. Dendrites bridge cracks, establish electrical conductivity, and aid in the jamming of granular media.

Molecular scale dendrites, known as dendrimers, aid in controlled adsorption, release of chemicals, ease of access to molecules by diffusion through bulk materials, and the formation of fractal-like hierarchical structures with branching structures for releasable adhesives [Svenson 2005] [Yu 2012] [Suh 2000].

Controlling geometric patterns of dendritic growth facilitates healing with the following: (1) *Crystal Orientation*: When dendrites grow out of a crystal, the underlying orientation controls the direction of growth [Sone 2007]. (20 *Mechanical Loading*: Mechanical loads induce a gradient of the internal stress distribution inside of a filament to alter the direction and location of new fiber growth. The effect occurs at the nanoscale with actin filament. New fibers grow by bending and spawn on the convex side of the

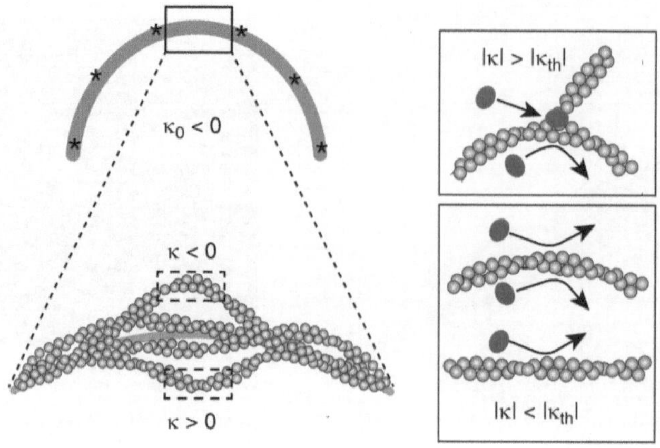

FIGURE 14.11 Twisting and bending of nanoscale actin filament spawns new filaments. (From [Risca 2012].)

curved section of bent fibers (Figure 14.11) [Risca 2012]. This leads to stress- or strain-adaptive filamentary growth response.

Active crack bridging is a fruitful line of self-healing development (Figure 14.12). Small-scale bridges are those that are smaller than the crack length. A large-scale bridge has dimensions on the order of the crack length [Tan 2012a].

Some methods of active bridging are as follows:

1. *Fluctuating Conditions Shepherd Particle Conglomeration into Dendrites:* Dendritic structures formed by the conglomeration of particles attaching particles to the end. Entropy may drive the attachment process. A complication is that the entropy landscape changes as each particle attaches to the chain. Periodically altering the environmental conditions injects free energy into the system. The free energy mobilizes the particles and makes them more susceptible to falling into an energy well by dissipative processes and attaching onto the end of the dendrite. Alternating magnetic fields act on suspended magnetic particles to produce dynamic self-assemblies of wire and arrays of spinners [Kokot 2015].
2. *Crack Surface Formation:* Crack surface free energy initiates dendrite.
3. *Electrochemical Conditions:* Crack surface chemistry and electrochemical environment in tight cracks promote dendrite formation.
4. *Stress Modulated:* Stress provides free energy for dendrite formation.
5. *Network Formation inside the Gel-like Material:* Complex liquids, colloidal suspensions, and gels are soft materials made of multiple interdispersed and organized constituents. Growth and formation of networks of internal filaments alters the strength and stiffness of the material, often in a tunable manner [Köhler 2011] [Karan 2016].

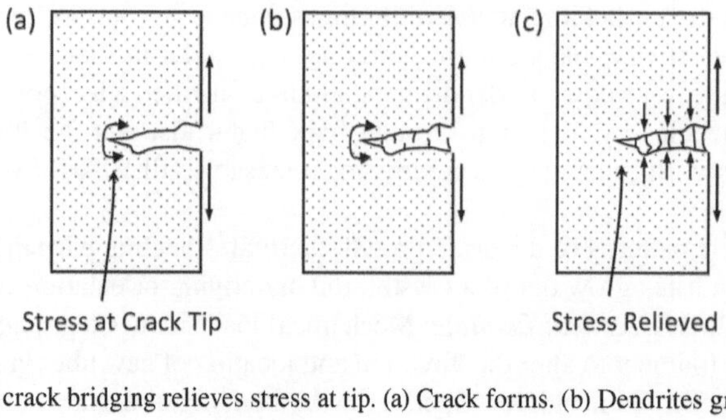

FIGURE 14.12 Dendrite crack bridging relieves stress at tip. (a) Crack forms. (b) Dendrites grow. (c) Dendrites bridge.

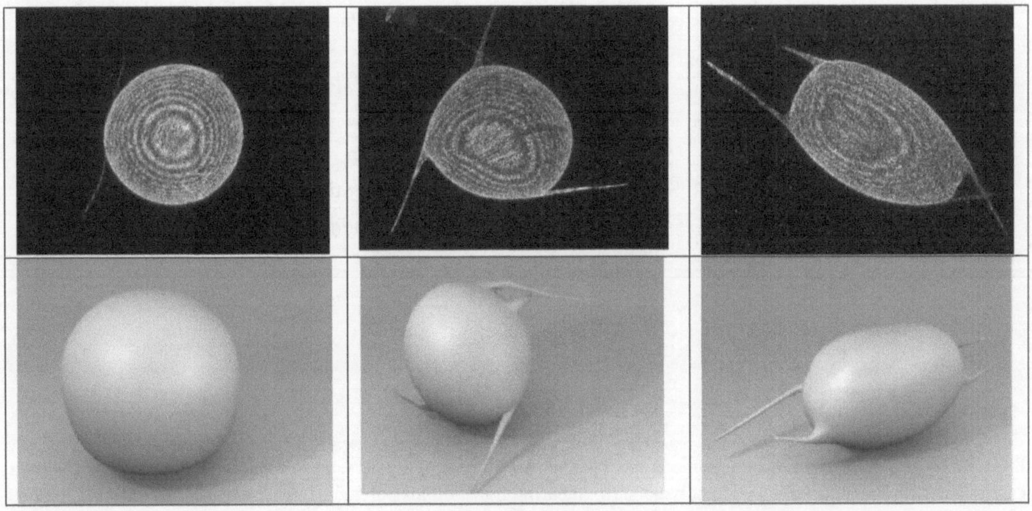

FIGURE 14.13 Formation of dendritic elements, that is, streaming filopodia-like protrusions, out of lipid vesicles at topological defect sites in nematic films consisting of microtubules with molecular motors. (From [Keber 2014].)

6. *Filopodia-like Protrusions:* Grow out of lipid vesicles by nematic films consisting of microtubules with molecular motors (Figure 14.13). The impetus of dendrite formation is topological defect that inevitably arises as a center when aligning fibers on the surface of a sphere.
7. *Molecular Self-assembly:* Chains of macromolecular structures form spontaneously as part of a thermal phase change. Small peptide molecules initially assemble into a lamellar plaque structure that upon cooling converts into macroscopic fibrils [Zhang 2010].

14.3 Healing Networks

14.3.1 Vascular Control

Vascular healing systems typically need to regulate the amount of flowing material to heal when and where needed. Table 14.8 lists some of the possibilities.

TABLE 14.8

Active Vascular and Fluid Transport Systems

Vascular System or Component	Product or Action	Potential for Self-healing	Reference
Vessel walls	Swelling and deswelling surface-attached polymer brush molecules	Regulate flow to mimic inflammation	[Gabriel 2007]
Flexible polymer vessel walls	External mechanical actuation alters flow	Regulate flow to mimic inflammation	[Gray 2012]
High-flow resistance of viscous fluid in narrow vessels	Thixotropic, non-Newtonian shear thinning	Easy flow for better material transport with narrow vessels	[Zhao 2004]
Surface with pre-patterned electrodes	Movement of droplets by selective electrowetting	Localized transport of healing droplets	[Ko 2014]
Liquids with permanent porosity due to molecular cages	Cage pores with potential chemical species selectivity and solvent-controlled bulk viscosity	Facile delivery of healing chemicals	[Giri 2015]

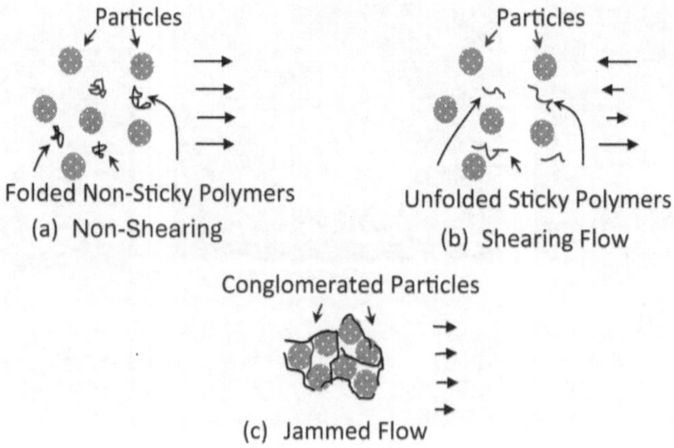

FIGURE 14.14 Shear-induced conglomeration of particles with sticky polymers. (a) Non-shearing flow with low viscosity carries particles and nonsticky folded polymers. (b) Shearing flow unfolds polymers to become sticky. (c) Sticky polymers conglomerate particles and jam flow.

14.4 Healing Tools

14.4.1 Jamming

Complex fluids containing sticky polymers and colloidal particles conglomerate in shear flows that mimic aspects of blood clotting (Figure 14.14) [Chen 2013a]. Active particles accentuate jamming by congealing and attaching to one another in response to specific cues. One possibility is sensitivity to deformation that alters the shape and sticking activity of molecules. Soft hydrogels bind and release synthetic microparticles coated with antigen bioproteins triggered by deformation [Galaev 2007].

14.4.2 Active Wicking

Wicks transport fluids through capillary actions. Active wicks provide enhanced functionality, such as increased transport and triggered responses. The wicking transport of material through spatially periodic structures at micro- and nanoscales benefits from periodic contractile actions that tune micro- and nanoscale flow dynamics to the spatial structure [Ahn 2015]. For example, oil-absorbing magnetic particles remove unwanted oil from bird feathers in a two-step process where the particles first absorb the oil and then magnetic fields remove the particles [Orbell 1999].

14.5 Coordinated Signaling and Repairs

14.5.1 Directed Assembly

Coordination accelerates and accentuates self-healing. Simple coordination uses damage-triggered repairs. More complicated coordination includes competing antagonistic and cooperating synergistic effects [Heuser 2015]. An overarching theme of self-healing in living biological systems is a damage-scale-dependent response with integration over length scales ranging from the macroscopic all the way down to the molecular. These natural actions inspire future directions for the control and coordination of human-built self-healing systems. Enabling technologies, that is, sensing, signaling, actuation, and repair, combine with holistic control systems.

14.5.2 Signaling

Advanced domain-specific signaling techniques may prove advantageous for engineered self-healing. Signaling sends information from a sender to a receiver in the form of physical quantities that differ from naturally occurring random configurations. Signaling in biological systems occurs through both biochemical and neurological pathways. In machines, structures, and systems, the signaling paths are electronic with optical, fluidic, and acoustoelastic pathways.

14.5.2.1 Damage-driven Signaling

Transducing and signaling damage are important components of self-repair processes. Many forms of damage growth comprise multiple competing and cooperating physical processes. This leads to multiple opportunities for chemical, electrical, mechanical, and optical damage-driven signaling (Table 14.9). Ideally such signals should identify the type, location, and extent of the damage. An example is cyclobutane-containing polymers of tricinnamates. Cracking of the material breaks the tricinnamates molecules into detectable fluorescent submolecules residing at the crack interface (Figure 14.15). Networks of molecular rotors transmit signals by a stimulated rotation of one molecule influencing neighboring particles to rotate, leading to a chain of rotations and long-distance transmission of a signal [Zhang 2016b].

14.5.2.2 Special Opportunities for Signaling

Multicellular organisms, including bacterial and biofilm communities, broadcast signals among the cells across the organism, such as by circulating chemical signals in a vascular system. Effective broadcasting benefits from high sensitivity and high selectivity. Biological systems solve the challenges of signaling through a variety of reactions, often nonlinear.

TABLE 14.9

Potential Damage Signaling Techniques

System or Component	Signaling Product or Action	Potential for Self-healing	References
Biological molecules, such as DNA, RNA, and proteins	Synthetic molecules using noncovalent binding of custom side link molecules	Precise signaling of state	[Hashidzume 2006]
Recognition-induced polymersomes	Microvesicles with a highly selective binding of diamidopyridine and thymine gold	Precise signaling of state	[Thibault 2002]
Dicyanocyclobutane mechanophores placed near to the center of polymer chains	Sonication releases cyanoacrylates	Polymerization promotes healing	[Kryger 2010]
Protein disulfide molecules	Mechanical stress stretches and unfolds molecules	Change in molecular configuration promotes chemical reactions	[Garcia 2009]
Flexible materials with mobile ionic component	Ionic gradients in bone due to cracks	Gradients drive healing agent into crack	[Yadav 2013]
Ionic polyelectrolyte brushes tethered onto surfaces	Compression of the brushes alters chemistry	Chemical signaling based on compression, for example, bruising	[Azzaroni 2006]
Flexoelectric liquid crystals	Deformation of elastomeric matrix containing liquid crystals	Material that sustains large deformations and produce electrical signals	[Chambers 2009]
Materials laden with mechanophores	Sonication with ultrasonic pressure waves deforms or splits molecules internal to structure	Initiation of chemical reaction, for example, precursors for the cyanoacrylate polymers	[Cravotto 2007] [Kryger 2010]

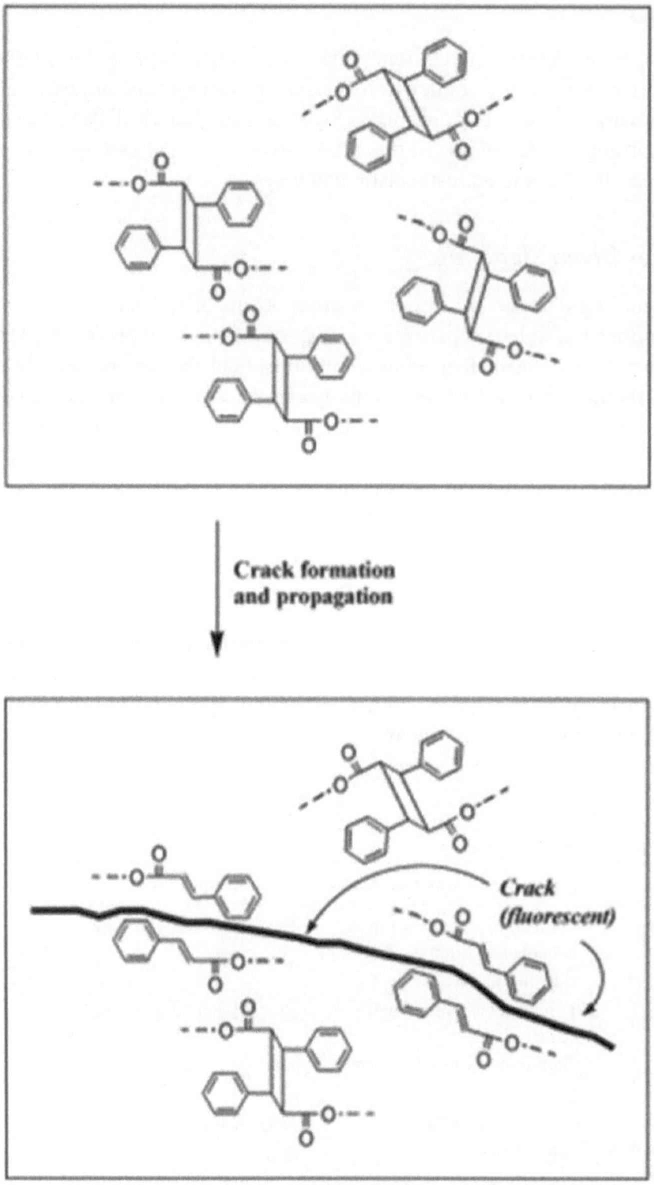

FIGURE 14.15 Cracking breaks tricinnamate molecules into fluorescent submolecules. (From [Cho 2008].)

14.5.2.3 Repair Initiated by Chemical and Molecular Scale Signaling

Chemical signaling transmits information through the presence and interaction of molecules. Chemical signaling requires (1) the production and/or release of specific chemicals by a transmitter, (2) the transport of the chemicals from the sender to the receiver, and (3) reception and recognition by the receiver. Most research to date focuses on unraveling the complicated nature of chemical signals in biological systems. Chemical species that are unique to the signal being transmitted enhance specificity and sensitivity. Sometimes relaxing specificity has the advantage of promoting the use of crosstalk between multiple signals and receptors. Wounds in plants cause internal water depletion stresses and increased susceptibility to fungal and insect attack. As part of a coordinated response, wound-signaling chemicals, such as oligopeptide systemin, oligosaccharides, abscisic acid, and jasmonates, all induce damage-amount-sensitive responses that regulate wound healing, water control, and anti-infestation response [Leon 2001] [Wasilewska 2008].

Plug and play chemistry uses molecules with complicated structural forms to produce highly specific binding reactions, typically through non-covalent binding. While relatively simple in concept, executing plug-and-play chemistry is sophisticated. Many methods use organic biochemical molecules with patterns of weak hydrogen bonds between hydrophobic and hydrophilic regions [Ilhan 2001]. The abstraction of networks theory simplifies understanding some of the complicated interactions.

Chemical amplification increases sensitivity. Biological systems routinely increase the response to signals with chemical amplification. An example is the competing coagulation and anticoagulation processes with nonlinear trigger thresholds that regulate blood clotting following injury and in healthy systems [Mann 2009]. A relatively mature, yet far less sophisticated, human-engineered chemical amplification technology is the chemically amplified photoresists used in microlithography. UV light cleaves molecules in resists to produce acids that catalyze further reactions [Ito 1985].

Mechanophores are molecules that react to mechanical stimulus, often by scission into functionally active groups. A goal is for mechanical damage to activate a mechanophore, which in turn initiates a catalyzed cascade of healing reactions. Sonication with ultrasonic pressure waves provides sufficient controlled and localized energy to drive mechanophore reactions [Cravotto 2007]. Sufficiently high-energy densities cause cavitation and bubble collapse in liquids leading to localized heating that stimulates reactions. An alternative mechanism has polymers with specific pendant shapes that deform under sonication and focus the motion to break specific bonds in the molecule. Dicyanocyclobutane is a potentially viable healing-response-initiating mechanophore. It cleaves under sonication into cyanoacrylate monomers that are precursors for the cyanoacrylate polymers, including superglues [Kryger 2010].

14.5.2.4 Bioelectric and Ionic Signaling and Computing

Voltage potentials and charge movements are alternative to chemistry for signaling and control. Metal conductors are the prevalent signal path in synthetic systems. Biological systems often transport electric charges as ions through liquids, gels, and across membranes. The conventional repair of solid metallic conductors melts metals at high temperatures to rejoin breaks with repair that solidifies upon cooling. Electrochemical deposition, metal particle percolation, and diffusion are alternative methods to repair metal conductors at lower temperatures. Biological systems repair conductive paths through tissue growth and electrolytic rebalancing at low temperatures compatible with life.

The versatility of biological signaling systems motivates developing synthetic versions. Biological systems read and output chemical signals in a digital logic fashion, including bio-based orthogonal AND NAND gates. One manifestation is *Escherichia coli* with a hetero-regulation module from *Pseudomonas syringae* [Wang 2011a]. Receiving both of the input signals and conditioning the signals with promoters produces an output. This thread of development holds potential for increasingly complex and programmed interactions, such as the genetic reprogramming of bacteria to control robotic machinery and controllable oscillator circuits [Heyde 2015] [Stricker 2008].

14.6 Active and Remote Control Capsules

Active and remotely controlled capsules provide high-performance dispensing of healing liquids. Table 14.10 lists some of the possible techniques that with further development may apply active capsule techniques to engineered self-healing systems. Energy approaches describe many aspects of the mechanics of active capsules. The total free energy F of a membrane is as follows [Smith 2007]:

$$F = F_{elastic} + F_{bend} + F_{adhesion} \tag{14.1}$$

TABLE 14.10

Active Release Microencapsulation Methods with Potential for Use in Damage Signaling and Self-healing

Encapsulation System	Release of Signaling Product or Action	Potential for Self-healing	References
Micelles made with antagonistic oppositely charged block ionomers	Stimulus, such as temperature	Controlled release of healing agent	[Park 2007]
Capsules with lenticular walls and weak points	Weak meridional layer ruptures in shear	Mechanical distortion of controlled magnitude releases healing agent	[Holme 2012]
Capsules with smart stitches that hold walls in place	Chemical triggers release stitches and rupture capture	Chemically induced self-immolation rupture of capsules to release healing agent	[Esser 2010] [Gunawan 2015]
Ultrathin wrapping with thickness <100 nm	Asymmetric capsules with ultrathin walls	Asymmetric geometry may enable smart and facile release of healing agents	[Paulsen 2015]
Thermally sensitive capsule	Rupture of capsule fine-tuned to specific temperature	Control of dispensing healing agents	[Farra 2012] [Barani 2008]
Thermally sensitive capsule	Magnetic hysteresis heating of ferromagnetic particle in core	Rapid triggered release of healing agent	[Loiseau 2016]
Photolabile capsule	UV light with high-energy photons selectively breaks down capsules	Remote control triggered release of healing agent	[Kim 2010b] [Coleman 2011]
Photolabile capsule	Nanoparticles and dyes absorb specific wavelengths	Remote control triggered release of anticorrosion agents	[Skorb 2009]
Photolabile capsule	IR-sensitive capsule	IR penetrates deeper than UV to release healing agents	[Yavuz 2009]
Chemically sensitive nanocages	Palladium and platinum nanocages break down with specific chemical stimuli, for example, trimethylamine	Chemical remote control of release of healing agent	[Pirondini 2002]

Here $F_{elastic}$ is the free energy of in-plane elastic deformation. F_{bend} is the free energy of bending. $F_{adhesion}$ is the free energy of particles adhering to the surface and for the surface to adhere to itself. Different values of these free energy functions lead to different behaviors:

Change Shape: Controlling of the shape of the capsules, such as with elongated forms, may promote specific capsule packing patterns [Calle 2015].

Drug Delivery Systems: The mature technologies of drug delivery systems provide guidance [Chen 2011a]. Vesicles are hollow bilayer containers with hydrophilic cores. Micelles are single-layer containers with hydrophobic cores. Hydrogels entangle three-dimensional networks of molecules to carry molecules interstitially; protein-based systems use intramolecular self-assembly with hydrophobic cores. Payload-stabilized systems use molecule-to-molecule stabilization.

Engulf Material while Remaining as a Capsule: Molecular recognition methods promote capture within capsules of specific molecules, such as proteins. Vault nanocapsules with gold nanoparticle probes recognize specific proteins and capture them with shuttle methods [Goldsmith 2009].

Internal Organization: Increased functionalization, such as increasing durability with support provided to the capsule walls with oriented molecules linked by hydrogen bonds, increases the sophistication of wall rupturing triggers [Atwood 2002]. Gel-like biological tissue, including the protoplasm inside of cells, transmits mechanical forces from the exterior to the interior for sensing purposes through nanoscale-embedded struts and filamentary structures of myosin and actin [Luo 2013a]. A collection of stiff rods derived from actin filaments forms a band-pass

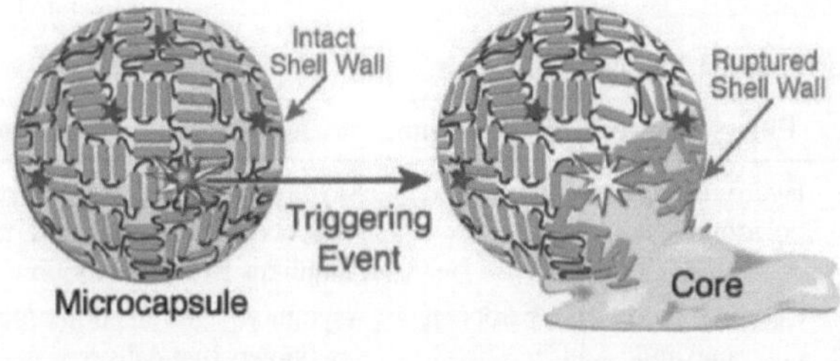

FIGURE 14.16 Self-immolative programmable microcapsule. (Reprinted with permission from [Esser 2010]. Copyright 2010, American Chemical Society.)

metamaterial structure that transmits specific frequencies of external motion to the interior of the cell [Shafrir 2002]. Fabricating gels with similar internal nanostructures produces similar mechanical behavior [Kouwer 2013].

Multi-capsule Interactions: Possible interactions include the breakup into a set of smaller capsules, capsules that merge with other capsules into a bigger capsule, and further clump together or form other patterned groupings of capsules.

Reactive Materials: Similar to microcapsules are micro- and nanoscale tubules that contain reactive materials, that is, healing agents, in the lumen. Natural halloysite clay nanotubes have lumen diameters on the order of 15 nm and lengths about 1,000 nm [Lvov 2013]. These tubes are easy to load with functional chemicals and to mix into polymers. The slow outflow of material through the ends controls delivery of the functional chemicals.

Release Material while Remaining as a Capsule: Microcapsules change surface permeability upon receipt of a signal, such as a change in pH, to release or absorb material, while remaining as an intact capsule [Ignacio 2013].

Smart Rupture: Biologic-derived molecular recognition methods provide highly specific activation for microcapsule wall breakdown and dissolution. An example is the use of two proteases in AND-gate logic to activate virus capsids [Judd 2014]. Many of these capsules use antagonistic effects to store free energy. A threshold exceeding stimulus destabilizes and ruptures the container (Figure 14.16).

14.6.1 Micro- and Nanorobots

Until recently, molecular robots have been the lore of science fiction. Technical advances are changing this paradigm to reality. Molecular biosystems inspire molecular robot technology. The healing of wounds to skin is an example. Keratinocyte cells on the periphery of wounds change their character when an injury breaches the skin. The cells stop differentiating and start migrating to fill the center of the wound [Henry 2003]. Molecular scale robots differ from ordinary chemistry by the appearance of control and information processing. The distinction is not crisp. Many actions in chemical thermodynamics take on information, and programmable control capabilities, especially when dealing with supramolecular chemistry.

Primary requirements of micro- and nanoscale robots are movement, actuation, sensing, control, and assembly. The mechanics scale down geometrically to submicron dimensions with inertia forces varying according to L^3 and actuator forces often scaling according to L^2. (Appendix D has more details on the effects of dimensional scaling.) Many small robots are similar to large robots. As the dimensions shrink even further, continuum approximations break down. Thermal oscillations and molecular forces play more dominant roles. It is at these smaller dimensions that robot designs may differ dramatically from the macroscale versions.

Microscale swimming robots must solve the problem of overcoming effectively large viscous forces. A biomimetic approach is with artificial flagella, similar to those in bacteria. Undulating irreversible motions for propulsion are simple in concept, but technically challenging to implement. Artificial flagella

TABLE 14.11

Energy Harvesting Microrobots with Engines for Actuation

Robot Type	Energy-harvesting Technique	Engine and/or Actuation	Reference
Microscopic swimmers	Hydrogen peroxide released by platinum catalyst	Asymmetric energy absorption propels swimming in pattern that amplifies Brownian motion	[Yamamoto 2013]
Microscopic swimmers	Carbon dioxide from carbonate in tranexamic acid	Asymmetric gas ejection propels swimmers that deliver therapeutics into the vasculature of wounds	[Baylis 2015]
Microscopic swimmers	Hydrolytic reaction of surfactant	Generation and asymmetric motion of lipophiles propels droplets	[Toyota 2009]
Microscopic swimmers	Carbonic anhydrase-functionalized micromotor using CO_2 hydration	Micromotor propels CO_2 scrubbers	[Uygun 2015]
Magnetic dipole microbeads	External fluctuating and rotating magnetic field	Beads assemble into controllable shapes, such as helices	[Cheang 2016]
Magnetic dipole microrods	External fluctuating and rotating magnetic field	Rods rotate and break up blood clots by accelerating therapeutics	[Cheng 2014b]
Molecular ring moving on a thread molecule	Alternating Coulombic attraction and repulsion due to alternating oxidation and reduction states	Tetracationic rings ratchet along a polymethylene chain	[Chuyang 2015]

with soft magnetic heads attached to helical tails, or snake-like magnetic nanoparticle arrangements, twist and wiggle so as to move under the power and control of external oscillating magnetic fields [Zhang 2009a] [Dreyfus 2005]. Ultrasonic acoustic excitation propels nanorods in living cells. Another method uses asymmetric 300 nm long gold nanorods, with motion bias-inducing concave and convex ends that translate and spin with 4 MHz excitation [Wang 2014].

Actuation requires an engine to convert available energy into desired motion. Table 14.11 lists some methods of harvesting free energy from the environment or working fluid by microrobots.

Molecular robots require structural elements, actuators, and controllers. DNA-based and supramolecular approaches provide large molecules as components in molecular robots. Small-molecule systems are also viable. Table 14.12 lists some possibilities, including molecular variants of conventional mechanisms, walkers with programmable paths, and robots that transport material under control.

14.6.2 Robot Teams, Crystals, Flocks, Swarms, and Mobs

Some tasks are easier with teams of simple robots than with individual sophisticated robots. Teams coordinate actions to build and repair structures, much as social insects build hives [Augugliaro 2014].

Individual robots use their bodies as structural elements and assemble with other robots to form permanent or temporary components of collective structures [Rus 2001]. The topology of the linkage determines whether collective structures are kinematic mechanisms, such as a chain, or effectively rigid bodies, as a tetrahedron. Collective robots build new structures, rebuild existing structures, and bridge patches across damage. Conversely, a collective structure disintegrates on cue for beneficial apoptotic applications. An example is clumps of therapeutic nanoparticles that conglomerate or disintegrate for the specific application of dissolving clots and blockages in blood vessels or stopping hemorrhages [Korin 2012].

Molecular scale structures, such as microtubules in living cells, impose directional and nonhomogeneous pathways for active molecules to follow. Such structures bias the direction and amount of material transport to avoid following the diffusion dynamics of Brownian motion. If the molecular structures provide runways for rapid directional motion, then positive feedback causes the transport to self-organize into a truncated Lévy walk that favors the efficient transport of material [Chen 2015b].

TABLE 14.12
Molecular Robot Techniques

Robot Structure	Actuation	Control	Reference
Molecular linkage formed by base pairing of DNA	Disassembly and reassembly in different cation concentrations	Click chemistry with DNA base pairing	[Gerling 2015]
DNA walker with spider geometry	Catalytic capture and release of legs on surface	Track of available catalytic sites on surface guides motion	[Lund 2010]
Walking DNA molecular robot	Catalytic capture and release of legs on surface	Molecular track with switches controls direction of movement	[Muscat 2011]
Ring molecules encircle and move on larger ring molecules	Sequence of chemical reactions as small ring moves on big ring	Kinematic constraint of motion around ring	[Wilson 2016]
Rotary DNA molecular motor with feet	Feet with DNA bases selectively catalyze to block and unblock react for motion along track	Internal and external DNA codes program switching movement along tracks	[Wickham 2012]
DNA-based container robot that attaches to surface	DNA aptamers latch container open or closed	Two distinct antigen molecules with AND-gate logic opens latch	[Douglas 2012]
DNA-based container and reactor robot swarms with molecular operations inside animals	Protein-specific opening of gate changes shape of container to release molecules	Interaction between different robot types creates logic-based release of molecules into animals	[Amir 2014]
Simple mechanisms – four-bar and slider crank	Stiff double-strand DNA combined with flexible single strand	DNA inputs open and close with zippered energy release	[Marras 2014]
Supramolecular assembly with gold atoms and adatoms on gold surface	Pulsed voltage at electronic resonances exerts work by tunneling of current intensity	Supramolecular structure guides rotation or translation	[Ohmann 2015]
Small-molecule robot arm that picks, moves, and places molecular cargo	Hydrazone-based rotary switch with protonated change in configuration	Selective protonation determines switching direction	[Kassem 2016]
Organic molecule with four paddle-wheel molecular appendages	Electrostatic actuation of isomeric configuration changes	Four-wheel motion for rolling along a surface	[Kudernac 2011]

Actively controlled droplets clean and heal surfaces. Droplets of two miscible liquids, such as propylene glycol and water, influence the behavior and motion of neighboring droplets by affecting surface tension through vapor pressure and evaporation [Cira 2015]. Surface tension propels droplets of room temperature liquid metal (eutectic indium gallium) in a mollusk-mimicking motion that digests aluminum flakes available in the ambient environment as fuel [Zhang 2015b]. Structured light spatiotemporally activates photoactive liquid crystal elastomers to control swimming and locomotion of soft robots [Palagi 2016]. Modular soft robots are rare but offer intriguing possibilities for future development.

14.7 Improved Tools for Repair

14.7.1 Remote Delivery of Energy for Repair with Ultrasound

Ultrasonic elastic waves deliver energy to remote locations for molecular and thermal healing. It is possible to direct and focus the waves on specific regions. Certain molecules have a sensitivity to ultrasound that differs from that to thermal, light, and electrical stimuli. Scission by ultrasound breaks covalent bonds to form new chemical species. A polymer system based on isomers of benzocyclobutene undergoes an isomer-independent ultrasonically induced ring opening reaction [Hickenboth 2007]. Ultrasound induces reactions that seed further reactions. Proof-of-concept demonstrations, that is, ultrasound-induced color changes, confirm the validity of remotely triggering cascades of healing reactions [Davis 2009].

14.8 Self-assembly

Self-assembly combines physical assembly of simple components with information regarding structural form to make more complicated structures. The information flow in self-assembly processes ranges from being entirely directed from a high level to being unsupervised with spontaneous free-form assembly, with many cases of hybridized directed and free-form control.

14.8.1 Directed Self-assembly

Directing the assembly of components to flow into a repair uses resources efficiently. The traditional approach to directed repair is human in the loop, with manual or machine-aided repairs. Directed repair extends to robots, such as those that inject structural foams, assemble, and repair structures [Revzen 2011].

A physical artifact, such as a template, or a data bank, such as DNA in biological systems, houses the information needed for self-assembly (Table 14.13). Templates have the advantage of quickly transducing the plan into a physical embodiment but are limited in the delivery of complex forms. Data banks provide more complex information but require encoding and decoding apparatus for transduction into physical forms.

An advantage of templates is the possibility of mass production of structural units with the number of structural units increasing linearly with time, especially if the template releases from the newly assembled structure and recycles for making additional parts. If a new part also acts as a template for a copy of itself, then the growth rate switches from linear to exponential. Self-replication of useful parts remains a technical challenge that has yet to be fully achieved. A demonstration at the molecular level is self-replication using seven-tile DNA motifs [Wang 2011b].

TABLE 14.13

Selected Reports of Templated Self-assembled Systems

Scale	Template	Assembly	References
Molecular	Gold, Au(111)	$Mn_{12}(acetate)_{16}$ with hybrid 3D effects on top of 2D template	[Saywell 2010]
Molecular	DNA	Templated assembly of molecular structures with orthogonal click bonds	[Yin 2014b]
Molecular	Supramolecular orthogonal noncovalent bonds	Functionalized polymer structures	[Hofmeier 2005]
Molecular	Alkyl-bridged di(8-hydroxyquinoline) ligands; metal–ligand preorganization	Templated assembly of helicate triple-stranded molecular structures using size-selective cations	[Wood 2015]
Molecular	Liquid crystal topological defects	Amphiphiles	[Chen 2017]
Microscale to mesoscale	Prepatterned solder bumps	Joining of electrical contacts and components with eutectic Bi-Sn solder in glycerol	[Morris 2007]
Microscale to mesoscale	Prepatterned surface	Assembly of micro-objects on pre-patterned surface followed by fixing step	[Jonas 2002]
Microscale to mesoscale	Prepatterned surface	Assembly of colloidal crystals and capsules by colloidal assembly on patterned surfaces and shapes	[van Blaaderen 1997] [Cayre 2004]
Molecular to macroscale	Biohybrid hydrogel with living fibroblasts dumping vascular growth factor into microchannels	Patterned growth of blood vessels	[Jeong 2012b]
Molecular to mesoscale	Diffusion of chemicals into gradient patterns through microchannel walls	Patterned assembly of free-swimming bacteria through chemotaxis	[Ahmed 2010]

14.8.2 Free-form Self-assembly

Free-form self-assembly builds structures without a template. Table 14.14 lists techniques [Whitesides 2002b].

TABLE 14.14

Free-form Self-assembled Systems and Components

Scale	Technique	Static versus Dynamic	References
Molecular	DNA into polyhedra with hierarchical and fractal structures assembly	Static	[He 2010b] [Shang 2015]
Molecular	Numerically designed protein assemblies	Static	[King 2014]
Molecular	pH-programmed sorting of two-component gelator assembly	Static	[Morris 2013]
Molecular to nanoscale	Programmed molecular recognition with DNA nanostructure geometry	Static	[Woo 2011]
Molecular to nanoscale	Ambiphillic organic molecules into nanotubes	Static	[Bayer 2011]
Molecular to microscale	Kinetic manipulation of block co-polymers for different nanostructure shapes and assemblies	Static	[Cui 2007]
Molecular	Programmed arrays of donor/receptor hydrogen bonds for site-specific binding	Static	[Beijer 1998]
Molecular	Helical dimerized polymer chains using hydrogen bonds and controlled hydrophobicity	Static	[Hirschberg 2000]
Molecular to nano	DNA click binding assembly of gold nanoparticles into clusters	Static	[Mirkin 1996]
Molecular to nanoscale	Hierarchical molecular assembly up through micelles by avoiding kinetic traps	Static	[Gröschel 2012]
Molecular to nanoscale	Assembly of linear structures using particles with controllable hydrophobic patches	Static	[Lee 2014]
Molecular to mesoscale	Custom DNA glue selectively assembles meso-scale components	Static	[Qi 2013]
Nanoscale to microscale	Assembly of variety of microlattices from binary nanoparticles	Static	[Shevchenko 2006]
Microscale to mesoscale	Triblock Janus particles in colloidal suspension form Kagome lattices and more complicated patterns	Static	[Chen 2011b] [Mao 2013]
Nanoscale to microscale	Hierarchical assembly of colloidal octapod nanopod crystals into chains and superlattice sheets	Static	[Miszta 2011]
Micro	Biohybrid self-assembly of muscle cells on microactuator	Static	[Xi 2005]
Molecular to microscale	Formation of asters and swirls of microtubules by molecular motos	Dynamic	[Ndlec 1997]
Molecular to microscale	Swirling collection motion of concentrated F-actin filaments	Dynamic	[Schaller 2010]
Nanoscale to microscale	Complex filamentary colloidal structures resulting from dynamic electric fields	Dynamic	[Leunissen 2009] [Demortière 2014] [Sapozhnikov 2003]
Nanoscale to microscale	Gold nanoparticles assemble into patterns and arrays on liquid surface in response to dynamic surfactant stimulus	Dynamic	[Sashuk 2013]
Microscale to macroscale	Asters of magnetic particles in slow oscillating magnetic fields with controlled manipulation	Dynamic	[Snezhko 2012]
Microscale to macroscale	Topological defects leading to organized structures	Dynamic	[Shi 2013]
Microscale to mesoscale	Janus particles assemble into tubes under oscillating magnetic fields	Dynamic	[Yan 2012]
Microscale to mesoscale	ATP-powered microtubules reptate and assemble into hexagonal arrays of vortices	Dynamic	[Sumino 2012]

14.8.3 Hybrid-directed and Free-form Self-assembly

Hybrid-directed and free-form self-assembly forms coherent structures from incoherent arrangements (Table 14.15). Random agitation brings different components together in different orientations, but only specific sides of components fit together and assemble into a structure. Multiple methods provide site-specific joining. Surface tension and compatible wettable surfaces adhere parts together in the hybrid assembly of electrical components (Figure 14.17) [Morris 2007] [Srinivasan 2001]. Lock and key methods provide highly specific joining capabilities (Figure 14.18).

Janus particles are microparticles with a dipolar distribution of functional properties, such as one side being hydrophilic and the other hydrophobic. Dipolar interactions produce orientation-specific particle binding and conglomeration. Complicated and controllable interactions between Janus particles lead to emergent collective patterns and behaviors [de Gennes 1992]. Asymmetric nanoparticles interlock and self-assemble into a variety of shapes, including capsules [Evers 2016].

TABLE 14.15

Selected Hybrid-directed and Spontaneous Self-assembled Systems

Scale	Process	Assembly	References
Molecular	DNA	DNA polyhedral and bricks	[Iinuma 2014] [Ke 2012]
Molecular	Size selectivity	Triple-stranded helicate-type complexes	[Albrecht 2002]
Molecular to microscale	pH-induced assembly of peptide	Nanostructured fibrous scaffold followed by hydroxyapatite mineralization	[Hartgerink 2001]
Molecular to microscale	Switching of electrolytic growth conditions	Controlled hierarchical assembly of carbonate silica nanostructures	[Noorduin 2013]
Molecular to microscale	Lock and key	Molecular binding and assembly of microparticles in colloidal system	[Sacanna 2010]
Molecular to microscale	Hydrophobic and hydrophilic patterning	Assembly of arrays of micropieces onto surfaces	[Bohringer 2003]
Molecular to microscale	Complementary DNA combined with complementary shapes in microparticles	Brick and mortar assembly of microstructures	[O'Brien 2015]
Microscale to mesoscale	Mesoscale parts self-assembly through selective microscale patterns and surface tension	Microelectronic assembly of discrete components, for example, LEDs	[Zheng 2006]
Microscale to mesoscale	Movement of identical microrobots with simple latching rules	Tuning of rules directs robots to swarm into polygonal shapes	[Arbuckle 2009]
Macroscale	Electronic sensing and control of latching parts together using local graph grammars	Tiling assemblies of geometric shapes containing electric circuit boards	[Klavins 2007]
Molecular to nanoscale	Elastin-like peptide membranes	Shape modified by peptide amphiphiles, including self-healing	[Inostroza 2015]
Molecular to microscale	Molecular cages of hydrophobic or hydrophilic cages oligomeric silsesquioxane assemble into tetrahedra	Tetrahedra use differences in hydrophobic and hydrophilic behavior to guide selective assembly into microscale structures	[Huang 2015a]
Molecular to macroscale	Hybrid mix of mussel and bacterial proteins	Hierarchical assembly of strong underwater adhesive	[Zhong 2014]

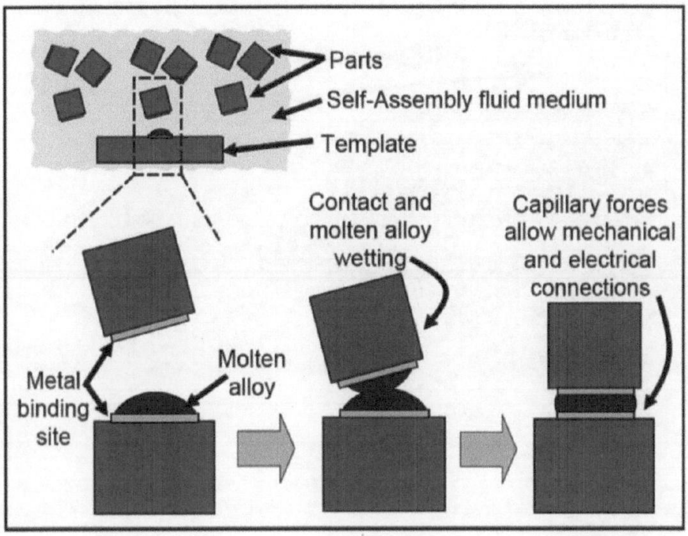

FIGURE 14.17 Microscale self-assembly using surface tension and capillary forces as sticky components. (From [Morris 2007].)

Biological tissue regeneration is sophisticated hybrid-directed and free-form self-assembly. Regeneration combines cell growth, differentiation, and spatial configuration – often through cooperating and antagonistic interactions (Figure 14.19). DNA and protein networks provide much of the information directing assembly, often through reaction–diffusion kinetics. Synthesizing materials with regenerative capabilities may use similar molecular building blocks and bind them together in specific multiple length scale arrangements.

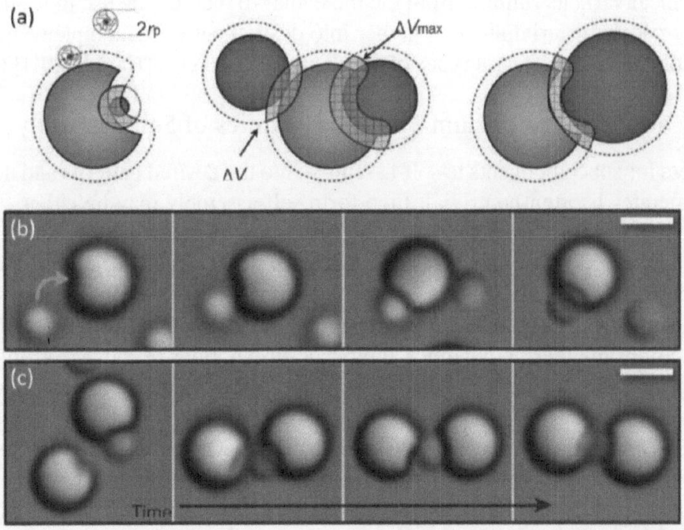

FIGURE 14.18 Lock and key selective binding of microparticles with selective molecular binding drives coordinated colloidal assembly. (From [Sacanna 2010].)

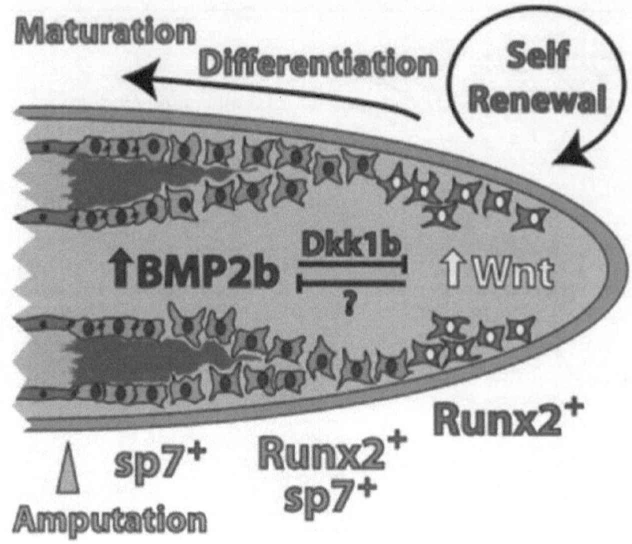

FIGURE 14.19 Coordinated cell replication, differentiation, and maturation in regeneration of bone following limb amputation in a zebrafish through antagonistic activities of the Wnt/b-catenin that promotes epithelial to mesenchymal transformation and bone morphogenetic protein (BMP) that promotes osteoblast differentiation. (From [Stewart 2014].)

14.8.4 Self-assembly for Healing and Functional Structures

Self-healing with self-assembly requires materials and components. Molecules with potential for healing often have excess free energy, which renders them long-term unstable. On-demand techniques increase stability by manufacturing relatively unstable healing molecules when needed from long-term stable feedstocks.

A challenge for self-assembly is to join crystalline structures with polymeric materials. Plug and play methods bind supramolecular assemblies in specific orientations onto the face of a lattice [So 2009a]. DNA works well for molecular scale plug and play self-assembly. DNA binds selectively with orthogonal matching. Hierarchical variants running from the molecular to macroscale use molecular scale recognition to adhere macroscale material pieces together into desired patterns. Complementary DNA nucleobases selectively join blobs of different types of acrylamide-based gels [Harada 2011] (Figure 14.20).

14.8.5 Entropy and Nonequilibrium Thermodynamics of Self-assembly

Designing a process for subcomponents to self-assemble into the desired patterns and not into messy and worthless conglomerates is often nontrivial. Free-form self-assembly may be either static or dynamic. Static structures assemble into a fixed state, such as crystals. Dynamic structures simultaneously add and remove material in nonequilibrium processes that maintain stable overall structural forms. Entropy and nonequilibrium thermodynamics describe some of the overall behavior of these processes. Equilibrium thermodynamics methods achieve a desired pattern with a system that falls into a state of minimum free energy. Nonequilibrium thermodynamics approaches use notions of causal entropic forces in which microstates emerge from macrostates in an adaptive manner that mimics aspects of cognitive adaptive behavior [Wissner 2013].

Dynamic dissipative structures respond to small disturbances by healing and converging back to the same form [Prigogine 1981]. Strong perturbations may move structures to different stable forms [Winfree 2006]. Living cells and hurricanes are examples. An interpretation of this behavior is that energy dissipation drives the structure into fractal subspaces of the possible phase space with processes that have negative Lyapunov exponents [Fialkowski 2006]. Another is systems that encircle non-Hermitian degeneracies, known as exceptional points [Berry 2004]. Many cases have a steady input of energy combined with a dissipation sink.

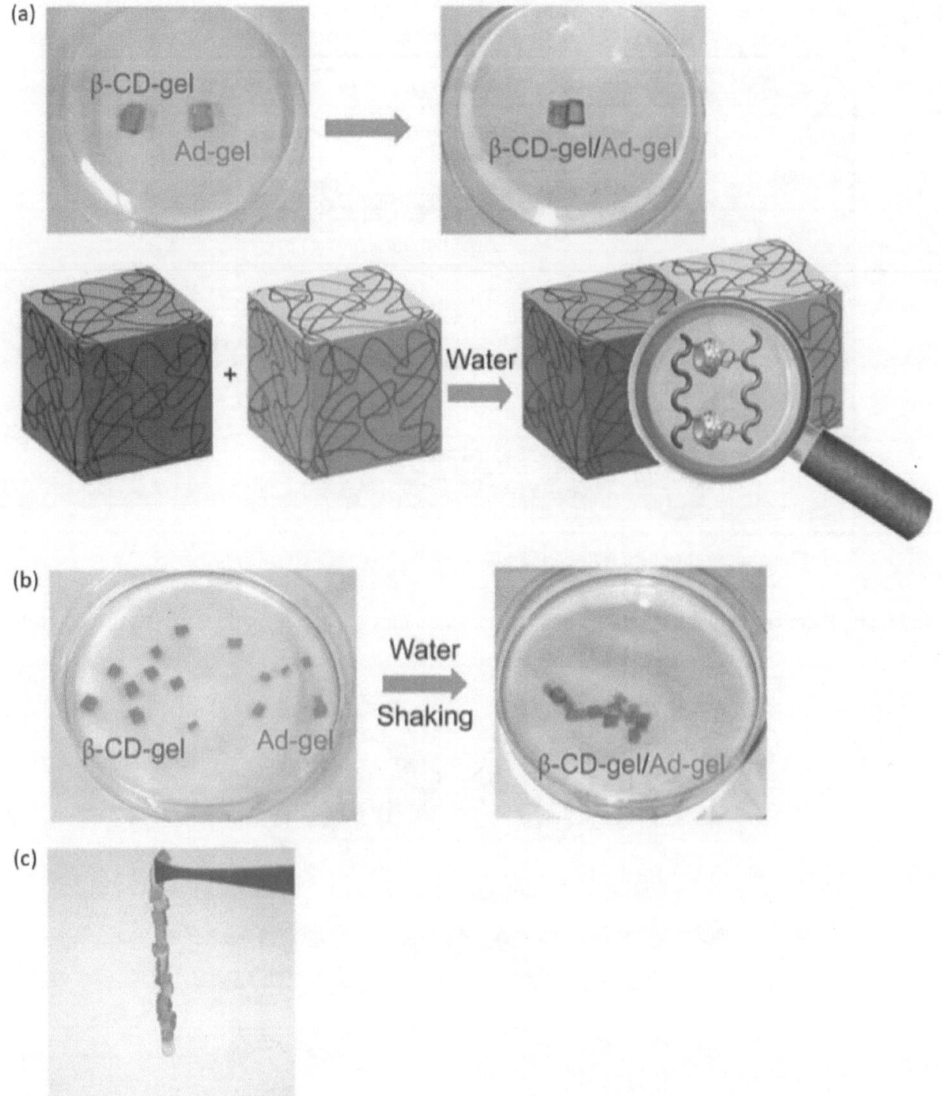

FIGURE 14.20 Molecular scale to macroscale recognition for directed self-assembly of macroscale acrylamide-based gels functionalized with different pairs of DNA nucleobases. (From [Harada 2011].)

Hierarchical structural forms appear when the assembly occurs on small length scales, which then assembles into larger units in a cascade of length scales up to the final assembly. The details of the assembly steps may vary with length scale, but the effect is to funnel the range of possible assembly configurations into the final form (Figure 14.21). Assembly is a reduction in the number of degrees of freedom in a process that maximizes the entropy of the entire system, but perhaps not at the local level [Onsager 1949].

Designing a material system to assemble into a desired pattern from first principles based on the geometry of the original shapes is complicated. Sometimes, it is possible to catalog possible shapes. Figure 14.22 shows part of a catalog of two-dimensional and three-dimensional shapes, including polyhedral and networks, based on square molecular building blocks and variations in process parameters. It enables self-assembly design for larger structural forms [Eddaoudi 2002].

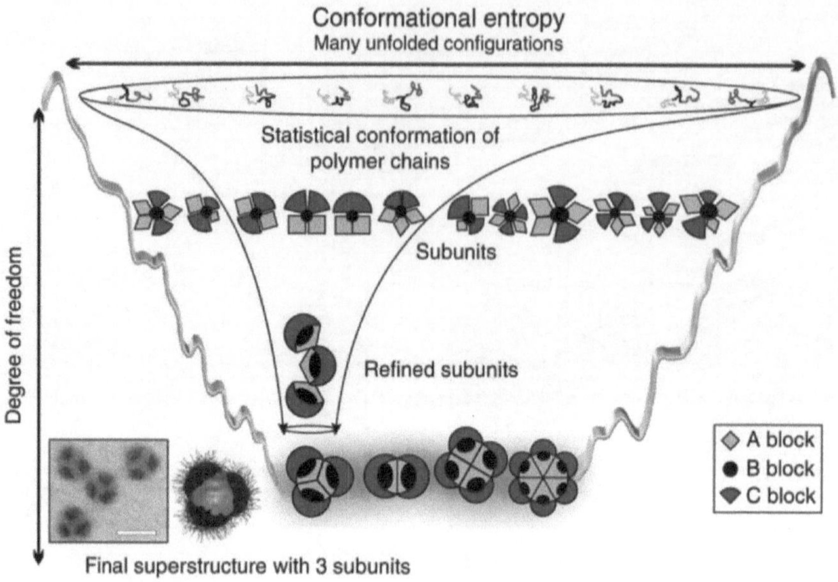

FIGURE 14.21 Hierarchical funnel of self-assembly reduces the number of degrees of freedom. (From [Gröschel 2012].)

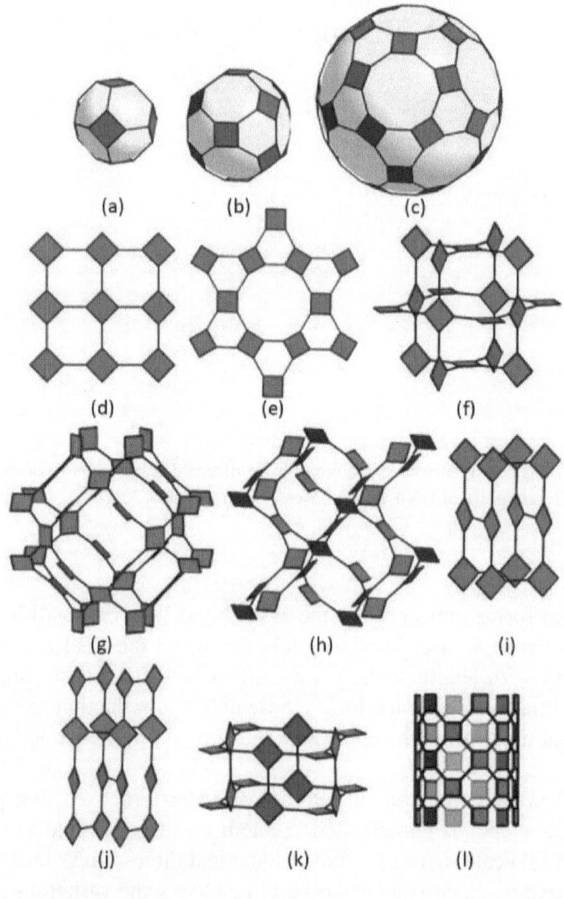

FIGURE 14.22 Catalog of shapes and networks formed from square molecular building blocks. (From [Eddaoudi 2002]. Copyright 2002, National Academy of Sciences, USA.]

14.9 Biohybrid Structures and Systems

14.9.1 Biohybrid Tissue

The incredible functionalization and versatility present in biological tissues opens wide ranges of biohybrid applications for self-healing.

14.9.1.1 Metastability

Many biomolecules are metastable, which implies that they react with other biomolecules at relatively small activation energies, and are easily damaged and self-repaired. Prepositioning repair-assisting molecules near to potential damage sites helps with mitigation. Photolyase UV damage to DNA creates the photoproduct (6-4PP) between two adjacent pyrimidine rings. Unchecked, the formation of (6-4PP) may be fatal. Flavoenzymes harvest energy from blue light and drive palliative reactions to repair the damage and restore vitality [Li 2010a]. The repair uses a template to provide configurational information with a sister chromatid that primes the repair with 3′ single strands of DNA to identify the proper template location [McAleenan 2013].

14.9.1.2 Assembly and Patterned Growth

Molecules extracted from formerly living tissue play key roles in material assembly for repair. The bacteria-derived protein Mrns6 guides forming high-quality magnetic nanoparticle crystals onto surfaces with feature sizes in tens of microns [Galloway 2012]. Chitosan is a polysaccharide derived from shrimp and other crustacean shells. The molecular structure consists of mers with NH_2 appendages that change hydrophobicity in response to pH and other stimuli, such as temperature. These changes affect hydrogen-bond linking among the polymer chains at the molecular scale and the density and overall size at the macroscale. High pH leads to shrinking. This thermal and pH-sensitive stimuli-responsive behavior fuels applications ranging from coagulating sealers with flocculation of polyaniline colloidal particles to shrinking fiber reinforcement in cementitious materials, such as concrete, that activate with high pH during cure [Cruz 2007] [Kim 2017].

Patterned assembly techniques extend to the growth of layered and structured materials. Controlled switching of pH and alcohol baths creates multilayer gels with onion-like structures, confirming that switching chemical environments creates patterned growth [Ladet 2008].

14.9.1.3 Synthesis and Synthetic Tissue

Many methods of developing biomimetic and biohybrid materials are tedious and time-consuming trial and error approaches. Quicker methods are appearing. One family of methods integrates RNA-seq to tease out favorable protein sequences and create customized materials, such as self-healing bio-elastomeric membranes from proteins that mimic the mechanically robust byssal threads of mollusks [Guerette 2013].

Large-scale biohybrid systems benefit from vasculature. It is generally easier to insert vessels during initial material system assembly than with a post-fabrication step. A viable approach is to build vascular scaffolding, such as with a 3D printer, and then seeding it with endothelial cells [Miller 2012]. The cells in and near the vessel walls thrive under suitable conditions, including angiogenesis (formation of new vessels), perivascular interactions with cells near to the vessels, and a prothrombic behavior (promotion of blood clotting) when exposed to inflammatory stimuli [Zheng 2012b].

Structural configurations in living systems are exceptionally complicated. Nonetheless, creating organisms from scratch that produce specific tasks is a reality. A major milestone was building viable organisms by synthesizing DNA for bacteria lacking a nucleus [Gibson 2010]. Computer-based methods are now able to manage the information to a level of sophistication that enables designing reconfigurable organisms [Kriegman 2020].

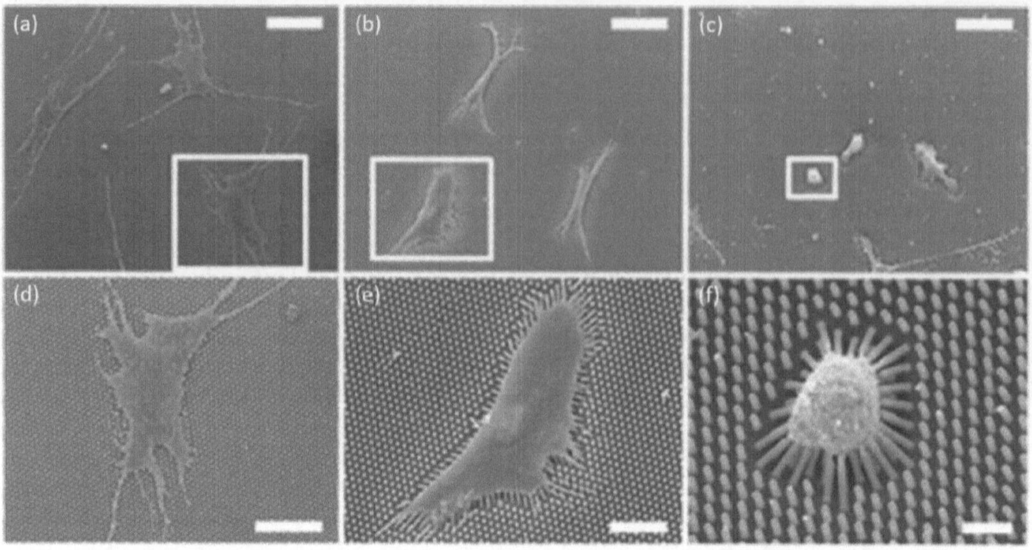

FIGURE 14.23 Effect of different micropillar compliances on different behaviors of human mesenchymal stem cell growth and differentiation scale bars: 100 μm (a–c), 50 μm (d), 30 μm (e), and 10 μm (f). (From [Fu 2010].)

14.9.1.4 Biohybrid-induced Healing

Living bio-tissue spontaneously heals damage on scales that range from the molecular to macroscale organs. How to get biohybrid material to induce healing? Some amphibians regenerate entire limbs [Brockes 1997]. Mammals largely lack such capabilities, except for the tips of the limbs, where intact nail tissue contains stem cells that orchestrate and regenerate the tips of severed fingers and toes [Takeo 2013].

14.9.1.5 Modulated Growth

Stress modulates the growth of biological tissue. The elastic compliance of microtextured elastomeric pillar array substrates modulate the growth and differentiation of human mesenchymal stem cells [Fu 2010]. Different compliances of the micropillars alter differentiation behaviors (Figure 14.23).

14.9.1.6 Active Materials

Living biological tissue is an extraordinary, sophisticated form of active matter. Genetic engineering programs organisms to exude chemicals that act as self-healing agents and scaffolds [Magennis 2014]. Programmed DNA-coated nanoparticles switch states to induce phase transitions in superlattices and colloids [Zhang 2015e]. Byssal thread biochemistry, especially the use of nitrocatecholamines, creates biocompatible underwater self-curing, self-healing, surface-reactive, and photodegradable materials [Shafiq 2012].

Biohybrid actuators provide systemic integration and control beyond that available with ordinary actuators. Incorporating living cells with biomechanical engines into actuators can be fairly simple – harvest and place cells of cellular components into compliant structures. The mechanisms that propel single cell organisms to move also propel micro-objects and micromachines. A system based on a metal-organic framework collects and organizes motions, as in the hydrophobic peptide molecular motor shown in Figure 14.24 [Ikezoe 2012]. Algae with flagella act as micro-oxen to transport small objects, such as polystyrene microspheres, in response to light stimuli [Weibel 2005]. Mobile gliding bacteria can turn rotors in micrometer-sized sweeps (Figure 14.25) [Hiratsuka 2006].

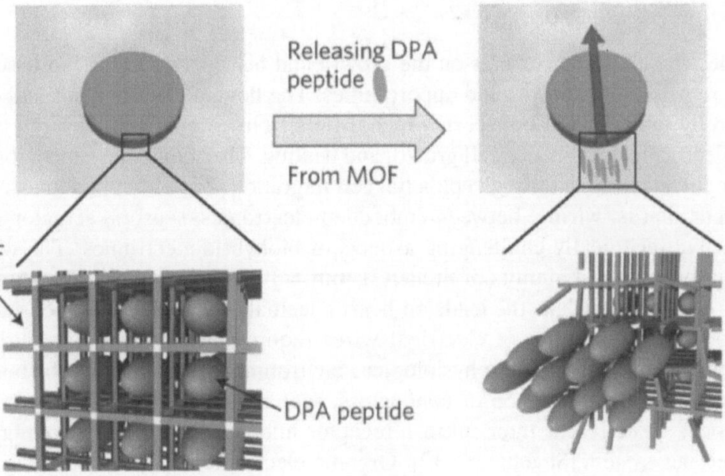

FIGURE 14.24 Organized metal peptide molecular motor biohybrid actuator. (From [Ikezoe 2012].)

Cells that self-organize into synchronized contractile units produce macroscale contractions. An example is cardiomyocytes that beat with collective frequencies on the order of 1.5 Hz. Similar to heart muscles, such actuators need supplemental restoring forces to return to an extended position for another contraction cycle [Chan 2012]. Bacillus spores are shrunken, desiccated, and relatively inert states of bacteria. When hydrated, the outer cortex of the bacillus spores swells with a mechanical energy density of 10 MJ/m^3 – a value two orders of magnitude greater than synthetic alternates. Genetic modifications create strains of bacillus with cortex material that is more active than natural variants [Chen 2014].

Active fluids stiffen, relax, and coagulate under control. Coagulation is particularly useful for flooding-type self-sealing electrical cable insulation where a breach in the outer insulation layer causes the sealing fluid to flow out of the gap and prevent the ingress of noxious fluids, such as water. Typical systems use polyisobutylene (PIB) as a gummy and viscous fluid. Triggered cross-linkable PIB may provide coagulating functionality [Desai 2014].

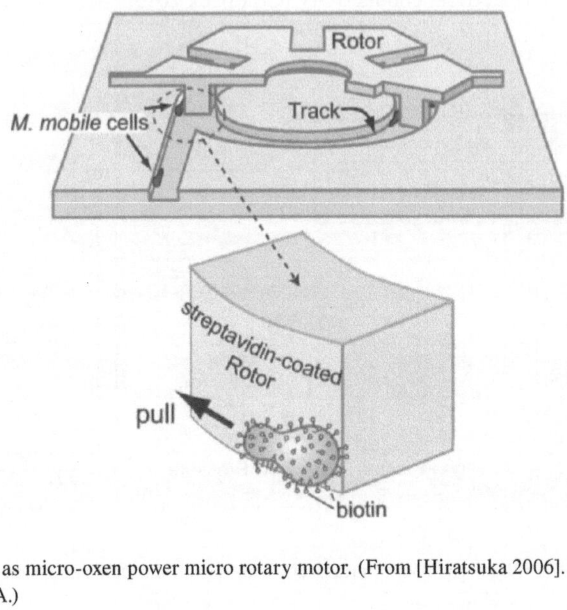

FIGURE 14.25 Bacteria as micro-oxen power micro rotary motor. (From [Hiratsuka 2006]. Copyright [2006], National Academy of Sciences, USA.)

14.9.1.7 Biohybrid Electronics

The role of electric fields and currents on the growth and behavior of living bio tissue is long studied yet has many remaining unknowns and opportunities. The flow of electric fields and currents in living systems generally follows ionic transport, which differs from conventional electron transport in metal conductors. Electric fields influence cell growth and healing. The autonomic generation of electric fields in human skin directs healing through epithelial cell migration [Zhao 2009] [Gumus 2010].

Interconnects, that is, wiring, between embedded electronics, sensors, actuators, and power supplies are vital but technically challenging aspects of biohybrid electronics. The wires are of small diameter (usually measured in units of French, 1 mm = 3 French), flexible, biocompatible, and durable. Some wire failures, such as the leads on heart pacemakers, have serious consequences [Rordorf 2013]. The long-term durability of electrical wires requires both biocompatibility and the ability to survive breakdown in the harsh physiological environment. Possibilities for boosting durability include specialized coatings on top of conductors, such as metal oxide shells on nanowires [Zhou 2014]. Electrodes coated with interleukin-1 receptor antagonists prevent scarring when placed in the central nervous system [Shacham 2012]. Organic electronics, made of conductive polymers that transport both electrons and ions, are alternatives with potential for biocompatibility and miniaturization [Owens 2010]. An interesting possibility is to exploit tunneling nanotubes that provide long-distance cell-to-cell electrical signaling [Wang 2010c]. Attaching graphene quantum dots to the surface of a single bacterial spore, such as *Bacillus subtilis*, creates an interface with electromechanical systems based on changes to the surface of the spore in response to changes in environmental conditions (Figure 14.26) [Sreeprasad 2015].

Multicellular biological tissue and entire organisms gain considerable functionality, including self-healing by heterogeneous structuring of properties at micro- and nanoscales. Can biohybrids provide such functionality? One possibility uses microbes to control the growth of the electrode material with a smart connectivity approach. At the molecular level, most proteins and biofilms are electrically insulating, but certain variants, including biofilms from *Geobacter sulfurreducens* and pilin nanofilaments, are conductive and may grow biohybrid electrical networks [Malvankar 2011]. Virus-derived templates grow gold and silver nanowires [Lee 2010b]. While not yet demonstrated for healing processes, the virus templates grow onto lithium anodes, something that is normally difficult to do with gold and silver. Adding surfactants improves growth performance. Most species of bacteria are electrically insulating, but some, such as *Shewanella oneidensis*, reduce metals through an extracellular transfer of electrons. Transferring this physiological mechanism to other bacteria, such as *E. coli*, produces hybrid bacteria that both thrive and produce electrical conductivity [Goldbeck 2013].

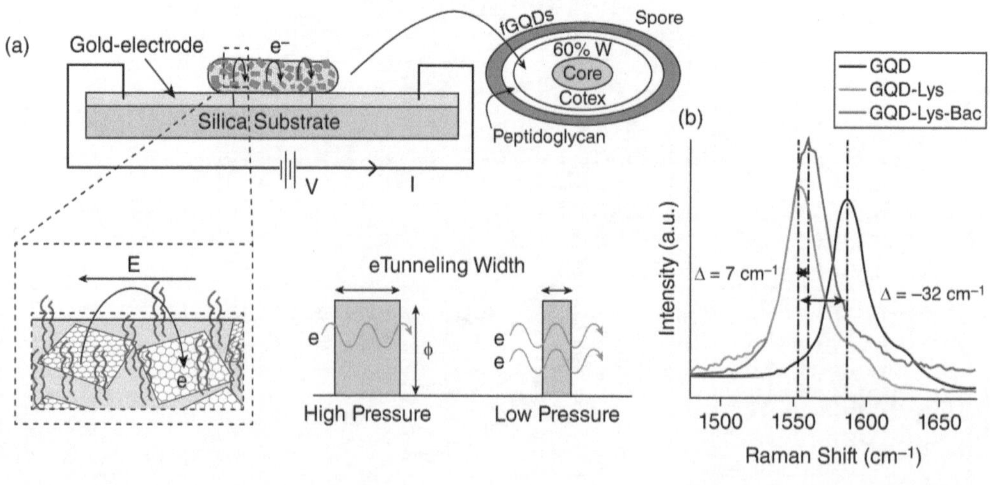

FIGURE 14.26 Functionalized bacteria spore interfaced with electromechanical system through quantum dots. (From [Sreeprasad 2015].)

14.9.1.8 Higher order Biomimicry

Multicellular life forms thrive with complicated and organized multiscale physical structures and interactions. It is natural to expect that engineered self-healing biological hybrid structures benefit from similar multiple-scale physical structures and interactions. Photosynthetic foam places many of the proteins and enzymes required for photosynthetic carbon fixation and sugar synthesis into an aspirating cellular matrix [Wendell 2010]. A key enabler of this is the use of the túngara frog surfactant protein Ranaspumin-2. The surfactant helps to concentrate lipid vesicles and coupled enzyme activity to boost the conversion efficiency of the photosynthetic activity to levels approaching 96%.

14.9.1.9 Complex Vascular Systems

Biological vascular systems are more than simple pipes. They actively manage the transport of materials through dilation, constriction, and diffusion. Higher order animals use a three-circuit system with fractal geometries of positive pumping (arterial), direct return (venous), and indirect cleanup return (lymphatic). Fabricating biohybrid vasculature often starts with scaffolding that approximates the correct geometry to support the seeding and growth of vascular endothelial cells [Leong 2013].

One potential area of future development for self-healing is a synthetic vascular system that includes a lymphatic system. In mammals, the lymphatic system is an open drainage system that removes material from tissues. The material being removed is detritus from healthy biological function, injury, and illness. The lymphatic system is also key to maintaining interstitial fluid levels and pressures. Damaged and defective lymphatic systems lead to disease and/or damage and death to tissue. An engineering application that mimics some of the action of lymphatic systems is conventional hydraulic power delivery and recirculating liquid lubricant systems in machinery. The working fluid, often with embedded detergents, dislodges, carries away, and filters out dirt and deleterious wear particles.

14.9.2 Biohybrid Controllers, Computers, and Quorum Sensing

Many aspects of biological life are computation and control. Organisms combine sensing of external conditions with internal states to make decisions and control actions to live, heal, and thrive. Such computation and control capabilities make superior artificial self-healing systems.

Molecular computers operate at a scale where the presence of specific chemicals causes reactions with measurable outputs. Different input chemicals produce different chemical outputs. One architecture mimics the operational steps and components of conventional computers and controllers, such as binary logic gates – NOT, AND, NAND, and N-IMPLY – in digital computers using synthetic molecular RNA–protein interactions inside living cells [Ausländer 2012], timescale control with genetic load-driver circuits [Mishra 2014].

Bacteria and other microbiota interact through chemical signaling, local feedback, and collective decision-making to form cooperative communities. In a sense, the individual cells vote. When enough cells vote to raise the chemical signals above a threshold value, the community responds collectively – a process known as quorum sensing.

Communities of microbiota are self-healing, highly resilient, and inspire engineering similar systems. A design constraint affecting both natural and artificial systems is the slow pace of direct diffusion of sensory and feedback chemicals which leads to desynchronization of behaviors over distances as small as 8 μm [Connell 2014]. Vascular systems help coordinate signaling, collective decision-making, and control. Experiments demonstrate the viability of controlling engineered biohybrid systems on spatial scales exceeding diffusion spatial limits by supplementing chemical signaling through rapid vascular transport. As an example, an arsenic-sensing system uses an array of up to 50 million small colonies of *E. coli* bacteria that interact by rapid gaseous transport to form a large colony. The frequency of quorum sensing synchronizes oscillations among the colonies. LED lighting indicators correspond with the presence and concentration of arsenic [Prindle 2012]. Communities of cells of different lineages that interact through activator and repressor signaling between the lineages may lead to more complicated and controllable behaviors, such as emergent oscillations [Chen 2015d].

14.10 Cognition, Self-reference, and Consciousness

Biological computers, that is, the brain in higher animals and the distributed coordinated intelligence in plants, are in many respects much more capable than human-built artificial computers. Biological computers are capable of cognition, self-reference, and the poorly understood process of consciousness, all while consuming orders of magnitude less energy than synthetic computers. There are many questions concerning intelligence, self-awareness, and consciousness. The present state of understanding is somewhat limited, but advances continue to be made on multiple fronts, and are largely beyond the scope of this discussion. The intent here is to discuss the possibilities of harnessing these powerful ideas and techniques for self-healing structures, machines, and systems.

Cognition involves thought processes that are more than reactive feedback control, and simple trigger response. Cognition implies additional considerations of sensing converted into meaning, that is, perception, memory, reaction, and anticipation [Haykin 2012]. With memory, when two nominally identical cognitive machines have different experiences, they exhibit different behaviors. At a higher level of cognition lies conscious systems.

What is consciousness? One of the greatest present-day scientific challenges is to explain the subjective experience of consciousness. This is a problem that has tantalized thinkers since the earliest days of philosophical and religious thought. Chalmers divides the problem of consciousness into two classes: the "easy problem" and the "hard problem" [Chalmers 1995]. The easy problem is the determination of physical processes that give rise to sensations. When a photon of light interacts with the retina in the eye, it triggers an electrochemical cascade of events that leads to the stimulus of specific regions in the brain. Tracing the pathway and understanding the electrochemical reactions, including automated signal processing along the way, is not trivial but appears to be scientifically achievable. The hard problem is to explain what happens when the brain perceives light and colors and an indication of the surrounding reality. Chalmers argues that this hard problem is not tractable in terms of present scientific tools.

Much research is presently underway attempting to use the scientific method to answer these questions, mostly through applications of the notion of multilevel hierarchies of function and information abstraction throughout a central nervous system, with self-consciousness appearing in those systems with higher degrees of complexity [Crick 1998] [Feinberg 2009] [Koch 2012]. Dimasio through detailed studies of the brain suggests that part of the explanation lies in that the brain and body are not separate. Key to cognitive perception and perhaps consciousness lies in the abstract projection of images of the body and surroundings as maps in the brain [Damasio 2010]. Sufficiently complex versions of this self-mapping process lead on a path to an understanding of self-consciousness. Nunez suggests that consciousness represents a form of yet undetermined superinformation [Nunez 2010]. Tononi developed a quantifiable set theory extension of Shannon's information theory to hypothesize that consciousness is the process of information integration [Tononi 2004]. Penrose suggests that the problem of consciousness aligns with some of the fundamental unresolved questions of quantum mechanics, such as the collapse of wave functions and the relation to quantum gravity [Penrose 2016].

Zombies are a gedanken-type tool used by philosophers to probe questions of consciousness. Zombies are fictitious entities with no subjective conscious experiences that give outward responses to external stimuli that match and are indistinguishable from the responses of a conscious being. In many respects, studying zombies replaces the religious question of whether someone has a soul. Zombies have experiences, memories, and autobiographies, and become cognitive. If they can have feelings, and subjective experiences, they may gain a soul. The spine and the digestive and immune systems of vertebrates possess many zombie-like characteristics. It remains an open (and perhaps fundamentally unresolvable) question as to whether an inanimate machine can move beyond a zombie-type existence into one with subjective conscious experiences [Kurzweil 2006]. In another gedanken-type exploration of the possible thought processes of bats, Nagel argues that subjective conscious experiences are impenetrable to those outside of the experience [Nagel 1974]. Humans can never truly know what it is like to be a bat because they cannot experience what a bat sees, hears, smells, and feels.

A counterargument is that there are many commonalities of some experiences. Bats feel pain as do apparently even the lowest order animals.

Nonetheless, within the context of these deep and perhaps impenetrable problems, there are practical situations where approaches based on cognitive self-referencing systems work for self-healing systems. The autotelic self-referencing approach to artificial intelligence leads to self-regulating robots [Steels 2004].

Computing systems and networks with high degrees of coordinated complexity and low capital costs are natural testbeds for developing self-aware machines and systems. Key attributes of self-aware systems are as follows [Agarwal 2009]:

1. *Introspective:* Observes itself, reflects on its behavior, and learns.
2. *Goal-oriented:* Tell the computer what you want. The job of the computer is to figure out how to get there.
3. *Adaptive:* The system analyzes the observations, computing the delta between the goal and observed state, and takes actions to optimize its behavior.
4. *Self-healing:* The system continues to function through faults and degrades gracefully.
5. *Approximate:* The system does not expend any more effort than necessary to meet goals.

An achievable goal is for a system to extend its survival even though there is the possibility of degraded performance.

An important subset of these topics is the role of pain. Pain is complicated with vastly different levels of sensitivity observed across individuals [Woolf 2010]. While progress has been made in identifying specific regions of the human brain and the nervous system associated with gating and pain intensity, understanding the subjective conscious and subconscious experience of pain remains a major unresolved challenge [Tölle 1999]. Nonetheless, pain seems to be necessary for the survival of animals. Pain helps with injury avoidance and taking actions that minimize reinjury. People with neurological diseases that suppress pain inadvertently injure themselves and do not survive very long [Cox 2006]. It remains unclear whether the sensation of pain is a useful contributor to healing. Other (primarily biochemical) signaling pathways are possible. One argument against the need for subjective feelings of pain lies with the plants, which may not feel pain in the subjective sense, yet have healing capabilities that often exceed that of animals.

Possible roles of pain related to self-healing and overall health management are as follows:

1. *Rapid Reaction:* Invoke rapid reactions that prevent serious injury
2. *Signaling:* Signal need for repair actions
3. *Behavior Modification:* Modify behavior following injury to avoid more damage
4. *Motivation:* Provide strong motivation to avoid dangerous conditions in the future
5. *Confirmation:* Confirm healing by disappearance of pain

Damage and repair are nonlinear memory-dependent processes. It is not surprising that pain is also nonlinear. A combination of interacting activation and inhibitory actions finely tunes pain sensations. A conceptual model for pain regulation is the Melzack–Wall pain gate (Figure 14.27). Such gates may be active at substantia gelatinosa and first-stage transmission cells in the central nervous system [Melzack 1965]. The gate combines large and small pain fibers with inhibitory and activation effects to regulate the passing of pain signals to the brain. Strong non-pain sensations on large fast neural pathways inhibit pain sensations that travel on smaller slow neural pathways with logic-based gating. In the absence of non-pain sensations to activate gating, the pain signals move onward for perception by the brain.

Attempts to create artificial sensations of pain in the control of autonomous machines include motivated learning [Starzyk 2012]. A combination of rewards and pain trains a machine to interact with an unknown environment. Pain acts as an event-driven bias of the learning process, with multiple pain centers.

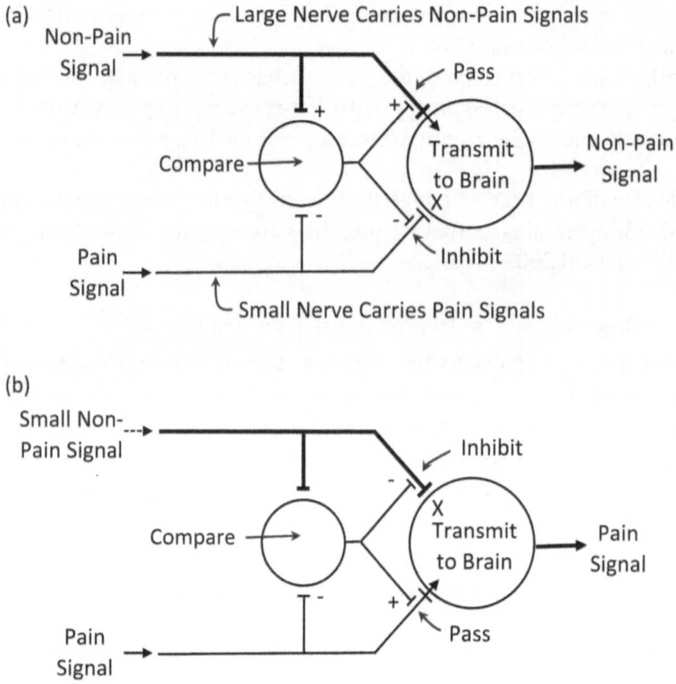

FIGURE 14.27 Melzack–Wall pain gate explains many aspects of pain perception. (a) Pain-inhibited – large non-pain sensations traveling along large fast neural pathways suppress the transmission of pain signals to the brain that travel along small and slow neural pathways. (b) Pain-transmitted – small non-pain sensations traveling along large fast neural pathways allow the gate to open for the transmission of pain signals to the brain. (Adapted from [Melzack 1965].)

The context of specific events affects and inhibits the degree of bias of particular pain paths in a manner inspired by the Melzack–Wall pain gate. Simulations indicate that such a learning system is superior to a rewards-only process.

Related to pain is the distinct sensation of itch [Bautista 2014] [Lee 2013b]. In mammals, acute itch sensations are histamine-dependent responses to noxious stimuli, such as mosquito bites, or can be histamine-independent, as in cases of chronic itch [Jeffry 2011]. From a biomimetic point of view, the advantage of separate itch and pain sensations is not clear but may lie in the use of overlapping fine and coarse measures of deleterious stimuli, along with different degrees of limbic interactions, that is, it is difficult to ignore pain, while most cases of itch are annoying, but not dominant, mental processes.

14.11 Special Opportunities and Applications

14.11.1 Very Large Structures

Earthquakes and related seismic events result from cracking and movement of subsurface geological structures. Pressure and heat-driven fusion of asperities heals the cracks, which may play a significant role in the occurrence rate and magnitude of earthquakes. Seismic observations are a viable method of assessing the healing of cracks along faults in the Earth's surface. Laboratory experiments indicate that the frequency content of seismic elastic waves increases with the amount of healing – an observation that may translate to other large structures [McLaskey 2012].

14.11.2 Self-cleaning Cities, Infrastructure, and Orbits

Poor sanitation is one of the most accurate indicators of urban poverty and health problems. Self-cleaning city infrastructure has the potential for large impact on prosperity and health for all segments of society. Self-cleaning city infrastructure runs from self-cleaning surfaces to sweepers and cleaners to urban-scale coordinated efforts, such as smart robotic trash pickup and disposal. An overall approach combines multimodality sensing, heterogeneous networking, data fusion and analysis, actuation, automated control, and mapping technologies to measure and promote urban sanitation and hygiene. It may be possible to take advantage of the *Rapid Cleaning Positive Feedback Hypothesis* where rapid cleaning alters physical and social dynamics such that it takes less overall effort to stay clean. An example is a storm sewer clog. Debris can catch and accumulate in a storm sewer. The debris restricts and alters the flow so that more debris accumulates and clogs form. A second example is graffiti. Rapid cleaning and/or repainting of graffiti discourages the placement of more graffiti through alteration of social dynamics and reduced satisfaction of graffiti artists/vandals. A potential consequence is for self-cleaning cities to be cleaner at lower cost.

Satellites in orbit about the earth tend to occupy specific orbital ranges for a variety of operational reasons. Inevitable accidents and uncontrolled trash release fills these favored orbits with high-velocity material that can do severe damage to vehicles in orbit. There is a serious need for systematic methods to clean these orbits.

14.11.3 Self-cleaning and Self-healing Feathers for Airplanes

Fixed wings on aircraft are highly effective in flight but are unable to take advantage of the nonlinear aerodynamics routinely used by birds with flapping and morphing wings. The original Wright brothers' airplanes used shape-morphing wings for steering and control. The advent of ailerons supplanted morphing wings and enabled building stronger yet rigid wings. The technology of morphing largely lay dormant for over a century. Recent advances are prompting a resurgence in morphing wing technology. A key technical challenge is how to make the wing surface sustain aerodynamic flows and carry loads while undergoing large shape changes. Solid materials (aluminum, composite, wood, etc.) are too stiff. Fabrics and elastomers may deform but lack stiffness. Articulated and reticulated underlying mechanical structures covered in engineered feathers may provide both stiffness and smooth aerodynamic form. Feathers need to be inherently lightweight and will benefit from self-healing and self-cleaning, perhaps with regenerative techniques.

14.11.4 Industrial Fluids

Many industrial processes and machines use fluids. Liquids with custom formulations heal and replenish the machinery as part of the process. How to make a liquid that heals machinery? One line of development mixes active particles into lubricating and cooling fluids. Passive microspheres mixed into drilling fluid enhances (reduces) the density of drilling fluid [Shinbach 2010]. A fluid mixed with collapsing micro balls provides more aggressive active density changes [Shim 2012].

Healing liquids should flow easily from reservoir to damaged sites for healing. The healing liquid could do more than just solidify into a plug. There is a need to condition and deal with the damaged site, such as by softening and breaking down, congealing and lysis, as needed to promote repair. Blood is a prime example of a high-performance congealing material. Upon receiving specific chemical signals, it forms clots consisting of complicated conglomerations of fibrin fibers, platelets, and other material. Fibrin has extraordinary mechanical properties. Fibrin fibers can undergo strains of up to 180% without permanent deformation and strains of 525% without rupture [Liu 2006a]. These strain levels are the highest known for proteins. It may be possible to engineer similar materials for artificial sealing compounds.

Additional directions for healing liquids formulate complex fluids with active microstructure. The functionality of the microcapsules may extend to the realm of mechanobiological mimicry by thrombotically glomming onto to one another when signaled and have controlled/tuned mechanical properties that alter the flow characteristics through microvessels [Merkel 2011].

14.11.5 Thermal Protection Systems

Thermal protection systems (TPSs) are vital components of multiple high-performance systems, such as spacecrafts that undergo hypersonic atmospheric reentry. Failure of a TPS component may cause catastrophic system-wide failure. Localized damage, such as from impact by debris or the incipient formation of hot spots, may heal through aggressive self-repairs. Possible healing actions include the following:

1. *Thermal Healing:* Repair the damage during a high-thermal loading cycle as with hypersonic atmospheric reentry. Heat drives the self-healing action during these transient events.
2. *Cold Healing:* Self-repair during a cool period before the next thermal load.
3. *Mitigation:* Survival-type mitigation actions during dangerous hypersonic maneuvers.
4. *Intumescent Healing:* Inject foaming intumescents as an uncured foam. Then, cure in place by the heat. Ceramics, such as SiC, with an internal blowing agent, such as polyurethane are a possibility [Stackpoole 2008].

14.11.6 Electrical Conductors

Multiple issues make self-healing of electrical conductors technically challenging opportunities. Most conductors are metallic – a relatively difficult material for autogenous repair because of the high melting point of metallic conductors. Additional challenges are the possibility of miswiring in cable bundles forming shorts or crossing signal connections. Conversely, the availability of electric current and power at the point of failure is an opportunity to drive self-healing.

The failure of electrical conductors is a major source of failure in modern engineered systems. Most electrical conductors are metallic. Failure of metal conductors results from mechanical stressors, such as fatigue due to excessive flexing, or trauma; electromechanical stressors, such as corrosion; electrical stressors, such as high-current surges and electromigration; and the failure of electrical solder contacts and conduction lines in printed circuit boards.

14.11.6.1 Self-assembling Percolating Conductors

Embedding metallic particles into an elastomer makes for a compliant solid conductor. If the mix loading of conductors is large enough, the individual particles contact one another and form conductive paths through the solid by percolation. Large deformations of the elastomer break the conductive paths, which reform and heal upon relaxation of the deformation. Certain geometric arrangements and distributions of the particles favor the formation and recovery of conductive paths, such as self-assembling networks of conductive nanoparticles in a polyurethane elastomer [Kim 2013b].

Applying an electric field to a conductive gel causes ionic components to migrate in a process called electrophoresis. The migration alters and controls the conduction paths set up by nanoparticle percolation [Nakanishi 2011].

CNTs are individually good conductors of electricity. High-junction resistivities hamper the use of percolation to form bulk conductive materials. It is possible to reduce the bulk resistance by an active growth process that exploits the Joule heating at junctions conducting electric currents to solder and deposit vaporized metals [Do 2013]. Magnetic particles suspended onto a surface formed by a liquid–air interface create a dynamic system with oscillating magnetic fields forming hydrodynamic vortices that cause the particles to move and conglomerate into snake-like patterns [Belkin 2007].

14.11.6.2 Conductive Liquids

It may be possible to repair metallic conductors using room temperature liquid conductors, such as mercury, eutectic gallium indium, and electrolytic conductors. An elastomeric tube filled with conductive liquid is a damage-mitigating conductor [Zhu 2013]. The tube stretches multiple times the original length while maintaining conductivity. The resistance changes with stretching and allows for using conductive liquid tubes as large deformation strain gages for applications, such as plethysmography (the measurement of volume changes in body parts) [Youdin 1976].

Using a liquid as a regenerator of conductive paths requires getting the liquid to stay in place at the damaged site and not migrating to undesired locations. It may be possible to heal with a two-step process: first, form the vessel wall and then flow the liquid conductor. Possible uses are liquid metal conductors as regenerative components in high current and heat applications, such as metal contacts in large DC motors, or large electrical switches. Gallium-based metals liquefy at room temperatures and then form oxide layers with high surface tension at metal–air interfaces. The application of small voltages on the order of 0.6 V removes the oxide layer, dramatically lowers the surface tension, and enables controlled movements of liquid metal droplets [Khan 2014] [Dickey 2014]. Microencapsulation stores and then releases electrically conductive liquids [Haishima 2001].

14.11.6.3 Whisker and Dendritic Growth

The growth of conductive dendritic structures is a potential means of restoring conductivity in damaged electrical wiring (Figure 14.28). One approach grows out with probing dendritic structures until a conductive path is achieved, and then rapidly grows in a reinforcing manner, possibly enhanced with chemical or electrochemical guidance.

How to grow and restore electrical conductivity:

1. *Dendrites:* Under the appropriate circumstances metal whiskers and dendrites grow out of a metal conductor, span across a gap, and form electrical contacts with other conductors. Perhaps the best documented case is the growth of tin whiskers out of lead-free solders. Tin whisker growth is normally undesirable. It causes electrical shorts and other problems in electronic packages, often at great expense [Karpov 2014] [Mathew 2011] [Fukuda 2006] [Han 2010]. Accentuating rather than suppressing tin whisker growth reconnects broken electrical wires and contacts. One method may be to use a protective coating, such as a tin–lead solder on top of tin wires. The tin–lead solder inhibits whisker growth. Lead-free solder promotes whisker growth. Cutting a coated wire exposes the tin and grows whiskers, perhaps outward, to contact another exposed tin surface. Adding rare earth elements and electric fields accelerates tin whisker growth [Dudek 2009] [Vasko 2015].
2. *Conductive Nanoparticles:* Form chains of conductive nanoparticles. Gold and silver nanoparticle chains, including those with an encapsulating filament, form nano peapod structures [Hu 2006].
3. *Electrolytic Nanowire Growth:* The flow of electrons through conductive media provides activity for growing or repairing conductive paths. One possibility is to deposit copper conductive connections out of a copper sulfate liquid bath. Another is a combination of light and electric fields that causes triarylamine to self-assemble spontaneously into conductive supramolecular nanowires between narrow submicron gaps [Faramarzi 2012].

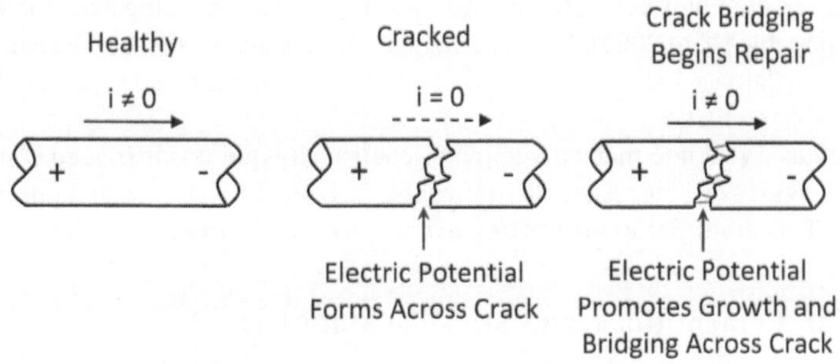

FIGURE 14.28 Crack bridging of conductors driven by electric potential.

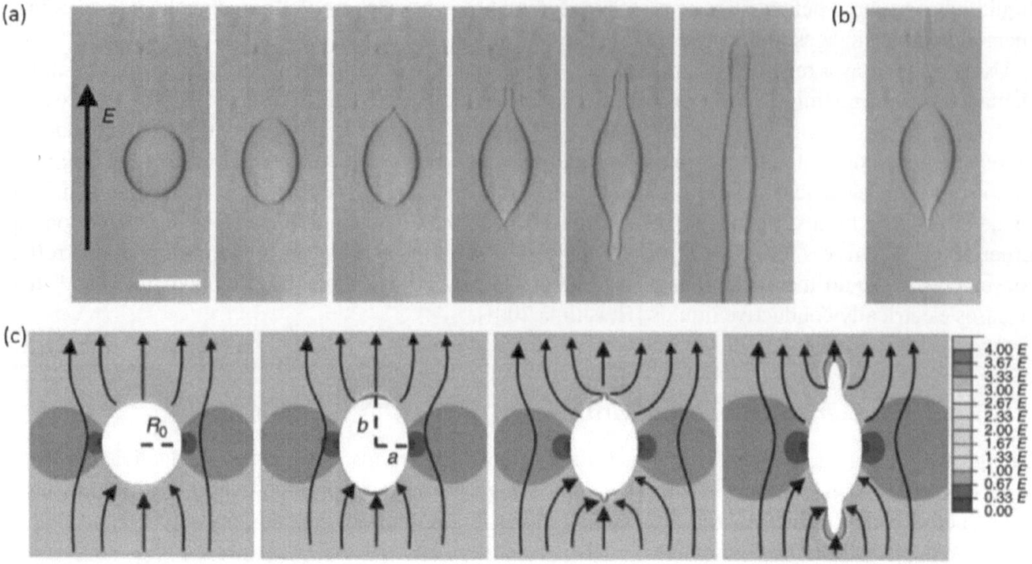

FIGURE 14.29 Stretching of a conductive liquid drop in a compliant dielectric by the application of a large electric voltage. (From [Wang 2012d].)

4. *Bacterial Growth:* Some bacteria that produce extracellular appendages to form electrically conductive nanowires. These nanowires may be healing agents for electrical contacts [El-Naggar 2010].
5. *Hierarchical Arrays:* Dendritic electrically conductive molecular molecules make effective electrical contacts between dissimilar materials, such as closely packed viologens that anchor onto gold surfaces [Kawauchi 2015].
6. *Electric Fields:* These stretch and control electrically conductive droplets to make an electrical connection (Figure 14.29) [Wang 2012d].

14.11.7 Electrochemical Chloride Extraction

Chlorides from deicing salts, ocean spray, and contaminated mixed ingredients migrate by diffusion through concrete, promote corrosion of the reinforcing steel, which swells to crack the concrete, and promote more chloride penetration and damage. Electromigration drives corrosion inhibitors into chloride-laden concrete [Pan 2008]. Electrochemical chloride extraction (ECE) is an active electromigration method that drives deleterious ions out of materials. The method places a steel or titanium mesh anode on the surface of the concrete and uses subsurface reinforcing steel as the cathode. Forced electrical currents cause the chloride ions to migrate to the surface and out of the concrete. Reduction of chloride load by over 60% and reversal of half-cell potential over the critical –200 mV threshold for 95% reduction of corrosion risk are possible [Kim 2007]. ECE has value in those situations where the chloride penetration is severe, but cracking damage has yet to occur. If cracking, delamination, and spalling damage are present, these need to be fixed first.

While often used as an ad hoc manually applied repair technique for distressed concrete, ECE may be very effective in aggressive corrosion environments, such as salt water, when built into the structure during fabrication. This is similar to the use of galvanic protection systems.

14.11.8 Granular, Cementitious, and Bricklike Materials

Remarkable physical materials at relatively low cost arise from the composite assembly of strong building blocks, such as bricks and stones, held together with weaker, more flexible, and geometrically accommodating materials, such as mortar or cement, and often strengthened with tension-based reinforcing material. These materials are low cost and easy to assemble into wide varieties of shapes while exhibiting exhibit high strength and mechanical properties. The matrix acts as a binder to hold the aggregate particles in place. Binders tend to be weak but are initially flowable to allow enveloping the aggregates before solidification. Biomimetics holds promise for these materials. Bone and abalone nacre are largely calcium carbonate but are remarkably tough [Du 2023].

Concrete provides mechanical strength in compression, largely through force transmission via a network of interacting stones (aggregate) and sand particles held together by cement. Ordinary Portland Cement (OPC) is a common cement in concrete. Polymer-based or polymer-inclusive concrete systems are increasingly being used in high-performance applications, where traditional ceramic-type cements are inadequate [Dinitz 2008]. The addition of polymers opens opportunities for creating more aggressive self-healing concretes.

Alkali-silica reaction (ASR) is a serious damage mode for concrete that may take years and decades to develop. The high pH of OPC interacts with some silica aggregates to form a gel that expands and causes the concrete to crack. At present, there are few remediation methods other than the use of aggregates that are not susceptible to ASR activity. An exception is the use of fly ash from coal plants in the concrete. Although the underlying mechanism is not entirely understood, large concrete structures built with aggregates known to be susceptible to ASR have largely avoided damage by the fortuitous inclusion of fly ash in the concrete mixes [Thomas 2012]. The fly ash intervenes to mitigate damage. Accentuating this effect could enable better preventative treatments and the use of more ASR-susceptible aggregates without jeopardy.

Appendix A: Acronyms and Abbreviations

1-D	One-dimensional
2-D	Two-dimensional
3-D	Three-dimensional
ACLM	Automated Contingency and Life Management
AFM	Atomic Force Microscope
AI	Artificial Intelligence
AIC	Adaptive Intelligent Controller
ASR	Alkali Silica Reaction
b.	*Bacillus*
BCC	Body-centered Cubic
BE SONOS	Bandgap-engineered Silicon Oxide–Nitride Oxide–Silicon
BTI	Bias Temperature Instability
CLASH	Cross-layered Accelerated Self-healing
CNT	Carbon Nanotubes
COD	Crack-opening Displacement
CP	Cathodic Protection
DA	Diels–Alder
DBTL	Dibutyltin Dilaurate
DC	Direct Current
DCB	Double-cantilever Beam
DCDC	Double-cleavage Drilled Compression
DCPD	Dicyclopentadiene
DEAP	Dielectric Electrically Active Polymer
DNA	Deoxy-ribonucleic Acid
DSC	Differential Scanning Calorimetry
e.	*Escherichia*
ECC	Engineered Cementitious Composite
ECE	Electrochemical Chloride Extraction
ECMO	Extracorporeal Membrane Oxygenator
EIG	Eutectic Indium Gallium
FCC	Face-centered Cubic
FLAIR	Fine Lightweight Aggregates as Internal Reservoirs
FMEA	Failure Mode Effects Analysis
FMECA	Failure Mode, Effects, and Criticality Analysis
FRP	Fiber-reinforced Polymer
GFCI	Ground-fault Current Interrupter
HA	Hierarchical Actuator
HCP	Hydrated Cement Paste
IDT	Indirect Tension
IFCS	Intelligent Flight Control System
ksi	Kilopounds per square inch
LSF	Living Snow Fences
MAC	Microarc Oxidation Coating
MEMS	Micro-electromechanical System
ML	Machine Learning
MOF	Metal–Organic Framework
MOV	Metal Oxide Varistor

OPC	Ordinary Portland Cement
OPH	Organo-phosphorous Hydrolase
PAN	Polyacrylonitrile
PCM	Phase Change Memory
PDES	Poly(diethylsiloxane)
PDMS	Polydimethylsiloxane
PFPE	Perfluoropolyether
PIB	Polyisobutylene
PLA	Poly(lactic acid)
PMMA	Poly(methyl methacrylate)
PUF	Poly(urea-formaldehyde)
rms	Root Mean Square
ROMP	Ring-opening Metathesis Polymerization
SLIPS	Slippery Liquid-infused Porous Surface
SMA	Shape Memory Alloy
SMAT	Surface Mechanical Attrition
SMM	Shape Memory Material
SMP	Shape Memory Polymer
STAR	Self-testing and Repairing
TBC	Thermal Barrier Coating
TPS	Thermal Protection System
UPV	Ultrasonic Pulse Velocity
UPy	2-Ureido-4[1*H*]-pyrimidone
UV	Ultraviolet
WWI	World War One
WWII	World War Two

Appendix B: Self-healing Patents

TABLE B.1
Selected Self-sealing Fuel Tank Patents

Inventor	Year	Patent Number	Title	Invention Technique
Ohnstad, Monk	2011	US 8,043,676	Sealing-reaction, Layer-effective, Stealth Liner for Synthetic Fuel Container	Synthetic fuel sealing, and constrained layer anisotropic swelling
Bennett, Ohnstad, Monk	2007	US 7,220,455	Material-selectable, Self-healing, Anti-leak Method for Coating Liquid Container	Method of making multilayer swelling elastomeric sealing
Grosvenor	1983	US 4,422,561	Fuel Tank Component	Flaps that plug holes
San Miguel	1978	US 4,088,240	Fuel Tank Leakage Fiber Flow Sealant	Wall-attached fibers flow into hole form scaffolding for plug
Cook	1974	US 3,801,425	Self Sealing Container	Liquid sealing layer solidifies by fuel solvent removal of liquefiers
Baker, Fogarty, Slagel	1972	US 3,698,587	Self Sealing Composite	Elastomeric foam layer provides resilience and shape rebound
Baker, Cook, Slagel	1971	US 3,577,306	Polymer Laminate to Prevent Severe Metal Petalling Damage	Fiber-reinforced elastomeric layer prevents petalling
Stedfeld	1968	US 3,379,336	Self-sealing Tank	Integrated wall flaps that plug pressure vessels
Boger	1951	US 2,715,085	Fuel Container	Nitrile layer separates gasoline from sealing layers
Gerke	1951	US 2,626,882	Self-healing Fuel Cell	Butyl rubber for low temperatures
Frolich	1950	US 2,497,123	Fuel Cell	Polymerized conjugated diolefin for low temperatures
Merrill	1948	US 2,440,965	Improved Tank for Hydrocarbon Fuels	Nylon barrier separates aromatic fuels from sealing layers
Dasher	1948	US 2,438,965	Self-sealing Fuel Tank	Foam layer mitigates hydraulic ram
Hershberger	1947	US 2,430,931	Laminated Fuel Tank	Polyvinyl acetal barrier separates aromatic fuels from sealing layers
Gray, De Weese	1947	US 2,421,613	Plastic Liner for Containers	Impermeable nylon layer separates fuel from active swelling sealing layers
Sanz	1942	US 2,411,116	Self-healing Container Forming Material	Multilayer rubber
Dasher, Crawford, Colley	1940	US 2,425,515	Self-sealing Fuel Tank	Layer of impervious rubber with solvent-swelling foam
Foges	1936	US 2,039,401	Automated Sealing Device	Adhesive layer sticks to projectile and drags into hole-forming plug
van Orman	1928	US 1,662,018	Leakproof Tank Cover	Gasoline-induced swelling of rubber layer
Friant	1922	US 1,436,985	Self Sealing Reservoir	Feathered array of flaps
Murdock	1920	US 1,349,290	War Aeroplane Fuel Tank	Rubber balls in compressed layer comingle with fuel and plug holes after bullet penetration

TABLE B.2
Selected Patents for the Manufacture of Self-sealing Fuel Tanks

Inventor	Year	Patent Number	Title	Invention
Hietz, Hill	1982	US 4,352,851	Void Filler Foam Fire Suppression System	Foam-filled cells mitigate damage
Villemain	1982	US 4,345,698	Aircraft Self-sealing Fuel Tank	Assembly using prefabricated self-sealing panels
Hollis Sr.	1976	US 3,969,563	Protective Wall Structure	Metal plate composite with rectilinear wire mesh for resilient lightweight wall fabrication for self-sealing tank
Bailey	1951	US 2,558,807	Method of Making Flexible Fuel Tanks	Spray on layering of material simplifies fabrication
Crawford	1948	US 2,466,811	Self-sealing Fuel Tank Construction	Multilayer fiber-reinforced polymer wall construction for self-sealing tank
Cunningham	1948	US 2,439,562	Fuel Tank	Stiff projectile resilient wall panel inserts that allow folding of tank for airplane insertion
Blanchard	1947	US 2,426,384	Manufacture of Flexible Containers	Breakable frame of paper and plaster of Paris as removable mandrel for layup of self-sealing tank
Crawford	1945	US 2,374,332	Self-sealing Fuel Tank	T-shaped mounting brackets for baffles inside self-sealing tank
Rethorst	1946	US 2,406,903	Fuel Container and Method of Making Same	Multilayer fabrication with chemical mastication of intermediate unvulcanized rubber layer
Eger	1946	US 2,401,626	Method for Forming Tanks	Collapsible and removable frame as a mandrel for multilayer self-sealing tank construction
Eger	1946	US 2,401,625	Method of Making Puncture Sealing Material	Method of precompressing elastomeric foam for puncture sealing
Imber	1921	US 1,392,892	Tank or Receptacle for Combustible or Other Liquids	Interior frame holding baffles supports construction of self-sealing tank on periphery

TABLE B.3
Selected Patents for the Manufacture of Self-sealing Pneumatic Tires

Inventor	Year	Patent Number	Title	Invention Technique
Semegen	1954	US 2,827,098	Article of Manufacture and Method for Making the Same	Layup process for acrylate sealing layer
Placentino	1953	US 2,632,492	Pneumatic Vehicle Tire Inner Tube	Layup process for foam-sealing layer in inner tubes using standard equipment
Iknayan	1943	US 2,332,913	Method of Manufacturing Puncture Sealing Inner Tube	Layup of compartmentalized self-healing inner tube
Carnahan	1942	US 2,283,183	Method of Producing Self Sealing Tubes	Layup procedure for self-sealing inner tube shield
Eberhard	1937	US 2,099,514	Puncture Sealing Inner Tube for Pneumatic Tires	Method of extruding and joining vulcanized and unvulcanized layers in an inner tube
Waber	1936	US 2,033,962	Punctureproof Tube and Method of Making Same	Vulcanizing accelerant to speed multilayer self-sealing tire tube manufacture
Ott	1931	US 1,818,349	Method of Manufacturing Inner Tubes for Pneumatic Tires	Procedure for creating unvulcanized healing layer in vulcanized inner tube
Armstrong	1919	US 1,311,738	Method of Making Puncture Proof Tire Tubes	Rolling layup of tube with self-sealing layer
Oberfelder	1917	US 1,244,236	Puncture Proof Tube and Process of Producing Same	Layup procedure for foamed tube with precompressed layer

TABLE B.4

Selected Patents and Published Applications on Self-healing Wire and Cable

Inventor	Year	Patent Number	Title	Invention
Alexander et al.	2013	US 8,409,479	Ceramifying Composition for Fire Protection	Environmental activation – heat causes silicone insulation with phosphate-enhanced strength to ceramify and survive fires
Jolley et al.	2012	US 20120321828	Self-healing Polymer Materials for Wire Insulation, Polyimides, Flat Surfaces, and Inflatable Structures	Flowable insulation – heat and/or microencapsulated solvent causes insulation to soften and flow into damage
Stanley et al.	2011	US 8,393,802	Cleanable and Self-cleaning Fiber Optic Connector	Self-cleaning fiber-optic connector – mechanism wipes contact paths clean with connector mating
Noddings KC	2011	US 8,050,527	Self-healing Optical Fiber Cable Assembly and Method of Making the Same	Flowable layer – heat-activated polymer seals cracks and defects in fiber-optic cables
Easter	2010	US 7,666,503	Self-healing Cables	Swelling insulation – water-swellable polymer gel plugs water leaks
Kauffman	2008	US 2008/ 0210453 Al	Water-soluble Polymer Coating for Use on Electrical Wiring	Environmental activation – water-soluble polymer can flow after damage to heal insulation
Huston and Esser	2007	US 7,302,145	Self-healing Cable Apparatus and Method	Swelling insulation – swelling material repairs insulation damage
Parrish	2007	US 7,285,306	Process for Self-repair of Insulation Material	Microencapsulated healing insulation – healing liquid for polyimide layers
Belli et al.	2006	US 7,087,842	Electric Cable Resistant to Water Penetration	Swelling insulation – water-swellable powder in expanded polymer blocks water leaks
Belli et al.	2003	US 6,664,476	Electrical Cable with Self-repairing Protection	Flooding insulation – reflowable dielectric polymer layer repairs insulation
Watkins Jr. and Morris Jr.	1999	US 5,862,030	Electrical Safety Device with Conductive Polymer Sensor	Mitigation – cable overheat and mechanical damage mitigation with distributed sensing system
Freeman CS, and Davis LE	1995	US 5,461,195	Filled Telecommunications Cable Having Temperature Stable Mutual Capacitance	Gel layer heals water-induced shorts in telecommunications cable
Priaroggia	1986	US 4,602,121	Oil-filled Electric Cable with Alternate Layers of Plastic and Paper Tape Insulation	Oil-filled electric power cable with paper insulation
Edwards	1974	US 3,844,860	Method of Making an Electric Power Cable	Flooding insulation – pre-swelling of paper tape wrap layer to aid making oil-filled power cable
Horton	1974	US 3,784,959	Self-healing Electrical Connector Means	Self-sealing connector – electrical contacts penetrate conformable dielectric
Zinser Jr. and Connelly	1973	US 3,775,548	Filled Telephone Cable	Flooding insulation – petroleum jelly and PIB flooding compound
Whitehead	1968	US 3,391,243	Enclosed Electric Power Transmission Conductor	Flooding insulation – gas-based vascular insulation
Kemp AR	1939	US 2,181,188	Insulated Cable	Flooding insulation – undersea cable using PIB for pressure equalization

TABLE B.5

Selected Self-sealing Tire Patents

Inventor	Year	Patent Number	Title	Invention Technique
Yamagiwa et al.	2008	US 7,316,253	Sealant for Prevention of Blowout, Tubeless Tire, and Tire Tube	Non-Newtonian sealant adapts to spinning tire dynamics
True and Glaser	2000	US 6,013,697	Tire Sealant Composition	Internal mixture of water, mica flakes, clay, and propylene glycol
Powell et al.	1980	US 4,186,042	Puncture Sealing Tire	Thin layer under tread flows into hole
Sweet and Kyriakides	1962	US 3,048,509	Puncture Sealing Means for Pneumatic Tires	Butadiene-styrene sealing layers enable retreading
Schutz	1959	US 2,905,229	Self-sealing Pneumatic Tire	Butadiene styrene sealing goo for wide temperature range
Iknayan	1956	US 2,756,801	Puncture-sealing Pneumatic Article	Modified butyl rubber-sealing goo for wide temperature range
Knill	1956	US 2,752,979	Puncture Sealing Inner Tube	Partially vulcanized sealing goo with fewer pores
Hardman	1953	US 2,657,729	Puncture Proof Tube and Sealing Material	Depolymerized rubber for high- temperature durable sealing goo
Tilton Jr.	1947	US 2,566,384	Self-sealing Article	Butyl rubber lightweight sealing layer
Loomis	1936	US 2,055,797	Inner Tube for Tires	Sealing layer that resists flow under centrifugal force
Foges	1936	US 2,039,401	Automated Sealing Device	Adhesive layer sticks to retracting puncture, dragging into hole, and forming plug
Knowlton	1934	US 1,977,281	Inner Tube and Method of Manufacture	Cellular sealing compartments
Richardson	1933	US 1,930,182	Valve Stemless Inner Tube, Self Sealing Section and the Like	Replaces valve stem with self-sealing tube and hypodermic needle for inflation
Landis	1931	US 1,156,155	Pneumatic Tire Tube	Mucalagenous layer for sealing puncture holes
Waber	1931	US 1,808,091	Pneumatic Tire Tube	Conformal self-sealing layer inner tube
Privett	1921	US 1,453,949	Inner Tube	Gummy inner tube with stiffener presses against tire
Jeffries	1918	US 1,285,719	Air Tube for Pneumatic Tires	Feathered flaps for sealing
Armstrong	1918	US 1,257,780	Plastic Composition	Rubber-oil plastic goo for sealing
Clare	1915	US 1,137,461	Self Healing Composition for Inner Tubes of Tires	Nonvulcanizable petroleum-based rubber sealant layer
Cochrane	1900	US 642,838	Pneumatic Tire	Precompressed sponge rubber layer

Appendix C: Supplemental Engineering Analysis and Science

C.1 Thermodynamics and Entropy

Thermodynamics and entropy are universal organizing principles for processes that combine stability, transformation, and information.

C.1.1 Classical Thermodynamics

Classical thermodynamics describes macroscopic transformations of isolated systems in terms of measurable complementary extensive and intensive physical properties [Callen 1960] [Fermi 1956]. Extensive quantities add and depend on the amount of material, such as volume V and mass M. Extensive properties of composite systems are the sum of the properties of the component subsystems. Intensive physical quantities are independent of the amount of material. Examples are temperature θ and pressure p. A fundamental relation among extensive quantities is as follows:

$$U = U(S, V, N_1, \ldots, N_r) \tag{C.1}$$

Here, U, S, V, and N_i are energy, entropy, volume, and molar quantity, respectively, of chemical species i. Small changes in the energy dU are as follows:

$$dU = \frac{\partial U}{\partial S} dS + \frac{\partial U}{\partial V} dV + \sum_{i=1}^{r} \frac{\partial U}{\partial N_i} dN_i \tag{C.2}$$

The partial derivatives of the energy U with respect to the extensive variables are intensive quantities, with common names.

$$\left(\frac{\partial U}{\partial S}\right)_{V,N_i} = \theta = \text{Temperature} \tag{C.3}$$

$$-\left(\frac{\partial U}{\partial V}\right)_{S,N_i} = p = \text{Pressure} \tag{C.4}$$

$$\left(\frac{\partial U}{\partial N_i}\right)_{S,V,N_{j\neq i}} = \mu_i = \text{Electrochemical potential} \tag{C.5}$$

If a system with potential energy $U(x_i)$ is in a stationary configuration at $x_i = x_{i0}$, then the system is at a point of equilibrium:

$$\left.\frac{\partial U}{\partial x_i}\right|_{x=x_0} = 0 \tag{C.6}$$

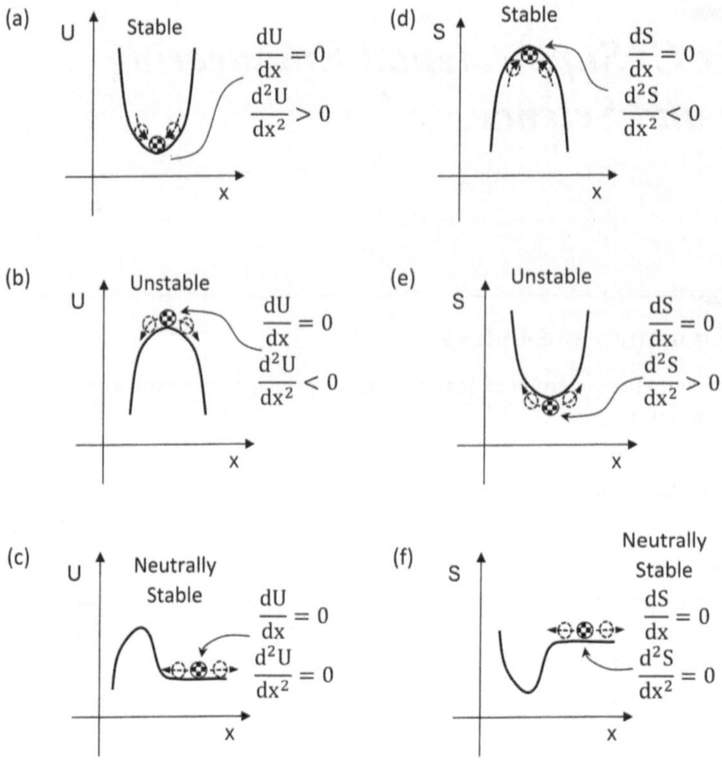

FIGURE C.1 Equilibrium point stability viewed in terms of energy with a ball rolling on a hill. (a) Stable. (b) Unstable. (c) Neutrally stable – equivalently as entropy in a thermodynamic system. (d) Stable. (e) Unstable. (f) Neutrally stable.

A ball rolling in a variable height gravitational landscape provides a conceptualization (Figure C.1). A ball is in a position of equilibrium at the bottom of a valley, on top of a hill, and on a flat. The valley position is stable, while the hilltop and flat points are respectively unstable and neutrally stable. A local quantification of stability depends on the curvature of the potential energy hill, that is, the sign of d^2U / dx^2 in a neighborhood of an equilibrium point (Figure C.1a–c). An alternative, but largely equivalent, point of view is that a thermodynamic system is stable when the entropy is at a local maximum (Figure C.1d–f).

Legendre transforms convert U (or S) with combinations of extensive and intensive variables to create useful equivalent representations:

$$h = U + PV \tag{C.7}$$

Small changes in h are as follows:

$$dh = dU - PdV - Vdp \tag{C.8}$$

At equilibrium, the entropy and molar concentrations are stationary in Eq. (C.2), that is, $dS = dN_i = 0$, and

$$dh = -Vdp \tag{C.9}$$

Enthalpy facilitates representations of constant pressure processes. Similar Legendre transforms facilitate examinations of constant temperature and volume, and constant temperature and pressure process with the Helmholtz free energy F and Gibbs free energy G:

$$F = U - \theta S \tag{C.10}$$

$$G = h - \theta S \tag{C.11}$$

At equilibrium,

$$dF = -Sd\theta - pdV \tag{C.12}$$

$$dG = VdP - Sd\theta \tag{C.13}$$

The application of energy, substance, heat, and work transforms a system from one state to another. Equations of state relate independent physical quantities during these transformations. An equation of state for a mechanically and chemically homogeneous fluid system is

$$f(p, V, \theta) = 0 \tag{C.14}$$

As a system moves from one state A to another state B, the work done W and heat transferred Q are as follows:

$$W_{AB} = \int_A^B dW = -\int_A^B p\,dV \tag{C.15}$$

$$Q_{AB} = \int_A^B dQ = \int_A^B \theta\,dS \tag{C.16}$$

Callen (1960) states that the fundamental problem in classical thermodynamics is to determine the movement of an isolated composite system following the removal of an internal constraint.

Being isolated, the total system does not exchange energy, substance, heat, and work through the system boundary, but it may exchange such quantities between subsystems (Figure C.2). The transformation of an isolated system following the release of an internal constraint maximizes entropy. A reservoir is a subsystem that is much larger than other subsystems. The large size of the reservoir causes specified thermodynamic properties to not change appreciably with the release of an internal constraint. The effect is to convert an isolated composite system into one where the small subsystem effectively acts as a nonisolated system with idealized external interactions with the large reservoir acting as the external environment.

The *first law of thermodynamics* for a quasi-static process that moves a system from state A to state B is

$$\Delta U_{AB} = W_{AB} + Q_{AB} \tag{C.17}$$

A cycle is a process that starts at A and then brings the system back to A, that is, A = B, $\Delta U_{AB} = \Delta U_{Cycle} = \Delta U_{AA}$. This leads to restating the first law of thermodynamics as the *conservation of energy*.

$$-W_{Cycle} = Q_{Cycle} \tag{C.18}$$

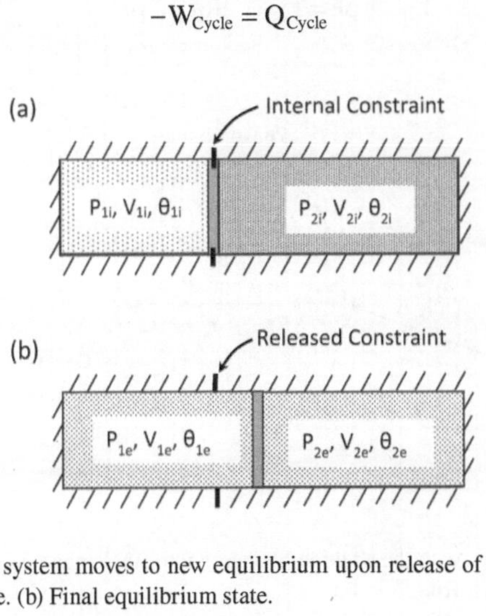

FIGURE C.2 Composite isolated system moves to new equilibrium upon release of internal thermodynamic constraint between subsystems. (a) Initial state. (b) Final equilibrium state.

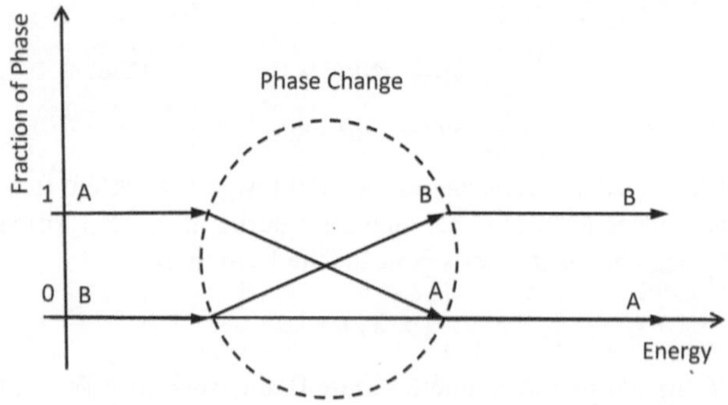

FIGURE C.3 Conversion of phase A into phase B by the addition of energy.

There are multiple equivalent statements of the second law of thermodynamics.

1. Heat flows from high temperatures to low temperatures.
2. No perpetual motion machines exist.
3. It is impossible to create a heat engine without a temperature difference.
4. Entropy increases in all processes.

An unstable energy versus state landscape causes homogeneous material systems to separate into two or more phases. Two phases coexist when each phase has the same average kinetic energy for each mechanical degree of freedom at the molecular scale. This appears at the macroscopic scale as both phases being at the same temperature. The addition, or removal, of energy from the combined two-phase system adds or removes potential energy, yet leaves the average kinetic energy, that is, temperature, of the two different phases equal. Conservation of energy converts one phase into the other (Figure C.3). Upon complete conversion of one phase into another, the material restabilizes in a single phase. Adding or removing more energy changes the temperature of the material (Figure C.4). More complicated phase transitions occur when multiple phases are present, such as water with three phases (gas, liquid, and solid) and mixtures of different chemical species, and metal alloys.

The concepts of classical thermodynamics extend to describing the behavior of continuous materials with an assumption that energy and entropy are continuous spatial and time varying quantities [Eringen 1980]. The first law of thermodynamics for a continuum (using indicial notation) is as follows:

$$\rho\dot{\varepsilon} = t_{kl}v_{l,k} + q_{k,k} + \rho h \tag{C.19}$$

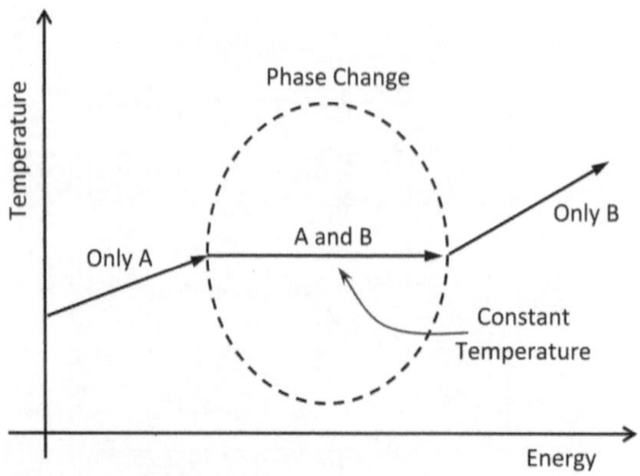

FIGURE C.4 Temperature changes with the addition of energy into single phase system (A or B), but is constant in two phase system (A and B) as it converts from A to B.

Here ρ is the mass density, $\dot{\varepsilon}$ is the rate of change of energy at a point, $v_{1,k}$ is the spatial gradient of velocity, t_{kl} is the stress tensor, q_k is the heat flux vector, and h is the internal heat generation. The second law of thermodynamics for a continuum is as follows:

$$\rho\gamma = \rho\left(\dot{\eta} - \frac{\dot{\varepsilon}}{\theta}\right) + \frac{1}{\theta} t_{kl} v_{1.k} + \frac{1}{\theta} q_k (\log\theta) v_{1k} \geq 0 \tag{C.20}$$

Here Υ is the total rate of entropy production and $\dot{\eta}$ is the local rate of entropy production.

Unstable homogeneous states may segregate into spatially heterogeneous forms with localized regions of different states and thermodynamic properties. Turing examined spatial localization of chemical species with competing diffusion and reaction effects [Turing 1952]. The goal was to explain the formation of patterns in biology, for example, spots on animals and the morphogenesis of embryos. The reaction–diffusion equations governing the spatial and temporal distribution of species X_i are as follows:

$$\frac{\partial X_i}{\partial t} = \nu(X_1, X_2, \ldots) + D_i \nabla^2 X_i \tag{C.21}$$

Here $\nu(X_i)$ is a spatially varying parameter, often a polynomial that introduces nonlinearity, and D_i is a diffusion coefficient. Perturbation methods linearize the equations for stability analysis and indicate the presence of bifurcation-type instabilities, known as *Turing bifurcations.* A characteristic feature of these bifurcations is the formation of spatially periodic patterns of chemical species concentrations (Figure C.5).

C.1.2 Statistical and Quantum Thermodynamics

The notions of equilibrium thermodynamics govern many physical processes. Largely beginning with the development of the kinetic theory of gases, it was recognized that matter consists of atoms and molecules and that all kinetic degrees of freedom, including – but not limited to –those of

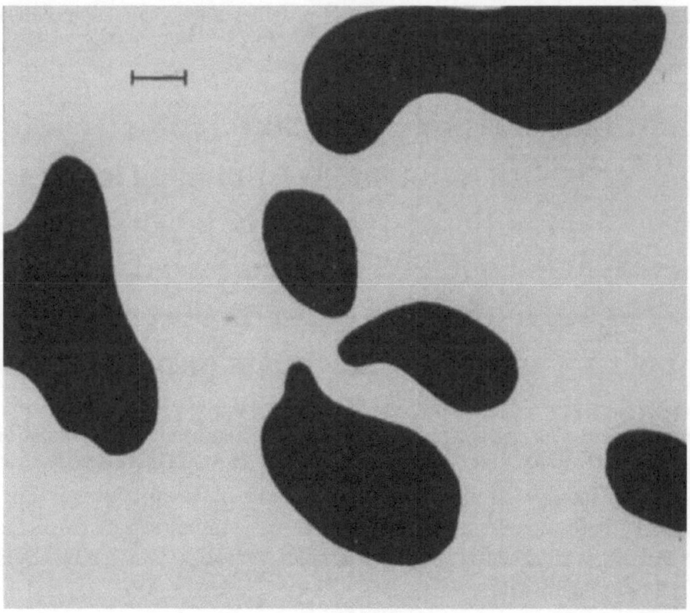

FIGURE C.5 Spatial patterns of concentration formed by Turing bifurcation in reaction–diffusion equations. (From [Turing 1952].)

atoms and molecules, are constantly in random motion. A result due to Einstein is that the random motions, that is, fluctuations δX_i, about most stable conditions of thermal equilibrium, follow Gaussian distributions.

$$p(\delta X_1, \delta X_2, \ldots, \delta X_N) = \Omega_1 \exp\left[\frac{1}{2k_B}\sum_{i=1}^{N}\sum_{j=1}^{N} S_{ij}\delta X_i \delta X_j\right] \tag{C.22}$$

$p(\delta_1, \delta_2, \ldots, \delta_N)$ is a multivariate probability distribution. Ω_1 is a normalization constant. N is the number of dynamical degrees of freedom. $k_B = 1.38 \times 10^{-23}$ J K^{-1} is Boltzmann's constant. S_{ij} are second-order coefficients in a power series expansion about equilibrium of the entropy $S(\delta_1, \delta_2, \ldots, \delta_N)$. A geometric interpretation is that the S_{ij} coefficients represent curvatures in the entropy surface, as shown in Figure C.1. Sharper curvatures represent increased dynamical stiffness, smaller fluctuations, and stability. Flatter curvatures represent decreased dynamical stiffness, larger fluctuations, and instability, such as when a system approaches a phase change.

Boltzmann's law indicates that when a system is in thermal equilibrium, the average kinetic energy $\langle E \rangle$ in each molecular degree of freedom is proportional to temperature:

$$\langle E \rangle = \frac{1}{2} k_B \theta \tag{C.23}$$

Higher temperatures increase the fluctuations of these degrees of freedom with increased mobility of molecules and atoms. The increased mobility manifests itself as increased diffusion rates. This allows chemical reactions to proceed that do not occur at lower temperatures.

The Gibbs entropy S_G of a system with moving components expressed in terms of the probability of the system being in state p_i is as follows:

$$S_G = -k_B \sum_i p_i \ln(p_i) \tag{C.24}$$

The von Neuman entropy S_{VN} is an extension into the quantum realm [Sagawa 2011]. The general state $|\psi\rangle$ of a quantum system is a superposition of N possible states $|\psi_i\rangle$ each with probability p_i. The density matrix ρ is

$$\rho = \sum_{i=1}^{N} p_i |\psi_i\rangle\langle\psi_i| \tag{C.25}$$

and

$$S_{VN} = -\mathrm{tr}(\rho \ln \rho) \tag{C.26}$$

tr(…) is the trace of a matrix, that is, sum of diagonals, and ln(…) is a matrix logarithm.

C.1.3 Nonequilibrium Thermodynamics

Describing nonequilibrium processes typically extends and modifies the formulations of equilibrium thermodynamics. Many aspects of nonequilibrium thermodynamics are principal unresolved fundamental problems of scientific inquiry. At the forefront is the issue of the direction of time. Newton's law of motion and almost all other fundamental laws of physics are symmetric with respect to time, that is, the laws appear the same if t is replaced by –t. For example, a linear single degree of freedom undamped oscillator has the equation of motion:

$$m\frac{d^2x}{dt^2} + kx = F(t) \tag{C.27}$$

Here m is the mass, x is the displacement, k is the spring stiffness, and F(t) is the applied external force. For the case of zero external force and nonzero initial conditions $x(0) = x_0$ and $dx(0)/dt = 0$, the motion is a sinusoid:

$$x(t) = x_0\cos(\omega t) \tag{C.28}$$

Here $\omega = (k/m)^{1/2}$ is the circular natural frequency. This solution is physically unrealizable because it corresponds to oscillating motion that does not stop. Macroscopic observation indicates unforced motions of oscillators due to nonzero initial conditions eventually dissipate and die out. A convenient mathematical model is to achieve physical realizability that introduces viscous damping, $c > 0$, to the equation of motion.

$$m\frac{d^2x}{dt^2} + c\frac{dx}{dt} + kx = F(t) \tag{C.29}$$

The response of the unforced system to the same initial conditions is an exponentially decaying sinusoid as t increases.

$$x(t) = x_0\exp\left(-\frac{ct}{2m}\right)\cos\left[\omega\left(1-\frac{c^2}{4mk}\right)t\right] \tag{C.30}$$

The distinction between the two cases is that the undamped oscillator has only even powers of derivatives of x in Eq. (C.27), whereas the damped oscillator has a mix of even and odd powers of derivatives in Eq. (C.29). Even power derivatives are symmetric with respect to a sign reversal in time. Odd power derivatives are antisymmetric. Reverse the direction of time in Eq. (C.28) and the motion does not change. Reverse the direction of time in Eq. (C.30) and the motion changes to exponentially growing oscillations.

Recognizing that dissipation is a universal process, Rayleigh formulated a set of phenomenological equations with first-order time derivatives of extensive physical quantities in the form of fluxes J_i, applied forces F_j, and coupling coefficients L_{ij}.

$$J_i = \sum_{j=1}^{N} L_{ij}F_j \tag{C.31}$$

The first-order time derivatives in the fluxes J_i introduce irreversibility and time-reversal asymmetry. Onsager assumed that the microscopic processes driving irreversible fluctuations are time-reversal symmetric. This led to the conclusion that the coupling coefficients L_{ij} show that the coupling coefficients are symmetric with respect to degree of freedom indices i and j.

$$L_{ij} = L_{ji} \tag{C.32}$$

Experiments confirm that these relations are valid for many irreversible processes operating near to equilibrium [Katchalsky 1965].

Active materials are examples of irreversible thermomechanical systems that derive action from free energy stored in the material. For example, an active gel consists of a fluid with an embedded network of active microfilaments [Prost 2015]. The stress tensor t_{ij} separates into passive t_{ij}^p and active t_{ij}^a components:

$$t_{ij} = t_{ij}^p + t_{ij}^a \tag{C.33}$$

The active stress represents the conversion chemical energy into mechanical energy by forces exerted among the embedded network of microfibers.

Most human-built and biological living systems distinguish themselves by an orderly nonequilibrium arrangement of constituents. In a discussion of the thermodynamics of living biological systems, Schrödinger introduced a normalized sign-reversed version of entropy called the negentropy, J(x) [Schrödinger 1944]. Other authors suggest that free energy may be a more appropriate term [Ho 2008]. In an infinite domain with a Gaussian distribution as the representation of a purely random case with minimum negentropy,

$$J(x) = H(x_{Gauss}) - H(x) \tag{C.34}$$

$H(x_{Gauss})$ is the entropy of the Gaussian variable with the same mean and variance as x. Properties of negentropy are as follows: (1) It is always nonnegative. (2) Gaussian distributions have zero negentropy (for infinite domains). (3) Negentropy of a Gaussian distribution is invariant under a linear transformation of coordinates. (4) Non-Gaussian variables have positive negentropy.

Dissipative structures are central to irreversible open systems that operate far from equilibrium [Prigogine 1981]. Local irreversible processes dissipate free energy while performing actions that stabilize and heal the structure (Figure C.6). Aeroelastic vortex shedding lock-in, hurricanes, cyclic chemical reactions, and living creatures are example stable dissipative systems.

Exceptional points are closely related to dissipative systems [Berry 2004]. Exceptional points are points in a 2D or higher parameter space for dynamic systems where the system operator is non-Hermitian with exceptional degenerate eigen-structures, that is, repeated eigenvalues with parallel instead of typically orthogonal eigenvectors. Exceptional points typically represent linearized operating points of nonlinear systems and describe a range of observed behaviors, such as the stabilization of lasers, microwaves, and bicycles [Kirillov 2018].

C.2 Logarithmic Measures and Information Theory

Shannon introduced entropy as a logarithmic measure of information and uncertainty in transmitted and received signals [Shannon 1949]. Logarithms transform multiplications into addition – a convenient property for probability-weighted representations of information. The analysis considers a signal with n possible outcomes and n associated probabilities $p_1, p_2, \ldots, p_n$. The signal has entropy H, with the following properties: (1) H changes continuously with changes in p_i. (2) An increase in the number of possible outcomes increases entropy. When all p_i are equal, that is, $p_i = 1/n$, then H increases monotonically with n. (3) Entropy sub-adds in sequential choices when weighted with the probability of the outcomes. These assumptions lead to the discrete variable formulation of H:

$$H = -K\sum_{i=1}^{n} p_i \ln(p_i) \tag{C.35}$$

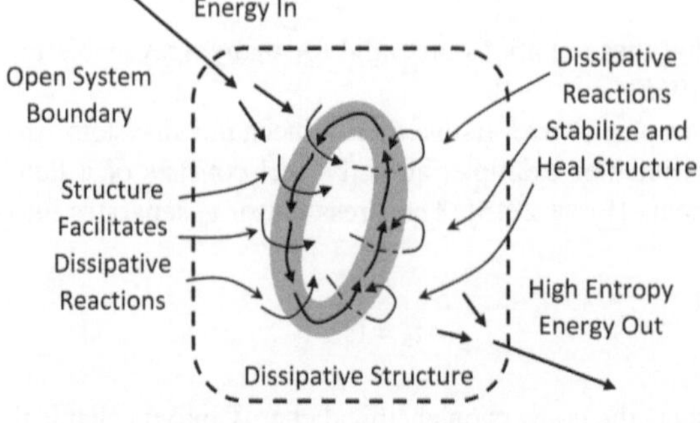

FIGURE C.6 Stable dissipative structure far from equilibrium facilitates reactions that stabilize and heal structure.

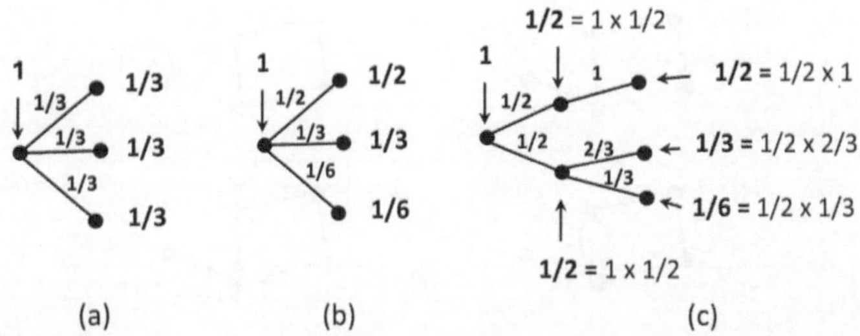

FIGURE C.7 Probabilities (in bold font) associated with three outcomes. (a) One choice and three outcomes with equal probabilities, H = 1.099. (b) One choice and three outcomes with unequal probabilities, H = 1.012. (c) Two successive choices and three outcomes with unequal probabilities that match b. Cases (b) and (c) have the same entropy.

For continuous variables,

$$H = -K\int p(x)\ln[p(x)]dx \tag{C.36}$$

K is a constant (often set = 1). The properties of entropy are as follows: (1) H = 0 if there is complete certainty in the outcome, that is, if one of the $p_i = 1$ and the rest of the $p_i = 0$. 2. H is a maximum if there is maximum uncertainty in the outcome, that is, $p_i = 1/n$ for all i. As an example, Figure C.7 illustrates different outcome probabilities affecting the entropy and invariance under sub-addition with three different cases, each with three outcomes.

Case a – One choice and three outcomes with equal probability of occurrence, that is, $p_1 = p_2 = p_3 = 1/3$:

$$H = -\left[\frac{1}{3}\ln\left(\frac{1}{3}\right) + \frac{1}{3}\ln\left(\frac{1}{3}\right) + \frac{1}{3}\ln\left(\frac{1}{3}\right)\right] = -(-0.366 - 0.366 - 0.366) = 1.099 \tag{C.37}$$

Case b – One choice and three outcomes with unequal probability of occurrence, that is, $p_1 = 1/2$, $p_2 = 1/3$, $p_3 = 1/6$.

$$H = -\left[\frac{1}{2}\ln\left(\frac{1}{2}\right) + \frac{1}{3}\ln\left(\frac{1}{3}\right) + \frac{1}{6}\ln\left(\frac{1}{6}\right)\right] = -(-0.347 - 0.366 - 0.299) = 1.012 \tag{C.38}$$

Case c – Two sequential choices with first choice probabilities, $p_{11} = 1/2$, $p_{21} = 1/2$, and local second choice probabilities, $p_{12} = 1$, $p_{22} = 2/3$, $p_{23} = 1/3$, leading to the same unequal global probabilities as Case b.

$$H = -\left\{\frac{1}{2}\ln\left(\frac{1}{2}\right) + \frac{1}{2}\left[\frac{2}{3}\ln\left(\frac{1}{3}\right) + \frac{1}{3}\ln\left(\frac{1}{6}\right)\right]\right\} = -[-0.347 + 0.5(-0.732 - 0.597)] = 1.012 \tag{C.39}$$

C.3 Graphs, Networks, and Percolation

Graph theory, network theory, and percolation modeling are interrelated mathematical tools based on abstractions of connectedness of entities. Networks are collections of interacting entities. The interconnections, number, and types of entities may change with time. The underlying theory of networks dates back to Euler. Much of the following describes more modern representations, with a tilt toward results with potential for self-healing [Lewis 2009] [Strogatz 2001] [Sporns 2004].

Connections enabling the flow of physical quantities and information among components and systems affect and often dominate the behavior of self-healing structures, machines, and systems. A judicious

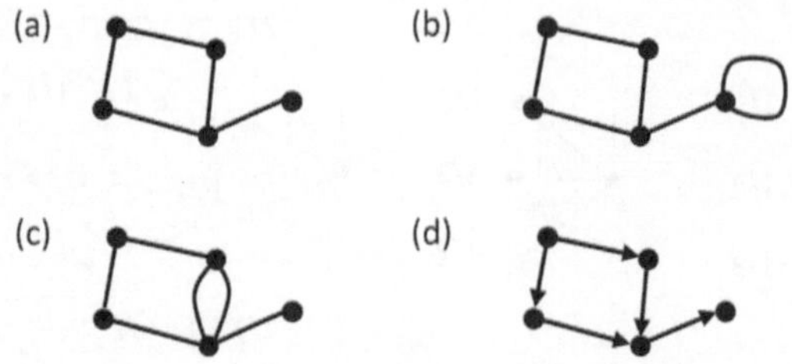

FIGURE C.8 Two-dimensional representations of various types of graphs. (a) Simple graph. (b) Non-simple graph with loop back edge. (c) Non-simple multigraph. (d) Directed graph.

choice of the level of abstraction enables graphs, networks, and percolation models to capture qualitative and quantitative essences of the behavior of interconnected systems of subsystems.

C.3.1 Graphs

A *graph* is a sparse binary description of connectedness between pairs of entities in a set. Graph theory literature refers to the entities as *vertices* and the connections as *edges*. Network literature often instead refers to the entities as *nodes* and the connections as *links*. A 2D graphical depiction of a graph draws the nodes as big dots and connects the nodes with line segments representing the edges.

C.3.1.1 Simple Graphs

Within this framework, a wide variety of distinct graphs and classes of graphs arise, most based on the patterns of interconnecting edges between vertices (Figure C.8). *Simple graphs* allow for the interconnections between pairs of vertices to be either connected by an edge or disconnected. Graphs with edge connections from one node back onto itself or with multiple edges between the same pairs of vertices are *non-simple* graphs. The former are *loop graphs* and the latter *multigraphs*. Graphs with edges that are directed rays are *directed graphs*. Figure C.9 shows some notable graph types.

The *degree* is the number of edges connecting to a vertex. The total number of distinct graph configurations $C_{n,r}$ of n vertices and r edges is as follows:

$$C_{n,r} = \binom{\binom{n}{2}}{r} \tag{C.40}$$

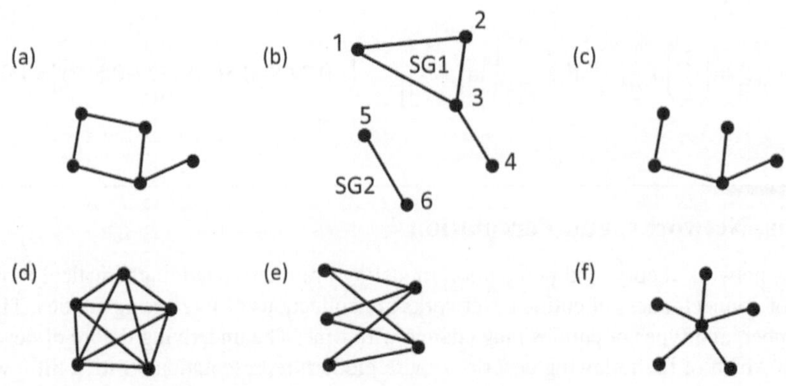

FIGURE C.9 Various notable graph types. (a) Connected. (b) Disconnected with vertices and subgraphs (SGi) labeled. (c) Tree. (d) Complete. (e) Bipartite. (f) Star.

with the combination operator defined as

$$\binom{n}{m} = \begin{cases} \dfrac{n!}{m!(n-m)!} & n > m \\ 0 & n < m \end{cases} \tag{C.41}$$

C.3.1.2 Random Graphs

Deterministic graphs have known configurations of vertices and edges. There are many situations where the arrangement of nodes and edges is random. Nondeterministic or random graphs prove useful in many applications. The modern study of random graph theory traces back to seminal papers by Erdős and Rényi [Erdős 1960]. An Erdős–Rényi graph has a fixed set of vertices, N, and fixed set of edges, k, with the distribution of edges being random with an independent and uniform distribution. The probability of an edge connecting two vertices is p, with $0 < p < 1$. The degree distribution P(k) is the probability of a node having degree k. The random assignment of edges to vertices, while avoiding multiple edges, leads to a binomial probability distribution [Newman 2001].

$$P(k) = \binom{N}{k} p^k (1-p)^{N-k} \tag{C.42}$$

When $p = z / N$ with N being large, the binomial distribution becomes a Poisson distribution:

$$P(k) = z^k \frac{\exp(-z)}{k!} \tag{C.43}$$

Many important networks have nodal distributions that depend on the presence and connectivity of other nodes, which lead to different network structures than the Erdős–Rényi graphs. Power law distributions of random network properties result when highly connected nodes tend to gain new connections more rapidly than nodes with lower connectivity [Blundell 1982] [Albert 2002] [Pastor-Satorras 2003] [Wu 2015]. The nodal degree distribution, P(k), for a power law distribution for a parameter q is as follows:

$$P(k) = k^{-q} \tag{C.44}$$

C.3.1.3 Quantitative Measures of Graph and Network Characteristics

Connectivity is the basis for multiple quantitative descriptions of graphs [Tsai 2001]. A *walk* is a sequence of alternating nodes and edges that begins and ends with a node. Walks include the possibility of backtracking. *Trails* are walks that do not repeat edges. A *circuit* is a walk with no repeated nodes, except the first and last node which coincide. A *connected graph* has a walk between every node. A *tree* is a connected graph with no circuits. A *spanning tree* connects all of the nodes in a graph. A *component* is a connected subset graph that does not connect with the rest of the graph.

Matrix methods assess the connectedness of graphs. Underlying many of these methods is the *Adjacency Matrix* [A] with elements A_{ij} that equal 1 when an edge connects vertices i and j, and 0 if they do not connect. For simple graphs with no loop edges, the diagonal elements equal zero. Figure C.9b shows a disconnected graph with 6 vertices, 5 edges, a subgraph SG1 (a.k.a. component) that is fully connected with 4 vertices and 4 edges, and subgraph SG2 with 2 vertices and 1 edge, respectively. $[A_G]$, $[A_{S1}]$, and $[A_{S2}]$ are the corresponding adjacency matrices. $[A_G]$ partitions into a form with the disconnected subgraphs on the diagonal. Raising an adjacency matrix to the power n leaves a matrix with entries corresponding to the number of possible walks of length n between nodes i and j. Note that the diagonals of $[A_{ij}]^n$ correspond to walks of length n that start and return to the same node:

$$[A_G]=\begin{bmatrix} 0 & 1 & 1 & 0 & 0 & 0 \\ 1 & 0 & 1 & 0 & 0 & 0 \\ 1 & 1 & 0 & 1 & 0 & 0 \\ 0 & 0 & 1 & 0 & 0 & 0 \\ 0 & 0 & 0 & 0 & 0 & 1 \\ 0 & 0 & 0 & 0 & 1 & 0 \end{bmatrix} \tag{C.45}$$

with

$$[A_{S1}]=\begin{bmatrix} 0 & 1 & 1 & 0 \\ 1 & 0 & 1 & 0 \\ 1 & 1 & 0 & 1 \\ 0 & 0 & 1 & 0 \end{bmatrix} \text{ and } [A_{S2}]=\begin{bmatrix} 0 & 1 \\ 1 & 0 \end{bmatrix} \tag{C.46}$$

$$[A_G]=\begin{bmatrix} A_{S1} & 0 \\ 0 & A_{S2} \end{bmatrix} \tag{C.47}$$

$$[A_{S1}]^3=\begin{bmatrix} 2 & 3 & 4 & 1 \\ 3 & 2 & 4 & 1 \\ 4 & 4 & 2 & 3 \\ 1 & 1 & 3 & 0 \end{bmatrix} \tag{C.48}$$

A complete graph, as in Figure C.9d, has L_c edges connecting all N nodes to all (N – 1) other nodes:

$$L_c=\frac{N(N-1)}{2} \tag{C.49}$$

In a network, some nodes may be more important than others. The arrangement and number of links reflect the importance and form the basis of node importance metrics.

Cluster Coefficient: An individual node u has a degree d(u). The neighborhood of node u consists of the d(u) nodes that connect to u. The neighboring nodes may also be interconnected, with L_{Nu} indicating the total number of links in the neighborhood for node u. The maximum possible L_{Nu} occurs when the neighborhood forms a complete graph:

$$L_{Nu\ max}=\frac{d(u)[d(u)-1]}{2} \tag{C.50}$$

The cluster coefficient of node u, Cc(u) is the ratio of the number of links in the neighborhood to the maximum number of links [Lewis 2010]:

$$Cc(u)=\frac{L_{Nu}}{L_{Nu\ max}}=\frac{2L_{Nu}}{d(u)[d(u)-1]} \tag{C.51}$$

The cluster coefficient of an entire graph, Cc(G) is the average cluster coefficient:

$$Cc(G)=\frac{1}{N}\sum_{i=1}^{N}Cc(u_i) \tag{C.52}$$

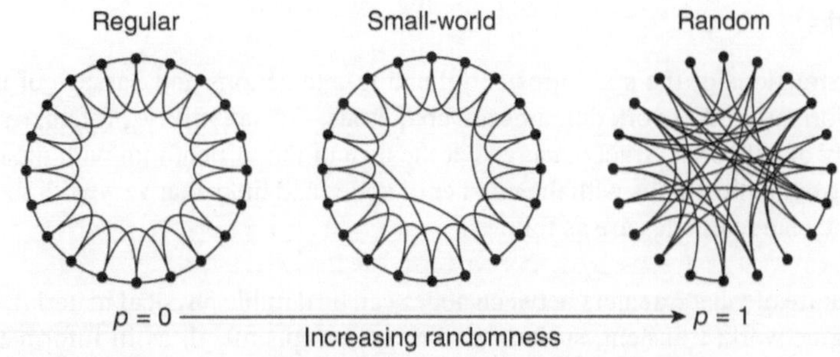

FIGURE C.10 Regular, small-world, and random graphs (From [Watts 1998].)

Sparseness versus being fully connected and the average maximum of the minimum path length between two nodes of connected graph leads to the notion of big and small worlds. Small-world graphs have cluster coefficients that are much higher than random graphs but have similar degree distribution histograms (Figure C.10) [Watts 1998].

Link efficiency of a graph, E(G), depends on the number of links m and the average path length, Avg_path_length(G):

$$E(G) = 1 - \frac{\text{avg\_path\_length}}{m} \tag{C.53}$$

Graph entropy I(G) measures the randomness of graphs. Larger entropy corresponds to more randomness. There are multiple definitions available [Mowshowitz 2012] [Dehmer 2011] [Li 2008c] [Wilhelm 2007]. Most variants use a version of the entropy measured on the nodal degree sequence distribution.

$$I(G) = -\sum_{i=1}^{\max d} h_i \ln(h_i) \tag{C.54}$$

When applied to networks, entropy quantifies the complexity and growth of networks, along with the spreading or contraction of information flows in directed graphs.

Entropy of Structural Organization: Graph models as abstractions of structures, machines, and systems enable entropy measures of the organizational and topological complexity of a structure. Possible measures drawn from molecular structures [Bonchev 1995] are as follows:

1. *Kolmogorov Complexity:* The length of the shortest string of 1's and 0's that a program can convert into a representation of the original structure, that is, the adjacency matrix. Asymmetric structures have a higher Kolmogorov complexity than symmetric variants of the same size.
2. *Shannon Entropy:* This is the probabilistic entropy measure due to Shannon adapted to structures by grouping structural features and attributes into classes and then tabulating the probabilities of finding structural components in the classes. Calculations based on the degree (number of edges per node) lead to branched structures having a higher entropy than cyclic structures.
3. *Topological Complexity*: Originally developed to assess the complexity of molecules, this metric, TC(G), tabulates the connected subgraphs in the original structural connectivity graph G. A_k is the tally of connected subgraphs with the number of edges, k. For a graph with n nodes and subgraphs with connected edge counts N_k, described by the adjacency matrix a_{ij} the topological complexity is

$$TC(G) = \sum_{k=0}^{n} A_k = \sum_{k=0}^{n} \sum_{i=0}^{N_k} \sum_{j=0}^{N_k} a_{ij} \tag{C.55}$$

C.3.2 Networks

Networks are abstractions of the structures coordinating interactions and passage of material, energy, forces, and/or information. Network domains are ubiquitous – social, utility, protein, transportation, etc. The modern study of networks largely emerged at the turn of the millennium with the representation of networks as being dynamic graphs with the number of nodes and links that vary with time [Lewis 2009]. Some features and characteristics are as follows:

1. *Nature:* Nature of what transfers between nodes can be durable physical material, such as water in a piping network; transient, such as electricity; or ephemeral, as in information or social interactions.
2. *Dynamism and Growth:* Behavior in the nodes and links, and the number of nodes and links vary with time and conditions. Networks often grow from the bottom-up, leaving a long-term imprint of early age development on network structures.
3. *Patterns:* Arrangements of subsystems and interconnections form patterns, including groups that act as subsystems. Patterns often repeat over space and timescales.
4. *Topology:* Global structure, including cycles, can have profound effect on network behavior.
5. *Nonlinear Influence:* The power of individual nodes depends in a nonlinear manner on the degree of the node, interconnection range (small versus large world), and topology.
6. *Stability:* Network structure may be stable with time or may be unstable and undergo large rapid changes.
7. *Emergence:* Large-scale structures with distinctive behaviors emerge in stable systems and presage instabilities.

C.3.2.1 Network Metrics, Taxonomy, and Topologies

Many aspects of networks are complicated. Qualitative descriptions tend to be taxonomic. More quantitative descriptions often use methods of statistical graph analysis.

Timescales of concern are as follows:

1. *Operational Timescales:* These derive from activity and interactions between entities. An example is electric power distribution networks that need to maintain frequency and phase stability with operational timescales ranging from milliseconds to days.
2. *Configurational Timescales:* These correspond to changes in topology of the network, that is, the addition and/or removal links and/or nodes.

Scale-free networks have degree sequence distribution P(k) that follow the power law form of Eq. (C.44) with the exponent $2 < q < 3$. A scale-free network has nodes with high degree numbers, but not many.

The properties of connected neighboring nodes have a large influence on overall network behavior [Aguirre 2013]. The pagerank algorithm quantifies the importance of nodes in networks describable by directed graphs. For a directed network with N nodes, adjacency matrix a_{ij}, $k_{out}(j)$ as the number of outgoing links at node j, and random jump reset parameter α, $0 \le \alpha \le 1$. The algorithm ranks the importance, $p_t(i)$, of node i, with an iterative, or equivalently eigenvalue, formula [Brin 1998].

$$p_t(i) = \frac{1-\alpha}{N} + \alpha \sum_{j=1}^{N} \frac{A_{ij} p_{t-1}(j)}{k_{out}(j)} \tag{C.56}$$

The pagerank formula ranks important nodes as those with more nodes pointing to it, with a weighting based on the importance of the pointing nodes. The rank increases if an important node points to the node, but is reduced by the fraction of outgoing nodes. The reset parameter α allows for random jumps in the referrals with a small α corresponding to a larger probability of random jumps.

Pagerank is stable in scale-free networks but is less stable and subject to perturbation in other network architectures [Ghoshal 2011].

C.3.2.2 Networks with Weighted Connections

Conventional graph theory uses a binary measure of connectedness between vertices and tallies items such as the path length between interconnections, but generally does not provide a description of the strength and type of a connection. Weighted graph techniques include the strength of the connections, at the expense of increased complexity. Some weighted connected graph and network techniques are as follows:

Bond Graphs: Directed graphs with specialized symbols to represent the flow of energy and related quantities in systems [Blundell 1982].

Petri Nets: Directed graphs with added information that enable them to change state with a sequential progression of events [Murata 1989]. The underlying structure is a directed, weighted, and bipartite graph. The bipartite nature separates the nodes into places and transitions with edges connecting the places to the transitions. The states evolve with time in a manner based on the network topology, structural properties, and predefined rules.

Weighted Networks: The strength of the weight between each node is given a value w_{ij}. In an unweighted network, w_{ij} is either 0 or 1. In a weighted network, w_{ij} is usually a positive number corresponding to the capacity or the amount of transport across the link. The degree of weighting of the links between clusters and overlapping clusters is a characteristic of different classes of complex networks. Many cases, such as cell phone and social and neural networks have strong links between nodes with overlapping neighborhood clusters [Pajevic 2012].

C.3.2.3 Networks with Static Topologies and Dynamic Nodal Properties

Dynamic networks change with time. A subclass is the networks with nodal properties that change with time but have fixed configurations of nodes and connections. Denoting $n_i(t)$ as the values of a property of N nodes, i = 1, 2, …, N, a general description of the evolution of a fixed node dynamical system is as follows:

$$\frac{dn_i}{dt} = W\left[n_i(t)\right] + \sum_{j=1}^{N} A_{ij} Q\left[n_i(t), n_j(t)\right] \tag{C.57}$$

Here W is the interaction of nodal properties with themselves. $A_{ij}Q$ describes cross-interactions between nodes. The nature of W and $A_{ij}Q$ affects stability in manners that capture the stability of a wide range of phenomena, including population dynamics, protein–protein interactions, and queuing theory [Barzel 2013]. A simple version for discrete time denotes $\{n(t)\}$ as a state vector and [M] as an $N \times N$ state transition matrix with the positive property $M_{ij} \geq 0$, and a unit time step [Aguirre 2013].

$$\{n(t+1)\} = [M]\{n(t)\} \tag{C.58}$$

In the long term, the system state converges to the eigenvector $\{u_1\}$ corresponding to the largest eigenvalue of [M]:

$$\lim_{t\to\infty}\{n(t)\} = \{u_1\} \tag{C.59}$$

A consequence of positive nature of [M] is that the components of the eigenvector $\{u_1\}$ are also positive, that is, $u_{1i} \geq 0$, i = 1, 2, …, N. A form of [M] tied to network connectivity is

$$[M] = (R - \mu)[I] + \frac{\mu}{S}[A] \tag{C.60}$$

Here R > 1, 0 < μ ≤ 1, and S > 0 are scalar constants representing the dynamics of the system nodal values, and [I] and [A] are identity and adjacency matrices, respectively. In Eq. (C.60), the adjacency matrix [A] and the transition matrix [M] differ by a matrix proportional to the identity matrix. The result is that [A] and [M] share the same eigenvectors, and the eigenvalues differ by a constant. The component value u_{1i} is the *centrality* of node i and is a measure of the importance of node i corresponding to λ_1. Centrality describes networks with propagating properties, such as epidemics and rumors.

Network stability is an important consideration. Dynamical systems, including those described by network models, can be stable or unstable against small perturbations. Classical methods useful with generic systems, such as equilibrium points near basins and hilltops and Lyapunov functions, also address many stability questions related to networks.

Based on the concepts of linear systems, the stability of networks of synchronized oscillators should favor random interconnections. However, observations of real-world networks, such as electric power, ecosystems, and nervous systems indicate that more regular arrangements favor stability. The disconnect between the model predicted by the linear model, as shown in Figure C.1, and observations led to an expanded notion of stability as similar to a ball rolling down in a basin filled with viscous fluid (Figure C.11) [Menck 2013]. The bigger the volume in the basin, the more robust the stability becomes. The nonlinear basin approach helps to explain the increased stability of regular over irregular topologies in synchronized networks.

The convergence of complex networks to stable states often depends on the appearance of multi-scale organizational structures. Network motifs based on scalable 3 and 4 node structures inside of large complex networks exploit the inherent stability of certain motifs to stabilize large networks [Angulo 2015]. Modular network structures also enable rapid evolutionary adaptation to external changes in environmental conditions [Clune 2013].

A wide class of stable systems fluctuate about an equilibrium point due to external and internal disturbances. Changes in system parameters normally alter the location of the equilibrium point in a smooth manner. Occasionally, a change in system parameters leads to a large jump in the position of the equilibrium points, in phenomena known as tipping point instabilities or fold bifurcations. Detecting these instabilities before they occur has advantages. The self-organization and slower recovery of spatially distributed fluctuations portends such instabilities [Moon 2015] [Dai 2013a].

Are some network topologies more adaptive than others? An examination of the role of topology in enzyme networks and adaptability establishes that a network requires a minimum of three nodes for adaptation [Ma 2009]. More complicated adaptive networks use three node network motifs as subunits. Stable adaptive networks are inherently dissipative, that is, they continuously absorb and dissipate energy to remain at a stable state. This leads to a trade-off of energy, speed, and accuracy [Lan 2012].

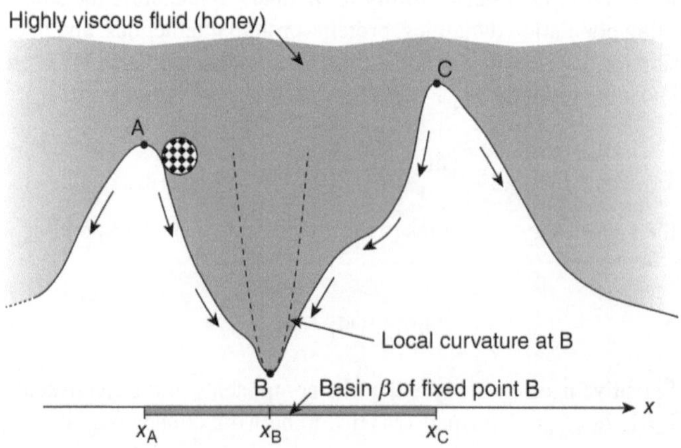

FIGURE C.11 Basin stability uses a nonlinear and nonlocal metric that is more robust than linear local curvature at equilibrium points as measures of stability. (From [Menck 2013].)

C.3.2.4 Control of Networks

Networks behave as dynamical systems when the nodes interact with one another and vary with time, possibly through the application of external control forces. In linear state space systems notation, the state of each node evolves according to a first-order differential equation:

$$\left\{\frac{dx(t)}{dt}\right\} = [A]\{x(t)\} + [B]\{u(t)\} \tag{C.61}$$

Here {x(t)} is an $N \times 1$ column vector representing the state of each node. [A] is an $N \times N$ state transition matrix. [B] is an $N \times M$ control matrix. {u(t)} is an $M \times 1$ control vector, where $M \leq N$. Networks are controllable when a suitable combination of inputs {u(t)} can control the value of every node [Kailath 1979]. The network is controllable if the $N \times NM$ controllability matrix [C] is of full rank:

$$[C] = [B \quad AB \quad A^2B \quad \cdots \quad A^{N-1}B] \tag{C.62}$$

Determining the rank of [C] for large systems is tedious, if not intractable. The following is a procedure for shortening the calculation: (1) Directed graphs provide abstracted representations of most networks. (2) The controllability versus noncontrollability of a network is the same as that of the abstracted system based on the equivalent directed graph. (3) Establishing the controllability of the equivalent directed graph network is simpler and requires $O(N^{1/2}L)$ steps, with L as the number of links [Liu 2011c]. Nonetheless, the size of the matrices in large-scale complex networks largely prevents using the rank of the controllability matrix to determine the controlling nodes. An alternative is network statistics. Calculations based on nodal degree distributions indicate that dense and homogeneous networks are easier to control than those with a heterogeneous structure [Liu 2011c]. A counterintuitive result is that the controlling nodes are not necessarily those with the highest number nodal degrees. The average degree $\langle k \rangle$ of the nodes is an important parameter. Controllable networks with $\langle k_{cr} \rangle$ above a critical value of approximately 9 tend to split into two forms – centralized and distributed control [Jia 2013].

Controlling the edges (links) between nodes in a directed network by correlating the input and output edges for specific nodes is an alternative [Nepusz 2012]. Edge control opens the possibility of robust controllability by edge removal.

C.3.2.5 Networks with Dynamic Topologies

The behavior of a network often depends on how it was built. Many networks form through dynamic building processes that sequentially add and remove nodes and links. In spite of the seemingly wide variety of possible dynamical behaviors in networks, there tends to be commonalities and universal behaviors. For example, biological systems use networks of regulatory signaling and control pathways as integral features of life. At the cellular level, networks that express genes through the production of proteins drive metabolism. These networks are complicated, stochastic, and temporally varying. Temporal Logic Analysis abstracts gene regulatory networks into discrete state space systems with structures that vary with time and enforce liveness by pruning configurations that converge to static states [Batt 2007].

C.3.2.6 Percolation

Percolation is a specialized application of network theory that describes cases where changes in parameters regarding near neighbor connections leads to dramatic changes in long range effects. The term "percolation" derives from an original problem concerning whether a liquid can percolate through a solid containing small pores, holes, and passages. Percolation occurs when there is sufficient connectivity between the holes. Changes to underlying features of construction can change the qualitative behavior of percolation. Many percolation phenomena change with behaviors reminiscent of phase changes. Below a critical value of the parameter, p, the connectivity is localized, and the path lengths of connections remain finite with a probability of 1. Above the critical value, the connectivity becomes long range and the path lengths become infinite with a probability of 1.

Percolation as a field of study largely began in 1957 with seminal work of Broadbent and Hammersley examining 2D lattices [Broadbent 1957] [Grimmett 1999]. A small number of relatively simple cases have known analytic solutions. Most other cases are intractable and instead rely upon numerical Monte Carlo–type simulations. Pike and Seager outlined a broader framework for percolation problems in which the locations of the constituents may be random, and the possible connectivity depends on the geometry and orientation of the constituents [Pike 1974]. The analysis uses a lattice of points with random connections between neighboring points and assumes that the near neighbor connections have a probability p of being connected and probability p–1 of being disconnected. When p = 0, there are no connections. Conversely, when p = 1, all neighboring points are connected. If there is a connected path between opposite sides, the lattice percolates. A goal is to determine which conditions of connective probabilities and lattice geometries cause the lattice to percolate (Figure C.12). Table C.1 shows the percolation threshold probability for lattices with random interconnections between neighboring nodes in selected regular lattices.

Dynamic networks may be steady and stable or may exhibit rapidly changing and complicated behaviors that occasionally result from relatively modest changes in structural assembly rules. For example, percolation modeling describes electrical conductivity in solids containing conductive nanowires in an insulating matrix. Numerical simulations indicate that restrictions on the orientation of wires affects macroscale conductivity, such as by the introduction of anisotropy and the identification of states where partially random wire orientation provides superior conductivity over fully random or fully ordered states [Jagota 2015].

Classical percolation theory considers cases of short-distance nearest-neighbor interconnections of nodes arranged in a fixed lattice. If the interconnections expand to longer distances, then it is possible

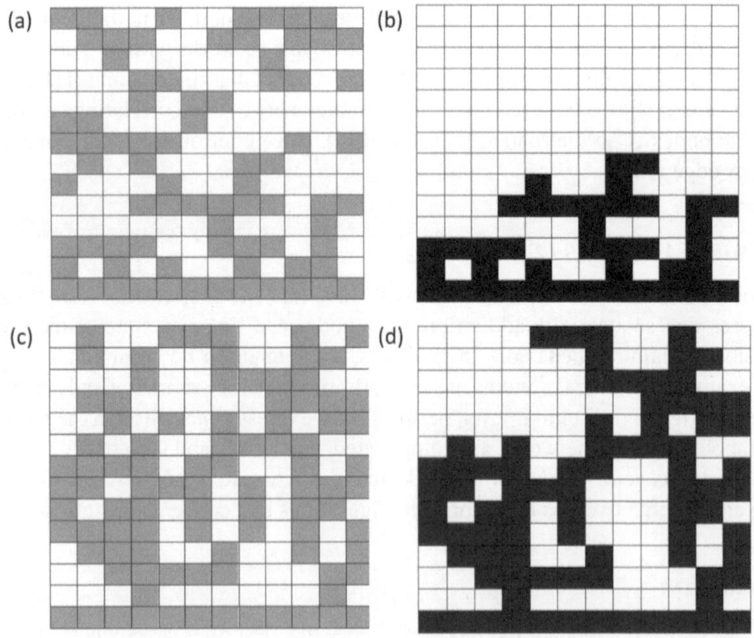

FIGURE C.12 Two-dimensional percolation based on nearest-neighbor connectivity. (a) 12 × 12 square lattice with random arrangement of dark and light cell types, with p = 0.40 dark cells. (b) Connectivity of p = 0.40 dark cells with the bottom edge. Dark cell ratio is less than critical value of 0.593 and connectivity does not percolate from bottom to top. (c) 12 × 12 square lattice with random arrangement of dark and light cell types, with p = 0.65 dark cells. (d) Connectivity of p = 0.65 dark cells with the bottom edge. Dark cell ratio exceeds critical value of 0.593 and connectivity percolates from bottom to top. (Adapted from [Peanasky 1991].)

TABLE C.1
Percolation Thresholds in Selected Regular Lattices [Pike 1974]

Lattice Type	Probability of Connection, p_c
Square	0.59275
Triangular	0.50000
Simple cubic	0.3117
Body-centered cubic	0.245

for discontinuous, that is, explosive, growth of interactions with slight changes in the value of the probability of connection [Boettcher 2012]. Networks that grow according to an Erdős–Rényi model exhibit a continuous phase-change connectivity percolation, whereas networks with different growth models, such as product or sum, produce different behaviors, with small changes in the growth rules, sometimes lead to big changes in the percolation behavior [Achlioptas 2009] [Radicchi 2013]. Another example of explosive growth occurs with the addition of a single link in percolation processes in finite systems. A competitive, rather than purely random, process governs the addition of links [Nagler 2011]. The process is continuous but can be quick enough to appear discontinuous [da Costa 2010].

c.4 Mechanics of Continua and Granular Systems

C.4.1 Non-Newtonian Fluids

Non-Newtonian fluids resist shear deformation with strengths that are not proportional to the rate of shear deformation. Useful in many descriptions is the rate of deformation tensor, **D**, defined as the symmetric portion of the velocity gradient tensor [Malvern 1969]:

$$\mathbf{D} = \begin{bmatrix} \frac{\partial v_x}{\partial x} & \frac{1}{2}\left(\frac{\partial v_x}{\partial y} + \frac{\partial v_y}{\partial x}\right) & \frac{1}{2}\left(\frac{\partial v_x}{\partial z} + \frac{\partial v_z}{\partial x}\right) \\ \frac{1}{2}\left(\frac{\partial v_x}{\partial y} + \frac{\partial v_y}{\partial x}\right) & \frac{\partial v_y}{\partial y} & \frac{1}{2}\left(\frac{\partial v_y}{\partial z} + \frac{\partial v_z}{\partial y}\right) \\ \frac{1}{2}\left(\frac{\partial v_x}{\partial z} + \frac{\partial v_z}{\partial x}\right) & \frac{1}{2}\left(\frac{\partial v_x}{\partial z} + \frac{\partial v_z}{\partial x}\right) & \frac{\partial v_z}{\partial z} \end{bmatrix} \tag{C.63}$$

Here v_x, v_y, and v_z are respectively the x, y, and z components of the flow velocity. Off-diagonal terms in the rate of deformation tensor, **D**, quantify the rate of shear deformation.

An experimental setup that isolates shear rate effects uses cross-shaped converging–diverging flow, as shown in Figure C.13. Typical experimental setups permit optical and other instrumental access from the top (z direction). The nominal velocity field is

$$\vec{v} = xS\vec{i} + yS\vec{j} + 0\vec{k} = v_x\vec{i} + v_y\vec{j} + v_z\vec{k} \tag{C.64}$$

with the corresponding **D** as follows:

$$\mathbf{D} = \begin{bmatrix} S & 0 & 0 \\ 0 & -S & 0 \\ 0 & 0 & 0 \end{bmatrix} \tag{C.65}$$

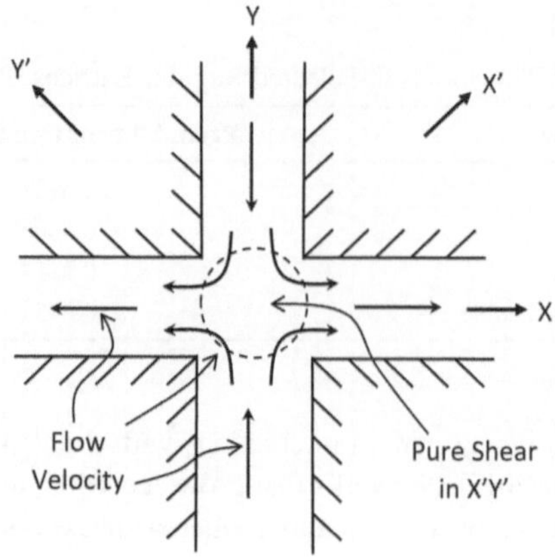

FIGURE C.13 Crossed flow channels produce flow with controllable shear rate. (Adapted from [Keller 1985].)

Transforming **D** to a coordinate frame rotated 45° about the z-axis leads to a rate of deformation tensor that represents pure shear deformation in the x′–y′ plane.

$$\mathbf{D} = \mathbf{D} = \begin{bmatrix} 0 & \frac{S}{2} & 0 \\ \frac{S}{2} & 0 & 0 \\ 0 & 0 & 0 \end{bmatrix} \tag{C.66}$$

Shear rate is an important factor in the use of coagulation processes to stop leaks. Some reasons are as follows: (1) High shear rates are characteristic of the flow through a leaking hole. (2) High shear rates provide available mechanical energy to convert the fluid from a liquid to a gelatinous state. (3) Plugging the leak reduces the rate of or eliminates entirely the shear deformation, which reduces or terminates the coagulation process. The practical use of shear rate–activated coagulation balances the kinematics of shear deformation with the kinetics of gelation processes.

C.4.1.1 Granular Mechanics and Jamming

The mechanics of loading and movement of granular material are distinctively nonlinear, leading to properties rarely seen in continuous media. Occasionally with minimal visual precursors, smooth flows quickly jam, and seemingly stable structures quickly avalanche. These nonlinearities interact over multiple length scales ranging from microscopic up through macroscopic.

Zero temperature is the term for when the kinetic energy of the granules is zero and the system is in a static or quasi-static state. For static jammed states, the equations of motion are those of static equilibrium but are indeterminate. There is a need for additional constitutive relations between stress and deformation. Granules interact with one another by forces that repulse when in close contact and vanish once the separation distance exceeds a threshold defined by the effective radii of the granules. A general relation for the force between particles i and j in terms of potential energy $V(r_{ij})$ is as follows [O'Hern 2003]:

$$V(r_{ij}) = \begin{cases} \epsilon\left(1 - r_{ij}/\sigma_{ij}\right)^{\alpha}/\alpha & \text{for} \quad r_{ij} < \sigma_{ij} \\ 0 & \text{for} \quad r_{ij} \geq \sigma_{ij} \end{cases} \tag{C.67}$$

Here ϵ is an energy scaling factor, r_{ij} is the center-to-center distance of the particles, and σ_{ij} is the sum of the radii of the two particles. Different values of the exponent α gives rise to different types of

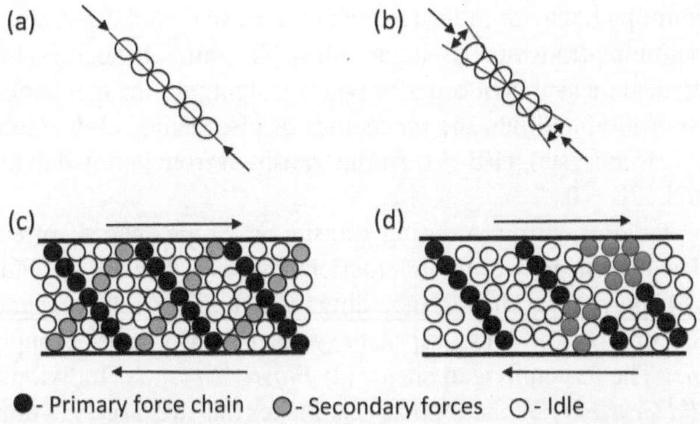

FIGURE C.14 Alignment of jammed or interlocking particles leading to directional stress. (a) Chain of rigid particles carries load on axis. (b) Chain of compressible particles carries on-axis and transverse loads. (c) Idealized grid of jammed particles in shear. (d) Nonuniform array of jammed particles in shear. (Adapted from [Cates 1998].)

interactions. If $\alpha = 2$, the interactions are those of a linear spring. If $\alpha = 3/2$, the interactions are those of a stiffening nonlinear spring. If $\alpha = 5/2$, the interactions are those of Hertz contact forces. When the granules are hard and essentially rigid, then the relation is one of either contact or noncontact, often supplemented with ad hoc assumptions concerning relations between the stresses.

A granular system in a jammed state forms a network of transmitted forces that carry shear stresses. Compression stresses dominate jammed states, which favors using a pressure tensor p_{ij} as the negative of the stress tensor σ_{ij}, that is, $p_{ij} = -\sigma_{ij}$, to represent stresses. A unidirectional constitutive model superposes an isotropic pressure term P and stresses aligned with a jammed chain of particles pointing in the direction n_i (Figure C.14) [Cates 1998].

$$p_{ij} = P\delta_{ij} + \Lambda n_i n_j \tag{C.68}$$

Such a jammed state is fragile. It cannot sustain significant shear or transverse loads relative to n_i. Three non-colinear chains of particles in directions n_i, m_i, and l_i stabilize a material into 3D stiffness with pressure tensor p_{ij}.

$$p_{ij} = \Lambda_1 n_i n_j + \Lambda_2 m_i m_j + \Lambda_3 l_i l_j \tag{C.69}$$

Studies of the network in a jammed granular medium reveal mechanical behaviors that differ from ordinary homogeneous solids. An example is the very noisy transmission of acoustic waves with the formation of spectra that span multiple frequency decades and have a spectral response of 1/fa with f as the frequency and a ≈ 2.2 [Liu 1992]. The transition from a flowing unjammed state into a jammed state is in many respects similar to a phase transition in materials. The phase diagram depends on volume density ρ, temperature T, and applied shear stress σ (Figure C.15a) [Olsson 2007] [Heussinger 2009] [Majmudar 2007].

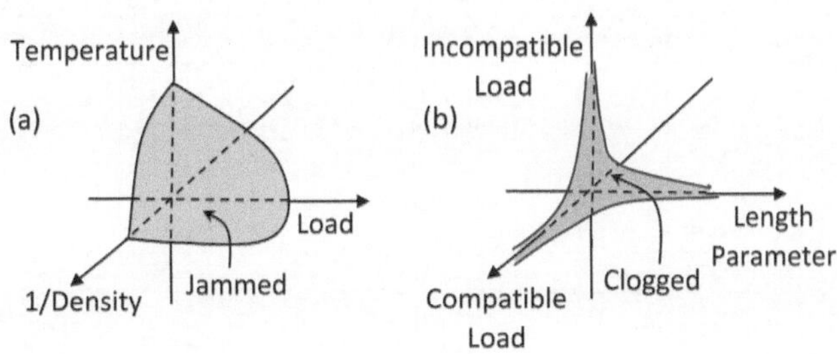

FIGURE C.15 Flow phase diagrams for granular and particulate media. (a) Jamming and (b) Clogging (Adapted from [Liu 1998] [Trappe 2001] [Zuriguel 2014].)

Detailed analysis of jamming behavior indicates the presence of complicated nonlinear scale-dependent dynamics near to the jamming transition [Banigan 2013]. The ratio of shear modulus G to bulk modulus B as a function of statistical measures of order provides insight into the mechanics of granular material near to the jamming transition and into the mechanics of disordered solids [Goodrich 2014]. For rigid spheres packed into a perfect crystal, G/B = 1. At the transition from jammed to unjammed, the material loses shear stiffness and G/B = 0.

Particle shape has a large effect on jamming transitions. Some shapes soften the aggregate, while other shapes stiffen. The nonlinearity of the interactions makes it difficult to predict behavior from first principles. A study of 145 polyhedral shapes, including the Platonic, Archimedean, Catalan, and Johnson solids along with zonohedra, prisms, and antiprisms, yielded four primary groupings of packing behavior: (1) *Liquid Crystals:* The assembly can shear. (2) *Plastic Crystals*: Individual granules can rotate. (3) *Disordered.* (4) *Solid Crystals* [Damasceno 2012]. Shapes that are largely symmetric tend to assemble into solid and plastic crystals. More asymmetric polyhedra form disordered states. Some flat elongated polyhedra form liquid crystals. Numerical models using the discrete element method capture some of the essential behavior of these granular healing systems [Herbst 2008]. Vertex models of cell-to-cell adhesion describe liquid to rigid transitions in single layers of biological tissue [Bi 2015]. Numerical methods that use nonlinear techniques, such as artificial evolution, are effective in exploring this design space [Bi 2011] [Miskin 2013].

Sand piles and sandstone are closely related cases of granular mechanics (Figure C.16). When the directions of the principal stresses in a sand pile are at a constant angle with respect to the central vertical axis, a continuum of internal stress arches forms [Wittmer 1996]. The stresses at the bottom depend on the diagonal length of the above arch, leading to a minimum value at the center. This stress flow leads to the interesting case of sandstone landforms, including arches, pedestal rocks, and pillars [Bruthans 2014]. Stress induced by gravity increases the interlocking of sand grains, which makes portions of the stone in compression more resilient to erosion and weathering. Subsequent erosion of regions with lesser compressive stress transfers the gravity load to the regions with higher compressive stress. The positive feedback of stress and erosion leads to the elongated arch shapes and pillars, in a manner somewhat similar to Wolff's law for the formation of bone structures (Figure C.17).

The flow of granular material through a constriction depends on relative size. A loose constriction leads to free flow that in many respects appears fluidic. A tight constriction causes jamming. In between the flow is intermittent with bursts of flow mixed between slowdowns. Oscillatory forcing, such as

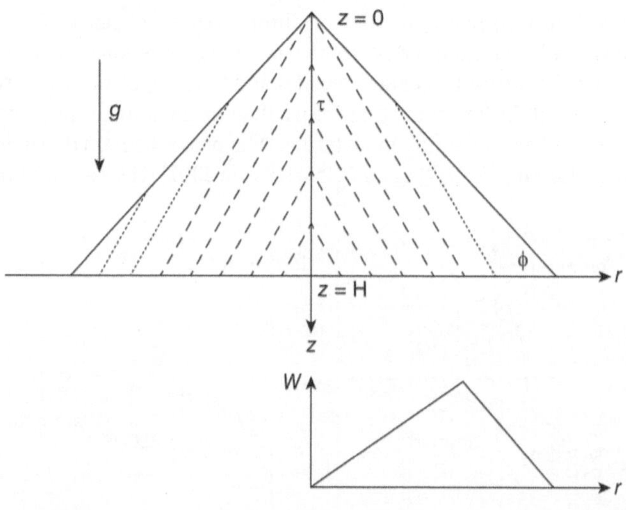

FIGURE C.16 Stress forms inside a sandpile as a set of nested arches with a minimum weight on the bottom at the center. (From [Wittmer 1996].)

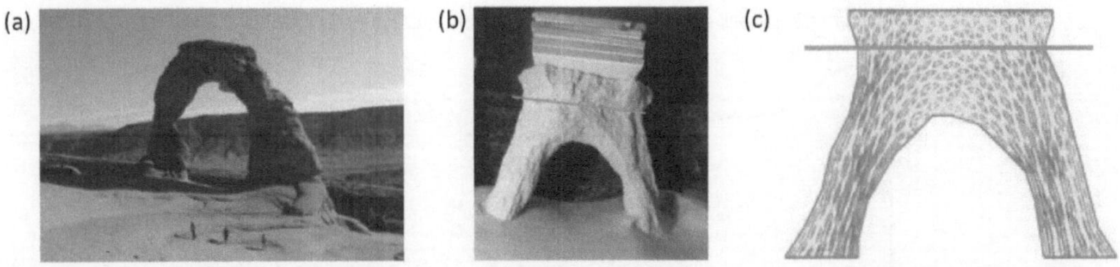

FIGURE C.17 Jamming processes lead to the formation of sandstone arches through positive feedback that localizes jamming and erosion. (a) Natural stone arch. (b) Laboratory demonstration of arch formation. (c) Numerical model. (From [Bruthans 2014].)

shaking, increases the rate of intermittent flow. Such behavior seems to be universal and spans inanimate phenomena, such as grains of salt in a shaker, to living creatures, such as sheep going into a pen. The flowing parameter Φ is a measure of intermittency defined in terms of the average clog time $\langle t_c \rangle$ and the average flow time $\langle t_f \rangle$ as follows:

$$\Phi = \frac{\langle t_f \rangle}{\langle t_f \rangle + \langle t_c \rangle} \quad (C.70)$$

Clogging occurs when $\Phi = 0$, free-flowing when $\Phi > 1$, and intermittent with jamming when $0 < \Phi < 1$. The intermittent behavior of the flow is random. The tail of the cumulative probability distribution follows the power law $\tau^{-\alpha}$ with power law distributions for the normalized time τ. Capturing this behavior is a 3D phase diagram with length parameter α, incompatible load parameter (promotes jamming), and reciprocal of a load parameter (promotes flow) as the axes (Figure C.15b) [Zuriguel 2014]. Attraction between particles accentuates jamming phenomena in a manner similar to applied stress [Trappe 2001].

The sticky colloid model captures a wide range of jamming behavior of colloidal particles held together by a polymer that activates under shear deformation. Originally posited for intercellular adhesion in biology, the model assumes that long-range electrodynamic forces cause particles to clump together while short-range electrostatic forces separate the particles, leaving them trapped close to one another (Figure C.18) [Bell 1984].

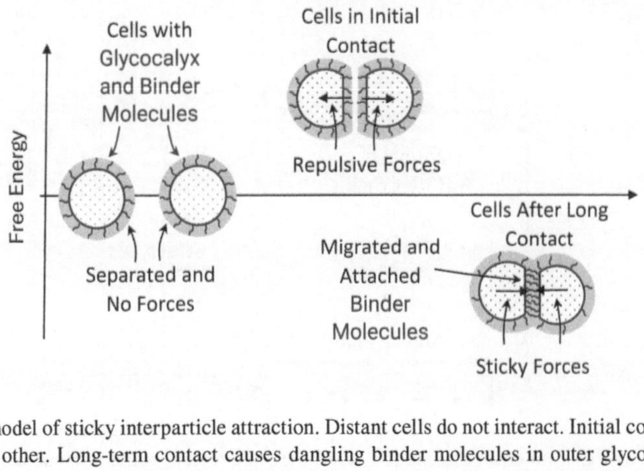

FIGURE C.18 Bell model of sticky interparticle attraction. Distant cells do not interact. Initial contact raises free energy and cells repulse each other. Long-term contact causes dangling binder molecules in outer glycocalyx layer to migrate, bind, and form attractive sticky forces with a free energy minimum. (Adapted from [Bell 1984].)

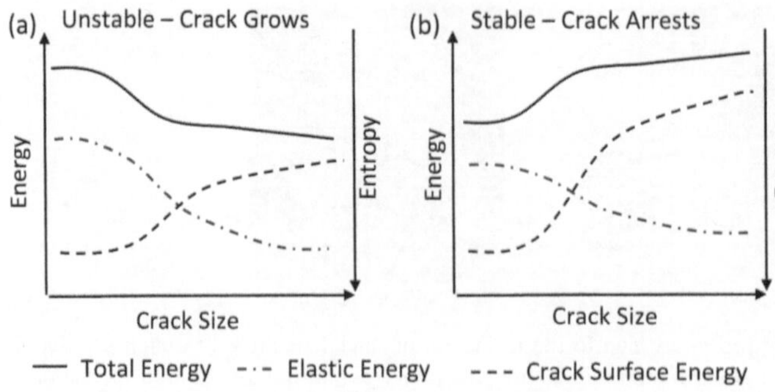

FIGURE C.19 Energy minimization and entropy maximization interpretation of crack growth. (a) Low crack surface energy promotes crack growth. (b) High crack surface energy arrests crack growth.

C.4.1.2 Crack Tip Mechanics

Fracture is a complicated multiscale phenomenon. Internal macroscopic stresses fracture solids by breaking bonds between atoms and molecules in an organized manner. A crack will grow if there is sufficient mechanical energy available to supply the energy needed to form a new fracture surface (Figure C.19). Materials that require large amounts of energy to form cracks are tough and fracture-resistant. Conversely, brittle materials require less energy to form crack surfaces and fracture easily.

The *stress intensity factor* K_i is a constant in a semiempirical model, inspired by the mechanics of Figure C.19, for determining whether a crack is stable or will grow [Boresi 2002]. K_i combines geometry, crack mode, and applied stress, with many formulas for particular situations tabulated, with the units being stress × $(\text{length})^{1/2}$. The subscript i denotes crack mode with typical values being I (opening), II (in-plane shear), and III (tearing or out-of-plane shear). K_I for a mode I elliptical, also known as Griffith crack, of width 2a in an infinite plate with nominal stress σ is as follows (Figure C.20):

$$K_I = \sigma\sqrt{\pi a} \tag{C.71}$$

Standardized experiments measure the critical stress intensity factor K_{ic} for a particular material at a specified temperature. Cracks are unstable and grow if $K_i > K_{ic}$. Conversely, cracks are stable and do not grow if $K_i < K_{ic}$.

The J-integral is a path-independent line integral in a 2D plane that quantifies the strain energy near to a crack tip in planar elements with 2D geometry [Rice 1968] (Figure C.21). The J-integral vanishes when calculated over a closed contour of a simply connected region. The difference in the integrated values along sub-contours, that is, Γ through an elastic region far from the crack tip and Γ_t through a

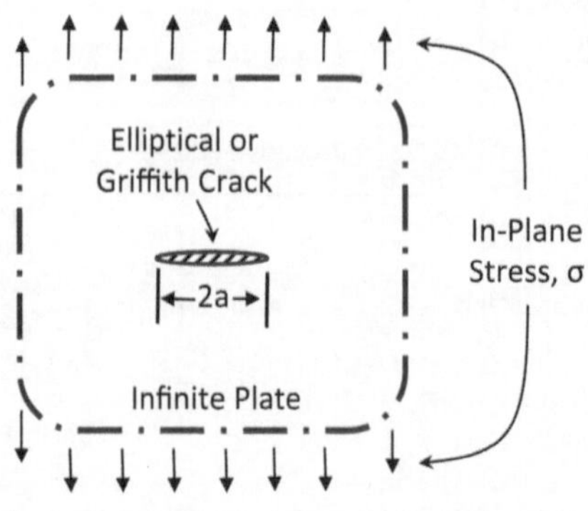

FIGURE C.20 Elliptical or Griffith crack of width 2a in infinite plate with nominal in-plane uniaxial stress σ, perpendicular to crack line.

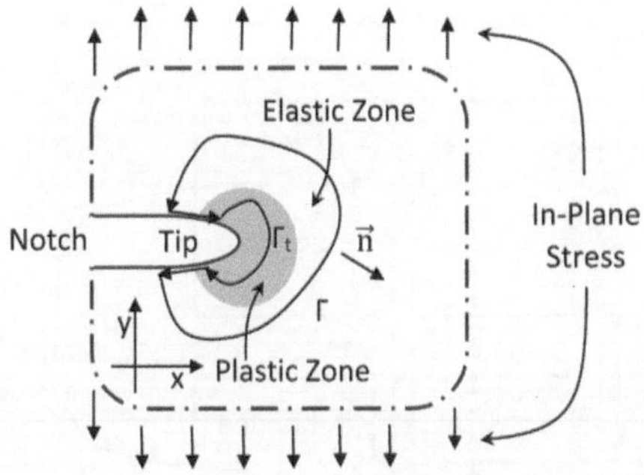

FIGURE C.21 Contour of invariant J-integral contour that surrounds crack tip. (Adapted from [Rice 1968].)

plastic region near to the crack tip, provides information regarding the propensity of crack growth and stress intensity factors.

$$J = \int_{\Gamma} W dy - \vec{T} \cdot \frac{\partial \vec{u}}{\partial x} ds \tag{C.72}$$

Here W is the strain energy density. $\vec{T} = \sigma_{ij} n_j$ is the stress (or traction) vector with unit normal vector $\vec{n}$. $\vec{u}$ is the displacement.

Bridging across a crack with material and/or mechanical devices to impose tensile closing stresses reduces the intensity and growth at the crack tip. Hardened particles, fibers, growing dendrites, and elastic springs are possible methods for bridging cracks. Figure C.22 shows the case of bridging with linear springs. As the gap of the crack opens, the springs pull harder to close the crack. Calculations with J-integrals find that bridging reduces the crack stress intensity factor by the ratio $\lambda = K / K_m$. K is the unbridged stress intensity factor. K_m is the bridging-modified stress intensity factor [Budiansky 1988]. Reduction factors in excess of 11 are reported.

The mechanics of fiber action across a crack are complicated, with debonding, friction, and stretching [Bennett 2008] [Seshadri 2007]. A model of debonding mechanics appears in Figure C.23. For simplicity, pullout is considered only in the bottom block, which separates from the top block by a gap opening Δ. The fiber debonds when the developed force F exceeds the debond strength F_D. As the fiber debonds, the free length L increases, which decreases the stiffness k of the fiber. If the bonding strength is too large, no debonding will occur. The opening crack causes large, localized strain in the fiber and it breaks. If the bonding strength is too weak, pullout occurs, and the fiber fails to deliver load across the crack. It is possible to tune the strength of the bonds with surface treatments to the fibers [Li 1998b] [Lin 1997].

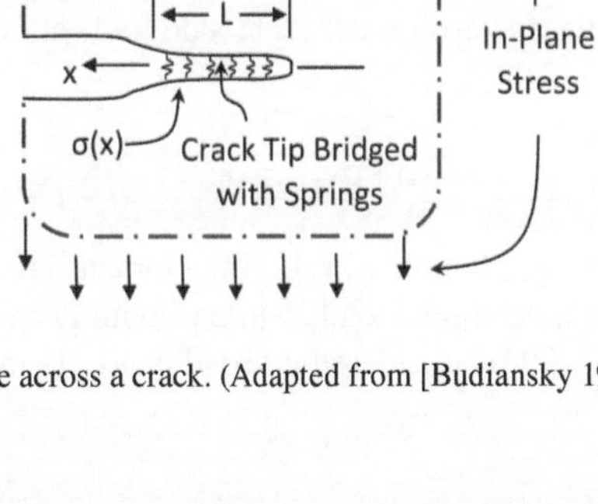

FIGURE C.22 Linear springs bridge across a crack. (Adapted from [Budiansky 1988].)

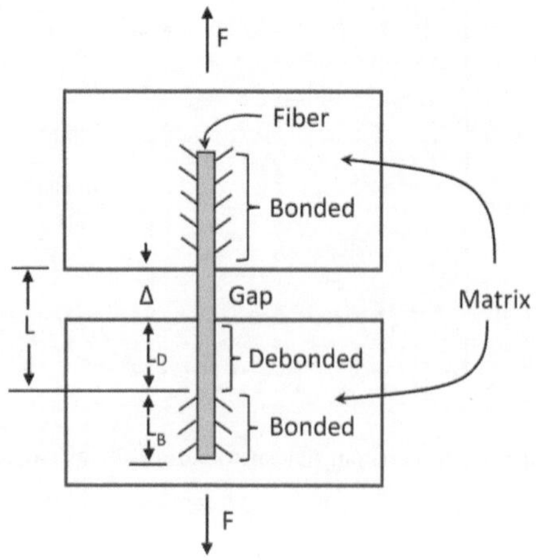

FIGURE C.23 Fiber pullout geometry.

Meso-, micro-, and nanoscale heterogeneity affect crack propagation by both tip blunting and crack bridging. Bone is a heterogeneous material with properties optimized to reduce dynamic fracture propagation. Bones use a combination of mineralized collagen ligament crack bridging and time-dependent crack blunting [Nalla 2005]. A semiempirical model for crack-bridging stress uses the following formula [Foote 1986]:

$$\frac{\sigma_{br}}{\sigma_{max}} = \left(1 - \frac{X}{L}\right)^{p} \tag{C.73}$$

σ_{br} is the bridging stress. σ_{max} is the maximum stress occurring at the crack tip. X is the distance behind the crack tip. L is a length parameter. p is a curve shape parameter determined from compliance data.

C.4.2 Growing Materials

Materials grow by accretion on the outside and by addition to the inside. Growth can be relatively simple, as in the case of macroscopic deposition of bulk materials, or quite complicated as in the molecular to macroscopic organization of biological tissue. From a macroscopic point of view, growth appears to violate macroscopic conservation laws – mass, energy, and entropy production. While it is not possible to violate these conservation laws, microscopic processes, especially those that involve transport into and out of the solid, along with microscopic organization produce what often appears to be a nonconservative process.

C.4.2.1 Growth Models

Continuum models of material growth generally differ from conventional continuum solid mechanics through modifications to the conservation of mass, that is, continuity, equations. A generalized continuity equation with added mass replaces zero on the right side with ρc. c is the mass growth rate factor [Mase 2010].

$$\dot{\rho} + \rho v_{k,k} = \rho c \tag{C.74}$$

Here $V_{k,k}$ is the trace of the velocity gradient tensor. Living biomaterials differ from nonliving materials in that they can grow and adapt to mechanical loads while providing exceptional strength and functionality [Thompson 1992]. Naumov (1994) considered a broad class of simple accretion growth problems based on the assumptions: (1) Material adds to the exterior of the body. Internal interstitial material addition does not occur. (2) The material and growth process are continuous. The body and surface forces act on the body as it grows. It is possible to model some aspects of growing materials by adapting conventional methods [Rausch 2014]. The addition of material alters both the mass and elastic nature of solids. Decomposing the deformation gradient tensor **F** into a product of an elastic-dependent tensor $\mathbf{F}^e$ and a growth tensor $\mathbf{F}^g$ simplifies the analysis, yet leaves the model capable of describing growth as a thermodynamically consistent dissipative system [Loret 2010].

$$\mathbf{F} = \mathbf{F}^e \cdot \mathbf{F}^g \tag{C.75}$$

A probabilistic and antagonistic model of growth by accretion of a surface of height h(x, y, t) is as follows [Kardar 1986]:

$$\frac{\partial h}{\partial t} = \vartheta \nabla^2 h + \frac{\lambda}{2}(\nabla h)^2 + \eta(x,y,t) \tag{C.76}$$

On the right side, the first term, $\vartheta \nabla^2 h$, represents surface tension relaxation. The second term, $\frac{\lambda}{2}(\nabla h)^2$, represents nonlinear growth that favors local variations in height. The third term, $\eta(x,y,t)$, is a zero mean Gaussian noise process.

Many biological structures adapt to environmental loads and stresses by adding and removing material. Wood and bone are well-studied examples. Hsu proposed a linear load adaptive model in which the applied stress results in anisotropic growth, perhaps similar to that exhibited by wood growth in trees [Hsu 1968]. Bone follows the remodeling process qualitatively described by Wolff's law in which loaded portions add material and get stronger, while unloaded portions resorb material and get weaker. Cowin and Hegedus proposed a load adaptive growth model for bone in which bone is a porous solid that adapts to load by changing the amount of porosity, but not the overall dimensions [Cowin 1976]. Adding material in response to loading increases the thickness of the pore walls and strengthens that bone. Carter quantified the concept by decomposing the stress state into the distortional octahedral, S_i, and mean dilatational, D_i, stress invariants and arguing that the distortional stresses promote bone growth, and compressive (negative sign) dilatational stresses promote bone resorption. The model formulates the stress index I for a set of c variant discrete loading conditions, with n_i being the number of load cycles per variant, and the empirical dilatational parameter k [Carter 1987].

$$I = \sum_{i=1}^{c} n_i (S_i + kD_i) \tag{C.77}$$

C.4.2.2 Mechanics of Wound Healing

Epidermal wound healing in biological systems moves cells, grows new cells, grows supporting vascular and neural structures, synthesizes extracellular matrix proteins, and forms scar tissue.

Simple models of wound repair track the concentration, C(x, y, z, t), of growth factor (GF) which promotes cell proliferation and healing (Figure C.24). A 1D model of the decay, diffusion, and production of GF is as follows:

$$\frac{\partial C}{\partial t} - D\nabla^2 C + \lambda C = PS(x,y,z) \tag{C.78}$$

Here D is the diffusion coefficient. λ is a decay factor. P is a production rate factor. S(x, y, z) is a source term which usually takes on values of 1 in active growth regions, and 0 elsewhere. Similar models apply in 2D and radial symmetry domains [Adam 1999] [Arnold 1999]. Note that Eq. (C.78) is a linearized nonhomogeneous form of the reaction–diffusion model in Eq. (C.21).

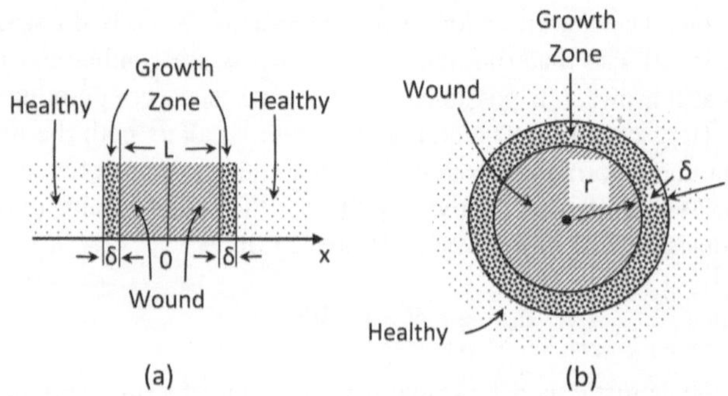

FIGURE C.24 Simple wound geometries. (a) 1D. (b) Circular with radial symmetry.

In wound healing, the rate of ingrowth normal to the boundary v_n is assumed to be of the following form:

$$v_n = (\alpha + \beta\kappa) H(c - \theta) \tag{C.79}$$

Here α and β are parameters such that $\alpha \geq 0$, $\beta \geq 0$, and $\alpha + \beta > 0$. κ is the local curvature of the wound. H is the Heaviside step function. Numerical simulations with this model indicate that healing occurs faster in regions of high local curvature with a large dependence on the choice of parameters α and β [Javierre 2009].

Control of healing is an antagonistic interaction of stimulation and inhibition with a nonlinear dependence on cell population density. The production of chemical stimulants requires living cells, with more cells being able to produce more stimulant. Cells actively produce stimulant at the low cellular population densities characteristic of an unhealed wound. The rate of stimulation wanes at the higher densities corresponding to healthier or healed conditions. A functional form for activation is

$$f(n) = \frac{\lambda c_0 n}{n_0} \cdot \left(\frac{n_0^2 + \alpha^2}{n^2 + \alpha^2} \right) \tag{C.80}$$

and for inhibition

$$f(n) = \frac{\lambda c_0 n}{n_0} \tag{C.81}$$

Here λ, c_0, and α are rate constants. n is the cellular density. n_0 is the density corresponding to a healthy steady state [Sherratt 1991].

A high priority is to close the wound quickly to establish and maintain an epidermal barrier between living tissue and the external environment. Epithelial bridges can span gaps to maintain tension across the gaps with actin filaments while facilitating cellular migration underneath [Vedula 2014].

C.5 Gaussian and Bessel Beams

A beam is a field with an intensity that takes on a maximal value along an axis of propagation and diminishes to zero in transverse directions [Durnin 1987a]. The ideal beam travels in a straight line, maintains a coherent phase front, and does not spread out. Such ideal behavior does not occur in practice. The underlying wave nature of many beams causes them to spread and distort in a process known as diffraction. Huygens' principle is a geometric framework for describing many aspects of diffraction (Figure C.25) [Born 1999]. The concept is that each point in a continuum converts an incoming disturbance into an outgoing spherical wave. Superposing waves emanating from neighboring points form macroscale wavefronts. Rays are directed lines normal to the wavefront.

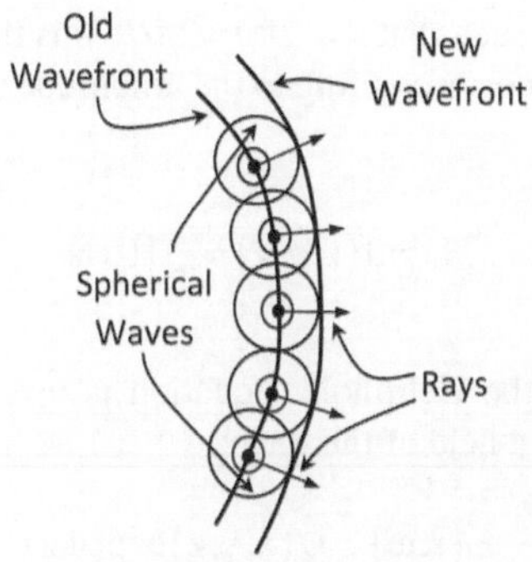

FIGURE C.25 Huygens' principle with spherical waves emanating from points along a wavefront form envelope that is new wavefront. Rays are normal to wavefront.

Waves are propagating disturbances in a field. Mathematical models range from simpler scalar and geometric representations to more sophisticated vector and quantum variants. The wave equation for propagation of a continuous and second-order differentiable scalar function $E(\vec{r},t)$ in a linear isotropic homogeneous medium with no sources:

$$\left(\nabla^2 - \frac{1}{c^2}\frac{\partial^2}{\partial t^2}\right)E(\vec{r},t) = 0 \tag{C.82}$$

Here c is the wave speed. t is the time. $\vec{r} = \vec{r}(x,y,z) = \vec{r}(\rho,\theta,z)$ is a position vector in Cartesian and cylindrical coordinates, respectively, with $\rho = (x^2 + y^2)$ and $\tan(\theta) = y / x$.

Diffraction effects cause a Gaussian beam to spread as it propagates. A beam of uniform intensity launches cleanly through a circular aperture and turns into a Gaussian beam [Verdeyen 1989].

Steady monochromatic light, such as from an ideal laser, separates as a product of harmonic oscillations with stationary spatial distributions. Complex numbers provide a convenient representation.

$$E(\vec{r},t) = \exp(-i\omega t)H(\vec{r}) = E(\vec{r};\omega) \tag{C.83}$$

Here $i = \sqrt{-1}$; $\omega = 2\pi f$ is the circular frequency; f is the frequency; T = 1/f is the period; $H(\vec{r})$ is a complex-valued function of position; $E(\vec{r};\omega)$ indicates that the field is now a complex function of the spatial coordinates with the frequency as a fixed parameter.

The intensity $I(\vec{r})$ is the time-average of the amplitude of the field at a point. The intensity of a harmonically oscillating field is as follows:

$$I(\vec{r}) = \frac{1}{T}\int_0^T \mathrm{Re}\left[E(\vec{r},t)\right]^2 dt = \frac{1}{2}H(\vec{r})H^*(\vec{r}) \tag{C.84}$$

The assumption of harmonic standing waves allows for converting the time derivatives into algebraic terms of k = (ω/c). The result is a conversion of a space-time differential equation into a simpler space-algebra differential equation, that is, the Helmholtz equation.

$$(\nabla^2 + k^2)E(\vec{r};\omega) = 0 \tag{C.85}$$

A plane wave solution to the Helmholtz equation is in many respects a specialized beam with infinite transverse spread. The amplitude is equal for all points in a transverse plane. A plane wave propagating in the z-direction is as follows:

$$E(\vec{r},t) = \exp\left[i(kz - \omega t)\right]H_0 \tag{C.86}$$

Here k is the angular wavenumber, such that $k = 2\pi\upsilon = 2\pi/\lambda$. υ is the wavenumber. λ is the wavelength. H_0 is a constant (possibly complex) corresponding to the amplitude of the wave. The intensity of the field has a constant value I_0:

$$I_0 = I(x,y,z) = \frac{1}{2} H_0 H_0^* \tag{C.87}$$

The paraxial approximation to the Helmholtz equation is useful for describing beams [Kogelnik 1966]. The approximation assumes a field of the form:

$$E(\vec{r};\omega) = \psi(x,y,z)\exp(ikz) \tag{C.88}$$

Substitution of Eq. (C.88) into the Helmholtz equation leads to

$$\left(\frac{\partial^2\psi}{\partial x^2} + \frac{\partial^2\psi}{\partial y^2} + \frac{\partial^2\psi}{\partial z^2} + ik\frac{\partial\psi}{\partial z}\right)\exp(ikz) = 0 \tag{C.89}$$

Key assumptions are that the spatial term $\psi(x,y,z)$ varies slowly enough in the direction of propagation, z, for the second derivative $\partial^2\psi / \partial z^2$ to be negligible. The result is the paraxial differential equation, with a form similar to the time-varying Schrödinger equation:

$$\frac{\partial^2\psi}{\partial x^2} + \frac{\partial^2\psi}{\partial y^2} + i2k\frac{\partial\psi}{\partial z} = 0 \tag{C.90}$$

Solutions to the paraxial equation include beams with Gauss profiles, and those with self-healing diffraction properties, that is, Airy, Bessel, and orbital angular momentum (OAM) beams.

Recasting the Helmholtz differential equation into an integral equation provides qualitative insights and quantitative tools for calculating diffraction and near-field effects. The technique begins with two scalar fields: $E(\vec{r};\omega)$ and $E'(\vec{r};\omega)$. Both fields are continuous and of second order differentiable on and inside an enclosing surface S, with unit normal vector $\vec{n}$ as positive pointing inward (some authors reverse the direction of $\vec{n}$). Green's second identity applies.

$$\iiint_V \left(E\nabla^2 E' - E'\nabla^2 E\right) dv = -\iint_S \left(E\frac{\partial E'}{\partial n} - E'\frac{\partial E}{\partial n}\right) dS \tag{C.91}$$

If both E and E′ are valid solutions to the Helmholtz equation, (C.85), the left side of Eq. (C.91) vanishes by direct substitution, forcing the right side to also vanish:

$$\iint_S \left(E\frac{\partial E'}{\partial n} - E'\frac{\partial E}{\partial n}\right) dS = 0 \tag{C.92}$$

Using E as the unknown field to be determined, E′ as a singular integrating kernel, and S as a surface with defined conditions in Eq. (C-92) forms the basis of a useful integral equation with the geometry of Figure C.26. A suitable E′ is as follows:

$$E' = \frac{e^{ikr}}{r} \tag{C.93}$$

Here r is the distance from a fixed point P(x, y, z) to the variable position point Q(ξ, η, ζ), with

$$r = \left[(x-\xi)^2 + (y-\eta)^2 + (z-\zeta)^2\right]^{1/2} \tag{C.94}$$

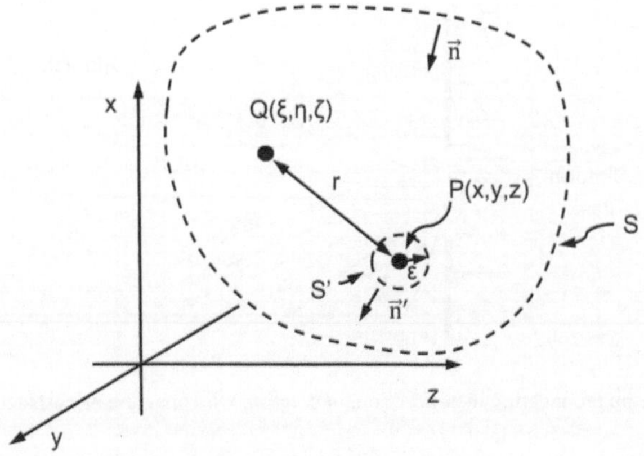

FIGURE C.26 Geometry of Green's integral of Helmholtz equation.

E′ is a solution to the Helmholtz equation everywhere, except at the local origin P, which has a 1/r singularity. Using spherical coordinates (r, θ, φ) with an origin shifted to P, the Laplace operator ∇^2 becomes

$$\nabla^2 = \left[\frac{1}{r^2}\frac{\partial}{\partial r}\left(r^2\frac{\partial}{\partial r}\right) + \frac{1}{r^2\sin\varphi}\frac{\partial}{\partial \varphi}\left(\sin\varphi\frac{\partial}{\partial \varphi}\right) + \frac{1}{r^2\sin^2\varphi}\frac{\partial^2}{\partial \theta^2}\right] \tag{C.95}$$

Since E′ is isotropic, the derivatives with respect to θ and φ vanish. Substitution verifies E′ as a solution of the Helmholtz equation throughout the volume with $r \neq 0$.

$$\left(\nabla^2 + k^2\right)E' = \left[\frac{1}{r^2}\frac{\partial}{\partial r}\left(r^2\frac{\partial}{\partial r}\right) + k^2\right]\frac{e^{ikr}}{r} = 0 \text{ for } r \neq 0 \tag{C.96}$$

The Green's surface integral in Eq. (C.92) avoids the point P where E′ is singular with r = 0 by using a small spherical surface S′ of radius ε and unit inward normal $\vec{n}'$ (Figure C.26).

$$\iint_{S+S'}\left[E\frac{\partial}{\partial n}\left(\frac{e^{ikr}}{r}\right) - \frac{e^{ikr}}{r}\frac{\partial E}{\partial n}\right]dS = 0 \tag{C.97}$$

On S′, r = ε, $\frac{\partial r}{\partial n'} = 1$, Ω is a solid angle so that $dS = \varepsilon^2 d\,\Omega$ and

$$\iint_{S}\left[E\frac{\partial}{\partial n}\left(\frac{e^{ikr}}{r}\right) - \frac{e^{ikr}}{r}\frac{\partial E}{\partial n}\right]dS = -\iint_{S'}\left[E\frac{e^{ik\varepsilon}}{\varepsilon}\left(ik - \frac{1}{\varepsilon}\right) - \frac{e^{ik\varepsilon}}{\varepsilon}\frac{\partial E}{\partial r}\right]\varepsilon^2\, d\Omega \tag{C.98}$$

Terms linear in ε vanish as the size of the S′ sphere shrinks. The remaining terms with E and $e^{ik\varepsilon}$ being smooth on S′ go to E(P) and 1, respectively. Combined with a total solid angle of 4π this forms the integral theorem of Helmholtz and Kirchoff (aka Kirchoff's diffraction formula):

$$E(P) = \frac{1}{4\pi}\iint_{S}\left[E\frac{\partial}{\partial n}\left(\frac{e^{ikr}}{r}\right) - \frac{e^{ikr}}{r}\frac{\partial E}{\partial n}\right]dS \tag{C.99}$$

A standard method of producing a structured beam traveling in the +z-direction begins with a beam of known properties propagating from z = −∞ and intersecting with an aperture surface at the x–y plane (Figure C.27).

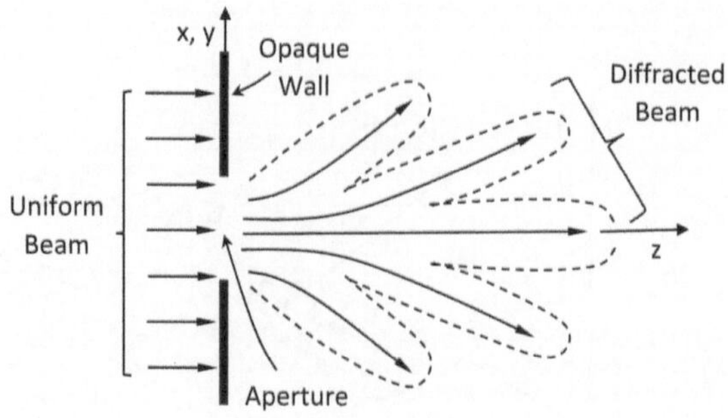

FIGURE C.27 Uniform beam propagating in z-direction, interacting with aperture or obstruction at z = 0 and diffracting for z > 0.

An assumed form for E on S_A is

$$E(x,y,0) = \psi(x,y,z)e^{ikz}\Big|_{z=0} \tag{C.100}$$

This boundary condition approximates many fields that produce beams that propagate in the +z-direction. It is assumed that the integral vanishes on a surface with z > 0 and distant from the origin S_B. This corresponds to the assumption that no waves come from the +z-direction. An examination of Figure C.28 indicates that on the z = 0 plane,

$$\frac{\partial r}{\partial n} = \frac{-z}{r} = -\cos(\theta_{ns}) \tag{C.101}$$

Applying Eq. (C.101) to Eq. (C.100),

$$\frac{\partial E}{\partial n} = \frac{-iEkz}{r} \tag{C.102}$$

The surface integral Eq. (C.99) reduces to

$$E(P) = \frac{1}{4\pi}\iint_{S_A} E\frac{e^{ikr}}{r}\left(i2k - \frac{1}{r}\right)\frac{-z}{r}dS \tag{C.103}$$

The action of Eq. (C.103) is diffraction. The portion of waves from z < 0 that pass through the aperture S_A superpose to form diffracted patterns of field and intensity for z > 0.

Direct integration of Eq. (C.103) is possible in only a small number of cases. Many approximate methods apply in regions where P is multiple wavelengths distant from the x–y plane and the lateral

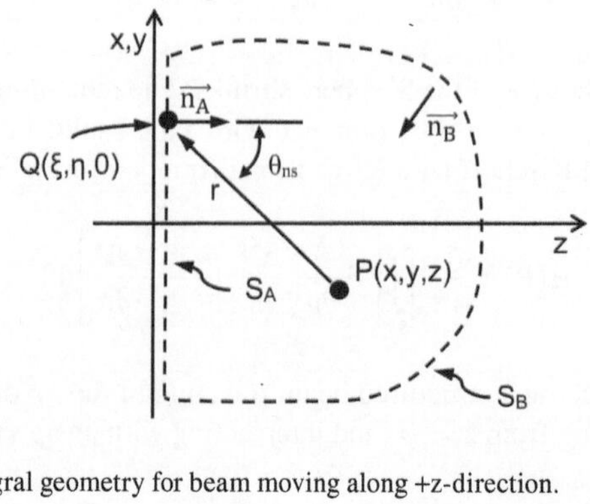

FIGURE C.28 Diffraction integral geometry for beam moving along +z-direction.

position offsets are small, that is, $r > \lambda = 2\pi/k$ and $\theta_{ns} \ll 1$. A large value of r reduces the bracketed term in Eq. (C.103):

$$i2k - \frac{1}{r} \approx i2k \tag{C.104}$$

On S_A,

$$r = z\left[1 + \left(\frac{x-\xi}{z}\right)^2 + \left(\frac{y-\eta}{z}\right)^2\right]^{1/2} \tag{C.105}$$

Expanding in a binomial series with $\theta_{ns} \ll 1$,

$$r = z\left\{1 + \frac{1}{2}\left[\left(\frac{x-\xi}{z}\right)^2 + \left(\frac{y-\eta}{z}\right)^2\right] - \frac{1}{8}\left[\left(\frac{x-\xi}{z}\right)^2 + \left(\frac{y-\eta}{z}\right)^2\right]^2 + \cdots\right\} \tag{C.106}$$

The successive terms in Eq. (C.106) decrease in size and decrease more rapidly as z increases. Truncating converts the Helmholtz–Kirchoff integral into a form of a Fourier transform, known as the Fraunhofer approximation:

$$E(x,y,z) = \frac{k}{2\pi iz}\exp\left\{ik\left[z + \left(\frac{x^2+y^2}{2z}\right)\right]\right\}\iint_{S_A} E(\xi,\eta,0)\exp\left[-i\frac{k}{z}(x\xi + y\eta)\right]d\xi\, d\eta \tag{C.107}$$

Including an additional higher order term leads to the Fresnel approximation. The Fraunhofer approximation works well for effects that lie multiple wavelengths distant from the x–y plane. The Fresnel approximation is more complicated but yields results for fields that are closer to the x–y plane.

Many beams have circular symmetry. Using circular coordinate transform relations

$$\begin{aligned} x &= R\cos\theta \quad & y &= R\sin\theta \\ \xi &= \rho\cos\varphi \quad & \eta &= \rho\sin\varphi \end{aligned} \tag{C.108}$$

leads to the Fraunhofer approximation in cylindrical coordinates (Figure C.29).

$$E(R,\theta,z) = \frac{k}{2\pi iz}\exp\left\{ik\left[z + \frac{R^2}{2z}\right]\right\}\int_0^{\rho}\int_0^{2\pi} E(\rho,\varphi,0)\exp\left[-i\frac{kR}{z}\rho\cos(\varphi-\theta)\right]d\varphi\rho\, d\rho \tag{C.109}$$

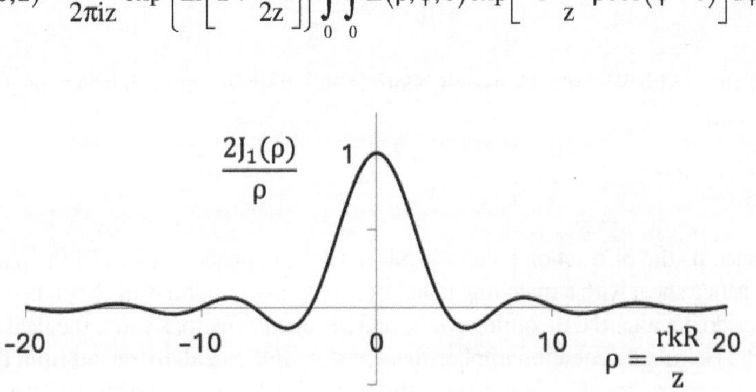

FIGURE C.29 Normalized signal strength versus radial position for Fraunhofer approximation of transmission through a circular aperture

The diffraction formula for cylindrical symmetry becomes a Hankel transformation with use of the Jacobi integral formula [Hildebrand 1976].

$$J_n(\beta) = \int_0^{2\pi} \exp\left[i(\beta\sin\theta - n\theta)\right]d\theta \tag{C.110}$$

$J_n(\beta)$ is the Bessel function of the first kind of order n.

$$E(R,z) = \frac{k}{iz}\exp\left\{ik\left[z + \frac{R^2}{2z}\right]\right\}\int_0^{\infty} E(\rho)J_0\left(\frac{kR\rho}{z}\right)\rho\, d\rho \tag{C.111}$$

$$E(R,\theta,z) = E(R,z) \text{ and } E(\rho,\varphi,0) = E(\rho).$$

A cylindrical Gaussian beam propagating in the z-direction and arriving at the z = 0 plane creates the S_A boundary condition in the form of a Gauss function with polar symmetry.

$$E_G(\rho,0) = E_0 \exp\left(\frac{-a^2\rho^2}{2}\right) \tag{C.112}$$

The Hankel transform of a Gauss function produces another Gauss function, typically with different scaling.

$$\int_0^{\infty} \exp\left(\frac{-a^2\rho^2}{2}\right)J_0(\beta\rho)\,\rho\, d\rho = \frac{1}{a^2}\exp\left[-\frac{1}{2}\left(\frac{\beta}{a}\right)^2\right] \tag{C.113}$$

This result confirms that a beam with Gauss-shape intensity at z = 0 propagates for z > 0 with $\beta = kR/z$ is another Gaussian.

$$E_G(R,z) = \frac{kE_0}{iza^2}\exp\left\{ik\left[z + \frac{R^2}{2z}\right]\right\}\exp\left[-\frac{1}{2}\left(\frac{kR}{za}\right)^2\right] \tag{C.114}$$

With the intensity distribution $I_G(R, z)$,

$$I_G(R,z) = E(R,z)E^*(R,z) = \frac{k^2E_0E_0^*}{z^2a^4}\exp\left[-\left(\frac{kR}{za}\right)^2\right] \tag{C.115}$$

The downrange centerline intensity, $I_G(0,z)$, is

$$I_G(0,z) = \frac{k^2E_0E_0^*}{z^2a^4} \tag{C.116}$$

A definition of the width W_G of a Gaussian beam is the diameter at which the intensity drops to $1/e^2$ of the centerline value.

$$W_G = \frac{za2\sqrt{2}}{k} \tag{C.117}$$

Figure C.27 concerns the diffraction pattern produced by an opaque object and the pattern formed by a complementary opaque sheet with a matching hole. A primary assumption is the boundary conditions where transparent regions do not alter the incoming waves and the opaque regions cause the field to vanish at z = 0. The linearity of the Helmholtz equation implies that the sum of the field diffracted from the opaque object, U_D, and the complementary hole U_C should equal that of the illuminating undisturbed wavefield U_I, that is,

$$U_D + U_C = U_I \tag{C.118}$$

The aperture function, aper(x, y), describes ideal apertures:

$$\text{aper}(x,y)=\begin{cases}1,(x,y)\in\text{transparent region}\\0,(x,y)\in\text{opaque region}\end{cases}\tag{C.119}$$

The aperture boundary condition imposed by a planar surface with an aperture at z = 0 onto a scalar wave propagating in the +z-direction is

$$E(x,y,0)=\text{aper}(x,y)H(x,y)e^{ikz}\Big|_{z=0}\tag{C.120}$$

The circ(r) function is a circular aperture function.

$$\text{circ}(\rho)=\begin{cases}1,\rho<1\\0,\rho>1\end{cases}\tag{C.121}$$

The incidence of a cylindrically symmetric wavefront onto a circular aperture of radius r leads to the boundary conditions:

$$E_C(\rho,0)=\text{circ}(r)H(\rho)e^{ikz}\Big|_{z=0}\tag{C.122}$$

For a plane wave, the S_A boundary condition is

$$E_C(\rho,0)=\text{circ}(r)E_0\,e^{ikz}\Big|_{z=0}\tag{C.123}$$

and

$$E_C(R,z)=\frac{k}{iz}\exp\left\{ik\left[z+\frac{R^2}{2z}\right]\right\}\int_0^\infty \text{circ}(r)E_0J_0\left(\frac{kR\rho}{z}\right)\rho\,d\rho\tag{C.124}$$

Integrating with the formula

$$\int J_0(\rho)\rho\,d\rho=\rho J_1(\rho)+C\tag{C.125}$$

leads to a cylindrical beam with shape $E_c(R, z)$:

$$E_C(R,z)=\frac{rE_0}{iR}\exp\left\{ik\left[z+\frac{R^2}{2z}\right]\right\}J_1\left(\frac{rkR}{z}\right)\tag{C.126}$$

The intensity distribution I(R,z) of the circular aperture beam is as follows:

$$I_C(R,z)=\frac{E_0E_0^*}{2}\left[\frac{r}{R}J_1\left(\frac{rkR}{z}\right)\right]^2\tag{C.127}$$

Applying L'Hôpital's rule to a series expansion of the Bessel function gives the centerline intensity $I_c(0, z)$:

$$I_C(0,z)=\left[\frac{1}{8}E_0E_0^*r^4k^2\right]\frac{1}{z^2}\tag{C.128}$$

A definition of the width W_c of the circular aperture beam is the diameter associated with the first zero in intensity:

$$W_C=\left(\frac{7.662}{rk}\right)z\tag{C.129}$$

Equations (C.116), (C.117), (C.128), and (C.129) indicate that both the Gauss and circular aperture beams have a centerline intensity that drops off as $1/z^2$ and a width that increases with z, that is, the beams spread out and weaken as they propagate. Since both beams exhibit similar spreading behavior, a question arises as to whether this is a common feature of all beams, or do beam shapes exist that do not change shape with propagation? The plane wave is a counter example. It starts out with a lateral spread that goes to infinity and does not spread or change with propagation. Durnin discovered a wider class of beams with intensity distributions that do not change with propagation [Durnin 1987a, 1987b]. These beams, known as *diffraction-free* or *propagation-invariant*, have the general form:

$$E(x,y,z\geq 0,t)=\exp\left[i(\beta z-\omega t)\right]\int_0^{2\pi} A(\varphi)\exp\left[i\alpha(x\cos\varphi+y\sin\varphi)\right]d\varphi \tag{C.130}$$

with the constraint

$$\beta^2+\alpha^2=\left(\frac{\omega}{c}\right)^2 \tag{C.131}$$

$A(\varphi)$ is an arbitrary complex integrable function. Direct substitution of Eq. (C.130) verifies that the diffraction-free beam is a solution to the Helmholtz equation. A simplification that fixes $A(\varphi)$ as a constant creates a subset of solutions, known as Bessel beams, with field E_B:

$$E_B(x,y,z\geq 0,t)=\exp\left[i(\beta z-\omega t)\right]\int_0^{2\pi} \exp\left[i\alpha(x\cos\varphi+y\sin\varphi)\right]d\varphi \tag{C.132}$$

Converting into cylindrical coordinates produces a form that readily integrates.

$$E_B(x,y,z\geq 0,t)=\exp\left[i(\beta z-\omega t)\right]\int_0^{2\pi} \exp\left\{i\alpha\left[R\cos(\varphi-\theta)\right]\right\}d\varphi \tag{C.133}$$

$$E_B(x,y,z\geq 0,t)=\exp\left[i(\beta z-\omega t)\right]J_0(\alpha R) \tag{C.134}$$

These are Bessel beams, which have an intensity distribution I_B that is invariant with propagation:

$$I_B=\frac{1}{2}\left|J_0(\alpha R)\right|^2 \tag{C.135}$$

If $\alpha = 0$, then $J_0(\alpha R) = 1$, and the Bessel beam reverts to the special case of a plane wave.

Bessel beams are axially symmetric self-reconstructing, that is, self-healing, systems. Airy beams are a class of diffraction-free beams that differ from the Durnin class of beams, and typically have bilateral symmetry. The shape of the diffraction pattern does not change with propagation, but the position does [Berry 1979]. The derivation makes use of the paraxial approximation, as in Eq. (C.90), for small-angle optical beam propagation, which leads to a 1D normalized Schrödinger equation [Siviloglou 2007].

$$i\frac{\partial\psi}{\partial\xi}+\frac{1}{2}\frac{\partial^2\psi}{\partial s^2}=0 \tag{C.136}$$

$s = x/x_0$ is a dimensionless transverse dimension. $\xi = z/kx_0^2$ is the dimensionless dimension in the direction of propagation. Solutions take the following form:

$$\psi(s,\xi)=Ai\left[s-\left(\frac{\xi}{2}\right)^2\right]\exp\left[i\left(\frac{s\xi}{2}-\frac{\xi^3}{12}\right)\right] \tag{C.137}$$

Ai(x) is the Airy function defined as a solution to the Airy differential equation:

$$\frac{d^2y}{dx^2}-xy=0 \tag{C.138}$$

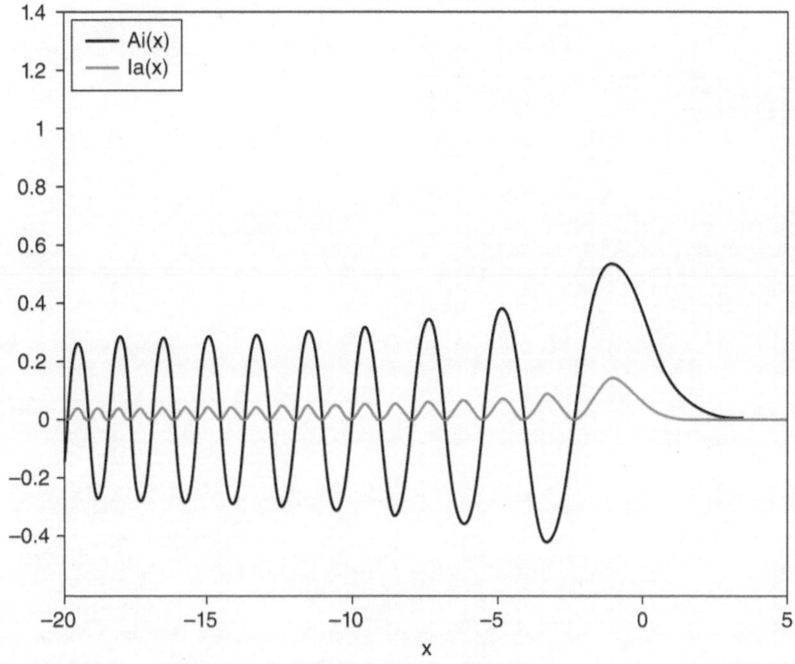

FIGURE C.30 Airy function and corresponding intensity $I_A(s,\xi)$.

The intensity pattern, $I_A(s, \xi)$, formed by the 1D Airy beam is as follows (Figure C.30):

$$I_A(s,\xi) = \frac{1}{2}\psi(s,\xi)\psi^*(s,\xi) = \frac{1}{2}\left\{Ai\left[s - \left(\frac{\xi}{2}\right)^2\right]\right\}^2 \tag{C.139}$$

The intensity pattern remains invariant with propagation in the ξ-direction, but the location shifts by $(\xi/2)^2$ in the s-direction. This movement has the curious property that the beam appears to move transversely as it propagates, even though there is no transverse influence.

Many authors denote the Bessel and Airy beams as being diffraction-free because the intensity distribution does not change with respect to propagation, except for perhaps with a superposed translation, rotation, or scaling. The phrase "diffraction-free" is a misnomer. The physical process of diffraction still occurs but results in an intensity pattern that does not change. Some authors prefer the phrase "propagation invariant."

The Bessel beams, including plane waves, are physically unrealizable. Like the plane wave, the required amount of energy in the beam is infinite due to the amplitude of Bessel functions dropping a off as $R^{-1/2}$ for large R [Hildebrand 1976]. It is possible to create optical beams with approximately invariant distributions over finite ranges. An example is the axicon cylindrical lens element [McLeod 1954]. Airy beams have finite energy [Besieris 2007]. Other propagation-invariant beam shapes are possible, including hollow beams with dark centers and hexagonal symmetry [Han 2015b].

Appendix D: Dimensional Scaling of Physical Effects

D.1 Dimensional Scaling

The behavior of physical phenomena depends heavily on the dimensions of the interactions. Some phenomena change little with size, others dramatically. Simple scaling rules often provide a good explanation. The first step establishes the dimensional units of interest, usually length or time, and occasionally others. The US National Institute of Standards and Technology recognizes six fundamental dimensional units to describe physical quantities:

1. Length (L)
2. Time (T)
3. Luminous intensity (I)
4. Temperature (T)
5. Mass (M)
6. Amount of substance (A)

These units provide a compact representation of physical units as follows [Hamilton 1996]:

$$Q = L^a T^b I^c K^d M^e A^f \tag{D.1}$$

The exponents a–f are usually integers or rational fractions. Equation (D.1) is useful for describing the effect of a change in scale of a dimensional unit, when the same physical phenomena dominate at both scales.

An example is the mechanics of animals. The strength of muscle and bones, S, largely depends on the cross-sectional area of limbs and torsos, and scales as the square of length L:

$$S \sim L^2 \tag{D.2}$$

The weight W depends on the volume of the animal:

$$W \sim L^3 \tag{D.3}$$

Small animals, such as ants, can lift several times their weight. Larger animals, such as humans, often struggle to lift a fraction of their body weight. Small animals, for example, insects and hummingbirds, tend to move quickly and with agility. Big animals are nominally sluggish. Those that can move quickly with agility, for example, the hippopotamus and bear, can do so only for short durations. An explanation considers the motion of an animal as the translation of a particle of lumped mass. Newton's second law of motion relates applied force and motion:

$$F = m\frac{d^2s}{dt^2} = ma \tag{D.4}$$

Here F is the applied force, m is the mass, s is the displacement, t is the time, and a is the acceleration. Integrating the acceleration twice with respect to time for constant force and mass, following quiescent initial conditions, leads to

$$s = \frac{1}{2}at^2 \tag{D.5}$$

and

$$F = ma = \frac{2ms}{t^2} \tag{D.6}$$

A conclusion is that moving a mass with constant acceleration over a specified distance and time implies the dimensional scaling relation:

$$F \sim L^4T^{-2} \tag{D.7}$$

Different physical circumstances can lead to different dimensional scaling effects for applied forces. For example, surface tension forces depend on the length of the contact area. Fluid pressures forces depend on the applied area, that is, the length squared. Gravity and inertial forces scale according to the amount of mass, that is, length cubed.

When the dimensional scales run over a wide range, it is often useful to use logarithms, exponents, or power scaling. Many manufacturers of physical devices scale their model sizes according to a power law rule [Pahl 1988]. Similar examples are denominations of circulating currency, the distribution of wealth in most economic systems, and the decibel scale in acoustic measurements.

The Buckingham PI theorem outlines a nonunique choice of dimensionless parameters to be used for maintaining dimensional similitude. Practical situations limit the ability to match dimensional similitude across the entire range of variables. For example, in aeroelastic wind tunnel testing, it is nearly impossible to satisfy both Reynolds number and reduced aeroelastic frequency scaling. Selecting which dimensionless parameters to match typically is an ad hoc process. It is possible to extend approximate similitude matching techniques by using specific criteria, such as energy dependence, to emphasize the more active variables in similitude [Kittirungsi 2008].

D.2 Trimmer Vectors

The Trimmer vector is a compact representation of the effect of different scaling exponent relations [Trimmer 1989]. The original use was to aid in the design of micro-electromechanical systems. The standard Trimmer vector, {T}, is a 4 × 1 column vector. Row i describes the effect of scaling a base dimension d raised to the ith power:

$$\{T\} = \begin{bmatrix} T_1 \\ T_2 \\ T_3 \\ T_4 \end{bmatrix} = \begin{bmatrix} T(d^1) \\ T(d^2) \\ T(d^3) \\ T(d^4) \end{bmatrix} \tag{D.8}$$

The algebra of Trimmer vectors follows the rules: (1) Raising {T} to a power, raises the individual components to that power. (2) Multiplying {T} by a scalar, such as d^m, multiplies each component by d^m:

$$\{T\}^n d^m = \begin{bmatrix} T_1^n d^m \\ T_2^n d^m \\ T_3^n d^m \\ T_4^n d^m \end{bmatrix} \tag{D.9}$$

A Trimmer force scaling vector $\{L^F\}$ represents the effects of forces that scale according to an integer power of length L.

$$\{L^F\} = \begin{bmatrix} L_1^F \\ L_2^F \\ L_3^F \\ L_4^F \end{bmatrix} = \begin{bmatrix} L^1 \\ L^2 \\ L^3 \\ L^4 \end{bmatrix} \tag{D.10}$$

Some examples:

$$L^F{}_1 = L^1 = \text{Liquid surface tension along a perimeter} \tag{D.11}$$

$$L^F{}_2 = L^2 = \text{Pressure over area} \tag{D.12}$$

$$L^F{}_3 = L^3 = \text{Inertia force acting on volume} \tag{D.13}$$

$$L^F{}_4 = L^4 = \text{Centrifugal body force with constant angular velocity} \tag{D.14}$$

Algebraic manipulations of $\{L^F\}$ produce vectors that represent the scaling of other physical effects. An example from solid mechanics is the deflection of a cantilever beam with a length-scale-dependent load applied to the end (Figure D.1). The dimensionless length-normalized deflection δ is

$$\delta = \frac{d}{L} = \frac{F}{L\left(3EI/L^3\right)} \sim FL^{-2} \tag{D.15}$$

Here d, L, b, h, I, and E are the end deflection, length, width, height, area moment of inertia, and Young's modulus, respectively.

The Trimmer vector for deflection, $\{\delta\}$, is $F \sim \{L^F\}$.

$$\{\delta\} = \begin{bmatrix} \delta_1 \\ \delta_2 \\ \delta_3 \\ \delta_4 \end{bmatrix} = \{L^F\}L^{-2} = \begin{bmatrix} L^{-1} \\ L^0 \\ L^1 \\ L^2 \end{bmatrix} \tag{D.16}$$

An example from rigid body dynamics is the acceleration a of a uniform density mass m, under constant force F.

$$a = \frac{F}{m} \tag{D.17}$$

The acceleration length scaling Trimmer vector $\{a\}$ becomes

$$\{a\} = \begin{bmatrix} a_1 \\ a_2 \\ a_3 \\ a_4 \end{bmatrix} = \{L^F\}L^{-3} = \begin{bmatrix} L^{-2} \\ L^{-1} \\ L^0 \\ L^1 \end{bmatrix} \tag{D.18}$$

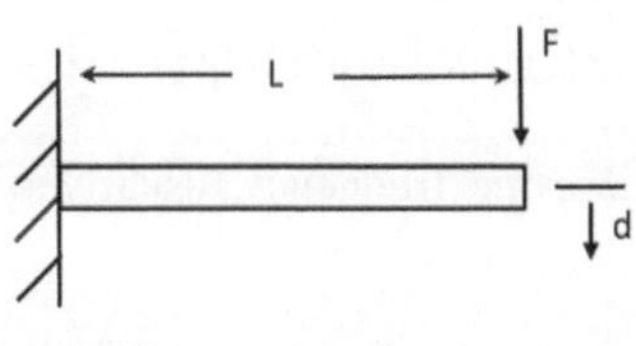

FIGURE D.1 Length-scale-dependent force F applied to the end of cantilever beam of length L with deflection d.

The time t required to move a rigid body through a distance s with constant acceleration is

$$t = \left[\frac{2s}{a}\right]^{1/2} \tag{D.19}$$

The associated Trimmer vector {t} for constant acceleration maneuvers is

$$\{t\} = \begin{bmatrix} t_1 \\ t_2 \\ t_3 \\ t_4 \end{bmatrix} = \left[\frac{L^1 L^3}{\{L^F\}}\right]^{1/2} = \begin{bmatrix} L^{3/2} \\ L^1 \\ L^{1/2} \\ L^0 \end{bmatrix} \tag{D.20}$$

The power density π for a constant acceleration move scales according to

$$\pi = \frac{P}{V} = \frac{Fs}{tV} \tag{D.21}$$

P, V, F, s, and t are mechanical power, volume, force, distance, and time of movement, respectively. The associated length scale Trimmer vector {π} for power density is

$$\{\pi\} = \begin{bmatrix} \pi_1 \\ \pi_2 \\ \pi_3 \\ \pi_4 \end{bmatrix} = \frac{\left[L^F\right]\left[L^1\right]}{\left[L^2\right]\left[L^{-F}\right]^{1/2}\left[L^3\right]} = \frac{\left[L^F\right]^{3/2}}{\left[L^4\right]} = \begin{bmatrix} L^{-5/2} \\ L^{-1} \\ L^{1/2} \\ L^2 \end{bmatrix} \tag{D.22}$$

D.3 Multi-scale Interactions

Multi-scale interactions drive many phenomena. Linear time-invariant systems do not alter the frequency content of input signals and generally do not support multi-scale interactions. Nonlinear interactions transfer energy from one time or spatial scale to another. Fractional dimensional scaling techniques take advantage of combined effects that scale with different dimensional powers.

D.3.1 Nonlinear Harmonic Distortion and Mixing

Harmonic analysis and balancing methods explain many aspects of multi-scale phenomena in nonlinear systems. The proportional system is a linear time-invariant system that converts an input, x(t), into an output, y(t), through the stiffness k.

$$y_L(t) = kx(t) \tag{D.23}$$

If the input to a linear system is a single sinusoid, expressed in complex numbers with cc denoting complex conjugate, that is,

$$x(t) = X_1\left[\exp(i2\pi f_1 t) + cc\right] \tag{D.24}$$

then the output is a sinusoid at the same frequency, usually with an amplitude and phase shift (Figure D.2a).

$$y_L(t) = kX_1\left[\exp(i2\pi f_1 t) + cc\right] \tag{D.25}$$

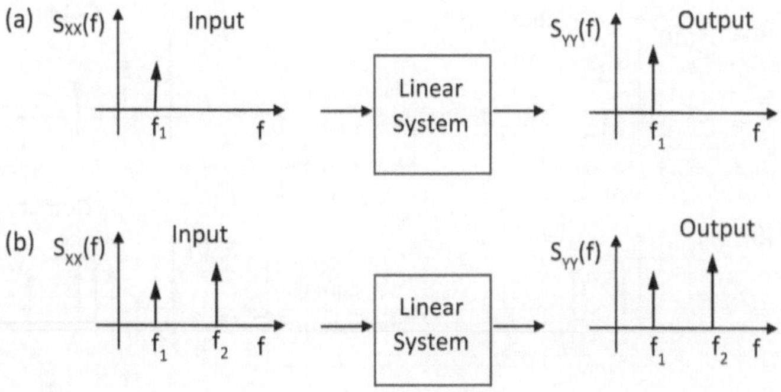

FIGURE D.2 Spectra of input and output signals of linear system. (a) Single-frequency input with single-frequency output. (b) Two frequency input with two frequency output.

When the input to a linear system is the sum of two sinusoids, that is,

$$x(t) = X_1[\exp(i2\pi f_1 t)] + X_2[\exp(i2\pi f_2 t)] + cc \tag{D.26}$$

there is no mixing or alteration of the frequency content of input signal by the linear system. The output is the sum of two sinusoids at the same frequencies as the input (Figure D.2b).

$$y_L(t) = k\{X_1[\exp(i2\pi f_1 t)] + X_2[\exp(i2\pi f_2 t)] + cc\} \tag{D.27}$$

Nonlinear systems do not follow the frequency superposition shown in Figure D.2 and instead create signals $y_N(t)$ at frequencies that multiply and mix those of the inputs. Cubic nonlinearities are perhaps more common, but quadratic nonlinearities demonstrate many of these features with simpler algebra.

$$y_N(t) = kx(t)^2 \tag{D.28}$$

The quadratic nonlinearity converts a single-frequency input, as in Eq. (D.31), into an output containing a component at twice the original frequency and a DC component. The result, sometimes termed harmonic distortion, when combined with a linear term appears in Figure D.3a.

$$y_N(t) = k\{X_1^2[\exp(i4\pi f_1 t) + 1 + cc]\} \tag{D.29}$$

The quadratic nonlinearity converts a double-frequency input, as in Eq. (D.26), into an output with multiple mixed harmonics, as shown in Figure D.3b.

$$\begin{aligned} y_N(t) = k[\![& X_1^2\{\exp(i4\pi f_1 t) + 1\} + X_1 X_2\{\exp[i2\pi(f_1 + f_2)t] + \exp[i2\pi(f_2 - f_1)t]\} \\ & + X_2^2\{\exp(i4\pi f_2 t) + 1\} + cc \end{aligned} \tag{D.30}$$

Adding feedback to a nonlinear system converts a single-input signal initially into one with a harmonic, as shown in Figure D.3b. Passing through the feedback loop mixes to form a cascade of higher harmonics to produce a frequency comb, as shown in Figure D.3c.

D.3.2 Branching and Fractal Scaling

Simple length scaling suggests that for biological systems, volume-based phenomena, such as overall energy consumption, should depend on L^3; and area-based phenomena, such as strength, should depend on L^2. Observations indicate that many phenomena, such as metabolic rate, vary according to $M^{3/4}$, with

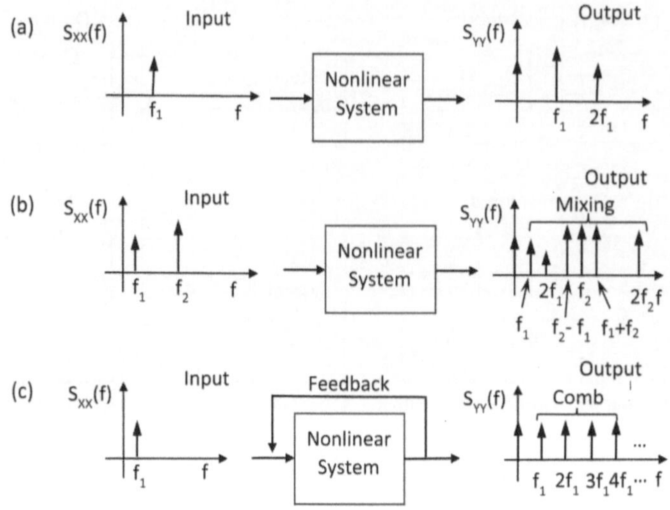

FIGURE D.3 Spectra of input and output signals of system with quadratic nonlinearity. (a) Single-frequency input with harmonic output at twice the input frequency. (b) Two-frequency input with harmonic mixing of frequencies as output. (c) Single-frequency input with feedback produces cascade of harmonics, that is, a frequency comb.

M as the total mass. Fractal and hierarchical organization of biological systems following rules based on minimizing energy consumption while filling volumes is an explanation for these 3/4 power allometric scaling laws [West 1997].

D.4 Selected Scaling Examples

D.4.1 Fluid Flow in Small Pipes

The nature of the flow of viscous fluids in pipes depends on dimensional scaling, especially at smaller dimensions (Figure D.4). The Hagen–Poiseuille law describes laminar flow in circular pipes based on the following assumptions:

1. The pipe is a long and slender cylinder with a smooth circular bore and constant diameter.
2. Fluid is incompressible, has uniform density, is linear viscous, that is, Newtonian, and does not slip at the pipe wall.
3. No external forces, for example, gravity.
4. Pressure varies with axial position but is constant across a circular section.
5. The flow is laminar and axially symmetric.

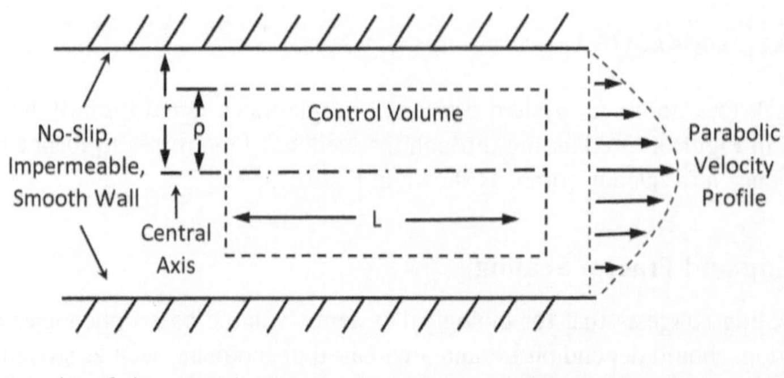

FIGURE D.4 Cross section of pipe, control volume, and resulting parabolic velocity profile in small-diameter pipe.

The analysis considers a cylindrical control volume in the pipe with radius ρ and length L. The absence of external forces leaves pressure and shear stress as the active forces. Since the flow is steady, the net force on the control volume is zero.

Steady flows, pipe-axis cylindrical symmetry and assumptions 1–4 imply that the pressure P depends on axial position x, but not with radial position ρ and angular position θ.

$$P(x,\ \rho,\ \theta) = P(x) \tag{D.31}$$

The kinematic assumptions imply that the velocity u, velocity gradient du/dρ, and shear stress τ depend on radial position ρ, but are independent of axial and angular position.

$$u(x,\ \rho,\ \theta) = u(\rho) \tag{D.32}$$

$$\tau(x,\ \rho,\ \theta) = \tau(\rho) \tag{D.33}$$

Viscous shear stresses develop for a Newtonian fluid:

$$\tau = \mu \frac{du}{d\rho} \tag{D.34}$$

Since the flow is steady, inertia forces vanish. Integrating the pressure and shear stress on the control volume surface produces a net of zero forces. In the radial direction, the control volume cylinder side wall pressures being radially symmetric for a given axial position integrate to zero. In the axial direction, the end face pressures P_1 and P_2 are uniform with radial position and the net end-face forces are the product of the pressure and face area. The control volume side wall axial stress arises due to viscous shear, producing a term equal to the product of shear stress and the circumferential surface area of the control volume. Summing the axial forces,

$$(P_1 - P_2)\pi\rho^2 + \tau 2\pi\rho L = 0 \tag{D.35}$$

Inserting Newtonian viscosity produces a differential equation:

$$\frac{du}{d\rho} = -\frac{(P_1 - P_2)\rho}{2\mu L} \tag{D.36}$$

Integrating from the central axis out to radial position ρ and applying a no slip boundary condition $u(r) = 0$ at the pipe wall produces a velocity field with an axisymmetric paraboloidal shape.

$$u(\rho) = \frac{(P_1 - P_2)}{4\mu L}\left(r^2 - \rho^2\right) \tag{D.37}$$

The volumetric flow rate Q is the integral of the velocity over a pipe section:

$$Q = \frac{(P_1 - P_2)}{8\mu L}\pi r^4 \tag{D.38}$$

The average volumetric flux q is

$$q = \frac{Q}{A} = \frac{(P_1 - P_2)}{8\mu L} r^2 \tag{D.39}$$

The pressure difference ΔP is

$$\Delta P = P_1 - P_2 = \frac{8Q\mu L}{\pi r^4} = \frac{8q\mu L}{r^2} \tag{D.40}$$

This result is the Hagen–Poiseuille law. The total flow rate depends on the fourth power of the tube diameter, which leads to the conclusion that smaller diameter tubes severely restrict flow rates.

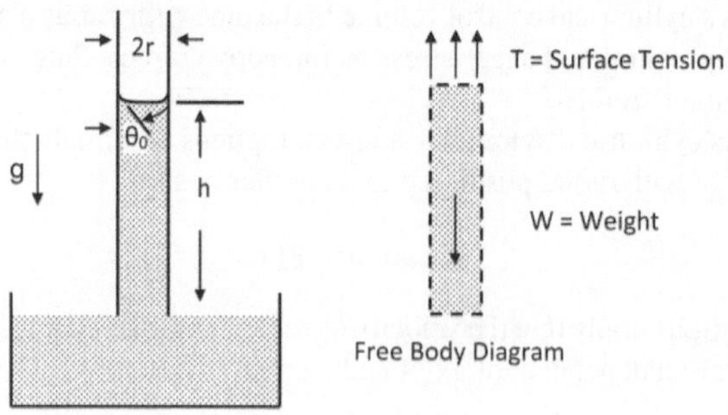

FIGURE D.5 Surface tension draws fluid into capillary to height h against gravity reaching static equilibrium.

D.4.2 Capillary and Wicking Effects

Capillary flows and wicking occur when tension forces form on the surface of liquids in contact with solid bodies. The influence of surface tension forces on free surfaces and lines of contact increases with reduction in feature dimensions in comparison with volume-dependent gravity and inertia forces. Hollow tubes, fiber bundles, porous solids, and textured surfaces all exhibit distinctive behaviors when contacting liquids. Surface tension forces draw liquids into tubes or onto the outside of fibers by capillary or wicking forces. A liquid that wets a solid surface pulls with a force per unit contact length γ, forming an angle of contact θ, relative to the surface. The net surface tension T exerted on a liquid in a tube of radius r is the product of the tension per unit length and the length of the contact line (Figure D.5).

$$T = 2\pi r\gamma\cos(\theta) \tag{D.41}$$

Gravity, with acceleration g, pulls down on the fluid column with mass density ρ. The height at static equilibrium is h.

$$h = \frac{2\gamma\cos(\theta)}{\rho g r} \tag{D.42}$$

Solids imbibe fluids with a surface-tension-activated process known as wicking. The Washburn model of wicking assumes that capillary surface tension, pressure differentials, gravity, and viscosity combine to induce fluid to flow quasi-statically into the solid [Washburn 1921]. The model combines capillary surface tension forces with Hagen–Poiseuille laminar viscous flow mechanics. Pressure differentials and gravity forces superpose to promote or suppress the flow, based on the specific circumstances. An illustrative case is the wicking of fluid into a capillary tube by surface tension, with negligible gravity and differential pressure effects. When balanced against viscosity, the surface tension draws the liquid into the tube such that static equilibrium does not occur. Instead, a quasi-static situation forms where the liquid column h(t) grows with time, as does the viscous force (Figure D.6).

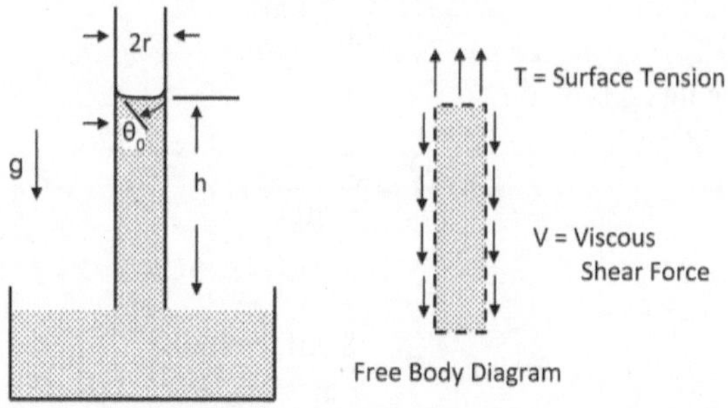

FIGURE D.6 Washburn model of capillary wicking. Surface tension draws the fluid into the tube. Viscous forces resist the flow. Gravity and inertia forces are negligible.

The Washburn model assumes that the surface tension force for the capillary acts as an effective differential pressure in Eq. (D.40), and the effective pipe length L equals the time-varying length of the fluid column h(t).

$$\Delta P = \frac{2\pi r\gamma\cos(\theta)}{\pi r^2} \tag{D.43}$$

The volumetric flow rate Q is

$$Q = \pi r^2 \frac{dh}{dt} \tag{D.44}$$

The Washburn-modified form of the Hagen–Poiseuille law is

$$\pi r^2 \frac{dh}{dt} = \frac{2\gamma\cos(\theta)\pi r^3}{8\mu h} \tag{D.45}$$

The height as a function of time is

$$h(t) = \left[\frac{r\gamma\cos(\theta)t}{2\mu}\right]^{1/2} \tag{D.46}$$

V(t) is the volume of liquid absorbed into a capillary.

$$V(t) = \pi r^2 h(t) = \pi\left[\frac{r^5\gamma\cos(\theta)t}{2\mu}\right]^{1/2} \tag{D.47}$$

The volume of liquid absorbed into a porous surface with an array of N holes of diameter r_i is

$$V(t) = \pi\left[\frac{\gamma\cos(\theta)t}{2\mu}\right]^{1/2}\left(\sum_{i=1}^{N} r_i^{5/2}\right) \tag{D.48}$$

Darcy's law is a continuum model of the flow through a porous medium based on mass and force balance.

$$\vec{q} = -\frac{\kappa}{\mu}(\nabla p - \rho\vec{g}) \tag{D.49}$$

Here, $\vec{q}$ is the velocity of flow expressed as a volumetric flux. κ is the fluid permeability. μ is the viscosity. p is the pressure. ρ is the mass density. $\vec{g}$ is the acceleration due to gravity [Bird 1960]. An advantage of Darcy's law is the ease of insertion into 3D numerical models of liquid imbibition into porous and swellable solids [Masoodi 2010].

Other geometries, such as patterned arrays of nanopillars or cracks, promote wicking in controllable directions (Figure D.7) [Zhou 2011] [Gardner 2014].

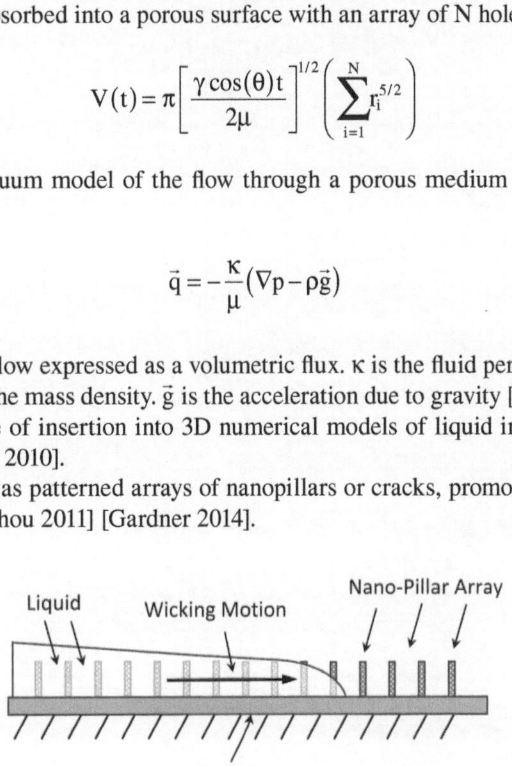

FIGURE D.7 Nanopillar array wicking of fluids. (Adapted from [Zhou 2011].)

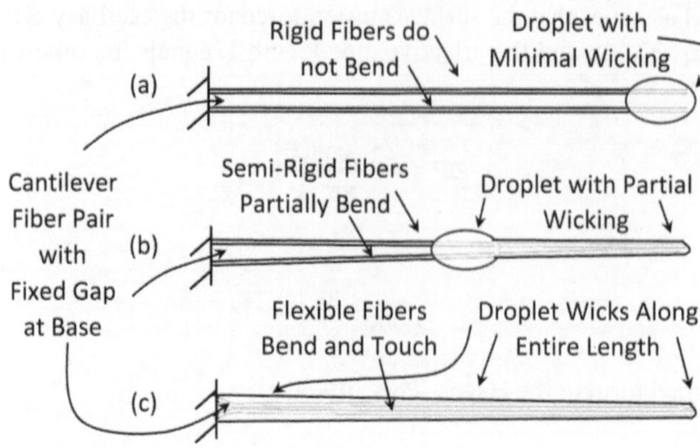

FIGURE D.8 Combined effect of droplet spreading and fiber coalescence due to wetting and fiber stiffness. (a) Stiff fibers bend little and droplet wicks to free end. (b) Moderately flexible fibers that bend with droplet that partially spreads to free end. (c) Flexible fibers bend easily and coalesce as droplet spreads along entire length. (Adapted from [Duprat 2012].)

An additional, but common, complication arises when liquids wet flexible microfeatures, such as fiber arrays. An example is for wetting to cause the fibers to bend and coalesce, while simultaneously causing an impinging droplet to spread (Figure D.8) [Duprat 2012].

References

[Abbott 1999] Abbott S, Ralston J, Reynolds G, Hayes R. (1999) "Reversible Wettability of Photoresponsive Pyrimidine-Coated Surfaces" Langmuir, 15, 26, 8923–8928, doi:10.1021/la990558o

[Abdul 2008] Abdul-Aziz A, Abumeri G, Garg M, Young PG. (2008) "Structural Testing of a Nickel Based Superalloy Metal Foam via NDT and Finite Element Analysis" Materials Evaluation, 66, 9, 949–954

[Abdullayev 2013] Abdullayev E, Abbasov V, Tursunbayeva A, Portnov V, Ibrahimov H, Mukhtarova G, Lvov Y. (2013) "Self-Healing Coatings Based on Halloysite Clay Polymer Composites for Protection of Copper Alloys" ACS Applied Materials & Interfaces, 5, 10, 4464–4471, doi:10.1021/am400936m

[Abe 2006] Abe O, Ohwa Y, Kuranobu YI. (2006) "Possibility of Enhanced Strength and Self-Recovery of Surface Damages of Ceramics Composites under Oxidative Conditions" Journal of the European Ceramic Society, 26, 4–5, 689–695, doi:10.1016/j.jeurceramsoc.2005.06.020

[Abu 2012] Abu Al-Rub R, Darabi M. (2012). "A Thermodynamic Framework for Constitutive Modeling of Time- and Rate-Dependent Materials. Part I: Theory" International Journal of Plasticity, 34, 61–92, doi:10.1016/j.ijplas.2012.01.002

[Acar 2014] Acar H, Çınar S, Thunga M, Kessler MR, Hashemi N, Montazami R. (2014) "Study of Physically Transient Insulating Materials as a Potential Platform for Transient Electronics and Bioelectronics" Advanced Functional Materials, 24, 26, 4135–4143, doi:10.1002/adfm.201304186

[Achal 2009] Achal V, Mukherjee A, Basu PC, Reddy MS. (2009) "Lactose Mother Liquor as an Alternative Nutrient Source for Microbial Concrete Production by *Sporosarcina pasteurii*" Journal of Industrial Microbiology and Biotechnology, 36, 433–438, doi 10.1007/s10295-008-0514-7

[Achlioptas 2009] Achlioptas D, D'Souza RM, Spencer J. (2009) "Explosive Percolation in Random Networks" Science, 323, 1453–1455, doi:10.1126/science.1167782

[ACI 2010] ACI Committee (2010) "Importance of Modulus of Elasticity in Surface Repair Materials" ACI 364.5T-10, TechNote, American Concrete Institute, Farmington Hills, MI

[Adam 1999] Adam JA. (1999) "A Simplified Model of Wound Healing (with Particular Reference to the Critical Size Defect)" Mathematical and Computer Modelling 30, 5–6, 23–32, doi:10.1016/S0895-7177(99)00145-4

[Agarwal 2006] Agarwal S, Horst S, Bognitzki M. (2006) "Electrospinning of Fluorinated Polymers: Formation of Superhydrophobic Surfaces" Macromolecular Materials and Engineering, 291, 592–601

[Agarwal 2009] Agarwal A, Miller J, Eastep J, Wentziaff D, Kasture H. (2009) "Self-Aware Computing" Final Technical Report, AFRL-RI-RS-TR-2009-161, Massachusetts Institute of Technology, Defense Advanced Research Projects Agency

[Ageev 2014] Ageev PG, Molchanov AA. (2014) "Plasma Source for Generating Nonlinear, Wide-Band, Periodic, Directed, Elastic Oscillations and a System and Method for Stimulating Wells, Deposits and Boreholes Using the Plasma Source" US Patent Application 20140027110

[Ageorges 2001] Ageorges C, Ye L, Hou M. (2001) "Advances in Fusion Bonding Techniques for Joining Thermoplastic Matrix Composites: A Review" Composites Part A: Applied Science and Manufacturing, 32, 839–857.

[Agrawal 2013] Agrawal A, Chipara AC, Shamoo Y, Patra PK, Carey BJ, Ajayan PM, Chapman WG, Verduzco R. (2013) "Dynamic Self-Stiffening in Liquid Crystal Elastomers" Nature Communications, 4, 1739, doi:10.1038/ncomms2772

[Aguirre 2013] Aguirre J, Papo D, Buldú JM. (2013) "Successful Strategies for Competing Networks" Nature Physics, 9, 230–234, doi:10.1038/nphys2556

[Ahangari 2014] Ahangari MG, Fereidoon A, Jahanshahi M, Sharifi N. (2014) "Effect of Nanoparticles on the Micromechanical and Surface Properties of Poly(urea-formaldehyde) Composite Microcapsules" Composites Part B: Engineering, 56, 450–455, doi:10.1016/j.compositesb.2013.08.071

[Ahmed 2010] Ahmed T, Shimizu TS, Stocker R. (2010) "Bacterial Chemotaxis in Linear and Nonlinear Steady Microfluidic Gradients" Nano Letters, 10, 3379–3385, doi:10.1021/nl101204e

[Ahmed 2015] Ahmed AS, Ramanujan RV. (2015) "Magnetic Field Triggered Multicycle Damage Sensing and Self Healing" Scientific Reports, 5, 13773, doi:10.1038/srep13773

[Ahn 2010] Ahn TH, Kishi T. (2010) "Crack Self-Healing Behavior of Cementitious Composites Incorporating Various Mineral Admixtures" Journal of Advanced Concrete Technology, 8, 2, 171–186

[Ahn 2014] Ahn BK, Lee DW, Israelachvili JN, Waite JH. (2014) "Surface-Initiated Self-Healing of Polymers in Aqueous Media" Nature Materials, 13, 867–872, doi:10.1038/nmat4037

[Ahn 2015] Ahn S, Lee SJ. (2015) "Optimum Periodicity of Repeated Contractile Actions Applied in Mass Transport" Scientific Reports, 5, 7800, doi:10.1038/srep07800

[Aindow 2010] Aindow M, Alpay SP, Liu Y, Mantese JV, Senturk BS. (2010) "Base Metal Alloys with Self-Healing Native Conductive Oxides for Electrical Contact Materials" Applied Physics Letters, 97, 152103, doi:10.1063/1.3499369

[Aissa 2012] Aissa B, Haddad E, Jamroz W, Hassani S, Farahani RD, Merle PG, Therriault D. (2012) "Micromechanical Characterization of Single-Walled Carbon Nanotube Reinforced Ethylidene Norbornene Nanocomposites for Self-Healing Applications" Smart Materials and Structures, 21, 105028, doi:10.1088/0964-1726/21/10/105028

[Aïssa 2012] Aïssa B, Therriault D, Haddad E, Jamroz W. (2012) "Self-Healing Materials Systems: Overview of Major Approaches and Recent Developed Technologies" Advances in Materials Science and Engineering, 854203, doi:10.1155/2012/854203

[Akimoto 2001] Akimoto H, Kanazawa T, Yamada M, Matsuda S, Shonaike GO, Murakami A. (2001) "Impact Fracture Behavior of Ethylene Ionomer and Structural Change after Stretching" Journal of Applied Polymer Science, 81, 7, 1712–1720, doi:10.1002/app.1603

[Akoglu 2009] Akoglu A, Sreeramareddy A, Josiah JG. (2009) "FPGA Based Distributed Self Healing Architecture for Reusable Systems" Cluster Computing, 12, 3, 269–284, doi:10.1007/s10586-009-0082-2

[Alaimo 2013] Alaimo A, Milazzo A, Orlando C, Messineo A. (2013) "Numerical Analysis of Piezoelectric Active Repair in the Presence of Frictional Contact Conditions" Sensors 13, 4, 4390–4403, doi:10.3390/s130404390

[Albert 2002] Albert R, Barabasi AL. (2002) "Statistical Mechanics of Complex Networks" Reviews of Modern Physics, 74, 47

[Albrecht 2002] Albrecht M, Blau O, Fröhlich R. (2002) "Size-Selectivity" in the Template-Directed Assembly of Dinuclear Triple-Stranded Helicates" Proceedings of the National Academy of Sciences of the United States of America, 99, 8, 4867–4872, doi:10.1073/pnas.062600799

[Aldea 2000] Aldea C, Song W, Popovics JS, Shah SP. (2000) "Extent of Healing of Cracked Normal Strength Concrete" Journal of the Materials in Civil Engineering, 12, 92–96

[Aldridge 2010] Aldridge M, Shankar C, Zhen C, Sui L, Kieffer J, Caruso M, Moore J. (2010) "Combined Experimental and Simulation Study of the Cure Kinetics of DCPD" Journal of the Composite Materials, 44, 2605–2618, doi:10.1177/0021998310373044

[Alessandri 2010] Alessandri I. (2010) "Writing, Self-Healing, and Self-Erasing on Conductive Pressure-Sensitive Adhesives" Small, 6, 1679–1685

[Alexander 2013] Alexander G, Cheng YB, Burford RP, Shanks R, Mansouri J, Barber KW, Rodrigo PD, Preston C. (2013) "Ceramifying Composition for Fire Protection" US Patent 8,409,479

[Alexeev 2008] Alexeev A, Uspal WE, Balazs AC. (2008) "Harnessing Janus Nanoparticles to Create Controllable Pores in Membranes" ACS Nano, 2, 1117–1122

[Aliev 2009] Aliev AE, Oh J, Kozlov ME, Kuznetsov AA, Fang S, Fonseca AF, Ovalle R, Lima MD, Haque MH, Gartstein YN, Zhang M, Zakhidov AA, Baughman RH. (2009) "Giant-Stroke, Superelastic Carbon Nanotube Aerogel Muscles" Science, 323, 5921, 1575–1578, doi:10.1126/science.1168312

[Aliko 2015] Aliko-Benítez A, Doblaré M, Sanz-Herrera JA. (2015) "Chemical-Diffusive Modeling of the Self-Healing Behavior in Concrete" International Journal of Solids and Structures, 69–70, 392–402, doi:10.1016/j.ijsolstr.2015.05.011

[Alizadeh 2012] Alizadeh A, Yamada M, Li R, Shang W, Otta S, Zhong S, Ge L, Dhinojwala A, Conway KR, Bahadur V, Vinciquerra AJ, Stephens B, Blohm ML. (2012) "Dynamics of Ice Nucleation on Water Repellent Surfaces" Langmuir 28, 6, 3180–3186, doi:10.1021/la2045256

[Allison 2012] Allison DB, Bozeman T, MacDonald M, Sevre AK, Webb E. (2012) "Use of Swellable Material in an Annular Seal Element to Prevent Leakage in a Subterranean Well" US Patent 8,127,859

[Almedia 2007] Almedia E, Diamantino TC, deSousa O. (2007) "Marine Paints: The Particular Case of Antifouling Paints" Progress in Organic Coatings, 59, 1, 2–20, doi:10.1016/j.porgcoat.2007.01.017

[Alshalif 2022] Alshalif AF, Juki MI, Tajarudin HA, et al. (2022) "Optimisation of Self-Healing of Bio-Foamed Concrete Bricks Pores Using *Bacillus tequilensis* under Different Temperature and CO_2 Curing Conditions" Scientific Reports, 12, 2682, doi:10.1038/s41598-022-05659-0

[Altman 2012] Altman RB, Terry DS, Zhou Z, Zheng Q, Geggier P, Kolster RA, Zhao Y, Javitch JA, Warren JD, Blanchard SC. (2012) "Cyanine Fluorophore Derivatives with Enhanced Photostability" Nature Methods, 9, 1, 68–71, doi:10.1038/nmeth.1774

[Alvarado 2013] Alvarado J, Sheinman M, Sharma A, MacKintosh FC, Koenderink GH. (2013) "Molecular Motors Robustly Drive Active Gels to a Critically Connected State" Nature Physics, 9, 591–597, doi:10.1038/nphys2715

[Aly 2012] Aly SA, Kamal AE, Al-Kofahi OM. (2012) "Network Protection Codes: Providing Self-Healing in Autonomic Networks Using Network Coding" Computer Networks, 56, 1, 99–111, doi:10.1016/j.comnet.2011.08.013

[Amamoto 2011] Amamoto Y, Kamada J, Otsuka H, Takahara A, Matyjaszewski K. (2011) "Repeatable Photoinduced Self-Healing of Covalently Cross-Linked Polymers through Reshuffling of Trithiocarbonate Units" Angewandte Chemie International Edition, 50, 1660–1663, doi:10.1002/anie.201003888

[Amendola 2009] Amendola V, Dini D, Polizzi S, Shen J, Kadish KM, Calvete MJ, Hanack M, Meneghetti M. (2009) "Self-Healing of Gold Nanoparticles in the Presence of Zinc Phthalocyanines and Their Very Efficient Nonlinear Absorption Performances" Journal of Physical Chemistry C, 113, 8688–8695, doi:10.1021/jp810921w

[Amin 2000] Amin M. (2000) "Toward Self-Healing Infrastructure Systems" IEEE Computer, 33, 8, 44–53, doi:10.1109/2.863967

[Amin 2001] Amin M. (2001) "Toward Self Healing Energy Infrastructure Systems" IEEE Computer Applications in Power, 14, 1, 20–28, doi:10.1109/67.893351

[Amin 2005] Amin SM, Wollenberg BF. (2005) "Toward a Smart Grid" IEEE Power & Energy Magazine, 3, 5, 34–41

[Amir 2014] Amir Y, Ben-Ishay E, Levner D, Ittah S, Abu-Horowitz A, Bachelet I. (2014) "Universal Computing by DNA Origami Robots in a Living Animal" Nature Nanotechnology, 9, 353–357, doi:10.1038/nnano.2014.58

[Anand 2012] Anand S, Paxson AT, Dhiman R, Smith JD, Varanasi KK. (2012) "Enhanced Condensation on Lubricant-Impregnated Nanotextured Surfaces" ACS Nano, 6, 11, 10122–10129, doi:10.1021/nn303867y

[Anderegg 1942] Anderegg FO. (1942) "Autogenous Healing in Mortars Containing Lime" ASTM Bulletin, May 1942, p 22

[Anderson 1986] Anderson RA (1986) "Mechanical Stress in a Dielectric Solid from a Uniform Electric Field" Physical Review B, 33, 2, 1303–1307

[Anderson 2008] Anderson JM, Rodriguez A, Chang DT. (2008) "Foreign Body Reaction to Biomaterials" Seminars in Immunology, 20, 86–100, doi:10.1016/j.smim.2007.11.004

[Anderson 2015] Anderson BR, Gunawidjaja R, Eilers H. (2015) "Self-Healing Organic-Dye-Based Random Lasers" Optics Letters, 40, 4, 577–580, doi:10.1364/OL.40.000577

[Ando 2005] Ando K, Furusawa K, Takahashi K, Sato S. (2005) "Crack-Healing Ability of Structural Ceramics and a New Methodology to Guarantee the Structural Integrity Using the Ability and Proof-Test" Journal of the European Ceramic Society, 25, 549–558, doi:10.1016/j.jeurceramsoc.2004.01.027

[Andreeva 2008] Andreeva DV, Fix D, Mohwald H, Shchukin DG. (2008) "Self-Healing Anticorrosion Coatings Based on pH-Sensitive Polyelectrolyte/Inhibitor Sandwich Like Nanostructures" Advanced Materials, 20, 14, 2789–94, doi:10.1002/adma.200800705

[Angelo 2013] Angelo C, Selejan P. (2013) "Technologies of the Self Healing Grid" Proceedings CIRED, 22nd International Conference and Exhibition on Electricity Distribution, Stockholm, Sweden, doi:10.1049/cp.2013.0968

[Anguiano 2007] Anguiano-Morales M, Martínez A, Iturbe-Castillo MD, Chávez-Cerda S, Alcalá-Ochoa N. (2007) "Self-Healing Property of a Caustic Optical Beam" Applied Optics, 46, 34, 8284–8290, doi:10.1364/AO.46.008284

[Angulo 2015] Angulo MT, Liu YY, Slotine JJ. (2015) "Network Motifs Emerge from Interconnections that Favour Stability" Nature Physics, 11, 848–852, doi:10.1038/nphys3402

[Antonini 2013] Antonini C, Bernagozzi I, Jung S, Poulikakos D, Marengo, M. (2013) "Water Drops Dancing on Ice: How Sublimation Leads to Drop Rebound" Physical Review Letters, 111, 014501, doi:10.1103/PhysRevLett.111.014501

[Appavoo 2003] Appavoo J, Hui K, Soules CA, Wisniewski RW, Da Silva DM, Krieger O, Auslander MA, Edelsohn DJ, Gamsa B, Ganger GR, McKenney P, Ostrowski M, Rosenburg B, Stumm M, Xenidis J. (2003) "Enabling Autonomic Behavior in Systems Software with Hot Swapping" IBM Systems Journal, 42, 1, 60–76

[Appel 2015] Appel EA, Tibbitt MW, Webber MJ, Mattix BA, Veiseh O, Langer R. (2015) "Self-Assembled Hydrogels Utilizing Polymer–Nanoparticle Interactions" Nature Communications, 6, 6295, doi:10.1038/ncomms7295

[Aragon 2007] Aragon AM, Hansen CJ, Wu W, Geubelle PH, Lewis J, White SR. (2007) "Computational Design and Optimization of a Biomimetic Self-Healing/Cooling Material" Proceedings of SPIE, Vol. 6526, Behavior and Mechanics of Multifunctional and Composite Materials, 65261G, doi:10.1117/12.717064

[Aramaki 2002] Aramaki K. (2002) "Self-Healing Mechanism of an Organosiloxane Polymer Film Containing Sodium Silicate and Cerium (III) Nitrate for Corrosion of Scratched Zinc Surface in 0.5 M NaCl" Corrosion Science, 44, 1621–1632.

[Arbuckle 2009] Arbuckle DJ, Requicha AA. (2009) "Self-Assembly and Self-Repair of Arbitrary Shapes by a Swarm of Reactive Robots: Algorithms and Simulations" Autonomous Robots, 28, 197–211, doi:10.1007/s10514-009-9162-7

[Arnold 1999] Arnold JS, Adam JA. (1999) "A Simplified Model of Wound Healing II: The Critical Size Defect in Two Dimensions" Mathematical and Computer Modelling 30, 11–12, 47–60, doi:10.1016/S0895-7177(99)00197-1

[Arora 2007] Arora WJ, Smith HI, Barbastathis G. (2007) "Membrane Folding by Ion Implantation Induced Stress to Fabricate Three-Dimensional Nanostructures" Microelectronic Engineering, 84, 1454–1458

[Arroyoa 2014] Arroyoa M, DeSimone A. (2014) "Shape Control of Active Surfaces Inspired by the Movement of Euglenids" Journal of the Mechanics and Physics of Solids, 62, 99–112, doi:10.1016/j.jmps.2013.09.017

[Arson 2012] Arson C, Xu H, Chester FM. (2012) "On the Definition of Damage in Time-Dependent Healing Models for Salt Rock" Géotechnique Letters, 2, 2, 67–71, doi:10.1680/geolett.12.00013

[Artus 2012] Artus GR, Seeger S. (2012) "Scale-Up of a Reaction Chamber for Superhydrophobic Coatings Based on Silicone Nanofilaments" Industrial & Engineering Chemistry Research, 51, 6, 2631–2636, doi:10.1021/ie202129z

[Arunchandran 2012] Arunchandran C, Ramya S, George RP, Mudali UK. (2012) "Self-Healing Corrosion Resistive Coatings Based on Inhibitor Loaded TiO_2 Nanocontainers" Journal of the Electrochemical Society, 159, 11, C552–C559, doi:10.1149/2.020212jes

[Astier 2009] Astier Y, Uzun O, Stellacci F. (2009) "Electrophysiological Study of Single Gold Nanoparticle/a-Hemolysin Complex Formation: A Nanotool to SlowDown ssDNA through the a-Hemolysin Nanopore" Small, 5, 1273–1278

[ASTM 2008a] ASTM (2008) "Standard Practice for Surface Wettability of Coatings, Substrates and Pigments by Advancing Contact Angle Measurement" D 7334 – 08, ASTM International, West Conshohocken, PA, USA

[ASTM 2008b] ASTM (2008) "Standard Test Method for Surface Wettability and Absorbency of Sheeted Materials Using an Automated Contact Angle Tester" D 5725 – 99 (Reapproved 2008), ASTM International, West Conshohocken, PA, USA

[ASTM 2013] ASTM. (2013) "Standard Specification for Flooding Compounds for Telecommunications Wire and Cable" D4730 – 13, ASTM International, West Conshohocken, PA, USA

[Atwood 2002] Atwood JL, Barbour LJ, Jerga A. (2002) "Organization of the Interior of Molecular Capsules by Hydrogen Bonding" Proceedings of the National Academy of Sciences of the United States of America, 99, 8, 4837–4841, doi:10.1073/pnas.082659799

[Augugliaro 2014] Augugliaro F, Lupashin S, Hamer M, Male C, Hehn M, Mueller MW, Willmann JS, Gramazio F, Kohler M, D'Andrea. (2014) "The Flight Assembled Architecture Installation: Cooperative Construction with Flying Machines" IEEE Control Systems Magazine, 34, 4, 46–64, doi:10.1109/MCS.2014.2320359.

[Ausländer 2012] Ausländer S, Ausländer D, Müller M, Wieland M, Fussenegger M. (2012) "Programmable Single-Cell Mammalian Biocomputers" Nature, 487, 123–127, doi:10.1038/nature11149

[Avizienis 1971] Avizienis A, Gilley GC, Mathur FP, et al. (1971) "The STAR (Self-Testing and Repairing) Computer: An Investigation of the Theory and Practice of Fault-Tolerant Computer Design" IEEE Transactions on Computers, C-20, 1312–1321. doi:10.1109/T-C.1971.223133

[Avramova 1993] Avramova N. (1993) "Study of the Healing Process of Polymers with Different Chemical Structure and Chain Mobility" Polymer 34, 1904–1907, doi:10.1016/0032-3861(93)90433-B

[Azimi 1996] Azimi HR, Pearson RA, Hertzberg RW. (1996) "Fatigue of Hybrid Epoxy Composites: Epoxies Containing Rubber and Hollow Glass Spheres" Polymer Engineering & Science, 36, 18, 2352–2365, doi:10.1002/pen.10633

[Azimi 2013] Azimi G, Dhiman R, Kwon HM, Paxson AT, Varanasi KK. (2013) "Hydrophobicity of Rare-Earth Oxide Ceramics" Nature Materials 12, 315–320, doi:10.1038/nmat3545

[Azizinamini 2014] Azizinamini A, Power EH, Myers GF, Ozyildirim HC, Kline E, Whitmore DW, Mertz DR. (2014) "Design Guide for Bridges for Service Life" Transportation Research Board, Washington DC, SHRP 2 Report S2-R19A-RW-2

[Azzaroni 2006] Azzaroni O, Trappmann B, van Rijn P, Zhou F, Kong B, Huck WT. (2006) "Mechanically Induced Generation of Counterions inside Surface-Grafted Charged Macromolecular Films: Towards Enhanced Mechanotransduction in Artificial Systems" Angewandte Chemie International Edition, 45, 44, 7440–7443, doi:10.1002/anie.200602666

[Bachmeier 2002] Bachmeier KL, Williams AE, Warmington JR, Bang SS. (2002) "Urease Activity in Microbiologically-Induced Calcite Precipitation" Journal of Biotechnology 93, 2, 171–181

[Bae 2005] Bae WS, Lestage DJ, Proia M, Heinhorst S, Urban MW. (2005) "Film Formation from Colloidal Dispersions Stabilized by Sugar Derivatives and Their Controllable Release for Selective Protein Adsorption" Biomacromolecules, 6, 2615–2621

[Bagheri 1996] Bagheri R, Pearson RA. (1996) "Role of Particle Cavitation in Rubber Toughened Epoxies: 1 Microvoid Toughening" Polymer, 37, 20, 4529–4538, doi:10.1016/0032-3861(96)00295-9

[Bagherifard 2015] Bagherifard S, Hickey DJ, de Luca AC, Malheiro VN, Markaki AE, Guagliano M, Webster TJ. (2015) "The Influence of Nanostructured Features on Bacterial Adhesion and Bone Cell Functions on Severely Shot Peened 316L Stainless Steel" Biomaterials, 73, 185–197, doi:10.1016/j.biomaterials.2015.09.019

[Bahder 1976] Bahder G. Eager Jr. GS, Silver DA. (1976) "Extruded Solid Dielectric High Voltage Cable Resistant to Electro-Chemical Trees" US Patent 3,943,271

[Bai 2010] Bai XM, Voter AF, Hoagland RG, Nastasi M, Uberuaga BP. (2010) "Boundaries via Interstitial Emission Efficient Annealing of Radiation Damage Near Grain" Science 327, 1631–1634, doi:10.1126/science.1183723

[Bai 2014a] Bai T, Liu S, Sun F, Sinclair A, Zhang L, Shao Q, Jiang S. (2014) "Zwitterionic Fusion in Hydrogels and Spontaneous and Time-Independent Self-Healing under Physiological Conditions" Biomaterials, 35, 13, 3926–3933, doi:10.1016/j.biomaterials.2014.01.077

[Bai 2014b] Bai Y, Chen Y, Wang Q, Wang T. (2014) "Poly(Vinyl Butyral) Based Polymer Networks with Dual-Responsive Shape Memory and Self-Healing Properties" Journal of Materials Chemistry A, 2, 9169–9177, doi:10.1039/C4TA00856A

[Bai 2015] Bai D, Zhang D, Zhang X, Chen G. (2015) "Origin of Self-Preservation Effect for Hydrate Decomposition: Coupling of Mass and Heat Transfer Resistances" Scientific Reports, 5, 14599, doi:10.1038/srep14599

[Bailey 2014] Bailey MM, Heddleston JM, Davis J, Staymates JL, Walker AR. (2014) "Functionalized, Carbon Nanotube Material for the Catalytic Degradation of Organophosphate Nerve Agents" Nano Research, 7, 3, 390–398, doi:10.1007/s12274-014-0405-3

[Bailin 1987] Bailin LJ, Agarwala VS. (1987) "Development of Microencapsulated DNBM Quaternary Ammonium Inhibitors for Paints" Proceedings of the 1987 Trl-Service Conference on Corrosion, Air Force Wright Aeronautical Laboratories, Materials Laboratory, US Air Force Academy, CO, AFWAL-TR-87-4139, Vol. II, pp 111–127

[Baker 1971] Baker RE, Cook RL, Slagel EC. (1971) "Polymer Laminate to Prevent Severe Metal Petalling Damage" US Patent 3,577,306

[Baker 1972] Baker RE, Fogarty RL, Slagel EC. (1972) "Self Sealing Composite" US Patent 3,698,587

[Balazs 2007] Balazs AC. (2007) "Modeling Self-Healing Materials" Materials Today, 10, 9, 18–23

[Baldi 2003] Baldi A, Gu Y, Loftness PE, Siegel RA, Ziaie B. (2003) "A Hydrogel-Actuated Environmentally Sensitive Microvalve for Active Flow Control" Journal of Microelectromechanical Systems, 12, 5, 613–621, doi:10.1109/JMEMS.2003.818070

[Baldwin 2012] Baldwin SP, Saltzman WM. (1998) "Materials for Protein Delivery in Tissue Engineering" Advanced Drug Delivery Reviews, 33, 71–86, doi:10.1016/s0169-409x(98)00021-0

[Ball 2023] Ball P. (2023) "Materials with Agency" Nature Materials, 22, 272–272, doi:10.1038/s41563-023-01501-8

[Ballauff 2007] Ballauff M, Lu Y. (2007) "'Smart' nanoparticles: Preparation, Characterization and Applications" Polymer, 48, 1815–1823

[Balsells 1993] Balsells PJ. (1993) "Canted Coil Spring in Length Filled with an Elastomer" US Patent 5,203,849

[Bang 2010] Bang SS, Lippert JJ, Yerra U, Mulukutla S, Ramakrishnan V. (2010) "Microbial Calcite, a Bio-Based Smart Nanomaterial in Concrete Remediation" International Journal of Smart and Nano Materials, 1, 1, 28–39, doi:10.1080/19475411003593451

[Banigan 2013] Banigan EJ, Illich MK, Stace-Naughton DJ, Egolf DA. (2013) "The Chaotic Dynamics of Jamming" Nature Physics, 9, 288–292, doi:10.1038/nphys2593

[Bar 2012] Bar-Cohen Y, Bao X, Sherrit S, Badescu M, Lih SS. (2012) "High-Speed Transport of Fluid Drops and Solid Particles via Surface Acoustic Waves" NASA Tech Briefs, Technical Support Package for NPO-46252, Jet Propulsion Laboratory, Pasadena, CA

[Barani 2008] Barani H, Montazer M. (2008) "A Review on Applications of Liposomes in Textile Processing" Journal of the Liposome Research, 18, 249–262, doi:10.1080/08982100802354665

[Barbe 2007] Barbe CJ Bartlett J. (2007) "Controlled Release Ceramic Particles, Compositions Thereof, Processes of Preparation and Methods of Use" US Patent 7,258,874

[Barbero 2005] Barbero EJ, Greco F, Lonetti P. (2005) "Continuum Damage Healing Mechanics with Application to Self-Healing Composites" International Journal of Damage Mechanics, 14, 51–81

[Bar-Cohen 2006] Bar-Cohen Y. (2006) "Biomimetics: Using Nature to Inspire Human Innovation" Bioinspiration & Biomimetics, 1, 1–12, doi:10.1088/1748-3182/1/1/P01

[Barth 2002] Barth, Jr. JE, Dreibelbis JH, Nelson EA, Anand DL, Pomichter G, Jakobsen P, Nelms MR, Leach J, Belansek GM. (2002) "Embedded DRAM Design and Architecture for the IBM 0.11-μm ASIC Offering" IBM Journal of Research and Development, 46, 6, 675–689

[Barthlott 1997] Barthlott W, Neinhuis C. (1997) "Purity of the Sacred Lotus, or Escape from Contamination in Biological Surfaces" Planta, 202, 1–8

[Bartolo 2006] Bartolo D, Bouamrirene F, Verneuil É, Buguin A, Silberzan P, Moulinet S. (2006) "Bouncing or Sticky Droplets: Impalement Transitions on Superhydrophobic Micropatterned Surfaces" Europhysics Letters, 74, 2, 299–305, doi:10.1209/epl/i2005-10522-3

[Barton 1952] Barton FD. (1952) "Stop-Leak Preparation" US Patent 2,580,719

[Barzel 2013] Barzel B, Barabási AL. (2013) "Universality in Network Dynamics" Nature Physics 9, 673–681, doi:10.1038/nphys2741

[Bassik 2010] Bassik N, Brafman A, Zarafshar AM, Jamal M, Luvsanjav D, Selaru FM, Gracias DH. (2010) "Enzymatically Triggered Actuation of Miniaturized Tools" Journal of the American Chemical Society, 132, 16314–16317

[Batt 2007] Batt G, Belta C, Weiss R. (2007) "Temporal Logic Analysis of Gene Networks under Parameter Uncertainty" IEEE Transactions on Automatic Control, 53, Special, 215–229, doi:10.1109/TAC.2007.911330

[Battocchi 2006] Battocchi D, Simoes AM, Tallman DE, Bierwagen GP. (2006) "Electrochemical Behaviour of a Mg-Rich Primer in the Protection of Al Alloys" Corrosion Science, 48, 1292–1306, doi:10.1016/j.corsci.2005.04.008

[Bauer 2009] Bauer G, Nellesen A, Sengespeick A, Speck T. (2009) "Fast Self-Repair Mechanisms in Plants: Biological Lattices as Role Models for the Development of Biomimetic Self-Healing, Mechanically Loaded Polymers" 6th Plant Biomechanics Conference – Cayenne, November 16–21

[Baumann 2018] Baumann A, Sánchez-Ferrer A, Jacomine L, Martinoty P, Le Houerou V, Ziebert F, Kulić IM. (2018) "Motorizing Fibres with Geometric Zero-Energy Modes" Nature Materials, 17, 523–527, doi:10.1038/s41563-018-0062-0

[Baumgartl 2008] Baumgartl J, Mazilu M, Dholakia K. (2008) "Abstract Optically Mediated Particle Clearing Using Airy Wavepackets" Nature Photonics, 2, 675–678, doi:10.1038/nphoton.2008.201

[Bautista 2014] Bautista DM, Wilson SR, Hoon MA. (2014) "Why We Scratch an Itch: The Molecules, Cells and Circuits of Itch" Nature Neuroscience, 17, 175–182, doi:10.1038/nn.3619

[Bayer 2011] Bayer IS, Biswas A. (2011) "Organic Nanotubes" Vacuum Technology & Coating, 12, 7, 18–25

[Bayer 2013] Bayer IS, Biswas A. (2013) "Self-Cleaning Surfaces" Vacuum Technology & Coating, 14, 5, 14–17

[Baylis 2015] Baylis JR, Yeon JH, Thomson MH, Kazerooni A, Wang X, St. John AE, Lim EB, Chien D, Lee A, Zhang JQ, Piret JM, Machan LS, Burke TF, White NJ, Kastrup CJ. (2015) "Self-Propelled Particles that Transport Cargo through Flowing Blood and Halt Hemorrhage" Science Advances, 1, 9, e1500379, doi:10.1126/sciadv.1500379

[Bazilevsky 2007] Bazilevsky AV, Yarin AL, Megaridis CM. (2007) "Co-Electrospinning of Core–Shell Fibers Using a Single-Nozzle Technique" Langmuir, 23, 5, 2311–2314, doi:10.1021/la063194q

[Bazin 1967] Bazin P, Saunier JB. (1967) "Deformability, Fatigue and Healing Properties of Asphalt Mixes" Proceedings of the Second International Conference on the Structural Design of Asphalt Pavements, Ann Arbor, MI, USA, p. 553–69.

[Beard 1971] Beard RV. (1971) "Failure Accommodation in Linear Systems through Self-Reorganization" Doctoral Dissertation, Aeronautics and Astronautics, MIT

[Beck 2003] Beck JB, Rowan SJ. (2003) "Multistimuli, Multiresponsive Metallo-Supramolecular Polymers" Journal of the American Chemical Society, 125, 13922

[Behl 2010] Behl M, Lendlein A. (2010) "Triple-Shape Polymers" Journal of Materials Chemistry, 20, 3335–3345, doi:10.1039/B922992B

[Behl 2013] Behl M, Kratz K, Noechel U, Sauter T, Lendlein A. (2013) "Temperature-Memory Polymer Actuators" Proceedings of the National Academy of Sciences of the United States of America, 110, 31, 12555–12559, doi:10.1073/pnas.1301895110

[Beiermann 2009] Beiermann BA, Keller MW, Sottos NR (2009) "Self-Healing Flexible Laminates for Resealing of Puncture Damage" Smart Materials and Structures, 18, 8, 1–7

[Beijer 1998] Beijer FH, Sijbesma RP, Kooijman H, Spek AL, Meijer EW. (1998) "Strong Dimerization of Ureidopyrimidones via Quadruple Hydrogen Bonding" Journal of the American Chemical Society, 120, 6761–6769

[Bejan 2006] Bejan A, Lorente S, Wang KM. (2006) "Networks of Channels for Self-Healing Composite Materials" Journal of the Applied Physics, 100, 033528, doi:/10.1063/1.2218768

[Belenkova 2012] Belenkova TL, Rimmerman D, Mentovich E, Gilon H, Hendler N, Richter S, Markovich G. (2012) "UV Induced Formation of Transparent Au-Ag Nanowire Mesh Film for Repairable OLED Devices" Journal of Materials Chemistry, 22, 24042–24047, doi:10.1039/C2JM35291E

[Belkin 2007] Belkin M, Snezhko A, Aranson IS, Kwok WK. (2007) "Driven Magnetic Particles on a Fluid Surface: Pattern Assisted Surface Flows" Physical Review Letters, 99, 158301, doi:10.1103/PhysRevLett.99.158301

[Belkin 2015] Belkin A, Hubler A Bezryadin A. (2015) "Self-Assembled Wiggling Nano-Structures and the Principle of Maximum Entropy Production" Scientific Reports, 5, 8323, doi:10.1038/srep08323

[Bell 1984] Bell GI, Dembo M, Bongrand P. (1984) "Cell adhesion. Competition between Nonspecific Repulsion and Specific Bonding" Biophysical Journal, 45, 1051–1064, doi:10.1016/S0006-3495(84)84252-6

[Bellan 2009] Bellan LM, Singh SP, Henderson PW, Porri TJ, Craighead HG, Spector JA. (2009) "Fabrication of an Artificial 3-Dimensional Vascular Network Using Sacrificial Sugar Structures" Soft Matter, 5, 7, 1354–1357, doi:10.1039/B819905A

[Belli 2006] Belli S, Bareggi A, Veggetti P, Castellani L, Pozzati G, Balconi L. (2006) "Electric Cable Resistant to Water Penetration" US Patent 7,087,842

[Bencherif 2012] Bencherif SA, Sands RW, Bhatta D, Aranya P, Verbeke CS, Edwards DA, Mooney DJ. (2012) "Injectable Preformed Scaffolds with Shape-Memory Properties" Proceedings of the National Academy of Sciences of the United States of America, 109, 48, 19590–19595, doi:10.1073/pnas.1211516109

[Bennett 1932] Bennett CE. (1932) "Cable System" (1932) US Patent 1,969,721

[Bennett 1934] Bennett CE. (1934) "Electric Cable System" US Patent 1,975,075

[Bennett 2006] Bennett IJ, Sloof WG. (2006) "Modeling the Influence of Reactive Elements on the Work of Adhesion between a Thermally Grown Oxide and a Bond Coat Alloy" Materials and Corrosion, 57, 3, 223–229, doi:10.1002/maco.200503928

[Bennett 2008] Bennett JA, Young RJ (2008) "A Strength Based Criterion for the Prediction of Stable Fibre Crack-Bridging" Composites Science and Technology 68, 1282–1296

[Bentz 2009] Bentz DP, Peltz MA, Snyder KA, Davis JM. (2009) "VERDiCT: Viscosity Enhancers Reducing Diffusion in Concrete Technology" Concrete International, 31, 1, 31–36

[Bererton 2002] Bererton C, Khosla P. (2002) "An Analysis of Cooperative Repair Capabilities in a Team of Robots" Proceedings 2002, IEEE International Conference on Robotics and Automation, Washington, DC

[Beres 1997] Beres W, Koul AK, Thamburaj R. (1997) "A Tapered Double-Cantilever-Beam Specimen Designed for Constant-K Testing at Elevated Temperatures" Journal of Testing and Evaluation, 25, 6, 536–542

[Bergman 2007] Bergman SD, Wudl F. (2007) "Mendable Polymers" Journal of Materials Chemistry, 2008, 18, 41–62, doi:10.1039/b713953p

[Bergmeister 2015] Bergmeister H, Seyidova N, Schreiber C, Strobl M, Grasl C, Walter I, Messner B, Baudis S, Fröhlich S, Marchetti-Deschmann M, Griesser M, di Franco M, Ksrsaki M, Liska R, Schima H. (2015) "Biodegradable, Thermoplastic Polyurethane Grafts for Small Diameter Vascular Replacements" Acta Biomaterialia, 11, 1, 104–113, doi:10.1016/j.actbio.2014.09.003

[Bernal 2015] Bernal RA, Aghaei A, Lee S, Ryu S, Sohn K, Huang J, Cai W, Espinosa H. (2015) "Intrinsic Bauschinger Effect and Recoverable Plasticity in Pentatwinned Silver Nanowires Tested in Tension" Nano Letters, 15, 1, 139–146, doi:10.1021/nl503237t

[Bernikowicz 1998] Bernikowicz P. (1998) "Terminator III: Design of Tin-based Biomimetic Self-Healing Alloy Tensile Specimens" TMS Student Awards Program, Northwestern University, Evanston, USA, http://www.tms.org/students/winners/bernikowicz/bernikowicz.html

[Berry 1979] Berry MV, Balazs NL. (1979) "Nonspreading Wave Packets" American Journal of Physics, 47, 3, 264–267

[Berry 2004] Berry MV. (2004) "Physics of Nonhermitian Degeneracies" Czechoslovak Journal of Physics, 54, 10, 1039–1047, doi:10.1023/B:CJOP.0000044002.05657.04

[Berton 2007] Berton S, Infanti S, Castellano MG, Hikosaka H. (2007) "Self-Centring Capacity of Seismic Isolation Systems" Structural Control and Health Monitoring, 14, 895–914

[Besieris 2007] Besieris IM, Shaarawi AM. (2007) "A Note on an Accelerating Finite Energy Airy Beam" Optics Letters, 32, 2447–2249

[Bettye 1999] Bettye L. Smith BL, Tilman E. Schaeffer TE, Viani M, Thompson JB, Frederick NA, Kindt J, Belcher A, Stuckyk GD, Morse DE, Hansma PK. (1999) "Molecularmechanistic Origin of the Toughness of Natural Adhesives, Fibres and Composites" Nature 399, 761–763, doi:10.1038/21607

[Beveridge 1983] Beveridge TJ, Meloche JD, Fyfe WS, Murray RG. (1983) "Diagenesis of Metals Chemically Complexed to Bacteria: Laboratory Formation of Metal Phosphates, Sulfides, and Organic Condensates in Artificial Sediments" Applied and Environmental Microbiology, 45, 3, 1094–1108

[Bhagavath 1998] Bhagavath VK. (1998) "Self-Healing Configuration for Delivering Data Services on a Hybrid Fiber-Coaxial (HFC) Network" US Patent 5,835,125

[Bhasin 2011] Bhasin A, Palvadi S, Little DN. (2011) "Influence of Aging and Temperature on Intrinsic Healing of Asphalt Binders," Transportation Research Record 2207, Transportation Research Board, Washington, DC, pp. 70–78

[Bhattacharya 2008] Bhattacharya SK, Kshirsagar A. (2008) "First Principle Study of Free and Surface Terminated CdTe Nanoparticles" European Physical Journal D, 48, 355–364, doi:10.1140/epjd/e2008-00114-3

[Bi 2011] Bi D, Zhang J, Chakraborty B, Behringer RP. (2011) "Jamming by Shear" Nature, 480, 355–358, doi:10.1038/nature10667

[Bi 2015] Bi D, Lopez JH, Schwarz JM, Manning ML. (2015) "A Density-Independent Rigidity Transition in Biological Tissues" Nature Physics, 11, 1074–1079, doi:10.1038/nphys3471

[Bielawski 2007] Bielawski CW, Grubbs RH. (2007) "Living Ring-Opening Metathesis Polymerization" Progress in Polymer Science, 32, 1–29, doi:10.1016/j.progpolymsci.2006.08.006

[Binder 2007] Binder WH, Petraru L, Roth T, Groh PW, Pálfi V, Keki S, Ivan B. (2007) "Magnetic and Temperature-Sensitive Release Gels from Supramolecular Polymers" Advanced Functional Materials, 17, 1317–1326, doi:10.1002/adfm.200601084

[Bindschadler 2007] Bindschadler M, McGrath JL. (2007) "Sheet Migration by Wounded Monolayers as an Emergent Property of Single-Cell Dynamics" Journal of Cell Science, 120, 876–884, doi:10.1242/jcs.03395

[Bird 1960] Bird RB, Stewart WE, Lightfoot EN. (1960) Transport Phenomena, Wiley, New York

[Bird 2013] Bird JC, Dhiman R, Kwon HM, Varanasi KK. (2013) "Reducing the Contact Time of a Bouncing Drop" Nature, 503, 385–388, doi:10.1038/nature12740

[Bixler 2012] Bixler GD, Bhushan B. (2012) "Bioinspired Rice Leaf and Butterfly Wing Surface Structures Combining Shark Skin and Lotus Effects" Soft Matter, 8, 11271–11284, doi:10.1039/C2SM26655E

[Biyani 2013] Biyani MV, Foster EJ, Weder C. (2013) "Light-Healable Supramolecular Nanocomposites Based on Modified Cellulose Nanocrystals" ACS Macro Letters, 2, 3, 236–240, doi:10.1021/mz400059w

[Blackman 2002] Blackman BR, Hadavinia H, Kinloch AJ, Paraschi M, Williams JG. (2002) "The Calculation of Adhesive Fracture Energies in Mode I: Revisiting the Tapered Double Cantilever Beam (TDCB) Test" Engineering Fracture Mechanics, 70, 233–248

[Blackmore 2006] Blackmore RD, Lepola WM. (2006) "Inflatable Heating Device for In-Situ Repair of Conduit and Method for Repairing Conduit" US Patent 7,052,567

[Blackshire 2004] Blackshire JL, Dosser L, Hix K. (2004) "Laser Processing of Micro-Cracks for Structural Life Extension" Proc SPIE Vol. 5392-26, Testing, Reliability, and Application of Micro- and Nano-Material Systems II

[Blaiszik 2008] Blaiszik BJ, Sottos NR, White SR. (2008) "Nanocapsules for Self-Healing Materials" Composites Science and Technology, 68, 978–986

[Blaiszik 2012] Blaiszik BJ, Kramer SL, Grady ME, McIlroy DA, Moore JS, Sottos NR, White SR. (2012) "Autonomic Restoration of Electrical Conductivity" Advanced Materials, 24, 398–401, doi:10.1002/adma.201102888

[Bleay 2001] Bleay SM, Loader CB, Hawyes VJ, Humberstone L, Curtis PT. (2001) "A Smart Repair System for Polymer Matrix Composites" Composites Part A: Applied Science and Manufacturing, 32, 1767–1776

[Blossey 2003] Blossey R. (2003) "Self-Cleaning Surfaces: Virtual Realities" Nature Materials, 2, 301–306

[Blundell 1982] Blundell A. (1982) "Bond Graphs for Modeling Engineering Systems", Ellis Horwood, Chichester

[Boesen 2011] Boesen MR, Keymeulen D, Madsen J, Lu T, Chao TH. (2011) "Integration of the Reconfigurable Self-Healing eDNA Architecture in an Embedded System" IEEE Aerospace Conference Proceedings 04/2011, doi:10.1109/AERO.2011.5747477

[Boettcher 2012] Boettcher S, Singh V, Ziff RM. (2012) "Ordinary Percolation with Discontinuous Transitions" Nature Communications, 3, 787, doi:10.1038/ncomms1774

[Bohn 2004] Bohn HF, Federle W. (2004) "Insect Aquaplaning: Nepenthes Pitcher Plants Capture Prey with the Peristome, a Fully Wettable Water-Lubricated Anisotropic Surface" Proceedings of the National Academy of Sciences of the United States of America, 101, 39, 14138–14143, doi:10.1073/pnas.0405885101

[Bohringer 2003] Bohringer KF. (2003) "Surface Modification and Modulation in Microstructures: Controlling Protein Adsorption, Monolayer Desorption, and Micro-Self-Assembly" Journal of Micromechanics and Microengineering, 13, 4, S1–S10

[Boiko 2004] Boiko YM, Bach A, Lyngaae-Jørgensen J. (2004) "Self-Bonding in an Amorphous Polymer below the Glass Transition: A T-Peel Test Investigation" Journal of Polymer Science. Part B: Polymer Physics, 42, 1861–1867, doi:10.1002/polb.20077

[Bommavaram 2009] Bommavaram RR, Bhasin A, Little DN. (2009) "Use of Dynamic Shear Rheometer to Determine the Intrinsic Healing Properties of Asphalt Binders" Transportation Research Board, 88th Annual Meeting, Washington, DC, paper 09-2083

[Bonchev 1995] Bonchev D. (1995) "Kolmogorov's Information, Shannon's Entropy, and Topological Complexity of Molecules" Bulgarian Chemical Communications, 28, 567–582

[Boncheva 2003] Boncheva M, Whitesides GM. (2003) "Self-Healing Systems Having a Design Stimulated by the Vertebrate Spine" Angewandte Chemie International Edition, 42, 23, 2644–2647, doi:10.1002/anie.200351010

[Bond 2007] Bond I, Trask R, Williams G, Williams H. (2007) "Autonomic Self-Healing and Damage Visualization in Fibre Reinforced Polymer Composites" Structural Health Monitoring, FK Chang (ed.), DEStech, Lancaster

[Bongard 2006] Bongard J, Zykov V, Lipson H. (2006) "Resilient Machines through Continuous Self-Modeling" Science, 314, 5802, 1118–1121

[Bongiovanni 2015] Bongiovanni D, Hu Y, Wetzel B, Robles RA, González GM, Marti-Panameño EA, Chen Z, Morandotti R. (2015) "Efficient Optical Energy Harvesting in Self-Accelerating Beams" Scientific Reports, 5, 13197, doi:10.1038/srep13197

[Bonnaure 1982] Bonnaure FP, Huibers AH, Boonders A. (1982) "A Laboratory Investigation of the Influence of Rest Periods on the Fatigue Characteristics of Bituminous Mixes" Journal of the Association of Asphalt Paving Technologists, 51, 104–128

[Bor 2010] Bor TC, Warnet L, Akkerman R, de Boer A. (2010) "Modeling of Stress Development during Thermal Damage Healing in Fiber-Reinforced Composite Materials Containing Embedded Shape Memory Alloy Wires" Journal of Composite Materials, 44, 2547

[Boresi 2002] Boresi AP, Schmidt RJ. (2002) *Advanced Mechanics of Materials*, 6th ed., Wiley

[Bose 2013] Bose S, Pal S, Natarajan B, Scoglio C M, Das S, Schulz NN. (2012) "Analysis of Optimal Reconfiguration of Shipboard Power Systems" IEEE Transactions on Power Systems, 27, 1, 189–197, doi:10.1109/TPWRS.2011.2163948

[Bose 2015] Bose RK, Hohlbein N, Garcia SJ, Schmidt AM, van der Zwaag S. (2015) "Relationship between the Network Dynamics, Supramolecular Relaxation Time and Healing Kinetics of Cobalt Poly(Butyl Acrylate) Ionomers" Polymer 69, 228–232, doi:10.1016/j.polymer.2015.03.049

[Bosia 2015] Bosia F, Merlino M, Pugno NM. (2015) "Fatigue of Self-Healing Hierarchical Soft Nanomaterials: The Case Study of the Tendon in Sportsmen" Journal of Materials Research, 30, 1, 2–9, doi:10.1557/jmr.2014.335

[Bouchal 1998] Bouchal Z, Wagner J, Chlup M. (1998) "Self-Reconstruction of a Distorted Nondiffracting Beam" Optics Communications, 151, 4–6, 207–211

[Boukerche 2006] Boukerche A, Werner R, Pazzi N, Araujo RB. (2006) "Fault-Tolerant Wireless Sensor Network Routing Protocols for the Supervision of Context-Aware Physical Environments" Journal of Parallel and Distributed Computing, 66, 586–599

[Bowers 2013] Bowers SM, Sengupta K, Dasgupta K, Parker BD, Hajimiri A. (2013) "Integrated Self-Healing for mm-Wave Power Amplifiers" IEEE Transactions on Microwave Theory and Techniques, 61, 3, 1301–1315, doi:10.1109/TMTT.2013.2243750

[Braiman 2007] Braiman-Wiksman L, Solomonik I, Spira R, Tennenbaum T. (2007) "Novel Insights into Wound Healing Sequence of Events" Toxicologic Pathology, 35, 6, 767–779, doi:10.1080/01926230701584189.

[Brammer 2009] Brammer KS, Choi C, Oh S, Cobb CJ, Connelly LS, Loya M, Kong SD, Jin S. (2009) "Antibiofouling, Sustained Antibiotic Release by Si Nanowire Templates" Nano Letters, 9, 10, 3570–3574, doi:10.1021/nl901769m

[Brancart 2014] Brancart J, Scheltjens G, Muselle T, Van Mele B, Terryn H, Van Assche G. (2014) "Atomic Force Microscopy-Based Study of Self-Healing Coatings Based on Reversible Polymer Network Systems" Journal of Intelligent Materials Systems and Structures, 25, 40–46, doi:10.1177/1045389X12457100

[Brandon 2011] Brandon EJ, Vozoff M, Kolawa EA, Studor GF, Lyons F, Keller MW, Beiermann B, White SR, Sottos NR, Curry MA, Banks DL, Brocato R, Zhou L, Jung S, Jackson TN, Champaigne K. (2011) "Structural Health Management Technologies for Inflatable/Deployable Structures: Integrating Sensing and Self-Healing" Acta Astronautica, 68, 883–903

[Braun 2010] Braun PV, Cho SH, White SR. (2010) "Self-Healing Coating System" US Patent 7,723,405

[Brin 1998] Brin S, Page L. (1998) "The Anatomy of a Large-Scale Hypertextual Web Search Engine" Computer Networks and ISDN Systems, 30, 107–117, doi:10.1016/S0169-7552(98)00110-X

[Broadbent 1957] Broadbent S, Hammersley J. (1957) "Percolation Processes I. Crystals and Mazes" Proceedings of the Cambridge Philosophical Society 53, 629–641, doi:10.1017/S0305004100032680

[Brochu 2011] Brochu AB, Craig SL, Reichert WM. (2011) "Self-Healing Biomaterials" Journal of Biomedical Materials Research Part A, 96A, 2, 492–506, doi:10.1002/jbm.a.32987

[Brockes 1997] Brockes JP. (1997) "Amphibian Limb Regeneration: Rebuilding a Complex Structure" Science, 276, 5309, 81–87, doi:10.1126/science.276.5309.81.

[Brodbeck 2012] Brodbeck L, Wang L, Iida F. (2012) "Robotic Body Extension Based on Hot Melt Adhesives" 2012 IEEE International Conference on Robotics and Automation (ICRA), 4322–4327, doi:10.1109/ICRA.2012.6225258

[Broky 2008] Broky J, Siviloglou GA, Dogariu A, Christodoulides DN. (2008) "Self-Healing Properties of Optical Airy Beams" Optics Express, 16, 17, 12880–12891, doi:10.1364/OE.16.012880

[Brostow 2002] Brostow W, Bujard B, Cassidiy PE, Hagg HE, Montemartini PE. (2002) "Effects of Fluoropolymer Addition to an Epoxy on Scratch Depth and Recovery" Materials Research Innovations, 6, 7–12, doi:10.1007/s10019-002-0161-y

[Broutman 1971] Broutman LJ, Sahu S. (1971) "The Effect of Interfacial Bonding on the Toughness of Glass Filled Polymers" Materials Science and Engineering, 8, 2, 98–107, doi:10.1016/0025-5416(71)90087-5

[Brown 2002] Brown EN, Sottos NR, White SR. (2002) "Fracture Testing of a Self-Healing Polymer Composite" Experimental Mechanics, 42, 4, 372–379

[Brown 2006] Brown EN, White SR, Sottos NR. (2006) "Fatigue Crack Propagation in Microcapsule-Toughened Epoxy" Journal of Materials Science, 41, 6266–6273, doi 10.1007/s10853-006-0512-y

[Brown 2011a] Brown EN. (2011) "Use of the Tapered Double-Cantilever Beam Geometry for Fracture Toughness Measurements and Its Application to the Quantification of Self-Healing" Journal of Strain Analysis, 46, 3, 167–186, doi:10.1177/0309324710396018

[Brown 2011b] Brown PS, Berson A, Talbot EL, Wood TJ, Schofield WCE, Bain CD, Badyal JP. (2011) "Impact of Picoliter Droplets on Superhydrophobic Surfaces with Ultralow Spreading Ratios" Langmuir 27, 22, 13897–13903, doi:10.1021/la203329n

[Brugués 2014] Brugués A, Anon E, Conte V, Veldhuis JH, Gupta M, Colombelli J, Muñoz JJ, Brodland GW, Ladoux B, Trepat X. (2014) "Forces Driving Epithelial Wound Healing" Nature Physics 10, 683–690, doi:10.1038/nphys3040

[Brumley 2007] Brumley D, Newsome J, and Song D. (2007) "Sting: An End-to-End Self-Healing System for Defending against Internet Worms" Advances in Information Security, 27, 147–70, doi:10.1007/978-0-387-44599-1_7

[Brunsveld 2001] Brunsveld L, Folmer BJ, Meijer EW, Sijbesma RP. (2001) "Supramolecular Polymers" Chemical Reviews, 101, 4071–4097, doi:10.1021/cr990125q

[Bruthans 2014] Bruthans J, Soukup J, Vaculikova J, Filippi M, Schweigstillova J, Mayo AL, Masin D, Kletetschka G, Rihosek J. (2014) "Sandstone Landforms Shaped by Negative Feedback between Stress and Erosion" Nature Geoscience, 7, 597–601, doi:10.1038/ngeo2209

[Brynko 1961] Brynko C. (1961) "Process of Making Dual-Walled Oil Containing Capsules" US Patent 2,969,331

[Bucher 1994] Bucher JP, Santesson L, Kern K. (1994) "Thermal Healing of Self-Assembled Organic Monolayers: Hexane- and Octadecanethiol on Au(111) and Ag(111)" Langmuir, 10, 4, 979–983, doi:10.1021/la00016a001

[Buchheit 2003] Buchheit RG, Guan H, Mahajanam S, Wong F. (2003) "Active Corrosion Protection and Corrosion Sensing in Chromate-Free Organic Coatings" Progress in Organic Coatings, 47, 174–182

[Buchwalter 1996] Buchwalter SL, Kosbar LL. (1996) "Cleavable Epoxy Resins: Design for Disassembly of a Thermoset" Journal of Polymer Science. Part A, Polymer Chemistry, 34, 249–260 doi:10.1002/(SICI)1099-0518(19960130)34:2<249::AID-POLA11>3.0.CO;2-Q

[Budiansky 1988] Budiansky B, Amazigo JC, Evans AG (1988) "Small-Scale Crack Bridging and the Fracture Toughness of Particulate-Reinforced Ceramics" Journal of the Mechanics and Physics of Solids, 36,167–187

[Bulinski 1998] Bulinski AT, Crinel JP, Noirhomme B, Densley RJ, Bamji S. (1998) "Polymer Oxidation and Water Treeing" IEEE Transactions on Dielectrics and Electrical Insulation, 5, 4, 558–570

[Bunker 2007] Bunker BC, Huber DL, Kushmerick JG, Dunbar T, Kelly M, Matzke C, Cao J, Jeppesen JO, Perkins J, Flood AH, Stoddart JF. (2007) "Switching Surface Chemistry with Supramolecular Machines" Langmuir, 23, 31–34

[Burattini 2009] Burattini S, Colquhoun HM, Fox JD, Friedmann D, Greenland BW, Harris PJ, Hayes W, Mackayc ME, Rowan SJ. (2009) "A Self-Repairing, Supramolecular Polymer System: Healability as a Consequence of Donor–Acceptor P–P Stacking Interactions" Chemical Communications, 2009, 6717–6719, doi:10.1039/b910648k

[Burken 2006] Burken JJ, Williams-Hayes P, Kaneshige JT, Stachowiak SJ. (2006) "Reconfigurable Control with Neural Network Augmentation for a Modified F-15 Aircraft" NASA/TM-2006-213678, National Aeronautics and Space Administration, Dryden Flight Research Center, Edwards, California 93523-0273

[Burnworth 2011] Burnworth M, Tang L, Kumpfer JR, Duncan AJ, Beyer FL, Fiore GL, Rowan SJ, Weder C. (2011) "Optically Healable Supramolecular Polymers" Nature, 472, 334–337, doi:10.1038/nature09963

[Burr 2010] Burr GW, Breitwisch MJ, Franceschini M, Garetto D, Gopalakrishnan K, Jackson B, Kurdi B, Lam C, Lastras LA, Padilla A, Rajendran BR, Raoux S, Shenoy RS. (2010) "Phase Change Memory Technology" Journal of Vacuum Science & Technology B, 28, 223–262, doi:10.1116/1.3301579

[Burris 2006] Burris DL, Sawyer WG. (2006) "Improved Wear Resistance in Alumina-PTFE Nanocomposites with Irregular Shaped Nanoparticles" Wear, 260, 915–918

[Burton 2006] Burton DS, Gao X, Brinson LC. (2006) "Finite Element Simulation of a Self-Healing Shape Memory Alloy Composite" Mechanics of Materials, 38, 5–6, 525–537, doi:10.1016/j.mechmat.2005.05.021

[Butler 1997] Butler KM. (1997) "Physical Modeling of Intumescent Fire Retardant Polymers" in Polymeric Foams Science and Technology, ACS Symposium Series 669, KC Khemani (ed.), Chapter 15, 214–230, American Chemical Society

[Butler 2004] Butler-Purry KL, Sarma ND. (2004) "Self-Healing Reconfiguration for Restoration of Naval Shipboard Power Systems" IEEE Transactions on Power Systems, 19, 2, 754–762, doi:10.1109/TPWRS.2003.821431

[Calle 2015] Calle LM, Li WN, Buhrow JW, Perusich SA, Jolley ST, Gibson TL, Williams MK. (2015) "Elongated Microcapsules and Their Formation" US Patent 9,108,178

[Callen 1960] Callen H. (1960) Thermodynamics: An Introduction to the Physical Theories of Equilibrium Thermostatics & Irreversible Thermodynamics, John Wiley & Sons Inc.

[Callies 2005] Callies M, Quéré D. (2005) "On Water Repellency" Soft Matter, 1, 55–61, doi:10.1039/B501657F

[Camino 1989] Camino G, Costa L. (1989) "Intumescent Fire-Retardant Systems" Polymer Degradation and Stability, 23, 4, 359–376

[Campaignolle 1997] Campaignolle X, Crolet JL. (1997) "Method for Studying Stabilization of Localized Corrosion on Carbon Steel by Sulfate-Reducing Bacteria" Corrosion, 53, 6, 440–447

[Canadell 2011] Canadell J, Goossens H, Klumperman H. (2011) "Self-Healing Materials Based on Disulfide Links" Macromolecules, 44, 2536–2541, doi:10.1021/ma2001492

[Cannon 1976] Cannon SL, McKenna GB, Statton WO. (1976) "Hard-Elastic Fibers: A Review of a Novel State for Crystalline Polymers" Journal of Polymer Science: Macromolecular Reviews, 11, 209

[Canpolat 1996] Canpolat M, Pekcan O. (1996) "Healing and Photon Diffusion during Sintering of High-T Latex Particles" Journal of Polymer Science: Part B: Polymer Physics, 34, 691–698

[Cao 2009] Cao L, Jones AK, Sikka VK, Wu J, Gao D. (2009) "Anti-Icing Superhydrophobic Coatings" Langmuir, 25, 21, 12444–12448, doi:10.1021/la902882b,

[Cao 2017] Cao Y, Morrissey TG, Acome E, Allec SI, Wong BM, Keplinger C, Wang C. (2017) "A Transparent, Self-Healing, Highly Stretchable Ionic Conductor" Advanced Materials, 29, 10, 1605099, doi:10.1002/adma.201605099

[Capadona 2012] Capadona JR, Tyler DJ, Zorman CA, Rowan SJ, Weder C. (2012) "Mechanically Adaptive Nanocomposites for Neural Interfacing" MRS Bulletin, 37, 6, 581–589, doi:10.1557/mrs.2012.97

[Capozzi 1958] Capozzi L. (1958) "Method of Packaging Epoxy Resins" US Patent 2,862,616

[Cardoso 2002] Cardoso RJ, Shukla A. (2002) "Effect of Particle Size and Surface Treatment on Constitutive Properties of Polyester–Cenosphere Composites" Journal of Materials Science, 37, 603–613

[Carey 2011] Carey BJ, Patra PK, Ci L, Silva GG, Ajayan PM. (2011) "Observation of Dynamic Strain Hardening in Polymer Nanocomposites" ACS Nano, 5, 4, 2715–2722, doi:10.1021/nn103104g

[Carlson 2006a] Carlson HC, Goretta KC. (2006) "Basic Materials Research Programs at the US Air Force Office of Scientific Research" Materials Science and Engineering B: Solid State Materials for Advanced Technology, 132, 2–7

[Carlson 2006b] Carlson JA, English JM, Coe DJ. (2006) "A Flexible, Self-Healing Sensor Skin" Smart Materials and Structures, 15, N129–N135, doi:10.1088/0964-1726/15/5/N05

[Carmona 2007] Carmona H, Kun F, Andrade J, Herrmann H. (2007). "Computer Simulation of Fatigue under Diametrical Compression" Physical Review E, 75, 4, 046115, doi:10.1103/PhysRevE.75.046115

[Carneiro 2012] Carneiro J, Tedim J, Fernandes SC, Freire CS, Silvestre AJ, Gandini AA, Ferreira MG, Zheludkevich ML. (2012) "Chitosan-Based Self-Healing Protective Coatings Doped with Cerium Nitrate for Corrosion Protection of Aluminum Alloy 2024" Progress in Organic Coatings, 75, 1–2, 8–13, doi:10.1016/j.porgcoat.2012.02.012

[Caron 2012] Caron M, Mills I. (2012) "Planning and Execution of Tele-Robotic Maintenance Operations on the ISS" SpaceOps 2012 Conference, Stockholm, Sweden, doi:10.2514/6.2012-1272635

[Carpenter 2006] Carpenter SH, Shen S. (2006) "A Dissipated Energy Approach to Study HMA Healing in Fatigue" Paper No. 06-0913, 86th Annual Meeting of the Transportation Research Board, Washington, DC

[Carter 1987] Carter DR. (1987) "Mechanical Loading History and Skeletal Biology" Journal of Biomechanics, 20, 11–12, 1095–1109, doi:10.1016/0021-9290(87)90027-3

[Caruso 2007] Caruso MM, Delafuente DA, Ho V, Moore JS, Sottos NR, White SR. (2007) "Solvent-Promoted Self-Healing Materials" Macromolecules, 40, 8830–8832.

[Caruso 2009a] Caruso MM, Davis DA, Shen Q, Odom SA, Sottos NR, White SR, Moore JS. (2009) "Mechanically-Induced Chemical Changes in Polymeric Materials" Chemical Reviews, 109, 5755–5798

[Caruso 2009b] Caruso MM, Schelkopf SR, Jackson AC, Landry AM, Braun PV, Moore JS. (2009) "Microcapsules Containing Suspensions of Carbon Nanotubes" Journal of Materials Chemistry, 6093–6096, doi:10.1039/b910673a

[Casado 2007] Casado CM, Poncela AV, Lorenzana A. (2007) "Adaptive Tuned Mass Damper for the Construction of Concrete Piers" Structural Engineering International, 17, 3, 252–255

[Casciati 2007] Casciati S. (2007) "Thermal Treatment Optimization for Cu-Based Shape Memory Alloys" Proceedings of the First International Conference on Self-Healing Materials, 18-20 April 2007, Noordwijk aan Zee, The Netherlands

[Cases 1997] Cases JM, Berend I, Francois M, Uriot JP, Michot LJ, Thomas F. (1997) "Mechanism of Adsorption and Desorption of Water Vapor by Homoionic Montmorillonite: 3. The Mg^{2+}, Ca^{2+}, Sr^{2+} and Ba^{2+} Exchanged Forms" Clays and Clay Minerals, 45, 1, 8–21

[Cassie 1944] Cassie AB, Baxter S. (1944) "Wettability of Porous Surfaces" Transactions of the Faraday Society, 40, 546–551

[Castano 2002] Castano A, Behar A, Will PM. (2002) "The Conro Modules for Reconfigurable Robots" IEEE/ASME Transactions on Mechatronics, 7, 4, 403–409, doi:10.1109/TMECH.2002.806233

[Castellucci 2009] Castellucci M. (2009) "Resistive Heating for Self-Healing Materials Based on Ionomeric Polymers" MS Thesis, Mechanical Engineering, Virginia Polytechnic Institute and State University

[Castle 2014] Castle T, Cho Y, Gong X, Jung E, Sussman DM, Yang S, Kamien RD. (2014) "Making the Cut: Lattice Kirigami Rules" Physical Review Letters, 113, 245502, doi:10.1103/PhysRevLett.113.245502

[Cates 1998] Cates ME, Wittmer JP, Bouchaud JP, Claudin P. (1998) "Jamming, Force Chains, and Fragile Matter" Physical Review Letters, 81, 9, 1841–1844

[Cayre 2004] Cayre OJ, Noble PF, Paunov VN. (2004) "Fabrication of Novel Colloidosome Microcapsules with Gelled Aqueous Cores" Journal of Materials Chemistry, 14, 3351–3355, doi:10.1039/B411359D

[Celestine 2015] Celestine AN, Sottos NR, White SR. (2015) "Autonomic Healing of PMMA Via Microencapsulated Solvent" Polymer, 69, 241–248, doi:10.1016/j.polymer.2015.03.072

[Chalmers 1995] Chalmers D (1995) The Conscious Mind: In Search of a Fundamental Theory, Oxford University Press, Oxford

[Chambers 2009] Chambers M, Verduzco R, Gleeson JT, Sprunt S, Jakli A. (2009) "Flexoelectricity of a Calamitic Liquid Crystal Elastomer Swollen with a Bent-Core Liquid Crystal" Journal of Materials Chemistry, 19, 7909–7913, doi:10.1039/b911652d

[Chan 2012] Chan V, Park K, Collens MB, Kong H, Saif TA, Bashir R (2012) "Development of Miniaturized Walking Biological Machines" Scientific Reports, 2, 857, doi:10.1038/srep00857

[Chang 1983] Chang JM, Aklonis JJ. (1983) "A Photothermal Reversibly Crosslinkable Polymer System" Journal of Polymer Science: Polymer Letters Edition, 21, 12, 999–1004, doi:10.1002/pol.1983.130211206

[Chang 2007] Chang DP, Dolbow JE, Zauscher S. (2007) "Switchable Friction of Stimulus-Responsive Hydrogels" Langmuir, 23, 250–257

[Cheang 2016] Cheang UK, Meshkati F, Kim H, Lee K, Fu HC, Kim MJ. (2016) "Versatile Microrobotics Using Simple Modular Subunits" Scientific Reports, 6, 30472, doi:10.1038/srep30472

[Checco 2013] Checco A, Rahman A, Black CT. (2013) "Robust Superhydrophobicity in Large-Area Nanostructured Surfaces Defined by Block-Copolymer Self Assembly" Advanced Materials, 26, 6, 886–891, doi:10.1002/adma.201304006

[Chen 1999] Chen W, Fadeev AY, Hsieh MC, Öner D, Youngblood J, McCarthy TJ. (1999) "Ultrahydrophobic and Ultralyophobic Surfaces: Some Comments and Examples" Langmuir, 15, 3395–3399, doi:10.1021/la990074s

[Chen 2002a] Chen CS, Lin CH, Tsai HY. (2002) "A Rule-Based Expert System with Colored Petri Net Models for Distribution System Service Restoration" IEEE Transactions on Power Systems, 17, 4, 1073–1080, doi:10.1109/TPWRS.2002.804947

[Chen 2002b] Chen X, Dam MA, Ono K, Mal A, Shen H, Nutt SR, Sheran K, Wudl F. (2002) "A Thermally Remendable Cross-Linked Polymeric Material" Science, 295, 1698–1702

[Chen 2007] Chen S, Zheng J, Li L, Jiang S. (2005) "Strong Resistance of Phosphorylcholine Self-Assembled Monolayers to Protein Adsorption: Insights into Nonfouling Properties of Zwitterionic Materials" Journal of the American Chemical Society, 127, 41, 14473–14478, doi:10.1021/ja054169u

[Chen 2010] Chen W, Güler G, Kuruvilla E, Schuster GB, Chiu HC, Riedo E. (2010) "Development of Self-Organizing, Self-Directing Molecular Nanowires: Synthesis and Characterization of Conjoined DNA–2,5-Bis(2-thienyl)pyrrole Oligomers" Macromolecules, 43, 9, 4032–4040, doi:10.1021/ma100409u

[Chen 2011a] Chen M, McDaniel JR, MacKay JA, Chilkoti A. (2011) "Nanoscale Self-Assembly for Delivery of Therapeutics and Imaging Agents" Technology and Innovation, 13, 5–25, doi:10.3727119498241IXI3003853539948

[Chen 2011b] Chen Q, Bae SC, Granick S. (2011) "Directed Self-Assembly of a Colloidal Kagome Lattice" Nature, 469, 381–384, doi:10.1038/nature09713

[Chen 2012] Chen Y, Kushner AM, Williams GA, Guan Z. (2012) "Multiphase Design of Autonomic Self-Healing Thermoplastic Elastomers" Nature Chemistry, 4, 467–472, doi:10.1038/nchem.1314

[Chen 2013a] Chen H, Fallah MA, Huck V, Angerer JI, Reininger AJ, Schneider SW, Schneider MF, Alexander-Katz A. (2013) "Blood-Clotting-Inspired Reversible Polymer–Colloid Composite Assembly in Flow" Nature Communications, 4, 1333, doi:10.1038/ncomms2326

[Chen 2013b] Chen Y, Simms R, Koh C, Lopp G, Roque R. (2013). "Development of a Test Method for Evaluation and Quantification of Healing in Asphalt Mixture" Road Materials and Pavement Design, 14, 4, 901–920, doi:10.1080/14680629.2013.844196

[Chen 2013c] Chen Z, Wang G, Xu Z, Li H, Dhôtel A, Zeng XC, Chen B, Saiter JM, Tan L. (2013) "Metal–Organic Frameworks Capable of Healing at Low Temperatures" Advanced Materials, 25, 42, 6106–6111, doi:10.1002/adma.201302471

[Chen 2013d] Chen C, Peters K, Li Y. (2013) "Self-Healing Sandwich Structures Incorporating Interfacial Layer with Vascular Network" Smart Materials and Structures, 22, 025031, doi:10.1088/0964-1726/22/2/025031

[Chen 2014] Chen X, Mahadevan L, Driks A, Sahin O. (2014) "Bacillus Spores as Building Blocks for Stimuli-Responsive Materials and Nanogenerators" Nature Nanotechnology 9, 137–141, doi:10.1038/nnano.2013.290

[Chen 2015a] Chen J, Li K, Wu S, Liu J, Liu K, Fan Q. (2017) "Durable Anti-Icing Coatings Based on Self-Sustainable Lubricating Layer" ACS Omega, 2, 5, 2047–2054, doi:10.1021/acsomega.7b00359

[Chen 2015b] Chen K, Wang B, Granick S. (2015) "Memoryless Self-Reinforcing Directionality in Endosomal Active Transport within Living Cells" Nature Materials, 14, 589–593, doi:10.1038/nmat4239

[Chen 2015c] Chen P, Xu Y, He S, Sun X, Pan S, Deng J, Chen D, Peng H. (2015) "Hierarchically Arranged Helical Fibre Actuators Driven by Solvents and Vapours" Nature Nanotechnology, 10, 12, 1077–1083, doi:10.1038/nnano.2015.198

[Chen 2015d] Chen Y, Kim JK, Hirning AJ, Josić K, Bennett MR. (2015) "Emergent Genetic Oscillations in a Synthetic Microbial Consortium" Science, 349, 6251, 986–9. doi:10.1126/science.aaa3794.

[Chen 2015e] Chen Y, Yu KY, Liu Y, Shao S, Wang H, Kirk MA, Wang J, Zhang X. (2015) "Damage-Tolerant Nanotwinned Metals with Nanovoids under Radiation Environments" Nature Communications, 6, 7036, doi:10.1038/ncomms8036

[Chen 2017] Chen J, Li K, Wu S, Liu J, Liu K, Fan Q. (2017) "Durable Anti-Icing Coatings Based on Self-Sustainable Lubricating Layer" ACS Omega, 2, 5, 2047–2054, doi:10.1021/acsomega.7b00359

[Chen 2022] Chen K, Liu M, Wang F, et al. (2022) "Highly Transparent, Self-Healing, and Self-Adhesive Double Network Hydrogel for Wearable Sensors" Frontiers in Bioengineering and Biotechnology, 10, 846401, doi:10.3389/fbioe.2022.846401

[Cheng 2007a] Cheng CJ, Chu LY, Ren PW, Zhang J, Hu L. (2007) "Preparation of Monodisperse Thermo-Sensitive Poly(*N*-Isopropylacrylamide) Hollow Microcapsules" Journal of Colloid and Interface Science, 313, 2, 383–388, doi:10.1016/j.jcis.2007.05.004

[Cheng 2007b] Cheng G, Zhang Z, Chen S, Bryers JD, Jiang S. (2007) "Inhibition of Bacterial Adhesion and Biofilm Formation on Zwitterionic Surfaces" Biomaterials, 28, 29, 4192–4199, doi:10.1016/j.biomaterials.2007.05.041

[Cheng 2014a] Cheng NG, Gopinath A, Wang L, Iagnemma K, Hosoi AE. (2014) "Thermally Tunable, Self-Healing Composites for Soft Robotic Applications" Macromolecular Materials and Engineering, 299, 11, 1279–1284, doi:10.1002/mame.201400017

[Cheng 2014b] Cheng R, Huang W, Huang L, Yang B, Mao L, Jin K, ZhuGe Q, Zhao Y. (2014) "Acceleration of Tissue Plasminogen Activator-Mediated Thrombolysis by Magnetically Powered Nanomotors" ACS Nano, 8, 8, 7746–7754, doi:10.1021/nn5029955

[Cheng 2016] Cheng Y, Todri-Sanial A, Yang J, Zhao W. (2016) "Alleviating through-Silicon-Via Electromigration for 3-D Integrated Circuits Taking Advantage of Self-Healing Effect" IEEE Transactions on Very Large Scale Integration (VLSI) Systems, 24, 3310–3322, doi:10.1109/TVLSI.2016.2543260

[Cheng 2018] Cheng C, Li J, Yang F, et al. (2018) "Renewable Eugenol-Based Functional Polymers with Self-Healing and High Temperature Resistance Properties" Journal of Polymer Research, 25, 1–13, doi:10.1007/s10965-018-1460-3

[Chien 2012] Chien C, Tang A, Hsiao F, Chang MF. (2012) "Dual-Control Self-Healing Architecture for High-Performance Radio SoCs" IEEE Design & Test of Computers, 29, 6, 40–51, doi:10.1109/MDT.2012.2213571

[Chino 2001] Chino K, Ashiura M. (2001) "Thermoreversible Cross-Linking Rubber Using Supramolecular Hydrogen Bonding Networks" Macromolecules, 34, 26, 9201–9204, doi:10.1021/ma011253v

[Chipara 2005] Chipara M, Wooley K. (2005) "Molecular Self-Healing Processes in Polymers" Materials Research Society Symposium Proceedings, 851, NN4.3.1

[Chirikjian 1996] Chirikjian GS, Pamecha A, Ebert-Uphoff I. (1996) "Evaluating Efficiency of Self-Reconfiguration in a Class of Modular Robots" Journal of Robotic Systems, 13, 5, 317–338

[Chluba 2015] Chluba C, Ge W, Lima de Miranda R, Strobel J, Kienle L, Quandt E, Wuttig M. (2015) "Ultralow-Fatigue Shape Memory Alloy Films" Science, 348, 6238, 1004–1007, doi:10.1126/science.1261164

[Cho 1995] Cho BR, Kardos JL. (1995) "Consolidation and Self-Bonding in Poly(Ether Ether Ketone) (PEEK)" Journal of Applied Polymer Science, 56, 11, 1435–1454, doi:10.1002/app.1995.070561106

[Cho 2006] Cho SH, Andersson HM, White SR, Sottos NR, Braun PV. (2006) "Polydimethylsiloxane-Based Self-Healing Materials" Advanced Materials, 18, 997–1000, doi:10.1002/adma.200501814

[Cho 2008] Cho SY, Kim JG, Chung CM. (2008) "A Fluorescent Crack Sensor Based on Cyclobutane-Containing Crosslinked Polymers of Tricinnamates" Sensors and Actuators B, 134, 822–825

[Choi 1992] Choi SR, Tikare V. (1992) "Crack-Healing Behavior of Hot-Pressed Silicon Nitride Due to Oxidation" Scripta Metallurgica et Materialia, 26, 8, 1263–1268, doi:10.1016/0956-716X(92)90574-X

[Choi 2010] Choi E, Cho SC, Hu JW, Park T, Chung YS. (2010) "Recovery and Residual Stress of SMA Wires and Applications for Concrete Structures" Smart Materials and Structures, 19, 094013 doi:10.1088/0964-1726/19/9/094013

[Chong 2010] Chong A, Renninger WH, Christodoulides DN, Wise FW. (2010) "Airy-Bessel Wave Packets as Versatile Linear Light Bullets" Nature Photonics, 4, 103–106, doi:10.1038/nphoton.2009.264

[Chou 2021] Chou T-C, Huang S-Y, Chen P-J, et al. (2021) "Electrical and Reliability Investigation of Cu-to-Cu Bonding with Silver Passivation Layer in 3-D Integration" IEEE Transactions on Components, Packaging and Manufacturing Technology, 11, 36–42, doi:10.1109/TCPMT.2020.3037365

[Christensen 2013] Christensen DJ, Schultz UP, Stoy K. (2013) "A Distributed and Morphology-Independent Strategy for Adaptive Locomotion in Self-Reconfigurable Modular Robots" Robotics and Autonomous Systems, 61, 1021–1035, https://doi.org/10.1016/j.robot.2013.05.009

[Chu 1995] Chu MC, Sato S, Kobayashi Y, Ando K. (1995) "Damage Healing and Strengthening Behavior in Intelligent Mullite/SiC Ceramics" Fatigue & Fracture of Engineering Materials & Structures, 18, 9, 1019–1029

[Chujo 1993] Chujo Y, Sada K, Naka A, Nomura R, Saegusa T. (1993) "Synthesis and Redox Gelation of Disulfide-Modified Polyoxazoline" Macromolecules, 26, 5, 883–887

[Chulikavit 2023] Chulikavit N, Wang C, Huynh T, et al. (2023) "Fireproofing Flammable Composites Using Mycelium: Investigating the Effect of Deacetylation on the Thermal Stability and Fire Reaction Properties of Mycelium" Polymer Degradation and Stability, 215, 110419, doi:10.1016/j.polymdegradstab.2023.110419

[Chung 2004a] Chung CM, Roh YS, Cho SY, Kim JG. (2004) "Crack Healing in Polymeric Materials via Photochemical [2+2] Cycloaddition" Chemistry of Materials, 16, 21, 3982–3984, doi:10.1021/cm049394+

[Chung 2004b] Chung DD. (2004) "Self-Heating Structural Materials" Smart Materials and Structures, 13, 562–565, doi:10.1088/0964-1726/13/3/015

[Chung 2007] Chung KK, Schumacher JF, Sampson EM, Burne RA, Antonelli PJ, Brennan AB. (2007) "Impact of Engineered Surface Microtopography on Biofilm Formation of *Staphylococcus aureus*" Biointerphases, 2, 89–94, doi:10.1116/1.2751405

[Chuyang 2015] Chuyang C, McGonigal PR, Schneebeli ST, Li H, Vermeulen NA. (2015) "An Artificial Molecular Pump" Nature Nanotechnology, 10, 6, 547–553, doi:10.1038/nnano.2015.96

[Cira 2015] Cira NJ, Benusiglio A, Prakash M. (2015) "Vapour-Mediated Sensing and Motility in Two-Component Droplets" Nature, 519, 446–450, doi:10.1038/nature14272

[Cline 1968] Cline DC, Harr GB. (1968) "Shipping Container" US Patent 3,416,692

[Clune 2013] Clune J, Mouret J-B, Lipson H. (2013) "The Evolutionary Origins of Modularity" Proceedings of the Royal Society B, 280, 20122863, doi:10.1098/rspb.2012.2863

[Cochrane 1900] Cochrane AP. (1900) "Pneumatic Tire" US Patent 642,838

[Coleman 1973] Coleman LB, Cohen MJ, Sandman DJ, Yamagishi FG, Garito AF, Heeger AJ. (1973) "Superconducting Fluctuations and the Peierls Instability in an Organic Solid" Solid State Communications, 12, 1125

[Coleman 2011] Coleman AC, Beierle JM, Stuart MC, Maciá B, Caroli G, Mika JT, van Dijken DJ, Chen J, Browne WR, Feringa BL. (2011) "Light-Induced Disassembly of Self-Assembled Vesicle-Capped Nanotubes Observed in Real Time" Nature Nanotechnology, 6, 9, 547–552

[Collichio 1957] Collichio WJ. (1957) "Flagged Bristle and Brush Made from Same" US Patent 2,812,530

[Comesaña 2015] Comesaña R, Lusquiños F, del Val J, Quintero F, Riveiro A, Boutinguiza M, Jones JR, Hill RG, Poua J. (2015) "Toward Smart Implant Synthesis: Bonding Bioceramics of Different Resorbability to Match Bone Growth Rates" Scientific Reports, 5, 10677, doi: 10.1038/srep10677

[Connell 2007] Connell L. (2007) "Crack Repair System Stops Water in Its Tracks" Concrete International, May

[Connell 2014] Connell JL, Kim J, Shear JB, Bard AJ, Whiteley M. (2014) "Real-Time Monitoring of Quorum Sensing in 3D-Printed Bacterial Aggregates Using Scanning Electrochemical Microscopy" Proceedings of the National Academy of Sciences of the United States of America, 111, 51, 18255–18260, doi:10.1073/pnas.1421211111

[Cook 1972] Cook RL. (1972) "Self-Sealing Structure for Use as a Fluid Barrier in Containers" US Patent 3,662,904

[Cook 1974] Cook RL. (1974) "Self-Sealing Container" US Patent 3,801,425

[Cordier 2008] Cordier P, Tournilhac F, Soulie´-Ziakovic C, Leibler L. (2008) "Self-Healing and Thermoreversible Rubber from Supramolecular Assembly" Nature, 451, 977–980, doi:10.1038/nature06669

[Corten 2009] Corten CC, Urban MW. (2009) "Repairing Polymers Using an Oscillating Magnetic Field" Advanced Materials, 21, 5011–5015, doi:10.1002/adma.200901940.

[Corvasce 2006] Corvasce FG, Colantonio D, Luigi L. (2006) "Tire with Tread Having Degradable Tread Filler" US Patent 7,581,575

[Cosco 2006] Cosco S, Ambrogi V, Musto P, Carfagna C. (2006) "Urea formaldehyde Microcapsules Containing an Epoxy Resin: Influence of Reaction Parameters on the Encapsulation Yield" Macromolecular Symposia, 234, 184–92

[Couairona 2007] Couairona A, Mysyrowicz A. (2007) "Femtosecond Filamentation in Transparent Media" Physics Reports, 441, 47–189

[Coulson 2000] Coulson SR, Woodward I, Badyal JP, Brewer SA. (2000) "Super-Repellent Composite Fluoropolymer Surfaces" Journal of Physical Chemistry B, 104, 37, 8836–8840, doi:10.1021/jp0000174

[Cowin 1976] Cowin SC, Hegedus DH. (1976) "Bone Remodeling I: Theory of Adaptive Elasticity" Journal of Elasticity, 6, 3, 313–326

[Cox 2006] Cox JJ, Reimann F, Nicholas AK, Thornton G, Roberts E, Springell K, Karbani G, Jafri H, Mannan J, Raashid Y, Al-Gazali L, Hamamy H, Valente EM, Gorman S, Williams R, McHale DP, Wood JN, Gribble FM, Woods CG. (2006) "An SCN9A Channelopathy Causes Congenital Inability to Experience Pain" Nature, 444, 894–898, doi:10.1038/nature05413

[Cranford 2012] Cranford SW, Tarakanova A, Pugno NM, Buehler MJ. (2012) "Nonlinear Material Behaviour of Spider Silk Yields Robust Webs" Nature, 482, 72–76

[Craven 1969] Craven JM. (1969) "Cross-Linked Thermally Reversible Polymers Produced from Condensation Polymers with Pendant Furan Groups Cross-Linked with Maleimides" US Patent 3,435,003

[Cravotto 2007] Cravotto G, Cintas P. (2007) "Forcing and Controlling Chemical Reactions with Ultrasound" Angewandte Chemie International Edition, 46, 29, 5476–5478, doi:10.1002/anie.200701567

[Creczynski 2009] Creczynski-Pasa TB, Millone MA, Munford ML, Lima VR, Vieira TO, Benitez GA, Pasa AA, Salvarezza RC, Vela ME. (2009) "Self-Assembled Dithiothreitol on Au Surfaces for Biological Applications: Phospholipid Bilayer Formation" Physical Chemistry Chemical Physics, 11, 1077–1084, doi:10.1039/b811964c

[Crichton 1999] Crichton SN, Tomozawa M, Hayden JS, Suratwala TI, Campbell JH. (1999) "Subcritical Crack Growth in a Phosphate Laser Glass" Journal of the American Ceramic Society, 82, 11, 3097–3104, doi:10.1111/j.1151-2916.1999.tb02208.x

[Crick 1998] Crick F, Koch C. (1998) "Consciousness and Neuroscience" Cerebral Cortex, 8, 2, 97–107

[Crum 2004] Crum V, Homan D, Bortner R. (2004) "Certification Challenges for Autonomous Flight Control Systems" AIAA 2004-5257, AIAA Guidance Navigation and Control Conference, Providence, RI

[Cruz 2007] Cruz-Silva R, Arizmendi L, Del-Angel M, Romero-Garcia J. (2007) "pH- and Thermosensitive Polyaniline Colloidal Particles Prepared by Enzymatic Polymerization" Langmuir, 23, 8–12

[Cui 2007] Cui H, Chen Z, Zhong S, Wooley KL, Pochan DJ. (2007) "Block Copolymer Assembly via Kinetic Control" Science, 317, 647

[Cui 2015] Cui J, Daniel D, Grinthal A, Lin K, Aizenberg J. (2015) "Dynamic Polymer Systems with Self-Regulated Secretion for the Control of Surface Properties and Material Healing" Nature Materials, 14, 790–795, doi:10.1038/nmat4325

[Cully 2015] Cully A, Clune J, Tarapore D, Mouret JB. (2015) "Robots that Can Adapt Like Animals" Nature, 521, 503–507, doi:10.1038/nature14422

[Ćurčić 1996] Ćurčić S, Özveren CS, Crowe L, Lo PK. (1996) "Electric Power Distribution Network Restoration: A Survey of Papers and a Review of the Restoration Problem" Electric Power Systems Research, 35, 73–86, doi:10.1016/0378-7796(95)00991-4

[Cushman 2012] Cushman AJ, Fehrman BC, Gruenig SD, Korde UA. (2012) "Experiments on the Focusing and Use of Acoustic Energy to Accelerate Polymer Healing" Proc SPIE, Vol. 8341, Active and Passive Smart Structures and Integrated Systems, Sodano HA (ed.), doi:10.1117/12.914756

[Cusson 2009] Cusson D, Qian SY. (2009) "Ten-Year Field Evaluation of Corrosion-Inhibiting Systems in Concrete Bridges Barrier Walls" ACI Materials Journal, 106, 3, 291–300

[D'Angelo 2010] D'Angelo MM, Gallman J, Johnson V, Garcia E, Tai J, Russell Young R. (2010) "N^{+3} Small Commercial Efficient and Quiet Transportation for Year 2030–2035" NASA/CR–2010-216691, National Aeronautics and Space Administration, Langley Research Center, Hampton, VA

[D'Hollander 2009] D'Hollander S, van Assche G, van Mele V, du Prez F. (2009) "Novel Synthetic Strategy toward Shape Memory Polyurethanes with a Well-Defined Switching Temperature" Polymer, 50, 19, 4447–4454, doi:10.1016/j.polymer.2009.07.021

[da Costa 2010] da Costa RA, Dorogovtsev SN, Goltsev A V, Mendes JF. (2010) "Explosive Percolation Transition Is Actually Continuous" Physical Review Letters, 105, 255701, doi:10.1103/PhysRevLett.105.255701

[Dai 2013a] Dai L, Korolev KS, Gore J. (2013) "Slower Recovery in Space Before Collapse of Connected Populations" Nature, 496, 355–358, doi:10.1038/nature12071

[Dai 2013b] Dai Q, Wang Z, Rosli M, Hasan M. (2013) "Investigation of Induction Healing Effects on Electrically Conductive Asphalt Mastic and Asphalt Concrete Beams through Fracture-Healing Tests" Construction and Building Materials, 49, 729–737, doi:10.1016/j.conbuildmat.2013.08.089

[Dakai 1999] Dakai L, Mingshun H, Baoqi T, Hao Q. (1999) "Research on Some Problems about Optical Fiber Embedded in Carbon Fiber Smart Structures" Proceedings of 2nd International Workshop on Structural Health Monitoring, Stanford University, F-K Chang (ed.), Stanford, CA September 8–10, Technomic Publishing, Lancaster

[Damasceno 2012] Damasceno PF, Engel M, Glotzer SC. (2012) "Predictive Self-Assembly of Polyhedra into Complex Structures" Science, 337, 6093, 453–457, doi:10.1126/science.1220869

[Damasio 2010] Damasio A. (2010) Self Comes to Mind, Pantheon, New York

[Dang 2005] Dang HP, Hosono M, Kagami N, Okada K, Sharma A, Smith GJ, Sri-Jayantha SM. (2005) "Self-Monitoring and Self-Healing of Frequency Specific Position Errors in a Data Storage Device" US Patent 6,937,422

[Daniel 2001] Daniel JS, Kim YR. (2001) "Laboratory Evaluation of Fatigue Damage and Healing of Asphalt Mixtures" Journal of Materials in Civil Engineering, 13, 6, 434–440

[Daoud 2004] Daoud WA, Xin JH. (2004) "Nucleation and Growth of Anatase Crystallites on Cotton Fabrics at Low Temperatures" Journal of the American Ceramic Society, 87, 953–955, doi:10.1111/j.1551-2916.2004.00953.x

[Darlington 2001] Darlington AB, Dat JF, Dixon MA. (2001) "The Biofiltration of Indoor Air: Air Flux and Temperature Influences the Removal of Toluene, Ethylbenzene, and Xylene" Environmental Science & Technology, 35, 240–246

[Dasher 1940] Dasher PJ, Crawford RA, Colley RS. (1940) "Self-Sealing Fuel Tank" US Patent 2,425,515

[Dasher 1948] Dasher PJ. (1948) "Self-Sealing Fuel Tank" US Patent, 2,438,965

[Dave 2009] Dave R, Terry DS, Munro JB, Blanchard SC. (2009) "Mitigating Unwanted Photophysical Processes for Improved Single-Molecule Fluorescence Imaging" Biophysical Journal, 96, 2371–2381, doi:10.1016/j.bpj.2008.11.061

[David 2007] David FM, Campbell RH. (2007) "Building a Self-Healing Operating System" Third IEEE International Symposium on Dependable, Autonomic and Secure Computing, R Sterritt, X Li, X Zou, M Hinchey, Y Dai (eds.), pp 3–10, doi:10.1109/DASC.2007.22

[Davie 1964] Davie EW, Ratnoff OD. (1964) "Waterfall Sequence for Intrinstic Blood Clotting" Science, 145, 1310–1312

[Davies 1948] Davies JM, Shipman JJ, Davidson WL. (1948) "Self-Sealing Fuel Tank Construction" US Patent 2,446,815

[Davies 2014] Davies E, Müller KH, Wong WC, Pickard CJ, Reid DG, Skepper JN, Duer MJ. (2014) "Citrate Bridges between Mineral Platelets in Bone" Proceedings of the National Academy of Sciences of the United States of America, 111, 14, E1354–E1363, doi:10.1073/pnas.1315080111

[Davis 2009] Davis DA, Hamilton A, Yang J, Cremar LD, Van Gough D, Potisek SL, Ong MT, Braun PV, Martínez TJ, White SR, Moore JS, Sottos NR. (2009) "Force-Induced Activation of Covalent Bonds in Mechanoresponsive Polymeric Materials" Nature, 459, 68–72, doi:10.1038/nature07970

[Davoudi 2014] Davoudi ZM, Kandjani AE, Bhatt AI, Kyratzis IL, O'Mullane AP, Bansal V. (2014) "Hybrid Antibacterial Fabrics with Extremely High Aspect Ratio Ag/AgTCNQ Nanowires" Advanced Functional Materials, 24, 8, 1047–1053, doi:10.1002/adfm.201302368

[de Crevoisier 1999] de Crevoisier G, Fabre P, Corpart JM, Leibler L. (1999) "Switchable Tackiness and Wettability of a Liquid Crystalline Polymer" Science, 285, 1246–1249.

[De Geest 2007] De Geest BG, Sanders NN, Sukhorukov GB, Demeester J, De Smedt SC. (2007) "Release Mechanisms for Polyelectrolyte Capsules" Chemical Society Reviews, 36, 636–649, doi:10.1039/B600460C

[de Gennes 1967] de Gennes PG. (1971) "Reptation of a Polymer Chain in the Presence of Fixed Obstacles" Journal of Chemical Physics, 55, 572–579, doi:10.1063/1.1675789

[de Gennes 1971] de Gennes PG. (1971) "Reptation of a Polymer Chain in the Presence of Fixed Obstacles" Journal of Chemical Physics, 55, 572–579, doi:10.1063/1.1675789

[de Gennes 1983] de Gennes PG. (1983) "Entangled Polymers," Physics Today, 36, 6, 33–39

[de Gennes 1985] de Gennes PG. (1985) "Wetting: Statics and Dynamics" Reviews of Modern Physics, 57, 3, 827–861, doi:/10.1103/RevModPhys.57.827

[de Gennes 1992] de Gennes PG. (1992) "Soft Matter" Reviews of Modern Physics, 64, 645–648, doi:10.1103/RevModPhys.64.645

[de Lucca 1987] de Lucca Freitas LL, Stadler R. (1987) "Thermoplastic Elastomers by Hydrogen Bonding. 3. Interrelations between Molecular Parameters and Rheological Properties" Macromolecules, 20, 10, 2478–2485, doi:10.1021/ma00176a027

[de Mestrai 1961] de Mestrai G. (1961) "Separable Fastening Device" US Patent 3,009,235

[De Muynck 2008] De Muynck W, Debrouwer D, De Belie N, Verstraete W. (2008) "Bacterial Carbonate Precipitation Improves the Durability of Cementitious Materials" Cement and Concrete Research, 38, 7, 1005–1014, doi:10.1016/j.cemconres.2008.03.005

[de Smet 2008] de Smet CA, Kunz J. (2008) "Rehabilitation of Bridges with Concrete Overlays" Proceedings of IABMAS 2008 Bridge Maintenance, Safety, Management, Health Monitoring and Informatics, HM Koh, D Frangopol (eds), pp 2901–2908

[de Vries 1994] de Vries MJ, Nasta M, Grinder CR, Claus RO, Masri SF. (1994) "Optical Fibers for Sensing, Healing and Reinforcement" SPIE Vol. 2191, Smart Sensing, Processing, and Instrumentation, JS Sirkis (ed.)

[De Vries 1997] De Vries J, Polder RB. (1997) "Hydrophobic Treatment of Concrete" Construction and Building Materials, 11, 4, 259–265, doi:10.1016/S0950-0618(97)00046-9

[De Windt 2003] De Windt W, Boon N, Siciliano SD, Verstraete W. (2003) "Cell Density Related H_2 Consumption in Relation to Anoxic Fe(0) Corrosion and Precipitation of Corrosion Products by *Shewanella oneidensis* MR-1" Environmental Microbiology, 5, 1192–1202, doi:10.1046/j.1462-2920.2003.00527.x

[Decker 2007] Decker MJ, Halbach CJ, Nam CH, Wagner NJ, Wetzel ED. (2007) "Stab Resistance of Shear Thickening Fluid (STF)-Treated Fabrics" Composites Science and Technology, 67, 3–4, 565–578.

[Degang 1999] Degang C, Xinmao G, Baoqi T, Hongdong H, Huaiqiang X. (1999) "The Investigation of Health Monitoring and Strength of Adaptive Composite Structures" Proceedings of 2nd International Workshop on Structural Health Monitoring, F-K Chang (ed.), Stanford University, Stanford, CA, September 8–10, Technomic Publishing, Lancaster, pp 921–925

[DeGraef 2005] DeGraef B, DeWindt W, Dick J, Verstraete W, DeBelie N. (2005) "Cleaning of Concrete Fouled by Lichens with the Aid of *Thiobacilli*" Materials and Structures, 38, 284, 875–882, doi:10.1617/14254

[Degtyar 2015] Degtyar E, Mlynarczyk B, Fratzl P, Harrington MJ. (2015) "Self-Healing Response in Supramolecular Polymers Based on Reversible Zinc–Histidine Interactions" Polymer, 69, 255–263, doi:10.1016/j.polymer.2015.03.030

[Dehmer 2011] Dehmer M, Mowshowitz A. (2011) "Generalized Graph Entropies" Complexity, 17, 45–50, doi:10.1002/cplx.20379

[DeJong 2010] DeJong JT, Mortensen BM, Martinez BC, Nelson DC. (2010) "Bio-Mediated Soil Improvement" Ecological Engineering, 36, 2, 197–210, doi:10.1016/j.ecoleng.2008.12.029

[Dementsov 2008] Dementsov A, Privman V. (2008) "Three-Dimensional Percolation Modeling of Self-Healing Composites" Physical Review E: Statistical, Nonlinear, and Soft Matter Physics, 78, 2, 021104

[Demetriou 2011] Demetriou MD, Launey ME, Garrett G, Schramm JP, Hofmann DC, Johnson WL, Ritchie RO. (2011) "A Damage-Tolerant Glass" Nature Materials, 10, 123–128, doi:10.1038/nmat2930

[Demortière 2014] Demortière A, Snezhko A, Sapozhnikov MV, Becker N, Proslier T, Aranson IS. (2014) "Self-Assembled Tunable Networks of Sticky Colloidal Particles" Nature Communications, 5, 3117, doi:10.1038/ncomms4117

[Deng 2006] Deng J, Ai J, Gao M, Luo C. (2006) "A Real-Time Engineering Software System for Failure Detection and Isolation of Self-Repairing Flight Control System" Proceedings of SPIE, 6358, doi:10.1117/12.718209

[Deng 2010] Deng G, Tang C, Li F, Jiang H, Chen Y. (2010) "Covalent Cross-Linked Polymer Gels with Reversible Sol–Gel Transition and Self-Healing Properties" Macromolecules, 43, 1191–1194, doi:10.1021/ma9022197

[Deng 2015] Deng CC, Brooks WL, Abboud KA, Sumerlin BS. (2015) "Boronic Acid-Based Hydrogels Undergo Self-Healing at Neutral and Acidic pH" ACS Macro Letters, 4, 2, 220–224, doi:10.1021/acsmacrolett.5b00018

[Derfus 2007] Derfus AM, Maltzahn GV, Harris TJ, Duza KS, Ruoslahti VE, Bhatia SN. (2007) "Remotely Triggered Release from Magnetic Nanoparticles" Advanced Materials, 19, 3932–3936, doi:10.1002/adma.200700091

[Desai 2014] Desai S, Boden M. (2014) "Crosslinkable Polyisobutylene-Based Polymers and Medical Devices Containing the Same" US Patent 8,529,934

[Deshpande 2011] Deshpande RD, Li J, Cheng YT, Verbrugge MW. (2011) "Liquid Metal Alloys as Self-Healing Negative Electrodes for Lithium Ion Batteries" Journal of Electrochemical Society, 158, 8, A845–A849, doi:10.1149/1.3591094

[DesRoches 2004] DesRoches R, McCormick J, Delemont M. (2004) "Cyclic Properties of Superelastic Shape Memory Alloy Wires and Bars" Journal of Structural Engineering, 130, 1, 38, doi:10.1061/(ASCE)0733-9445(2004)

[Dettre 1964] Dettre RH, Johnson RE. (1964) "Contact Angle Hysteresis II. Contact Angle Measurements on Rough Surfaces" Advances in Chemistry, 43, 136–144, doi:10.1021/ba-1964-0043.ch008

[Devanathan 2008] Devanathan R, Weber WJ. (2008) "Dynamic Annealing of Defects in Irradiated Zirconia-Based Ceramics" Journal of Materials Research, 23, 3, 593–597

[D'Heurle 1971] D'Heurle FM. (1971) "Electromigration and Failure in Electronics: An Introduction" Proceedings of IEEE, 59, 1409–1418

[Dhir 1973] Dhir RK, Sangha CM, Munday JG. (1973) "Strength and Deformation Properties of Autogenously Healed Mortars," ACI Journal, 70, 231–236

[Di Prima 2007] Di Prima MA, Lesniewski M, Gall K, McDowell DL, Sanderson T, D Campbell D. (2007) "Thermo-Mechanical Behavior of Epoxy Shape Memory Polymer Foams" Smart Materials and Structures, 16, 2330–2340, doi:10.1088/0964-1726/16/6/037

[Diamond 1999] Diamond SL. (1999) "Engineering Design of Optimal Strategies for Blood Clot Dissolution" Annual Review of Biomedical Engineering, 1, 427–461

[Diamond 2005] Diamond J. (2005) Guns, Germs and Steel, Norton, ISBN 978-0393061314

[Dickey 2014] Dickey MD. (2014) "Emerging Applications of Liquid Metals Featuring Surface Oxides" ACS Applied Materials & Interfaces, 6, 21, 18369–18379, doi:10.1021/am5043017

[Diegelmann 2004] Diegelmann RF, Evans MC. (2004) "Wound Healing: An Overview of Acute, Fibrotic and Delayed Healing" Frontiers in Bioscience, 9, 283–289

[Dikić 2012] Dikić T, Ming W, van Benthem RA, Esteves AC, de With G. (2012) "Self-Replenishing Surfaces" Advanced Materials, 24, 3701–3704, doi:10.1002/adma.201200807

[Ding 2009] Ding Z, Srivastava SK, Cartes DA, Suryanarayanan S. (2009) "Dynamic Simulation-Based Analysis of a New Load Shedding Scheme for a Notional Destroyer-Class Shipboard Power System" IEEE Transactions on Industry Applications, 45, 3, 1163–1174, doi:10.1109/TIA.2009.2018965

[Ding 2012] Ding X, Yang C, Lim TP, Hsu LY, Engler AC, Hedrick JL, Yang YY. (2012) "Antibacterial and Antifouling Catheter Coatings Using Surface Grafted Peg-B-Cationic Polycarbonate Diblock Copolymers" Biomaterials, 33, 28, 6593–6603, doi:10.1016/j.biomaterials.2012.06.001

[Ding 2015] Ding F, Wu S, Wang S, Xiong Y, Li Y, Li B, Deng H, Du Y, Xiao L, Shi X. (2015) "A Dynamic and Self-Crosslinked Polysaccharide Hydrogel with Autonomous Self-Healing Ability" Soft Matter, 11, 3971–3976, doi:10.1039/C5SM00587F

[Dinitz 2008] Dinitz AM, Bang W. (2008) "The Use of Polymer Concrete Materials for Construction, Maintenance, Rehabilitation and Preservation of Concrete and Steel Orthotropic Bridge Decks" Bridge Maintenance, Safety, Management, Health Monitoring and Informatics, HM Koh, D Frangopol (eds), CRC Press, ISBN 978-0-415-46844-2, pp 2874–2879

[Dirienzo 1920] Dirienzo JM, Dirienzo JA. (1920) "Inner Tube" US Patent 1,339,948

[Diu 2014] Diu T, Faruqui N, Sjöström T, Lamarre B, Jenkinson HF, Su B, Ryadnov MG. (2014) "Cicada-Inspired Cell-Instructive Nanopatterned Arrays" Scientific Reports, 4, 7122, doi:10.1038/srep07122

[Do 2013] Do JW, Estrada D, Xie X, Chang NN, Mallek J, Girolami GS, Rogers JA, Pop E, Lyding JW. (2013) "Nanosoldering Carbon Nanotube Junctions by Local Chemical Vapor Deposition for Improved Device Performance" Nano Letters, 13, 12, 5844–5850, doi:10.1021/nl4026083

[Doe 2005] Doe TA, Burgess CJ, Philbrook K, Hauter A. (2005) "Advancing Technology for Cable Durability" Structure, 22–24

[Döhler 2015] Döhler D, Peterlik H, Binder WH. (2015) "A Dual Crosslinked Self-Healing System: Supramolecular and Covalent Network Formation of Four-Arm Star Polymers" Polymer, 69, 264–273, doi:10.1016/j.polymer.2015.01.073

[Douglas 2009] Douglas SM, Dietz H, Liedl T, Hogberg B, Graf F, Shih WM. (2009) "Self-Assembly of DNA into Nanoscale Three-Dimensional Shapes" Nature, 459, 414–418, doi:10.1038/nature08016

[Douglas 2012] Douglas SM, Bachelet I, Church GM. (2012) "A Logic-Gated Nanorobot for Targeted Transport of Molecular Payloads" Science, 335, 6070, 831–834, doi:10.1126/science.1214081

[Drechsler 2009] Drechsler U, Singh W, Sharma A. (2009) "Functionalization of Polymers with Reactive Species Having Bond-Stabilized Decontamination Activity" US Patent Application 20090012204

[Drexler 1981] Drexler KE. (1981) "Molecular Engineering: An Approach to the Development of General Capabilities for Molecular Manipulation" Proceedings of the National Academy of Sciences of the United States of America, 78, 9, 5275–5278

[Dreyfus 2005] Dreyfus R, Baudry J, Roper ML, Fermigier M, Stone HA, Bibette J. (2005) "Microscopic Artificial Swimmers" Nature, 437, 862–865, doi:10.1038/nature04090

[Dry 1992a] Dry CM. (1992) "Passive Smart Materials for Sensing and Actuation." Proceedings: Conference on Recent Advances in Adaptive and Sensory Materials and Their Applications, CA Rogers, RC Rogers (eds), VPI &SU, Blacksburg, Virginia, April 27–29, Technomic, Lancaster, England, pp 207–223

[Dry 1992b] Dry CM. (1992) "Smart Materials Which Sense, Activate and Repair Damage; Hollow Porous Fibres in Composites Release Chemicals from Fibers for Self-Healing, Damage Prevention and/or Dynamic Control" Proceedings of 1st European Conference on Smart Structures and Materials (Glasgow), B Culshaw, PT Gardiner, A McDonach (eds), pp 367–370, EOS/SPIE and IOP Publishing, Ltd.

[Dry 1992c] Dry C. (1992) "Passive Tuneable Fibers and Matrices" International Journal of Modern Physics B, 6, 15–16, 2763–2771

[Dry 1993a] Dry C. (1993) "Passive Smart Materials for Sensing and Actuation" Journal of Intelligent Materials Systems and Structures, 4, 420–425

[Dry 1993b] Dry CM, Sottos NR. (1993) "Passive Smart Self-Repair in Polymer Matrix Composite Materials" SPIE 1916, Smart Structures and Materials 1993: Smart Materials

[Dry 1994] Dry C. (1994) "Matrix Cracking Repair and Filling Using Active and Passive Modes for Smart Timed Release of Chemicals from Fibers into Cement Matrices" Smart Mater Struc 3 118–123

[Dry 1996a] Dry C. (1996) "Self-Repairing, Reinforced Matrix Materials" US Patent 5,561,173

[Dry 1996b] Dry C, McMillan W. (1996) "Three-Part Methylmethacrylate Adhesive System as an Internal Delivery System for Smart Responsive Concrete" Smart Materials and Structures, 5, 3, 297–300

[Dry 1996c] Dry C. (1996) "Cementitious Materials" US Patent 5,575,841

[Dry 1996d] Dry C. (1996) "Procedures Developed for Self-Repair of Polymer Matrix Composite Materials." Composite Structures 35, 263–269

[Dry 1996e] Dry C. (1996) "Smart Earthquake-Resistant Materials: Using Time-Released Adhesives for Damping, Stiffening, and Deflection Control" Proceedings of SPIE 2779, 3rd International Conference on Intelligent Materials and 3rd European Conference on Smart Structures and Materials, PF Gobin, J Tatibouet (eds), doi:10.1117/12.237085

[Dry 1997] Dry CM. (1997) "Self-Repairing, Reinforced Matrix Materials" US Patent 5,660,624

[Dry 1998] Dry CM. (1998) "Smart-Fiber-Reinforced Matrix Composites" US Patent 5,803,963

[Dry 2000] Dry CM. (2000) "Three Designs for the Internal Release of Sealants, Adhesives, and Waterproofing Chemicals into Concrete to Reduce Permeability" Cement and Concrete Research, 30, 1969–1977

[Dry 2001a] Dry C. (2001) "In-Service Repair of Highway Bridges and Pavements by Internal Time-Release Repair Chemicals" Final Report for Highway IDEA Project 37, Transportation Research Board, NW, Washington, DC

[Dry 2001b] Dry CM. (2001) "Self-Repairing, Reinforced Matrix Materials" US Patent 6,261,360

[Dry 2003a] Dry C, Corsaw M, Bayer E. (2003) "A Comparison of Internal Self-Repair with Resin Injection in Repair of Concrete" Journal of Adhesion Science and Technology, 17, 1, 79–89, doi:10.1163/15685610360472457

[Dry 2003b] Dry C. (2003) "Self Repairing of Composites" SPIE, Vol. 5055-54, Smart Electronics, MEMS, BioMEMS, and Nanotechnology

[Dry 2008] Dry C. (2008) "Self Repairing Composites for Airplane Components" Sensors and Smart Structures Technologies for Civil, Mechanical, and Aerospace Systems 2008, M Tomizuka (ed.), Proc SPIE, Vol. 6932, doi:10.1117/12.776497

[Du 2011] Du X, He J. (2011) "A Self-Templated Etching Route to Surface-Rough Silica Nanoparticles for Superhydrophobic Coatings" ACS Applied Materials & Interfaces, 3, 4, 1269–1276, doi:10.1021/am200079w

[Du 2012] Du PY, Wu JY, Hsu TH, Lee MH, Wang TY, Cheng HY, Lai EK, Lai SC, Lung HL, Kim SB, BrightSky MJ, Zhu Y, Mittal S, Cheek R, Raoux S, Joseph EA, Schrott A, Li J, Lam C. (2012) "The Impact of Melting during Reset Operation on the Reliability of Phase Change Memory" 2012 IEEE International Reliability Physics Symposium (IRPS), 6C.2.1–6C.2.6, doi:10.1109/IRPS.2012.6241872

[Du 2023] Du F, Alghamdi S, Yang J, Huston D, Tan T. (2023) "Interfacial Mechanical Behavior in Nacre of Red Abalone and Other Shells: A Review" ACS Biomaterials Science & Engineering, 9, 7, 3843–3859, doi:acsbiomaterials.2c00080

[Duan 2005] Duan ZH, Cai Z, Yu J. (2005) "Fault Diagnosis and Fault Tolerant Control for Wheeled Mobile Robots under Unknown Environments: A Survey" Proceeding of IEEE International Conference on Robotics and Automation, pp 3428–3433, doi:10.1109/ROBOT.2005.1570640

[Duan 2008] Duan WH, Quek ST, Wang Q. (2008) "Finite Element Analysis of the Piezoelectric-Based Repair of a Delaminated Beam" Smart Materials and Structures, 17, 015017, doi:10.1088/0964-1726/17/01/015017

[Dubiel 2002] Dubiel M, Hsu CH, Chien CC, Mansfield F, Newman DK. (2002) "Microbial Iron Respiration Can Protect Steel from Corrosion" Applied and Environmental Microbiology, 68, 3, 1440–1445, doi:10.1128/AEM.68.3.1440-1445.2002

[Dubreuil 2003] Dubreuil F, Elsner N, Fery A. (2003) "Elastic Properties of Polyelectrolyte Capsules Studied by Atomic-Force Microscopy and RICM" European Physical Journal E 12, 215–221, doi:10.1140/epje/i2003-10056-0

[Dudek 2009] Dudek MA, Chawla N. (2009) "Mechanisms for Sn Whisker Growth in Rare Earth-Containing Pb-Free Solders" Acta Materialia, 57, 15, 4588–4599, doi:10.1016/j.actamat.2009.06.031

[Duenas 2007] Duenas T, Jha A, Lee W, Bortolin R, Mal A, Ooi TK, Corder A. (2007) "Structural Health Monitoring with Self-Healing Morphing Skins" Structural Health Monitoring 2007, FK Chang (ed.), DEStech, Lancaster

[Duffy 1998] Duffy DC, McDonald JC, Schueller JA, Whitesides GM. (1998) "Rapid Prototyping of Microfluidic Systems in Poly(dimethylsiloxane)" Analytical Chemistry, 70, 4974–4984, doi:10.1021/ac980656z

[Duki 2011] Duki SF, Kolmakov GV, Yashin VV, Kowalewski T, Matyjaszewski K, Balazs AC. (2011) "Modeling the Nanoscratching of Self-Healing Materials" Journal of Chemical Physics, 134, 084901, doi:10.1063/1.3556744

[Dumitru 2013] Dumitru TA. (2013) "Fission-Track Geochronology" in Quaternary Geochronology: Methods and Applications, JS Noller, JM Sowers, WR Lettis (eds), American Geophysical Union, pp 131–155

[Duprat 2012] Duprat C, Protiére S, Beebe AY, Stone HA. (2012) "Wetting of Flexible Fibre Arrays" Nature, 482, 510–513, doi:10.1038/nature10779

[Durnin 1987a] Durnin J, Miceli Jr. JJ, Eberly JH. (1987) "Diffraction-Free Beams" Physical Review Letters, 58, 15, 1499–1501

[Durnin 1987b] Durnin J. (1987) "Exact Solutions for Nondiffracting Beams. I. The Scalar Theory" Journal of the Optical Society of America, 4, 4, 651–654

[Dussan 1985] Dussan V. EB. (1985) "On the Ability of Drops or Bubbles to Stick to Non-Horizontal Surfaces of Solids. Part 2. Small Drops or Bubbles Having Contact Angles of Arbitrary Size" Journal of Fluid Mechanics, 151, 1, doi:10.1017/S0022112085000842

[Dyke 1996] Dyke SJ, Jr. BFS, Sain MK, Carlson JD (1996) Modeling and control of magnetorheological dampers for seismic response reduction. Smart Materials and Structures, 5, 565, doi:10.1088/0964-1726/5/5/006

[Eapen 2005] Eapen KC, Patton ST, Smallwood SA, Phillips BS, Zabinski JS. (2005) "MEMS Lubricants Based on Bound and Mobile Phases of Hydrocarbon Compounds: Film Deposition and Performance Evaluation" Journal of Microelectromechanical Systems, 14, 954–960

[Ebbing 2002] Ebbing MH, Villa MJ, Valpuesta JM, Prados P, de Mendoza J. (2002) "Resorcinarenes with 2-Benzimidazolone Bridges: Self-Aggregation, Self-Assembled Dimeric Capsules, and Guest Encapsulation" Proceedings of the National Academy of Sciences of the United States of America, 99, 8, 4962–4966, doi:10.1073/pnas.072652599

[Eckelmeyer 1946] Eckelmeyer Jr. EH. (1946) "The Story of the Self-Sealing Tank" US Naval Institute Proceedings, 72, 2, 205–219

[Eckhoff 2001] Eckhoff A, Lucena A, Medelius P, Perotti J. (2001) "Advanced Data Acquisition Technology Self-Healing, Self-Calibrating Signal Conditioner" Attachment to Disclosure of Invention and New Technology Report KSC-12301, NASA, Kennedy Space Center

[Eddaoudi 2002] Eddaoudi M, Kim J, Vodak D, Sudik A, Wachter J, O'Keeffe M, Yaghi OM. (2002) "Geometric Requirements and Examples of Important Structures in the Assembly of Square Building Blocks" Proceedings of the National Academy of Sciences of the United States of America, 99, 8, 4900–4904, doi:10.1073/pnas.082051899

[Edmiston 2013] Edmiston PL. (2013) "Modified Sol–Gel Derived Sorbent Material and Method for Using the Same" US Patent Application 20130012379

[Edvardsen 1999] Edvardsen C. (1999) "Water Permeability and Autogenous Healing of Cracks in Concrete," ACI Materials Journal, 96, 6, 448–455

[Edwards 1967] Edwards SF. (1967) "The Statistical Mechanics of Polymerized Material" Proceedings of the Physical Society of London, 92, 9–16, doi:10.1088/0370-1328/92/1/303

[Edwards 2002] Edwards D, Hubbs W, Hovater M. (2002) "Capacitors Could Help Protect Against Hypervelocity Impacts" Technical Support Package, NASA Tech Briefs MFS-31995-1

[Egloffstein 2001] Egloffstein TA. (2001) "Natural Bentonites: Influence of the Ion Exchange and Partial Desiccation on Permeability and Self-Healing Capacity of Bentonites Used in GCL" Geotextiles and Geomembranes, 19, 7, 427–444, doi:10.1016/S0266-1144(01)00017-6

[Ehsani 2010] Ehsani M. (2010) "FRP Super Laminates: Transforming the Future of Repair and Retrofit with FRP" Concrete International, March, 49–53

[Eichel 1959] Eichel HJ. (1959) "Adhesive Tape Containing Pressure-Rupturable Capsules" US Patent 2,907,682

[Eisenberg 1990] Eisenberg A, Hird B, Moore RB. (1990) "A New Multiplet-Cluster Model for the Morphology of Random Ionomers" Macromolecules, 23, 18, 4098–4107, doi:10.1021/ma00220a012

[El-Ganainy 2018] El-Ganainy R, Makris KG, Khajavikhan M, et al. (2018) "Non-Hermitian Physics and PT Symmetry" Nature Physics, 14, 11–19. doi:10.1038/nphys4323

[Elliott 2000] Elliott C, Heile B. (2000) "Self-Organizing, Self-Healing Wireless Networks" 2000 IEEE International Conference on Personal Wireless Communications, Hyderabad, 355–362, doi:10.1109/ICPWC.2000.905836

[El-Naggar 2010] El-Naggar MY, Wanger G, Leung KM, Yuzvinsky TD, Southam G, Yang J, Lau WM, Nealson KH, Gorby YA. (2010) "Electrical Transport along Bacterial Nanowires from *Shewanella oneidensis* MR-1" Proceedings of the National Academy of Sciences of the United States of America, 107, 42, 18127–18131, doi:10.1073/pnas.1004880107

[Elsadig 2009] Elsadig M, Abdullah A. (2009) "Biological Inspired Intrusion Prevention and Self-Healing System for Network Security Based on Danger Theory" International Journal of Video & Image Processing and Network Security, 9, 9, 1–15

[England 2013] England JL. (2013) "Statistical Physics of Self-Replication" Journal of Chemical Physics, 139, 121923, doi:10.1063/1.4818538

[Enos 1998] Enos DG, Guilbert CR, Norman CF, Kehr JA, Boyer CE. (1998) "Improving the Damage Tolerance, and Extending the Service Life of Fusion-Bonded Epoxy Coatings" Proceedings of International Conference on Corrosion and Rehabilitation of Reinforced Concrete Structures, Orlando, FL

[Eppel 2014] Eppel S, Portnoy M. (2014) "Reversible Multistep Synthesis with Equilibrium Properties Based on a Selection-Oriented Process with a Repetitive Sequence of Steps" Journal of Physical Chemistry B, 118, 32, 9733–9744, doi:10.1021/jp5051645

[Erbel 2007] Erbel R, Di Mario C, Bartunek J, Bonnier J, de Bruyne B, Eberli FR, Erne P, Haude M, Heublein B, Horrigan M, Ilsley C, Böse D, Koolen J, Lüscher TF, Weissman N, Waksman R. (2007) "Temporary Scaffolding of Coronary Arteries with Bioabsorbable Magnesium Stents: A Prospective, Non-Randomised Multicentre Trial" Lancet, 369, 9576, 1869–1875, doi:10.1016/S0140-6736(07)60853-8

[Erbil 2003] Erbil HY, Demirel AL, Avci Y, Mert O. (2003) "Transformation of a Simple Plastic into a Superhydrophobic Surface" Science 299, 5611, 1377–1380, doi:10.1126/science.1078365

[Erdős 1960] Erdős P, Rényi A. (1960) "On the Evolution of Random Graphs" Publicationes Mathematicae 6, 290–297

[Ericsson 1921] Ericsson HC. (1921) "Anti-Explosive and Non-Inflammable Gasolene Tank" US Patent 1,381,175

[Eringen 1980] Eringen AC. (1980) Mechanics of Continua, 2nd edition, Krieger Pub Co

[Esser 2005] Esser B, Huston D, Spencer G, Burns D, Kahn E. (2005) "Active Self-Healing Wire Insulation" SPIE 5762-02, Smart Structures and Materials: Industrial and Commercial Applications of Smart Structures Technologies, San Diego, CA.

[Esser 2010] Esser-Kahn AP, Sottos NR, White SR, Moore JS. (2010) "Programmable Microcapsules from Self-Immolative Polymers" Journal of the American Chemical Society, 132, 10266–10268. doi:10.1021/ja104812p

[Evers 2016] Evers CH, Luiken JA, Bolhuis PG, Kegel WK. (2016) "Self-Assembly of Microcapsules via Colloidal Bond Hybridization and Anisotropy" Nature 534, 364–368, doi:10.1038/nature17956

[Extrand 1995] Extrand CW, Kumagai Y. (1995) "Liquid Drops on an Inclined Plane: The Relation between Contact Angles, Drop Shape, and Retentive Force" Journal of Colloid and Interface Science, 170, 2, 515–521, doi:10.1006/jcis.1995.1130

[Extrand 2006] Extrand CW. (2006) "Designing for Optimum Liquid Repellency" Langmuir, 22, 1711–1714

[Faghihnejad 2014] Faghihnejad A, Feldman KE, Yu J, Tirrell MV, Israelachvili JN, Hawker CJ, Kramer EJ, Zeng H. (2014) "Adhesion and Surface Interactions of a Self-Healing Polymer with Multiple Hydrogen-Bonding Groups" Advanced Functional Materials, 24, 16, 2322–2333, doi:10.1002/adfm.201303013

[Fahrbach 2010] Fahrbach FO, Simon P, Rohrbach A. (2010) "Microscopy with Self-Reconstructing Beams" Nature Photonics, 4, 780–785, doi:10.1038/NPHOTON.2010.204

[Fall 2001] Fall R. (2001) "Puncture Reversal of Ethylene Ionomers: Mechanistic Studies" Masters Thesis, Chemistry, Virginia Polytechnic Institute and State University

[Fan 2014] Fan JA, Yeo WH, Su Y, Hattori Y, Lee W, Jung SY, Zhang Y, Liu Z, Cheng H, Falgout L, Bajema M, Coleman T, Gregoire D, Larsen RJ, Huang Y, Rogers JA. (2014) "Fractal Design Concepts for Stretchable Electronics" Nature Communications, 5, 3266, doi:10.1038/ncomms4266

[Fang 2013] Fang Y, Wang CF, Zhang ZH, Shao H, Chen S. (2013) "Robust Self-Healing Hydrogels Assisted by Cross-Linked Nanofiber Networks" Scientific Reports, 3, 2811, doi:10.1038/srep02811

[Faramarzi 2012] Faramarzi V, Niess F, Moulin E, Maaloum M, Dayen JF, Beaufrand JB, Zanettini S, Doudin B, Giuseppone N. (2012) "Light-Triggered Self-Construction of Supramolecular Organic Nanowires as Metallic Interconnects" Nature Chemistry, 4, 485–490, doi:10.1038/nchem.1332

[Farra 2012] Farra R, Sheppard NF, McCabe L, Neer RM, Anderson JM, Santini Jr. JT, Cima MJ, Langer R. (2012) "First-in-Human Testing of a Wirelessly Controlled Drug Delivery Microchip" Science Translational Medicine, 4, 122, 122ra21, doi:10.1126/scitranslmed.3003276

[Fay 1999] Fay FF, Carraway R, Ikebe M, Walker J. (1999) "Photosensitive Caged Macromolecules" US Patent 5,998,580

[Fay 2005] Fay F, Linossier I, Langlois V, Haras D, Vallee-Rehel K. (2005) "SEM and EDX Analysis: Two Powerful Techniques for the Study of Antifouling Paints" Progress in Organic Coatings, 54, 216–223, doi:10.1016/j.porgcoat.2005.05.005

[Feinberg 2009] Feinberg TE. (2009) From Axons to Identity: Neurological Explorations of the Nature of the Self, W W Norton, New York

[Felton 2014] Felton S, Tolley M, Demaine E, Rus D, Wood R. (2014) "A Method for Building Self-Folding Machines" Science, 345, 6197, 644–646, doi:10.1126/science.1252610

[Feng 2015] Feng S, Wang S, Tao Y, Shang W, Deng S, Zheng Y, Hou Y. (2015) "Radial Wettable Gradient of Hot Surface to Control Droplets Movement in Directions" Scientific Reports, 5, 10067, doi:10.1038/srep10067

[Ferl 2007] Ferl J, Ware J, Cadogan D, Yavorsky J. (2007) "Self-Healing Technology for Gas Retention Structures and Space Suit Systems" SAE Technical Paper, 2007-01-3211, doi:10.4271/2007-01-3211

[Fermi 1956] Fermi E (1956) *Thermodynamics*, Dover Publications

[Ferrara 2014] Ferrara L, Krelani V, Carsana M. (2014) "A "Fracture Testing" Based Approach to Assess Crack Healing of Concrete with and without Crystalline Admixtures" Construction and Building Materials, 68, 535–551, doi:10.1016/j.conbuildmat.2014.07.008

[Ferraris 1973] Ferraris J, Cowan DO, Walatka V, Perlstein JH. (1973) "Electron Transfer in a New Highly Conducting Donor–Acceptor Complex" Journal of the American Chemical Society, 3, 95, 3, doi:10.1021/ja00784a066

[Ferris 1992] Ferris FG, Stehmeier LG (1992) "Bacteriogenic Mineral Plugging" US Patent 5,143,155

[Fery 2004] Fery A, Dubreuil F, Moehwald H. (2004) "Mechanics of Artificial Microcapsules" New Journal of Physics, 6, 18

[Feynman 1960] Feynman R. (1960) "There's Plenty of Room at the Bottom" Engineering and Science, 23, 5, 22–36

[Fialkowski 2006] Fialkowski M, Bishop KJ, Klajn R, Smoukov SK, Campbell CJ, Grzybowski BA. (2006) "Principles and Implementations of Dissipative (Dynamic) Self-Assembly" Journal of Physical Chemistry B, 110, 6, 2482–2496, doi:10.1021/jp054153q

[Fiore 2013] Fiore GL, Rowan SJ, Weder C. (2013) "Optically Healable Polymers" Chemical Society Reviews, 42, 7278–7288, doi:10.1039/C3CS35471G

[Fischer 2010a] Fischer H. (2010) "Self-Repairing Material Systems: A Dream or a Reality?" Natural Science, 2, 8, 873–901, doi:10.4236/ns.2010.28110

[Fischer 2010b] Fischer HR, Tempelaars K, Kerpershoek A, Dingemans T, Iqbal M, van Lonkhuyzen H, Iwanowsky B, Semprimoschnig C. (2010) "Development of Flexible LEO-Resistant PI Films for Space Applications Using a Self-Healing Mechanism by Surface-Directed Phase Separation of Block Copolymers" ACS Applied Materials & Interfaces, 2, 8, 2218–2225, doi:10.1021/am100223v

[Fisher 2006] Fisher B, Szylakowski G, Easter MR. (2006) "Fault Protected Electrical Cable" US Patent Application 20060105616

[Fleet 2015] Fleet EJ, Zhang Y, Hayes SA, Smith PJ. (2015) "Inkjet Printing of Self-Healing Polymers for Enhanced Composite Interlaminar Properties" Journal of Materials Chemistry A, 3, 2283–2293, doi:10.1039/C4TA05422A

[Flores 2013] Flores EI, Friswell MI, Xia Y. (2013) "Variable Stiffness Biological and Bio-Inspired Materials" Journal of Intelligent Materials Systems and Structures, 24, 5, 529–540, doi:10.1177/1045389X12461722

[Fock 1988] Fock J, Hahn G, Holzer G, Radisch H. (1988) "Transparent Coating of Flexible, Elastic Polyurethane for Transparent Glass or Synthetic Resin Substrates" US Patent 4,745,152

[Fong 2001] Fong H, Vaia RA, Sanders JH, Lincoln D, Vreugdenhil AJ, Liu W, Bultman J, Chen C. (2001) "Self-Passivation of Polymer-Layered Silicate Nanocomposites" Chemistry of Materials, 13, 4123–4129

[Fontana 1978] Fontana MG, Greene ND. (1978) Corrosion Engineering, 2nd edition, Mc-Graw-Hill

[Foote 1986] Foote RM, Mai YM, Cotterell B. (1986) "Crack Growth Resistance Curves in Strain-Softening Materials" Journal of the Mechanics and Physics of Solids, 34, 593–607, doi:10.1016/0022-5096(86)90039-6

[Förster 2002] Förster S, Plantenberg T. (2002) "From Self-Organizing Polymers to Nanohybrid and Biomaterials" Angewandte Chemie International Edition, 41, 688–714

[Fortney 2007] Fortney PJ, Shahrooz BM, Rassati GA. (2007) "Large-Scale Testing of a Replaceable "Fuse" Steel Coupling Beam" Journal of Structural Engineering, 133, 12, 1801–1807, doi:10.1061/(ASCE)0733-9445(2007)133:12(1801)

[Foster 2011] Foster M, Nakashima R. (2011) "Japan Nuclear Plant Plugs Highly Radioactive Leak" Associated Press, April 5, 2011

[Foster 2014] Foster PC, Mlot NJ, Lin A, Hu DL. (2014) "Fire Ants Actively Control Spacing and Orientation within Self-Assemblages" Journal of Experimental Biology, 217, 2089–2100, doi:10.1242/jeb.093021

[Frahm 1911] Frahm H. (1911) "Device for Damping Vibrations of Bodies" US Patent 989,958

[Francken 1979] Francken L. (1979) "Fatigue Performance of a Bituminous Road Mix under Realistic Test Conditions," Transportation Research Record 712, Transportation Research Board, Washington, DC, pp. 30–36

[Fratzl 2007] Fratzl P. (2007) "Biomimetic Materials Research: What Can We Really Learn from Nature's Structural Materials?" Journal of the Royal Society Interface, 4, 637–642, doi:10.1098/rsif.2007.0218

[Fredrickson 1996] Fredrickson JK, Gorby YA. (1996) "Environmental Processes Mediated by Iron Reducing Bacteria" Current Opinion in Biotechnology, 7, 3, 287–294, doi:10.1016/S0958-1669(96)80032-2

[Frei 2013] Frei R, McWilliam R, Derrick B, Purvis A, Tiwari A, Serugendo GD. (2013) "Selfhealing and Self-Repairing Technologies" International Journal of Advanced Manufacturing Technology, 69, 5–8, 1033–1061, doi:10.1007/s00170-013-5070-2

[Freitas 1982] Freitas, Jr. RA, Gilbreath WP (1982) "Replicating Systems Concepts: Self-Replicating Lunar Factory and Demonstration" Chapter 5, Advanced Automation for Space Missions, Proceedings of the 1980 NASA/ASEE Summer Study, NASA, Scientific and Technical Information Branch, Conference Publication 2255, US Government Printing Office, Washington, DC

[Frenkel 2015] Frenkel D. (2015) "Order through Entropy" Nature Materials, 14, 9–12, doi:10.1038/nmat4178

[Frère 2008] Frère W, Danicher L, Gramain P. (1998) "Preparation of Polyurethane Microcapsules by Interfacial Polycondensation" European Polymer Journal, 34, 193–199

[Friant 1922] Friant E. (1922) "Self Sealing Reservoir" US Patent 1,436,985

[Frisken 1993] Frisken SJ. (1993) "Light-Induced Optical Waveguide Uptapers" Optics Letters, 18, 1035–1037

[Fritzler 2014] Fritzler T, Azarian MH, Pecht MG. (2014) "Scintillation Conditioning of Tantalum Capacitors with Manganese Dioxide Cathodes" IEEE Transactions on Device and Materials Reliability, 14, 2, 630–638

[Fry 1976] Fry PF. (1976) "A Review of the Analysis of Hydrodynamic Ram," AFFDL-TR-75-102, DTIC ADA031996, Air Force Flight Dynamics Laboratory, Wright-Patterson Air Force Base, Ohio

[Fu 2007] Fu Q, Rao GV, Ward TL, Lu Y, Lopez GP. (2007) "Thermoresponsive Transport through Ordered Mesoporous Silica/PNIPAAm Copolymer Membranes and Microspheres" Langmuir, 23, 170–174

[Fu 2010] Fu J, Wang YK, Yang MT, Desai RA, Yu X, Liu Z, Chen CS. (2010) "Mechanical Regulation of Cell Function with Geometrically Modulated Elastomeric Substrates" Nature Methods, 7, 9, 733–736

[Fu 2017] Fu F, Chen Z, Zhao Z, Wang H, Shang L, Gu Z, Zhao Y. (2017) "Bio-Inspired Self-Healing Structural Color Hydrogel" Proceedings of the National Academy of Sciences, 114, 23, 5900–5905, doi:10.1073/pnas.1703616114

[Fugon 2012] Fugon D, Chen C, Peters K. (2012) "Self-Healing Sandwich Composite Structures" Proceedings of SPIE, Vol. 8345, Sensors and Smart Structures Technologies for Civil, Mechanical, and Aerospace Systems, M Tomizuka, C-B Yun, J P Lynch (eds.), doi:10.1117/12.915165

[Fujishima 1972] Fujishima A, Honda K. (1972) "Electrochemical Photolysis of Water at a Semiconductor Electrode" Nature, 238, 37–38, doi:10.1038/238037a0

[Fujishima 2000] Fujishima A, Rao TN, Tryk DA. (2000) "Titanium Dioxide Photocatalysis" Journal of Photochemistry and Photobiology C: Photochemistry Reviews 1, 1–21

[Fukuda 2006] Fukuda Y, Osterman M, Pecht M. (2006) "The Effect of Annealing on Tin Whisker Growth" IEEE Transactions on Electronics Packaging Manufacturing, 29, 4, 252–258, doi:10.1109/TEPM.2006.887390

[Fürstner 2005] Fürstner R, Barthlott W, Neinhuis C, Walzel P. (2005) "Wetting and Self-Cleaning Properties of Artificial Superhydrophobic Surfaces" Langmuir, 21, 956–961

[Gabriel 2007] Gabriel S, Duwez AS, Jérôme R, Jérôme C. (2007) "Thermoresponsive Coatings Strongly Adhering to (Semi)conducting Surfaces" Langmuir, 23, 159–161

[Gaharwar 2014] Gaharwar AK, Avery RK, Assmann A, Paul A, McKinley GH, Khademhosseini A, Olsen BD. (2014) "Shear-Thinning Nanocomposite Hydrogels for the Treatment of Hemorrhage" ACS Nano, 8 (10), 9833–9842, doi:10.1021/nn503719n

[Galaev 2007] Galaev IY, Dainiak MB, Plieva F, Mattiasson B. (2007) "Effect of Matrix Elasticity on Affinity Binding and Release of Bioparticles: Elution of Bound Cells by Temperature-Induced Shrinkage of the Smart Macroporous Hydrogel" Langmuir, 23, 35–40

[Galloway 2012] Galloway JM, Bramble JP, Rawlings AE, Burnell G, Evans SD, Staniland SS. (2012) "Biotemplated Magnetic Nanoparticle Arrays" Small, 8, 2, 204–208

[Galvin 1964] Galvin GD, Naylor H. (1964) "Effect of Lubricants on the Fatigue of Steel and Other Metals" Proceedings of the Institution of Mechanical Engineers, 179, 857–875

[Ganek 2003] Ganek AG, Corbi TA. (2003) "The Dawning of the Autonomic Computing Era" IBM Systems Journal 42, 1, 5–18

[Gao 2001] Gao KW, Qiao LJ, Chu WY. (2001) "In Situ TEM Observation of Crack Healing in α-Fe" Scripta Materialia, 44, 7, 1055–1059, doi:10.1016/S1359-6462(01)00671-6

[Gao 2007] Gao X, Yan X, Yao X, Xu L, Zhang K, Zhang J, Yang B, Jiang L. (2007) "The Dry-Style Antifogging Properties of Mosquito Compound Eyes and Artificial Analogues Prepared by Soft Lithography" Advanced Materials, 19, 2213, doi:10.1002/adma.200601946

[Gao 2008] Gao L, Fadeev AY, McCarthy TJ. (2008) "Superhydrophobicity and Contact-Line Issues" MRS Bulletin, 33, 747–751

[Gao 2010a] Gao J, Suo J. (2010) "Proposal of Self-Healing Coatings for Nuclear Fusion Applications" Surface & Coatings Technology, 204, 23, 3876–3881, doi:10.1016/j.surfcoat.2010.05.007

[Gao 2010b] Gao Y, Huang Y, Feng S, Gu G, Qing FL. (2010) "Novel Superhydrophobic and Highly Oleophobic PFPE-Modified Silica Nanocomposite" Journal of Materials Science, 45, 2, 460–466

[Garagnani 1925] Garagnani M. (1925) "Fuel Tank for Aeroplanes and the Like" US Patent 1,535,462

[Garcés 2002] Garcés-Chávez V, McGloin D, Melville H, et al. (2002) "Simultaneous Micromanipulation in Multiple Planes Using a Self-Reconstructing Light Beam" Nature, 419, 145–147, doi:10.1038/nature01007

[Garcia 2009] Garcia A, Schlangen E, van de Ven M, Liu Q. (2009) "Electrical Conductivity of Asphalt Mortar Containing Conductive Fibers and Fillers" Construction and Building Materials, 23, 10, 3175–3181

[Garcia 2009] Garcia-Manyes S, Liang J, Szoszkiewicz R, Kuo TL, Fernández JM. (2009) "Force-Activated Reactivity Switch in a Bimolecular Chemical Reaction" Nature Chemistry, 1, 236–242, doi:10.1038/nchem.207

[Garcia 2010] Garcia ME, Lin Y, Sodano HA. (2010) "Autonomous Materials with Controlled Toughening and Healing" Journal of Applied Physics, 108, 093512, doi:10.1063/1.3499351

[García 2011] García SJ, Fischer HR, van der Zwaag S. (2011) "A Critical Appraisal of the Potential of Self Healing Polymeric Coatings" Progress in Organic Coatings, 72, 3, 211–221, doi:10.1016/j.porgcoat.2011.06.016

[García 2012] García, Á. (2012). "Self-Healing of Open Cracks in Asphalt Mastic" Fuel, 93, 264–272, doi:10.1016/j.fuel.2011.09.009

[García 2014] García JM, Jones GO, Virwani K, McCloskey BD, Boday DJ, ter Huurne G, Horn HW, Coady DJ, Bintaleb AM, Alabdulrahman AM, Alsewailem F, Almegren HA, Hedrick JL. (2014) "Recyclable, Strong Thermosets and Organogels via Paraformaldehyde Condensation with Diamines" Science, 344, 6185, 732–735, doi:10.1126/science.1251484

[Gardner 1966] Gardner GL. (1966) "Manufacturing Encapsulated Products" Chemical Engineering Progress, 62, 4, 87–91

[Gardner 2014] Gardner D, Jefferson A, Hoffman A, Lark R. (2014) "Simulation of the Capillary Flow of an Autonomic Healing Agent in Discrete Cracks in Cementitious Materials" Cement and Concrete Research, 58, 35–44, doi:10.1016/j.cemconres.2014.01.005

[Garlan 2002] Garlan D, Schmerl B. (2002) "Model-Based Adaptation for Self-Healing Systems" Proceedings of the First Workshop on Self-Healing Systems, Proceeding WOSS '02, pp 27–32, Charleston, SC, ACM, New York, NY, doi:10.1145/582128.582134

[Garnett 2012] Garnett EC, Cai W, Cha JJ, Mahmood F, Connor ST, Christoforo MG, Cui Y, McGehee MD, Brongersma ML. (2012) "Self-Limited Plasmonic Welding of Silver Nanowire Junctions" Nature Materials, 11, 241–249, doi:10.1038/nmat3238

[Geng 2013] Geng F, Ma R, Nakamura A, Akatsuka K, Ebina Y, Yamauchi Y, Miyamoto N, Tateyama Y, Sasaki T. (2013) "Unusually Stable ~100-Fold Reversible and Instantaneous Swelling of Inorganic Layered Materials" Nature Communications, 4, 1632, doi:10.1038/ncomms2641

[Genzer 2008] Genzer J, Marmur A. (2008) "Biological and Synthetic Self-Cleaning Surfaces" MRS Bulletin, 33, 742–746

[Geppert 2004] Geppert L. (2004) "Electrical Fuse Lets Chips Heal Themselves" IEEE Spectrum, https://spectrum.ieee.org/electrical-fuse-lets-chips-heal-themselves

[Gerling 2015] Gerling T, Wagenbauer KF, Neuner AM, Dietz H. (2015) "Dynamic DNA Devices and Assemblies Formed by Shape-Complementary, Non-Base Pairing 3D Components" Science, 347, 6229, 1446–1452, doi:10.1126/science.aaa5372

[Gheneim 2002] Gheneim R, Perez-Berumen C, Gandini A. (2002) "Diels–Alder Reactions with Novel Polymeric Dienes and Dienophiles: Synthesis of Reversibly Cross-Linked Elastomers" Macromolecules, 35, 19, 7246–7253, doi:10.1021/ma020343c

[Ghezzo 2010] Ghezzo F, Smith DR, Starr TN, Perram T, Starr AF, Darlington TK, Baldwin RK, Oldenburg SJ. (2010) "Development and Characterization of Healable Carbon Fiber Composites with a Reversibly Cross Linked Polymer" Journal of Composite Materials, 44, 1587–1603

[Ghiorse 1984] Ghiorse WC. (1984) "Biology of Iron-and Manganese-Depositing Bacteria" Annual Review of Microbiology, 38, 515–550, doi:10.1146/annurev.mi.38.100184.002503

[Ghosh 2007] Ghosh D, Sharman R, Rao HR, Upadhyaya S. (2007) "Self-Healing Systems: Survey and Synthesis" Decision Support Systems, 42, 2164–2185

[Ghosh 2009a] Ghosh B, Urban MW. (2009) "Self-Repairing Oxetane-Substituted Chitosan Polyurethane Networks" Science, 323, 5920, 1458–1460

[Ghosh 2009b] Ghosh SK. (2009) Self-Healing Materials: Fundamentals, Design Strategies, and Applications, Wiley-VCH, Weinheim, Germany

[Ghoshal 2011] Ghoshal G, Barabási AL. (2011) "Ranking Stability and Super-Stable Nodes an Complex Networks" Nature Communications, 2, 394, doi:10.1038/ncomms1396

[Gibson 1997] Gibson LJ Ashby MF. (1997) "Cellular Solids Structure and Properties", 2nd edition. Cambridge University Press, Cambridge, UK

[Gibson 2010] Gibson DG, Glass JI, Lartigue C, Noskov VN, Chuang RY, Algire MA, Benders GA, Montague MG, Ma L, Moodie MM, Merryman C, Vashee S, Krishnakumar R, Assad-Garcia N, Andrews-Pfannkoch C, Denisova EA, Lei Young L, Qi ZQ, Segall-Shapiro TH, Calvey CH, Parmar PP, Hutchison III CA2, Smith HO, Venter JC. (2010) "Creation of a Bacterial Cell Controlled by a Chemically Synthesized Genome" Science, 329, 5987, 52–56, doi:10.1126/science.1190719

[Gilmore 1972] Gilmore JP, Mckern RA. (1972) "A Redundant Strapdown Inertial Reference Unit (SIRU)" Journal of Spacecraft, 9, 1, 39–47

[Gilmore 2003] Gilmore DA. (2003) "Apparatus and Method for Self-Tuning a Piano" US Patent 6,559,369

[Gilpin 2008] Gilpin K, Kotay K, Rus D, Vasilescu I (2008) "Miche: Modular Shape Formation by Self-Disassembly" International Journal of Robotics Research, 27, 3–4, 345–372, doi:10.1177/0278364907085557

[Girand 2005] Girand S, Bourbigot S, Rochery M, Vroman I, Tighzert L, Delobel R, Pouch F. (2005) "Flame Retarded Polyurea with Microencapsulated Ammonium Phosphate for Textile Coating" Polymer Degradation and Stability, 88, 1, 106–113, doi:10.1016/j.polymdegradstab.2004.01.028

[Giri 2015] Giri N, Del Pópolo MG, Melaugh G, Greenaway RL, Rätzke K, Koschine T, Pison L, Costa Gomes MF, Cooper AI, James SL. (2015) "Liquids with Permanent Porosity" Nature, 527, 216–220, doi:10.1038/nature16072

[Glogowski 2006] Glogowski E, Tangirala R, Russell TP, Emrick T. (2006) "Functionalization of Nanoparticles for Dispersion in Polymers and Assembly in Fluids" Journal of Polymer Science. Part A, Polymer Chemistry, 44, 5076–86, doi:10.1002/pola.21598

[Gobbini 2016] Gobbini E, Cassani C, Villa M, Bonetti D, Longhese MP. (2016) "Functions and Regulation of the MRX Complex at DNA Double-Strand Breaks" Microbial Cell, 3, 8, 329–337, doi:10.15698/mic2016.08.517.

[Godowsky 1955] Godowsky L. (1955) "Mixed Packet Photographic Emulsions" US Patent 2,698,794

[Gohar 1995] Gohar Y, Parker R, Rebut PH. (1995) "ITER Blanket Designs" Fusion Engineering and Design, 27, 52–61

[Goldbeck 2013] Goldbeck CP, Jensen HM, TerAvest MA, Beedle N, Appling Y Hepler M, Cambray G, Mutalik V, Angenent LT, Ajo-Franklin CM. (2013) "Tuning Promoter Strengths for Improved Synthesis and Function of Electron Conduits in *Escherichia coli*" ACS Synthetic Biology, 2, 150–159, doi:10.1021/sb300119v

[Goldsmith 2009] Goldsmith LE, Pupols M, Kickhoefer VA, Rome LH, Monbouquette HG. (2009) "Utilization of a Protein "Shuttle" to Load Vault Nanocapsules with Gold Probes and Proteins" ACS Nano, 3, 10, 3175–3183

[Goldstein 2009] Goldstein SC, Mowry TC, Campbell JD, Ashley-Rollman MP, De Rosa M, Funiak S, Hoburg JF, Karagozler ME, Kirby B, Lee P, Pillai P, Reid JR, Stancil DD, Weller MP. (2009) "Beyond Audio and Video: Using Claytronics to Enable Pario" AI Magazine, 2009, 29–45

[Gollapudi 1995] Gollapudi UK, Knutson CL, Bang SS, Islam MR. (1995) "A New Method for Controlling Leaching through Permeable Channels" Chemosphere, 30, 4, 695–705, doi:10.1016/0045-6535(94)00435-W

[González 2011] González-García Y, Mola JM, De Graeve I, Scheltjens G, Van Mele B, Terryn H. (2011) "A Combined Mechanical, Microscopic and Local Electrochemical Evaluation of Self-Healing Properties of Shape-Memory Polyurethane Coatings" Electrochimica Acta, 56, 26, 9619–9626, doi:10.1016/j.electacta.2011.03.081

[Goodell 2012] Goodell R, Wisotzkey S, Carr H, Fyfe E. (2012) "Status of Program to Develop Low Cost Application of Carbon Fiber to Strengthen Water and Other Pipes" Proceedings of SPIE, Vol. 8347, Nondestructive Characterization for Composite Materials, Aerospace Engineering, Civil Infrastructure, and Homeland Security, AL Gyekenyesi, TY Yu, PJ Shull, AA Diaz, HF Wu, eds.

[Goodrich 2014] Goodrich CP, Liu AJ, Nagel SR. (2014) "Solids between the Mechanical Extremes of Order and Disorder" Nature Physics, 10, 578–581, doi:10.1038/nphys3006

[Goodwin 2008] Goodwin AL, Calleja M, Conterio MJ, Dove M T Evans J S Keen DA, Peters L, Tucker MG. (2008) "Colossal Positive and Negative Thermal Expansion in the Framework Material $Ag_3[Co(CN)6]$" Science, 319, 5864, 794–797, doi:10.1126/science.1151442

[Gosvami 2015] Gosvami NN, Bares JA, Mangolini F, Konicek AR, Yablon DG, Carpick RW. (2015) "Tribology. Mechanisms of Antiwear Tribofilm Growth Revealed In Situ by Single-Asperity Sliding Contacts" Science 348, 6230, 102–106, doi:10.1126/science.1258788

[Götzen 2003] Götzen N, Cross AR, Ifjuc PG, Rapoff AJ. (2003) "Understanding Stress Concentration about a Nutrient Foramen" Journal of Biomechanics, 36, 1511–1521

[Gould 2003] Gould P. (2003) "Smart Clean Surfaces" Materials Today, 6, 6, 44

[Goyal 2012] Goyal A, Swaminathan M, Chatterjee A, Howard DC, Cressler JD. (2012) "A New Self-Healing Methodology for RF Amplifier Circuits Based on Oscillation Principles" IEEE Transactions on Very Large Scale Integration (VLSI) Systems, 20, 10, 1835–1848, doi:10.1109/TVLSI.2011.2163953

[Graff 2002] Graff A, Sauer M, Van Gelder P, Meier W. (2002) "Virus-Assisted Loading of Polymer Nanocontainer" Proceedings of the National Academy of Sciences of the United States of America, 99, 8, 5064–5068, doi:10.1073/pnas.062654499

[Gramelspacher 1939] Gramelspacher CU. (1939) "One-Piece Tire" US Patent 2,241,593

[Grande 2012] Grande AM, Coppi S, Di Landro L, Sala G, Giacomuzzo C, Francesconi A, Rahman MA. (2012) "An Experimental Study of the Self-Healing Behavior of Ionomeric Systems under Ballistic Impact Tests" Proceedings of SPIE, Vol. 8342, Behavior and Mechanics of Multifunctional Materials and Composites, NC Goulbou rne, Z Ounaies (eds), doi:10.1117/12.911870

[Granger 2007] Granger S, Loukili A, Pijaudier-Cabot G, Chanvillard G. (2007) "Experimental Characterization of the Self-Healing of Cracks in an Ultra High Performance Cementitious Material: Mechanical Tests and Acoustic Emission Analysis" Cement and Concrete Research, 37, 519–527

[Gray 1984] Gray RJ. (1984) "Autogenous Healing of Fibre/Matrix Interfacial Bond in Fibre-Reinforced Mortar" Cement and Concrete Research, 14, 3, 315–317, doi:10.1016/0008-8846(84)90047-4

[Gray 2012] Gray BL. (2012) "Microfluidics on Compliant Substrates: Recent Developments in Foldable and Bendable Devices and System Packaging" Proceedings of SPIE, Vol. 8344, Nanosensors, Biosensors, and Info-Tech Sensors and Systems, VK Varadan (ed.)

[Green 1942] Green BK. (1942) "Coating for Paper" US Patent 2,299,693

[Greenberg 2000] Greenberg CB, Harris CS, Korthuis V, Kutilek LA, Singleton DE, Szanyi J, Thiel JP. (2000) "Photocatalytically-Activated Self-Cleaning Article and Method of Making Same" US Patent 6,027,766

[Grigoriev 2009] Grigoriev DO, Kohler K, Skorb E, Shchukin DG, Mohwald H. (2009) "Polyelectrolyte Complexes as a 'Smart' Depot for Self-Healing Anticorrosion Coatings" Soft Matter 5, 7, 1426–1432, doi:10.1039/b815147d

[Grimmett 1999] Grimmett G (1999) Percolation, Springer, Berlin, Germany

[Grindy 2015] Grindy SC, Learsch R, Mozhdehi D, Cheng J, Barrett DG, Guan Z, Messersmith PB, Holten-Andersen N. (2015) "Control of Hierarchical Polymer Mechanics with Bioinspired Metal-Coordination Dynamics" Nature Materials, 14, 1210–1216, doi:10.1038/nmat4401

[Grondahl 2013] Grondahl CM, Smith RL, Dudley JC. (2013) "Film Riding Pressure Actuated Leaf Seal Assembly" US Patent 8,474,827

[Gröschel 2012] Gröschel AH, Schacher FH, Schmalz H, Borisov OV, Zhulina EB, Walther A, Müller AH. (2012) "Precise Hierarchical Self-Assembly of Multicompartment Micelles" Nature Communications, 1–10, doi:10.1038/ncomms1707

[Groten 2013] Groten J, Rühe J. (2013) "Surfaces with Combined Microscale and Nanoscale Structures: A Route to Mechanically Stable Superhydrophobic Surfaces?" Langmuir, 29, 11, 3765–3772, doi:10.1021/la304641q

[Grosvenor 1983] Grosvenor RL, Fox M. (1983) "Fuel Tank Component" US Patent 4,422,561, Dec 27

[Grubbs 2004] Grubbs RH. (2004) "Olefin Metathesis" Tetrahedron, 60, 7117–7140

[Gu 2003] Gu ZZ, Uetsuka H, Takahashi K, Nakajima R, Onishi H, Fujishima A, Sato O. (2003) "Structural Color and the Lotus Effect" Angewandte Chemie International Edition, 42, 894–897

[Guadagno 2013] Guadagno L, Raimondo M, Naddeo C, Mariconda A, Corvino R, Longo P, Vittoria V, Russo S, Iannuzzo G. (2013) "Process for Preparing Self-Healing Composite Materials of High Efficiency for Structural Applications" US Patent 8,481,615

[Guan 2004] Guan Z, Roland JT, Bai JZ, Ma SX, McIntire TM, Nguyen M. (2004) "Modular Domain Structure: A Biomimetic Strategy for Advanced Polymeric Materials" Journal of the American Chemical Society, 126, 7, 2058–2065, doi:10.1021/ja039127p

[Gudipati 2005] Gudipati CS, Finlay JA, Callow JA, Callow ME, Wooley KL. (2005) "The Antifouling and Fouling-Release Performance of Hyperbranched Fluoropolymer (HBFP)–Poly(ethylene glycol) (PEG) Composite Coatings Evaluated by Adsorption of Biomacromolecules and the Green Fouling Alga Ulva" Langmuir, 21, 7, 3044–3053, doi:10.1021/la048015o

[Guerette 2013] Guerette PA, Hoon S, Seow Y, Raida M, Masic A, Wong FT, Ho VH, Kong KW, Demirel MC, Pena-Francesch A, Amini S, Tay GZ, Ding D, Miserez A. (2013) "Accelerating the Design of Biomimetic Materials by Integrating RNA-Seq with Proteomics and Materials Science" Nature Biotechnology, 31, 908–915, doi:10.1038/nbt.2671

[Guérin 2003] Guérin G, Mauger F, Prud'homme RE. (2003) "The Adhesion of Amorphous Polystyrene Surfaces below T_g" Polymer, 44, 24, 7477–7784, doi:10.1016/j.polymer.2003.09.018

[Guldin 2013] Guldin S, Kohn P, Stefik M, Song J, Divitini G, Ecarla F, Ducati C, Wiesner U, Steiner U. (2013) "Self-Cleaning Antireflective Optical Coatings" Nano Letters, 13, 11, 5329–5335, doi:10.1021/nl402832u

[Gumus 2010] Gumus A, Califano JP, Wan AM, Huynh J, Reinhart-King CA, Malliaras GG. (2010) "Control of Cell Migration Using a Conducting Polymer Device" Soft Matter, 6, 5138–5142, doi:10.1039/B923064E

[Gunawan 2015] Gunawan ST, Kempe K, Bonnard T, Cui J, Alt K, Law LS, Wang X, Westein E, Such GK, Karlheinz PK, Hagemeyer CE, Caruso F. (2015) "Multifunctional Thrombin-Activatable Polymer Capsules for Specific Targeting to Activated Platelets" Advanced Materials, 27, 35, 5153–5157, doi:10.1002/adma.201502243

[Gunes 2010] Gunes IS, Pérez-Bolivar C, Cao F, Jimenez GA, Anzenbacher Jr. P, Jana SC. (2010) "Analysis of Non-Covalent Interactions between the Nanoparticulate Fillers and the Matrix Polymer as Applied to Shape Memory Performance" Journal of Materials Chemistry, 20, 3467–3474, doi:10.1039/B922027E

[Guo 2004] Guo C, Feng L, Zhai J, Wang G, Song Y, Jiang L, Zhu D. (2004) "Large-Area Fabrication of a Nanostructure-Induced Hydrophobic Surface from a Hydrophilic Polymer" Chemphyschem, 5, 5, 750–753, doi:10.1002/cphc.200400013

[Guo 2017] Guo X, Stan MR. (2017) "Implications of Accelerated Self-Healing as a Key Design Knob for Cross-Layer Resilience" Integration, 56, 167–180, doi:10.1016/j.vlsi.2016.10.008

[Gupta 2006] Gupta S, Zhang Q, Emrick T, Balazs AC, Russell TP. (2006) "Entropy-Driven Segregation of Nanoparticles to Cracks in Multilayered Composite Polymer Structures" Nature Materials, 5, 3, 229–233, doi:10.1038/nmat1582

[Habault 2013] Habault D, Zhang H, Zhao Y. (2013) "Light-Triggered Self-Healing and Shape-Memory Polymers" Chemical Society Reviews, 42, 7244e56. doi:10.1039/C3CS35489J

[Haechler 2023] Haechler I, Ferru N, Schnoering G, et al. (2023) "Transparent Sunlight-Activated Antifogging Metamaterials" Nature Nanotechnology, 18, 137–144, doi:10.1038/s41565-022-01267-1

[Haegermann 1930] Haegermann G. (1930) "Lime Exudations on Concrete under Water" Chemical Abstracts, 24, 5127

[Hager 2010] Hager MD, Greil P, Leyens C, van der Zwaag S, Schubert US. (2010) "Self-Healing Materials" Advanced Materials, 22, 5424–5430, doi:10.1002/adma201003036

[Haishima 2001] Haishima T. (2001) "Process for Producing Microencapsulated Electroconductive Filler" US Patent 6,255,138

[Hakimi 2013] Hakimi O, Mouthuy PA, Carr A. (2013) "Synthetic and Degradable Patches: An Emerging Solution for Rotator Cuff Repair" International Journal of Experimental Pathology, 94, 4, 287–292, doi:10.1111/iep.12030

[Haldar 2015] Haldar U, Bauri K, Li R, Faust R, De P. (2015) "Polyisobutylene-Based pH-Responsive Self-Healing Polymeric Gels" ACS Applied Materials & Interfaces, 2015, 7, 16, 8779–8788, doi:10.1021/acsami.5b01272

[Hall 1978] Hall CA. (1978) "Self-Sealing Fuel Lines" US Patent 4,216,803

[Ham 2010] Ham MH, Choi JH, Boghossian AA, Jeng ES, Graff RA, Heller DA, Chang AC, Mattis A, Bayburt TH, Grinkova YV, Zeiger AS, Van Vliet KJ, Hobbie EK, Sligar SG, Wraight CA, Strano MS. (2010) "Photoelectrochemical Complexes for Solar Energy Conversion that Chemically and Autonomously Regenerate" Nature Chemistry, 2, 929–936, doi:10.1038/nchem.822

[Hamada 2003] Hamada K, Kawano F, Asaoka K. (2003) "Shape Recovery of Shape Memory Alloy Fiber Embedded Resin Matrix Smart Composite after Crack Repair" Dental Materials Journal, 22, 2, 160–167

[Hamdy 2011] Hamdy AS, Doench I, Möhwald H. (2011) "Assessment of a One-Step Intelligent Self-Healing Vanadia Protective Coatings for Magnesium Alloys in Corrosive Media" Electrochimica Acta, 56, 5, 2493–2502, doi:10.1016/j.electacta.2010.11.103

[Hamilton 1996] Hamilton B. (1996) "A Compact Representation of Units" Hewlett-Packard Technical Report, HPL-96-61

[Hamilton 2009] Hamilton AR, Sottos NR, White SR. (2010) "Self-Healing of Internal Damage in Synthetic Vascular Materials" Advanced Materials, 22, 5159–5163, doi:10.1002/adma.201002561

[Han 2010] Han S, Osterman M, Pecht M. (2010) "Electrical Shorting Propensity of Tin Whiskers" IEEE Transactions on Electronics Packaging Manufacturing, 33, 3, 205–211, doi:10.1109/TEPM.2010.2053377

[Han 2015a] Han B, Wang Y, Dong S, Zhang L, Ding S, Yu X, Ou J. (2015) "Smart Concretes and Structures: A Review" Journal of Intelligent Materials Systems and Structures, 26, 11, 1303–1345, doi:10.1177/1045389X15586452

[Han 2015b] Han Y, Liu C. (2015) "Propagation-Invariant Hollow Beams with Hexagonal Symmetry" Optics Communications, 284, 9, 2264–2267, doi:10.1016/j.optcom.2010.12.034

[Hanley 1941] Hanley AJ. (1941) "Insulating Material and Method of Manufacture" US Patent 2,249,275

[Hannant 1983] Hannant DJ, Keer JG. (1983). "Autogenous Healing of Thin Cement Based Sheets" Cement and Concrete Research, 13, 3, 357–365, doi:10.1016/0008-8846(83)90035-2

[Hansen 2009] Hansen CJ, Wu W, Toohey KS, Sottos NR, White SR, Lewis JA. (2009) "Self-Healing Materials with Interpenetrating Microvascular Networks" Advanced Materials, 21, 4143–4147, doi:10.1002/adma.200900588

[Hao 2014] Hao Q, Pang Y, Zhao Y, Zhang J, Feng J, Yao S. (2014) "Mechanism of Delayed Frost Growth on Superhydrophobic Surfaces with Jumping Condensates: More Than Interdrop Freezing" Langmuir, 30, 51, 15416–15422, doi:10.1021/la504166x

[Harada 2011] Harada A, Kobayashi R, Takashima Y, Hashidzume A, Yamaguchi H. (2011) "Macroscopic Self-Assembly through Molecular Recognition" Nature Chemistry, 3, 34–37, doi:10.1038/nchem.893

[Harr 1971] Harr GB. (1971) "Self-Sealing Fuel Tank" US Patent 3,563,846

[Harreld 2004] Harreld JH, Wong MS, Hansma PK, Morse DE, Stucky GD. (2004) "Self-Healing Organosiloxane Materials Containing Reversible and Energy-Dispersive Crosslinking Domains" US Patent 6,783,709

[Harrington 2011] Harrington MJ, Razghandi K, Ditsch F, Guiducci L, Rueggeberg M, Dunlop JW, Fratzl P, Neinhuis C, Burgert I. (2011) "Origami-like Unfolding of Hydro-Actuated Ice Plant Seed Capsules" Nature Communications, 2, 1–7, doi:10.1038/ncomms1336

[Harris 2004] Harris KM, Rajagopalan M. (2004) "Self Healing Polymers in Sports Equipment" US Patent 6,794,472

[Hart 2013] Hart LR, Harries JL, Greenland BW, Colquhoun HM, Hayes W. (2013) "Healable Supramolecular Polymers" Polymer Chemistry, 4, 18, 4860–4870, doi:10.1039/C3PY00081H

[Hartgerink 2001] Hartgerink JD, Beniash E, Stupp SI. (2001) "Self-Assembly and Mineralization of Peptide-Amphiphile Nanofibers" Science, 294, 5547, 1684–1688, doi:10.1126/science.1063187

[Hashidzume 2006] Hashidzume A, Tomatsu I, Harada A. (2006) "Interaction of Cyclodextrins with Side Chains of Water Soluble Polymers: A Simple Model for Biological Molecular Recognition and Its Utilization for Stimuli-Responsive Systems" Polymer, 47, 17, 6011–6027, doi:10.1016/j.polymer.2006.06.021

[Hattori 2009] Hattori D, Chen Y, Matthews BJ, Salwinski L, Sabatti C, Grueber WB, Zipursky SL. (2009) "Robust Discrimination between Self and Non-Self Neurites Requires Thousands of Dscam1 Isoforms" Nature, 461, 644–648, doi:10.1038/nature08431

[Hauser 2012] Hauser RG, Abdelhadi R, McGriff D, Retel LK (2012) "Deaths Caused by the Failure of Riata and Riata St Implantable Cardioverter-Defibrillator Leads" Heart Rhythm, 9, 1227–1235, doi:10.1016/j.hrthm.2012.03.048

[Hautakangas 2008] Hautakangas S, Schut H, van Dijk NH, Rivera Díaz del Castillo PEJ, van der Zwaag S. (2008) "Self Healing of Deformation Damage in Underaged Al-Cu-Mg Alloys" Scripta Materialia, 58, 719–722

[Hawkes 2010] Hawkes E, An B, Benbernou NM, Tanaka H, Kim S, Demaine ED, Rus D, Wood RJ. (2010) "Programmable Matter by Folding" Proceedings of the National Academy of Sciences of the United States of America, 107, 28, 12441–12445, doi:10.1073/pnas.0914069107

[Hayashida 2003] Hayashida K, Tsuge S, Ohtani H. (2003) "Flame Retardant Mechanism of Polydimethylsiloxane Material Containing Platinum Compound Studied by Analytical Pyrolysis Techniques and Alkaline Hydrolysis Gas Chromatography" Polymer, 44, 5611–5616

[Hayes 2007a] Hayes SA, Zhang W, Branthwaite M, Jones FR. (2007) "Self-Healing of Damage in Fibre-Reinforced Polymer-Matrix Composites" Journal of the Royal Society Interface, 4, 381–387 doi:10.1098/rsif.2006.0209

[Hayes 2007b] Hayes SA, Jones FR, Marshiya K, Zhang W. (2007) "A Self-Healing Thermosetting Composite Material" Composites: Part A, 38, 1116–1120

[Hayes 2012] Hayes TP, Saia J, Trehan A. (2012) "The Forgiving Graph: A Distributed Data Structure for Low Stretch under Adversarial Attack" Distributed Computing, 25, 261–278, doi:10.1007/s00446-012-0160-1

[Haykin 2012] Haykin S. (2012) Cognitive Dynamic Systems: Perception-Action Cycle, Radar, and Radio, Cambridge University Press

[He 2008] He X, Shi X. (2008) "Chloride Permeability and Microstructure of Portland Cement Mortars Incorporating Nanomaterials" Proceedings of TRB 2008 Annual Meeting, Washington, DC, Paper No. 08-1041

[He 2009] He X, Shi X. (2009) "Self-Repairing Coating for Corrosion Protection of Aluminum Alloys" Progress in Organic Coatings, 65, 1, 37–43, doi:10.1016/j.porgcoat.2008.09.003.

[He 2010a] He S, van Dijk NH, Schut H, Van Der Zwaag S. (2010) "Thermally Activated Precipitation at Deformation-Induced Defects in Fe-Cu and Fe-Cu-B-N Alloys Studied by Positron Annihilation Spectroscopy" Physical Review B, 81, 094103, doi:10.1103/PhysRevB.81.094103

[He 2010b] He Y, Ye T, Su M, Zhang C, Ribbe AE, Jiang W, Mao C. (2010) "Hierarchical Self-Assembly of DNA into Symmetric Supramolecular Polyhedra" Nature, 452, 198–201, doi:10.1038/nature06597

[He 2011] He L, Fullenkamp DE, Rivera JG, Messersmith PB. (2011) "pH Responsive Self-Healing Hydrogels Formed by Boronate–Catechol Complexation" Chemical Communications, 47, 7497, doi:10.1039/c1cc11928a

[He 2015] He Y, Liao S, Jia H, Cao Y, Wang Z, Wang Y. (2015) "A Self-Healing Electronic Sensor Based on Thermal-Sensitive Fluids" Advanced Materials, 27, 31, 4622–4627, doi:10.1002/adma.201501436

[Healy 1994] Healy JW. (1994) "Fuel Dispensing Hose Breakaway Assembly" US Patent 5,297,574

[Hearn 1998] Hearn N. (1998) "Self-Sealing, Autogenous Healing and Continued Hydration: What Is the Difference?" Materials and Structures, 31, 8, 563–567

[Heikenfeld 2008] Heikenfeld J, Dhindsa M. (2008) "Electrowetting on Superhydrophobic Surfaces: Present Status and Prospects" Journal of Adhesion Science and Technology, 22, 319–334, doi:10.1163/156856108X295347

[Heitz 1978] Heitz RM, Hill F. (1978) "Self-Sealing Fuel Line Assembly" US Patent 4,115,616

[Hemmelgarn 2014] Hemmelgarn CD, Margraf TW, Havens DE, Reed JL, Snyder LW, Louderbaugh A, Dietsch BA. (2014) "Composite Self-Healing System" US Patent Application 20140186476

[Henry 2003] Henry G, Li W, Garner W, Woodley DT. (2003) "Migration of Human Keratinocytes in Plasma and Serum and Wound Re-Epithelialisation" Lancet, 361, 574–576, doi:10.1016/S0140-6736(03)12510-X

[Herbert 2013] Herbert EN, Li VC. (2013) "Self-Healing of Engineered Cementitious Composites in the Natural Environment" Materials, 6, 2831–2845, doi:10.3390/ma6072831

[Herbst 2008] Herbst O, Luding S. (2008) "Modelling Particulate Self-Healing Materials and Application to Uni-Axial Compression" International Journal of Fracture, 154, 87–103, doi:10.1007/s10704-008-9299-y

[Herbst 2012] Herbst F, Seiffert S, Binder WH. (2012) "Dynamic Supramolecular Poly(isobutylene)s for Self-Healing Materials" Polymer Chemistry, 3, 3084–3092, doi:10.1039/C2PY20265D

[Hering 2001] Hering A. (2001) "The Proof Is in the Fire" Chemical Innovation, 31, 5, 17–21

[Hermanson 2001] Hermanson KD, Lumsdon SO, Williams JP, Kaler EW, Velev OD. (2001) "Dielectrophoretic Assembly of Electrically Functional Microwires from Nanoparticle Suspensions" Science, 294, 5544, 1082–1086, doi:10.1126/science.1063821

[Herrera 2009] Herrera LK, Videla HA. (2009) "Role of Iron-Reducing Bacteria in Corrosion and Protection of Carbon Steel" International Biodeterioration and Biodegradation, 63, 891–895, doi:10.1016/j.ibiod.2009.06.003

[Hesp 2007] Hesp SA, Iliuta S, Shirokoff JW. (2007) "Reversible Aging in Asphalt Binders" Energy Fuels, 21, 2, 1112–1121, doi:10.1021/ef060463b

[Heuser 2015] Heuser T, Steppert AK, Lopez CM, Zhu B, Walther A. (2015) "Generic Concept to Program the Time Domain of Self-Assemblies with a Self-Regulation Mechanism" Nano Letters, 15, 4, 2213–2219, doi:10.1021/nl5039506

[Heussinger 2009] Heussinger C, Barrat JL. (2009) "Jamming Transition as Probed by Quasistatic Shear Flow" Physical Review Letters, 102, 218303, doi:10.1103/PhysRevLett.102.218303

[Heyde 2015] Heyde KC, Ruder WC. (2015) "Exploring Host–Microbiome Interactions Using an In Silico Model of Biomimetic Robots and Engineered Living Cells" Scientific Reports, 5, 11988, doi:10.1038/srep11988

[Hickenboth 2007] Hickenboth CR, Moore JS, White SR, Sottos NR, Baudry J, Wilson SR. (2007) "Biasing Reaction Pathways with Mechanical Force" Nature, 446, 423–427, doi:10.1038/nature05681

[Hietz 1982] Hietz RM, Hill FM. (1982) "Void Filler Foam Fire Suppression System" US Patent 4,352,851

[Higaki 2006] Higaki Y, Otsuka H, Takahara A. (2006) "A Thermodynamic Polymer Cross-Linking System Based on Radically Exchangeable Covalent Bonds" Macromolecules, 39, 2121–2125

[Hikasa 2004] Hikasa A, Sekino T, Hayashi Y, Rajagopalan R, Niihara K. (2004) "Preparation and Corrosion Studies of Self-Healing Multi-Layered Nano Coatings of Silica and Swelling Clay" Materials Research Innovations, 8, 84–88

[Hildebrand 1976] Hildebrand F (1976) Advanced Calculus for Applications, 2nd edition. Prentice-Hall, Englewood Cliffs, NJ

[Hiratsuka 2006] Hiratsuka Y, Miyata M, Tada T, Uyeda TQ. (2006) "A Microrotary Motor Powered by Bacteria" Proceedings of the National Academy of Sciences of the United States of America, 103, 37, 13618–13623, doi:10.1073/pnas.0604122103

[Hirohata 2008] Hirohata M, Kim YC. (2008) "Behavior under Compressive Loads of Steel Structural Members Repaired by Heating and Pressing" Proceedings of IABMAS 2008, Bridge Maintenance, Safety, Management, Health Monitoring and Informatics, HM Koh, D Frangopol (eds), pp 2858–2865

[Hirschberg 2000] Hirschberg JH, Brunsveld L, Ramzi A, Vekemans JA, Sijbesma RP, Meijer EW. (2000) "Helical Self-Assembled Polymers from Cooperative Stacking of Hydrogen-Bonded Pairs" Nature, 407, 167–170, doi:10.1038/35025027

[Ho 1998] Ho CT. (1998) "Reactive Two-Part Polyurethane Compositions and Optionally Self-Healable and Scratch-Resistant Coatings Prepared Therefrom" US Patent 5,798,409

[Ho 2004] Ho W, Yu JC, Lin J, Yu J, Li P. (2004) "Preparation and Photocatalytic Behavior of MoS_2 and WS_2 Nanocluster Sensitized TiO_2" Langmuir, 20, 14, 5865–5869, doi:10.1021/la049838g

[Ho 2008] Ho MW. (2008) "The Rainbow and the Worm: The Physics of Organisms", 3rd edition, World Scientific

[Ho 2022] Ho PS, Hu CK, Gall M, Sukharev V. (2022) "Electromigration in Metals Fundamentals to Nano-Interconnects", Cambridge University Press

[Hoare 2009] Hoare T, Santamaria J, Goya GF, Irusta S, Lin D, Lau S, Padera R, Langer R, Kohane DS. (2009) "A Magnetically Triggered Composite Membrane for On-Demand Drug Delivery" Nano Letters, 9, 10, 3651–3657

[Hobbs 2000] Hobbs GK. (2000) Accelerated Reliability Engineering: HALT and HASS, Wiley, Chichester

[Hodzic 2016] Hodzic A, Smith PJ. (2016) "Composites" US Patent Application 20160167352

[Hof 2002] Hof F, Rebek Jr. J. (2002) "Molecules within Molecules: Recognition through Self-Assembly" Proceedings of the National Academy of Sciences of the United States of America, 99, 8 4775–4777, doi:10.1073/pnas.042663699

[Höfener 2014] Höfener M, Schüppstuhl T. (2014) "A Method for Increasing the Accuracy of 'On-Workpiece' Machining with Small Industrial Robots for Composite Repair" Production Process, 8, 6, 701–709, doi:10.1007/s11740-014-0570-y

[Hoffman 2014] Hoffman MD, Van Hove AH, Benoit DS. (2014) "Degradable Hydrogels for Spatiotemporal Control of Mesenchymal Stem Cells Localized at Decellularized Bone Allografts" Acta Biomaterialia, 10, 8, 3431–3441, doi:10.1016/j.actbio.2014.04.012.

[Hofmeier 2005] Hofmeier H, Schubert US. (2005) "Combination of Orthogonal Supramolecular Interactions in Polymeric Architectures" Chem Communications, 2423–2432, doi:10.1039/B419171D

[Hofstadter 1999] Hofstadter D (1999) Gödel, Escher, Bach: An Eternal Golden Braid. Basic Books

[Hohlbein 2015] Hohlbein N, Shaaban A, Bras AR, Pyckhout-Hintzen W, Schmidt AM. (2015) "Self-Healing Dynamic Bond-Based Rubbers: Understanding the Mechanisms in Ionomeric Elastomer Model Systems" Physical Chemistry Chemical Physics, 17, 32, 21005–21017, doi:10.1039/c5cp00620a

[Holme 2012] Holme MN, Fedotenko IA, Abegg D, Althaus J, Babel L, Favarger F, Reiter R, Tanasescu R, Zaffalon PL, Ziegler A, Müller B, Saxer T, Zumbuehl A. (2012) "Shear-Stress Sensitive Lenticular Vesicles for Targeted Drug Delivery" Nature Nanotechnology, 7, 536–543, doi:10.1038/NNANO.2012.84

[Holmes 2007] Holmes DP, Crosby AJ. (2007) "Snapping Surfaces" Advanced Materials, 19, 3589-3593, doi:10.1002/adma.200700584

[Holmes 2010] Holmes J. (2010) "Development of an Automated Pavement Crack Sealing System" GDOT Research Project No. 2047, Department of Transportation, Georgia, USA

[Holt 1947] Holt HS. (1947) "Fuel Container" US Patent 2,422,239

[Holten 2011] Holten-Andersen N, Harrington MJ, Birkedal H, Leed BP, Messersmith PB, Lee KY, Waite JH. (2011) "pH-Induced Metal-Ligand Cross-Links Inspired by Mussel Yield Self-Healing Polymer Networks with Near-Covalent Elastic Moduli" Proceedings of the National Academy of Sciences of the United States of America, 108, 7, 2651–2655, doi/10.1073/pnas.1015862108

[Holzman 1973] Holzman L, Powanda T, Tracy J. (1973) "Permanent Press Fabric Resin from Crotonylidenediurea Glyoxal Formal Dehyde Aminoplast Material" US Patent 3,764,263

[Homma 2009] Homma D, Mihashi H, Nishiwaki T. (2009) "Self-Healing Capability of Fiber Reinforced Cementitious Composites" Journal of Advanced Concrete Technology, 7, 2, 217–228, doi:10.3151/jact.7.217

[Hong 2013] Hong G, Zhang H, Lin Y, Chen Y, XuY, Weng W, Xia H. (2013) "Mechanoresponsive Healable Metallosupramolecular Polymers" Macromolecules, 46, 21, 8649–8656, doi:10.1021/ma4017532

[Hong 2023] Hong Z-J, Liu D, Hu H-W, et al. (2023) "Low-Temperature Hybrid Bonding with High Electromigration Resistance Scheme for Application on Heterogeneous Integration" Applied Surface Science, 610, 155470, doi:10.1016/j.apsusc.2022.155470

[Hook 2012] Hook AL, Chang CY, Yang J, Luckett J, Cockayne A, Atkinson S, Mei Y, Bayston R, Irvine DJ, Langer R, Anderson DG, Williams P, Davies MC, Alexander MR. (2012) "Combinatorial Discovery of Polymers Resistant to Bacterial Attachment" Nature Biotechnology, 30, 9, 868–875

[Hoover 1947] Hoover WR. (1947) "Fuel Tank" US Patent 2,429,688

[Hosford 2003] Hosford JP. (2003) "Pumps and Filter Assemblies with Application to Aquatic and Other Environments" US Patent 6,520,752

[Hou 2011] Hou Y, Liu CC, Sun K, Zhang P, Liu S, Mizumura D. (2011) "Computation of Milestones for Decision Support during System Restoration" IEEE Transactions on Power Systems, 26, 3, 1399–1409, doi:10.1109/TPWRS.2010.2089540

[Hou 2014] Hou F, Shen WM. (2014) "Graph-Based Optimal Reconfiguration Planning for Self-Reconfigurable Robots" Robotics and Autonomous Systems, 62, 7, 1047–1059, doi:10.1016/j.robot.2013.06.014

[Hou 2015] Hou X, Hu Y, Grinthal G, Khan M, Aizenberg J. (2015) "Liquid-Based Gating Mechanism with Tunable Multiphase Selectivity and Antifouling Behaviour" Nature, 519, 70–73, doi:10.1038/nature14253

[Hougie 2004] Hougie C. (2004) "The Waterfall-Cascade and Autoprothrombin Hypotheses of Blood Coagulation: Personal Reflections from an Observer" Journal of Thrombosis and Haemostasis, 2, 1225–1233.

[House 1966] House PA. (1966) "Self-Sealing Construction for Space Vehicles" US Patent 3,291,333

[Howard 1952] Howard AM, Meyer LS. (1952) "Bulletproof Fuel Tank" US Patent 2,601,525

[Hsiao 2011] Hsiao E, Bradley LC, Kim SH. (2011) "Improved Substrate Protection and Self-Healing of Boundary Lubrication Film Consisting of Polydimethylsiloxane with Cationic Side Groups" Tribology Letters, 41, 33–40, doi:10.1007/s11249-010-9679-0

[Hsieh 2001] Hsieh HC, Yang TJ, Lee S. (2001) "Crack Healing in Poly(methyl methacrylate) Induced by Co-Solvent of Methanol and Ethanol" Polymer, 42, 3, 1227–1241, doi:10.1016/S0032-3861(00)00407-9

[Hsu 1968] Hsu FH. (1968) "The Influence of Mechanical Loads on the Form of a Growing Elastic Body" Journal of Biomechanics, 1, 4, 303–311, doi:10.1016/0021-9290(68)90024-9

[Hu 2006] Hu MS, Chen HL, Shen CH, Hong LS, Huang BR, Chen KH, Chen LC. (2006) "Photosensitive Gold-Nanoparticle-Embedded Dielectric Nanowires" Nature Materials, 5, 2, 102–106, doi:10.1038/nmat1564

[Hu 2010] Hu J, Chen S. (2010) "A Review of Actively Moving Polymers in Textile Applications" Journal of Materials Chemistry, 20, 3346–3355, doi:10.1039/B922872A

[Huang 2007] Huang RF, Chen CH, Wu CW. (2007) "Economic Aspects of Memory Built-In Self-Repair" IEEE Design & Test of Computers, 24, 2, 164–172, doi:10.1109/MDT.2007.41

[Huang 2009] Huang JH, Kim J, Agrawal N, Sudarsan AP, Maxim JE, Jayaraman A, Ugaz VM. (2009) "Rapid Fabrication of Bio-inspired 3D Microfluidic Vascular Networks" Advanced Materials, 21, 35, 3567–3571

[Huang 2010] Huang CY, Trask RS, Bond IP. (2010) "Characterization and Analysis of Carbon Fibre-Reinforced Polymer Composite Laminates with Embedded Circular Vasculature" Journal of the Royal Society Interface, 7, 49, 1229–1241, doi:10.1098/rsif.2009.0534

[Huang 2012] Huang H, Ye G. (2012) "Simulation of Self-Healing by Further Hydration in Cementitious Materials" Cement and Concrete Composites, 34, 4, 460–467, doi:10.1016/j.cemconcomp.2012.01.003

[Huang 2015a] Huang M, Hsu CH, Wang J, Mei S, Dong X, Li Y, Li M, Liu H, Zhang W, Aida T, Zhang WB, Yue K, Cheng SZ. (2015) "Selective Assemblies of Giant Tetrahedra via Precisely Controlled Positional Interactions" Science, 348, 6233, 424–428, doi:10.1126/science.aaa2421.

[Huang 2015b] Huang W, Besar K, Zhang Y, Yang S, Wiedman G, Liu Y, Guo W, Song J, Hemker K, Hristova K, Kymissis IJ, Katz HE. (2015) "A High-Capacitance Salt-Free Dielectric for Self-Healable, Printable, and Flexible Organic Field Effect Transistors and Chemical Sensor" Advanced Functional Materials, 3745–3755, doi:10.1002/adfm.201404228

[Huang 2016] Huang L, Tan L, Zheng W. (2016) "Renovated Comprehensive Multilevel Evaluation Approach to Self-Healing of Asphalt Mixtures." International Journal of Geomechanics, 16, 1, B4014002, doi:10.1061/(ASCE)GM.1943-5622.0000451

[Huber 2005] Huber A, Hinkley JA. (2005) "Impression Testing of Self-Healing Polymers" NASA/TM-2005-213532, NASA, Langley Research Center, Hampton, VA

[Hull 1969] Hull DR, Piccard JA. (1969) "Self-Healing Reinforced Concrete Structures and Process for the Preparation Thereof" US Patent 3,466,822

[Hunt 2014] Hunt S, McKay TG, Anderson IA. (2014) "A Self-Healing Dielectric Elastomer Actuator" Applied Physics Letters, 104, 113701, doi:10.1063/1.4869294

[Huovinen 2014] Huovinen E, Takkunen L, Korpela T, Suvanto M, Pakkanen TT, Pakkanen TA. (2014) "Mechanically Robust Superhydrophobic Polymer Surfaces Based on Protective Micropillars" Langmuir, 30, 5, 1435–1443, doi:10.1021/la404248d

[Hur 2015] Hur SM, Thapar V, Ramírez-Hernández A, Khaira G, Segal-Peretz T, Rincon-Delgadillo PA, Li W, Müller M, Nealey PF, de Pablo JJ. (2015) "Molecular Pathways for Defect Annihilation in Directed Self-Assembly" Proceedings of the National Academy of Sciences of the United States of America, 112, 46, 14144–14149, doi:10.1073/pnas.1508225112

[Hura 2013] Hura GL, Tsai CL, Claridge SA, Mendillo ML, Smith JM, Williams GJ, Mastroianni AJ, Alivisatos AP, Putnam CD, Kolodner RD, Tainer JA. (2013) "DNA Conformations in Mismatch Repair Probed in Solution by X-Ray Scattering from Gold Nanocrystals" Proceedings of the National Academy of Sciences of the United States of America, 110, 43, 17308–17313, doi:10.1073/pnas.1308595110

[Hurley 2011a] Hurley DA, Huston DR (2011) "Coordinated Sensing and Active Repair for Self-Healing" Smart Materials and Structures, 20, 025010. doi:10.1088/0964-1726/20/2/025010

[Hurley 2011b] Hurley DA, Huston DR. (2011) "Self-Sealing Pneumatic Pressure Vessel with Passive and Active Methods" ASME 2011 Pressure Vessels and Piping Conference, American Society of Mechanical Engineers Digital Collection, pp 107–112, doi:10.1115/PVP2011-58008

[Hurst 2005] Hurst SJ, Ammerlaan JA, McCurdy RJ. (2005) "Coatings on Substrates" US Patent 6,840,061

[Hüsken 2009] Hüsken G, Hunger M, Brouwers HJ. (2009) "Experimental Study of Photocatalytic Concrete Products for Air Purification" Building and Environment, 44, 12, 2463–2474, doi:10.1016/j.buildenv.2009.04.010

[Huston 1998] Huston D, Spillman Jr. W, Drzewiczewski S, Eid B. (1998) "Position Stabilization of a Guyed Antenna Tower" Proceedings of SPIE 3329, Smart Structures and Materials 1998: Smart Structures and Integrated Systems, July 27, doi:10.1117/12.316881

[Huston 2005] Huston D, Esser B, Spencer G, Kahn E. (2005) "Hierarchical Actuator Systems" Proceedings of SPIE 5762, Smart Structures and Materials 2005: Industrial and Commercial Applications of Smart Structures Technologies, pp 311–319, doi:10.1117/12.607220

[Huston 2010] Huston D. (2010) Structural Sensing Health Monitoring and Prognosis, Taylor and Francis, Boca Raton, ISBN-13: 978-0750309196

[Huston 2011] Huston DR, Hurley DA. (2011) "Health Assessment and Coordination of Self-Sealing Structures" 3rd International Conference on Self-Healing Materials, Bath, UK, June 27–29, I Bond and R Varley (eds)

[Huston 2013] Huston D, Burns D, Razinger J, Seal R. (2013) "Actuated Wound Sensing, Closing and Healing in Flexible Sheets Using Functional Macro Cells" ICSHM 2013: Proceedings of the 4th International Conference on Self-Healing Materials, Ghent, Belgium, pp 177–180, http://resolver.tudelft.nl/uuid:85f21e19-dc57-4380-bd6b-d1c1b55c23e5

[Ichimura 2000] Ichimura K, Oh SK, Nakagawa M. (2000) "Light-Driven Motion of Liquids on a Photoresponsive Surface" Science, 288, 5471, 1624–1626

[Ignacio 2013] Ignacio-de Leon PA, Zharov I. (2013) "SiO_2@Au Core–Shell Nanospheres Self-Assemble to Form Colloidal Crystals that Can Be Sintered and Surface Modified to Produce pH-Controlled Membranes" Langmuir, 29, 11, 3749–3756, doi:10.1021/la304069x

[Iinuma 2014] Iinuma R, Ke Y, Jungmann R, Schlichthaerle T, Woehrstein JB, Yin P. (2014) "Polyhedra Self-Assembled from DNA Tripods and Characterized with 3D DNA-PAINT" Science, 344, 6179, 65–69, doi:10.1126/science.1250944

[Ikezoe 2012] Ikezoe Y, Washino G, Uemura T, Kitagawa S, Matsui H. (2012) "Autonomous Motors of a Metal–Organic Framework Powered by Reorganization of Self-Assembled Peptides at Interfaces" Nature Materials, 11, 1081–1085, doi:10.1038/nmat3461

[Ilhan 2001] Ilhan F, Gray M, Rotello VM. (2001) "Reversible Side Chain Modification through Noncovalent Interactions: 'Plug and Play' Polymers" Macromolecules, 34, 2597–2601

[Imaizumi 2001] Imaizumi K, Ohba T, Ikeda Y, Takeda K. (2001) "Self-Repairing Mechanism of Polymer Composite" Materials Science Research International, 7, 249–253

[Imato 2012] Imato K, Nishihara M, Kanehara T, Amamoto Y, Takahara A, Otsuka H. (2012) "Self-Healing of Chemical Gels Cross-Linked by Diarylbibenzofuranone-Based Trigger-Free Dynamic Covalent Bonds at Room Temperature" Angewandte Chemie International Edition, 51, 1138–1142, doi:10.1002/anie.201104069

[Indirli 2008] Indirli M, Castellano MG. (2008) "Shape Memory Alloy Devices for the Structural Improvement of Masonry Heritage Structures" International Journal of Architectural Heritage, 2, 2, 93–119, doi:10.1080/15583050701636258

[Inostroza 2015] Inostroza-Brito KE, Collin E, Siton-Mendelson O, Smith KH, Monge-Marcet A, Ferreira DS,Pérez Rodríguez R, Alonso M, Rodríguez-Cabello JC, Reis RL, Sagués F, Botto L, Bitton R, Azevedo HS, Mata A. (2015) "Co-Assembly, Spatiotemporal Control and Morphogenesis of a Hybrid Protein-Peptide System" Nature Chemistry, 7, 897–904, doi:10.1038/nchem.2349

[Ionov 2010] Ionov L. (2010) "Actively-Moving Materials Based on Stimuli-Responsive Polymers" Journal of Materials Chemistry, 20, 3382–3390, doi:10.1039/B922718K

[Irie 2001] Irie M, Kobatake S, Horichi M. (2001) "Reversible Surface Morphology Changes of a Photochromic Diarylethene Single Crystal by Photoirradiation" Science, 291, 5509, 1769–1772, doi:10.1126/science.291.5509.1769

[Irvine 2012] Irvine WT, Bowick MJ, Chaikin PM. (2012) "Fractionalization of Interstitials in Curved Colloidal Crystals" Nature Materials, 11, 948–951, doi:10.1038/NMAT3429

[Isaacs 2013] Isaacs B, Lark R, Jefferson T, Davies R, Dunn S. (2013) "Crack Healing of Cementitious Materials Using Shrinkable Polymer Tendons" Structural Concrete, 14, 2, 138–147, doi:10.1002/suco.201200013

[Ismail 2002] Ismail KM, Gehrig T, Jayaraman A, Wood TK, Trandem K, Arps PJ, Earthman JC. (2002) "Corrosion Control of Mild Steel by Aerobic Bacteria under Continuous Flow Conditions" Corrosion, 58, 5, 417–423

[Ito 1985] Ito H, Willson CG, Frechet JM. (1985) "Positive- and Negative-Working Resist Compositions with Acid Generating Photoinitiator and Polymer with Acid Labile Groups Pendant from Polymer Backbone" US Patent 4,491,628

[Ivanova 2013] Ivanova EP, Hasan J, Webb HK, Gervinskas G, Juodkazis S, Truong VK, Wu AH, Lamb RN, Baulin VA, Watson GS, Watson JA, Mainwaring DE, Crawford RJ. (2013) "Bactericidal Activity of Black Silicon" Nature Communications, 4, 2838, doi:10.1038/ncomms3838

[Iyer 2015] Iyer BV, Yashin VV, Balazs AC. (2015) "Harnessing Biomimetic Catch Bonds to Create Mechanically Robust Nanoparticle Networks" Polymer, 69, 310–320, doi:10.1016/j.polymer.2015.01.015

[Jabbari 1993] Jabbari E, Peppas NA. (1993) "Use of ATR-FTIR to Study Interdiffusion in Polystyrene and Poly(vinyl methyl ether)" Macromolecules, 26, 2175–2186, doi:10.1021/ma00061a006

[Jackson 2010] Jackson AC, Bartelt JA, Marczewski K, Sottos NR, Braun PV. (2010) "Silica-Protected Micron and Sub-Micron Capsules and Particles for Self-Healing at the Microscale" Macromolecular Rapid Communications, 31, 1–7, doi:10.1002/marc.201000468

[Jacobsen 1996] Jacobsen S, Marchand J, Boisvert L. (1996) "Effect of Cracking and Healing on Chloride Transport in OPC Concrete" Cement and Concrete Research, 26, 6, 869–881

[Jadhav 2010] Jadhav RS, Hundiwale DG, Mahulikar PP. (2010) "Synthesis and Characterization of Phenol-Formaldehyde Microcapsules Containing Linseed Oil and Its Use in Epoxy for Self-Healing and Anticorrosive Coating" Journal of Applied Polymer Science, 119, 2911–2916, doi:10.1002/app.33010.

[Jagota 2015] Jagota M, Tansu N. (2015) "Conductivity of Nanowire Arrays under Random and Ordered Orientation Configurations" Scientific Reports, 5, 10219, doi:10.1038/srep10219

[Jau 2011] Jau WC. (2011) "Self-Curing Concrete" US Patent 8,016,939

[Javierre 2009] Javierre E, Vermolen FJ, Vuik C, van der Zwaag S. (2009) "A Mathematical Analysis of Physiological and Morphological Aspects of Wound Closure" Journal of Mathematical Biology, 59, 605–630, doi:10.1007/s00285-008-0242-7

[Jefferson 1998] Jefferson GN. (1998) "Device for Neutralizing and Preventing Formation of Scale and Method" US Patent 5,738,766

[Jefferson 2010] Jefferson A, Joseph C, Lark R, Isaacs B, Dunn S, Weager B. (2010). "A New System for Crack Closure of Cementitious Materials Using Shrinkable Polymers" Cement and Concrete Research, 40, 795–801, doi:10.1016/j.cemconres.2010.01.004

[Jeffry 2011] Jeffry J, Kim S, Chen ZF. (2011) "Itch Signaling in the Nervous System" Physiology, 26, 286–292, doi:10.1152/physiol.00007.2011

[Jensen 1962] Jensen EH, Wagner JG. (1962) "Coacervation Process for Encapsulation of Lipophilic Materials" US Patent 3,069,370

[Jensen 2013] Jensen OM. (2013) "Use of Superabsorbent Polymers in Concrete" Concrete International, 35, 1, 48–52

[Jeong 2012a] Jeong C, Choi CH. (2012) "Single-Step Direct Fabrication of Pillar-on-Pore Hybrid Nanostructures in Anodizing Aluminum for Superior Superhydrophobic Efficiency" ACS Applied Materials & Interfaces, 4, 2, 842–848, doi:10.1021/am201514n

[Jeong 2012b] Jeong JH, Chan V, Cha C, Zorlutuna P, Dyck C, Hsia KJ, Bashir R, Kong H. (2012) "'Living' Microvascular Stamp for Patterning of Functional Neovessels: Orchestrated Control of Matrix Property and Geometry" Advanced Materials, 24, 58–63, doi:10.1002/adma.201103207

[Ji 2021] Ji Y, Li X, Zhou L, Chen A. (2021) "Analytical Study on the Pressure Caused by Two Spherical Projectiles Penetrating a Liquid-Filled Container" European Journal of Mechanics B: Fluids, 88, 103–122, doi:10.1016/j.euromechflu.2021.03.003

[Jia 2013] Jia T, Liu YY, Csóka E, Pósfai M, Slotine JJ, Barabási AL. (2013) "Emergence of Bimodality in Controlling Complex Networks" Nature Communications, 4, 2002, doi:10.1038/ncomms3002

[Jiang 2004] Jiang C, Markutsya S, Pikus Y, Tsukruk VV. (2004) "Freely Suspended Nanocomposite Membranes as Highly Sensitive Sensors" Nature Materials, 3, 721–728, doi:10.1038/nmat1212

[Jiang 2014] Jiang Z, Li W, Yuan Z. (2014) "Influence of Mineral Additives and Environmental Conditions on the Self-Healing Capabilities of Cementitious Materials" Cement and Concrete Composites, 57, 116–127, doi:10.1016/j.cemconcomp.2014.11.014

[Jin 2011] Jin HJ, Weissmüller J. (2011) "A Material with Electrically Tunable Strength and Flow Stress" Science, 332, 1179–1182, doi:10.1126/science.1202190

[Jin 2012] Jin H, Mangun CL, Stradley DS, Moore JS, Sottos NR, White SR. (2012) "Self-Healing Thermoset Using Encapsulated Epoxy-Amine Healing Chemistry" Polymer, 53, 2, 581–587

[Jisr 2005] Jisr RM, Rmaile HH, Schlenoff JB. (2005) "Hydrophobic and Ultrahydrophobic Multilayer Thin Films from Perfluorinated Polyelectrolytes" Angewandte Chemie International Edition, 44, 782, doi:10.1002/ange.200461645

[Johanson 2008] Johanson U, Punning A, Kruusmaa M, Aabloo A. (2008) "Self Healing Properties of Cu-Pt Coated Ionic Polymer Actuators" Proceedings of SPIE, Vol. 6927, Electroactive Polymer Actuators and Devices (EAPAD) 2008, Y. Bar-Cohen (ed.), doi:10.1117/12.775770

[John 2010] John M, Li G. (2010) "Self-Healing of Sandwich Structures with a Grid Stiffened Shape Memory Polymer Syntactic Foam Core" Smart Materials and Structures, 19, 075013, doi:10.1088/0964-1726/19/7/075013

[Johnson 1964] Johnson Jr. RE, Dettre RH. (1964) "Contact Angle Hysteresis I: Study of an Idealized Rough Surface" Contact Angle, Wettability, and Adhesion, Advances in Chemistry, Vol. 43, Chapter 7, pp 112–135, American Chemical Society, FM Fowkes (ed.), doi:10.1021/ba-1964-0043.ch007

[Johnson 2000] Johnson DH, White EI, D'Angelo Jr. JJ, Dicks D, Decker AL. (2000) "Wiring System Diagnostic Techniques for Legacy Aircraft" Fourth Joint NASA/FAA/DOD Conference on Aging Aircraft, St. Louis, MO

[Jonas 2002] Jonas U, del Campo A, Krüger C, Glasser G, Boos D. (2002) "Colloidal Assemblies on Patterned Silane Layers" Proceedings of the National Academy of Sciences of the United States of America, 99, 8, 5024–5027, doi:10.1073/pnas.072685399

[Jones 2006] Jones AS, Rule JD, Moore JS, White SR, Sottos NR. (2006) "Catalyst Morphology and Dissolution Kinetics of Self-Healing Polymers" Chemistry of Materials, 18, 1312–1317

[Jones 2007] Jones AS, Rule JD, Moore JS, Sottos NR, White SR. (2007) "Life Extension of Self-Healing Polymers with Rapidly Growing Fatigue Cracks" Journal of the Royal Society Interface, 4, 395–403, doi:10.1098/rsif.2006.0199, Published online 19 December

[Jones 2009] Jones F, Hayes SA. (2009) "A Self-Healing Composite Material Comprising a Fibre-Reinforced Polymeric Matrix, Wherein the Polymeric Matrix Comprises a Thermosetting Polymer and a Thermoplastic Polymer" US Patent Application 20090015272

[Jonkers 2010] Jonkers HM, Thijssen A, Muyzer G, Copuroglu O, Schlangen E. (2010) "Application of Bacteria as Self-Healing Agent for the Development of Sustainable Concrete" Ecological Engineering, 36, 2, 230–235

[Jonkers 2011] Jonkers HM. (2011) "Bacteria-Based Self-Healing Concrete" HERON, 56, 1/2, 1–12

[Jorgensen 2004] Jorgensen MW, Ostergaard EH, Lund HH. (2004) "Modular ATRON: Modules for a Self-Reconfigurable Robot" Proceedings. 2004 IEEE/RSJ International Conference on Intelligent Robots and Systems, 2, 2068–2073, doi:10.1109/IROS.2004.1389702

[Joseph 2008] Joseph C, Jefferson AD, Isaacs B, Lark R, Gardner D. (2008) "Experimental and Numerical Study of the Fracture and Self-Healing of Cementitious Materials" Magazine of Concrete Research, 62, 11, 831–843, doi:10.1680/macr.2010.62.11.831

[Jradi 2008] Jradi S, Soppera O, Lougnot D. (2008) "Fabrication of Polymer Waveguides between Two Optical Fibers Using Spatially Controlled Light-Induced Polymerization" Applied Optics, 47, 22, 3987–3993, doi:10.1364/AO.47.003987

[Jud 1979] Jud K, Kausch HH. (1979) "Load Transfer through Chain Molecules after lnterpenetration at Interfaces," Polymer Bulletin, 1, 697–707

[Judd 2014] Judd J, Ho ML, Tiwari A, Gomez EJ, Dempsey C, Van Vliet K, Igoshin OA, Silberg JJ, Agbandje-McKenna M, Suh J. (2014) "Tunable Protease-Activatable Virus Nanonodes" ACS Nano, 8, 5, 4740–4746, doi:10.1021/nn500550q

[Jung 1997] Jung D, Hegeman A, Sottos NR, Geubelle PH, White SR. (1997) "Self-Healing Composites Using Embedded Microspheres" American Society of Mechanical Engineers, MD-80, 265–275

[Jung 2010] Jung DH, Jeong HM, Kim BK. (2010) "Organic–Inorganic Chemical Hybrids Having Shape Memory Effect" Journal of Materials Chemistry, 20, 3458–3466, doi:10.1039/B922775J

[Jungbauer 2008] Jungbauer S, Gao H, Spatz JP, Kemkemer R. (2008) "Two Characteristic Regimes in Frequency-Dependent Dynamic Reorientation of Fibroblasts on Cyclically Stretched Substrates" Biophysical Journal, 95, 7, 3470–3478, doi:10.1529/biophysj.107.128611

[Kagami 2001] Kagami M, Yamashita T, Ito H. (2001) "Light-Induced Self-Written Three-Dimensional Optical Waveguide" Applied Physics Letters, 79, 8, 1079–1081

[Kailath 1979] Kailath T. (1979) Linear Systems, Prentice-Hall, Englewood Cliffs

[Kaitz 2015] Kaitz JA, Lee OP, Moore JS. (2015) "Depolymerizable Polymers: Preparation, Applications, and Future Outlook" MRS Communications, 5, 191–204, doi:10.1557/mrc.2015.28

[Kalkowski 2015] Kalkowski MK, Waters TP, Rustighi E. (2015) "Removing Surface Accretions with Piezo-Excited High-Frequency Structural Waves" Proceedings of SPIE, Vol. 9431, Active and Passive Smart Structures and Integrated Systems, 94311T, doi:10.1117/12.2087048

[Kallappa 2005] Kallappa P, Hailu H. (2005) "Automated Contingency and Life Management for Integrated Power and Propulsion Systems" Proceedings of GT2005, ASME Turbo Expo 2005: Power for Land, Sea and Air, GT2005-68587, pp 647–664, Reno-Tahoe, NV

[Kamada 2010] Kamada J, Koynov K, Corten C, Juhari A, Yoon JA, Urban MW, Balazs AC, Matyjaszewski K. (2010) "Redox Responsive Behavior of Thiol/Disulfi Defunctionalized Star Polymers Synthesized via Atom Transfer Radical Polymerization" Macromolecules, 43, 4133–4139

[Kammakakam 2019] Kammakakam I, O'Harra KE, Dennis GP, Jackson EM, Bara JE. (2019) "Self-Healing Imidazolium-Based Ionene-Polyamide Membranes: An Experimental Study on Physical and Gas Transport Properties" Polymer International, 68, 6, 1123–1129, doi:10.1002/pi.5802

[Kamphaus 2008] Kamphaus JM, Rule JD, Moore JS, Sottos NR, White SR. (2008) "A New Self-Healing Epoxy with Tungsten (VI) Chloride Catalyst" Journal of the Royal Society: Interface 5, 95–103, doi:10.1098/rsif.2007.1071.

[Kan 2010] Kan LL, Shi HS, Sakulich AR, Li VC. (2010) "Self-Healing Characterization of Engineered Cementitious Composite Materials" ACI Materials Journal, 107, 6, 617–624

[Kaneshige 2007] Kaneshige J, Krishnakumar K. (2007) "Artificial Immune System Approach for Air Combat Maneuvering" Proceedings of SPIE 6560, Intelligent Computing: Theory and Applications, 656009, doi:10.1117/12.718892

[Kansal 2007] Kansal A, Hsu J, Zahedi S, Srivastava MB. (2007) "Power Management in Energy Harvesting Sensor Networks" ACM Transactions on Embedded Computing Systems, 6, 4, 32–38, doi:10.1145/1274858.1274870

[Karan 2016] Karan CK, Bhattacharjee M. (2016) "Self-Healing and Moldable Metallogels as the Recyclable Materials for Selective Dye Adsorption and Separation" ACS Applied Materials & Interfaces, 8, 8, 5526–5535, doi:10.1021/acsami.5b09831

[Kardar 1986] Kardar M, Parisi G, Zhang YC. (1986) "Dynamic Scaling of Growing Interfaces" Physical Review Letters, 56, 9, 889–892

[Karki 2012] Karki K, Epstein E, Cho JH, Jia Z, Li T, Picraux ST, Wang C, Cumings J. (2012) "Lithium-Assisted Electrochemical Welding in Silicon Nanowire Battery Electrodes" Nano Letters, 12, 3, 1392–1397, doi:10.1021/nl204063u

[Karpov 2014] Karpov VG. (2014) "Electrostatic Theory of Metal Whiskers" Physical Review Applied, 1, 044001, doi:10.1103/PhysRevApplied.1.044001

[Karrfalt 1998] Karrfalt HA. (1998) "Multicomponent Self-Sealing Seam Tape" US Patent 5,843,552

[Kashi 2009] Kashi KB, Croes K, Pavlacky D, Garty J, Gelling VJ, Croll SG, Webster DC, Li W, Hintze P, Calle LM, Buhrow J, Curran J. (2009) "Responsive Microcapsules for Corrosion Inhibition" DOD Corrosion Conference, https://www.dau.edu/cop/cpc/documents/responsive-microcapsules-corrosion-inhibition

[Kashiwagi 2005] Kashiwagi T, Du F, Douglas JF, Winey KI, Harris Jr. RH, Shields JR. (2005) "Nanoparticle Networks Reduce the Flammability of Polymer Nanocomposites" Nature Materials, 4, 12, 928–933

[Kassab 2006] Kassab GS. (2006) "Scaling Laws of Vascular Trees: Form and Function" American Journal of Physiology: Heart and Circulatory Physiology, 290, H894–H903, doi:10.1152/ajpheart.00579.2005

[Kassem 2016] Kassem S, Lee AT, Leigh DA, Markevicius A, Solà J. (2016) "Pick-Up, Transport and Release of a Molecular Cargo Using a Small-Molecule Robotic Arm" Nature Chemistry, 8, 138–143, doi:10.1038/nchem.2410

[Katchalsky 1965] Katchalsky AK, Curran PF (1965) "Nonequilibrium Thermodynamics in Biophysics", 1st edition, Harvard University Press, Cambridge, MA

[Katifori 2010] Katifori E, Gergely J. Szöllősi GJ, Magnasco MO. (2010) "Damage and Fluctuations Induce Loops in Optimal Transport Networks" Physical Review Letters, 104, 048704, doi:10.1103/PhysRevLett.104.048704

[Kausch 1983] Kausch HH. (1983) "The Nature of Defects and Their Role in Large Deformation and Fracture of Engineering Thermoplastics" Pure Applied Chemistry, 55, 833–844

[Kavitha 2009] Kavitha AA, Singha, NK. (2009) "'Click Chemistry' In Tailor-Made Polymethacrylates Bearing Reactive Furfuryl Functionality: A New Class of Self-Healing Polymeric Material" ACS Applied Materials & Interfaces, 1, 7, 1427–1436, doi:10.1021/am900124c

[Kawagoe 1997] Kawagoe M, Nakanishi M, Qiu J, Morita M. (1997) "Growth and Healing of a Surface Crack in Poly(methyl methacrylate) under Case II Diffusion of Methanol" Polymer, 38, 5969–5975, doi:0.1016/S0032-3861(97)00153-5

[Kawaguchi 2004] Kawaguchi T, Pearson RA. (2004) "The Moisture Effect on the Fatigue Crack Growth of Glass Particle and Fiber Reinforced Epoxies with Strong and Weak Bonding Conditions Part 1. Macroscopic Fatigue Crack Propagation Behavior" Composites Science and Technology, 64, 1981–1989

[Kawauchi 2015] Kawauchi T, Oguchi Y, Nagai K, Iyoda T. (2015) "Conical Gradient Junctions of Dendritic Viologen Arrays on Electrodes" Scientific Reports, 5, 11122, doi:10.1038/srep11122

[Ke 2011] Ke Y, Ong LL, Shih WM, Yin P. (2012) "Three-Dimensional Structures Self-Assembled from DNA Bricks" Science, 338, 6111, 1177–1183, doi:10.1126/science.1227268

[Keber 2014] Keber FC, Loiseau E, Sanchez T, DeCamp SJ, Giomi L, Bowick MJ, Marchetti MC, Dogic Z, Bausch AR. (2014) "Topology and Dynamics of Active Nematic Vesicles" Science, 345, 6201, 1135–1139, doi:10.1126/science.1254784

[Keller 1985] Keller A, Odell JA. (1985) "The Extensibility of Macromolecules in Solution: A New Focus for Macromolecular Science" Colloid & Polymer Science, 263, 181–201

[Keller 2006a] Keller K, Pfeiffer E, Ullmann T, Ritter H. (2006) "Safety Improvements in TPS Development" Proceedings of 5th European Workshop on Thermal Protection Systems and Hot Structures, Noordwijk, The Netherlands, K, Fletcher (ed.)

[Keller 2006b] Keller MW, Sottos NR. (2006) "Mechanical Properties of Microcapsules Used in a Self-Healing Polymer" Experimental Mechanics, 46, 725–733, doi:10.1007/s11340-006-9659-3

[Keller 2007] Keller MW, White SR, Sottos NR. (2007) "A Self-Healing Poly(dimethyl siloxane) Elastomer" Advanced Functional Materials, 17, 2399–2404, doi:10.1002/adfm.200700086

[Kemell 2006] Kemell M, Färm E, Leskelä M, Ritala M. (2006) "Transparent Superhydrophobic Surfaces by Self-Assembly of Hydrophobic Monolayers on Nanostructured Surfaces" Physica Status Solidi A, 203, 6, 1453–1458, doi:10.1002/pssa.200566127

[Kendig 2003] Kendig M, Hon M, Warren L. (2003) "'Smart' Corrosion Inhibiting Coatings" Progress in Organic Coatings, 47, 183–189

[Kenneth 1956] Kenneth RL, Floyd O. (1956) "Autogenous Healing of Cement Paste" Journal of the American Concrete Institute, 52, 6, 52–63

[Kersey 2007] Kersey FR, Loveless DM, Craig SL. (2007) "A Hybrid Polymer Gel with Controlled Rates of Cross-Link Rupture and Self-Repair" Journal of the Royal Society: Interface, 4, 373–380

[Kessler 2002] Kessler MR, White SR. (2002) "Cure Kinetics of the Ring-Opening Metathesis Polymerization of Dicyclopentadiene" Journal of Polymer Science, Part A: Polymer Chemistry, 40, 2373–2383

[Kessler 2003] Kessler MR, Sottos NR, White SR. (2003) "Self-Healing Structural Composite Materials" Composites: Part A, 34, 743–753

[Khan 2014] Khan MR, Eaker CB, Bowden EF, Dickey MD. (2014) "Giant and Switchable Surface Activity of Liquid Metal via Surface Oxidation" Proceedings of the National Academy of Sciences of the United States of America, 111, 39, 14047–14051, doi:10.1073/pnas.1412227111

[Khushalani 2008] Khushalani S, Solanki J, Schulz NN. (2008) "Optimized Restoration of Combined AC/DC Shipboard Power Systems Including Distributed Generation and Islanding Techniques" Electrical Power Systems Research, 78, 9, 1528–1536

[Killian 2011] Killian CE, Metzler RA, Gong Y, Churchill TH, Olson IC, Trubetskoy V, Christensen MB, Fournelle JH, De Carlo F, Cohen S, Mahamid J, Scholl A, Young A, Doran A, Wilt FH, Coppersmith SN, Gilbert P. (2011) "Self-Sharpening Mechanism of the Sea Urchin Tooth" Advanced Functional Materials, 21, 682–690, doi:10.1002/adfm.201001546

[Kim 1983] Kim YH, Wool RP. (1983) "A Theory of Healing at a Polymer–Polymer Interface" Macromolecules, 16, 1115–1120

[Kim 1994] Kim KD, Sperling LH, Klein A, Hammouda B. (1994) "Reptation Time, Temperature, and Cosurfactant Effects on the Molecular Interdiffusion Rate during Polystyrene Latex Film Formation" Macromolecules, 27, 6841–6850, doi:10.1021/ma00101a024

[Kim 1996] Kim HJ, Lee KJ, Lee HH. (1996) "Healing of Fractured Polymers by Interdiffusion" Polymer, 37, 20, 4593–4597, doi:10.1016/0032-3861(96)00304-7

[Kim 2001] Kim HS, Khamis MA. (2001) "Fracture and Impact Behaviours of Hollow Micro-Sphere/Epoxy Resin Composites" Composites Part A: Applied Science and Manufacturing, 32, 1311–1317

[Kim 2003a] Kim YR, Little DN, Lytton RL. (2003) "Fatigue and Healing Characterization of Asphalt Mixes" Journal of Materials in Civil Engineering (ASCE), 15, 75–83, doi:10.1061/(ASCE)0899-1561(2003)15:1(75)

[Kim 2003b] Kim YW, Ando K, Chu MC. (2003) "Crack-Healing Behavior of Liquid-Phase Sintered Silicon Carbide Ceramics" Journal of the American Ceramic Society, 86, 465–470, doi:10.1111/j.1151-2916.2003.tb03322.x

[Kim 2004a] Kim S, Park H, Choi W. (2004) "Comparative Study of Homogeneous and Heterogeneous Photocatalytic Redox Reactions: $PW_{12}O_{40}^{3-}$ vs TiO_2" Journal of Physical Chemistry B, 108, 6402

[Kim 2004b] Kim WJ, Yoon YH, Kim IS, Kim KR. (2004) "Development of Self-Diagnosis Concrete for Damages" Advanced Smart Materials and Smart Structures Technology, FK Chang, CB Yun, BF Spencer, Jr. (eds), Destech, Lancaster

[Kim 2006a] Kim B, Roque R. (2006) "Evaluation of Healing Property of Asphalt Mixture" Paper No. 06-0596, 86th Annual Meeting of the Transportation Research Board, Washington, DC

[Kim 2006b] Kim S, Lorente S, Bejan A. (2006) "Vascularized Materials: Tree-Shaped Flow Architectures Matched Canopy to Canopy" Journal of Applied Physics, 100, 6, 063525, doi:/10.1063/1.2349479

[Kim 2007] Kim J, Lee H, Edwards R, Jacobsen SK. (2007) "Electrochemical Chloride Extraction Method to Control and Mitigate Corrosion in Rebar Embedded in Concrete," Transportation Research Record No. 1991, 78–85. doi:10.3141/1991-10

[Kim 2009] Kim J, Lee HJ, Kim YR, Kim HB. (2009) "A Drainage System for Mitigating Moisture Damage to Bridge Deck Pavements" The Baltic Journal of Road and Bridge Engineering, 4, 4, 168–176, doi:10.3846/1822-427X.2009.4.168-176

[Kim 2010a] Kim DH, Viventi J, Amsden JJ, Xiao J, Vigeland L, Kim YS, Blanco JA, Panilaitis B, Frechette ES, Contreras D, Kaplan DL, Omenetto FG, Huang Y, Hwang KC, Zakin MR, Litt B, Rogers JA. (2010) "Dissolvable Films of Silk Fibroin for Ultrathin Conformal Bio-Integrated Electronics" Materials, 9, 511–517, doi:10.1038/nmat2745

[Kim 2010b] Kim MS, Gruneich J, Jing H, Diamond SL. (2010) "Photo-Induced Release of Active Plasmid from Crosslinked Nanoparticles: *O*-Nitrobenzyl/Methacrylate Functionalized Polyethyleneimine" Journal of Materials Chemistry, 20, 3396–3403, doi:10.1039/B922613C

[Kim 2012a] Kim J, Hanna JA, Byun M, Santangelo CD, Hayward RC. (2012) "Designing Responsive Buckled Surfaces by Halftone Gel Lithography" Science, 335, 1201, doi:10.1126/science.1215309

[Kim 2012b] Kim P, Wong TS, Alvarenga J, Kreder MJ, Adorno-Martinez WE, Aizenberg J. (2012) "Liquid-Infused Nanostructured Surfaces with Extreme Anti-Ice and Anti-Frost Performance" ACS Nano, 6, 8, 6569–6577, doi:10.1021/nn302310q

[Kim 2013a] Kim SJ, Kim J, Moon MW, Lee KR, Kim HY. (2013) "Experimental Study of Drop Spreading on Textured Superhydrophilic Surfaces" Physics of Fluids, 25, 092110, doi:10.1063/1.4821985

[Kim 2013b] Kim Y, Zhu J, Yeom B, Di Prima M, Su X, Kim JG, Yoo SJ, Uher C, Kotov NA. (2013) "Stretchable Nanoparticle Conductors with Self-Organized Conductive Pathways" Nature, 500, 59–63, doi:10.1038/nature12401

[Kim 2015a] Kim B, Minisi M, McFarlane S. (2015) "Limited Range Projectile" US Patent 9,121,679

[Kim 2015b] Kim DH, Jung MC, Cho SH, Kim SH, Kim HY, Lee HJ, Oh KH, Moon MW. (2015) "UV-Responsive Nano-Sponge for Oil Absorption and Desorption" Scientific Reports, 5, 12908, doi:10.1038/srep12908

[Kim 2015c] Kim YS, Liu M, Ishida Y, Ebina Y, Osada M, Sasaki T, Hikima T, Takata M, Aida T. (2015) "Thermoresponsive Actuation Enabled by Permittivity Switching in an Electrostatically Anisotropic Hydrogel" Nature Materials, 14, 1002–1007, doi:10.1038/nmat4363

[Kim 2016a] Kim CL, Kim DE. (2016) "Self-Healing Characteristics of Collagen Coatings with Respect to Surface Abrasion" Scientific Reports, 6, 20563, doi:10.1038/srep20563

[Kim 2016b] Kim Y, Macfarlane RJ, Jones MR, Mirkin CA. (2016) "Transmutable Nanoparticles with Reconfigurable Surface Ligands" Science, 351, 6273, 579–582, doi:10.1126/science.aad2212

[Kim 2017] Kim ES, Lee JK, Lee PC, Huston DR, Tan T, Al-Ghamdi S. (2017) "Reinforced Cementitious Composite with In Situ Shrinking Microfibers" Smart Materials and Structures, 26, 3, 03LT01, doi:10.1088/1361-665X/aa5a2e

[Kim 2018] Kim S-M, Jeon H, Shin S-H, Park S-A, Jegal J, Hwang SY, Oh DX, Park J. (2018) "Superior Toughness and Fast Self-Healing at Room Temperature Engineered by Transparent Elastomers" Advanced Materials, 30, 1705145, doi:10.1002/adma.201705145

[Kim 2023] Kim S, Hsiao YH, Lee Y, Zhu W, Ren Z, Niroui F, Chen Y. (2023) "Laser-Assisted Failure Recovery for Dielectric Elastomer Actuators in Aerial Robots" Science Robotics, 8, 76, eadf4278, doi:10.1126/scirobotics.adf4278

[King 2008] King Jr. CR, Sekar D, Bakir MS, Dang B, Pikarsky J, Meindl JD. (2008) "3D Stacking of Chips with Electrical and Microfluidic I/O Interconnects" IEEE 58th Electronic Components and Technology Conference, doi:10.1109/ECTC.2008.4549941

[King 2014] King NP, Bale JB, Sheffler W, McNamara DE, Gonen S, Gonen T, Yeates TO, Baker D. (2014) "Accurate Design of Co-Assembling Multi-Component Protein Nanomaterials" Nature, 510, 103–108, doi:10.1038/nature13404

[Kinlen 2002] Kinlen PJ, Ding Y, Silverman DC. (2002) "Corrosion Protection of Mild Steel Using Sulfonic and Phosphonic Acid-Doped Polyanilines" Corrosion, 58, 6, 490–497

[Kirillov 2018] Kirillov O. (2018) "Locating the Sets of Exceptional Points in Dissipative Systems and the Self-Stability of Bicycles" Entropy, 20, 502, doi:10.3390/e20070502

[Kirk 2009] Kirk JG, Naik S, Moosbrugger JC, Morrison DJ, Volkov D, Sokolov I. (2009) "Self-Healing Epoxy Composites Based on the Use of Nanoporous Silica" International Journal of Fracture, 159, 101–102, doi 10.1007/s10704-009-9375-y

[Kishi 2012] Kishi T, Ahn TH, Hosoda A, Kobayashi K. (2012) "Cement Admixture and Cement Composition and Concrete Containing the Cement Admixture" US Patent 8,105,433

[Kittirungsi 2008] Kittirungsi B. (2008) "A Scaling Methodology for Dynamic Systems: Quantification of Approximate Similitude and Use in Multiobjective Design" PhD Dissertation, Mechanical Engineering, University of Michigan

[Klajn 2009] Klajn R, Wesson PJ, Bishop KJ, Bartosz A. (2009) "Writing Self-Erasing Images Using Metastable Nanoparticle Inks" Angewandte Chemie International Edition, 48, 7035–7039, doi:10.1002/anie.200901119

[Klarbring 2003] Klarbring A, Petersson J, Torstenfelt B, Karlsson M. (2003) "Topology Optimization of Flow Networks" Computer Methods in Applied Mechanics and Engineering, 192, 3909–3932

[Klavins 2007] Klavins E. (2007) "Programmable Self-Assembly" IEEE Control Systems, 27, 4, 43–56, doi:10.1109/MCS.2007.384126

[Knowlton 1934] Knowlton AS. (1934) "Inner Tube and Method of Manufacture" US Patent 1,977,281

[Ko 2014] Ko H, Lee J, Kim Y, Lee B, Jung CH, Choi JH, Kwon OS, Shin K. (2014) "Active Digital Microfluidic Paper Chips with Inkjet-Printed Patterned Electrodes" Advanced Materials, 26, 15, 2335–2340, doi:10.1002/adma.201305014

[Kobayashi 1995] Kobayashi N, Yoshioki H, Yoshida K, Sagawa K, Ishihara S. (1995) "Intumescent Fire-Resistant Coating, Fire-Resistant Material, and Process for Producing the Fire-Resistant Material" US Patent 5,401,793

[Koch 2009] Koch K, Bhushan B, Jung YC, Barthlott W. (2009) "Fabrication of Artificial Lotus Leaves and Significance of Hierarchical Structure for Superhydrophobicity and Low Adhesion" Soft Matter, 5, 1386–1393, doi:10.1039/B818940D

[Koch 2012] Koch C. (2012) Consciousness: Confessions of a Romantic Reductionist, The MIT Press, Cambirdge, MA

[Koene 2009] Koene BE, Own SH, Taushanoff RS. (2009) "Self Healing Superhydrophobic Coatings for Corrosion Protection" DOD Corrosion Conference, DJ Dunmire (ed.), Washington, DC

[Koerner 2004] Koerner H, Price G, Pearce NA, Alexander M, Vaia RA. (2004) "Remotely Actuated Polymer Nanocomposites: Stress-Recovery of Carbon-Nanotube-Filled Thermoplastic Elastomers" Nature Materials, 3, 2, 115–120

[Koester 2008] Koester KJ, Ager JW, Ritchie RO. (2008) "The True Toughness of Human Cortical Bone Measured with Realistically Short Cracks" Nature Materials, 7, 672–677, doi:10.1038/nmat2221

[Kogelnik 1966] Kogelnik H, Li T. (1966) "Laser Beams and Resonators" Applied Optics, 5, 10, 1550–1567

[Koh 2011] Koh SJ, Li T, Zhou J, Zhao X, Hong W, Zhu J, Suo Z. (2011) "Mechanisms of Large Actuation Strain in Dielectric Elastomers" Journal of Polymer Science Part B: Polymer Physics, 49, 504–515

[Köhler 2011] Köhler S, Schaller V, Andreas R. Bausch AR. (2011) "Structure Formation in Active Networks" Nature Materials, 10, 462–468, doi:10.1038/NMAT3009

[Kokot 2015] Kokot G, Piet D, Whitesides GM, Aranson IS, Snezhko A. (2015) "Emergence of Reconfigurable Wires and Spinners via Dynamic Self-Assembly" Scientific Reports, 5, 9528, doi:10.1038/srep09528

[Koleva 2009] Koleva DA, Guo Z, van Breugel K, de Wit JH. (2009) "Conventional and Pulse Cathodic Protection of Reinforced Concrete: Electrochemical Behavior of the Steel Reinforcement after Corrosion and Protection" Materials and Corrosion, 60, 5, 344–354.

[Kolmakov 2009] Kolmakov GV, Matyjaszewski K, Balazs AC. (2009) "Harnessing Labile Bonds between Nanogel Particles to Create Self-Healing Materials" ACS Nano, 3, 885–892.

[Kolmschlag 2004] Kolmschlag G, Steurer A. (2004) "Extruded Polytetrafluoroethylene Foam" US Patent 6,683,255

[Kolsky 1959] Kolsky H. (1959) "Fractures Produced by Stress Waves" Fracture, Proceedings of International Conference on Atomic Mechanisms of Fracture, BL Averbach, DK Felbeck, GT Hahn, DA Thomas (eds), April 12–16, pp 281–295, Technology Press, MIT, Swampscott, MA

[Konstantin 2007] Konstantin G. Kornev KG, Burstyn H, Kamath Y. (2007) "Electroimpregnation of Yarns and Fabrics with Nonwetting Liquids" Journal of Applied Physics, 101, 114901, doi:10.1063/1.2740333

[Koos 2011] Koos E, Willenbacher N. (2011) "Capillary Forces in Suspension Rheology" Science, 331, 897–898, doi:10.1126/science.1199243

[Korin 2012] Korin N, Kanapathipillai M, Matthews BD, Crescente M, Brill A, Mammoto T, Ghosh K, Jurek S, Bencherif SA, Bhatta D, Coskun AU, Feldman CL, Wagner DD, Ingber. (2012) "Shear-Activated Nanotherapeutics for Drug Targeting to Obstructed Blood Vessels" Science, 337, 738, doi:10.1126/science.1217815

[Korkmaz 2012] Korkmaz S, Ali NB, Smith IF. (2012) "Configuration of Control System for Damage Tolerance of a Tensegrity Bridge" Advanced Engineering Informatics, 26, 145–155

[Korouš 2000] Korouš Y, Chu MC, Nakatani M, Ando K. (2000) "Crack Healing Behavior of SiC Ceramics" Journal of the American Ceramic Society, 83, 11, 2788–2792, doi:10.1111/j.1151-2916.2000.tb01632.x

[Kotrotsos 2021] Kotrotsos A. (2021) "An Innovative Synergy between Solution Electrospinning Process Technique and Self-Healing of Materials: A Critical Review" Polymer Engineering & Science, 61, 5–21, doi:10.1002/pen.25559

[Kötteritzsch 2015] Kötteritzsch J, Hager M, Schubert U. (2015) "Tuning the Self-Healing Behavior of One-Component Intrinsic Polymers" Polymer, 69, 321–329, doi:10.1016/j.polymer.2015.03.027

[Kouhi 2012] Kouhi M, Mohebbi A, Mirzaei M. (2012) "Evaluation of the Corrosion Inhibition Effect of Micro/Nanocapsulated Polymeric Coatings: A Comparative Study by Use of EIS and Tafel Experiments and the Area under the Bode Plot" Research on Chemical Intermediates, 39, 2048–2062, doi 10.1007/s11164-012-0736-1

[Kousourakis 2008] Kousourakis A, Bannister MK, Mouritz AP. (2008) "Tensile and Compressive Properties of Polymer Laminates Containing Internal Sensor Cavities" Composites Part A: Applied Science and Manufacturing, 39, 9, 1394–1403.

[Kouwer 2013] Kouwer PH, Koepf M, Le Sage VA, Jaspers M, van Buul AM, Eksteen-Akeroyd ZH, Woltinge T, Schwartz E, Kitto HJ, Hoogenboom R, Picken SJ, Nolte RJ, Mendes E, Rowan AE. (2013) "Responsive Biomimetic Networks from Polyisocyanopeptide Hydrogels" Nature, 493, 651–655, doi:10.1038/nature11839

[Kowalski 2010] Kowalski D, Ueda M, Ohtsuka T. (2010) "Self-Healing Ion-Permselective Conducting Polymer Coating" Journal of Materials Chemistry, 20, 7630–7633, doi:10.1039/C0JM00866D

[Kraft 1927] Kraft HT. (1927) "Fuel Tank" US Patent 1,779,397

[Kratz 2012] Kratz K, Narasimhan A, Tangirala R, Moon SC, Revanur R, Kundu S, Kim HS, Crosby AJ, Russell TP, Emrick T, Kolmakov G, Balazs AC. (2012) "Probing and Repairing Damaged Surfaces with Nanoparticle-Containing Microcapsules" Nature Nanotechnology, 7, 87–90, doi:10.1038/nnano.2011.235

[Kretinin 2014] Kretinin AV, Cao Y, Tu JS, et al. (2014) "Electronic Properties of Graphene Encapsulated with Different Two-Dimensional Atomic Crystals" Nano Letters, 14, 3270–3276, doi:10.1021/nl5006542

[Kriegman 2020] Kriegman S, Blackiston D, Levin M, Bongard J. (2020) "A Scalable Pipeline for Designing Reconfigurable Organisms" Proceedings of the National Academy of Sciences of the United States of America, 117, 4, 1853–1859, doi:10.1073/pnas.1910837117

[Kringos 2011] Kringos N, Schmets A, Scarpas A. (2011) "Towards an Understanding of the Self-Healing Capacity of Asphaltic Mixtures" HERON, 56, 1/2, 45–74

[Krupenkin 2004] Krupenkin TN, Taylor JA, Schneider TM, Yang S. (2004) "From Rolling Ball to Complete Wetting: The Dynamic Tuning of Liquids on Nanostructured Surfaces" Langmuir, 20, 3824–3827

[Kryger 2010] Kryger MJ, Ong MT, Odom SA, Sottos NR, White SR, Martinez TJ, Moore JS. (2010) "Masked Cyanoacrylates Unveiled by Mechanical Force" Journal of the American Chemical Society, 132, 13, 4558–4559, doi:10.1021/ja1008932

[Kuang 2008] Kuang Y, Ou J. (2008) "Self-Repairing Performance of Concrete Beams Strengthened Using Superelastic SMA Wires in Combination with Adhesives Released from Hollow Fibers" Smart Materials and Structures, 17, 025020 doi:10.1088/0964-1726/17/2/025020

[Kuang 2023] Kuang X, Rong Q, Belal S, et al. (2023) "Self-Enhancing Sono-Inks Enable Deep-Penetration Acoustic Volumetric Printing" Science, 382, 1148–1155, doi:10.1126/science.adi1563

[Kudernac 2011] Kudernac T, Ruangsupapichat N, Parschau M, Maciá B, Katsonis N, Harutyunyan SR, Ernst KH, Feringa BL. (2011) "Electrically Driven Directional Motion of a Four-Wheeled Molecule on a Metal Surface" Nature, 479, 208–211

[Kuksenok 2007] Kuksenok O, Yashin VY, Balazs AC. (2007) "Mechanically Induced Chemical Oscillations and Motion in Responsive Gels" Soft Matter, 3, 1138–1144, doi:10.1039/b707393c

[Kumar 2008] Kumar A, Stephenson LD. (2008) "Self-Healing Coatings Using Microcapsules to Suppress Lead Dust" US Patent 7,342,057

[Kumar 2010] Kumar UN, Kratz K, Wagermaier W, Behl M, Lendlein A. (2010) "Non-Contact Actuation of Triple-Shape Effect in Multiphase Polymer Network Nanocomposites in Alternating Magnetic Field" Journal of Materials Chemistry, 20, 3404–3415, doi:10.1039/B923000A

[Kuramoto 1994] Kuramoto N, Hayashi K, Nagai K. (1994) "Thermoreversible Reaction of Diels–Alder Polymer Composed of Difurufuryladipate with Bismaleimidodi Phenylmethane" Journal of Polymer Science Part A: Polymer Chemistry, 32, 2501–2504

[Kurzweil 2006] Kurzweil R. (2006) The Singularity Is Near: When Humans Transcend Biology, Penguin Books, New York

[Kushner 2009] Kushner AM, Vossler JD, Williams GA, Guan Z. (2009) "A Biomimetic Modular Polymer with Tough and Adaptive Properties" Journal of the American Chemical Society, 131, 8766–8768, doi:10.1021/ja9009666

[Kutzer 2008] Kutzer MD, Armand M, Scheidt DH, Lin E, Chirikjian GS. (2008) "Toward Cooperative Team-diagnosis in Multi-robot Systems" International Journal of Robotics Research, 27, 9, 1069–1090, doi:10.1177/0278364908095700

[Kwok 2007] Kwok N, Hahn HT. (2007) "Resistance Heating for Self-healing Composites" Journal of Composite Materials, 41, 13, 1635–1654, doi:10.1177/0021998306069876

[Kwon 2012] Kwon E, Kuczera R. (2012) "Self-Repairing Boot for a Constant Velocity Joint" US Patent 8,088,015

[Laczynski 2002] Laczynski M, Jayaraman S, Craychee T, Miyasaka C, Tittmann B. (2002) "Study of the Feasibility of Healing Delaminations in an AS4/PEEK Plate" SPIE Vol. 4704, Proceedings of NDE and Health Monitoring of Aerospace Materials and Civil Infrastructures, A Geyekenyesi, S Shepard, D Huston, A Aktan, P Schull (eds)

[Ladet 2008] Ladet S, David L, Domard A. (2008) "Multi-Membrane Hydrogels" Nature, 452, 6

[Lafont 2014] Lafont U, Moreno-Belle C, van Zeijl H, van der Zwaag S. (2014) "Self-Healing Thermally Conductive Adhesives" Journal of Intelligent Materials Systems and Structures, 25, 1, 67–74, doi:10.11 77/1045389X13498314

[Lafuma 2003] Lafuma A, Quéré D. (2003) "Superhydrophobic States" Nature Materials, 457–460

[Lahann 2003] Lahann J, Mitragotri S, Tran TN, Kaido H, Sundaram J, Choi IS, Hoffer S, Somorjai GA, Langer R. (2003) "A Reversibly Switching Surface" Science 299, 371 doi:10.1126/science.1078933

[Lai 2013] Lai YK, Pan F, Fuchs H, Chi LF. (2013) "In Situ Surface-Modification-Induced Superhydrophobic Patterns with Reversible Wettability and Adhesion" Advanced Materials, 25, 1682–1686, doi:10.1002/adma.201203797

[Lam 2011] Lam WA, Chaudhuri O, Crow A, Webster KD, Li TD, Kita A, Huang J, Fletcher DA. (2011) "Mechanics and Contraction Dynamics of Single Platelets and Implications for Clot Stiffening" Nature Materials, 10, 61–66, doi:10.1038/nmat2903

[Lamaka 2007] Lamaka SV, Zheludkevich ML, Yasakau KA, Serra R, Poznyak SK, Ferreira MG. (2007) "Nanoporous Titania Interlayer as Reservoir of Corrosion Inhibitors for Coatings with Self-Healing Ability" Progress in Organic Coatings, 58, 2–3, 127–135, doi:10.1016/j.porgcoat.2006.08.029

[Lan 2012] Lan G, Sartori P, Neumann S, Sourjik V, Tu Y. (2012) "The Energy-Speed-Accuracy Trade-Off in Sensory Adaptation" Nature Physics, 8, 422–428, doi:10.1038/NPHYS2276

[Landis (1915] Landis NR. (1915) "Vehicle Tire" US Patent 1,156,155

[Lang 2014] Lang N, Pereira MJ, Lee Y, Friehs I, Vasilyev NV, Feins EN, Ablasser K, O'Cearbhaill ED, Xu C, Fabozzo A, Padera R, Wasserman S, Freudenthal F, Ferreira LS, Langer R, Karp JM, del Nido PJ. (2014) "A Blood-Resistant Surgical Glue for Minimally Invasive Repair of Vessels and Heart Defects" Science Translational Medicine, 6, 218, 218ra6, doi:10.1126/scitranslmed.3006557

[Lanzara 2009] Lanzara G, Yoon Y, Liu H, Peng S, Lee WI. (2009) "Carbon Nanotube Reservoirs for Self-Healing Materials" Nanotechnology, 20, 335704, doi:10.1088/0957-4484/20/33/335704

[Larin 2006] Larin GE, Bernklau N, Kessler MR, DiCesare JC. (2006) "Rheokinetics of Ring-Opening Metathesis Polymerization of Norbornene-Based Monomers Intended for Self-Healing Applications" Polymer Engineering & Science, 46, 12, 1804–1811, doi:10.1002/pen.20655

[Larson 2023] Larson NM, Mueller J, Chortos A, et al. (2023) "Rotational Multimaterial Printing of Filaments with Subvoxel Control" Nature, 613, 682–688, doi:10.1038/s41586-022-05490-7

[Lau 2003] Lau KK, Bico J, Teo KB, Chhowalla M, Amaratunga GA, Milne WI, McKinley GH, Gleason KK. (2003) "Superhydrophobic Carbon Nanotube Forests" Nano Letters, 3, 12, 1701–1705, doi:10.1021/nl034704t

[Lau 2012] Lau GK, Chua SL, Shiau LL, Tan AW. (2012) "Self-Clearing Dielectric Elastomer Actuators Using Charcoal-Powder Electrodes" Proc SPIE, Vol. 8340, Electroactive Polymer Actuators and Devices (EAPAD), Y Bar-Cohen (ed.), doi:10.1117/12.915493

[Laver 2009] Laver R. (2009) "Defining and Measuring State of Good Repair" Transportation Research Board, Annual Meeting CD-ROM, Paper No. 09-2556, Washington, DC

[Law 2014] Law JB, Ng AM, He AY, Low HY. (2014) "Bioinspired Ultrahigh Water Pinning Nanostructures" Langmuir, 30, 1, 325–331, doi:10.1021/la4034996

[Lednicer 2003] Lednicer D. (2003) "A Capsule History" American Heritage of Invention and Technology, 19, 2, 50–55

[Lee 2001] Lee CH, Hall DV, Perkowski MA, Jun DS. (2001) "Self-Repairable GALs" Journal of Systems Architecture, 47, 119–135, doi:10.1016/S1383-7621(00)00061-8

[Lee 2002] Lee YJ, Hirose S. (2002) "Three-Legged Walking for Fault-Tolerant Locomotion of Demining Quadruped Robots" Advanced Robotics, 16, 5, 415–426, doi:10.1163/156855302601829l8

[Lee 2003] Lee YC, Hong YP, Lee HY, Kim H, Jung YJ, Ko KH, Jung HS, Hong KS. (2003) "Photocatalysis and Hydrophilicity of Doped TiO_2 Thin Films" Journal of Colloid and Interface Science, 267, 127, doi:10.1016/S0021-9797(03)00603-9

[Lee 2004a] Lee JY, Buxton GA, Balazs AC. (2004) "Using Nanoparticles to Create Self-Healing Composites" Journal of Chemical Physics, 121, 5531, doi:10.1063/1.1784432

[Lee 2004b] Lee SH. (2004) "Photocatalytic Nanocomposites Based on TiO_2 and Carbon Nanotubes" Doctoral Dissertation, University of Florida

[Lee 2005] Lee JC, Kim YJ. (2005) "Calculation of the Interruption Process of a Self-Blast Circuit Breaker" IEEE Transactions on Magnetics, 41, 5, 1592–1595

[Lee 2006] Lee JK, Hong SJ, Liu X, Yoon SH. (2004) "Characterization of Dicyclopentadiene and 5-Ethylidene-2-Norbornene as Self-Healing Agents for Polymer Composite and Its Microcapsules" Macromolecular Research, 12, 478–483, doi:10.1007/BF03218430

[Lee 2007] Lee JY, Lim DP, Lim DS. (2007) "Tribological Behavior of PTFE Nanocomposite Films Reinforced with Carbon Nanoparticles" Composites: Part B, 38, 810–816

[Lee 2008] Lee K, Moses M, Chirikjian GS. (2008) "Robotic Self-Replication in Structured Environments: Physical Demonstrations and Complexity Measures" International Journal of Robotics Research, 27, 3–4, 387–401, doi:10.1177/0278364907084982

[Lee 2010a] Lee HJ, Owens JR. (2010) "Motion of Liquid Droplets on a Superhydrophobic Oleophobic Surface" Journal of Materials Science, 46, 69–76, doi:10.1007/s10853-010-4810-z

[Lee 2010b] Lee YJ, Lee Y, Oh D, Chen T, Ceder G, Belcher AM. (2010) "Biologically Activated Noble Metal Alloys at the Nanoscale: For Lithium Ion Battery Anodes" Nano Letters, 10, 2433–2440, doi:10.1021/nl1005993

[Lee 2012a] Lee JH, Veysset D, Singer JP, Retsch M, Saini G, Pezeril T, Nelson KA, Thomas EL. (2012) "High Strain Rate Deformation of Layered Nanocomposites" Nature Communications, 3, 1164, 1–9, doi:10.1038/ncomms2166

[Lee 2012b] Lee KK, Ryoo JD, Joo BS. (2012) "Faultless Protection Methods in Self-Healing Ethernet Ring Networks" ETRI Journal, 34, 816–826, doi:10.4218/etrij.12.1812.0102

[Lee 2013a] Lee B, Chen Y, Fu D, Yi HT, Czelen K, Najafov H, Podzorov V. (2013) "Trap Healing and Ultralow-Noise Hall Effect at the Surface of Organic Semiconductors" Nature Materials, 12, 1125–1129, doi:10.1038/nmat3781

[Lee 2013b] Lee C-H, Sugiyama T, Kataoka A, Kudo A, Fujino F, Chen YW, Mitsuyama Y, Nomura S, Yoshioka T. (2013) "Analysis for Distinctive Activation Patterns of Pain and Itchy in the Human Brain Cortex Measured Using Near Infrared Spectroscopy (NIRS)" PLoS One, 8, 10, e75360, doi:10.1371/journal.pone.0075360

[Lee 2013c] Lee SG, Ham DS, Lee DY, Bong H, Cho K. (2013) "Transparent Superhydrophobic/Translucent Superamphiphobic Coatings Based on Silica–Fluoropolymer Hybrid Nanoparticles" Langmuir, 29, 48, 15051–15057

[Lee 2013d] Lee WJ, Weber B, Feltrin G, Czaderski C, Motavalli M, Leinenbach C. (2013) "Stress Recovery Behaviour of an Fe–Mn–Si–Cr–Ni–VC Shape Memory Alloy Used for Prestressing" Smart Materials and Structures, 22, 12, doi:10.1088/0964-1726/22/12/125037

[Lee 2014] Lee HY, Shin SH, Drews AM, Chirsan AM, Lewis SA, Bishop KJ. (2014) "Self-Assembly of Nanoparticle Amphiphiles with Adaptive Surface Chemistry" ACS Nano, 8, 10, 9979–9987, doi:10.1021/nn504734v

[Lee 2015] Lee HS. (2015) "Application of the Viscoelastic Continuum Damage Mechanics to Asphalt Mixtures under Indirect Tensile Load" Journal of Engineering Mechanics, 141, 04014154, doi:10.1061/(ASCE)EM.1943-7889.0000876

[Legrain 2014] Legrain A, Janson TG, Berenschot JW, Abelmann L, Tas NR. (2014) "Controllable Elastocapillary Folding of Three-Dimensional Micro-Objects by Through-Wafer Filling" Journal of Applied Physics, 115, 214905, doi:10.1063/1.4878460

[Lei 2014] Lei ZQ, Xiang HP, Yuan YJ, Rong MZ, Zhang MQ. (2014) "Room-Temperature Self-Healable and Remoldable Cross-linked Polymer Based on the Dynamic Exchange of Disulfide Bonds" Chemistry of Materials, 26, 6, 2038–2046, doi:10.1021/cm4040616

[LeMieux 2007] LeMieux MC, Peleshanko S, Anderson KD, Tsukruk VV. (2007) "Adaptive Nanomechanical Response of Stratified Polymer Brush Structures" Langmuir, 23, 265–273, doi:10.1021/la061723k

[Lendlein 2002] Lendlein A, Kelch S. (2002) "Shape-Memory Effect: From Temporary Shape to Permanent Shape" Angewandte Chemie International Edition, 41, 12, 2034–2057, doi:10.1002/chin.200240251.

[Leng 2009] Leng J, Lu H, Liu Y, Huang WM, Du S. (2009) "Shape-Memory Polymers: A Class of Novel Smart Materials" MRS Bulletin, 34, 848–855

[Lenhardt 2010] Lenhardt JM, Ong MT, Choe R, Evenhuis CR, Martinez TJ, Craig SL. (2010) "Trapping a Diradical Transition State by Mechanochemical Polymer Extension" Science, 329, 1057–1060, doi:10.1126/science.1193412

[Leon 2001] Leon J, Rojo E, Sanchez-Serrano J. (2001) "Wound Signalling in Plants" Journal of Experimental Botany, 52, 354, 1–9. doi:10.1093/jexbot/52.354.1

[Leong 2013] Leong MF, Toh JK, Du C, Narayanan K, Lu HF, Lim TC, Wan AC, Ying JY. (2013) "Patterned Prevascularised Tissue Constructs by Assembly of Polyelectrolyte Hydrogel Fibres" Nature Communications, 4, 2353, doi:10.1038/ncomms3353

[Lepech 2006] Lepech MD, Li VC. (2006) "Long Term Durability Performance of Engineered Cementitious Composites" Restoration of Buildings and Monuments, 12, 2, 119–132

[Leseman 2007] Leseman ZC, Koppaka SB, Mackin TJ. (2007) "A Fracture Mechanics Description of Stress-Wave Repair in Stiction-Failed Microcantilevers: Theory and Experiments" Journal of Microelectromechanical Systems, 16, 4

[Leslie 2014] Leslie DC, Waterhouse A, Berthet JB, Valentin TM,Watters AL, Jain A, Kim P, Hatton BD, Nedder A, Donovan K, Super EH, Howell C, Johnson CP, Vu TL, Bolgen DE, Rifai S, Hansen AR, Aizenberg M, Super M, Aizenberg J, Ingber DE. (2014) "A Bioinspired Omniphobic Surface Coating on Medical Devices Prevents Thrombosis and Biofouling" Nature Biotechnology, 32, 1134–1140, doi:10.1038/nbt.3020

[Leunissen 2009] Leunissen ME, Vutukuri HR, van Blaaderen A. (2009) "Directing Colloidal Self-Assembly with Biaxial Electric Fields" Advanced Materials, 21, 30, 3116–3120, doi:10.1002/adma.200900640

[Lewis 2006] Lewis JA. (2006) "Direct Ink Writing of 3D Functional Materials" Advanced Functional Materials, 16, 17, 2193–2204, doi:10.1002/adfm.200600434

[Lewis 2009] Lewis TG. (2009) Network Science Theory and Practice, Wiley, Hoboken

[Li 1998a] Li VC, Lim YM, Chan YW. (1998) "Feasibility Study of a Passive Smart Self-Healing Cementitious Composite" Composites Part B, 29B, 819–827

[Li 1998b] Li VC, Wu HC. (1998) "Control of Interface Properties between Fiber/Cementitious Material Using Plasma Treatment" US Patent 5,788,760

[Li 2001] Li XZ, Li FB, Yang CL, Ge WK. (2001) "Photocatalytic Activity of WO_x-TiO_2 under Visible Light Irradiation" Journal of Photochemistry and Photobiology A: Chemistry, 141, 2–3, 209–217, doi:10.1016/S1010-6030(01)00446-4

[Li 2006] Li Y, Liang GZ, Xie JQ, Gou J, Li L. (2006) "Thermal Stability of Microencapsulated Epoxy Resins with Poly(urea-formaldehyde)" Polymer Degradation and Stability, 91, 2300–2306, doi:10.1016/j.polymdegradstab.2006.04.026

[Li 2007a] Li L, Li Q and Zhang F (2007) "Behavior of Smart Concrete Beams with Embedded Shape Memory Alloy Bundles" Journal of Intelligent Material Systems and Structures, 18, 1003–1014

[Li 2007b] Li XM, Reinhoudt D, Crego-Calama M. (2007) "What Do We Need for a Superhydrophobic Surface? A Review on the Recent Progress in the Preparation of Superhydrophobic Surfaces" Chemical Society Reviews, 36, 1350–1368, doi:10.1039/B602486F

[Li 2008a] Li G, John M. (2008) "A Self-Healing Smart Syntactic Foam under Multiple Impacts" Composites Science and Technology, 68, 3337–3343, doi:10.1016/j.compscitech.2008.09.009

[Li 2008b] Li H, Wang R, Hu H, Liu W. (2008) "Surface Modification of Self-Healing Poly(urea-formaldehyde) Microcapsules Using Silane-Coupling Agent" Applied Surface Science, 255, 5, 1, 1894–1900, doi:10.1016/j.apsusc.2008.06.170

[Li 2008c] Li J, Wang BH, Wang WX, Zhou T. (2008) "Network Entropy Based on Topology Configuration and Its Computation to Random Networks" Chinese Physics Letters, 25, 11, 4177–4180

[Li 2008d] Li P, Zhao Z, Hong X, Yu H. (2008) "Study on the Intelligent Self-Healing Fiber Optic Microbend Sensors Based on the Photocurable Material" SPIE 7267, Smart Materials V, 7267–7217, doi:10.1117/12.810686

[Li 2009a] Li VC, Yang EH. (2009) "Engineered Self-Healing Cementitious Composites" US Patent 7,572,501

[Li 2009b] Li W. (2009) "Evaluation of Hybrid Binder for Dense and Open-Graded Asphalt Mixtures" PhD Dissertation, University of Florida, Gainesville, FL

[Li 2010a] Li J, Liu Z, Tan C, Guo X, Wang L, Sancar A, Zhong D. (2010) "Dynamics and Mechanism of Repair of Ultraviolet Induced (6-4) Photoproduct by Photolyase" Nature, 466, 12, 887–890, doi:10.1038/nature09192

[Li 2010b] Li X, Du X, He J. (2010) "Self-Cleaning Antireflective Coatings Assembled from Peculiar Mesoporous Silica Nanoparticles" Langmuir, 26, 16, 13528–13534, doi:10.1021/la1016824

[Li 2012a] Li G, Meng H, Hu J. (2012) "Healable Thermoset Polymer Composite Embedded with Stimuli-Responsive Fibres" Journal of the Royal Society Interface, 9, 3279–3287, doi:10.1098/rsif.2012.0409

[Li 2012b] Li G. (2012) "A Shape Memory Polymer Based Self-Healing Sealant for Expansion Joint" Final Report for Highway IDEA Project 142, Transportation Research Board, Washington, DC

[Li 2012c] Li S, Song G, Kwakernaak K, van der Zwaag S, Sloof WG. (2012) "Multiple Crack Healing of a Ti_2AlC Ceramic" Journal of the European Ceramic Society, 32, 1813–1820, doi:10.1016/j.jeurceramsoc.2012.01.017

[Li 2012d] Li Y, Chen SS, Wu MC, Sun JQ. (2012) "Polyelectrolyte Multilayers Impart Healability to Highly Electrically Conductive Films" Advanced Materials, 24, 4578–4582, doi:10.1002/adma.201201306

[Li 2013a] 1641 Li W, Jiang Z, Yang Z, Zhao N, Yuan W (2013) "Self-Healing Efficiency of Cementitious Materials Containing Microcapsules Filled with Healing Adhesive: Mechanical Restoration and Healing Process Monitored by Water Absorption" PLoS One, 8, 11, e81616, doi:10.1371/journal.pone.0081616

[Li 2013b] Li G, Wie JJ, Nguyen NA, Chung WJ, Kim ET, Char K, Mackay ME, Pyun J. (2013) "Synthesis, Self-Assembly and Reversible Healing of Supramolecular Perfluoropolyethers" Journal of Polymer Science Part A: Polymer Chemistry, 51, 17, 3598–3606, doi:10.1002/pola.26777

[Li 2015] Li Q, Fuks G, Moulin E, Maaloum M, Rawiso M, Kulic I, Foy JT, Giuseppone N. (2015) "Macroscopic Contraction of a Gel Induced by the Integrated Motion of Light-Driven Molecular Motors" Nature Nanotechnology, 10, 161–165, doi:10.1038/nnano.2014.315

[Libeskind 1995] Libeskind-Hadas R, Shrivastava N, Melhem RG, Liu CL. (1995) "Optimal Reconfiguration Algorithms for Real-Time Fault-Tolerant Processor Arrays" IEEE Transactions on Parallel and Distributed Systems, 6, 5, 498–510, doi:10.1109/71.382318

[Liedl 2010] Liedl T, Hogberg B, Tytell J, Ingber DE, Shih WM. (2010) "Self-Assembly of Three-Dimensional Prestressed Tensegrity Structures from DNA" Nature Nanotechnology, 5, 520–524, doi:10.1038/NNANO.2010.107

[Lim 2008] Lim KY. (2008) "A Development of Repair Mechanism and Control Technologies for Bottom Part of Concrete Bridge" Proceedings of IABMAS 2008 Bridge Maintenance, Safety, Management, Health Monitoring and Informatics, HM Koh, D Frangopol (eds), pp 1231–1237

[Lim 2014] Lim K, Durand M, Baudelet M, Richardson M. (2014) "Transition from Linear- to Nonlinear-Focusing Regime in Filamentation" Scientific Reports, 4, 7217, doi:10.1038/srep07217

[Lima 2011] Lima MD, Fang S, Lepró X, Lewis C, Ovalle-Robles R, Carretero-González J, Castillo-Martínez E, Kozlov ME, Oh J, Rawat N, Haines CS, Haque MH, Aare V, Stoughton S, Zakhidov AA, Baughman RH. (2011) "Biscrolling Nanotube Sheets and Functional Guests into Yarns" Science, 7, 331,6013, 51–55, doi:10.1126/science.1195912

[Lima 2012] Lima MD, Li N, Jung de Andrade M, Fang S, Oh J, Spinks GM, Kozlov ME, Haines CS, Suh D, Foroughi J, Kim SJ, Chen Y, Ware T, Shin MK, Machado LD, Fonseca AF, Madden JD, Voit WE, Galvão DS, Baughman RH. (2012) "Electrically, Chemically, and Photonically Powered Torsional and Tensile Actuation of Hybrid Carbon Nanotube Yarn Muscles" Science, 338, 928–932, doi:10.1126/science.1226762

[Lin 1990] Lin CB, Lee SB, Liu KS. (1990) "Methanol-Induced Crack Healing in Poly(methyl methacrylate)" Polymer Engineering & Science, 30, 1399–406, doi:10.1002/pen.760302109

[Lin 1997] Lin Z, Li VC. (1997) "Crack Bridging in Fiber Reinforced Cementitious Composites with Slip-Hardening Interfaces" Journal of the Mechanics and Physics of Solids, 45, 5, 763–787, doi:10.1016/S0022-5096(96)00095-6

[Lin 2012] Lin J, Dellinger J, Genevet P, Cluzel B, de Fornel F, Capasso F. (2012) "Cosine-Gauss Plasmon Beam: A Localized Long-Range Nondiffracting Surface Wave" Physical Review Letters 109, 093904, doi:10.1103/PhysRevLett.109.093904

[Lin 2014] Lin J, Cretu O, Zhou W, Suenaga K, Prasai D, Bolotin KI, Cuong NT, Otani M, Okada S, Lupini AR, Idrobo JC, Caudel D, Burger A, Ghimire NJ, Yan J, Mandrus DG, Pennycook SJ, Pantelides ST. (2014) "Flexible Metallic Nanowires with Self-Adaptive Contacts to Semiconducting Transition-Metal Dichalcogenide Monolayers" Nature Nanotechnology, 9, 436–442, doi:10.1038/nnano.2014.81

[Lin 2016] Lin CA, Tsai ML, Wei WR, Lai KY, He JH. (2016) "Packaging Glass with a Hierarchically Nanostructured Surface: A Universal Method to Achieve Self-Cleaning Omnidirectional Solar Cells" ACS Nano, 10, 1, 549–555, doi:10.1021/acsnano.5b05564

[Ling 2011] Ling J, Rong MZ, Zhang MQ. (2011) "Coumarin Imparts Repeated Photochemical Remendability to Polyurethane" Journal of Materials Chemistry, 21, 18373–18380, doi:10.1039/C1JM13467A

[Little 2001] Little DN, Lytton RL, Williams AD, Chen CW. (2001). "Microdamage Healing in Asphalt and Asphalt Concrete, Volume I: Microdamage and Microdamage Healing, Project Summary Report" FHWA-RD-98-144, Texas Transportation Institution, College Station, TX.

[Little 2006] Little BJ, Lee JS, Ray RI. (2006) "Diagnosing Microbiologically Influenced Corrosion: A State-of-the-art Review" Corrosion 62, 11, 1006–1017

[Liu 1987] Liu X, Lee JK, Kessler MR. (2011) "Microencapsulation of Self-Healing Agents with Melamine-Urea-Formaldehyde by the Shirasu Porous Glass (SPG) Emulsification Technique" Macromolecular Research, 19, 10, 1056–1061, doi:10.1007/s13233-011-1009-3

[Liu 1992] Liu CH, Nagel SR. (1992) "Sound in Sand" Physical Review Letters, 68, 2301–2305, doi:/10.1103/PhysRevLett.68.2301

[Liu 1996] Liu KK, Williams DR, Briscoe BJ (1996) "Compressive Deformation of a Single Microcapsule" Physical Review D, 54, 6, 6673–6680, doi:10.1103/PhysRevE.54.6673

[Liu 1998] Liu AJ, Nagel SR. (1998) "Jamming Is Not Just Cool Any More" Nature, 396, 21–22, doi:10.1038/23819

[Liu 2001] Liu WD, Neuenhoffer A, Ghosn M, Moses F. (2001) "Redundancy in Highway Bridge Substructures" NCHRP Report 458, National Academy Press, Washington, DC, 2001

[Liu 2004] Liu H, Feng L, Zhai J, Jiang L, Zhu D. (2004) "Reversible Wettability of a Chemical Vapor Deposition Prepared ZnO Film between Superhydrophobicity and Superhydrophilicity" Langmuir 20, 14, 5659–5661, doi:10.1021/la036280o

[Liu 2006a] Liu W, Jawerth LM, Sparks EA, Falvo MR, Hantgan RR, Superfine R, Lord ST and Guthold M. (2006) "Fibrin Fibers Have Extraordinary Extensibility and Elasticity" Science, 313, 634

[Liu 2006b] Liu YL, Hsieh CY. (2006) "Crosslinked Epoxy Materials Exhibiting Thermal Remendablility and Removability from Multifunctional Maleimide and Furan Compounds" Journal of Polymer Science. Part A: Polymer Chemistry, 44, 2, 905–913, doi:10.1002/pola.21184

[Liu 2007a] Liu TJ. (2007) "Fracture Mechanics and Crack Contact Analyses of the Active Repair of Multi-Layered Piezoelectric Patches Bonded on Cracked Structures" Theoretical and Applied Fracture Mechanics, 47, 2, 120–132, doi:10.1016/j.tafmec.2006.11.004

[Liu 2007b] Liu YL, Chen YW. (2007) "Thermally Reversible Cross-Linked Polyamides with High Toughness and Self-Repairing Ability from Maleimide and Furan-Functionalized Aromatic Polyamides" Macromolecular Chemistry and Physics, 208, 224–232

[Liu 2008a] Liu CK, Yang TJ, Shen JS, Lee S. (2008) "Some Recent Results on Crack Healing of Poly(methyl methacrylate)" Engineering Fracture Mechanics, 75, 4876–4885, doi:10.1016/j.engfracmech.2008.06.014

[Liu 2008b] Liu HA, Gnade BE, Balkus Jr. KJ. (2008) "A Delivery System for Self-Healing Inorganic Films" Advanced Functional Materials, 18, 3620–3629, doi:10.1002/adfm.200701470

[Liu 2008c] Liu WN, Sun X, Khaleel MA. (2008) "Predicting Young's Modulus of Glass/Ceramic Sealant for Solid Oxide Fuel Cell Considering the Combined Effects of Aging, Micro-Voids and Self-Healing" Journal of Power Sources, 185, 2, 1193–1200, doi:10.1016/j.jpowsour.2008.07.017

[Liu 2011a] Liu N, Sun G, Zhu J. (2011) "Photo-Induced Self-Cleaning Functions on 2-Anthraquinone Carboxylic Acid Treated Cotton Fabrics" Journal of Materials Chemistry, 21, 15383, doi:10.1039/c1jm12805a

[Liu 2011b] Liu Q, García A, Schlangen E, van de Ven M. (2011) "Induction Healing of Asphalt Mastic and Porous Asphalt Concrete" Construction and Building Materials, 25, 9, 3746–3752, doi:10.1016/j.conbuildmat.2011.04.016

[Liu 2011c] Liu YY, Slotine JJ, Barabási AL (2011) "Controllability of Complex Networks" Nature, 473, 167–173, doi:10.1038/nature10011

[Liu 2012a] Liu K, Cheng C, Cheng Z, Wang K, Ramesh R, Wu J. (2012) "Giant-Amplitude, High-Work Density Microactuators with Phase Transition Activated Nanolayer Bimorphs" Nano Letters, 12, 6302–6308, doi:10.1021/nl303405g

[Liu 2012b] Liu Q, Schlangen E., van de Ven M., van Bochove G., van Montfort J. (2012). "Evaluation of the Induction Healing Effect of Porous Asphalt Concrete through Four Point Bending Fatigue Test." Construction and Building Materials, 29, 403–409, doi:10.1016/j.conbuildmat.2011.10.058a

[Liu 2012c] Liu F, Li F, Deng G, Chen Y, Zhang B, Zhang J, Liu CY. (2012) "Rheological Images of Dynamic Covalent Polymer Networks and Mechanisms behind Mechanical and Self-Healing Properties" Macromolecules, 45, 3, 1636–1645, doi:10.1021/ma202461eb

[Liu 2013] Liu K, Cheng C, Suh J, Tang-Kong R, Fu D, Lee S, Zhou J, Chua LO, Wu J. (2013) "Powerful, Multifunctional Torsional Micromuscles Activated by Phase Transition" Advanced Materials, 26, 11, 1746–1750, doi:10.1002/adma.201304064

[Liu 2014] Liu TL, Kim CJ. (2014) "Turning a Surface Superrepellent Even to Completely Wetting Liquids" Science 28, 346, 6213, 1096–1100, doi:10.1126/science.1254787

[Liu 2024] Liu F, Terakawa T, Long S, Komori M. (2024) "Rigid-Foldable Cylindrical Origami with Tunable Mechanical Behaviors" Scientific Reports, 14, 145, doi:10.1038/s41598-023-50353-4

[Loiseau 2016] Loiseau L, de Boiry AQ, Niedermair F, Albrecht G, Rühs PA, Studart AR. (2016) "Explosive Raspberries: Controlled Magnetically Triggered Bursting of Microcapsules" Advanced Functional Materials, 26, 22, 4007–4015, doi: adfm.201504656

[Lokuge 2007] Lokuge I, Wang X, Bohn PW, Seitz F. (2007) "Temperature-Controlled Flow Switching in Nanocapillary Array Membranes Mediated by Poly(*N*-isopropylacrylamide) Polymer Brushes Grafted by Atom Transfer Radical Polymerization" Langmuir 23, 305–311

[Long 2015] Long Y, Liu C, Zhao B, Song K, Yang G, Tung CH. (2015) "Bio-Inspired Controlled Release through Compression–Relaxation Cycles of Microcapsules" NPG Asia Materials 7, e148, doi:10.1038/am.2014.114

[Loos 1994] Loos AC, Li MC. (1994) "Non-Isothermal Autohesion Model for Amorphous Composites" Journal of Thermoplastic Composite Materials, 7, 280–310, doi:10.1177/089270579400700401

[López 2016] López AB, de la Cal JC, Asua JM. (2016) "Highly Hydrophobic Coatings from Waterborne Latexes" Langmuir, 32, 30, 7459–7466, doi:10.1021/acs.langmuir.6b01072

[Lorente 2009] Lorente S, Bejan A. (2009) "Vascularized Smart Materials: Designed Porous Media for Self-Healing and Self Cooling" Journal of Porous Media, 12, 1, 1–18

[Loret 2010] Loret B, Simões FM. (2010) "Elastic-Growing Tissues: A Growth Rate Law that Satisfies the Dissipation Inequality" Mechanics of Materials, 42, 8, 782–796, doi:10.1016/j.mechmat.2010.06.001

[Lorret 2009] Lorret O, Francová D, Waldner G, Stelzer N. (2009) "W-Doped Titania Nanoparticles for UV and Visible-Light Photocatalytic Reactions" Applied Catalysis B: Environmental, 91, 1–2, 7, 39–46, doi:10.1016/j.apcatb.2009.05.005

[Lovley 1997] Lovley DR, Coates JD. (1997) "Bioremediation of Metal Contamination" Current Opinion in Biotechnology, 8, 3, 285–289, doi:10.1016/S0958-1669(97)80005-5

[Low 2012] Low SH, Tan AW, Shiau LL, Lau GK. (2012) "Actuated Strains in Excess of 100% in Dielectric Elastomer Actuators Using Silver Film Electrodes" Proceedings of SPIE, Vol. 8340, Electroactive Polymer Actuators and Devices (EAPAD), Y Bar-Cohen (ed.), doi:10.1117/12.915410

[Loy 2002] Loy DA, Wheeler DR, Russick EM, McElhanon JR, Sanders RS. (2002) "Method of Making Thermally Removable Epoxies" US Patent, 6,337,384

[Lu 2013] Lu PK, Mahapatra MK. (2013) "Barium Oxide, Calcium Oxide, Magnesia, and Alkali Oxide Free Glass" US Patent 8,541,327

[Lu 2015] Lu Y, Sathasivam S, Song J, Crick CR, Carmalt CJ, Parkin IP. (2015) "Robust Self-Cleaning Surfaces that Function When Exposed to Either Air or Oil" Science, 347, 6226, 1132–1135, doi:10.1126/science.aaa0946

[Lubelli 2011] Lubelli B, Nijland TG, van Hees RP. (2011) "Self-Healing of Lime Based Mortars: Microscopy Observations on Case Studies" HERON Vol. 56 No. 1/2

[Lucas 2001] Lucas PJ, Taylor AJ. (2001) "Stable, Foamed Caulk and Sealant Compounds and Methods of Use Thereof" US Patent 6,284,077

[Luding 2008] Luding S, Suiker A. (2008) "Self-Healing of Damaged Particulate Materials through Sintering" Philosophical Magazine, 88, 28–29, 3445–3457

[Lue 2012] Lue HT, Du PY, Chen CP, Chen WC, Hsieh CC, Hsiao YH, Shih YH, Lu CY. (2012) "Radically Extending the Cycling Endurance of Flash Memory (to >100 M Cycles) by Using Built-in Thermal Annealing to Self-Heal the Stress-Induced Damage" 2012 IEEE International Electron Devices Meeting (IEDM), pp 9.1.1–9.1.4, San Francisco, CA, doi:10.1109/IEDM.2012.6479008

[Lueke 2001] Lueke JS, Ariaratnam ST. (2001) "Rehabilitation of Underground Infrastructure Utilizing Trenchless Pipe Replacement" ASCE Practice Periodical on Structural Design and Construction, 6, 1, 25–34.

[Lulevich 2004] Lulevich VV, Andrienko D, Vinogradova OI (2004) "Elasticity of Polyelectrolyte Multilayer Microcapsules" Journal of Chemical Physics, 120, 8, 3822–3826, doi:10.1063/1.1644104

[Lumley 2002] Lumley RN, Morton AJ, Polmear IJ. (2002) "Enhanced Creep Performance in an Al–Cu–Mg–Ag Alloy through Underageing" Acta Materialia, 50, 3597–3608

[Lund 2010] Lund K, Manzo AJ, Dabby N, Michelotti N, Johnson-Buck A, Nangreave J, Taylor S, Pei R, Stojanovic MN, Walter NG, Winfree E, Yan H. (2010) "Molecular Robots Guided by Prescriptive Landscapes" Nature, 465, 206–210, doi:10.1038/nature09012

[Luo 2001] Luo J, Hepel M. (2001) "Photoelectrochemical Degradation of Naphthol Blue Black Diazo Dye on WO_3 Film Electrode" Electrochimica Acta, 46, 2913–2922

[Luo 2013a] Luo T, Mohan K, Iglesias PA, Robinson DN. (2013) "Molecular Mechanisms of Cellular Mechanosensing" Nature Materials, 12, 1064–1071, doi:10.1038/nmat3772

[Luo 2013b] Luo X, Mather PT. (2013) "Shape Memory Assisted Self-Healing Coating" ACS Macro Letters, 2, 2, 152–156, doi:10.1021/mz400017x

[Lutterman 2009] Lutterman DA, Surendranath Y, Nocera DG. (2009) "A Self-Healing Oxygen-Evolving Catalyst" Journal of the American Chemical Society, 131, 11, 3838–3839, doi:10.1021/ja900023k

[Lv 2014] Lv Z, Chen H, Yuan H. (2014) "Analytical Solution on Dosage of Self-Healing Agents in Cementitious Materials: Long Capsule Model" Journal of Intelligent Materials Systems and Structures, 25, 1, 47–57, doi:10.1177/1045389X12457250

[Lvov 2013] Lvov Y, Abdullayev E. (2013) "Functional Polymereclay Nanotube Composites with Sustained Release of Chemical Agents" Progress in Polymer Science, 38, 10–11, 1690–1719, doi:10.1016/j.progpolymsci.2013.05.009

[Lyon 2009] Lyon LA, Meng Z, Singh N, Sorrell CD, St. John A. (2009) "Thermoresponsive Microgel-Based Materials" Chemical Society Reviews, 38, 865–874, doi:10.1039/B715522K

[Lytton 2000] Lytton RL. (2000) "Characterizing Asphalt Pavement for Performance" Transportation Research Record 1723, Transportation Research Board, Washington DC, pp. 5–16.

[Lytton 2001] Lytton RL, Chen CW, Little DN. (2001) "Microdamage Healing in Asphalt and Asphalt Concrete, Volume III: A Micromechanics Fracture and Healing Model for Asphalt Concrete" FHWA-RD-98-143, US Department of Transportation, Federal Highway Administration, McLean, VA.

[Ma 2005] Ma M, Mao Y, Gupta M, Gleason KK, Rutledge GC. (2005) "Superhydrophobic Fabrics Produced by Electrospinning and Chemical Vapor Deposition" Macromolecules, 38, 23, 9742–9748, doi:10.1021/ma0511189

[Ma 2006] Ma M, Hill RM. (2006) "Superhydrophobic Surfaces" Current Opinion in Colloid & Interface Science, 11, 193–202

[Ma 2009] Ma W, Trusina A, El-Samad H, Lim WA, Tang C. (2009) "Defining Network Topologies that Can Achieve Biochemical Adaptation" Cell, 138, 4, 760–773, doi:10.1016/j.cell.2009.06.013

[Ma 2014] Ma H, Qian S, Zhang Z. (2014) "Effect of Self-Healing on Water Permeability and Mechanical Property of Medium-Early-Strength Engineered Cementitious Composites" Construction and Building Materials, 68, 92–101, doi:10.1016/j.conbuildmat.2014.05.065

[Macdonald 2006] Macdonald DD. (2006) "Reflections on the History of Electrochemical Impedance Spectroscopy" Electrochimica Acta 51, 1376–1388, doi:10.1016/j.electacta.2005.02.107

[Macfarlane 1964] Macfarlane RG. (1964) "An Enzyme Cascade in the Blood Clotting Mechanism, and Its Function as a Biochemical Amplifier" Nature, 202, 498–499, doi:10.1038/202498a0

[Mackel 2007] Mackel MJ, Sanchez S, Kornfield JA. (2007) "Humidity-Dependent Wetting Properties of High Hysteresis Surfaces" Langmuir, 23, 3–7

[Maddocks 1925] Maddocks FT. (1925) "Fractured Concrete Test Specimens Heal and Show Increased Compressive Strength When Retested" California Highways, 2, 8, 11

[Madsen 1998] Madsen, FT. (1998) "Clay Mineralogical Investigations Related to Nuclear Waste Disposal" Clay Minerals, 33, 1, 109–129, doi:10.1180/000985598545318

[Magennis 2014] Magennis EP, Fernandez-Trillo F, Sui C, Spain SG, Bradshaw DJ, Churchley D, Mantovani G, Winzer K, Alexander C. (2014) "Bacteria-Instructed Synthesis of Polymers for Self-Selective Microbial Binding and Labelling" Nature Materials, 13, 748–755, doi:10.1038/nmat3949

[Mahadevan 2005] Mahadevan L, Rica S. (2005) "Self-Organized Origami" Science, 307, 5716, 1740, doi:10.1126/science.1105169

[Maiti 2006] Maiti S, Shankar C, Geubelle PH, Kieffer J. (2006) "Continuum and Molecular-Level Modeling of Fatigue Crack Retardation in Self-Healing Polymers" Journal of Engineering Materials and Technology, 128, 595–602, doi:10.1115/1.2345452

[Maitra 2014] Maitra T, Antonini C, Tiwari MK, Mularczyk A, Imeri Z, Schoch P, Poulikakos D. (2014) "Supercooled Water Drops Impacting Superhydrophobic Textures" Langmuir, 30, 36, 10855–10861, doi:10.1021/la502675a

[Majdandzic 2014] Majdandzic A, Podobnik B, Buldyrev SV, Kenett DY, Havlin S, Stanley HE. (2014) "Spontaneous Recovery in Dynamical Networks" Nature Physics, 10, 34–38, doi:10.1038/nphys2819

[Majmudar 2007] Majmudar TS, Sperl M, Luding S, Behringer RP. (2007) "Jamming Transition in Granular Systems" Physical Review Letters, 98, 058001, doi:10.1103/PhysRevLett.98.058001

[Malvankar 2011] Malvankar NS, Vargas M, Nevin KP, Franks AE, Leang C, Kim BC, lnoue K, Mester T, Covalla SF, Johnson JP. Rotello VM, Tuominen MT, Lovley DR. (2011) "Tunable Metallic-Like Conductivity in Microbial Nanowire Networks" Nature Nanotechnology, 6, 573–579, doi:10.1038/nnano.2011.119

[Malvern 1969] Malvern LE. (1969) Introduction to the Mechanics of a Continuous Medium Prentice-Hall, Englewood Cliffs

[Manfredi 2014] Manfredi E, Cohades A, Richard I, Michaud V. (2014) "Assessment of Solvent Capsule-Based Healing for Woven E-Glass Fibre-Reinforced Polymers" Smart Materials and Structures, 24, 015019, doi:10.1088/0964-1726/24/1/015019

[Mangat 1987] Mangat PS, Gurusamy K. (1987) "Permissible Crack Widths in Steel Fibre Reinforced Rnarine Concrete" Materials and Structures, 20, 5, 338–347

[Mange 2000] Mange D, Sipper M, Stauffer A, Tempesti G. (2000) "Toward Robust Integrated Circuits: The Embryonics Approach" Proceedings of IEEE, 88, 4, 516–543, doi:10.1109/5.842998

[Mann 2009] Mann KG, Orfeo T, Butenas S, Undas A, Brummel-Ziedins K. (2009) "Blood Coagulation Dynamics in Haemostasis" Hämostaseologie, 29, 1, 7–16

[Mann 2010] Mann KG. (2010) "Taking the Thrombin 'Fork'" Arteriosclerosis, Thrombosis, and Vascular Biology, 30, 1293–1299

[Manna 2001] Manna A, Imae T, Aoi K, Okada M, Yogo T. (2001) "Synthesis of Dendrimer-Passivated Noble Metal Nanoparticles in a Polar Medium: Comparison of Size between Silver and Gold Particles" Chemistry of Materials, 13, 1674–1681, doi:10.1021/cm000416b

[Manuel 2008] Manuel MV. (2008) "Principles of Self-Healing in Metals and Alloys: An Introduction" in Self-Healing Materials: Fundamentals, Design Strategies, and Applications (ed S. K. Ghosh), Wiley-VCH Verlag GmbH & Co. KGaA, Weinheim, Germany, doi:10.1002/9783527625376.ch8

[Mao 2013] Mao X, Chen Q, Granick S. (2013) "Entropy Favours Open Colloidal Lattices" Nature Materials, 12, 217–222, doi:10.1038/nmat3496

[Marcus 1971] Marcus HL, Sih GC. (1971) "A Crackline-Loaded Edge-Crack Stress Corrosion Specimen" Engineering Fracture Mechanics, 3, 4, 453–461, doi:10.1016/0013-7944(71)90058-0

[Mariconda 2015] Mariconda A, Longo P, Agovino A, Guadagno L, Sorrentino A, Raimondo M. (2015) "Synthesis of Ruthenium Catalysts Functionalized Graphene Oxide for Self-Healing Applications" Polymer, 69, 330–342, doi:/10.1016/j.polymer.2015.04.048

[Marmur 2004] Marmur A. (2004) "The Lotus Effect: Superhydrophobicity and Metastability" Langmuir, 20, 3517–3519, doi:10.1021/la036369u

[Marras 2014] Marras AE, Zhou L, Su HJ, Castro CE. (2014) "Programmable Motion of DNA Origami Mechanisms" Proceedings of the National Academy of Sciences of the United States of America, 112, 3, 713–718, doi:10.1073/pnas.1408869112

[Martin 1996] Martin T. (1996) "Process for Rehabilitation of Sewer Collection System Structures" US Patent 5,538,755

[Martin 1997] Martin P. (1997) "Wound Healing: Aiming for Perfect Skin Regeneration" Science, 276, 75–81.

[Martin 2010] Martin PM. (2010) "Functional Biomaterials: Self Healing Biological Materials" Vacuum Technology & Coating, 11, 8, 6–11

[Martin 2013a] Martin PM. (2013) "Self-Healing Thin Films I: Metallized and Conductive Materials" Vacuum Technology & Coating, 14, 4, 34–39

[Martin 2013b] Martin PM. (2013) "Self-Healing Thin Films II: Polymers and Polymer Composites" Vacuum Technology & Coating, 14, 5, 39–43

[Martin 2015] Martin DC. (2015) "Molecular Design, Synthesis, and Characterization of Conjugated Polymers for Interfacing Electronic Biomedical Devices with Living Tissue" MRS Communications, 5, 2, 131–153, doi:10.1557/mrc.2015.17

[Maruyama 1989] Maruyama S, Viskantat R, Aiharat T. (1989) "Active Thermal Protection System Against Intense Irradiation" Journal of Thermophysics, 3, 4, 389–394

[Mase 2010] Mase G, Smelser R, Mase G. (2010) Continuum Mechanics for Engineers, 3rd edition, Taylor and Francis

[Masoodi 2010] Masoodi R. (2010) "Modeling Imbibition of Liquids into Rigid and Swelling Porous Media" PhD Dissertation, Engineering, University of Wisconsin-Milwaukee

[Masoud 2011] Masoud H, Alexeev A. (2011) "Harnessing Synthetic Cilia to Regulate Motion of Microparticles" Soft Matter, 7, 8702, doi:10.1039/c1sm05423f

[Masuda 1973] Masuda S. (1973) "Apparatus for Electric Field Curtain of Contact Type" US Patent 3,778,678

[Mather 2014] Mather P, Luo X. (2014) "Stimuli-Responsive Product" US Patent 8,683,798

[Mathew 2011] Mathew S, Wang W, Osterman M, Pecht M. (2011) "Assessment of Solder-Dipping as a Tin Whisker Mitigation Strategy" IEEE Transactions on Components, Packaging and Manufacturing Technology, 1, 6, 957–963

[Matsuda 2019] Matsuda T, Kawakami R, Namba R, et al (2019) "Mechanoresponsive Self-Growing Hydrogels Inspired by Muscle Training" Science, 363, 504–508, doi:10.1126/science.aau9533

[Matsuzaki 2003] Matsuzaki A, Yamaji T, Yamashita M. (2003) "Development of a New Organic Composite Coating for Enhancing Corrosion Resistance of 55% Al-Zn Alloy Coated Steel Sheet" Surface and Coatings Technology, 655–657, doi:10.1016/S0257-8972(03)00110-5.

[Mauldin 2007] Mauldin TC, Rule JD, Sottos NR, White SR, Moore JS. (2007) "Self-Healing Kinetics and the Stereoisomers of Dicyclopentadiene" Journal of the Royal Society: Interface, 4, 389–393, doi:10.1098/rsif.2006.0200

[Maunder 2007] Maunder AL, Bareggi A, Balconi L, Dell'Anna G, Pozzati G, Belli S. (2007) "Electrical Cable with Self-Repairing Protection and Apparatus for Manufacturing the Same" US Patent 7,204,896

[Mauricio 2012] Mauricio MR, Guilherme MR, Kunita MH, Muniz EC, Rubira AF. (2012) "Designing Nanostructured Microspheres with Well-Defined Outlines by Mixing Carboxyl-Functionalized Amylose and Magnetite via Ultrasound" Chemical Engineering Journal, 189, 456–463, doi:10.1016/j.cej.2012.02.044

[Maxey 2012] Maxey C, Creech G, Raman S, Rockway J, Groves K, Quach T, Orlando L, Mattamana A. (2012) "Mixed-Signal SoCs with In Situ Self-Healing Circuitry" IEEE Design & Test of Computers, 29, 6, 27–39, doi:10.1109/MDT.2012.2226014

[Mazumder 2005] Mazumder MK, Sims RA, Wilson JD, Biris AS. (2005) "Transparent Self-Cleaning Dust Shield" US Patent 6,911,593

[Mazumder 2007] Mazumder MK, Sharma R, Biris AS, Zhang J, Calle C, Zahn M (2007): Self-Cleaning Transparent Dust Shields for Protecting Solar Panels and Other Devices, Particulate Science and Technology, 25, 1, 5–20, doi:10.1080/02726350601146341

[McAleenan 2013] McAleenan A, Clemente-Blanco A, Cordon-Preciado V, Sen N, Esteras M, Jarmuz A, Aragón L. (2013) "Post-Replicative Repair Involves Separase-Dependent Removal of the Kleisin Subunit of Cohesin" Nature, 493, 250–254, doi:10.1038/nature11630

[McElhanon 2002] McElhanon JR, Russick EM, Wheeler DR, Loy DA, Aubert JH. (2002) "Removable Foams Based on an Epoxy Resin Incorporating Reversible Diels–Alder Adducts" Journal of Applied Polymer Science, 85, 1496–1502, doi 10.1002/app.10753

[McGarel 1987] McGarel JO, Wool RP. (1987) "Craze Growth and Healing in Polystyrene" Journal of Polymer Science B, 25, 12, 2541–2560

[McLaren 2014] McLaren M, Mhlanga T, Padgett MJ, Roux FS, Forbes A. (2014) "Self-Healing of Quantum Entanglement after an Obstruction" Nature Communications, 5, 3248, doi:10.1038/ncomms4248

[McLaskey 2012] McLaskey GC, Thomas AM, Glaser SD, Nadeau RM. (2012) "Fault Healing Promotes High-Frequency Earthquakes in Laboratory Experiments and on Natural Faults" Nature, 491, 101–104, doi:10.1038/nature11512

[McLaughlin 1948] McLaughlin L. (1948) "Self-Sealing Container" US Patent 2,439,366

[McLeod 1954] McLeod JH. (1954) "The Axicon: A New Type of Optical Element" JOSA, 44, 8, 592–597, doi:10.1364/JOSA.44.000592

[Medelius 2002] Medelius PJ Eckhoff AJ Angel LR Perotti JM. (2002) "Advanced Self-Calibrating, Self-Repairing Data Acquisition System" US Patent 6,462,684

[Medelius 2013] Medelius PJ, Gibson TL, Lewis ME. (2013) "Method of Fault Detection and Rerouting" US Patent 8,593,153

[Mehta 2008] Mehta V, Frecker M, Lesieutre G. (2008) "Contact-Aided Compliant Mechanisms for Morphing Aircraft Skin" Proceedings of SPIE, Vol. 6926, Modeling, Signal Processing, and Control for Smart Structures, D.K. Lindner (ed.), doi:10.1117/12.773599

[Mehta 2009] Mehta NK, Bogere MN. (2009) "Environmental Studies of Smart/Self-Healing Coating System for Steel" Progress in Organic Coatings, 64, 419–428

[Meier 1962] Meier DJ, Kahn A. (1962) "New Curing Agents for Polyepoxides and Method for Their Preparation" US Patent 3,018,258

[Meier 2000] Meier W. (2000) "Polymer Nanocapsules" Chemical Society Reviews, 29, 295–303, doi:10.1039/a809106d

[Meiller 1999] Meiller M, Roche AA, Sautereau H. (1999) "Tapered Double Cantilever Beam Test Used as a Practical Adhesion Test for Metal/Adhesive/Metal Systems" Journal of Adhesion Science and Technology, 13, 7, 773–788

[Melzack 1965] Melzack R, Wall PD. (1965) "Pain Mechanisms: A New Theory" Science, 150, 3699, 971–979, doi:10.1126/science.150.3699.971

[Menck 2013] Menck PJ, Heitzig J, Marwan N, Kurths J. (2013) "How Basin Stability Complements the Linear-Stability Paradigm" Nature Physics, 9, 89–92, doi:10.1038/NPHYS2516

[Meng 2010] Meng H, Hu J. (2010) "A Brief Review of Stimulus-Active Polymers Responsive to Thermal, Light, Magnetic, Electric, and Water/Solvent Stimuli" Journal of Intelligent Materials Systems and Structures, 21, 9, 859–885, doi:10.1177/1045389X10369718

[Meng 2010] Meng LM, Yuan YC, Rong MZ, Zhang MQ. (2010) "A Dual Mechanism Single-Component Self-Healing Strategy for Polymers" Journal of the Materials Chemistry, 20, 29, 6030–6038, doi:10.1039/c0jm00268b

[Mercier 1896] Mercier P. (1896) "Material for Protecting Vessels, Recepticles, &c." US Patent 561,905

[Merkel 2011] Merkel TJ, Jones SW, Herlihy KP, Kersey FR, Shields AR, Napier M, Luft JC, Wug H, Zamboni WC, Wang AZ, Bear JE, DeSimone JM. (2011) "Using Mechanobiological Mimicry of Red Blood Cells to Extend Circulation Times of Hydrogel Microparticles" Proceedings of the National Academy of Sciences of the United States of America, 108, 2, 586–591, doi:10.1073/pnas.1010013108

[Merle 2014] Merle P, Guntzburger Y, Haddad E, Hoa SV, Thatte G. (2014) "Self Healing Composite Material and Method of Manufacturing Same" US Patent 8,865,798

[Meure 2009] Meure S, Wu DY, Furman S. (2009) "Polyethylene-*co*-Methacrylic Acid Healing Agents for Mendable Epoxy Resins" Acta Materialia, 57, 4312–4320, doi:10.1016/j.actamat.2009.05.032

[Meyer 2014] Meyer SN, Amoyel M, Bergantiños C, de la Cova C, Schertel C, Basler K, Johnston LA. (2014) "An Ancient Defense System Eliminates Unfit Cells from Developing Tissues during Cell Competition" Science, 346, 6214, 1258236, doi:10.1126/science.1258236

[Meyers 2013] Meyers MA, McKittrick J, Chen PY. (2013) "Structural Biological Materials: Critical Mechanics–Materials Connections" Science, 339, 6121, 773–779, doi:10.1126/science.1220854

[Miao 1995] Miao S, Wang M, Schreyer H. (1995) "Constitutive Models for Healing of Materials with Application to Compaction of Crushed Rock Salt" Journal of Engineering Mechanics, 121, 10, 1122–1129, doi:10.1061/(ASCE)0733-9399(1995)121:10(1122)

[Miao 2015] Miao T, Fenn SL, Charron PN, Oldinski RA. (2015) "Self-Healing and Thermoresponsive Dual-Cross-Linked Alginate Hydrogels Based on Supramolecular Inclusion Complexes" Biomacromolecules, 16, 12, 3740–3750, doi:10.1021/acs.biomac.5b00940

[Miccichè 2008] Miccichè F, Fischer H, Varley R, van der Zwaag S. (2008) "Moisture Induced Crack Filling in Barrier Coatings Containing Montmorillonite as an Expandable Phase" Surface and Coatings Technology, 202, 3346

[Mihashi 2012] Mihashi H, Nishiwaki T. (2012) "Development of Engineered Self-Healing and Self-Repairing Concrete-State-of-the-Art Report" Journal of Advanced Concrete Technology, 10, 5, 170–184, doi:10.3151/jact.10.170

[Miljkovic 2013] Miljkovic N, Preston DJ, Enrigh R, Wang EN. (2013) "Electric-Field-Enhanced Condensation on Superhydrophobic Nanostructured Surfaces" ACS Nano, 7, 12, 11043–11054, doi:10.1021/nn404707j

[Miller 2012] Miller JS, Stevens KR, Yang MT, Baker BM, Nguyen DH, Cohen DM, Toro E, Chen AA, Galie PA, Yu X, Chaturvedi R, Bhatia SN, Chen CS. (2012) "Rapid Casting of Patterned Vascular Networks for Perfusable Engineered Three-Dimensional Tissues" Nature Materials, 11, 768–774, doi:10.1038/NMAT3357

[Mimitou 2008] Mimitou EP, Symington LS. (2008) "Sae2, Exol and Sgs1 Collaborate in DNA Double-Strand Break Processing" Nature, 455, 770–774, doi:10.1038/nature07312

[Minakuchi 2014] Minakuchi S, Sun D, Takeda N. (2014) "Hierarchical System for Autonomous Sensing-Healing of Delamination in Large-Scale Composite Structures" Smart Materials and Structures, 23, 115014, doi:10.1088/0964-1726/23/11/115014

[Minko 2003] Minko S, Müller M, Motornov M, Nitschke M, Grundke K, Stamm M. (2003) "Two-Level Structured Self-Adaptive Surfaces with Reversibly Tunable Properties" Journal of the American Chemical Society, 125, 13, 3896–3900, doi:10.1021/ja0279693

[Mirkin 1996] Mirkin C, Letsinger R, Storhoff J. (1996) "A DNA-Based Method for Rationally Assembling Nanoparticles into Macroscopic Materials" Nature, 382, 607–609, doi:10.1038/382607a0

[Mishchenko 2010] Mishchenko L, Hatton B, Bahadur V, Taylor JA, Krupenkin T, Aizenberg J. (2010) "Design of Ice-free Nanostructured Surfaces Based on Repulsion of Impacting Water Droplets" ACS Nano, 4, 12, 7699–7707, doi:10.1021/nn102557p

[Mishra 2014] Mishra D, Rivera PM, Lin A, Del Vecchio D, Weiss R. (2014) "A Load Driver Device for Engineering Modularity in Biological Networks" Nature Biotechnology, 32, 1268–1275, doi:10.1038/nbt.3044

[Miskin 2013] Miskin MZ, Jaeger HM. (2013) "Adapting Granular Materials through Artificial Evolution" Nature Materials, 12, 326–331, doi:10.1038/nmat3543

[Miszta 2011] Miszta K, de Graaf J, Bertoni G, Dorfs D, Brescia R, Marras S, Ceseracciu L, Cingolani R, van Roij R, Dijkstra M, Manna L. (2011) "Hierarchical Self-Assembly of Suspended Branched Colloidal Nanocrystals into Superlattice Structures" Nature Materials, 10, 872–876, doi:10.1038/nmat3121

[Miu 1997] Miu KN, Chiang HD, Yuan B, Darling G. (1997) "Fast Service Restoration for Large-Scale Distribution Systems with Priority Customers and Constraints" IEEE, Proceedings of 20th International Conference on Power Industry Computer Applications, Vol. 13, pp 789–795, doi:10.1109/PICA.1997.599366

[Miyashita 2015] Miyashita S, Guitron S, Ludersdorfer M, Sung C, Rus D. (2015) "An Untethered Miniature Origami Robot that Self-Folds, Walks, Swims, and Degrades" 2015 IEEE International Conference on Robotics and Automation (ICRA), 1490–1496, doi:10.1109/ICRA.2015.7139386

[Miyazaki 2010] Miyazaki E, Tagawa M, Yokota K, Yokota R, Kimoto Y, Ishizawa J. (2010) "Investigation into Tolerance of Polysiloxane-Block-Polyimide Film against Atomic Oxygen" Acta Astronautica, 66, 5–6, 922–928, doi:10.1016/j.actaastro.2009.09.002

[Mladenov 2008] Mladenov M, Mock M, Grosspietsch KE. (2008) "Fault Monitoring and Correction in a Walking Robot Using LMS Filters" IEEE 2008 International Workshop on Intelligent Solutions in Embedded Systems, Regensburg, pp 1–10

[Mohammadi 1993] Mohammadi N, Klein A, Sperling LH. (1993) "Polymer Chain Rupture and the Fracture Behavior of Glassy Polystyrene" Macromolecules, 26, 5, 1019–1026, doi:10.1021/ma00057a022

[Molchanov 2007] Molchanov VS, Philippova OE, Khokhlov AR. (2007) "Self-Assembled Networks Highly Responsive to Hydrocarbons" Langmuir, 23, 105–111

[Moll 2010] Moll JL, White SR, Sottos NR. (2010) "A Self-Sealing Fiber-Reinforced Composite" Journal of Composite Materials, 44, 2573–2585. doi:10.177/0021998306356605

[Mondal 2022] Mondal P, Cohen SM (2022) "Self-Healing Mixed Matrix Membranes Containing Metal–Organic Frameworks" Chemical Science, 13, 12127–12135, doi:10.1039/D2SC04345A

[Monier 2015] Monier B, Gettings M, Gay G, Mangeat T, Schott S, Guarner A, Magali S. (2015) "Apico-Basal Forces Exerted by Apoptotic Cells Drive Epithelium Folding" Nature, 518, 245–248, doi:10.1038/nature14152

[Monk 2007] Monk RA, Ohnstad TS, Henry JJ. (2007) "Projectile Barrier Method for Sealing Liquid Container" US Patent 7,229,673

[Montarnal 2009] Montarnal D, Tournilhac F, Hidalgo M, Couturier JL, Leibler L. (2009) "Versatile One-Pot Synthesis of Supramolecular Plastics and Self-Healing Rubbers" Journal of the American Chemical Society, 131, 7966–7967

[Montemor 2012] Montemor MF, Snihirova DV, Taryba MG, Lamaka SV, Kartsonakis IA, Balaskas AC, Kordas GC, Tedim J, Kuznetsova A, Zheludkevich ML, Ferreira MG. (2012) "Evaluation of Self-Healing Ability in Protective Coatings Modified with Combinations of Layered Double Hydroxides and Cerium Molibdate Nanocontainers Filled with Corrosion Inhibitors" Electrochimica Acta, 60, 31–40, doi:10.1016/j.electacta.2011.10.078

[Montero 2015] Montero de Espinosa L, Fiore GL, Weder C, Foster EJ, Simon YC. (2015) "Healable Supramolecular Polymer Solids" Progress in Polymer Science, 49–50, 60–78, doi:10.1016/j.progpolymsci.2015.04.003

[Mookhoek 2010] Mookhoek SD, Mayo SC, Hughes AE, Furman SA, Fischer HR, van der Zwaag S. (2010) "Applying SEM-Based X-Ray Microtomography to Observe Self-Healing in Solvent Encapsulated Thermoplastic Materials" Advanced Energy Materials, 12, 3, 228–234, doi:10.1002/adem.200900289

[Moon 2015] Moon H, Lu TC. (2015) "Network Catastrophe: Self-Organized Patterns Reveal Both the Instability and the Structure of Complex Networks" Scientific Reports, 5, 9450, doi:10.1038/srep09450

[Moore 2009] Moore JS, Rule JD, White SR, Sottos NR, Brown EN. (2009) "Wax Particles for Protection of Activators, and Multifunctional Autonomically Healing Composite Materials" US Patent 7,566,747

[Morariu 2009] Morariu S, Lămătic IE, Bercea M. (2009) "Rheological Behavior of Smectite Aqueous Dispersions" Revue Roumaine de Chimie, 54, 11–12, 975–980

[Morris 2007] Morris CJ. (2007) "Microscale Self-Assembled Electrical Contacts" Report Number ARL-TR-4298, US Army Research Laboratory, Adelphi, MD

[Morris 2013] Morris KL, Chen L, Raeburn J, Sellick OR, Cotanda P, Paul A, Griffiths PC, King SM, O'Reilly RK, Serpell LC, Adams DJ. (2013) "Chemically Programmed Self-Sorting of Gelator Networks" Nature Communications, 4, 1480, doi:10.1038/ncomms2499

[Mostovoy 1976] Mostovoy S, Mai YW. (1976) "Contoured Double Cantilever Beam Specimens for Fracture Toughness Measurement of Adhesive Joints" Journal of Materials Science, 11, 11, 2154–2157

[Motoku 1999] Motoku M, Vaidya UK, Janowski GM. (1999) "Parametric Studies on Self-Repairing Approaches for Resin Infused Composites Subjected to Low Velocity Impact" Smart Materials and Structures, 8, 623–638

[Motornov 2007] Motornov M, Sheparovych R, Tokarev I, Roiter Y, Minko S. (2007) "Nonwettable Thin Films from Hybrid Polymer Brushes Can Be Hydrophilic" Langmuir 23, 13–19

[Motuku 1999] Motuku M, Vaidya UK, Janowski GM. (1999) "Parametric Studies on Self-Repairing Approaches for Resin Infused Composites Subjected to Low Velocity Impact" Smart Materials and Structures, 8, 623–638

[Mougey 1919] Mougey HC, Jacobs JM. (1919) "Fuel Tank" US Patent 1,298,080

[Mowshowitz 2012] Mowshowitz A, Dehmer M. (2012) "Entropy and the Complexity of Graphs Revisited" Entropy 14, 3, 559–570, doi:10.3390/e14030559

[Moyano 2014] Moyano DF, Saha K, Prakash G, Yan B, Kong H, Yazdani M, Rotello VM. (2014) "Fabrication of Corona-Free Nanoparticles with Tunable Hydrophobicity" ACS Nano, 8, 7, 6748–6755, doi:10.1021/nn5006478

[Muljadi 1998] Muljadi E, Forsyth T, Butterfield CP. (1998) "Soft-Stall Control versus Furling Control for Small Wind Turbine Power Regulation" Presented at Windpower '98 Bakersfield, CA, https://www.nrel.gov/docs/legosti/old/25100.pdf

[Müller 2015] Müller U, Gindl-Altmutter W, Konnerth J, Maier GA, Keckes J. (2015) "Synergy of Multi-Scale Toughening and Protective Mechanisms at Hierarchical Branch-Stem Interfaces" Scientific Reports, 5, 14522, doi:10.1038/srep14522

[Murata 1989] Murata T. (1989) "Petri Nets: Properties, Analysis and Applications" Proceedings of IEEE, 77, 4, 541–580

[Murata 2001] Murata S, Yoshida E, Kurokawa H, Tomita K, Kokaji S. (2001) "Self-Repairing Mechanical Systems" Autonomous Robots, 10, 1, 7–21, doi:10.1023/A:1026540318188

[Murdock 1919] Murdock GJ. (1919) "Self Sealing Fuel Tank" US Patent 1,312,745

[Murdock 1921] Murdock GJ. (1921) "Self Puncture Sealing Covering for Fuel Containers" US Patent 1,386,791

[Murphy 2008] Murphy EB, Bolanos E, Shaffner-Hamann C, Wudl F, Nutt SR, Auad ML. (2008) Synthesis and Characterization of a Single-Component Thermally Remendable Polymer Network: Staudinger and Stille Revisited" Macromolecules, 41, 5203–5209, doi:10.1021/ma800432g

[Murphy 2009] Murphy EB, Wudl F. (2009) "The World of Smart Healable Materials" Progress in Polymer Science, 35, 1–2, 223–251, doi:10.1016/j.progpolymsci.2009.10.006

[Murray 1926] Murray CD. (1926) "The Physiological Principle of Minimum Work: I. The Vascular System and the Cost of Blood Volume" Proceedings of the National Academy of Sciences of the United States of America, 12, 207–214

[Muscat 2011] Muscat RA, Bath J, Turberfield AJ. (2011) "A Programmable Molecular Robot" Nano Letters, 11, 982–987, doi:10.1021/nl1037165

[Musil 1987] Musil I, Holzer G, Raedisch H. (1987) "Coating Layer with Self-Repairing Properties for Panes Exposed to Atmospheric Agents" US Patent 4,657,796

[Muzenski 2015] Muzenski S, Flores-Vivian I, Sobolev K. (2015) "Durability of Superhydrophobic Engineered Cementitious Composites" Construction and Building Materials, 81, 291–297, doi:10.1016/j.conbuildmat.2015.02.014

[Na 2014] Na JH, Evans AA, Bae J, Chiappelli MC, Santangelo CD, Lang RJ, Hull TC, Hayward RC. (2014) "Programming Reversibly Self-Folding Origami with Micropatterned Photo-Crosslinkable Polymer Trilayers" Advanced Materials, 27, 1, 79–85, doi:10.1002/adma.201403510

[NAE 2010] National Academy of Engineering. (2010) "Interim Report on Causes of the Deepwater Horizon Oil Rig Blowout and Ways to Prevent Such Events" Committee for the Analysis of Causes of the Deepwater Horizon Explosion, Fire, and Oil Spill to Identify Measures to Prevent Similar Accidents in the Future, Washington, DC, ISBN: 0-309-16382-X

[Nagaya 2006] Nagaya K, Ikai S, Chiba M, Chao X. (2006) "Tire with Self-Repairing Mechanism" JSME International Journal, Series C, 49, 379–384, doi:10.1299/jsmec.49.379

[Nagel 1974] Nagel T. (1974) "What Is It Like to Be a Bat?" The Philosophical Review, 83, 435–450

[Nagler 2011] Nagler J, Levina A, Timme M. (2011) "Impact of Single Links in Competitive Percolation" Nature Physics, 7, 265–270

[Nair 2008] Nair KP, Breedveld V, Weck M. (2008) "Complementary Hydrogen-Bonded Thermoreversible Polymer Networks with Tunable Properties" Macromolecules, 41, 3429–3438.

[Nair 2014] Nair V, Dave B. (2014) "Anti-Reflective and Anti-Soiling Coatings with Self-Cleaning Properties" US Patent 8,864,897

[Nakahata 2011] Nakahata M, Takashima Y, Yamaguchi H, Harada A. (2011) "Redox-Responsive Self-Healing Materials Formed from Host–Guest Polymers" Nature Communications 2, 511, doi:10.1038/ncomms1521

[Nakanishi 2011] Nakanishi H, Walker DA, Bishop KJ, Wesson PJ, Yan Y, Soh S, Swaminathan S, Grzybowski BA. (2011) "Dynamic Internal Gradients Control and Direct Electric Currents within Nanostructured Materials" Nature Nanotechnology, 6, 740–746, doi:10.1038/nnano.2011.165

[Nalla 2005] Nalla RK, Kruzic JJ, Kinney JH, Ritchie RO. (2005) "Mechanistic Aspects of Fracture and R-Curve Behavior in Human Cortical Bones" Biomaterials, 26, 217–231

[Napp 2006] Napp N, Burden S, Klavins E (2006) "The Statistical Dynamics of Programmed Robotic Self-Assembly. Proceedings of the IEEE International Conference on Robotics and Automation, Orlando, FL, USA, pp 1469–1476, doi:10.1109/ROBOT.2006.1641916

[Narayanan 2013] Narayanan B, Weeks SL, Jariwala BN, Macco B, Weber JW, Rathi SJ, van de Sanden MC, Sutter P, Agarwal S, Ciobanu CV. (2013) "Carbon Monoxide-Induced Reduction and Healing of Graphene Oxide" Journal of Vacuum Science & Technology A, 31, 040601, doi:10.1116/1.4803839

[Narkis 1982] Narkis M, Bell JP. (1982) "An Unusual Visual Microcracking/Healing Phenomenon in Polycarbonate at Room Temperature" Journal of Applied Polymer Science, 27, 8, 2809–2814, doi:10.1002/app.1982.070270807

[NASA 2007] NASA. (2007) "Two Devices for Removing Sludge from Bioreactor Wastewater" NASA Tech Briefs Lyndon B. Johnson Space Center, Houston, TX

[NASA 2012] NASA. (2012) "Self-Cleaning Particulate Prefilter Media" NASA Tech Briefs, September, pp 20–21, https://ntrs.nasa.gov/citations/20120014115

[Nath 2014] Nath RK, Zain MF, Kadhum AA, Kisha AB. (2014) "An Investigation of $LiNbO_3$ Photocatalyst Coating on Concrete Surface for Improving Indoor Air Quality" Construction and Building Materials, 54, 348–353, doi:10.1016/j.conbuildmat.2013.12.072

[Nayak 2013] Nayak BK, Caffrey PO, Speck CR, Gupta MC. (2013) "Superhydrophobic Surfaces by Replication of Micro/Nano-Structures Fabricated by Ultrafast-Laser-Microtexturing" Applied Surface Science, 266, 27–32, doi:10.1016/j.apsusc.2012.11.052

[Ndlec 1997] Ndlec FJ, Surrey T, Maggs AC, Leibler S. (1997) "Self-Organization of Microtubules and Motors" Nature 389, 305–308, doi:10.1038/38532

[Neinhuis 1997] Neinhuis C, Barthlott C. (1997) "Characterization and Distribution of Water-Repellent, Self-Cleaning Plant Surfaces" Annals of Botany, 79, 667–677

[Nejad 2015] Nejad HB, Robertson JM, Mather PT. (2015) "Interwoven Polymer Composites via Dual-Electrospinning with Shape Memory and Self-Healing Properties" MRS Communications, 5, 2, 211–221, doi:10.1557/mrc.2015.39

[Nemat 2004] Nemat-Nasser S, Vakilamirkhizi A, Plaisted TA, Santos C, Starr AF, Nemat-Nassar SC. (2004) "Composites with Negative Refractive Index, Thermal, Self-Healing and Self-Sensing Functionality" Advanced Smart Materials and Smart Structures Technology, FK Chang, CB Yun, BF Spencer, Jr. (eds), Destech, Lancaster, 2004

[Nepal 2008] Nepal D, Balasubramanian S, Simonian AL, Davis VA. (2008) "Strong Antimicrobial Coatings: Single-Walled Carbon Nanotubes Armored with Biopolymers" Nano Letters, 8, 7, 1896–1901, doi:10.1021/nl080522t

[Nepusz 2012] Nepusz T, Vicsek T. (2012) "Controlling Edge Dynamics in Complex Networks" Nature Physics, 8, 568–573, doi:10.1038/NPHYS2327

[Neumann 1966] Neumann JV, Burks AW. (1966) "Theory of Self-Reproducing Automata" University of Illinois Press

[Neuser 2013] Neuser S, Michaud V. (2013) "Effect of Aging on the Performance of Solvent-Based Self-Healing Materials" Polymer Chemistry, 4, 4993–4999, doi:10.1039/c3py00064h

[Neves 2001] Neves BR, Salmon ME, Troughton Jr. EB, Russell PE. (2001) "Self-Healing on OPA Self-Assembled Monolayers" Nanotechnology, 12, 285, doi:10.1088/0957-4484/12/3/315

[Neville 2002] Neville AM. (2002) "Autogenous Healing: A Concrete Miracle?" Concrete International, 24, 11, 76–82

[Neville 2008] Neville JL, Larsen JL, Peterson SW, Bandi M, Lockstedt A, Cariveau PT, White AC, Griffo A, Siracki Michael. (2008) "Self Relieving Seal" US Patent 7,387,178

[Newman 2001] Newman ME, Strogatz SH, Watts DJ. (2001) "Random Graphs with Arbitrary Degree Distributions and Their Applications" Physical Review E, 64, 026118, doi:10.1103/PhysRevE.64.026118

[Nguyen 1987] Nguyen HV, Padmanabhan S, Desisto WJ, Bose A. (1987) "Sessile Drops on Nonhorizontal Solid Substrates" Journal of Colloid and Interface Science, 115, 2, 410–416

[Nielsen 2010] Nielsen C, Weizman O, Nemat-Nasser S. (2010) "Characterization of Healable Polymers" Proceedings of SPIE, Vol. 7644, 76441B, Behavior and Mechanics of Multifunctional Materials and Composites Z Ounaies, J Li (eds)

[Nishiwaki 2006] Nishiwaki T, Mihashi H, Jang BK, Miura K. (2006) "Development of Self-Healing System for Concrete with Selective Heating around Crack" Journal of Advanced Concrete Technology, 4, 2, 267–275.

[Niu 2005] Niu W, O'Sullivan C, Rambo BM, Smith MD, Lavigne JJ. (2005) "Self-Repairing Polymers: Poly(dioxaborolane)s Containing Trigonal Planar Boron" Chemical Communications, 4342–4344, doi:10.1039/b504634c

[Nji 2010] Nji J, Li G. (2010) "A Self-Healing 3D Woven Fabric Reinforced Shape Memory Polymer Composite for Impact Mitigation" Smart Materials and Structures, 19, 035007 doi:10.1088/0964-1726/19/3/035007

[Noguchi 2001] Noguchi H, Takasu M. (2001) "Self-Assembly of Amphiphiles into Vesicles: A Brownian Dynamics Simulation" Physical Review E, 64, 041913, doi:10.1103/PhysRevE.64.041913

[Noimark 2014] Noimark S, Allan E, Parkin IP. (2014) "Light-Activated Antimicrobial Surfaces with Enhanced Efficacy Induced by a Dark-Activated Mechanism" Chemical Science, 5, 2216–2223, doi:10.1039/C3SC53186D

[Noorduin 2013] Noorduin WL, Grinthal A, Mahadevan L, Aizenberg J. (2013) "Rationally Designed Complex, Hierarchical Microarchitectures" Science, 340, 6134, 832–837, doi:10.1126/science.1234621

[Norris 2011] Norris CJ, Bond IP, Trask RS. (2011) "The Role of Embedded Bioinspired Vasculature on Damage Formation in Self-Healing Carbon Fibre Reinforced Composites" Composites Part A: Appl Sci Manuf, 42, 6, 639–648, doi:10.1016/j.compositesa.2011.02.003

[Nosonovsky 2009] Nosonovsky M, Amano R, Lucci JM, Rohatgi PK. (2009) "Physical Chemistry of Self-Organization and Self-Healing in Metals" Physical Chemistry Chemical Physics, 11, 41, 9530–9536, doi:10.1039/b912433k

[Nosonovsky 2012] Nosonovsky M, Rohatgi PK. (2012) "Biomimetics in Materials Science: Self-Healing, Self-Lubricating, and Self-Cleaning Materials" Chapter 2, 25–51, Springer Series in Materials Science 152, doi:10.1007/978-1-4614-0926-7_2

[Nowak 2002] Nowak AP, Breedveld V, Pakstis L, Ozbas B, Pine DJ, Pochan D, Deming TJ. (2002) "Rapidly Recovering Hydrogel Scaffolds from Self-Assembling Diblock Copolypeptide Amphiphiles" Nature, 417, 424–428

[Nunez 2010] Nunez PL. (2010) Brain, Mind, and the Structure of Reality, Oxford

[Nyberg 1977] Nyberg DD. (1977) "Heat-Shrinkable Laminate" US Patent 4,035,534

[O'Brien 2015] O'Brien MN, Jones MR, Lee B, Mirkin CA. (2015) "Anisotropic Nanoparticle Complementarity in DNA-Mediated Co-Crystallization" Nature Materials, 14, 833–839 doi:10.1038/nmat4293

[O'Connor 1980] O'Connor K, Wool R. (1980) "Optical Studies of Void Formation and Healing in Styrene-Isoprene-Styrene Block Copolymers" Journal of Applied Physics, 51, 10, 5075–5079

[O'Harra 2020a] O'Harra K, Sadaba N, Irigoyen M, Ruipérez F, Aguirresarobe R, Sardon H, Bara J. (2020) "Nearly Perfect 3D Structures Obtained by Assembly of Printed Parts of Polyamide Ionene Self-Healing Elastomer" ACS Applied Polymer Materials, 2, 11, 4352–4359, doi:10.1021/acsapm.0c00799

[O'Harra 2020b] O'Harra KE, Kammakakam I, Noll DM, Turflinger EM, Dennis GP, Jackson EM, Bara JE. (2020) "Synthesis and Performance of Aromatic Polyamide Ionenes as Gas Separation Membranes" Membranes, 10, 3, 51, doi:10.3390/membranes10030051

[O'Hern 2003] O'Hern CS, Silbert LE, Liu AJ, Nagel SR. (2003) "Jamming at Zero Temperature and Zero Applied Stress: The Epitome of Disorder" Physical Review E, 68, 011306, doi:10.1103/PhysRevE.68.011306

[Odom 2010] Odom SA, Caruso MM, Finke AD, Prokup AM, Ritchey JA, Leonard JH, White SR, Sottos NR, Moore JS. (2010) "Restoration of Conductivity with TTF-TCNQ Charge-Transfer Salts" Advanced Functional Materials, 20, 1721–1727, doi:10.1002/adfm.201000159

[Oghenekevwe 2009] Oghenekevwe V, Redmond S, Hiltz M, Rembala R. (2009) "Human and Robotic Repair of a Solar Array Wing During ISS Assembly Mission 10A" Acta Astronautica, 65, 1717–1722

[Ogi 2014] Ogi S, Sugiyasu K, Manna S, Samitsu S, Takeuchi M. (2014) "Living Supramolecular Polymerization Realized through a Biomimetic Approach" Nature Chemistry, 6, 188–195, doi:10.1038/nchem.1849

[Ogisu 2003] Ogisu T, Shimanuki M, Kiyoshima S, Takaki J, Takeda N. (2003) "Damage Suppression System Using Embedded SMA (Shape Memory Alloy) Foils in CFRP Laminates" Proceedings of SPIE, Vol. 5054-23, Industrial and Commercial Applications of Smart Structures Technologies

[Ogurtani 2005] Ogurtani TO, Oren EE. (2005) "Irreversible Thermodynamics of Triple Junctions during the Intergranular Void Motion under the Electromigration Forces" International Journal of Solids and Structures, 42, 13, 3918–3952, doi:10.1016/j.ijsolstr.2004.11.013

[Oh 2008] Oh JK, Drumright R, Siegwart DJ, Matyjaszewski K. (2008) "The Development of Microgels/Nanogels for Drug Delivery Applications" Progress in Polymer Science, 33, 448–477

[Oh 2016] Oh JY, Rondeau-Gagné S, Chiu YC, Chortos A, Lissel F, Wang GJ, Schroeder BC, Kurosawa T, Lopez J, Katsumata T, Xu J, Zhu C, Gu X, Bae WG, Kim Y, Jin L, Chung JW, Tok JB, Bao Z. (2016) "Intrinsically Stretchable and Healable Semiconducting Polymer for Organic Transistors" Nature 539, 411–415, doi:10.1038/nature20102

[Ohkubo 2010] Ohkubo Y, Tsuji I, Onishi S, Ogawa K. (2010) "Preparation and Characterization of Super-Hydrophobic and Oleophobic Surface" Journal of Material Science, 45, 4963–4969, doi:10.1007/s10853-010-4362-2

[Ohmann 2015] Ohmann R, Meyer J, Nickel A, Echeverria J, Grisolia M, Joachim C, Moresco F, Cuniberti G. (2015) "Supramolecular Rotor and Translator at Work: On-Surface Movement of Single Atoms" ACS Nano, 9, 8, 8394–8400, doi:10.1021/acsnano.5b03131

[Ohnstad 2011] Ohnstad TS, Monk RA. (2011) "Sealing-Reaction, Layer-Effective, Stealth Liner for Synthetic Fuel Container" US Patent 8,043,676

[Ohouo 2013] Ohouo PY, Bastos de Oliveira FM, Liu Y, Ma CJ, Smolka MB. (2013) "DNA-Repair Scaffolds Dampen Checkpoint Signalling by Counteracting the Adaptor Rad9" Nature, 493, 120–124, doi:10.1038/nature11658

[Oku 2004] Oku T, Furusho Y. Takata T. (2004) "A Concept for Recyclable Cross-Linked Polymers: Topologically Networked Polyrotaxane Capable of Undergoing Reversible Assembly and Disassembly" Angewandte Chemie International Edition, 43, 966–969, doi:10.1002/anie.200353046

[Olevitch 1972] Olevitch A. (1972) "Self-Sealing Fuel Tank" US Patent 3,654,057

[Oliver 2014] Oliver R. (2014) "Particulate Capture from a High Energy Discharge Device" US Patent 8,790,434

[Olney 1994] Olney RD. (1994) "Vehicle Wheel Including Self-Inflating Mechanism" US Patent 5,355,924

[Olsson 2007] Olsson P, Teitel S. (2007) "Critical Scaling of Shear Viscosity at the Jamming Transition" Physical Review Letters, 99, 178001, doi:10.1103/PhysRevE.83.031307

[Olugebefola 2010] Olugebefola SC, Aragón AM, Hansen CJ, Hamilton AR, Kozola BD, Wu W, Geubelle PH, Lewis JA, Sottos NR, White SR. (2010) "Polymer Microvascular Network Composites" Journal of Composite Materials, 44, 2587–2603, doi:10.1177/0021998310371537

[Ono 2005] Ono M, Ishida W, Nakao W, Ando K, Mori S, Yokouchi M. (2005) "Crack-Healing Behavior under Stress, Strength and Fracture Toughness of Mullite/SiC Whisker Composite Ceramic" Journal of the Society of Materials Science, 54, 2, 207–214

[Onsager 1949] Onsager L (1949) "The Effects of Shape on the Interaction of Colloidal Particles" Annals of the New York Academy of Sciences, 51, 627–659, doi:10.1111/j.1749-6632.1949.tb27296.x

[Oosterom 2002] Oosterom M, Babuška R, Verbruggen HB. (2002) "Soft Computing Applications in Aircraft Sensor Management and Flight Control Law Reconfiguration" Transactions on Systems, Man, and Cybernetics C, 32, 125–39

[Orain 1980] Orain R, Heuser H, Ohlenforst H, Pelzer R. (1980) "Safety Window Comprising Self-Healing Polymeric Layer" US Patent 4,232,080

[Orbell 1999] Orbell JD, Tan EK, Coutts M, Bigger SW, Ngeh LN. (1999) "Cleansing Oiled Feathers – Magnetically" Marine Pollution Bulletin, 38, 3, 219–221, doi:10.1016/S0025-326X(98)00186-6

[Ormondroyd 1928] Ormondroyd J, Den Hartog JP. (1928) "The Theory of the Dynamic Vibration Absorber" Trans ASME APM-50-7, pp 9–22

[Örnek 2002] Örnek D, Jayaraman A, Syrett BC, Hsu CH, Mansfeld FB, Wood TK. (2002) "Pitting Corrosion Inhibition of Aluminum 2024 by Bacillus Biofilms Secreting Polyaspartate or γ-Polyglutamate" Applied Microbiology and Biotechnology, 58, 651–657, doi 10.1007/s00253-002-0942-7

[Ottelé 2011] Ottelé M, Perini K, Fraaij AL, Haasa EM, Raiteri R. (2011) "Comparative Life Cycle Analysis for Green Façades and Living Wall Systems" Energy and Buildings, 43, 12, 3419–3429, doi:10.1016/j.enbuild.2011.09.010

[Otten 2004] Otten A, Herminghaus S. (2004) "How Plants Keep Dry: A Physicist's Point of View" Langmuir, 20, 6, 2405–2408, doi:10.1021/la034961d

[Ou 2014] Ou R, Eberts K, Skandan G, Lee SP, Iezzi R, Eberly DE. (2014) "Self-Healing Polymer Nanocomposite Coatings for Use on Surfaces Made of Wood" US Patent 8,664,298

[Outwater 1969] Outwater JO, Gerry DJ. (1969) "On the Fracture Energy, Rehealing Velocity and Refracture Energy of Cast Epoxy Resin" Journal of Adhesion, 1, 290–298.

[Owens 2010] Owens RM, Malliaras GG. (2010) "Organic Electronics at the Interface with Biology" MRS Bulletin, 35, 449–456

[Öztürk 2015] Öztürk BÖ, Şehitoğlu SK. (2015) "Applications of Ruthenium Indenylidene Catalysts on ROMP-Based Self-Healing Epoxy Systems" Polymer, 69, 343–348, doi:10.1016/j.polymer.2015.03.037

[Pahl 1988] Pahl G, Beitz W, Wallace K. (1988) Engineering Design: A Systematic Approach, Springer-Verlag, New York

[Paine 2016] Paine K, Alazhari M, Sharma T, Cooper R, Heath A. (2016) "Design and Performance of Bacteria-Based Self-Healing Concrete" In: Jones MR, Newlands MD, Halliday JE, Csetenyi LJ, Zheng L, McCarthy MJ, Dyer TD (eds) The 9th International Concrete Conference, Dundee, pp 545–554.

[Pajevic 2012] Pajevic S, Plenz D. (2012) "The Organization of Strong Links in Complex Networks" Nature Physics, 8, 429–436, doi:10.1038/NPHYS2257

[Palagi 2016] Palagi S, Mark AG, Reigh SY, Melde K, Qiu T, Zeng H, Parmeggiani C, Martella D, Sanchez-Castillo A, Kapernaum N, Giesselmann F, Wiersma DS, Lauga E, Fischer P. (2016) "Structured Light Enables Biomimetic Swimming and Versatile Locomotion of Photoresponsive Soft Microrobots" Nature Materials, 15, 647–653, doi:/10.1038/nmat4569

[Paliwoda 2005] Paliwoda-Porebska G, Stratmann M, Rohwerder M, Potje-Kamloth K, Lu Y, Pich AZ, Adler H- J (2005) "On the Development of Polypyrrole Coatings with Self-Healing Properties for Iron Corrosion Protection" Corrosion Science, 47, 3216–3233, doi:10.1016/j.corsci.2005.05.057

[Palleau 2013a] Palleau E, Morales D, Dickey MD, Velev OD. (2013) "Reversible Patterning and Actuation of Hydrogels by Electrically Assisted Ionoprinting" Nature Communications, 4, 2257, doi:10.1038/ncomms3257

[Palleau 2013b] Palleau E, Reece S, Desai SC, Smith ME, Dickey MD. (2013) "Self-Healing Stretchable Wires for Reconfigurable Circuit Wiring and 3D Microfluidics" Advanced Materials, 25, 11, 1589–1592, doi:10.1002/adma.201203921

[Palvadi 2012] Palvadi S, Bhasin A, Little DN. (2012) "Method to Quantify Healing in Asphalt Composites by Continuum Damage Approach" Transportation Research Record, 2296, doi:10.3141/2296-09

[Pan 2008] Pan T, Shi X. (2008) "Assessment of Electrical Injection of Corrosion Inhibitor for Corrosion" Proceedings of Transportation Research Board 2008 Annual Meeting, Washington, DC, Paper No. 08-1973

[Pan 2013] Pan S, Kota AK, Mabry JM, Tuteja A. (2013) "Superomniphobic Surfaces for Effective Chemical Shielding" Journal of the American Chemical Society, 135, 578–581, dx.doi.org/10.1021/ja310517s

[Pandey 2014] Pandey R, Ghosh R, Han SP. (2014) "Apparatus and Method for Admitting New Devices in a Self-Healing, Self-Organizing Mesh Network" US Patent 8,885,548

[Pang 2005] Pang JW, Bond IP. (2005) "'Bleeding Composites': Damage Detection and Self-Repair Using a Biomimetic Approach" Composites: Part A, 36, 183–188

[Panse 2013] Panse D. (2013) "Burn Protective Materials" US Patent 8,383,528

[Pareek 2014] Pareek S, Shrestha KC, Suzuki Y, Omori T, Kainuma R, Araki Y. (2014) "Feasibility of Externally Activated Self-Repairing Concrete with Epoxy Injection Network and Cu-Al-Mn Superelastic Alloy Reinforcing Bars" Smart Materials and Structures, 23, 10, 105027, doi:10.1088/0964-1726/23/10/105027

[Park 1996] Park JH, Kassner TF. (1996) "CaO Insulator Coatings and Self-Healing of Defects on V-Cr-Ti Alloys in Liquid Lithium System" Journal of Nuclear Materials, 233–237, 476–481

[Park 2003a] Park G, Muntges DE, Inman DJ. (2003) "Self-Repairing Joints Employing Shape-Memory Alloy Actuators" Journal of Metals, 55, 33–37

[Park 2003b] Park JH, Sawada A, Kestle B, Rink D, Natesan K, Mattas RF. (2003) "Effect of Thin Y-O and Si-O Films on In-situ Formed CaO Coatings on V-4%Cr-4%Ti in Liquid 2.8 at.% Ca-Li" Materials Research Society Symposia Proceedings, 751, Z3.25.1–5

[Park 2004a] Park SE, Joo H, Kang JW. (2004) "Effect of Impurities in TiO_2 Thin Films on Trichloroethylene Conversion" Solar Energy Materials & Solar Cells, 83, 39, doi:10.1016/j.solmat.2004.02.012

[Park 2004b] Park W, Albright D, Addleston C, Won WK, Lee K, Chirikjian GS. (2004) "Robotic Self-Repair in a Semistructured Environment" Proceedings of Robosphere 2004, NASA Ames Research Center, Ames, CA, SP Colombano, ed.

[Park 2007] Park JS, Akiyama Y, Yamasaki Y, Kataoka K. (2007) "Preparation and Characterization of Polyion Complex Micelles with a Novel Thermosensitive Poly(2-isopropyl-2-oxazoline) Shell via the Complexation of Oppositely Charged Block Ionomers" Langmuir 23, 138–146

[Park 2009] Park JS, Kim HS, Hahn HT. (2009) "Healing Behavior of a Matrix Crack on a Carbon Fiber/Mendomer Composite" Composites Science and Technology, 69, 1082–1087

[Park 2012] Park KC, Choi HJ, Chang CH, Cohen RE, McKinley GH, Barbastathis G. (2012) "Nanotextured Silica Surfaces with Robust Superhydrophobicity and Omnidirectional Broadband Supertransmissivity" ACS Nano, 6, 5, 3789–3799, doi:10.1021/nn301112t

[Park 2016a] Park B, Choi YC, Cha SW, Choi SC. (2016) "Quantitative Evaluation of Self-Healing Efficiency in Cracked Cementitious Material Using Modified Absorption Test" HealCON-conference, 28–29, Delft, The Netherlands

[Park 2016b] Park CH, Lee SY, Hwang DS, Shin DW, Cho DH, Lee KH, Kim TW, Kim TW, Lee M, Kim DS, Doherty CM, Thornton AW, Hill AJ, Guiver MD, Lee YM. (2016) "Nanocrack-Regulated Self-Humidifying Membranes" Nature, 532, 480–483, doi:10.1038/nature17634

[Parke 2007] Parke TJ. (2007) "Self Cleaning Shredding Device Having Movable Cleaning Rings" US Patent 7,311,279

[Parkin 2005] Parkin IP, Palgrave RG. (2005) "Self-Cleaning Coatings" Journal of Materials Chemistry, 15, 1689–1695, doi:10.1039/b412803f

[Pastine 2009] Pastine SJ, Okawa D, Zettl A, Fréchet JM. (2009) "Chemicals on Demand with Phototriggerable Microcapsules" Journal of the American Chemical Society, 131, 13586–13587, doi:10.1021/ja905378v

[Pastor-Satorras 2003] Pastor-Satorras R, Rubi M, Diaz-Guilera A. (2003) Statistical Mechanics of Complex Networks, Springer-Verlag

[Patankar 2004] Patankar NA. (2004) "Mimicking the Lotus Effect: Influence of Double Roughness Structures and Slender Pillars" Langmuir, 20, 8209–8213

[Patel 2010] Patel AJ, Sottos NR, Wetzel ED, White SR. (2010) "Autonomic Healing of Low-Velocity Impact Damage in Fiber-Reinforced Composites" Composites: Part A, 41, 360–368

[Patrick 2012] Patrick JF, Sottos NR, White SR. (2012) "Microvascular Based Self-Healing Polymeric Foam" Polymer, 53, 19, 4231–4240, doi:10.1016/j.polymer.2012.07.021

[Paugh 2006] Paugh B, Sterry K. (2006) "Designing with Inflatable Seals" Machine Design, 78, 5, 98–105

[Paul 1976] Paul DR. (1976) "Polymers in Controlled Release Technology" in Controlled Release Polymeric Formulations DR Paul, FW Harris (eds), pp 1–14, ACS Symposium Series, American Chemical Society, Washington, DC, doi:10.1021/bk-1976-0033

[Paulsen 2015] Paulsen JD, Démery V, Santangelo CD, Russell TP, Davidovitch B, Menon N. (2015) "Optimal Wrapping of Liquid Droplets with Ultrathin Sheets" Nature Materials, 14, 1206–1209 doi:10.1038/nmat4397

[Paulusse 2004] Paulusse JM, Sijbesma RP. (2004) "Reversible Mechanochemistry of a Pd(II) Coordination Polymer" Angewandte Chemie International Edition, 43, 4460, doi:10.1002/anie.200460040

[Paxson 2013] Paxson AT, Varanasi KK. (2013) "Self-Similarity of Contact Line Depinning from Textured Surfaces" Nature Communications 4, 1492, doi:10.1038/ncomms2482

[Paz 1997] Paz Y, Heller A. (1997) "Photo-oxidatively Self-Cleaning Transparent Titanium Dioxide Films on Soda Lime Glass: The Deleterious Effect of Sodium Contamination and Its Prevention" Journal of Materials Research, 12, 2759–2766.

[Peairs 2003] Peairs DM, Park G, Inman DJ. (2003) "Practical Issues in Self-Repairing Bolted Joints" SPIE Vol. 5056-10 Smart Structures and Integrated Systems

[Peanasky 1991] Peanasky JS, Long JM, Wool RP. (1991) "Percolation Effects in Degradable Polyethylene–Starch Blends" Journal of Polymer Science: Polymer Physics, 29, 565–579

[Pecht 2010] Pecht MG, Gu J. (2010) "Prognostics and Health Management Implementation for Self Cognizant Electronic Products" US Patent Application 20100191503

[Pei 2014] Pei Z, Yang Y, Chen Q, Terentjev EM, Wei Y, Ji Y. (2014) "Mouldable Liquid-Crystalline Elastomer Actuators with Exchangeable Covalent Bonds" Nature Materials, 13, 36–41, doi:10.1038/nmat3812

[Peleshanko 2007] Peleshanko S, Anderson KD, Goodman M, Determan MD, Mallapragada SK, Tsukruk VV. (2007) "Thermoresponsive Reversible Behavior of Multistimuli Pluronic-Based Pentablock Copolymer at the Air–Water Interface" Langmuir, 23, 25–30

[Pelletier 2011] Pelletier M, Bose A. (2011) "Self-Mending Composites Incorporating Encapsulated Mending Agents" US Patent Application 20110316189

[Peng 2006] Peng S, Manohar R. (2006) "Self-Healing Asynchronous Arrays" 12th IEEE International Symposium Asynchronous Circuits and Systems, pp 45–55, Grenoble, France, doi:10.1109/ASYNC.2006.25

[Penrose 2016] Penrose R (2016) The Emperor's New Mind: Concerning Computers, Minds, and the Laws of Physics, Oxford University Press

[Perez 2004] Perez N. (2004) "Cathodic Protection" Electrochemistry and Corrosion Science, Chapter 8, 247–294, Kluwer Academic Publishers, Boston

[Perez 2012] Perez-Herrera RA, Fernandez-Vallejo M, Lopez-Amo M. (2012) "Robust Fiber-Optic Sensor Networks" Photonic Sensors, 2, 4, 366–380, doi:10.1007/s13320-012-0083-2

[Pescara 1944] Pescara MG. (1944) "Fuel Tank for Airplanes" US Patent 2,354,701

[Peskar 2015] Peskar J, Gunderson Olson ME, Entezarian M. (2015) "Impact Resistant Battery" US Patent 9,136,558

[Peterlik 2006] Peterlik H, Roschger P, Klaushofer K, Fratzl P. (2006) "From Brittle to Ductile Fracture of Bone" Nature Materials, 5, 52–55, doi:10.1038/nmat1545

[Peterson 2010] Peterson AM, Jensen RE, Palmese GR. (2010) "Room-Temperature Healing of a Thermosetting Polymer Using the Diels–Alder Reaction" ACS Applied Materials & Interfaces, 2, 4, 1141–1149, doi:10.1021/am9009378

[Petit 2007] Petit L, Bouteiller L, Brûlet A, Lafuma F, Hourdet D. (2007) "Responsive Hybrid Self-Assemblies in Aqueous Media" Langmuir, 23, 147–158

[Petrie 2000] Petrie EM. (2000) Handbook of Adhesives and Sealants McGraw-Hill, New York

[Peyratout 2003] Peyratout CS, Möhwald H, Dähne L. (2003) "Preparation of Photosensitive Dye Aggregates and Fluorescent Nanocrystals in Microreaction Containers" Advanced Materials, 15, 20, 1722–1726

[Pflaum 2017] Pflaum M, Kühn-Kauffeldt M, Schmeckebier S, Dipresa D, Chauhan K, Wiegmann B, Haug RJ, Schein J, Haverich A, Korossis S. (2017) "Endothelialization and Characterization of Titanium Dioxide-Coated Gas-Exchange Membranes for Application in the Bioartificial Lung" Acta Biomaterialia, 50, 510–521, doi:10.1016/j.actbio.2016.12.017

[Phadke 2012] Phadke A, Zhang C, Arman B, Hsu CC, Mashelkar RA, Lele AK, Tauber MJ, Arya G, Varghese S. (2012) "Rapid Self-Healing Hydrogels" Proceedings of the National Academy of Sciences of the United States of America, 109, 12, 4383–4388, doi:10.1073/pnas.1201122109

[Phillips 2010] Phillips DM, Baur JW. (2010) "A Granular Core for Self-Healing, Variable Modulus Sandwich Composites" Journal of the Composite Materials, 44, 22, 2527–2545, doi:10.1177/0021998310371531

[Piermattei 2009] sh287 Piermattei A, Karthikeyan S, Sijbesma RP. (2009) "Activating Catalysts with Mechanical Force" Nature Chemistry, 1, 133–137, doi:10.1038/nchem.167

[Piestun 1998] Piestun R, Shamir J. (1998) "Generalized Propagation-Invariant Wave Fields" Journal of the Optical Society of America A, 15, 12, 3039–3044

[Pike 1974] Pike GE, Seager CH. (1974) "Percolation and Conductivity: A Computer Study I*" Physical Review B, 10, 4, 1421–1434

[Pinori 2011] Pinori E, Berglin M, Brive LM, Hulander M, Dahlström M, Elwing H. (2011) "Multi-Seasonal Barnacle (Balanus Improvisus) Protection Achieved by Trace Amounts of o Macrocyclic Lactone (Ivermectin) Included In Rosin-Based Coatings" Biofouling: The Journal of Bioadhesion and Biofilm Research, 27, 9, 941–953, doi:10.1080/08927014.2011.616636

[Pirondini 2002] Pirondini L, Bertolini F, Cantadori B, Ugozzoli F, Massera C, Dalcanale E. (2002) "Design and Self-Assembly of Wide and Robust Coordination Cages" Proceedings of the National Academy of Sciences of the United States of America, 99, 8, 4911–4915, doi:10.1073/pnas.072612199

[Plaisted 2003] Plaisted TA, Amirkhizi AV, Arbelaez D, Nemat-Nasser SC, Nemat-Nasser S. (2003) "Self-Healing Structural Composites with Electromagnetic Functionality" SPIE, Vol. 5054-43 Industrial and Commercial Applications of Smart Structures Technologies

[Plaisted 2007] Plaisted TA, Nemat-Nasser S. (2007) "Quantitative Evaluation of Fracture, Healing and Re-Healing of a Reversibly Cross-Linked Polymer" Acta Materialia, 55, 5684–5696

[Pleydell 2002] Pleydell CP, David T, Smye SW, Berridge DC. (2002) "A Mathematical Model of Post-Canalization Thrombolysis" Physics in Medicine & Biology, 47, 209–234

[Półrolniczak 2021] Półrolniczak A, Katrusiak A. (2021) "Self-Healing Ferroelastic Metal–Organic Framework Sensing Guests, Pressure and Chemical Environment" Materials Advances, 2, 4677–4684; doi:10.1039/D1MA00111F

[Popkin 2016] Popkin G. (2016) "The Physics of Life" Nature, 529, 16–18, doi:10.1038/529016a

[Popov 2007] Popov P, Lagoudas DC. (2007) "A 3-D Constitutive Model for Shape Memory Alloys Incorporating Pseudoelasticity and Detwinning of Self-Accomodated Martensite" International Journal of Plasticity, 23, 10–11, 1679–1720.

[Potier 2014] Potier F, Guinault A, Delalande S, Sanchez C, Ribot F, Rozes L. (2014) "Nano-Building Block Based-Hybrid Organiceinorganic Copolymers with Self-Healing Properties" Polymer Chemistry, 5, 4474–4479, doi:10.1039/C4PY00172A

[Potters 1949] Potters S. (1949) "Self-Sealing Oil Tank" US Patent 2,482,366

[Powell 1980] Powell JA, Messerly JW, Shippy RL. (1980) "Puncture Sealing Tire" US Patent 4,186,042

[Prager 1981] Prager S, Tirrell M. (1981) "The Healing Process at Polymer–Polymer Interfaces" Journal of Chemical Physics, 75, 10, 5194–5198, doi:10.1063/1.441871

[Pramanik 2015] Pramanik NB, Nando GB, Singha NK. (2015) "Self-Healing Polymeric Gel via RAFT Polymerization and Diels–Alder Click Chemistry" Polymer, 69, 349–356, doi:10.1016/j.polymer.2015.01.023

[Prigogine 1981] Prigogine I (1981) From Being to Becoming: Time and Complexity in the Physical Sciences, W H Freeman & Co

[Prindle 2012] Prindle A, Samayoa P, Razinkov I, Danino T, Tsimring LS, Hasty J. (2012) "A Sensing Array of Radically Coupled Genetic 'Biopixels'" Nature, 481, 39–44, doi:10.1038/nature10722

[Privman 2007] Privman V, Dementsov A, Sokolov I. (2007) "Modeling of Self-Healing Polymer Composites Reinforced with Nanoporous Glass Fibers" Journal of Computational and Theoretical Nanoscience, 4, 1, 190–193

[Prost 2015] Prost J, Jülicher F, Joanny J-F. (2015) "Active Gel Physics" Nature Physics, 11, 111–117, doi:10.1038/nphys3224

[Prymak 2006] Prymak J, Staubli P, Blais P, Long B. (2006) "Scintillation Testing Reveals Self-Healing in Solid Electrolytic Capacitors" Proceedings of CARTS Europe 2006, Bad Homburg, Germany

[Pugno 2011] Pugno N, Abdalrahman T. (2011) "Modeling the Self-Healing of Biological or Bio-Inspired Nanomaterials" Journal of the International Space Elevator Consortium, 1, 1, 79–86

[Puretskiy 2012] Puretskiy N, Stoychev G, Synytska A, Ionov L. (2012) "Surfaces with Self-Repairable Ultrahydrophobicity Based on Self-Organizing Freely Floating Colloidal Particles" Langmuir, 28, 3679–3682, doi:10.1021/la204232g

[Puzder 2004] Puzder A, Williamson AJ, Gygi F, Galli G. (2004) "Self-Healing of CdSe Nanocrystals: First-Principles Calculations" Physical Review Letters, 92, 217401, doi:10.1103/PhysRevLett.92.217401

[Qi 2006] Qi K, Daoud WA, Xin JH, Mak CL, Tang W, Cheung WP. (2006) "Self-Cleaning Cotton" Journal of Materials Chemistry, 16, 4567–4574, doi:10.1039/b610861j

[Qi 2009] Qi H, Chen T, Yao L, Zuo T. (2009) "Micromachining of Microchannel on the Polycarbonate Substrate with CO_2 Laser Direct-Writing Ablation" Optics and Lasers in Engineering, 47, 5, 594–598, doi:10.1016/j.optlaseng.2008.09.004

[Qi 2013] Qi H, Ghodousi M, Du Y, Grun C, Bae H, Yin P, Khademhosseini A. (2013) "DNA-Directed Self-Assembly of Shape-Controlled Hydrogels" Nature Communications, 4, 2275, doi:10.1038/ncomms3275

[Qiao 2021] Qiao X, Midya B, Gao Z, et al. (2021) "Higher-Dimensional Supersymmetric Microlaser Arrays" Science, 372, 403–408, doi:10.1126/science.abg3904

[Qian 2009] Qian S, Zhou J, de Rooij MR, Schlangen E, Yea G, van Breugel K. (2009) "Self-Healing Behavior of Strain Hardening Cementitious Composites Incorporating Local Waste Materials" Cement & Concrete Composites, 31, 613–621

[Qian 2010] Qian SZ, Zhou J, Schlangen E. (2010) "Influence of Curing Condition and Precracking Time on the Self-Healing Behavior of Engineered Cementitious Composites" Cement & Concrete Composites, 32, 686–693, doi:10.1016/j.cemconcomp.2010.07.015

[Qin 2013] Qin Z, Buehler MJ. (2013) "Impact Tolerance in Mussel Thread Networks by Heterogeneous Material Distribution" Nature Communications, 4, 2187, doi:10.1038/ncomms3187

[Qiu 2011] Qiu J, van de Ven M, Wu S, Yu J, Molenaar A. (2011) "Investigating Self Healing Behaviour of Pure Bitumen Using Dynamic Shear Rheometer" Fuel, 90, 8, 2710–2720.

[Quattrociocchi 2014] Quattrociocchi W, Caldarelli G, Scala A (2014) "Self-Healing Networks: Redundancy and Structure" PLoS One, 9, 2, e87986, doi:10.1371/journal.pone.0087986

[Quéré 2003] Quéré D, Lafuma A, Bico J. (2003) "Slippy and Sticky Microtextured Solids" Nanotechnology, 14, 1109–1112, doi:S0957-4484(03)63599-3

[Rabiei 2013] Rabiei A, Sandukas S. (2013) "Processing and Evaluation of Bioactive Coatings on Polymeric Implants" Journal of Biomedical Materials Research Part A, 101A, 9, 2621–2629, doi:10.1002/jbm.a.34557

[Rabinovitch 2007] Rabinovitch O. (2007) "Piezoelectric Control of Edge Debonding in Beams Strengthened with Composite Materials: Part I – Analytical Modeling" Journal of Composite Materials, 41, 525–546, doi:10.1177/0021998306063790

[Rabone 2010] Rabone J, Yue YF, Chong SY, Stylianou KC, Bacsa J, Bradshaw D, Darling GR, Berry NG, Khimyak YZ, Ganin AY, Wiper P, Claridge JB, Rosseinsky MJ. (2010) "An Adaptable Peptide-Based Porous Material" Science, 329, 5995, 1053–1057, doi:10.1126/science.1190672

[Racicot 1997] Racicot R, Brown T, Yang SC. (1997) "Corrosion Protection of Aluminum Alloys by Double-Strand Polyaniline" Synthetic Metals, 85, 1–3, 1263–1264, doi:10.1016/S0379-6779(97)80232-9

[Radicchi 2013] Radicchi F, Arenas A. (2013) "Abrupt Transition in the Structural Formation of Interconnected Networks" Nature Physics, 9, 717–720, doi:10.1038/nphys2761

[Rädisch 1986] Rädisch H, Scholz W. (1986) "Transparent, Anti-Fogging Coating Comprised of Plastic Material Containing a Surface Active Agent" US Patent 4,609,688

[Rafsanjani 2017] Rafsanjani A, Bertoldi K. (2017) "Buckling-Induced Kirigami" Physical Review Letters, 118, 084301, doi:10.1103/PhysRevLett.118.084301

[Raghavan 1999] Raghavan J, Wool RP. (1999) "Interfaces in Repair, Recycling, Joining and Manufacturing of Polymers and Polymer Composites" Journal of Applied Polymer Science, 71, 775–785

[Rahaman 2014] Rahaman MN, Bal BS, Huang W. (2014) "Review: Emerging Developments in the Use of Bioactive Glasses for Treating Infected Prosthetic Joints" Materials Science & Engineering C: Materials for Biological Applications, 41, 224–231, doi:10.1016/j.msec.2014.04.055

[Rahmathullah 2009] Rahmathullah MA, Palmese GR. (2009) "Crack-Healing Behavior of Epoxy-Amine Thermosets" Journal of Applied Polymer Science, 113, 2191–2201, doi:10.1002/app.30152

[Raimondo 2013] Raimondo M, Guadagno L. (2013) "Healing Efficiency of Epoxy-Based Materials for Structural Applications" Polymer Composites, 34, 9, 1525–1532, doi:10.1002/pc.22539

[Rajagopalan 2007] Rajagopalan J, Han JH, Taher M, Saif A. (2007) "Plastic Deformation Recovery in Freestanding Nanocrystalline Aluminum and Gold Thin Films" Science, 315, 30

[Ramachandran 2013] Ramachandran S. (2013) "Systems and Techniques for Generating Bessel Beams" US Patent 8,358,888

[Ramaswamy 2010] Ramaswamy S. (2010) "The Mechanics and Statistics of Active Matter" Annual Review of Condensed Matter Physics, 1: 323–345, doi:10.1146/annurev-conmatphys-070909-104101

[Ramirez 2013] Ramirez AL, Kean ZS, Orlicki JA, Champhekar M, Elsakr SM, Krause WE, Craig SL. (2013) "Mechanochemical Strengthening of a Synthetic Polymer in Response to Typically Destructive Shear Forces" Nature Chemistry, 5, 757–761, doi:10.1038/nchem.1720

[Ramm 1998] Ramm W, Biscoping M. (1998) "Autogeneous Healing and Reinforcement Corrosion of Water-Penetrated Separation Cracks in Reinforced Concrete" Nuclear Engineering and Design, 179, 2, 191–200, doi:10.1016/S0029-5493(97)00266-5

[Rampaul 2003] Rampaul A, Parkin IP, O'Neill SA, DeSouza J, Mills A, Elliott N. (2003) "Titania and Tungsten Doped Titania Thin Films on Glass; Active Photocatalysts" Polyhedron, 22, 1, 35–44, doi:10.1016/S0277-5387(02)01333-5

[Rampf 2013] Rampf M, Speck O, Speck T, Luchsinger RH. (2013) "Investigation of a Fast Mechanical Self-Repair Mechanism for Inflatable Structures" International Journal of Engineering & Science, 63, 61–70

[Rao 2011] Rao Z, Inoue M, Matsuda M, Taguchi T. (2011) "Quick Self-Healing and Thermo-Reversible Liposome Gel" Colloids and Surfaces B: Biointerfaces, 82, 1, 196–202, doi:10.1016/j.colsurfb.2010.08.038

[Rasmussen 2012] Rasmussen L, Meixler LD, Gentile CA. (2012) "Considerations for Contractile Electroactive Materials and Actuators" Electroactive Polymer Actuators and Devices (EAPAD) 2012, Y Bar-Cohen (ed.), Proc SPIE, Vol. 8340, doi:10.1117/12.914988

[Ratna 2008] Ratna D, Karger-Kocsis J. (2008) "Recent Advances in Shape Memory Polymers and Composites: A Review" Journal of Materials Science, 43, 254–269, doi:10.1007/s10853-007-2176-7

[Raupach 2005] Raupach M, Wolff L. (2005) "Long-Term Durability of Hydrophobic Treatment on Concrete" Surface Coatings International Part B, Coatings Trans, 88, 2, 127–133

[Rausch 2014] Rausch MK, Kuhl E. (2014) "On the Mechanics of Growing Thin Biological Membranes" Journal of Mechanics and Physics of Solids, 63, 128–140, doi:10.1016/j.jmps.2013.09.015

[Reece 2002] Reece D, Waller J, Ware N, Sasse P. (2002) "Electrical Cable Having a Self-Sealing Agent and Method for Preventing Water from Contacting the Conductor" US Patent 6,359,231

[Reed 1994] Reed CW, Cichanowski SW. (1994) "The Fundamentals of Aging in HV Polymer-Film Capacitors" IEEE Transactions on Dielectrics and Electrical Insulation, 1, 5, 904–922, doi:10.1109/94.326658

[Reeder 2014] Reeder J, Kaltenbrunner M, Ware T, Arreaga-Salas D, Avendano-Bolivar A, Yokota T, Inoue Y, Sekino M, Voit W, Sekitani T, Someya T. (2014) "Mechanically Adaptive Organic Transistors for Implantable Electronics" Advanced Materials, 26, 29, 4967–4973, doi:10.1002/adma.201400420

[Rekondo 2014] Rekondo A, Martin R, Ruiz de Luzuriaga A, Cabañero G, Grande HJ, Odriozola I. (2014) "Catalyst-Free Room-Temperature Self-Healing Elastomers Based on Aromatic Disulfide Metathesis" Materials Horizons, 1, 237–240, doi:10.1039/C3MH00061C

[Rembala 2009] Rembala R, Ower C. (2009) "Robotic Assembly and Maintenance of Future Space Stations Based on the ISS Mission Operations Experience" Acta Astronautica, 65, 912–920, doi:10.1016/j.actaastro.2009.03.064

[Ren 2006] Ren F, Nah JW, Tu KN, Xiong B, Xu L, Pang JH. (2006) "Electromigration Induced Ductile-to-Brittle Transition in Lead-Free Solder Joints" Applied Physics Letters, 89, 141914, doi:10.1063/1.2358113

[Ren 2014] Ren W. (2014) "Wetting Transition on Patterned Surfaces: Transition States and Energy Barriers" Langmuir, 30, 10, 2879–2885, doi:10.1021/la404518q

[Renner 2011] Renner LD, Weibel DB. (2011) "Physicochemical Regulation of Biofilm Formation" MRS Bulletin, 36, 347–355, doi:10.1557/mrs.2011.65

[Reutenauer 2009] Reutenauer P, Buhler E, Boul PJ, Candau SJ, Lehn JM. (2009) "Room Temperature Dynamic Polymers Based on Diels–Alder Chemistry" Chemistry: A European Journal, 15, 8, 1893–1900, doi:10.1002/chem.200802145

[Revzen 2011] Revzen S, Bhoite M, Macasieb A, Yim M. (2011) "Structure Synthesis On-the-Fly in a Modular Robot" IEEE IROS 2011 Conference, San Francisco, CA

[Reynaert 1990] Reynaert EA. (1990) "Cable Splicing and Termination System" US Patent 4,943,685

[Rice 1968] Rice JR. (1968) "A Path Independent Integral and the Approximate Analysis of Strain Concentration by Notches and Cracks" Journal of Applied Mechanics, 35, 2, 379–386, doi:10.1115/1.3601206

[Richardson 1933] Richardson AJ. (1933) "Valve Stemless Inner Tube, Self Sealing Section and the Like" US Patent 1,930,182

[Richardson 1996] Richardson MO, Wisheart MJ. (1996) "Review of Low-Velocity Impact Properties of Composite Materials" Composites A, 27, 1123–31

[Ring 2012] Ring JD, Lindberg J, Mourka A, Mazilu M, Dholakia K, Dennis MR. (2012) "Auto-Focusing and Self-Healing of Pearcey Beams" Optics Express, 20, 17, 18955

[Rios 2008] Rios PF, Dodiuk H, Kenig S, S. McCarthy S, Dotan A. (2008) "Durable Ultra-Hydrophobic Surfaces for Self-Cleaning Applications" Polymers for Advanced Technologies, 19, 1684–1691

[Risca 2012] Risca VI, Wang EB, Chaudhuri O, Chia JJ, Geissler PL, Fletcher DA. (2012) "Actin Filament Curvature Biases Branching Direction" Proceedings of the National Academy of Sciences of the United States of America, 109, 8, 2913–2918, doi:10.1073/pnas.1114292109

[Ritchie 1988] Ritchie RO. (1988) "Mechanisms of Fatigue Crack Propagation in Metals, Ceramics and Composites: Role of Crack Tip Shielding" Materials Science and Engineering, A103, 15–28

[Rivadeneyra 1991] Rivadeneyra MA, Delgado R, Quesada E, Ramos-Cormenzana A, (1991) "Precipitation of Calcium Carbonate by Deleya Halophile in Media Containing NaCl as Sole Salt" Current Microbiology, 22, 185–90

[Rochon 1995] Rochon P, Batala E, Natanshon A. (1995) "Optically Induced Surface Gratings on Azoaromatic Polymer Films" Applied Physics Letters, 66, 136–139, doi:10.1063/1.113541

[Roddy 2011] Roddy CW. (2011) "Methods and Compositions Comprising a Dual Oil/Water-Swellable Particle" US Patent 7,934,554

[Rodriguez 2002] Rodriguez M, Sauerbrey R, Wille H, et al. (2002) "Triggering and Guiding Megavolt Discharges by Use of Laser-Induced Ionized Filaments" Optics Letters, 27, 772, doi:10.1364/OL.27.000772

[Rodriguez 2003] Rodriguez-Navarro C, Rodriguez-Gallego M, Ben Chekroun K, Gonzalez-Munoz MT. (2003) "Conservation of Ornamental Stone by *Myxococcus xanthus*–Induced Carbonate Biomineralization" Applied and Environmental Microbiology, 69, 4, 2182–2193

[Rodriguez 2011] Rodriguez ED, Luo X, Mather PT. (2011) "Shape Memory Miscible Blends for Thermal Mending" ACS Applied Materials & Interfaces, 3, 2, 152–161, doi:10.1021/am101012c

[Rohatgi 2013] Rohatgi PK. (2013) "Self Healing Metals and Alloys, Including Structural Alloys and Self-Healing Solders" US Patent 8,518,531

[Roig-Flores 2016] Roig-Flores M, Moscato S, Serna P, Ferrar L. (2015) "Self-Healing Capability of Concrete with Crystalline Admixtures in Different Environments" Construction and Building Materials, 86, 1–11, doi:10.1016/j.conbuildmat.2015.03.091

[Romanov 2003] Romanov AE. (2003) "Mechanics and Physics of Disclinations in Solids" European Journal of Mechanics A: Solids 22, 727–741, doi:10.1016/S0997-7538(03)00089-5

[Roostalu 2012] Roostalu U, Strähle U. (2012) "In Vivo Imaging of Molecular Interactions at Damaged Sarcolemma" Developmental Cell, 22, 515–529, doi 10.1016/j.devcel.2011.12.008

[Roque 2012] Roque R, Simms R, Chen Y, Koh C, Lopp G. (2012) "Development of a Test Method That Will Allow Evaluation and Quantification of the Effects of Healing on Asphalt Mixture" Final Report, University of Florida Project, No. 00084223, Gainesville, FL, pp 32611–6580

[Rordorf 2013] Rordorf R, Poggio L, Savastano S, Vicentini A, Petracci B, Chieffo E, Klersy C, Landolina M. (2013) "Failure of Implantable Cardioverter-Defibrillator Leads: A Matter of Lead Size?" Heart Rhythm, 10, 2, 184–190, doi:10.1016/j.hrthm.2012.10.017

[Rose 2014] Rose S, Prevoteau A, Elzière P, Hourdet D, Marcellan A, Leibler L. (2014) "Nanoparticle Solutions as Adhesives for Gels and Biological Tissues" Nature, 505, 382–385, doi:10.1038/nature12806

[Rossiter 2012] Rossiter J, Scarpa F, Takashima K, Walters P. (2012) "Design of a Deployable Structure with Shape Memory Polymers" Proceedings of SPIE, Vol. 8342, 83420Y, Behavior and Mechanics of Multifunctional Materials and Composites, NC Goulbourne, Z Ounaies, doi:10.1117/12.915476

[Rosthauser 1988] Rosthauser JW, Markusch PH. (1988) "Swellable Coating Compositions" US Patent 4,728,711

[Rothemund 2006] Rothemund PW. (2006) "Folding DNA to Create Nanoscale Shapes and Patterns" Nature, 440, 297–302, doi:10.1038/nature04586

[Rousseau 2010] Rousseau IA, Xie T. (2010) "Shape Memory Epoxy: Composition, Structure, Properties and Shape Memory Performances" Journal of Materials Chemistry, 20, 3431–3441, doi:10.1039/B923394F

[Roy 2015] Roy N, Bruchmann B, Lehn JM. (2015) "DYNAMERS: Dynamic Polymers as Self-Healing Materials" Chemical Society Reviews, 44, 3786–3807, doi:10.1039/C5CS00194C

[Rozynek 2014] Rozynek Z, Mikkelsen A, Dommersnes P, Fossum JO. (2014) "Electroformation of Janus and Patchy Capsules" Nature Communications, 5, 3945, doi:10.1038/ncomms4945

[Rtimi 2013] Rtimi S, Baghriche O, Pulgarina C, Lavanchy JC, Kiwi J. (2013) "Growth of TiO_2/Cu Films by HiPIMS for Accelerated Bacterial Loss of Viability" Surface and Coatings Technology, 232, 804–813, doi:10.1016/j.surfcoat.2013.06.102

[Ruan 2013] Ruan M, Li W, Wang B, Deng B, Ma F, Yu Z. (2013) "Preparation and Anti-Icing Behavior of Superhydrophobic Surfaces on Aluminum Alloy Substrates" Langmuir, 29, 27, 8482–8491, doi:10.1021/la400979d

[Rubenchik 2014] Rubenchik AM, Fedoruk MP, Turitsyn SK. (2014) "The Effect of Self-Focusing on Laser Space-Debris Cleaning" Light: Science & Applications, 3, e159, doi:10.1038/lsa.2014.40

[Rubenstein 2014] Rubenstein M, Cornejo A, Nagpal R. (2014) "Programmable Self-Assembly in a Thousand-Robot Swarm" Science, 345, 6198, 795–799, doi:10.1126/science.1254295

[Rule 2007] Rule JD, Sottos NR, White SR. (2007) "Effect of Microcapsule Size on the Performance of Self-Healing Polymers" Polymer, 48, 3520–3529

[Runyon 2004] Runyon MK, Johnson-Kerner BL, Ismagilov RF. (2004) "Minimal Functional Model of Hemostasis in a Biomimetic Microfluidic System" Angewandte Chemie International Edition, 43, 1531–1536 doi:10.1002/anie.200353428

[Rus 2001] Rus D, Vona M. (2001) "Crystalline Robots: Self-Reconfiguration with Compressible Unit Modules" Autonomous Robots, 10, 107–124

[Russell 1993] Russell TP, Deline VR, Dozier WD, Felcher GP, Agrawal G, Wool RP, Mays JW. (1993) "Direct Observation of Reptation at Polymer Interfaces" Nature, 1993, 365, 235–237, doi:10.1038/365235a0

[Sacanna 2010] Sacanna S, Irvine WT, Chaikin PM, Pine DJ. (2010) "Lock and Key Colloids" Nature, 464, 575–578, doi:10.1038/nature08906

[Sagawa 2011] Sagawa H, Yoshida N. (2011) Fundamentals of Quantum Information, World Scientific

[Saha 2007] Saha GK. (2007) "Software-Implemented Self-Healing System" CLEI Electronic Journal, 10, 2, 5

[Sahmaran 2007] Sahmaran M, Li M, Li VC. (2007) "Transport Properties of Engineered Cementitious Composites under Chloride Exposure" ACI Materials Journal, 303–310

[Şahmaran 2008] Şahmaran M, Keskin SB, Ozerkan G, Yaman IO. (2008) "Self-Healing of Mechanically-Loaded Self Consolidating Concretes with High Volumes of Fly Ash" Cement & Concrete Composites, 30, 872–879, doi:10.1016/j.cemconcomp.2008.07.001

[Sahrani 2008] Sahrani FK, Aziz M, Ibrahim Z, Yahya A. (2008) "Open Circuit Potential Study of Stainless Steel in Environment Containing Marine Sulphate-Reducing Bacteria" Sains Malaysiana, 37, 4, 359–364

[Saihi 2005] Saihi D, Vroman I, Giraud S, Bourbigot S. (2005) "Microencapsulation of Ammonium Phosphate with a Polyurethane Shell Part I: Coacervation Technique" Reactive and Functional Polymers, 64, 3, 127–138, doi:10.1016/j.reactfunctpolym.2005.05.004

[Sakai 2003] Sakai Y, Kitagawa Y, Fukuta T, Iiba M. (2003) "Experimental Study on Enhancement of Self-Restoration of Concrete Beams Using SMA Wire" Proccedings of SPIE, Vol. 5057-22 Smart Systems and Nondestructive Evaluation for Civil Infrastructures

[Salamone 1988] Salamone JC, Chung Y, Clough SB, Watterson AC. (1988) "Thermally Reversible, Covalently Crosslinked Polyphosphazenes" Journal of Polymer Science. Part A, Polymer Chemistry, 26, 2923–2939.

[Samadzadeh 2010] Samadzadeh M, Boura SH, Peikari M, Kasiriha SM, Ashrafi A. (2010) "A Review on Self -Healing Coatings Based on Micro/Nanocapsules" Progress in Organic Coatings 68, 159–164, doi:10.1016/j.porgcoat.2010.01.006

[Samie 2013] Samie M, Dragffy G, Pipe T, Perinpanayagam S. (2013) "Unicellular Self-Healing Electronic Array" Procedia CIRP 11 400-405, 2nd International Through-life Engineering Services Conference

[Samorí 2004] Samorí P, Müllen K, Rabe JP. (2004) "Molecular-Scale Tracking of the Self-Healing of Polycrystalline Monolayers at the Solid–Liquid Interface" Advanced Materials, 16, 1761–1765, doi:10.1002/adma.200400132

[San 1978] San Miguel. (1978) "Fuel Tank Leakage Fiber Flow Sealant" US Patent 4,088,240

[Sanada 2006] Sanada K, Yasuda I, Shindo Y. (2006) "Transverse Tensile Strength of Unidirectional Fibre-Reinforced Polymers and Self-Healing of Interfacial Debonding" Plastics, Rubber and Composites, 35, 2, 67–72, doi:10.1179/174328906X79914

[Sancar 2004] Sancar A, Lindsey-Boltz LA, Unsal-Kaçmaz K, Linn S. (2004) "Molecular Mechanisms of Mammalian DNA Repair and the DNA Damage Checkpoints" Annual Review of Biochemistry, 73, 39–85, doi:10.1146/annurev.biochem.73.011303.073723

[Sanchez 2004] Sanchez-Moral S, Garcia-Guinea J, Luque L. (2004) "Carbonation Kinetics in Roman-Like Lime Mortars" Materiales de Construction, 54, 275, 23–37, doi:10.3989/mc.2004.v54.i275.245

[Sanchez 2012] Sanchez T, Chen DT, DeCamp SJ, Heymann M, Dogic Z. (2012) "Spontaneous Motion in Hierarchically Assembled Active Matter" Nature, 491, 431–434, doi:10.1038/nature11591

[Santagata 2013] Santagata E, Baglieri O, Tsantilis L, Dalmazzo D. (2013) "Evaluation of Self Healing Properties of Bituminous Binders Taking into Account Steric Hardening Effects" Construction and Building Materials, 41, 60–67

[Sapozhnikov 2003] Sapozhnikov MV, Tolmachev YV, Aranson IS, Kwok WK. (2003) "Dynamic Self-Assembly and Patterns in Electrostatically Driven Granular Media" Physical Review Letters, 90, 114301, doi:10.1103/PhysRevLett.90.114301

[Sarangapani 2007] Sarangapani S, Kumar A, Thies C, Stephenson LD. (2007) "Self-Healing Coating and Microcapsules to Make Same" US Patent 7,192,993

[Sariola 2015] Sariola V, Pena-Francesch A, Jung H, Çetinkaya HJ, Pacheco C, Sitti M, Demirel MC. (2015) "Segmented Molecular Design of Self-Healing Proteinaceous Materials" Scientific Reports, 5, doi:10.1038/srep13482

[Sarro 1994] Sarro CA. (1994) "Fluid-Filled Electric Power Cable System with Two-Way Piston Check Valve" US Patent 5,280,131

[Sasaki 1998] Sasaki H, Shouji M. (1998) "Control of Hydrophobic Character of Super-Waterrepellent Surface by UV Irradiation" Chemistry Letters, 27, 4, 293–294, doi:10.1246/cl.1998.293

[Sashuk 2013] Sashuk V, Winkler K, Żywociński A, Wojciechowski T, Górecka E, Fiałkowski M. (2013) "Nanoparticles in a Capillary Trap: Dynamic Self-Assembly at Fluid Interfaces" ACS Nano, 7, 10, 8833–8839, doi:10.1021/nn403297f

[Sasse 2005] Sasse P, Nuckles K. (2005) "A Summary of Testing of Sureseal Cable" Southwire Company, Technology Center, Carrollton, GA

[Saurín 2015] Saurín N, Sanes J, Bermúdez MD. (2015) "Self-Healing of Abrasion Damage in Epoxy Resin–Ionic Liquid Nanocomposites" Tribology Letters, 8, 1, doi:10.1007/s11249-015-0490-9

[Sauvant 2008] Sauvant-Moynot V, Gonzalez S, Kittel J. (2008) "Self-Healing Coatings: An Alternative Route for Anticorrosion Protection" Progress in Organic Coatings, 63, 3, 307–315, doi:10.1016/j.porgcoat.2008.03.004

[Sawyer 2003] Sawyer WG, Freudenberg KD, Bhimaraj P, Schadler S. (2003) "A Study on the Friction and Wear Behavior of PTFE Filled with Alumina Nanoparticles" Wear, 254, 573–580

[Saxena 2007] Saxena A, Orchard ME, Zhang B, Vachtsevanos G, Tang L, Lee Y, Wardi Y. (2007) "Automated Contingency Management for Propulsion Systems" Proceedings of European Control Conference 2007, Kos, Greece, 3515–3522, WeC11.1, SG Tzafestas and P Antsaklis, eds.

[Saywell 2010] Saywell A, Magnano G, Satterley CJ, Perdigão LM, Britton AJ, Taleb N, Giménez-López M, Champness NR, O'Shea JN, Beton PH. (2010) "Self-Assembled Aggregates Formed by Single-Molecule Magnets on a Gold Surface" Nature Communications, 1, 75, doi:10.1038/ncomms1075

[Schaedler 2011] Schaedler TA, Jacobsen AJ, Torrents A, Sorensen AE, Lian J, Greer JR, Valdevit L, Carter WB. (2011) "Ultralight Metallic Microlattices" Science, 334, 962–965, doi:10.1126/science.1211649

[Schäfer 2015] Schäfer S, Kickelbick G. (2015) "Self-Healing Polymer Nanocomposites Based on Diels–Alder-Reactions with Silica Nanoparticles: The Role of the Polymer Matrix" Polymer, 69, 357–368, doi:10.1016/j.polymer.2015.03.017

[Schaller 2010] Schaller V, Weber C, Semmrich C, Frey E, Bausch AR. (2010) "Polar Patterns of Driven Filaments" Nature, 467, 73–77, doi:10.1038/nature09312

[Scheller 2014] Scheller M, Mills MS, Miri MA, Cheng W, Moloney JV, Kolesik M, Polynkin P, Christodoulides DN. (2014) "Externally Refuelled Optical Filaments" Nature Photonics, 8, 297–301 doi:10.1038/nphoton.2014.47

[Scherer 1941] Scherer RP. (1941) "Capsule and Method of Making the Same" US Patent 2,234,479

[Schiessl 1996] Schiessl P, Brauer N. (1996) "Influence of Autogenous Healing of Cracks on Corrosion of Reinforcement" Durability of Building Materials and Compoments, 7, 542–552

[Schlangen 2006] Schlangen E, ter Heide N, van Breugel K. (2006) "Crack Healing of Early Age Cracks in Concrete" M.S. Konsta-Goutos. (ed.). Measuring, Monitoring and Modeling Concrete Properties, 273–284. Springer, The Netherlands.

[Schmidt 2006] Schmidt AM. (2006) "Electromagnetic Activation of Shape Memory Polymer Networks Containing Magnetic Nanoparticles" Macromolecular Rapid Communications, 27, 1168–1172, doi:10.1002/marc.200600225

[Schniepp 2006] Schniepp HC, Saville DA, Aksay IA. (2006) "Self-Healing of Surfactant Surface Micelles on Millisecond Time Scales" Journal of the American Chemical Society, 12378–12379.

[Schoen 2010] Schoen DT, Schoen AP, Hu L, Kim HS, Heilshorn SC, Cui Y. (2010) "High Speed Water Sterilization Using One-Dimensional Nanostructures" Nano Letters, 10, 9, pp 3628–3632, doi:10.1021/nl101944e

[Schrock 2006] Schrock RR. (2006) "Multiple Metal–Carbon Bonds for Catalytic Metathesis Reactions (Nobel Lecture)" Angewandte Chemie International Edition, 45, 3748, doi:10.1002/anie.200600085

[Schrödinger 1944] Schrödinger E. (1944) What Is Life, Cambridge University Press, Cambridge

[Schumacher 2007] Schumacher JF, Carman ML, Estes TG, Feinberg AW, Wilson LH, Callow ME, Callow JA, Finlay JA, Brennan AB. (2007) "Engineered Antifouling Microtopographies: Effect of Feature Size, Geometry, and Roughness on Settlement of Zoospores of the Green Alga Ulva" Biofouling: The Journal of Bioadhesion and Biofilm Research, 23, 55–62, doi:10.1080/08927010601136957

[Schutz 1959] Schutz JM. (1959) "Self Sealing Pneumatic Tire" US Patent 2,905,229

[Schwartz 1970] Schwartz SS. (1970) "Self-Scaling Space Suit" US Patent 3,536,576

[Scott 2005] Scott TF, Schneider AD, Cook WD, Bowman CN. (2005) "Photoinduced Plasticity in Cross-Linked Polymers" Science, 308, 5728, 1615–1617, doi:10.1126/science.1110505

[Sedlacek 2003] Sedlacek J, Vostracky Z, Knobloch H, Schramm HH, Wiesinger C. (2003) "Optimization of High-Voltage Self-Blast Interrupters by Gas Flow and Electric Field Computations" IEEE Transactions on Power Delivery, 18, 4, 1228–1235

[Semprimosching 2006] Semprimosching C. (2006) "Enabling Self-Healing Capabilities: A Small Step to Bio-Mimetic Materials" European Space Agency Materials Report Number 4476, Noordwijk: European Space Agency

[Sen 2010] Sen R, Mullins G. (2010) "Underwater Fiber-Reinforced Polymer Repair of Corroding Piles Incorporating Cathodic Protection" Final Report for Highway IDEA Project 128, Transportation Research Board, NW, Washington, DC

[Serra 2012] Serra-Picamal X, Conte V, Vincent R, Anon E, Tambe DT, Bazellieres E, Butler JP, Fredberg JJ, Trepat X. (2012) "Mechanical Waves during Tissue Expansion" Nature Physics, 8, 628–634, doi:10.1038/nphys2355

[Seshadri 2007] Seshadri M, Saigal S. (2007) "Crack Bridging in Polymer Nanocomposites" Journal of Engineering Mechanics, 133, 8

[Seymour 2023] Seymour LM, Maragh J, Sabatini P, et al. (2023) "Hot Mixing: Mechanistic Insights into the Durability of Ancient Roman Concrete" Science Advances, 9, eadd1602. doi:10.1126/sciadv.add1602

[Shacham 2012] Shacham-Diamand Y, Mintz M, Taub A. (2012) "IL-1 Receptor Antagonist-Coated Electrode and Uses Thereof" US Patent Application 20140249396

[Shafiq 2012] Shafiq Z, Cui J, Pastor-Pérez L, San Miguel V, Gropeanu RA, Serrano C, del Campo A. (2012) "Bioinspired Underwater Bonding and Debonding on Demand" Angewandte Chemie International Edition, 51, 4332–4335, doi:10.1002/anie.201108629

[Shafrir 2002] Shafrir Y, Forgacs G. (2002) "Mechanotransduction through the Cytoskeleton" American Journal of Physiology-Cell Physiology, 282, 3, C479–C486, doi:10.1152/ajpcell.00394.2001

[Shah 2008] Shah PB, Brinker JC, Koene BE. (2008) "Hydrophobic Organic–Inorganic Hybrid Silane Coatings" US Patent Application 20080113188

[Shah 2015] Shah AA, Schultz B, Zhang W, Glotzer SC, Solomon MJ. (2015) "Actuation of Shape-Memory Colloidal Fibres of Janus Ellipsoids" Nature Materials, 14, 117–124, doi:10.1038/nmat4111

[Shajil 2012] Shajil N, Srinivasan SM, Santhanam M. (2012) "An Innovative Approach to Achieve Re-Centering and Ductility of Cement Mortar Beams through Randomly Distributed Pseudo-Elastic Shape Memory Alloy Fibers" Proceedings of SPIE, Vol. 8342, Behavior and Mechanics of Multifunctional Materials and Composites, NC Goulbourne, Z Ounaies (eds), doi:10.1117/12.915829

[Shan 2010] Shan L, Tan Y, Underwood S, Kim YR. (2010) "Application of Thixotropy to Analyze Fatigue and Healing Characteristics of Asphalt Binder" Transportation Research Record: Journal of the Transportation Research Board, 2179, 1, 85–92, doi:10.3141/2179-10

[Shang 2015] Shang J, Wang Y, Chen M, Dai J, Zhou X, Kuttner J, Hilt G, Shao X, Gottfried JM, Wu K. (2015) "Assembling Molecular Sierpiński Triangle Fractals" Nature Chemistry, 7, 389–393, doi:10.1038/nchem.2211

[Shanmuganathan 2010] Shanmuganathan K, Capadona JR, Rowan SJ, Weder C. (2010) "Biomimetic Mechanically Adaptive Nanocomposites" Progress in Polymer Science, 35, 1–2, 212–222, doi:10.1016/j.progpolymsci.2009.10.005

[Shannon 1949] Shannon CE, Weaver W. (1949) The Mathematical Theory of Communication, University of Illinois Press, Urbana and Chicago

[Sharon 2004] Sharon E, Marder M, Swinney H. (2004) "Leaves, Flowers and Garbage Bags: Making Waves" American Scientist, 92, 3, 254–261, doi:10.1511/2004.3.25

[Shchukin 2007] Shchukin DG, Möhwald H. (2007) "Surface-Engineered Nanocontainers for Entrapment of Corrosion Inhibitors" Advanced Functional Materials, 17, 1451–1458

[She 2008] She Z, Zhang B, Jin C, Feng Q, Xu Y. (2008) "Preparation and In Vitro Degradation of Porous Three-Dimensional Silk Fibroin/Chitosan Scaffold" Polymer Degradation and Stability, 93, 1316–1322, doi:10.1016/j.polymdegradstab.2008.04.001

[Shen 2002] Shen JS, Harmon JP, Lee S. (2002) "Thermally-Induced Crack Healing in Poly(methyl methacrylate)" Journal of Materials Research, 17, 1335–40

[Shen 2014] Shen S, Lu X, Zhang Y, Lytton R. (2014). "Fracture and Viscoelastic Properties of Asphalt Binders during Fatigue and Rest Periods." Journal of Testing and Evaluation, 42, 1, 20130030, doi:10.1520/JTE20130030

[Shen 2015] Shen Y, Tao J, Tao H, Chen S, Pan L, Wang T. (2015) "Anti-Icing Potential of Superhydrophobic Ti_6Al_4V Surfaces: Ice Nucleation and Growth" Langmuir, 31, 39, 10799–10806, doi:10.1021/acs.langmuir.5b02946

[Sheng 2009] Sheng X, Lee JK, Kessler MR. (2009) "Influence of Cross-Link Density on the Properties of ROMP Thermosets" Polymer, 50, 5, 1264–1269, doi:10.1016/j.polymer.2009.01.021

[Shenoy 2012] Shenoy VB, Gracias DH. (2012) "Self-Folding Thin-Film Materials: From Nanopolyhedra to Graphene Origami" MRS Bulletin, 37, 9, 847–854

[Shepherd 2011] Shepherd RF, Ilievski F, Choi W, Morin SA, Stokes AA, Mazzeo AD, Chen X, Wang M, Whitesides GM. (2011) "Multigait Soft Robot" Proceedings of the National Academy of Sciences of the United States of America, 108, 51, 20400–20403

[Sherman 1981] Sherman T. (1981) "On Connecting Large Vessels to Small the Meaning of Murray's Law" Journal of General Physiology, 78, 4, 431–453.

[Sherratt 1991] Sherratt JA, Murray JD. (1991) "Mathematical Analysis of a Basic Model for Epidermal Wound Healing" Journal of Mathematical Biology, 29, 389–404

[Shevchenko 2006] Shevchenko EV, Talapin DV, Kotov NA, O'Brien S, Murray CB. (2006) "Structural Diversity in Binary Nanoparticle Superlattices" Nature, 439, 55–59, doi:10.1038/nature04414

[Shi 2005] Shi C, Booth R. (2005) "Laboratory Development and Field Demonstration of Self-Sealing/Self-Healing Landfill" Waste Management, 25, 3, 231–238, doi:10.1016/j.wasman.2004.11.006

[Shi 2008] Shi D, Matsusaki M, Kaneko T, Akashi M. (2008) "Photo-Cross-Linking and Cleavage Induced Reversible Size Change of Bio-Based Nanoparticles" Macromolecules, 41, 8167–8172

[Shi 2011] Shi X, Xie N, Gong J. (2011) "Recent Progress in the Research on Microbially Influenced Corrosion: A Bird's Eye View through the Engineering Lens" Recent Patents on Corrosion Science, 1, 118–131

[Shi 2013] Shi X, Ma Y. (2013) "Topological Structure Dynamics Revealing Collective Evolution in Active Nematics" Nature Communications, 4, 3013, doi:10.1038/ncomms4013

[Shi 2015] Shi Y, Wang M, Ma C, Wang Y, Li X, Yu G. (2015) "A Conductive Self-Healing Hybrid Gel Enabled by Metal–Ligand Supramolecule and Nanostructured Conductive Polymer" Nano Letters, 15, 9, 6276–6281, doi:10.1021/acs.nanolett.5b03069

[Shiammoto 2005] Shiammoto A, Zhao H, Abè H. (2005) "Vibration Characteristics in a Smart Bridge Model Using Shape Memory Alloy Fiber Reinforced Composite" SPIE 5761 Active Materials: Behavior and Mechanics

[Shibuichi 1996] Shibuichi S, Onda T, Satoh N, Tsujii K. (1996) "Super Water-Repellent Surfaces Resulting from Fractal Structure" Journal of Physical Chemistry, 100, 50, 19512–19517, doi:10.1021/jp9616728

[Shigo 1984] Shigo AL. (1984) "Compartmentalization: A Conceptual Framework for Understanding How Trees Grow and Defend Themselves" Annual Review of Phytopathology, 22, 1, 189–214, doi:10.1146/annurev.py.22.090184.001201

[Shim 2012] Shim J, Perdigou C, Chen C, Bertoldi K, Reis PM. (2012) "Buckling-Induced Encapsulation of Structured Elastic Shells under Pressure" Proceedings of the National Academy of Sciences of the United States of America, 109, 16, 5978–5983, doi:10.1073/pnas.1115674109

[Shim 2013] Shim J, Shin JH, Lee IY, Choi D, Baek JW, Heo J, Park W, Leem JW, Yu JS, Jung WS5, Saraswat K, Park JH. (2013) "Effects of Point Defect Healing on Phosphorus Implanted Germanium N+/P Junction and Its Thermal Stability" Journal of Applied Physics, 114, 094515, doi:10.1063/1.4820580

[Shimada 2004] Shimada S, Takahasi Y, Sugino Y, Hara S, Yamamoto K J. (2004) "Autonomic Healing of a Pinhole in Polyethylene and Photografted Polyethylene-*g*-poly(hexyl methacrylate) Films" Polym Sci, Part B: Polym Phys, 42, 1705

[Shimamoto 2007] Shimamoto A, Zhao H, Azakami T. (2007) "Active Control for Stress Intensity of Crack-Tips under Mixed Mode by Shape Memory TiNi Fiber Epoxy Composites" Smart Materials and Structures, 16, N13–N21 doi:10.1088/0964-1726/16/3/N01

[Shin 2005] Shin ME. (2005) "Self-Healing Components in Robust Software Architecture for Concurrent and Distributed Systems" Science of Computer Programming, 57, 1, 27–44, doi:10.1016/j.scico.2004.10.003

[Shinbach 2010] Shinbach MP, Nwabunma D. (2010) "Drilling Fluid Containing Microspheres and Use Thereof" US Patent 7,767,629

[Shinya 2006] Shinya N, Kyono J, Laha K. (2006) "Self-Healing Effect of Boron Nitride Precipitation on Creep Cavitation in Austenitic Stainless Steel" Journal of Intelligent Material Systems and Structures, 17, 12, 1127–1133, doi:10.1177/1045389X06065238

[Shirrif 1930] Shirrif FW. (1930) "Food Product and Method of Preparing" US Patent 1,758,089

[Shivakumar 2004] Shivakumar K, Emmanwori L. (2004) "Mechanics of Failure of Composite Laminates with an Embedded Fiber Optic Sensor" Journal of Composite Materials, 38, 669–680

[Shivapooja 2013] Shivapooja P, Wang Q, Orihuela B, Rittschof D, López GP, Zhao X. (2013) "Bioinspired Surfaces with Dynamic Topography for Active Control of Biofouling" Advanced Materials, 25, 10, 1430–1434, doi:10.1002/adma.201203374

[Shrestha 2015] Shrestha KC, Saiidi MS, Cruz CA. (2015) "Advanced Materials for Control of Post-Earthquake Damage in Bridges" Smart Materials and Structures, 24, 2, 025035, doi:10.1088/0964-1726/24/2/025035

[Shuler 2010] Shuler Jr, RL. (2010) "Methods and Circuitry for Reconfigurable SEU/SET Tolerance" US Patent 7,859,292

[Si 2002] Si Z, Little DN, Lytton RL. (2002) "Evaluation of Fatigue Healing Effect of Asphalt Concrete by Pseudostiffness" Transportation Research Record 1789, Transportation Research Board, Washington, DC, 73–79, doi:10.3141/1789-08

[Siat 1997] Siat C, Bourbigot S, Le Bras M. (1997) "Thermal Behavior of Polyamide-6-Based Intumescent Formulations: A Kinetic Study" Polymer Degradation and Stability, 58, 303–313

[Siddiqi 2009] Siddiqi A, de Weck OL. (2009) "Reconfigurability in Planetary Surface Vehicles" Acta Astronautica, 64, 5–6, 589–601, doi:10.1016/j.actaastro.2008.10.010

[Sidorenko 1999] Sidorenko A, Minko S, Schenk-Meuser K, Duschner H, Stamm M. (1999) "Switching of Polymer Brushes" Langmuir, 15, 8349–8355

[Silverberg 2015] Silverberg JL, Na JH, Evans AA, Liu B, Hull TC, Santangelo CD, Lang RJ, Hayward RC, Cohen I. (2015) "Origami Structures with a Critical Transition to Bistability Arising from Hidden Degrees of Freedom" Nature Materials, 14, 389–393, doi:10.1038/nmat4232

[Simpson 2010] Simpson B, Nunnery G, Tannenbaum R, Kalaitzidou K. (2010) "Capture/Release Ability of Thermo-Responsive Polymer Particles" Journal of Materials Chemistry, 20, 3496–3501, doi:10.1039/B922972H

[Singer 1999] Singer AJ, Clark RA. (1999) "Cutaneous Wound Healing" New England Journal of Medicine, 341, 738–746, doi:10.1056/NEJM199909023411006

[Singh 2004] Singh A, Lee Y, Dressick W. (2004) "Self-Cleaning Fabrics for Decontamination of Organophosphorous Pesticides and Related Chemical Agents" Advanced Materials, 16, 23–24, doi:10.1002/adma.200400660

[Sisomphon 2011] Sisomphon K, Copuroglu O, Fraaij A. (2011) "Application of Encapsulated Lightweight Aggregate Impregnated with Sodium Monofluorophosphate as a Self-Healing Agent in Blast Furnace Slag Mortar" Heron, 56, 1 / 2, 13–32

[Siviloglou 2007] Siviloglou GA, Broky J, Dogariu A, Christodoulides D N. (2007) "Observation of Accelerating Airy Beams." Physical Review Letters, 99, 213901, doi:10.1103/PhysRevLett.99.213901

[Skelton 2001] Skelton RE, Adhikari R, Pinaud JP, Chan WJ, Helton W. (2001) "An Introduction to the Mechanics of Tensegrity Structures" Proceedings of 40th IEEE Conference on Decision and Controls, 4254–4259, doi:10.1109/CDC.2001.980861

[Skipor 2006] Skipor A, Scheifer S, Olson B. (2006) "Self-Healing Polymer Compositions" US Patent 7,108,914

[Skorb 2009] Skorb EV, Skirtach AG, Sviridov DV, Shchukin DG, Miihwald H. (2009) "Laser-Controllable Coatings for Corrosion Protection" ACS Nano, 3, 7, 1753–1760, doi:10.1021/nn900347x

[Skrabania 2007] Skrabania K, Kristen J, Laschewsky A, Akdemir O, Hoth A, Lutz JF. (2007) "Design, Synthesis, and Aqueous Aggregation Behavior of Nonionic Single and Multiple Thermoresponsive Polymers" Langmuir, 23, 84–93

[Slenski 2004] Slenski GA, Metzler Jr. PS. (2004) "Wired for Success Ensuring Aircraft Wiring Integrity Requires a Proactive Systems Approach" AMTIAC Quarterly, 8, 3

[Slocum 2013] Slocum AH, Kazem BI, Figueredo S. (2013) "Vortex-Induced Cleaning of Surfaces" US Patent Application 20130047978

[Slocum 2017] Slocum AH, Rojas FE. (2017) "Method and Apparatus for Bringing under Control an Uncontrolled Flow through a Flow Device" US Patent 9,719,331

[Sloof 2016] Sloof WG, Pei R, McDonald SA, Fife JL, Shen L, Boatemaa L, Farle AS, Yan K, Zhang X, van der Zwaag S, Lee PD, Withers PJ. (2016) "Repeated Crack Healing in MAX-Phase Ceramics Revealed by 4D in Situ Synchrotron X-Ray Tomographic Microscopy" Scientific Reports, 6, 23040, doi:10.1038/srep23040

[Small 2001] Small JH, Loy DA, Wheeler DR, McElhanon JR, Saunders RS. (2001) "Method of Making Thermally Removable Polymeric Encapsulants" US Patent 6,271,335

[Smith 1947] Smith WC, Haworth JP. (1947) "Puncture-Proof Fuel Cell US Patent 2,416,231

[Smith 2005] Smith KA, Tyagi S, Balazs AC. (2005) "Healing Surface Defects with Nanoparticle-Filled Polymer Coatings: Effect of Particle Geometry" Macromolecules, 38, 10138–10147

[Smith 2007] Smith KA, Jasnow D, Balazs AC. (2007) "Designing Synthetic Vesicles that Engulf Nanoscopic Particles" Journal of Chemical Physics, 127, 084703

[Smith 2013] Smith JD, Dhiman R, Anand S, Reza-Garduno E, Cohen RE, McKinley GH, Varanasi KK. (2013) "Droplet Mobility on Lubricant-Impregnated Surfaces" Soft Matter, 9, 1772–1780, doi:10.1039/C2SM27032C

[Smith 2014] Smith SW, Newman JA, Piascik RS, Glaessgen EH. (2014) "System for Repairing Cracks in Structures" US Patent 8,679,642

[Snelson 1960] Snelson K. (1960) "Continuous Tension, Discontinuous Compression Structures" US Patent 3,169,611

[Snezhko 2012] Snezhko A, Aranson IS. (2012) "Magnetic Manipulation of Self-Assembled Colloidal Asters" Nature Materials, 10, 698–703, doi:10.1038/NMAT3083

[Snoeck 2015] Snoeck D, De Belie N. (2015) "From Straw in Bricks to Modern Use of Microfibers in Cementitious Composites for Improved Autogenous Healing: A Review" Construction and Building Materials, 95, 774–787, doi:10.1016/j.conbuildmat.2015.07.018

[So 2009a] So CR, Kulp JL, Oren EE, Zareie H, Tamerler C, Evans JS, Sarikaya M. (2009) "Molecular Recognition and Supramolecular Self-Assembly of a Genetically Engineered Gold Binding Peptide on Au{111}" ACS Nano, 3, 6, 1525–1531, doi:10.1021/nn900171s

[So 2009b] So JH, Thelen J, Qusba A, Hayes GJ, Lazzi G, Dickey MD. (2009) "Reversibly Deformable and Mechanically Tunable Fluidic Antennas" Advanced Functional Materials, 19, 3632–3637, doi:10.1002/adfm.200900604

[Sohn 2006] Sohn KE, Perry R, Ober CK, Kramer EJ, Callow ME, Callow JA, Fischer DA. (2006) "Anti-Biofouling Properties of Comblike Block Copolymers with Amphiphilic Side Chains" Langmuir, 22, 11, 5075–5086, doi:10.1021/la052978l

[Solberg 2008] Solberg K, Dhakal RP, Bradley B, Mander JB, Luoman Li. (2008) "Seismic Performance of Damage-Protected Beam-Column Joints" ACI Structural Journal, 105-522, 205–210

[Solga 2007] Solga A, Cerman Z, Striffler BF, Spaeth M, Barthlott W. (2007) "The Dream of Staying Clean: Lotus and Biomimetic Surfaces" Bioinspiration & Biomimetics, 2, S126–S134 doi:10.1088/1748-3182/2/4/S02

[Somoza 2007] Somoza A, Macchi CE, Lumley RN, Polmear IJ, Dupasquier A, Ferragut R. (2007) "Role of Vacancies during Creep and Secondary Precipitation in an Underaged Al-Cu-Mg-Ag Alloy" Physica Status Solidi C, 4, 10, 3473–3476, doi:10.1002/pssc.200675737

[Sone 2007] Sone H, Mishima T, Miyachi A, Fukuda T, Hosaka S. (2007) "Growth Control of Self-Assembled $ErSi_2$ Nanowires on Si(001) and Si(110) Surfaces" Microelectronic Engineering, 84, 1491–1495

[Song 2008] Song GM, Pei YT, Sloof WG, Li SB, De Hosson JT, van der Zwaag S. (2008) "Oxidation Induced Crack Healing of Ti_3AlC_2 Ceramics" Scripta Materialia, 58, 13–16, doi:10.1016/j.scriptamat.2007.09.006

[Song 2011] Song YJ, Peters KJ. (2011) "A Self-Repairing Polymer Waveguide Sensor" Smart Materials and Structures, 20 065005 (12pp) doi:10.1088/0964-1726/20/6/065005

[Song 2013] Song YK, Jo YH, Lim YJ, Cho SY, Yu HC, Ryu BC, Lee SI, Chung CM. (2013) "Sunlight-Induced Self-Healing of a Microcapsule-Type Protective Coating" ACS Applied Materials & Interfaces, 5, 4, 1378–1384, doi:10.1021/am302728m

[Sonntag 2004] Sonntag P, Hoerner P, Cheymol A, Argy G, Riess G, Reiter G. (2004) "Biocide Squirting from an Elastomeric Tri-Layer Film" Nature Materials, 3, 311–315, doi:10.1038/nmat1113

[Soroushian 2001] Soroushian P, Ostowari K, Nossoni A, Chowdhury H. (2001) "Repair and Strengthening of Concrete Structures through Application of Corrective Posttensioning Forces with Shape Memory Alloys" Transportation Research Record, 1770, doi:10.3141/1770-03

[South 2010] South AB, Lyon LA. (2010) "Autonomic Self-Healing of Hydrogel Thin Films" Angewandte Chemie International Edition, 122, 779–783, doi:10.1002/ange.200906040

[Soutis 1999] Soutis C, Duan DM, Goutas P. (1999) "Compressive Behaviour of CFRP Laminates Repaired with Adhesively Bonded External Patches" Composite Structures, 45, 4, 289–301, doi:10.1016/S0263-8223(99)00033-1

[Spathi 2015] Spathi C, Young N, Heng JY, Vandeperre LJ, Cheeseman CR. (2015) "A Simple Method for Preparing Super-Hydrophobic Powder from Paper Sludge Ash" Materials Letters, 142, 80–83, doi:10.1016/j.matlet.2014.11.123

[Spears 2014] Spears Jr. MW, Herman ES, Gaulding JC, Lyon LA. (2014) "Dynamic Materials from Microgel Multilayers" Langmuir, 30, 22, 6314–6323, doi:10.1021/la403058t

[Sperling 1994] Sperling LH, Klein A, Sambasivam M, Kim KD. (1994) "Molecular Basis of Healing and Fracture at Polymer Interfaces" Polymers for Advanced Technologies, 5, 9, 453–472, doi:10.1002/pat.1994.220050901

[Sporns 2004] Sporns O, Chialvo DR, Kaiser M, Hilgetag CC. (2004) "Organization, Development and Function of Complex Brain Networks" TRENDS in Cognitive Sciences, 8, 9, doi:10.1016/j.tics.2004.07.008

[Spröwitz 2014] Spröwitz A, Moeckel R, Vespignan M, Bonardi S, Ijspeert AJ. (2014) "Roombots: A Hardware Perspective on 3D Self-Reconfiguration and Locomotion with a Homogeneous Modular Robot" Robotics and Autonomous Systems, 62, 7, 1016–1033, doi:10.1016/j.robot.2013.08.011

[Spruell 2003] Spruell SL, Robertson SM, Ware, Jr. JN. (2003) "Self-Sealing Electrical Cable Having a Finned Inner Layer" US Patent 6,573,456

[Spruell 2013] Spruell SL, Sasse PA GA, Nuckles KM, Reece D. (2013) "Self-Sealing Electrical Cable Using Rubber Resins" US Patent 8,470,108

[Sreeprasad 2015] Sreeprasad TS, Nguyen P, Alshogeathri A, Hibbeler L, Martinez F, McNeil N, Berry V. (2015) "Graphene Quantum Dots Interfaced with Single Bacterial Spore for Bio-Electromechanical Devices: A Graphene Cytobot" Scientific Reports, 5, 9138, doi:10.1038/srep09138

[Srinivasan 2001] Srinivasan U, Liepmann D, Howe RT. (2001) "Microstructure to Substrate Self-Assembly Using Capillary Forces" Journal of Microelectromechanical Systems, 10, 1, 17–24, doi:10.1109/84.911087

[Srivastava 2014] Srivastava S, Nykypanchuk D, Fukuto M, Halverson JD, Tkachenko AV, Yager KG, Gang O. (2014) "Two-Dimensional DNA-Programmable Assembly of Nanoparticles at Liquid Interfaces" Journal of the American Chemical Society, 136, 23, 8323–8332, doi:10.1021/ja501749b

[Stack 2011] Stack J. (2011) "Fuel Protection" Loglines, Defense Logistics Agency, Jan–Feb, 22–23

[Stackpoole 2008] Stackpoole M. (2008) "Ceramic Foams for Novel TPS Applications" NASA Tech Briefs, ARC-15260-1, NASA Ames Research Center, Moffett Field CA 94035

[Stahl 2008] Stahl Sr JP. (2008) "Self-Adjusting Intumescent Firestopping Apparatus" US Patent 7,373,761

[Starzyk 2012] Starzyk JA, Graham JT, Raif P, Tan AH. (2012) "Motivated Learning for the Development of Autonomous Systems" Cognitive Systems Research, 14, 10–25

[Stauffer 2009] Stauffer A, Rossier J. (2009) "Self-Testable and Self-Repairable Bio-Inspired Configurable Circuits" NASA/ESA Conference on Adaptive Hardware and Systems, IEEE, doi:10.1109/AHS.2009.19

[Steels 2004] Steels L. (2004) "The Autotelic Principle, in Embodied Artificial Intelligence" Lecture Notes in Computer Science, Vol. 3139, pp 231–242, Springer-Verlag, F Iida, R Pfeifer, L Steels, Y Kuniyoshi (eds), doi:10.1007/978-3-540-27833-7_17

[Stefani 2000] Stefani AM. (2000) "Observations on Efforts to Address Concerns about Aircraft Wiring" US Federal Aviation Administration, Washington, DC, Report No. AV-2001-004

[Stein 2003] Stein J, Truby K, Darkangelo-Wood C, Stein J, Gardner M, Swain G, Kavanagh C, Kovach B, Schultz M, Wiebe D, Holm E, Montemarano J, Wendt D, Smith C, Meyer A. (2003) "Silicone Foul Release Coatings: Effect of the Interaction of Oil and Coating Functionalities on the Magnitude of Macrofouling Attachment Strengths" Biofouling, 19, 71, doi:10.1080/0892701031000089525

[Stellman 2007] Stellman P, Buchner T, Arora WJ, Barbastathis G. (2007) "Dynamics of Nanostructured Origami" Journal of Microelectromechanical Systems, 16, 4, 932–949

[Stewart 2014] Stewart S, Gomez AW, Armstrong BE, Henner A, Stankunas K. (2014) "Sequential and Opposing Activities of Wnt and BMP Coordinate Zebrafish Bone Regeneration" Cell Reports, 6, 3, 482–498, doi:10.1016/j.celrep.2014.01.010

[Stocks 1999] Stocks-Fischer S, Galinat JK, Bang SS. (1999) "Microbiological Precipitation of $CaCO_3$," Soil Biology and Biochemistry, 31, 1563–1571.

[Stricker 2008] Stricker J, Cookson S, Bennett MR, Mather WH, Tsimring LS, Hasty J. (2008) "A Fast, Robust and Tunable Synthetic Gene Oscillator" Nature, 456, 516–519, doi:10.1038/nature07389

[Stubblefield 1998] Stubblefield MA, Yang C, Pang SS, Lea RH. (1998) "Development of Heat-Activated Joining Technology for Composite-to-Composite Pipe Using Prepreg Fabric" Polymer Engineering and Science, 38, 1, 143–149

[Stukalin 2013] Stukalin EB, Cai LH, Kumar NA, Leibler L, Rubinstein M. (2013) "Self-Healing of Unentangled Polymer Networks with Reversible Bonds" Macromolecules, 46, 18, 7525–7541, doi:10.1021/ma401111n

[Style 2015] Style RW, Boltyanskiy R, Allen B, Jensen KE, Foote HP, Wettlaufer JS, Dufresne ER. (2015) "Stiffening Solids with Liquid Inclusions" Nature Physics, 11, 82–87, doi:10.1038/nphys3181

[Subramanian 2004] Subramanian V, Wolf EE, Kamat PV. (2004) "Catalysis with TiO_2/Gold Nanocomposites. Effect of Metal Particle Size on the Fermi Level Equilibration" Journal of the American Chemical Society, 126, 4943, doi:10.1021/ja0315199

[Subramanyam 2013] Subramanyam SB, Rykaczewski K, Varanasi KK. (2013) "Ice Adhesion on Lubricant-Impregnated Textured Surfaces" Langmuir, 29, 44, 13414–13418, doi:10.1021/la402456c

[Sudduth 2006] Sudduth RD. (2006) "Analysis of the Maximum Tensile Strength of a Composite with Spherical Particulates" Journal of Composite Materials, 40, 4, 301–331

[Sugiyama 2008] Sugiyama R, Yamane K, Nakao W, Takahashi K, Ando K. (2008) "Effect of Difference in Crack-healing Ability on Fatigue Behavior of Alumina/Silicon Carbide Composites" Journal of Intelligent Materials Systems and Structures, 19, 411–415

[Suh 2000] Suh HJ, Bharathi P, Beebe DJ, Moore JS. (2000) "Dendritic Material as a Dry-Release Sacrificial Layer" Journal of Microelectromechanical Systems, 9, 2, 198–205

[Sukhotskaya 1983] Sukhotskaya SS, Mazhorova VP, Terekhin YN. (1983) "Effect of Autogenous Healing of Concrete Subjected to Periodic Freeze-Thaw Cycles" Journal of Hydrotechnical Construction, 17, 294–296, doi:10.1007/BF01427180

[Suleiman 2017] Suleiman AR, Nehdi ML. (2017) "Modeling Self-Healing of Concrete Using Hybrid Genetic Algorithm-Artificial Neural Network" Materials, 10, 2, 135, doi:10.3390/ma10020135

[Sullivan 1946] Sullivan DJ. (1946) "Fuel Container" US Patent 2,405,986

[Sultan 2001] Sultan C, Corless M, Skelton RE. (2001) "The Prestressabilty Problem of Tensegrity Structures: Some Analytical Solutions" International Journal of Solids and Structures, 38, 30–31, 5223–5252, doi:10.1016/S0020-7683(00)00401-7

[Sumino 2012] Sumino Y, Nagai KH, Shitaka Y, Tanaka D, Yoshikawa K, Chaté H, Oiwa K. (2012) "Large-Scale Vortex Lattice Emerging from Collectively Moving Microtubules" Nature, 483, 448–452, doi:10.1038/nature10874

[Sun 2001] Sun RD, Nakajima A, Fujishima A, Watanabe T, Hashimoto K. (2001) "Photoinduced Surface Wettability Conversion of ZnO and TiO_2 Thin Films" Journal of Physical Chemistry B, 105, 10, 1984–1990, doi:10.1021/jp002525j

[Sun 2011] Sun T, Qing G. (2011) "Biomimetic Smart Interface Materials for Biological Applications" Advanced Materials, 23, 12, H57–H77, doi:10.1002/adma.201004326

[Sun 2016] Sun Y, Meng Q, Qian M, Liu B, Gao K, Ma Y, Wen M, Zheng W. (2016) "Enhancement of Oxidation Resistance via a Self-Healing Boron Carbide Coating on Diamond Particles" Scientific Reports, 6, 20198, doi:10.1038/srep20198

[Sundaresan 2013] Sundaresan VB, Morgan A, Castellucci M. (2013) "Self-Healing of Ionomeric Polymers with Carbon Fibers from Medium-Velocity Impact and Resistive Heating," Smart Materials Research, 271546, doi:10.1155/2013/271546

[Sunder 1999] Sunder A, Krämer M, Hanselmann R, Mülhaupt R, Frey H. (1999) "Molecular Nanocapsules Based on Amphiphilic Hyperbranched Polyglycerols" Angewandte Chemie International Edition, 38, 23, 3552–3555

[Sundstrom 2015] Sundstrom G. (2015) "Assessment and Placement of Living Snow Fences to Reduce Highway Maintenance Costs and Improve Safety (Living Snow Fences)" Study No: 047-10, Report No. CDOT-2015-01, Colorado Department of Transportation, Denver, CO

[Suryanarayana 2008] Suryanarayana C, Rao KC, Kumar D. (2008) "Preparation and Characterization of Microcapsules Containing Linseed Oil and Its Use in Self-Healing Coatings" Progress in Organic Coatings, 63, 72–78, doi:10.1016/j.porgcoat.2008.04.008

[Suthakorn 2003] Suthakorn J, Kwan Y, Chirikjian GS. (2003) "A Semi-Autonomous Replicating Robotic System" Proceedings of 2003 IEEE International Conference on Intelligent Robots and Applications (CIRA), Kobe, Japan, pp. 776–781, doi:10.1109/CIRA.2003.1222279

[Suzuki 2011] Suzuki Y, Inou N, Kimura H, Koseki M. (2011) "Self-Reconfigurable Modular Robots Adaptively Transforming a Mechanical Structure: Algorithm for Adaptive Transformation to Load Condition" Journal of Robotics, 2011, 794251, doi:10.1155/2011/794251

[Suzuki 2015] Suzuki R, Weber CA, Frey E, Bausch AR. (2015) "Polar Pattern Formation in Driven Filament Systems Requires Non-Binary Particle Collisions" Nature Physics, 11, 839–843, doi:10.1038/nphys3423

[Svenson 2005] Svenson S, Tomalia DA. (2005) "Dendrimers in Biomedical Applications: Reflections on the Field" Advanced Drug Delivery Reviews, 57, 14, 2106–2129, doi:10.1016/j.addr.2005.09.018

[Swain 1996] Swain GW, Schultz MP. (1996) "The Testing and Evaluation of Non-Toxic Antifouling Coatings" Biofouling, 10, 1–3, 187–197, doi:10.1080/08927019609386279

[Sylvester 2006] Sylvester D, Blaauw D, Karl E. (2006) "ElastIC: An Adaptive Self-Healing Architecture for Unpredictable Silicon," IEEE Design Test Computing, 23, 6, 484–490, doi:10.1109/MDT.2006.145

[Syrett 2010] Syrett JA, Mantovani G, Barton WR, Price DP, Haddleton DM. (2010) "Self-Healing Polymers Prepared via Living Radical Polymerization" Polymer Chemistry, 1, 102–106, doi:10.1039/b9py00316a

[Szabó 2011] Szabó T, Molnár-Nagy L, János Bognár J, Nyikos L, Telegdi J. (2011) "Self-Healing Microcapsules and Slow Release Microspheres in Paints" Progress in Organic Coatings, 72, 1–2, 52–57, doi:10.1016/j.porgcoat.2011.03.014

[Szwarc 1956] Szwarc M. (1956) "'Living' Polymers" Nature, 178, 1168–1169

[Tadano 1989] Tadano K, Hirasawa E, Yamamoto H, Yano S. (1989) "Order–Disorder Transition of Ionic Clusters in Ionomers" Macromolecules, 22, 226–233

[Tadmor 2004] Tadmor R. (2004) "Line Energy and the Relation between Advancing, Receding, and Young Contact Angles" Langmuir, 20, 7659–7664

[Takahashi 2003] Takahashi K, Kim BS, Chu MC, Sato S, Ando K. (2003) "Crack-Healing Behavior and Static Fatigue Strength of Si_3N_4/SiC Ceramics Held under Stress at Temperature (800, 900, 1000 C)" Journal of the European Ceramic Society, 23, 1971–1978, doi:10.1016/S0955-2219(02)00428-4

[Takahashi 2010a] Takahashi M. (2010) "Self-Repairing Control via an Unstable Filter" Control Engineering Practice" 18, 2, 140–146, doi:10.1016/j.conengprac.2009.10.001

[Takahashi 2010b] Takahashi K, Park JS, Hahn HT. (2010) "An Addressable Conducting Network for Autonomic Structural Health Management of Composite Structures" Smart Materials and Structures, 19, 105023, doi:10.1088/0964-1726/19/10/105023

[Takeda 2003] Takeda K, Tanahashi M, Unno H. (2003) "Self-Repairing Mechanism of Plastics" Science and Technology of Advanced Materials, 4, 435–44.

[Takei 2010] Takei K, Takahashi T, Ho JC, Ko H, Gillies AG, Leu PW, Fearing RS, Javey A. (2010) "Nanowire Active-Matrix Circuitry for Low-Voltage Macroscale Artificial Skin" Nature Materials, 9, 821–826.

[Takeo 2013] Takeo M, Chou WC, Sun Q, Lee W, Rabbani P, Loomis C, Taketo MM, Ito M. (2013) "Wnt Activation in Nail Epithelium Couples Nail Growth to Digit Regeneration" Nature, 499, 228–232, doi:10.1038/nature12214

[Taleb 2013] Taleb NN, Douady R. (2013) "Mathematical Definition, Mapping, and Detection of (Anti) Fragility" Quantitative Finance, 13, 11, 1677–1689, doi:10.1080/14697688.2013.800219

[Tan 1996] Tan YJ, Bailey S, Kinsella B. (1996) "An Investigation of the Formation and Destruction of Corrosion Inhibitor Films Using Electrochemical Impedance Spectroscopy" Corrosion Science, 38, 1545–1561, doi:10.1016/0010-938X(96)00047-9

[Tan 2012a] Tan T, Santos SF, Savastano Jr. H, Soboyejo WO. (2012) "Fracture and Resistance-Curve Behavior in Hybrid Natural Fiber and Polypropylene Fiber Reinforced Composites" Journal of Materials Science, 47, 2864–2874, doi 10.1007/s10853-011-6116-1

[Tan 2012b] Tan Y, Shan L, Kim Y, Underwood B. (2012) "Healing Characteristics of Asphalt Binder" Construction and Building Materials, 27, 1, 570–577, doi:10.1016/j.conbuildmat.2011.07.006

[Tang 2014] Tang SY, Khoshmanesh K, Sivan V, Petersen P, O'Mullane AP, Abbotte D, Mitchell A, Kalantar-zadeha K. (2014) "Liquid Metal Enabled Pump" Proceedings of the National Academy of Sciences of the United States of America, 111, 9, 3304–3309, doi:10.1073/pnas.1319878111

[Taylor 1939] Taylor HF. (1939) "Vitamin Preparation and Method of Making Same" US Patent 2,183,053

[Tee 2012] Tee BC, Wang C, Allen R, Bao Z. (2012) "An Electrically and Mechanically Self-Healing Composite with Pressure- and Flexion-Sensitive Properties for Electronic Skin Applications" Nature Nanotechnology, 7, 825–832, doi:10.1038/nnano.2012.192

[Telecka 2016] Telecka A, Murthy S, Schneider L, Pranov H, Taboryski R. (2016) "Superhydrophobic Properties of Nanotextured Polypropylene Foils Fabricated by Roll-to-Roll Extrusion Coating" ACS Macro Letters, 9, 1034–1038 doi:10.1021/acsmacrolett.6b00550

[Termkhajornkit 2009] Termkhajornkit P, Nawa T, Yamashiro Y, Saito T. (2009) "Self-Healing Ability of Fly Ash-Cement Systems," Cement and Concrete Composites, 31, 3, 195–203

[Tesoro 1993] Tesoro GC, Sastri VR. (1993) "Polyimide Resins" US Patent 5,260,411

[Teverovsky 2021] Teverovsky A. (2021) "Breakdown and Self-Healing in Tantalum Capacitors" IEEE Transactions on Dielectrics and Electrical Insulation, 28, 663–671, doi:10.1109/TDEI.2020.009240

[Thacher 1919] Thacher SP. (1919) "Tank" US Patent 1,297,305

[Thakur 2015] Thakur VK, Kessler MR. (2015) "Self-Healing Polymer Nanocomposite Materials: A Review" Polymer, 69, 369–383, doi:10.1016/j.polymer.2015.04.086

[Therriault 2005] Therriault D, Shepherd RF, White SR, Lewis JA. (2005) "Fugitive Inks for Direct-Write Assembly of Three-Dimensional Microvascular Networks" Advanced Materials, 17, 4, 395–399, doi:10.1002/adma.200400481

[Thibault 2002] Thibault Jr. RJ, Galow TH, Turnberg ET, Gray M, Hotchkiss PH, Rotello VM. (2002) "Specific Interactions of Complementary Mono- and Multivalent Guests with Recognition-Induced Polymersomes" Journal of the American Chemical Society, 124, 51, 15249–15254, doi:10.1021/ja026418+

[Thieme 2001] Thieme M, Frenzel R, Schmidt S, Simon F, Hennig A, Worch H, Lunkwitz K, Scharnweber D. (2001) "Generation of Ultrahydrophobic Properties of Aluminium: A First Step to Self-Cleaning Transparently Coated Metal Surfaces" Advanced Energy Materials, 3, 9, 691–695, doi:10.1002/1527-2648(200109)3:9<691::AID-ADEM691>3.0.CO;2-8

[Thoma 2003] Thoma SG, Giunta RK, Stavig ME, Emerson JA, Morales AM. (2003) "Autonomic Healing of Epoxy Using Micro-Encapsulated Dicyclopentadiene" Sand2003-1590, Sandia National Laboratories

[Thomas 2012] Thomas M, Hooton RD, Rogers C, Fournier B. (2012) "50 Years Old and Still Going Strong" Concrete International, 34, 1, 35–40

[Thompson 1992] Thompson DW (1992) On Growth and Form: The Complete Revised Edition, Dover Publications, New York

[Thompson 2001] Thompson JB, Kindt JH, Drake B, Hansma HG, Morse DE, Paul K. Hansma PK. (2001) "Bone Indentation Recovery Time Correlates with Bond Reforming Time" Nature, 414, 773–775

[Thostenson 2006] Thostenson ET, Chou TW. (2006) "Carbon Nanotube Networks: Sensing of Distributed Strain and Damage for Life Prediction and Self Healing" Advanced Materials, 18, 2837–2841, doi:10.1002/adma.200600977

[Tian 2009] Tian Q, Yuan YC, Rong MZ, Zhang MQ. (2009) "A Thermally Remendable Epoxy Resin" Journal of Materials Chemistry, 19, 1289, doi:10.1039/b811938d

[Tiarks 2001] Tiarks F, Landfester K, Antonietti M. (2001) "Preparation of Polymeric Nanocapsules by Miniemulsion Polymerization" Langmuir, 17, 3, 908–918, doi:10.1021/la001276n

[Tillotson 2014] Tillotson BJ, Fischer BG. (2014) "Method and System for Attenuating Shock Waves via an Inflatable Enclosure" US Patent 8,677,881

[Tillotson 2015] Tillotson BJ. (2015) "Method and System for Shockwave Attenuation via Electromagnetic Arc" US Patent 8,981,261

[Timonen 2013] Timonen JV, Latikka M, Ikkala O, Ras RH. (2013) "Free-Decay and Resonant Methods for Investigating the Fundamental Limit of Superhydrophobicity" Nature Communications, 4, 2398, doi:10.1038/ncomms3398

[Timoshenko 1961] Timoshenko S, Gere JM (1961) Theory of Elastic Stability, 2nd edition, McGraw-Hill, New York

[Tinnefeld 2012] Tinnefeld P, Cordes T. (2012) "'Self-Healing' Dyes: Intramolecular Stabilization of Organic Fluorophores" Nature Methods, 9, 426–427

[Todd 2005] Todd EM, Quinn JR, Park T, Zimmerman SC. (2005) "Fidelity in the Supramolecular Assembly of Triply and Quadruply Hydrogen-Bonded Complexes" Israel Journal of Chemistry, 45, 381–389

[Toldy 2011] Toldy A, Szolnoki B, Marosi G. (2011) "Flame Retardancy of Fibre-Reinforced Epoxy Resin Composites for Aerospace Applications" Polymer Degradation and Stability, 96, 371–376, doi:10.1016/j.polymdegradstab.2010.03.021

[Toledo 2005] Toledo Filho RD, Ghavami K, Sanjuanc MA, England GL. (2005) "Free, Restrained and Drying Shrinkage of Cement Mortar Composites Reinforced with Vegetable Fibres" Cement & Concrete Composites, 27, 537–546

[Tölle 1999] Tölle TR, Kaufmann T, Siessmeier T, Lautenbacher S, Berthele A, Munz F, Zieglgänsberger W, Willoch F, Schwaiger M, Conrad B, Bartenstein P. (1999) "Region-Specific Encoding of Sensory and Affective Components of Pain in the Human Brain: A Positron Emission Tomography Correlation Analysis" Annals of Neurology, 45, 1, 40–47, doi:10.1002/1531-8249(199901)45:1<40::AID-ART8>3.0.CO;2-L

[Tomar 2008] Tomar V, Gungor MR, Maroudas D. (2008) "Current-Induced Stabilization of Surface Morphology in Stressed Solids" Physical Review Letters, 100, 036106

[Tomayko 2003] Tomayko JE. (2003) Story of Self-Repairing Flight Control Systems, NASA Dryden Flight Research Center, C Gelzer (ed.)

[Tomczak 2004] Tomczak S, Marchant D, Svejda S, Minton T, Brunsvold A. (2004) "Properties and Improved Space Survivability of POSS (Polyhedral Oligomeric Silsesquioxane) Polyimides" DARP A433, Air Force Research Laboratory (AFMC), Edwards AFB, CA, 93524–7680 USA

[Tong 2010] Tong XM, Zhang T, Yang MZ, Zhang Q. (2010) "Preparation and Characterization of Novel Melamine Modified Poly(urea-formaldehyde) Self-Repairing Microcapsules" Colloids and Surfaces A: Physicochemical and Engineering Aspects, 371, 1–3, 91–97, doi:10.1016/j.colsurfa.2010.09.009

[Tononi 2004] Tononi, G. (2004) "An Information Integration Theory of Consciousness" BMC Neuroscience, 5, 42 doi:10.1186/1471-2202-5-42

[Toohey 2007] Toohey KS, Sottos NR, Lewis JA, Moore JS, White SR. (2007) "Self-Healing Materials with Microvascular Networks" Nature Materials, 6, 581–585

[Toyota 2009] Toyota T, Maru N, Hanczyc MM, Ikegami T, Sugawara T. (2009) "Self-Propelled Oil Droplets Consuming 'Fuel' Surfactant" Journal of the American Chemical Society, 131, 14, 5012–5013, doi:10.1021/ja806689p

[Tran 2013] Tran PA, Webster TJ. (2013) "Antimicrobial Selenium Nanoparticle Coatings on Polymeric Medical Devices" Nanotechnology, 24, 15, doi:10.1088/0957-4484/24/15/155101

[Trappe 2001] Trappe V, Prasad V, Cipelletti L, Segre PN, Weitz DA. (2001) "Jamming Phase Diagram for Attractive Particles" Nature, 411, 772–775, doi:10.1038/35081021

[Trask 2007a] Trask RS, Williams GJ, Bond IP. (2007) "Bioinspired Self-Healing of Advanced Composite Structures Using Hollow Glass Fibres" Journal of the Royal Society Interface, 4, 363–371

[Trask 2007b] Trask RS, Williams HR, Bond IP. (2007) "Self-Healing Polymer Composites: Mimicking Nature to Enhance Performance" Bioinspiration & Biomimetics, 2, P1–P9 doi:10.1088/1748-3182/2/1/P01

[Trau 1995] Trau M, Sankaran S, Saville DA, Aksay IA. (1995) "Electric-Field-Induced Pattern Formation in Colloidal Dispersions" Nature, 374, 437–439

[Trimmer 1989] Trimmer WS. (1989) "Microrobots and Micromechanical Systems" Sensors and Actuators, 19, 267–287, doi:10.1016/0250-6874(89)87079-9

[Trotter 2000] Trotter JA, Tipper J, Lyons-Levy G, Chino K, Heuer AH, Liu Z, Mrksich M, Hodneland C, Dillmore WS, Koob TJ, Koob-Emunds MM, Kadler K, Holmes D. (2000) "Towards a Fibrous Composite with Dynamically Controlled Stiffness: Lessons from Echinoderms" Biochemical Society Transactions, 28, 4, 357–362

[True 2000] True GC, Glaser TJ. (2000) "Tire Sealant Composition" US Patent 6,013,697

[Truszkowski 2009] Truszkowski W, Hallock L, Rouff C, Karlin J, Rash J, Hinchey MG, Sterritt R. (2009) Autonomous and Autonomic Systems with Applications to NASA Intelligent Spacecraft Operations and Exploration Systems, Springer, Berlin, ISBN 978-1-84628-233-1

[Tsai 2001] Tsai LW. (2001) Mechanism Design Enumeration of Kinematic Structures According to Function, CRC Press, Boca Raton

[Tsai 2009] Tsai P, Pacheco S, Pirat C, Lefferts L, Lohse D. (2009) "Drop Impact upon Micro- and Nanostructured Superhydrophobic Surfaces" Langmuir, 25 (20),12293–12298, doi:10.1021/la900330q

[Tsangouri 2015] Tsangouri E, Aggelis D, Van Hemelrijck D. (2015) "Quantifying Thermoset Polymers Healing Efficiency: A Systematic Review of Mechanical Testing" Progress in Polymer Science, 49–50, 154–174, doi:10.1016/j.progpolymsci.2015.06.002

[Tuan 2008] Tuan CY. (2008) "Roca Spur Bridge: The Implementation of an Innovative Deicing Technology" Journal of Cold Regions Engineering, 22,1–15. doi:10.1061/(ASCE)0887-381X(2008)22:1(1)

[Turing 1952] Turing AM. (1952) "The Chemical Basis of Morphogenesis" Philosophical Transactions of the Royal Society B: Biological Sciences, 237, 37–72, doi:/10.1098/rstb.1952.0012

[Tuteja 2008] Tuteja A, Choi W, McKinley GH, Cohen RE, Rubne MF. (2008) "Design Parameters for Superhydrophobicity and Superoleophobicity" MRS Bulletin, 33, 752–758

[Tvergaard 1992] Tvergaard V. (1992) "Effect of Ductile Particle Debonding during Crack Bridging in Ceramics" International Journal of Mechanical Sciences, 34, 8, 635–649, doi:10.1016/0020-7403(92)90060-T

[Twede 2005] Twede D. (2005) "The Cask Age: The Technology and History of Wooden Barrels" Packaging Technology and Science, 18, 253–264, doi:10.1002/pts.696

[Tyagi 2004] Tyagi S, Lee JY, Buxton GA, and Balazs AC. (2004) "Using Nanocomposite Coatings to Heal Surface Defects" Macromolecules, 37, 24, 9160–9168, doi:10.1021/ma048773l

[Underwood 1970] Underwood TA, Wickersham Jr. WS, Sutton RW. (1970) "Self-Sealing Fuel Cell Wall" US Patent 3,509,016

[Urban 2009] Urban MW. (2009) "Stratification, Stimuli-Responsiveness, Self-Healing, and Signaling in Polymer Networks" Progress in Polymer Science, 34, 8, pp 679–687, doi:10.1016/j.progpolymsci.2009.03.004

[Urnes 2001] Urnes Sr J, Davidson R, Jacobson S. (2001) "A Damage Adaptive Flight Control System Using Neural Network Technology" Proceedings of IEEE on American Control Conference 2001, 4, 2907–2912, doi:10.1109/ACC.2001.946344

[Uygun 2015] Uygun M, Singh VV, Kaufmann K, Uygun DA, de Oliveira SD, Wang J. (2015) "Micromotor-Based Biomimetic Carbon Dioxide Sequestration: Towards Mobile Microscrubbers" Angewandte Chemie, 127, 44, 13092–13096, doi:10.1002/ange.201505155

[Vaccaro 2001] Vaccaro E, Waite JH. (2001) "Yield and Post-Yield Behaviour of Mussel Byssal Thread: A Self-Healing Biomolecular Material" Biomacromolecules, 2, 906–11

[Vaia 1999] Vaia RA, Price G, Ruth PN, Nguyen HT, Lichtenhan J. (1999) "Polymer Layered Silicate Nanocomposites as High Performance Ablative Materials" Applied Clay Science, 15, 67–92

[Vaiyapuri 2013] Vaiyapuri R, Greenland BW, Colquhoun HM, Elliott JM, Hayes W. (2013) "Molecular Recognition between Functionalized Gold Nanoparticles and Healable, Supramolecular, Polymer Blends: A Route to Property Enhancement" Polymer Chemistry, 2013, 4, 4902–4909, doi:10.1039/C3PY00086A

[Vakarelski 2012] Vakarelski IU, Patankar NA, Marston JO, Chan DY, Thoroddsen ST. (2012) "Stabilization of Leidenfrost Vapour Layer by Textured Superhydrophobic Surfaces" Nature, 489, 274–277, doi:10.1038/nature11418

[Van Belleghem 2017] Van Belleghem B, Van den Heede P, Van Tittelboom K, De Belie N. (2017) "Quantification of the Service Life Extension and Environmental Benefit of Chloride Exposed Self-Healing Concrete" Materials, 10, 1, 5, doi:10.3390/ma10010005

[van Blaaderen 1997] van Blaaderen A, Ruel R, Wiltzius P. (1997) "Template-Directed Colloidal Crystallization" Nature, 385, 321–324, doi:10.1038/385321a0

[van den Dungen 2010] van den Dungen ET, Loos B, Klumperman B. (2010) "Use of a Profluorophore for Visualization of the Rupture of Capsules in Self-Healing Coatings" Macromolecular Rapid Communications, 31, 625–628, doi:10.1002/marc.200900728

[van der Mee 2008] van der Mee MA, Goossens JG, van Duin M. (2008) "Thermoreversible Cross-Linking of Maleated Ethylene/Propylene Copolymers with Diamines and Amino-Alcohols" Polymer, 49, 5, 1239–1248, doi:10.1016/j.polymer.2008.01.031

[van der Velde 2013] van der Velde JH, Ploetz E, Hiermaier M, Oelerich J, de Vries JW, Roelfes G, Cordes T. (2013) "Mechanism of Intramolecular Photostabilization in Self-Healing Cyanine Fluorophores" ChemPhysChem, 14, 18, 4084–4093, doi:10.1002/cphc.201300785

[van der Wal 2007] van der Wal P, Steiner U. (2007) "Super-Hydrophobic Surfaces Made from Teflon" Soft Matter, 3, 426–429, doi:10.1039/b613947g [van der Zwaag 2009] van der Zwaag S, van Dijk NH, Jonkers HM, Mookhoek SD, Sloof WG. (2009) "Self-Healing Behaviour in Man-Made Engineering Materials: Bioinspired But Taking into Account Their Intrinsic Character" Philosophical Transactions of the Royal Society A, 367, 1689–1704, doi:10.1098/rsta.2009.0020

[Van Tittelboom 2010] Van Tittelboom K, de Belie N, de Muynck W, Verstraete W. (2010) "Use of Bacteria to Repair Cracks in Concrete" Cement and Concrete Research, 40, 1, 157–166, doi:10.1016/j.cemconres.2009.08.025

[Van Tittelboom 2011] Van Tittelboom K, De Belie N, Van Loo D, Jacobs P. (2011) "Self-Healing Efficiency of Cementitious Materials Containing Tubular Capsules Filled with Healing Agent" Cement & Concrete Composites 33, 497–505

[Vandeparre 2013] Vandeparre H, Liu Q, Minev IR, Suo Z, Lacour SP. (2013) "Localization of Folds and Cracks in Thin Metal Films Coated on Flexible Elastomer Foams" Advanced Materials, 25, 22, 3117–3121, doi:10.1002/adma.201300587

[Varghese 2006] Varghese S, Lele A, Mashelkar R. (2006) "Metal-Ion-Mediated Healing of Gels" Journal of Polymer Science: Part A: Polymer Chemistry, 44, 666–70, doi:10.1002/pola.21177

[Varley 2008] Varley RJ, van der Zwaag S. (2008) "Towards an Understanding of Thermally Activated Self-Healing of an Ionomer System during Ballistic Penetration" Acta Materialia, 56 5737–5750

[Varley 2010] Varley RJ, van der Zwaag S. (2010) "Autonomous Damage Initiated Healing in a Thermo-Responsive Ionomer" Polymer International, 59, 1031–1038, doi:10.1002/pi.2841

[Vasko 2015] Vasko AC, Grice CR, Kostic AD, Karpov VG. (2015) "Evidence of Electric-Field-Accelerated Growth of Tin Whiskers" MRS Communications, 5, 619–622, doi:10.1557/mrc.2015.64

[Vaughan 1908] Vaughan JW, Bryan WC. (1908) "Burglar Proof Safejacket" US Patent 888,052

[Vazquez 2005] Vazquez JM, Cano RJ, Jensen BJ, Weiser ES. (2005) "Polyimide Foams" US Patent 6,956,066

[Vedula 2014] Vedula SR, Hirata H, Nai MH, Brugués A, Toyama Y, Trepat X, Lim CT, Ladoux B. (2014) "Epithelial Bridges Maintain Tissue Integrity during Collective Cell Migration" Nature Materials, 13, 87–96, doi:10.1038/nmat3814

[Velnar 2009] Velnar T, Bailey T, Smrkolj V. (2009) "The Wound Healing Process: An Overview of the Cellular and Molecular Mechanisms" Journal of International Medical Research, 37, 1528–1542, doi:10.1177/147323000903700531

[Veraart 2012] Veraart AJ, Faassen EJ, Dakos V, van Nes EH, Lürling M, Scheffer M. (2012) "Recovery Rates Reflect Distance to a Tipping Point in a Living System" Nature, 481, 357–359, doi:10.1038/nature10723

[Verdeyen 1989] Verdeyen JT. (1989) Laser Electronics, 2nd edition, Prentice-Hall, Englewood Cliffs

[Vermolen 2006] Vermolen FJ, van Baaren E, Adam JA. (2006) "A Simplified Model for Growth Factor Induced Healing of Wounds" Mathematical and Computer Modelling, 44, 9–10, 887–898, Nov

[Videla 2009] Videla HA, Herrera LK. (2009) "Understanding Microbial Inhibition of Corrosion: A Comprehensive Overview" International Biodeterioration and Biodegradation, 63, 896–900

[Vilela 2017] Vilela D, Stanton MM, Parmar J, Sánchez S. (2017) "Microbots Decorated with Silver Nanoparticles Kill Bacteria in Aqueous Media" ACS Applied Materials & Interfaces, 9, 27, 22093–22100, doi:10.1021/acsami.7b03006

[Villares 2004] Villares Lenz Cesar D. (2004) "Self-Cleaning Drum Filter" US Patent 6,808,076

[Vincenzo 2009] Vincenzo A, Meneghetti M. (2009) "Self-Healing at the Nanoscale" Nanoscale, 1, 1, 74–88, doi:10.1039/b9nr00146h

[Viry 2010] Viry L, Mercader C, Miaudet P, Zakri C, Derré A, Kuhn A, Maugey M, Poulin P. (2010) "Nanotube Fibers for Electromechanical and Shape Memory Actuators" Journal of Materials Chemistry, 20, 3487–3495, doi:10.1039/B924430A

[Vogel 2013] Vogel N, Belisle RA, Hatton B, Wong TS, Aizenberg J. (2013) "Transparency and Damage Tolerance of Patternable Omniphobic Lubricated Surfaces Based on Inverse Colloidal Monolayers" Nature Communicatios, 4, 2176, doi:10.1038/ncomms3176

[Volynskii 2009] Volynskii AL, Bakeev NF. (2009) "Healing of Interfacial Surfaces in Polymer Systems" Polymer Science, Series A, 51, 10, pp. 1096–1126, doi:10.1134/S0965545X09100071

[Vorobyev 2015] Vorobyev AY, Guo C. (2015) "Multifunctional Surfaces Produced by Femtosecond Laser Pulses" Journal of Applied Physics, 117, 033103, doi:10.1063/1.4905616

[Vourdas 2007] Vourdas N, Tserepi A, Gogolides E. (2007) "Nanotextured Super-Hydrophobic Transparent Poly(methyl methacrylate) Surfaces Using High-Density Plasma Processing" Nanotechnology, 18, 125304, doi:10.1088/0957-4484/18/12/125304

[Voyiadjis 2011] Voyiadjis G, Shojaei A, Li G. (2011) "A Thermodynamic Consistent Damage and Healing Model for Self Healing Materials" International Journal of Plasticity, 27, 7, 1025–1044, doi:10.1016/j.ijplas.2010.11.002

[Vuillaume 2002] Vuillaume PY, Jonas AM, Laschewsky A. (2002) "Ordered Polyelectrolyte "Multilayers". 5. Photo-Cross-Linking of Hybrid Films Containing an Unsaturated and Hydrophobized Poly(diallylammonium) Salt and Exfoliated Clay" Macromolecules 2002, 35, 5004–5012

[Vuong 2019] Vuong VQ, Nishimoto Y, Fedorov DG, Sumpter BG, Niehaus TA, Irle S. (2019) "The Fragment Molecular Orbital Method Based on Long-Range Corrected Density-Functional Tight-Binding" Journal of Chemical Theory and Computation, 15, 5, 3008–3020, doi:10.1021/acs.jctc.9b00108

[Waber 1939] Waber JW. (1939) "Inner Tube" US Patent 2,161,490

[Wagener 1993] Wagener KB, Engle LP. (1993) "Thermally Reversible Polymer Linkages II. Linear Addition Polymers" Journal of Polymer Science Part A: Polymer Chemistry, 31, 865–875, doi:10.1002/pola.1993.080310404

[Wagner 1974] Wagner EF. (1974) "Autogenous Healing of Cracks in Cement-Mortar Linings for Gray-Iron and Ductile-Iron Water Pipe" Journal of American Water Works Association, 66, 6, 358–360.

[Waisman 2011] Waisman H, Montoya A, Betti R, Noyan I. (2011)" Load Transfer and Recovery Length in Parallel Wires of Suspension Bridge Cables" Journal of Engineering Mechanics, 137, 4, 227–237, doi:10.1061/(ASCE)EM.1943-7889.0000220

[Walker 2022] Walker A, Stankovic T. (2022) "Algorithmic Design of Origami Mechanisms and Tessellations" Communications Materials, 3, 1–8, doi:10.1038/s43246-022-00227-5

[Wang 1993] Wang YC, Winnik MA. (1993) "Energy-Transfer Study of Polymer Diffusion in Melt-Pressed Films of Poly(methylmethacrylate)" Macromolecules, 26, 12, 3147–3150

[Wang 1994] Wang PP, Lee S, Harmon J. (1994) "Ethanol-Induced Crack Healing in Poly(methyl mMethacrylate)" Journal of Polymer Science, Part B: Polymer Physics, 32, 1217–1227, doi:10.1002/polb.1994.090320709.

[Wang 1997] Wang R, Hashimoto K, Fujishima A, Chikuni M, Kojima E, Kitamura A, Shimohigoshi M, Watanabe T. (1997) "Light-Induced Amphiphilic Surfaces" Nature, 338, 431–432

[Wang 2002] Wang X. (2002) "Shape Memory Alloy Volume Fraction of Pre-Stretched Shape Memory Alloy Wire-Reinforced Composites for Structural Damage Repair" Smart Materials and Structures, 11, 4, 590–595, doi:10.1088/0964-1726/11/4/315

[Wang 2003] Wang H, Li Z. (2003) "The Shrinkage of Grain-Boundary Voids under Pressure" Metallurgical and Materials Transactions A, 34, 7, 1493–1500

[Wang 2004a] Wang H, Presuel F, Kelly RG. (2004) "Computational Modeling of Inhibitor Release and Transport from Multifunctional Organic Coatings" Electrochim Acta 49, 2, 239–255, doi:10.1016/j.electacta.2003.08.006

[Wang 2004b] Wang Q, Quek ST. (2004) "Repair of Delaminated Beams via Piezoelectric Patches" Smart Materials and Structures, 13, 1222–1229, doi:10.1088/0964-1726/13/5/026

[Wang 2007a] Wang KM, Lorente S, Bejan A. (2007) "Vascularization with Grids of Channels: Multiple Scales, Loops and Body Shapes" Journal of Physics D: Applied Physics, 40, 4740–4749 doi:10.1088/0022-3727/40/15/057

[Wang 2007b] Wang Y, Bolanos E, Wudl F, Hahn T, Kwok N. (2007) "Self-Healing Polymers and Composites Based on Thermal Activation" Dapino MJ (ed.) Proceedings of SPIE Behavior and Mechanics of Multifunctional and Composite Materials, 6526, 65261I.

[Wang 2009] Wang R, Li H, Hu H, He X, Liu W. (2009) "Preparation and Characterization of Self-Healing Microcapsules with Poly(urea-formaldehyde) Grafted Epoxy Functional Group Shell" Journal of Applied Polymer Science, 113, 3, 1501–1506, doi:10.1002/app.30001

[Wang 2010a] Wang HP, Yuan YC, Rong MZ, Zhang MQ. (2010) "Self-Healing of Thermoplastics via Living Polymerisation" Macromolecules, 43, 2, 595–598, doi:10.1021/ma902021v

[Wang 2010b] Wang Q, Mynar JL, Yoshida M, Lee E, Lee M, Okuro K, Kinbara K, Aida T. (2010) "High-Water-Content Mouldable Hydrogels by Mixing Clay and a Dendritic Molecular Binder" Nature, 463, doi:10.1038/nature08693

[Wang 2010c] Wang X, Veruki ML, Bukoreshtliev NV, Hartveit E, Gerdes HH. (2010) "Animal Cells Connected by Nanotubes Can be Electrically Coupled through Interposed Gap-Junction Channels" Proceedings of the National Academy of Sciences of the United States of America, 107, 40, 17194–17199, doi/10.1073/pnas.1006785107

[Wang 2011a] Wang B, Kitney RI, Joly N, Buck M. (2011) "Engineering Modular and Orthogonal Genetic Logic Gates for Robust Digital-Like Synthetic Biology" Nature Communications, 2, 508, 1–9, doi:10.1038/ncomms1516

[Wang 2011b] Wang T, Sha R, Dreyfus R, Leunissen ME, Maass C, Pine DJ, Chaikin PM, Seeman NC. (2011) "Self-Replication of Information-Bearing Nanoscale Patterns" Nature, 478, 225–228, doi:10.1038/nature10500

[Wang 2012a] Wang CC, Huang WM, Ding Z, Zhao Y, Purnawali H, Zheng LX, Fan H, He CB. (2012) "Rubber-Like Shape Memory Polymeric Materials with Repeatable Thermal-Assisted Healing Function" Smart Materials and Structures, 21, 11, 115010, doi:10.1088/0964-1726/21/11/115010

[Wang 2012b] Wang J, Lu C, Wang Q, Xiao P, Ke F, Bai Y, Shen Y, Wang Y, Chen B, Liao X, Gao H. (2012) "Self-Healing in Fractured GaAs Nanowires" Acta Materialia, 60, 15, 5593–5600, doi:10.1016/j.actamat.2012.07.013

[Wang 2012c] Wang J, van Tittelboom K, De Belie N, Verstraete W. (2012) "Use of Silica Gel or Polyurethane Immobilized Bacteria for Self-Healing Concrete" Construction and Building Materials, 26, 1, 532–540, doi:10.1016/j.conbuildmat.2011.06.054

[Wang 2012d] Wang Q, Suo Z, Zhao X. (2012) "Bursting Drops in Solid Dielectrics Caused by High Voltages" Nature Communications, 3, 1157. doi:10.1038/ncomms2178

[Wang 2012e] Wang Q, Wu N. (2012) "A Review on Structural Enhancement and Repair Using Piezoelectric Materials and Shape Memory Alloys" Smart Materials and Structures, 2, 1 013001 doi:10.1088/0964-1726/21/1/013001

[Wang 2012f] Wang T, Zhuang J, Lynch J, Chen O, Wang Z, Wang X, LaMontagne D, Wu H, Wang Z, Cao YC. (2012) "Self-Assembled Colloidal Superparticles from Nanorods" Science, 338, 6105, 358–363, doi:10.1126/science.1224221

[Wang 2013a] Wang C, Wu H, Chen Z, McDowell MT, Cui Y, Bao Z. (2013) "Self-Healing Chemistry Enables the Stable Operation of Silicon Microparticle Anodes for High-Energy Lithium-Ion Batteries" Nature Chemistry, 5, 1042–1048, doi:10.1038/nchem.1802

[Wang 2013b] Wang E, Desai MS, Lee SW. (2013) "Light-Controlled Graphene-Elastin Composite Hydrogel Actuators" Nano Letters, 13, 6, 2826–2830, doi:10.1021/nl401088b

[Wang 2013c] Wang Z, Zhu L, Li W, Liu H. (2013) "Rapid Reversible Superhydrophobicity-to-Superhydrophilicity Transition on Alternating Current Etched Brass" ACS Applied Materials & Interfaces, 5, 11, 4808–4814, doi:10.1021/am400299f

[Wang 2014] Wang W, Li S, Mair L, Ahmed S, Huang TJ, Mallouk TE. (2014) "Acoustic Propulsion of Nanorod Motors Inside Living Cells" Angewandte Chemie International Edition, 53, 12, 3201–3204, doi:10.1002/anie.201309629

[Wang 2015a] Wang M, Trlica C, Khan MR, Dickey MD, Adams JJ. (2015) "A Reconfigurable Liquid Metal Antenna Driven by Electrochemically Controlled Capillarity" Journal of Applied Physics, 117, 194901, doi:10.1063/1.4919605

[Wang 2015b] Wang N, Xiong D, Deng Y, Shi Y, Wang K. (2015) "Mechanically Robust Superhydrophobic Steel Surface with Anti-Icing, UV-Durability, and Corrosion Resistance Properties" ACS Applied Materials & Interfaces 7, 11, 6260–6272, doi:10.1021/acsami.5b00558

[Wang 2015c] Wang Q, Yao X, Liu H, Quéré D, Jiang L. (2015) "Self-Removal of Condensed Water on the Legs of Water Striders" Proceedings of the National Academy of Sciences of the United States of America, 112, 30, 9247–9252, doi:10.1073/pnas.1506874112

[Wang 2015d] Wang ZJ, Li QJ, Cui YN, Liu ZL, Ma E, Li J, Sun J, Zhuang Z, Dao M, Shan ZW, Suresh S. (2015) "Cyclic Deformation Leads to Defect Healing and Strengthening of Small-Volume Metal Crystals" Proceedings of the National Academy of Sciences of the United States of America, 112, 44, 13502–13507, doi:10.1073/pnas.1518200112

[Wang 2016] Wang L, McCarthy TJ. (2016) "Covalently Attached Liquids: Instant Omniphobic Surfaces with Unprecedented Repellency" Angewandte Chemie International Edition, 55, 1, 244–248, doi:10.1002/anie.201509385

[Wang 2018] Wang Y, Cheng Z, Liu Z, Kang H, Liu Y. (2018) "Cellulose Nanofibers/Polyurethane Shape Memory Composites with Fast Water-Responsivity" Journal of Materials Chemistry B, 2018, 6, 1668–1677, doi: 10.1039/C7TB03069J

[Ware 2005] Ware JN, Spruell SL. (2005) "Self-Sealing Electrical Cable Having a Finned or Ribbed Structure between Protective Layers" US Patent 6,914,193

[Washburn 1921] Washburn EW. (1921) "The Dynamics of Capillary Flow" Physical Review, XVII., 3, 273–283

[Washburn 1968] Washburn RM, Kremer PA, Beaucamp NA. (1968) "Epoxy Systems Containing Encapsulated Fracturable Curing Agents" US Patent 3,395,105, July 30

[Wasilewska 2008] Wasilewska A, Vlad F, Sirichandra C, Redko Y, Jammes F, Valon C, Freya NF, Leung J. (2008) "An Update on Abscisic Acid Signaling in Plants and More ..." Molecular Plant, 1, 2, 198–217

[Watson 1946] Watson FK. (1946) "Production of Shaped Structure from Proteins" US Patent 2,399,084

[Watts 1998] Watts DJ, Strogatz SH. (1998) "Collective Dynamics of 'Small-World' Networks" Nature, 393, 440–442, https://doi.org/10.1038/30918

[Weber 2015] Weber C, Liebig T, Gensler M, Pithan L, Bommel S, Bléger D, Rabe JP, Hech S, Kowarik S. (2015) "Light-Controlled "Molecular Zippers" Based on Azobenzene Main Chain Polymers" Macromolecules, 48, 5, 1531–1537, doi:10.1021/ma502551b

[Wechsatol 2005] Wechsatol W, Lorente S, Bejan A. (2005) "Tree-Shaped Networks with Loops" International Journal of Heat and Mass Transfer, 48, 3–4, 573–583, doi:10.1016/j.ijheatmasstransfer.2004.08.020

[Wei 2012] Wei B, Dai M, Yin P. (2012) "Complex Shapes Self-Assembled from Single-Stranded DNA Tiles" Nature, 485, 623–626, doi:10.1038/nature11075

[Wei 2014] Wei Z, Yang JH, Zhou J, Xu F, Zrínyi M, Dussault PH, Osada Y, Chen YM. (2014) "Self-Healing Gels Based on Constitutional Dynamic Chemistry and Their Potential Applications" Chemical Society Reviews, 43, 8114–8131, doi:10.1039/C4CS00219A

[Weibel 2005] Weibel DB, Garstecki P, Ryan D, DiLuzio WR, Mayer M, Seto JE, Whitesides GM. (2005) "Microoxen: Microorganisms to Move Microscale Loads" Proceedings of the National Academy of Sciences of the United States of America, 102, 34, 11963–11967, doi:10.1073/pnas.0505481102

[Weiser 2000] Weiser ES, Johnson TF, St Clair TL, Echigo Y, Kaneshiro H, Grimsley BW. (2000) "Polyimide Foams for Aerospace Vehicles" High Performance Polymers, 12, 1, doi:10.1088/0954-0083/12/1/301

[Welp 1999] Welp KA, Wool RP, Agrawal G, Satija SK, Pispas S, Mays J. (1999) "Direct Observation of Polymer Dynamics: Mobility Comparison between Central and End Section Chain Segments" Macromolecules, 32, 15, 5127

[Wen 2011] Wen L, Wang YM, Liu Y, Zhou Y, Guo LX, Ouyang JH, Jia DC. (2011) "EIS Study of a Self-Repairing Microarc Oxidation Coating" Corrosion Science, 53, 2, 618–623, doi:10.1016/j.corsci.2010.10.010

[Wendell 2010] Wendell D, Todd J, Montemagno C. (2010) "Artificial Photosynthesis in Ranaspumin-2 Based Foam" Nano Letters, 10, 3231–3236, doi:10.1021/nl100550k

[Wenzel 1936] Wenzel RN. (1936) "Resistance of Solid Surfaces to Wetting by Water" Industrial & Engineering Chemistry Research, 28, 8, 988–994, doi:10.1021/ie50320a024

[Werfel 2014] Werfel J, Petersen K, Nagpal R. (2014) "Designing Collective Behavior in a Termite-Inspired Robot Construction Team" Science, 343, 6172, 754–758, doi:10.1126/science.1245842

[Wertzberger 2010] Wertzberger BE, Steere JT, Pfeifer RM, Nensel MA, Latta MA, Gross SM. (2010) "Physical Characterization of a Self-Healing Dental Restorative Material" Journal of Applied Polymer Science, 118, 1, 428–434, doi:10.1002/app.31542

[Wessling 1996] Wessling B. (1996) "Corrosion Prevention with an Organic Metal (Polyaniline): Surface Ennobling, Passivation, Corrosion Test Results" Materials and Corrosion, 47, 8, 439–445, doi:10.1002/maco.19960470804

[West 1997] West GB, Brown JH, Enquist BJ. (1997) "A General Model for the Origin of Allometric Scaling Laws in Biology" Science, 276, 5309, 122–126, doi:0.1126/science.276.5309.122

[White 2001] White SR, Sottos NR, Geubelle PH, Moore JS, Kessler MR, Sriram SR, Brown EN, Viswanathan. (2001) "Autonomic Healing of Polymer Composites" Nature, 409, 794–797

[White 2003] White SR, Sottos NR, Geubelle PH, Moore JS, Sriram SR, Kessler MR, Brown EN. (2003) "Multifunctional Autonomically Healing Composite Material" US Patent 6,518,330

[White 2005] White SR, Sottos NR, Geubelle PH, Moore JS, Siriram SR, Kessler MR, Brown EN. (2005) "Multifunctional Autonomically Healing Composite Material" US Patent 6,858,659

[White 2009] White PJ, Yim M. (2009) "Reliable External Actuation for Full Reachability in Robotic Modular Self-Reconfiguration" International Journal of Robotics Research, 29, 5, 598–612, doi:10.1177/0278364909351942

[Whitehead 1968] Whitehead DL. (1968) "Enclosed Electric Power Transmission Conductor" US Patent 3,391,243

[Whitesides 2002a] Whitesides GM, Boncheva M. (2002) "Beyond Molecules: Self-Assembly of Mesoscopic and Macroscopic Components" Proceedings of the National Academy of Sciences of the United States of America, 99, 8, 4769–4774, doi:10.1073/pnas.082065899

[Whitesides 2002b] Whitesides GM, Grzybowski BA. (2002) "Self-Assembly at All Scales" Science, 295, 2418–2421

[Whitlow 1991] Whitlow SJ, Wool RP. (1991) "Diffusion of Polymers at Interfaces: A Secondary Ion Mass Spectroscopy Study" Macromolecules, 24, 5926–5938

[Wickham 2012] Wickham SF, Bath J, Katsuda Y, Endo M, Hidaka K, Sugiyama H, Turberfield AJ. (2012) "A DNA-Based Molecular Motor that Can Navigate a Network of Tracks" Nature Nanotechnology, 7, 169–173, doi:10.1038/nnano.2011.253

[Wietor 2009] Wietor JL, Dimopoulos A, Govaert LE, van Benthem RA, de With G, Sijbesma RP. (2009) "Preemptive Healing through Supramolecular Cross-Links" Macromolecules, 42, 17, 6640–6646, doi:10.1021/ma901174r

[Wiktor 2011] Wiktor V, Jonkers HM. (2011) "Quantification of Crack-Healing in Novel Bacteria-Based Self-Healing Concrete" Cement and Concrete Composites, 33, 763–770, doi:/10.1016/j.cemconcomp.2011.03.012

[Wiles 1980] Wiles DM, Carlsson DJ. (1980) "Photostabilisation Mechanisms in Polymers: A Review" Polymer Degradation and Stability, 3 (1980–81) 61–72

[Wilhelm 2007] Wilhelm T, Hollunder J. (2007) "Information Theoretic Description of Networks" Physica A, 385, 1, 385–396, doi:10.1016/j.physa.2007.06.029

[Wilkinson 1943] Wilkinson B, Douglas WD. (1943) "Petrol and Like Pipes for Use in Aircraft" US Patent 2,308,342

[Williams 2007a] Williams G, Trask RS, Bond IP. (2007) "A Self-Healing Carbon Fibre Reinforced Polymer for Aerospace Applications" Composites Part A: Applied Science and Manufacturing, 38, 1525–1532, doi:10.1016/j.compositesa.2007.01.013

[Williams 2007b] Williams HR, Trask RS, Bond IP. (2007) "Self-Healing Composite Sandwich Structures" Smart Materials and Structures, 16, 1198–1207, doi:10.1088/0964-1726/16/4/031

[Williams 2007c] Williams KA, Boydston AJ, Bielawski CW. (2007) "Towards Electrically Conductive, Self-Healing Materials" Journal of the Royal Society: Interface 4, 359–362

[Williams 2008a] Williams HR, Trask RS, Knights AC, Williams ER, Bond IP (2008) "Biomimetic Reliability Strategies for Self-Healing Vascular Networks in Engineering Materials" Journal of Royal Society Interface, 5, 735–747

[Williams 2008b] Williams HR, Trask RS, Weaver PM, Bond IP. (2008) "Minimum Mass Vascular Networks in Multifunctional Materials" Journal of the Royal Society: Interface, 5, 18, 55–65 doi:10.1098/rsif.2007.1022

[Wills 2001] Wills L, Kannan S, Sander S, Guler M, Heck B, Prasad JV, Schrage D, Vachtsevanos GV. (2001) "An Open Platform for Reconfigurable Control Systems" IEEE Control Systems Magazine, 21, 3, 49–64

[Wilson 1941] Wilson RF, Klingman WL. (1941) "Inner Tube" US Patent 2,237,245

[Wilson 1956] Wilson RF. (1956) "Fuel Cell" US Patent 2,754,992

[Wilson 2008] Wilson GO, Caruso MM, Reimer NT, White SR, Sottos NR, Moore JS. (2008) "Evaluation of Ruthenium Catalysts for Ring-Opening Metathesis Polymerization-Based Self-Healing Applications" Chemistry of Materials, 20, 3288–3297

[Wilson 2010] Wilson GO, Henderson JW, Caruso MM, Blaiszik BJ, McIntire PJ, Sottos NR, White SR, Moore JS. (2010) "Evaluation of Peroxide Initiators for Radical Polymerization-Based Self-Healing Applications" Journal of Polymer Science, Part A: Polymer Chemistry, 48, 2698–2708, doi:10.1002/pola.24053

[Wilson 2016] Wilson MR, Solà J, Carlone A, Goldup SM, Lebrasseur N, Leigh DA. (2016) "An Autonomous Chemically Fuelled Small-Molecule Motor" Nature, 534, 235–240, doi:10.1038/nature18013

[Winchester 1974] Winchester HF. (1974) "Shock and Fire Attenuating Fuel Tank" US Patent 3,787,279

[Winfree 2006] Winfree E. (2006) "Self-healing Tile Sets" Nanotechnology: Science and Computation, Part I, 55–78, J Chen, N Jonoska, G Rozenberg (eds), doi:10.1007/3-540-30296-4_4

[Wisner 2002] Wisner JA, Beer PD, Berry NG, Tomapatanage B. (2002) "Anion Recognition as a Method for Templating Pseudorotaxane Formation" Proceedings of the National Academy of Sciences of the United States of America, 99, 8, 4983–4986, doi:10.1073/pnas.062637999

[Wissner 2013] Wissner-Gross AD, Freer CE. (2013) "Causal Entropic Forces" Physical Review Letters, 110, 168702, doi:10.1103/PhysRevLett.110.168702

[Wittmer 1996] Wittmer JP, Claudin P, Cates ME, Bouchaud JP. (1996) "An Explanation for the Central Stress Minimum in Sand Piles" Nature, 382, 336–338, doi:10.1038/382336a0

[Wojtecki 2011] Wojtecki RJ, Meador MA, Rowan SJ. (2011) "Using the Dynamic Bond to Access Macroscopically Responsive Structurally Dynamic Polymers" Nature Materials, 10, 14–27, doi:10.1038/nmat2891

[Wong 2011] Wong TS, Kang SH, Tang SK, Smythe EJ, Hatton BD, Grinthal A, Aizenberg J. (2011) "Bioinspired Self-Repairing Slippery Surfaces with Pressure-Stable Omniphobicity" Nature, 477, 443–447, doi:10.1038/nature10447

[Wong 2015] Wong HS, Barakat R, Alhilali A, Saleh M, Cheeseman CR. (2015) "Hydrophobic Concrete Using Waste Paper Sludge Ash" Cement and Concrete Research, 70, 9–20, doi:10.1016/j.cemconres.2015.01.005

[Woo 2011] Woo S, Rothemund PW. (2011) "Programmable Molecular Recognition Based on the Geometry of DNA Nanostructures" Nature Chemistry, 3, 620–627, doi:10.1038/nchem.1070

[Wood 2015] Wood CS, Ronson TK, Belenguer AM, Holstein JJ, Nitschke JR. (2015) "Two-Stage Directed Self-Assembly of a Cyclic [3]Catenane" Nature Chemistry, 7, 354–358, doi:10.1038/nchem.2205

[Woodward 2003] Woodward I, Schofield WC, Roucoules V, Badyal JP. (2003) "Super-Hydrophobic Surfaces Produced by Plasma Fluorination of Polybutadiene Films" Langmuir, 19, 8, 3432–3438, doi:10.1021/la020427e

[Wool 1981] Wool RP, O'Connor KM. (1981) "A Theory of Crack Healing in Polymers" Journal of Applied Physics, 52, 5953–5963.

[Wool 1982] Wool RP, O'Connor KM. (1982) "Time Dependence of Crack Healing" Journal of Polymer Science: Polymer Letters, 20, 7–16

[Wool 2006] Wool RP. (2006) "Adhesion at Polymer–Polymer Interfaces: A Rigidity Percolation Approach" C. R. Chimie, 9, 25–44, doi:10.1016/j.crci.2005.04.008

[Wool 2008] Wool RP. (2008) "Self-Healing Materials: A Review" Soft Matter, 4, 400–418, doi:10.1039/b711716g

[Woolf 2010] Woolf CJ. (2010) "What Is This Thing Called Pain?" Journal of Clinical Investigation, 120, 11, 3742–3744, doi:10.1172/JCI45178.

[Woolley 2000] Woolley K, Martin P. (2000) "Conserved Mechanisms of Repair: From Damaged Single Cells to Wounds in Multicellular Tissues" BioEssays, 22, 911–919

[World Health Organization 2009] World Health Organization. (2009) "Basic Documents, 45th Ed." WHO Press, World Health Organization, Geneva, Switzerland

[Wouters 2009] Wouters M, Craenmehr E, Tempelaars K, Fischer H, Stroeks N, van Zanten J. (2009) "Preparation and Properties of a Novel Remendable Coating Concept" Progress in Organic Coatings, 64, 156–162

[Wrenn SM] Wrenn SM, Griswold ED, Uhl FE, Uriarte JJ, Park HE, Coffey AL, Dearborn JS, Ahlers BA, Deng B, Lam YW, Huston DR, Lee PC, Wagner DE, Weiss DJ. (2018) "Avian Lungs: A Novel Scaffold for Lung Bioengineering" PLoS One, 13, 6, e0198956. doi:10.1371/. PMID: 29949597

[Wu 1994] Wu T, Lee S. (1994) "Carbon Tetrachloride-Induced Crack Healing in Polycarbonate" Journal of Polymer Science, Part B: Polymer Physics, 3, 2055–2064.

[Wu 2004] Wu Y, Bekke M, Inoue Y, Sugimura H, Kitaguchi H, Liu C, Takai O. (2004) "Mechanical Durability of Ultra-Water-Repellent Thin Film by Microwave Plasma-Enhanced CVD" Thin Solid Films, 457, 122–127, doi:10.1016/j.tsf.2003.12.007

[Wu 2010a] Wu CY, Feng KM, Peng PC, Lin CY. (2010) "Three-Dimensional Mesh-Based Multipoint Sensing System with Self-Healing Functionality" IEEE Photonics Technology Letters, 22, 8, 565–567, doi:10.1109/LPT.2010.2042709

[Wu 2010b] Wu W, Hansen CJ, Aragón AM, Geubelle PH, White SR, Lewis JA. (2010) "Direct-Write Assembly of Biomimetic Microvascular Networks for Efficient Fluid Transport" Soft Matter, 6, 739–742, doi:10.1039/b918436h

[Wu 2011] Wu D, Long M. (2011) "Realizing Visible-Light-Induced Self-Cleaning Property of Cotton through Coating N-TiO_2 Film and Loading AgI Particles" ACS Applied Materials & Interfaces, 3, 4770–4774, doi: dx.doi.org/10.1021/am201251d

[Wu 2012a] Wu AS, Coppola AM, Sinnott MJ, Chou TW, Thostenson ET, Byun JH, Kim BS. (2012) "Sensing of Damage and Healing in Three-Dimensional Braided Composites with Vascular Channels" Composites Science and Technology, 72, 13, 1618–1626, doi:10.1016/j.compscitech.2012.06.012

[Wu 2012b] Wu ZL, Moshe M, Greener J, Therien-Aubin H, Nie Z, Sharon E, Kumacheva E. (2012) "Three-Dimensional Shape Transformations of Hydrogel Sheets Induced by Small-Scale Modulation of Internal Stresses" Nature Communications, 4, 1586, doi:10.1038/ncomms2549

[Wu 2015] Wu Z, Menichetti G, Rahmede C, Bianconi G. (2015) "Emergent Complex Network Geometry" Scientific Reports, 5, 10073, doi:10.1038/srep10073

[Wu 2020] Wu M, Yuan L, Jiang F, et al. (2020) "Strong Autonomic Self-Healing Biobased Polyamide Elastomers" Chemistry of Materials, 32, 8325–8332, doi:10.1021/acs.chemMater0c02169

[Wudl 2005] Wudl, F, Chen X. (2005) "Thermally Re-Mendable Cross-Linked Polymers" US Patent 6,933,361

[Wyatt 2012] Wyatt G, Zamora D, Smith D, Schroder S, Paudel D, Knight J, Kilberg D, Current D, Gullickson D, Taff S. (2012) "Economic and Environmental Costs and Benefits of Living Snow Fences: Safety, Mobility, and Transportation Authority Benefits, Farmer Costs, and Carbon Impacts" Research Project Final Report 2012-03, Minnesota Department of Transportation, USA

[Wynn's 2008] Wynn's Australia Pty Ltd. (2008) "Premium Radiator Stop-Leak Material Safety Data Sheet" Wetherill Park NSW, Australia

[Xi 2005] Xi J, Schmidt JJ, Montemagno CD. (2005) "Self-Assembled Microdevices Driven by Muscle" Nature Materials, 4, 180–184, doi:10.1038/nmat1308

[Xiao 2007] Xiao P, Venayagamoorthy GK, Corzine KA, Woodley R. (2007) "Self-Healing Control with Multifunctional Gate Drive Circuits for Power Converters" IEEE Indus App Soc Ann Mtg - IAS, pp. 1852–1858, doi:10.1109/07IAS.2007.282

[Xiao 2009] Xiao DS, Yuan YC, Rong MZ, Zhang MQ. (2009) "Self-Healing Epoxy Based on Cationic Chain Polymerization" Polymer, 50, 13, 2967–2975

[Xiao 2010] Xiao X, Xie T, Cheng YT. (2010) "Self-Healable Graphene Polymer Composites" Journal of Materials Chemistry, 20, 3508–3514, doi:10.1039/c0jm00307g

[Xiao 2015] Xiao X, Kong D, Qiu X, Zhang W, Liu Y, Zhang S, Zhang F, Hu Y, Leng J. (2015) "Shape Memory Polymers with High and Low Temperature Resistant Properties" Scientific Reports, 5, 14137, doi:10.1038/srep14137

[Xie 2008] Xie T, Xiao X. (2008) "Self-Peeling Reversible Dry Adhesive System" Chemistry of Materials, 20, 9, 2866–2868, doi:10.1021/cm800173c

[Xing 2013] Xing S, Jiang J, Pan T. (2013) "Interfacial Microfluidic Transport on Micropatterned Superhydrophobic Textile" Lab Chip, 13, 1937–1947, doi:10.1039/C3LC41255E

[Xing 2016] Xing L, Li Q, Zhang G, Zhang X, Liu F, Liu L, Huang Y, Wang Q. (2016) "Self-Healable Polymer Nanocomposites Capable of Simultaneously Recovering Multiple Functionalities" Advanced Functional Materials, 26, 20, 3524–3531, doi:10.1002/adfm.201505305

[Xiong 2015] Xiong W, Tang J, Zhu G, Han N, Schlangen E, Dong B, Wang X, Xing F. (2015) "A Novel Capsule-Based Self-Recovery System with a Chloride Ion Trigger" Scientific Reports, 5, 10866, doi:10.1038/srep10866

[Xu 2003] Xu H, Hampe EM, Rudkevich DM. (2003) "Applying Reversible Chemistry of CO_2 to Supramolecular Polymers" Chemical Communications, 2828–2829, doi:10.1039/B309580K

[Xu 2004] Xu J, Fogleman EA, Craig SL. (2004) "Structure and Properties of DNA-Based Reversible Polymers" Macromolecules, 37, 5, 1863–1870, doi:10.1021/ma035546v

[Xu 2007] Xu B, Toutanji HA, Gilbert JA, Biszick K. (2007) "Different Techniques and Methods of Self Healing" Structural Health Monitoring, FK Chang (ed.), DEStech, Lancaster, pp 507–515

[Xu 2010a] Xu B, Fu YQ, Ahmad M, Luo JK, Huang WM, Kraft A, Reuben R, Pei YT, Chen ZG, De Hossond JT. (2010) "Thermo-Mechanical Properties of Polystyrene-Based Shape Memory Nanocomposites" Journal of Materials Chemistry, 20, 3442–3448, doi:10.1039/B923238A

[Xu 2010b] Xu W, Li G. (2010) "Constitutive Modeling of Shape Memory Polymer Based Self-Healing Syntactic Foam" International Journal of Solids and Structures, 47, 9, 1306–1316, doi:10.1016/j.ijsolstr.2010.01.015

[Xu 2012] Xu F, Zhu Y. (2012) "Highly Conductive and Stretchable Silver Nanowire Conductors" Advanced Materials, 24, 37, 5117–5122, doi:10.1002/adma.201201886

[Xu 2013a] Xu GQ, Demkowicz MJ. (2013) "Healing of Nanocracks by Disclinations" Physical Review Letters, 111, 145501, doi:10.1103/PhysRevLett.111.145501

[Xu 2013b] Xu QF, Liu Y, Lin FJ, Mondal B, Lyons AM. (2013) "Superhydrophobic TiO_2–Polymer Nanocomposite Surface with UV-Induced Reversible Wettability and Self-Cleaning Properties" ACS Applied Materials & Interfaces, 18, 8915–8924, doi:10.1021/am401668y

[Xu 2015] Xu S, Yan Z, Jang K, Huang W, Fu H, Kim J, Wei Z, Flavin M, McCracken J, Wang R, Badea A, Liu Y, Xiao D, Zhou G, Lee J, Chung HU, Cheng H, Ren W, Banks A, Li X, Paik U, Nuzzo RG, Huang Y, Zhang Y, Rogers JA. (2015) "Assembly of Micro/Nanomaterials into Complex, Three-Dimensional Architectures By Compressive Buckling" Science, 347, 6218, 154–159, doi:10.1126/science.1260960

[Xue 2013] Xue L, Li HB, Brodsky EE, Xu ZQ, Kano Y, Wang H, Mori JJ, Si JL, Pei JL, Zhang W, Yang G, Sun ZM, Huang Y. (2013) "Continuous Permeability Measurements Record Healing Inside the Wenchuan Earthquake Fault Zone" Science, 340, 6140, 1555–1559, doi:10.1126/science.1237237

[Yadav 2013] Yadav V, Freedman JD, Grinstaff M, Sen A. (2013) "Bone-Crack Detection, Targeting, and Repair Using Ion Gradients" Angewandte Chemie International Edition, 52, 42, 10997–11001, doi:10.1002/anie.201305759

[Yamada 2001] Yamada K, Watanabe W, Toma T, Itoh K, Nishii J. (2001) "In Situ Observation of Photoinduced Refractive-Index Changes in Filaments Formed in Glasses by Femtosecond Laser Pulses" Optics Letters, 26, 19–26

[Yamagiwa 2008] Yamagiwa T, Tanaka A, Tanaka A, Fujiwara K, Deguchi N, Yamashita K. (2008) "Sealant for Prevention of Blowout, Tubeless Tire, and Tire Tube" US Patent 7,316,253

[Yamaguchi 2007] Yamaguchi M, Ono S, Terano M. (2007) "Self-Repairing Property of Polymer Network with Dangling Chains" Materials Letters, 61, 1396–1399

[Yamamoto 2013] Yamamoto D, Mukai A, Okita N, Yoshikawa K, Shioi A. (2013) "Catalytic Micromotor Generating Self-Propelled Regular Motion through Random Fluctuation" Journal of Chemical Physics, 139, 034705, doi:10.1063/1.4813791

[Yamasamit 2023] Yamasamit N, Sangkeaw P, Jitchaijaroen W, et al. (2023) "Effect of *Bacillus subtilis* on Mechanical and Self-Healing Properties in Mortar with Different Crack Widths and Curing Conditions" Scientific Reports, 13, 7844, doi:10.1038/s41598-023-34837-x

[Yan 2012] Yan J, Bloom M, Bae SC, Luijten E, Granick S. (2012) "Linking Synchronization to Self-Assembly Using Magnetic Janus Colloids" Nature, 491, 578–581, doi:10.1038/nature11619

[Yan 2013] Yan T, Xu S, Peng Q, Zhao L, Zhao X, Lei X, Zhang F. (2013) "Self-Healing of Layered Double Hydroxide Film by Dissolution/Recrystallization for Corrosion Protection of Aluminum" Journal of the Electrochemical Society, 160, 10, C480–C486, doi:10.1149/2.053310jes

[Yang 1998] Yang JM, Kim JH. (1998) "Fault-Tolerant Locomotion of the Hexapod Robot" IEEE Transactions on Systems, Man, and Cybernetics B, 28, 1, 109–116.

[Yang 2006] Yang C, Tartaglino U, Persson B. (2006) "Influence of Surface Roughness on Superhydrophobicity" Physical Review Letters, 97, 116103, doi:10.1103/PhysRevLett.97.116103

[Yang 2007] Yang H, Fang Z, Fu X, Tong L. (2007) "A Novel Glass Fiber-Supported Platinum Catalyst for Self-Healing Polymer Composites: Structure and Reactivity" Chinese Journal of Catalysis, 28, 11, 947–952, doi:10.1016/S1872-2067(07)60081-3

[Yang 2008] Yang J, Keller MW, Moore JS, White SR, Sottos NR. (2008) "Microencapsulation of Isocyanates for Self-Healing Polymers" Macromolecules, 41, 9650–9655

[Yang 2009a] Yang Y, Lepech MD, Yang EH, Li VC. (2009) "Autogenous Healing of Engineered Cementitious Composites under Wet–Dry Cycles" Cement and Concrete Research, 39, 382–390

[Yang 2009b] Yang Z, Hollar J, He X, Shi X. (2009) "Feasibility Investigation of Self-Healing Cementitious Composite Using Oil Core/Silica Gel Shell Passive Smart Microcapsules" SPIE Proceedings, Vol. 7493, Second International Conference on Smart Materials and Nanotechnology in Engineering, J Leng, AK Asundi, W Ecke (eds)

[Yang 2011] Yang Z, Wei Z, Le-ping L, Hong-mei W, Wu-jun L. (2011) "The Self-Healing Composite Anticorrosion Coating" Physics Procedia, 18, 216–221, doi:10.1016/j.phpro.2011.06.084

[Yang 2012] Yang B, Zhang Y, Zhang X, Tao L, Lia S, Wei Y. (2012) "Facilely Prepared Inexpensive and Biocompatible Self-Healing Hydrogel: A New Injectable Cell Therapy Carrier" Polymer Chemistry, 3, 3235–3238, doi:10.1039/C2PY20627G

[Yang 2013] Yang SY, Eoin D. O'Cearbhaill ED, Sisk GC, Park KM, Cho WK, Villiger M, Bouma BE, Pomahac B, Karp JM. (2013) "A Bio-Inspired Swellable Microneedle Adhesive for Mechanical Interlocking with Tissue" Nature Communications, 4, 1702, doi:10.1038/ncomms2715

[Yang 2015a] Yang H, Liang F, Chen Y, Wang Q, Qu X, Yang Z. (2015) "Lotus Leaf Inspired Robust Superhydrophobic Coating from Strawberry-Like Janus Particles" NPG Asia Materials 7, e176; doi:10.1038/am.2015.33

[Yang 2015b] Yang W, Sherman VR, Gludovatz B, Schaible E, Stewart P, Ritchie RO, Meyers MA. (2015) "On the Tear Resistance of Skin" Nature Communications, 6, 6649, doi:10.1038/ncomms7649

[Yang 2015c] Yang WJ, Tao X, Zhao T, Weng L, Kang EN, Wang L. (2015) "Antifouling and Antibacterial Hydrogel Coatings with Self-Healing Properties Based on a Dynamic Disulfide Exchange Reaction" Polymer Chemistry, 6, 7027–7035, doi:10.1039/C5PY00936G

[Yang 2016] Yang X, Liu X, Lu Y, Zhou S, Gao M, Song J, Xu W. (2016) "Controlling the Adhesion of Superhydrophobic Surfaces Using Electrolyte Jet Machining Techniques" Scientific Reports, 6, 23985, doi:10.1038/srep23985

[Yao 2000] Yao F, Ando K, Chu MC, Sato S. (2000) "Crack-Healing Behavior, High-Temperature and Fatigue Strength of SiC-Reinforced Silicon Nitride Composite" Journal of Materials Science Letters, 12, 19, 1081–1084.

[Yao 2002] Yao K, Tay FE. (2002) "Self-Mending of Microcracks in Barium Titanate Glass-Ceramic Thin Films with High Dielectric Constant" Journal of the American Ceramic Society, 85, 496–498, doi:10.1111/j.1151-2916.2002.tb00121.x

[Yao 2013] Yao X, Hu Y, Grinthal A, Wong TS, Mahadevan L, Aizenberg J. (2013) "Adaptive Fluid-Infused Porous Films with Tunable Transparency and Wettability" Nature Materials, 12, 529–534, doi:10.1038/nmat3598

[Yashin 2007] Yashin VV, Balazs AC. (2007) "Theoretical and Computational Modeling of Self-Oscillating Polymer Gels" Journal of Chemical Physics, 126, 124707, doi:10.1063/1.2672951

[Yasuda 2003] Yasuda M, Akao N, Hara N, Sugimoto K. (2003) "Self-Healing Corrosion Protection Ability of Composition-Gradient Al_2O_3·Nb Nanocomposite" Journal of the Electrochemical Society, 150, 10, B481–B487, doi:10.1149/1.1604787

[Yavuz 2009] Yavuz MS, Cheng Y, Chen J, Cobley CM, Zhang Q, Rycenga M, Xie J, Kim C, Song KH, Schwartz AG, Wang LV, Xia Y. (2009) "Gold Nanocages Covered by Smart Polymers for Controlled Release with Near-Infrared Light" Nature Materials, 8, 935–939, doi:10.1038/NMAT2564

[Yebra 2004] Yebra DM, Kiil S, Dam-Johansen K. (2004) "Antifouling Technology: Past, Present and Future Steps towards Efficient and Environmentally Friendly Antifouling Coatings" Progress in Organic Coatings, 50, 75–104, doi:10.1016/j.porgcoat.2003.06.001

[Yehia 2010] Yehia S, Host J. (2010) "Conductive Concrete for Cathodic Protection of Bridge Decks" ACI Materials Journal, 107, 6, 577–585

[Yen 2001] Yen GG, Feng W. (2001) "Winner Takes All Experts Network for Sensor Validation" ISA Trans, 40, 2, 99–110, doi:10.1016/S0019-0578(00)00047-1

[Yim 2007] Yim M, Shen WM, Salemi B, Rus D, Moll M, Lipson H, Klavins E, Chirikjian GS. (2007) "Modular Self-Reconfigurable Robot Systems [Grand Challenges of Robotics]" IEEE Robotics & Automation Magazine, 14, 1, 43–52, doi:10.1109/MRA.2007.339623

[Yin 2008] Yin T, Rong MZ, Wu J, Chen H, Zhang MQ. (2008) "Healing of Impact Damage in Woven Glass Fabric Reinforced Epoxy Composites" Composites: Part A 39, 1479–1487

[Yin 2011] Yin L, Zhu L, Wang Q, Ding J, Chen Q. (2011) "Superhydrophobicity of Natural and Artificial Surfaces under Controlled Condensation Conditions" ACS Applied Materials & Interfaces, 3, 4, 1254–1260, doi:10.1021/am200061t

[Yin 2014a] Yin L, Huang X, Xu H, Zhang Y, Lam J, Cheng J, Rogers JA. (2014) "Materials, Designs, and Operational Characteristics for Fully Biodegradable Primary Batteries" Advanced Materials, 26, 23, 3879–3884, doi:10.1002/adma.201306304

[Yin 2014b] Yin P, Choi HM, Calvert CR, Pierce NA. (2008) "Programming Biomolecular Self-Assembly Pathways" Nature, 451, 318–322, doi:10.1038/nature06451

[Yoon 2012] Yoon JA, Kamada J, Koynov K, Mohin J, Nicolaÿ R, Zhang Y, Balazs AC, Kowalewski T, Matyjaszewski K. (2012) "Self-Healing Polymer Films Based on Thiol–Disulfide Exchange Reactions and Self-Healing Kinetics Measured Using Atomic Force Microscopy" Macromolecules, 45, 1, 142–149, doi:10.1021/ma2015134

[Yoshida 1999] Yoshida E, Murata S, Tomita K, Kurokawa H, Kokaji S. (1999) "An Experimental Study on a Self-Repairing Modular Machine" Robotics and Autonomous Systems, 29, 1, 79–89, doi:10.1016/S0921-8890(99)00040-8

[Yoshie 2010] Yoshie N, Watanabe M, Araki H, Ishida K. (2010) "Thermo-Responsive Mending of Polymers Crosslinked by Thermally Reversible Covalent Bond: Polymers from Bisfuranic Terminated Poly(ethylene adipate) and Tris-Maleimide" Polymer Degradation and Stability, 95, 826–829.

[Yoshimitsu 2002] Yoshimitsu Z, Nakajima A, Watanabe T, Hashimoto K. (2002) "Effects of Surface Structure on the Hydrophobicity and Sliding Behavior of Water Droplets" Langmuir, 18, 15, 5818–5822, doi:10.1021/la020088p

[Youdin 1976] Youdin L, Reich T. (1976) "Mercury-in-Rubber (Whitney) Strain Gauge Temperature Compensation and Analysis of Error Caused by Temperature Drift" Annals of Biomedical Engineering, 4, 220–231

[Youngblood 2008] Youngblood JP, Sottos NR. (2008) "Bioinspired Materials for Self-Cleaning and Self-Healing" MRS Bulletin, 33, 732–741

[Yu 2006] Yu CH, Nagpal R. (2006) "A Self-adaptive Framework for Modular Robots in a Dynamic Environment: Theory and Applications" International Journal of Robotics Research, 30, 8, 1015–1036, doi:10.1177/0278364910384753

[Yu 2008] Yu X, Wang J, Zhang M, Yang L, Li J, Yang P, Cao D. (2008) "Synthesis, Characterization and Anticorrosion Performance of Molybdate Pillared Hydrotalcite/in Situ Created ZnO Composite as Pigment for Mg–Li Alloy Protection" Surface and Coatings Technology, 203, 3–4, 250–255, doi:10.1016/j.surfcoat.2008.08.074

[Yu 2012] Yu X, Liu Z, Janzen J, Chafeeva I, Horte S, Chen W, Kainthan RK, Kizhakkedathu JN, Brooks DE. (2012) "Polyvalent Choline Phosphate as a Universal Biomembrane Adhesive" Nature Materials, 11, 468–476, doi:10.1038/nmat3272

[Yuan 2006] Yuan L, Liang GZ, Xie JQ, Gou J, Li L. (2006) "Thermal Stability of Microencapsulated Epoxy Resins with Poly(urea-formaldehyde)" Polymer Degradation and Stability, 91, 2300–2306

[Yuan 2008a] Yuan W, Hu L, Yu Z, Lam T, Biggs J, Ha SM, Xi D, Chen B, Senesky MK, Grüner G, Pei Q. (2008) "Fault-Tolerant Dielectric Elastomer Actuators Using Single-Walled Carbon Nanotube Electrodes" Advanced Materials, 20, 621–625, doi:10.1002/adma.200701018

[Yuan 2008b] Yuan YC, Yin T, Rong MZ, Zhang MQ. (2008) "Self Healing in Polymers and Polymer Composites. Concepts, Realization and Outlook: A Review" eXPRESS Polymer Letters, 2, 4, 238–250, doi:10.3144/expresspolymlett.2008.29

[Yuan 2010] Yuan G, Li C, Satija SK, Karim A, Douglas JF, Han CC. (2010) "Observation of a Characteristic Length Scale in the Healing of Glassy Polymer Interfaces" Soft Matter, 6, 2153–2159, doi:10.1039/C002046J

[Yuan 2011] Yuan M, Lu J, Kong G, Che C. (2011) "Self Healing Ability of Silicate Conversion Coatings on Hot Dip Galvanized Steels" Surface and Coatings Technology, 205, 19, 4507–4513, doi:10.1016/j.surfcoat.2011.03.088

[Yuan 2012] Yuan Q, Xu Z, Yakobson BI, Ding F. (2012) "Efficient Defect Healing in Catalytic Carbon Nanotube Growth" Physical Review Letters, 108, 245505, doi:10.1103/PhysRevLett.108.245505

[Yue 2010] Yue ZQ, Fan S, Yang Z, Li L, Zhang L, Zhang Z. (2010) "A Hypothesis for Crack Free Interior Surfaces of Longyou Caverns Caved in Argillaceous Siltstone 2000 Years Ago" Frontiers of Architecture and Civil Engineering in China, 4, 2, 165–177, doi:10.1007/s11709-010-0018-1

[Yufa 2009] Yufa NA, Li J, Sibener SJ. (2009) "Diblock Copolymer Healing" Polymer, 50, 12, 2630–2634, doi:10.1016/j.polymer.2009.03.037

[Yuranova 2007] Yuranova T, Laub D, Kiwi J. (2007) "Synthesis, Activity and Characterization of Textiles Showing Self-Cleaning Activity under Daylight Irradiation" Catalysis Today, 122, 109–117

[Zackay 1967] Zackay VF, Parker ER. (1967) "TRIP Steel: Metal that Repairs Itself" Steel, 161, 73–76

[Zak 2007] Zak M. (2007) "Complexity for Survival of Livings" Chaos, Solitons and Fractals, 32, 1154–1167

[Zako 1999] Zako M, Takano N. (1999) "Intelligent Material Systems Using Epoxy Particles to Repair Microcracks and Delamination Damage in GFRP" Journal of Intelligent Materials Systems and Structures, 10, 836–841

[Zamora 2015] Zamora DS, Ogdahl E, Wyatt G, Smith DJ, Johnson G, Current D, Gullickson D. (2015) "Assessing the Use of Shrub-Willows for Living Snow Fences in Minnesota" Minnesota Department of Transportation, MN/RC 2015–46

[Zang 2013] Zang J, Ryu S, Pugno N, Wang Q, Tu Q, Buehler MJ, Zhao X. (2013) "Multifunctionality and Control of the Crumpling and Unfolding of Large-Area Graphene" Nature Materials, 12, 321–325, doi:10.1038/nmat3542

[Zavada 2015] Zavada SR, McHardy NR, Gordon KL, Scott TF. (2015) "Rapid, Puncture-Initiated Healing via Oxygen-Mediated Polymerization" ACS Macro Letters, 4, 8, 819–824, doi:10.1021/acsmacrolett.5b00315

[Zebrowski 2003] Zebrowski J, Prasad V, Zhang W, Walker LM, Weitz DA. (2003) "Shake-Gels: Shear-Induced Gelation of Laponite/PEO Mixtures" Colloids and Surfaces A: Physicochemical and Engineering Aspects, 213, 189–197

[Zeng 2013] Zeng F, Han Y, Yan ZC, Liu CY, Chen CF. (2013) "Supramolecular Polymer Gel with Multi Stimuli Responsive, Self-Healing and Erasable Properties Generated by Host–Guest Interactions" Polymer, 54, 26, 6929–6935, doi:10.1016/j.polymer.2013.10.048

[Zhai 2006] Zhai L, Berg MC, Cebeci FC, Kim Y, Milwid JM, Rubner MF, Cohen RE. (2006) "Patterned Superhydrophobic Surfaces: Toward a Synthetic Mimic of the Namib Desert Beetle" Nano Letters, 6, 6, 1213–1217

[Zhang 1989] Zhang H, Wool RP. (1989) "Concentration Profile for a Polymer–Polymer Interface. 1. Identical Chemical Composition and Molecular Weight" Macromolecules, 22, 7, 3018–3021, doi:10.1021/ma00197a024

[Zhang 1998] Zhang YH, Edwards L, Plumbridge WJ. (1998) "Crack Healing in a Silicon Nitride Ceramics" Journal of the American Ceramic Society, 81, 7, 1861–1868, doi:10.1111/j.1151-2916.1998.tb02558.x

[Zhang 2006] Zhang L, Zhou Z, Cheng B, DeSimmone JM, Samulski ET. (2006) "Superhydrophobic Behavior of a Perfluoropolyether Lotus-Leaf-like Topography" Langmuir, 22, 20, 8576–8580, doi:10.1021/la061400o

[Zhang 2007a] Zhang HL, Lorente S, Bejan A. (2007) "Vascularization with Trees that Alternate with Upside-Down Trees" Journal of Applied Physics, 101, 9, 094904, doi:10.1063/1.2723186

[Zhang 2007b] Zhang W, Sakalkar V, Koratkar N. (2007) "In Situ Health Monitoring and Repair In Composites Using Carbon Nanotube Additives" Applied Physics Letters, 91, 133102, doi:10.1063/1.2783970

[Zhang 2008] Zhang X, Xu L, Du S, Han W, Han J. (2008) "Crack-Healing Behavior of Zirconium Diboride Composite Reinforced with Silicon Carbide Whiskers" Scripta Materialia, 59, 11, 1222–1225, doi:10.1016/j.scriptamat.2008.08.013

[Zhang 2009a] Zhang L, Abbott JJ, Dong L, Kratochvil BE, Bell D, Nelson BJ. (2009) "Artificial Bacterial Flagella: Fabrication and Magnetic Control" Applied Physics Letters, 94, 064107, doi:10.1063/1.3079655

[Zhang 2009b] Zhang Y, Broekhuis AA, Picchioni F. (2009) "Thermally Self-Healing Polymeric Materials: The Next Step to Recycling Thermoset Polymers?" Macromolecules, 42, 1906

[Zhang 2010] Zhang S, Greenfield MA, Mata A, Palmer LC, Bitton R, Mantei JR, Aparicio C, de la Cruz MO, Stupp SI. (2010) "A Self-Assembly Pathway to Aligned Monodomain Gels" Nature Materials, 9, 594–601, doi:10.1038/NMAT2778

[Zhang 2011] Zhang YL, Wang JN, He Y, He Y, Xu BB, Wei S, Xiao FS. (2011) "Solvothermal Synthesis of Nanoporous Polymer Chalk for Painting Superhydrophobic Surfaces" Langmuir, 27, 20, 12585–12590, doi:10.1021/la2018264

[Zhang 2012a] Zhang L, Lü C, Li Y, Lin Z, Wang Z, Dong H, Wang T, Zhang X, Li X, Zhang J, Yang B. (2012) "Fabrication of Biomimetic High Performance Antireflective and Antifogging Film by Spin-Coating" Journal of Colloid Interface Science, 374, 1, 89–95, doi:10.1016/j.jcis.2012.01.051

[Zhang 2012b] Zhang YL, Yang B, Zhang XY, Xu LX, Tao L, Li SX, Wei Y. (2012) "A Magnetic Self-Healing Hydrogel" Chemical Communications, 48, 9305–9307, doi:10.1039/C2CC34745H

[Zhang 2012c] Zhang Z, Hu Y, Liu Z, Guo T. (2012) "Synthesis and Evaluation of a Moisture-Promoted Healing Copolymer" Polymer 53, 14, 2979–2990, doi:10.1016/j.polymer.2012.04.048

[Zhang 2013] Zhang J, Ke X, Gou G, Seidel J, Xiang B, Yu P, Liang WI, Minor AM, Chu Y, Van Tendeloo G, Ren X, Ramesh R. (2013) "A Nanoscale Shape Memory Oxide" Nature Communications, 4, 2768, doi:10.1038/ncomms3768

[Zhang 2014] Zhang H, Hao R, Jackson JK, Chiao M, Yu H. (2014) "Janus Ultrathin Film from Multi-Level Self-Assembly at Air–Water Interfaces" Chemical Communications, 50, 14843–14846, doi:10.1039/C4CC06798C

[Zhang 2015a] Zhang F, Jiang S, Wu S, Li Y, Mao C, Liu Y, Yan H. (2015) "Complex Wireframe DNA Origami Nanostructures with Multi-Arm Junction Vertices" Nature Nanotechnology, 10, 779–784, doi:10.1038/nnano.2015.162

[Zhang 2015b] Zhang J, Yao Y, Sheng L, Liu J. (2015) "Self-Fueled Biomimetic Liquid Metal Mollusk" Advanced Materials, 27, 16, 2648–2655

[Zhang 2015c] Zhang P, Li G. (2015) "Healing-on-Demand Composites Based on Polymer Artificial Muscle" Polymer, 64, 29–38, doi:10.1016/j.polymer.2015.03.022

[Zhang 2015d] Zhang X, He J. (2015) "Hydrogen-Bonding-Supported Self-Healing Antifogging Thin Films" Scientific Reports, 5, 9227, doi:10.1038/srep09227

[Zhang 2015e] Zhang Y, Pal S, Srinivasan B, Vo T, Kumar S, Gang O. (2015) "Selective Transformations between Nanoparticle Superlattices via the Reprogramming of DNA-Mediated Interactions" Nature Materials, 14, 840–847 doi:10.1038/nmat4296

[Zhang 2015f] Zhang Y, Rocco C, Karasu F, van der Ven LG, van Benthem RA, Allonas X, Croutxé-Barghorn C, Esteves AC. (2015) "UV-Cured Self-Replenishing Hydrophobic Polymer Films" Polymer, 69, 384–393, doi:10.1016/j.polymer.2015.02.036

[Zhang 2016a] Zhang H, Wang C, Zhu G, Zacharia NS. (2016) "Self-Healing of Bulk Polyelectrolyte Complex Material as a Function of pH and Salt" ACS Applied Materials & Interfaces, 8, 39, 26258–26265, doi:10.1021/acsami.6b06776

[Zhang 2016b] Zhang Y, Kersell H, Stefak R, Echeverria J, Iancu V, Perera UG, Li Y, Deshpande A, Braun KF, Joachim C, Rapenne G, Hla SW. (2016) "Simultaneous and Coordinated Rotational Switching of All Molecular Rotors in a Network" Nature Nanotechnology, 11, 8, doi:10.1038/nnano.2016.69

[Zhao 2004] Zhao Y, Beck JB, Rowan SJ, Jamieson AM. (2004) "Rheological Behavior of Shear-Responsive Metallo-Supramolecular Gels" Macromolecules, 37, 3529–3530

[Zhao 2009] Zhao M. (2009) "Electrical Fields in Wound Healing: An Overriding Signal that Directs Cell Migration" Seminars in Cell and Developmental Biology, 20, 6, 674–682, doi:10.1016/j.semcdb.2008.12.009

[Zhao 2010] Zhao X, Kim J, Cezar CA, Huebsch N, Lee K, Bouhadir K, Mooney DJ. (2010) "Active Scaffolds for On-Demand Drug and Cell Delivery" Proceedings of the National Academy of Sciences of the United States of America, 108, 1, 67–72, doi:10.1073/pnas.1007862108

[Zhao 2012] Zhao Y, Xu Z, Wang X, Lin T. (2012) "Photoreactive Azido-Containing Silica Nanoparticle/Polycation Multilayers: Durable Superhydrophobic Coating on Cotton Fabrics" Langmuir, 28, 6328–6335, doi:10.1021/la300281q

[Zheludkevich 2007] Zheludkevich ML, Shchukin DG, Yasakau KA, Möhwald H, Ferreira MG. (2007) "Anticorrosion Coatings with Self-Healing Effect Based on Nanocontainers Impregnated with Corrosion Inhibitor" Chemistry of Materials, 19, 402–411

[Zheng 2002] Zheng Y, Micic M, Mello SV, Mabrouki M, Andreopoulos FM, Konka V, Pham SM, Leblanc RM. (2002) "PEG-Based Hydrogel Synthesis via the Photodimerization of Anthracene Groups" Macromolecules, 35, 13, 5228–5234, doi:10.1021/ma012263z

[Zheng 2006] Zheng W, Chung J, Jacobs HO. (2006) "Fluidic Heterogeneous Microsystems Assembly and Packaging" Journal of Microelectromechanical Systems, 15, 4, 864–870, doi:10.1109/JMEMS.2006.878885

[Zheng 2012a] Zheng P, McCarthy TJ. (2012) "A Surprise from 1954: Siloxane Equilibration Is a Simple, Robust, and Obvious Polymer Self-Healing Mechanism" Journal of the American Chemical Society, 134, 4, 2024–2027, doi:10.1021/ja2113257

[Zheng 2012b] Zheng Y, Chen J, Craven M, Choi NW, Totorica S, Diaz-Santana A, Kermani P, Hempstead B, Fischbach-Teschl C, López JA, Stroock AD. (2012) "In Vitro Microvessels for the Study of Angiogenesis and Thrombosis" Proceedings of the National Academy of Sciences of the United States of America, 109, 24, 9342–9347, doi:pnas.1201240109

[Zhong 2008] Zhong S, Yao W. (2008) "Influence of Damage Degree on Self-Healing of Concrete" Construct ion and Building Materials, 22, 6, 1137–1142, doi:10.1016/j.conbuildmat.2007.02.006

[Zhong 2014] Zhong C, Gurry T, Cheng AA, Downey J, Deng Z, Stultz CM, Lu TK. (2014) "Strong Underwater Adhesives Made by Self-Assembling Multi-Protein Nanofibres" Nature Nanotechnology, 9, 858–866, doi:10.1038/nnano.2014.199

[Zhou 1999] Zhou Y, Kassab GS, Molloi S. (1999) "On the Design of the Coronary Arterial Tree: A Generalization of Murray's Law" Physics in Medicine & Biology, 44, 2929, doi:/10.1088/0031-9155/44/12/306

[Zhou 2011] Zhou J, Bronikowski M, Noca F, Sansom EB. (2011) "Nanoscale Wicking Methods and Devices" US Patent 8,021,967

[Zhou 2014a] Zhou W, Dai X, Fu TM, Xie C, Liu J, Lieber CM. (2014) "Long Term Stability of Nanowire Nanoelectronics in Physiological Environments" Nano Letters, 14, 3, 1614–1619, doi:10.1021/nl500070h

[Zhou 2014b] Zhou W, Li X, Lu J, Huang N, Chen L, Qi Z, Li L, Liang H. (2014) "Toughening Mystery of Natural Rubber Deciphered by Double Network Incorporating Hierarchical Structures" Scientific Reports, 16, 4, 7502, doi:10.1038/srep07502

[Zhou 2015] Zhou Z, Wang Z, Zhou Y, Pang S, Wang D, Xu H, Liu Z, Padture NP, Cui G. (2015) "Methylamine-Gas-Induced Defect-Healing Behavior of $CH_3NH_3PbI_3$ Thin Films for Perovskite Solar Cells" Angewandte Chemie International Edition, 54, 33, 9705–9709, doi:10.1002/anie.201504379

[Zhou 2018] Zhou KG, Vasu KS, Cherian CT, Neek-Amal M, Zhang JC, Ghorbanfekr-Kalashami H, Huang K, Marshall OP, Kravets VG, Abraham J, Su Y, Grigorenko AN, Pratt A, Geim AK, Peeters FM, Novoselov KS, Nair RR. (2018) "Electrically Controlled Water Permeation through Graphene Oxide Membranes" Nature, 559, 236–240, doi:10.1038/s41586-018-0292-y

[Zhou 2024] Zhou T, Wan X, Huang DZ, et al. (2024) "Ai-Aided Geometric Design of Anti-Infection Catheters" Science Advances, 10, eadj1741, doi:10.1126/sciadv.adj1741

[Zhu 2010] Zhu J, Hsu CM, Yu Z, Fan S, Cui Y. (2010) "Nanodome Solar Cells with Efficient Light Management and Self-Cleaning" Nano Letters, 10, 1979–1984, doi:10.1021/nl9034237

[Zhu 2012] Zhu W, Pao GM, Satoh A, Cumming G, Monaghan JR, Harkins TT, Bryant SV, Voss SR, Gardiner DM, Hunter T. (2012) "Activation of Germline-Specific Genes Is Required for Limb Regeneration in the Mexican Axolotl" Developmental Biology, 370, 42–51, doi:10.1016/j.ydbio.2012.07.021

[Zhu 2013] Zhu S, So JH, Mays R, Desai S, Barnes WR, Pourdeyhimi B, Dickey MD. (2013) "Ultrastretchable Fibers with Metallic Conductivity Using a Liquid Metal Alloy Core" Advanced Functional Materials, 23, 18, 2308–2314, doi:10.1002/adfm.201202405

[Zhu 2014] Zhu M, Rong MZ, Zhang MQ. (2014) "Self-Healing Polymeric Materials towards Non-Structural Recovery of Functional Properties" Polymer International, 63, 10, 1741–1749, doi:10.1002/pi.4723

[Zhu 2015] Zhu C, Arson C. (2015). "A Model of Damage and Healing Coupling Halite Thermo-mechanical Behavior to Microstructure Evolution." Geotechnical and Geological Engineering, 33, 2, 389–410, doi:10.1007/s10706-014-9797-9

[Zhu 2023]. Zhu D, Wang Z, Xie J, et al. (2023) "Metal–Organic Framework-Based Self-Healing Hydrogel Fiber Random Lasers" Nanoscale, 15, 10685–10692, doi:10.1039/D3NR00675A

[Zoldesi 2006] Zoldesi CI, van Walree CA. Imhof A. (2006) "Deformable Hollow Hybrid Silica/Siloxane Colloids by Emulsion Templating" Langmuir, 22, 4343–4352

[Zou 2012] Zou J, Roque R, Byron T. (2012) "Effect of HMA Ageing and Potential Healing on Top-Down Cracking Using HVS" Road Materials and Pavement Design, 13, 3, 518–533, doi:10.1080/14680629.2012.709177

[Zuiderduin 2007] Zuiderduin WCJ, Westzaan C, Huétink J, Gaymans RJ. (2003) "Toughening of Polypropylene with Calcium Carbonate Particles" Polymer, 44, 261–275

[Zuriguel 2014] Zuriguel I, Parisi DR, Hidalgo RC, Lozano C, Janda A, Gago PA, Peralta JP, Ferrer LM, Pugnaloni LA, Clément E, Maza D, Pagonabarraga I, Garcimartín A. (2014) "Clogging Transition of Many-Particle Systems Flowing through Bottlenecks" Scientific Reports, 4, 7324, doi:10.1038/srep07324

[Zykov 2007] Zykov V, Mytilinaios E, Desnoyer M, Lipson H. (2007) "Evolved and Designed Self-Reproducing Modular Robotics" IEEE Transactions on Robotics, 23, 2, 308–319, doi:10.1109/TRO.2007.894685

Index

C

D

M

N

T

Y

Z